GEOMETRY
Its Elements and Structure

J. Houston Banks
Alfred S. Posamentier
Robert L. Bannister

Webster Division
McGraw-Hill Book Company

New York · St. Louis · San Francisco · Dallas
Düsseldorf · Johannesburg · Kuala Lumpur
London · Mexico · Montreal · New Delhi · Panama
Rio de Janeiro · Singapore · Sydney · Toronto

Elements and Structure Series

Algebra: Its Elements and Structure, Book I, Second Edition
Geometry: Its Elements and Structure
Algebra: Its Elements and Structure, Book II, Second Edition

Editor **Irwin F. Harris**

Editing and Styling John Aliano

Text Design and Illustrations Ralph J. Kellner and Associates

Cover Design Leslie Adalman

Production William Weiss

Technical Illustrations J & R Services

Grateful acknowledgement is made for photographs from the following sources:

Brown Brothers 298

French Government Tourist Office 345

Greek National Tourist Office 345

New York Convention and Visitors Bureau 349

Bettman Archive 429, 502

Library of Congress Catalog Card Number: 77-38819

ISBN 07-003681-0

Preface To the Student

This book was written for you to read and enjoy. It is a logical development of a body of mathematical knowledge known as geometry. Included in this book are topics from plane geometry, solid geometry, coordinate geometry, vector geometry, and non-Euclidean geometry. At times the work may be easy; at times it may be challenging. At all times we hope that the work will be interesting and exciting.

How does one read and study a geometry book? You will find that each new concept in the book is carefully developed and then clarified with many examples. In addition, there are over 2,000 illustrations to help you visualize the conditions of the problems. Prior to each set of exercises, there are numerous examples with their solutions completely worked out in detail. Before attempting to start any assignment, you should read the text and study the sample solutions that precede the exercises. The Class Exercises that appear in each section will also help you to understand the concepts of the sections more clearly. Study this material with pencil at hand, supplying illustrations where they are not provided. Try the sample exercises yourself before you look at the solutions. If necessary, reread the section that precedes the exercises.

Within each chapter are inserts to be used as supplementary reading. These come under the headings A Look at the Past and Something to Think About. You will find the course far more interesting if you take the time to read and explore these ideas. Further supplementary material is given at the conclusion of selected chapters in the section entitled Mathematical Excursion. This material is definitely more challenging and has been prepared for those of you with special interests and abilities in mathematics.

Also included at the end of each chapter are a vocabulary list, a set of review exercises, a chapter test, and topics for suggested research. These should help you to organize and evaluate your knowledge before going on to new work.

The cumulative review at the close of every three chapters is a unique feature of this text. The reviews are designed for self-evaluation, and you should enjoy the experience of working through these.

The chapter-opening pages illustrate a few, but only a few, of the applications of geometry to everyday life. We hope you will think about these and try to name other situations in which geometry is used.

The study of mathematics can be an exciting and rewarding experience. We wish you luck as you begin this year's study of geometry.

The authors

Preface To the Teacher

As indicated in the preface to the students, *Geometry: Its Elements and Structure* has been prepared to present a logical development of a course in geometry that emphasizes basic concepts and understandings of many branches of the tree of geometry. It has been written with the reports and the recommendations of various commissions and curriculum groups in mind.

Each of the authors has a strong interest in secondary education and is either currently teaching, or has taught, at this level and is now involved in the training of secondary school teachers of mathematics. All are active members of the National Council of Teachers of Mathematics. Consequently, the material presented is felt to be realistic in terms of what high school students can be expected to absorb. Furthermore, there are an abundance of exercises, reviews, and tests for constant and ongoing reinforcement and evaluation. The considerable number of exercises allows you to spiral your homework assignments for optimum individualization.

In addition, class exercises are provided in those places where they will be helpful to the development and/or reinforcement of the lesson. Summaries are placed within the chapters where they will be of greatest use to the individual student.

Both the content and the development is based on the suggestions offered by classroom teachers and professors of mathematics. To provide for individual differences, all sets of exercises have been graded, with most having levels A, B, and C designated. Level A exercises consist of exercises that are essential and are within the grasp of all students. Level B exercises are more challenging but still within the grasp of most students, and level C exercises consist of problems that should serve to challenge your best students. This gradation of exercises, together with optional sections in the book, allows it to be used for classes of varying abilities.

End-of-chapter material is extensive and can be used in various ways, as indicated by the annotations on these pages. The Mathematical Excursions that appear at the end of selected chapters are yet another means of providing enrichment for gifted students, whereas the review exercises should prove helpful to slower students. Furthermore, a complete booklet of chapter tests is available as a supplementary aid.

Finally, we wish to take this opportunity to express our appreciation and gratitude to all the many people who have made this project possible. We are indebted to the reviewers from New York City, Arlington Heights, Illinois, and Columbia, Missouri for their helpful comments.

We welcome all comments, criticisms, and suggestions from both teachers and students alike. We hope that you will enjoy teaching from this book as much as we have enjoyed preparing it for you.

J. Houston Banks
Alfred S. Posamentier
Robert L. Bannister

About the Authors

J. Houston Banks is Professor of Mathematics and Chairman of the Department of Mathematics, George Peabody College, Nashville, Tennessee. He is a past member of the Board of Directors of the National Council of Teachers of Mathematics and has served as a consultant for elementary and secondary mathematics curriculum groups both domestically and internationally. Dr. Banks has written many articles for mathematical and scientific publications and has authored or co-authored arithmetic and algebra textbooks and professional books for teachers.

Alfred S. Posamentier is a member of the faculty of The City College of The City University of New York. He is a member of the National Council of Teachers of Mathematics and has been a program participant at various meetings sponsored by that group. Mr. Posamentier is the author of various journal articles and the author or co-author of mathematics books for high school students. A past consultant for various elementary and secondary school mathematics projects, he has extensive experience as a high school teacher. While teaching at Theodore Roosevelt High School, New York City, he created a new course in mathematical enrichment for high school students.

Robert L. Bannister is an instructor at Horton Watkins High School, Ladue, Missouri. Previous to this, he served as a mathematics editor and has taught at St. Louis University and at the University of Missouri at St. Louis. Mr. Bannister is a member of the National Council of Teachers of Mathematics and has assisted schools with the coordination of mathematics laboratory materials. He is currently engaged in research in mathematics curriculum development.

Contents

Learning Aids

A Look at the Past

Something to Think About

Mathematical Excursions

Summaries

1 Elements of Geometry

In other mathematics classes, you used points and lines to graph the solution sets of equations and inequalities. In these classes you also became familiar with the plane, and such geometric figures as the triangle, parallelogram, square, cube, pyramid, and sphere. As a result, you understand certain concepts basic to geometry. This year, your past experience will serve as your springboard to a deeper understanding of the compactness and completeness of this area of mathematics.

1-1 SETS

Sets are everyday concepts. A set of dishes, a swarm of bees, and a fold of sheep are examples of nonmathematical sets. In mathematics, we are concerned with sets whose members or **elements** are identifiable. Sets of this type are called **well-defined** sets.

An example of one of the many sets you studied in algebra is the set of natural or counting numbers. We customarily name sets with capital letters and write the elements of sets within braces. For example, if we let N represent the set of natural numbers, we write $N = \{1, 2, 3, \ldots\}$. How many elements are in N? The ellipsis dots (. . .) tell us that the obvious pattern begun in the set so far continues throughout the set. N is an **infinite** set. Is the set X, $X = \{1, 2, 3, \ldots, 99\}$, an infinite set? How many elements are in the set X?

Other sets you have studied include the whole numbers and the integers. We can form a correspondence between each set of numbers and a set of points of a line. These correspondences and graphings form a model that aids in showing the relationship that exists between these sets.

Set **Model**

$N = \{1, 2, 3, \ldots\}$

$W = \{0, 1, 2, 3, \ldots\}$

$I = \{\ldots, -3, -2, -1, 0, 1, 2, \ldots\}$

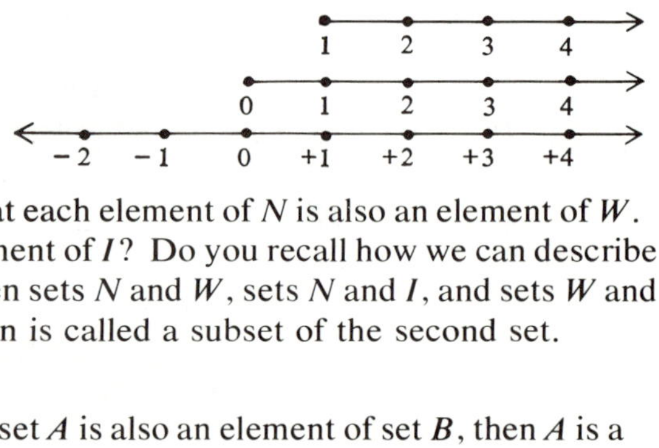

From these models we can see that each element of N is also an element of W. Is every element of N also an element of I? Do you recall how we can describe the relationship that exists between sets N and W, sets N and I, and sets W and I? In each case the first set given is called a subset of the second set.

D **Definition 1-1** If every element of set A is also an element of set B, then A is a **subset** of B. We denote this by $A \subseteq B$.

The set $\{1, 2, 3, 4, 5, 6, 7\}$ is a subset of the set of whole numbers. Is the set $\{2, 4, 6, 10\}$ a subset of the set $\{10, 2, 6, 4\}$? Every set is a subset of itself. The natural numbers and the counting numbers are two different names for the same set. We see that there are two types of subsets. The set of counting numbers is a proper subset of the set of whole numbers. It is not, however, a proper subset of the set of natural numbers.

D **Definition 1-2** If set A is a subset of set B, but set B contains at least one element not in set A, then A is a **proper subset** of B. We denote this by $A \subset B$.

Example 1 Show that $\{2, 3, 7\}$ is a subset of N.

Solution The numbers 2, 3, and 7 are all natural numbers. Hence,

$$\{2, 3, 7\} \subseteq N$$

Example 2 How can we show $N \subset W$?

Solution Every element of N is also an element of W. W, however, contains zero while N does not. Thus, N is a proper subset of W. $N \subset W$.

We have already said that the natural numbers and the counting numbers are two different names for the same set. The two sets have exactly the same members. When this occurs, we say that the sets are equal sets.

D **Definition 1-3** Set A and set B are **equal sets** if and only if the two sets have exactly the same elements. We denote the equality of A and B by $A = B$.

If $A = \{0, 1, 2\}$ and $B = \{5, 7, 9\}$, are A and B equal? Since they do not have the same elements, they are not equal. Because A and B have the same number of elements, we say that they are equivalent sets.

D **Definition 1-4** Set A and set B are **equivalent sets** if and only if the two sets have exactly the same number of elements.

If two sets are equal, are they equivalent? Are equivalent sets equal sets?

Class Exercises

Classify each of the following sets as finite or infinite:

1. $\{2, 4, 6, 10\}$
2. $\{3, 3.1, 4\}$
3. $\{\frac{4}{2}, \frac{6}{2}, \frac{8}{2}, \frac{10}{2}, \ldots\}$
4. $\{-3, -1, 1, 3, \ldots\}$
5. One of the sets above is a subset of another of the above sets. Name the set and its subset.

Determine which of the following are true and which are false:

6. $0 \subseteq \{0\}$
7. $\{a, b, c\} = \{c, b, a\}$
8. The set of triangles $\subseteq$ the set of geometric figures.
9. $\{-2, \pi, 7\} \subseteq$ the set of real numbers
10. {a, e, i, o, u} is equivalent to $\{7, 11, 15, 19, 23\}$
11. $\{\Delta, \square, \bigcirc\} \subseteq$ the set of geometric figures

Set Operations

We think of a line as a set of points. If two lines intersect, their intersection is a point common to both lines. When any two sets share common elements, we say that their intersection is the set of common elements.

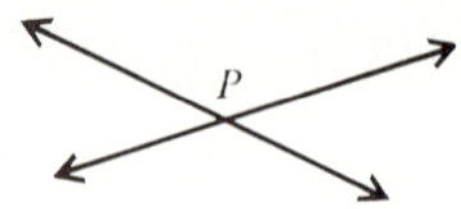

D **Definition 1-5** The **intersection** of sets A and B is the set whose elements are elements of both A and B. We denote the intersection of A and B by $A \cap B$.

Example 3 Find the intersection of $A = \{1, 3, 4, 9\}$ and $B = \{5, 7, 9, 11, \ldots\}$.

Solution $A \cap B = \{9\}$, since 9 is the only element of A that is also an element of B.

The set containing no members is the **empty set**, $\varnothing$. When the intersection of two sets is the empty set, the sets are called **disjoint sets**. The empty set is a subset of all sets.

Example 4 If $P = \{\text{a, b, y, z}\}$ and $Q = \{0, 1, 2, 7, 9\}$, find $P \cap Q$.

Solution There are no elements of P that are also elements of Q; their intersection contains no members, and hence $P \cap Q = \varnothing$.

We often think of geometric figures as the combination of various sets of points. If two lines intersect, the combination of the lines consists of all points of both lines. We refer to the set formed by combining the elements of the two sets as the union of the two sets.

D **Definition 1-6** The **union** of sets A and B is the set whose elements are elements of at least one of the sets, A and B. We denote the union of A and B by $A \cup B$.

If the sets are disjoint, it is easy to form the union. If the sets are not disjoint, the union is sometimes difficult to form. Why?

Example 5 Describe the set combining the points of the line through A and B and the points of circle O.

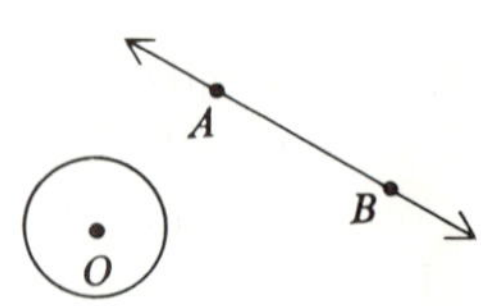

Solution The union of the two sets is the set of all the points of the line and of the circle.

Example 6 Describe the union of the set of points contained by figure $ABGH$ and the set of points contained by figure $CDFE$.

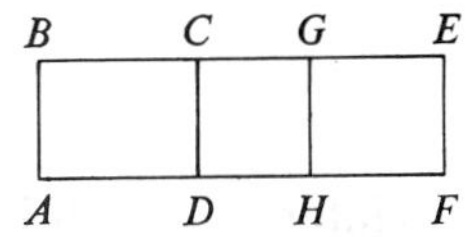

Solution The set must contain every point in the region bounded by figure $ABGH$ and every point in the region bounded by figure $CDFE$. The union of the two sets is the set of all points in the region bounded by figure $ABEF$.

In Example 6, the points bounded by figure $CDGH$ are contained in both region $ABGH$ and region $CDFE$. When forming the union of two nondisjoint sets we must consider how many times to count or list these common elements. We shall agree to list the common elements only once.

Example 7 If $A = \{0, 1, 2, 3, 4, 5, 6, 7, 8, 9\}$, $B = \{0, 2, 4, 6, 8\}$, and $C = \{1, 3, 5, 7, 9\}$, find $A \cap C$, $A \cap B$, $B \cup C$, and $B \cap C$.

Solution

$A \cap C = \{1, 3, 5, 7, 9\}$; that is, $A \cap C = C$

$A \cap B = \{0, 2, 4, 6, 8\}$; that is, $A \cap B = B$

$B \cup C = \{0, 2, 4, 6, 8, 1, 3, 5, 7, 9\}$; that is, $B \cup C = A$

$B \cap C = \varnothing$

EXERCISES

Classify each of the following sets as well-defined or not well-defined:

1. $\{1, 3, 5, 7, \ldots\}$
2. the set of beautiful girls in school
3. $\{5, 10, 15, \ldots\}$
4. the set of rational numbers
5. all n such that $\frac{n}{2}$ is an odd number
6. every third tall tree

List the members of each of the following sets:

7. the set of boys in your class
8. the set of whole numbers greater than 3 but less than 7
9. the set of even counting numbers less than 100
10. the set of four-sided triangles

11. List the subsets of the set $\{3, 5, 7\}$.
12. Given any two sets A and B, is the set $A \cap B$ a subset of set B?
13. Is set B always a proper subset of set $A \cup B$? Explain.

Let U = the set of natural numbers less than 25, $A = \{5, 10, 15, 20\}$, and $B = \{1, 3, 5, 7, 9, 11, 13, 15, 17, 19, 21, 23\}$. List the members of each of the following sets:

14. $A \cup B$
15. $A \cap B$
16. $A \cap A$
17. $U \cup B$
18. $(A \cup B) \cup U$
19. $(A \cap B) \cap U$

20. $U \cap B$ 21. $A \cup (B \cup U)$ 22. $A \cap (B \cap U)$
23. $U \cap (A \cup B)$ 24. $U \cap A$ 25. $\varnothing \cap U$
26. $(U \cap A) \cup (U \cap B)$ 27. $(U \cap A) \cup B$ 28. $(A \cup U) \cap \varnothing$

Describe the intersection of the sets of points for each of the following:

29. Figures $ABEF$ and $BCDE$

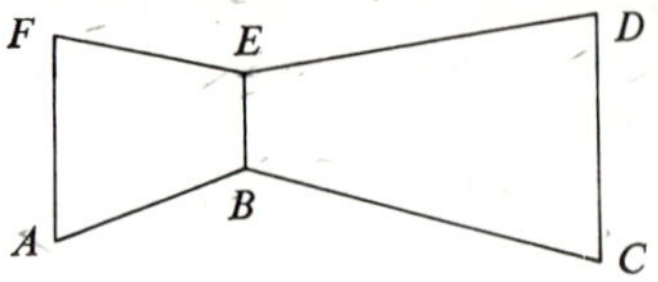

30. Circle O and line AD

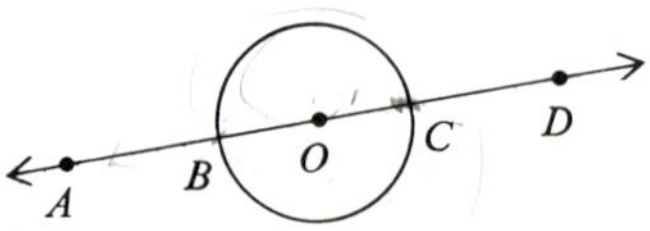

31. Lines UT, PQ, and SR

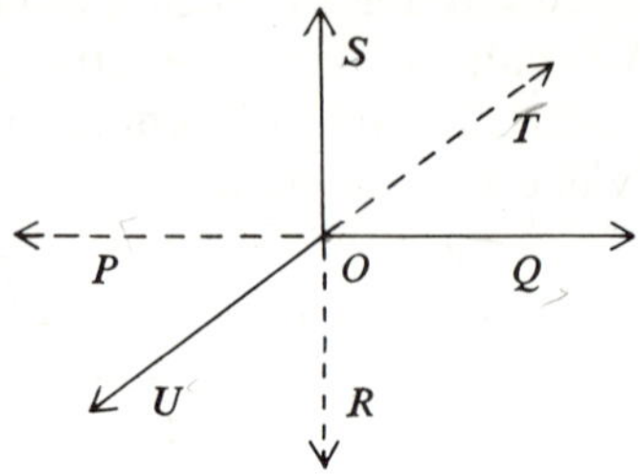

Describe the union of the sets of points in each of the following regions:

32. $ABCD$ and $CDEF$ 33. ABC and PBD 34. PQR and QRS

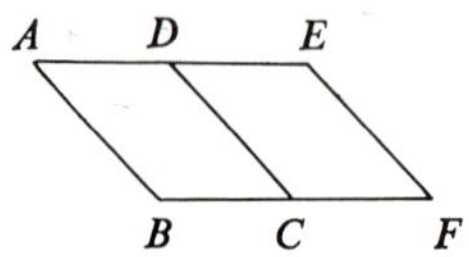

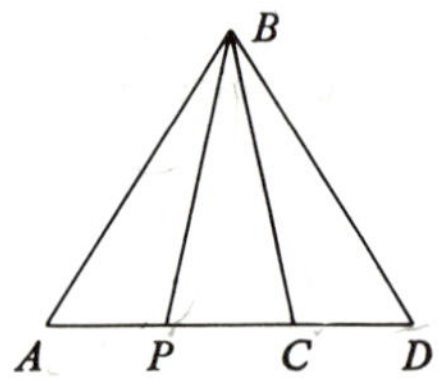

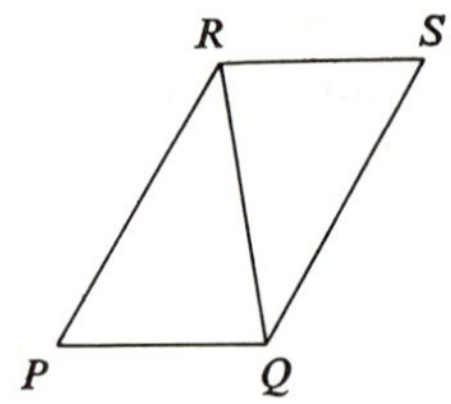

SOMETHING TO THINK ABOUT

Venn Diagrams

John Venn (1834–1923) devised a method for showing models of statements about sets. We call these models *Venn diagrams*. He represented a set by a circular region. By using several circles at a time, he showed the result of an operation, such as union or intersection, by shading. For example, $A \cap B$ can be shown in either of the following ways:

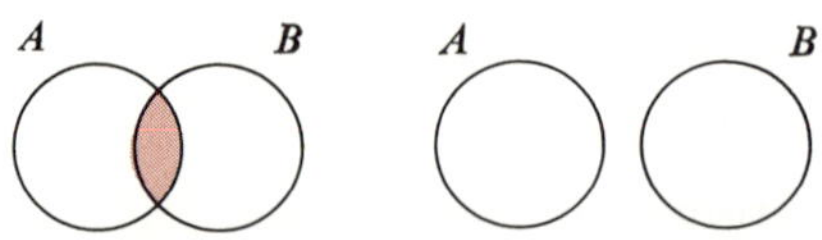

The intersection of sets A and B in the figure at left above is shown by the shaded region. In the figure on the right above, sets A and B are shown to be disjoint. Since their intersection is the empty set, there is no shading in the figure.

The union of the sets A and B can be shown in either of the following ways:

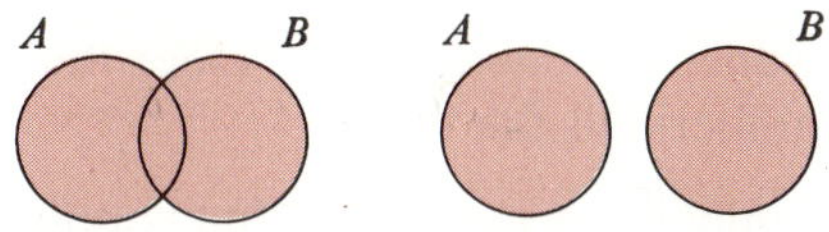

The set containing every element under consideration is called the **universal** set. We shall let U represent the universal set. To show a model of U, we use a closed region different from the regions used for other sets. We usually use a rectangular region for U, with circular regions inside it for its subsets. In the figure on the left below, set A is a subset of U. The figure at right below shows all the members of U that are not members of set A. This shaded set is called the *complement* of set A. We shall denote the complement of set A by A'.

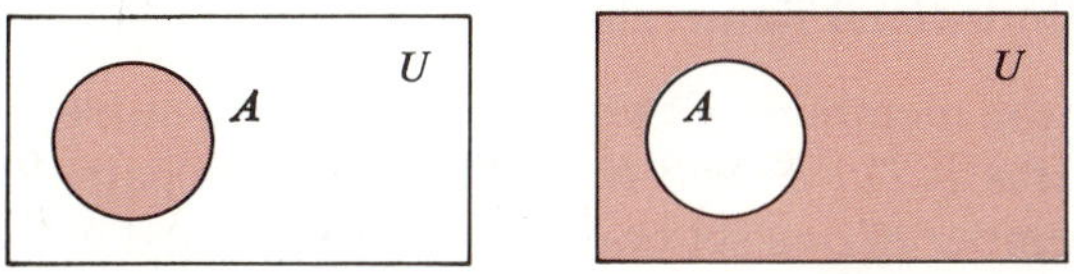

To show that a statement such as $A \cap B = B \cap A$ is true, we can construct a Venn diagram to represent each part of the statement and then show that the shaded regions are the same.

Use Venn diagrams to show that the following are true statements:

$$(A \cup B)' = A' \cap B' \qquad (A \cap B)' = A' \cup B'$$

These statements were first proved by Augustus De Morgan (1806–1871) and are therefore known as De Morgan's laws.

1-2 REAL NUMBERS

In the last section, we graphed counting numbers, whole numbers, and integers on a number line. Examine the following model. Do you see that there are points of the line that do not have integers corresponding to them? By using fractional parts, we can name some of these points with rational numbers.

Rational Numbers

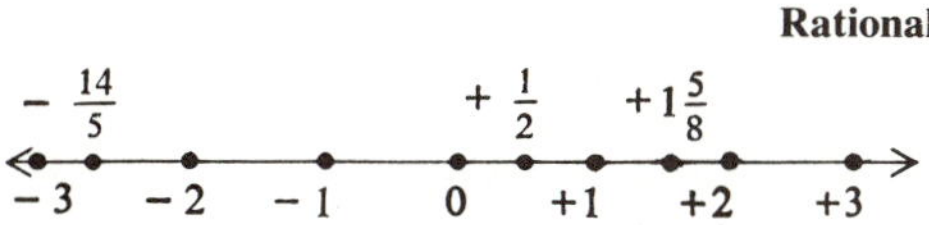

$\left\{\frac{a}{b}\text{: } a, b \text{ are integers; } b \neq 0\right\}$

The notation $\left\{\frac{a}{b}\text{: } a, b \text{ are integers; } b \neq 0\right\}$ is called **set-builder notation**. We use set-builder notation as our shorthand method of indicating sets. The notation above is read "the set of all numbers in the form $\frac{a}{b}$, such that a and b are integers and b does not equal zero." Does the set of points corresponding to the rational numbers fill the line?

There are some numbers we have used that cannot be represented by the ratio of two integers. Numbers such as $\sqrt{2}$, π, or 0.010010001 . . . are examples of numbers that do not fit the form $\frac{a}{b}$. These numbers belong to the set of **irrational numbers**, which together with the set of rational numbers forms the set of real numbers.

Definition 1-7 The set of **real numbers** is the set formed by the union of the set of rational numbers and the set of irrational numbers. All real numbers can be expressed with decimal numerals.

When we pair the members of two sets, we set up a correspondence between the two sets. If $A = \{a, b\}$ and $B = \{1, 2\}$, we can pair a with 1 and b with 2. When each member of one set corresponds to one and only one member of the second set, and each member of the second set corresponds to one and only one member of the first set, we say that there is a one-to-one correspondence between the sets. There is a one-to-one correspondence between the real numbers and the points of a line. To each real number there corresponds exactly one point of the line, and to each point of the line there corresponds exactly one real number. We say that the real numbers fill the line.

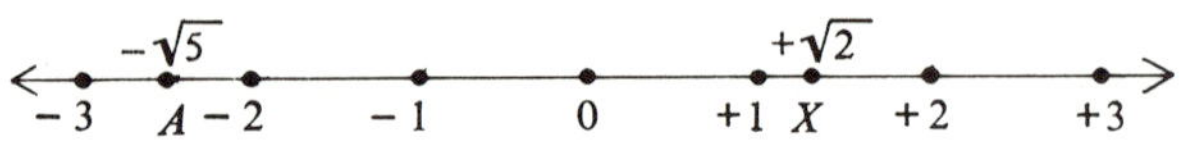

We use the entire line with all its points as a model for the real numbers. Each point is called the **graph** of the number associated with it. The number associated with the point is called the **coordinate** of the point. The point corresponding to 0 is called the **origin**. The point one unit to the right of the origin is the graph of the number 1. We can also say that 1 is the coordinate of the point one unit to the right of the origin. In the model above, the graph of $-\sqrt{5}$ is the point at A and the coordinate of the point at X is $\sqrt{2}$.

Order

Since the coordinates of the points are real numbers, we can use the line to show order. If a point lies to the left of a second point, then the coordinate of the first point is less than the coordinate of the second point. In the model above, $-\sqrt{5} < \sqrt{2}$, because A is to the left of X. We could also write $\sqrt{2} > -\sqrt{5}$. Since we are using the real numbers, the properties of real numbers you learned in algebra will apply to our work in geometry. The essential properties regarding equality and order are listed on page 605.

Absolute Value

The correspondence we have discussed so far is between the real numbers and the points of a line. There are many correspondences that are made between the real numbers and a given subset of the real numbers. Do you recall the correspondence called **absolute value**? The symbol we use for the absolute value of a number such as -5 is $|-5|$.

To find the absolute value of a real number, first determine if the number is positive, zero, or negative. If the number is positive or zero, the absolute value is the number itself. If the number is negative, the absolute value is the opposite of the number. Thus, $|-5| = 5$, $|-\sqrt{3}| = \sqrt{3}$, $|7| = 7$, $|-\pi| = \pi$, and $|0| = 0$. Formally, we write the following:

D **Definition 1-8** For $x \geq 0$, $|x| = x$; for $x < 0$, $|x| = -x$.

Example 1 Evaluate $|18 - 2|$.

Solution $|18 - 2| = |16|$. Since 16 is positive, the absolute value is positive. Therefore $|18 - 2| = 16$.

Example 2 Evaluate $|20 - 25|$.

Solution $|20 - 25| = |-5|$. Since -5 is negative, we use the opposite of -5. Therefore $|20 - 25| = -(-5) = 5$.

Example 3 Solve $|x - 1| = 3$.

Solution $x - 1$ could be positive or negative. (It is impossible for $x - 1$ to equal 0. Why?) If $x - 1$ is positive, then $|x - 1| = x - 1$. Thus, $x - 1 = 3$ or $x = 4$. If $x - 1$ is negative, then $|x - 1| = -(x - 1)$ and $-(x - 1) = 3$, or $x = -2$. The solution set is $\{-2, 4\}$.

Class Exercises

List the elements of each of these sets:

1. $\{x\colon\ x = 7\}$
2. $\{z\colon\ 2z = 25\}$
3. $\left\{p\colon\ \frac{p}{3} \text{ is an integer}\right\}$
4. $\left\{n\colon\ \frac{n}{2} \text{ is } \textit{not} \text{ an integer}\right\}$

Use set-builder notation to describe these sets:

5. $\{0, 1, 2, 3, 4\}$
6. the set of even integers
7. $\{5, 10, 15, 20, 25, \ldots\}$
8. $\{-2, -1, 0, +1, +2\}$

9. Order $-\frac{8}{7}$, $1\frac{1}{3}$, 0, -1, and 4.3 as they would appear on the number line.
10. Is the statement $-3.01 < -3 < -2.9$ true or false?

Evaluate each of the following:
11. $|16|$ 12. $|-22.5|$ 13. $|9-7|$ 14. $|11-7|$ 15. $|(-2)(-3)|$

Describe the solution set of each of the following:
16. $3x+1>16$ 17. $2x>3x+1$ 18. $|x|=2$ 19. $|x+\frac{1}{2}|=\frac{1}{2}$

Distance

Let the one-inch mark on a ruler correspond to the point of the number line whose coordinate is 1, when the left edge of the ruler corresponds to the origin. Can we assign a number to each point of the number line to represent the point's distance from the origin? Can this number be negative?

What is the distance from -5 to 0? Is it 5?
What is the distance from $-\frac{5}{2}$ to 0? Is it $\frac{5}{2}$?
What is the distance from 0 to 0? Is it 0?

Since distance cannot be negative, the absolute value of the coordinate of each point is the same as its distance from the origin. Thus, the distance from x to 0 is $|x|$. How do we find the distance between any two points? Is there only one number representing this distance? The following examples will help to answer these questions:

Example 4 Find the distance between the graphs of 5 and 3.

Solution Since the points on the number line are the same distance apart as the units on the ruler, we can use the ruler. The graph of 3 is to the left of the graph of 5, so place the edge of the ruler at the 3 and read the mark that corresponds to the 5 on the number line.

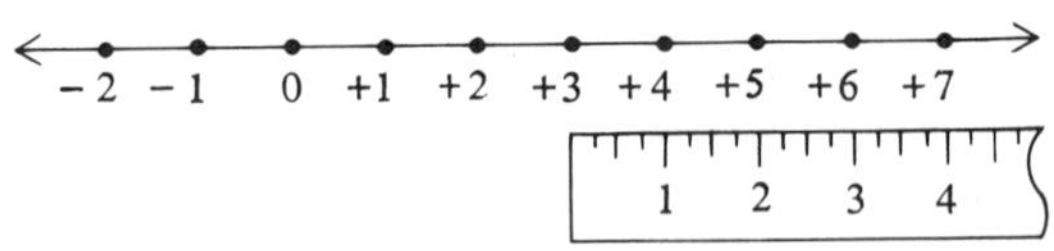

The distance is 2.
There is a second way of finding this distance.
We can use the difference between the coordinates of the points.
That is, $5-3=2$. Since distance is associated with absolute value, we can write $|5-3|=|2|=2$. The distance is 2.

Example 5 Find the distance between the graphs of -2 and 12.

Solution Using the first method, we place the edge of the ruler at -2. Since this point is to the left of the point 12, read the mark on the ruler that corresponds to the 12 on the number line. The distance is 14.

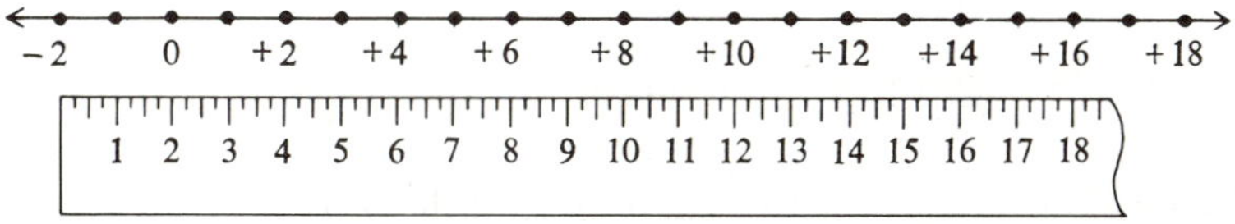

Using the second method, we find the absolute value of the difference between the coordinates. This difference is $|12 - (-2)| = |14|$. The distance is 14.

These examples lead us to make the following postulate and definition about distance. Postulates are statements which we assume but do not prove. We shall discuss postulates in greater detail in Chapter 2.

Postulate 1-1 **The Distance Postulate** To every pair of distinct points there corresponds a unique positive number.

Definition 1-9 The **distance** between two points A and B is the unique number corresponding to them. We denote the distance from A to B by AB.

Example 6 Using coordinates, find the distance between the graphs of points A and B, where A is -17 and B is -2.

Solution Use the absolute value of the difference between the coordinates.

$$\begin{aligned} AB &= |(-17) - (-2)| = |(-17) + 2| \\ &= |-15| \text{ or } 15 \end{aligned}$$

The distance is 15. Check this, using your ruler. Our discussion above leads to the following postulate:

Postulate 1-2 **The Ruler Postulate** A correspondence between the real numbers and the points of a line can be made such that:

1. to every real number there corresponds exactly one point of the line;
2. to every point of the line there corresponds exactly one real number;
3. the distance between any two points is the absolute value of the difference of the coordinates of the points.

The above information is useful in determining when one point of a line is between two other points of the line. If we have a line that passes through points A and B, and we know that there is a distinct point C of the line, how can we tell if C is between A and B? There are three possibilities for C:

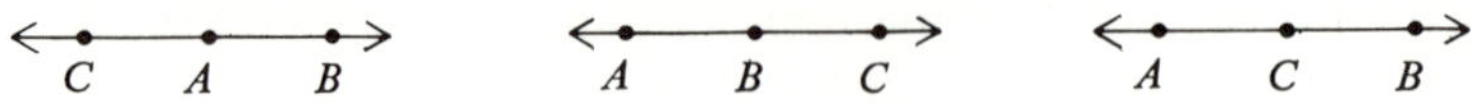

From the last figure, (C between A and B), we see that $AC + CB = AB$.

D

Definition 1-10 If points A, B, and C are distinct points of a line such that $AC + CB = AB$ then point C lies *between* points A and B.

EXERCISES

A

Use set-builder notation to describe these sets:

1. the whole numbers between 7 and 16
2. the even integers divisible by 5
3. the multiples of 5 greater than 25
4. the pairs of numbers whose sum is 17
5. any pair of numbers such that one is twice the other

Indicate which of the following are true and which are false:

6. If point A lies to the left of point B on the number line, then the coordinate of A is greater than the coordinate of B.
7. If $a > b$, then $a - 7 > b - 7$. 8. Since $5 > 2$, $5x > 2x$. 9. $|-7| = 7$
10. The distance between the graphs of (-5) and (3) is 2.
11. Since $AB + BC = AC$, and A, B, and C are distinct points of the line, B lies between A and C.
12. Order $-\frac{14}{5}$, 5, -4.1, 0, 2, $\frac{36}{6}$ as they would appear on the number line.

Evaluate each of the following:

13. $|8 + 2.1|$ 14. $-3 - |-2|$ 15. $|13| - |51|$
16. $-|(2)(-3)|$ 17. $|-3| + |2 - 3|$ 18. $|-2| \cdot |5|$

Solve each of the following:

19. $x - \frac{7}{2} \leq \frac{3}{2} + 2x$ 20. $2x - 3 < 5$ 21. $3x - 1 > 0$
22. $2(x - 1) > 8$ 23. $2(x - 3) < 5x + 1$ 24. $5x + 3 < x + 2$
25. $|2x| = 6$ 26. $|x - 1| = 14$ 27. $3|x - 5| = 9$
28. $|x| \geq 1$ 29. $|x| < 1$ 30. $2|x - \frac{1}{2}| = 3$

Name the property of equality or of order illustrated by each of the following:

31. If $2x = 6$, then $6 = 2x$. 32. Either $x = 6$ or $x \neq 6$.
33. If $3 > -2$ and $-2 > -5$, then $3 > -5$. 34. If $2x = 6$, then $\frac{1}{2}(2x) = \frac{1}{2}(6)$.
35. $AB = 0$, $AB < 0$, or $AB > 0$ 36. $XY = XY$
37. If $PQ - 7 = 0$, then $PQ - 7 + 7 = 0 + 7$. 38. If $2x > 4$, then $x > 2$.

B Point Q lies between points P and R of a line.

39. If $PR = 12$ and $QR = 3$, find PQ.
40. If $PQ = 5$ and $PR = 8$, find QR.
41. If $PQ = 8$ and $QR = 9$, find PR.
42. If $PR = 7$ and $QP = 2$, find QR.
43. If P, Q, and R lie on a line such that $PR = 12$ and $PQ = 15$, does Q lie between P and R?
44. M has the coordinate -3. Name the coordinate(s) of the point(s) of a line which are a distance of 7 from point M of the line.
45. If $x \ngtr 3$, what can you conclude?
46. If $3x > 2y$ and $2y > 28$, can you conclude that $x > \frac{28}{3}$?
47. Is it true that if $3 < 12$, then $3x < 12x$ when x is positive?

Solve each of the following:

48. $\frac{5}{2}(x - 1) \leq \frac{7}{2}(x + 1)$
49. $5(x - \frac{7}{5}) \leq 2(x + \frac{11}{2})$
50. $2|x - 1| < 5$
51. $5|x - \frac{1}{2}| < 11$

1-3 BASIC TERMS

"A square is a four-sided convex polygon with square corners." This definition attempts to define a square in terms of **square**. We call this a *circular definition*. If we are to know the meanings of words, we cannot use circular definitions. The definition above uses terms which have not been previously defined or accepted. We cannot understand a definition if we do not understand its terms. The above definition is a poor definition.

Definitions are vital to the organization of a logical system. They supply meaning to concepts. To be useful, however, they must be understandable. We shall discuss definitions in more detail in Chapter 2.

Point, *line*, and *plane* are terms basic to any course of geometry. We shall not define these words. Their meaning should be readily understood. Examine the following figures carefully. We shall use them to illustrate many basic terms. A more thorough study of these concepts is included in later chapters.

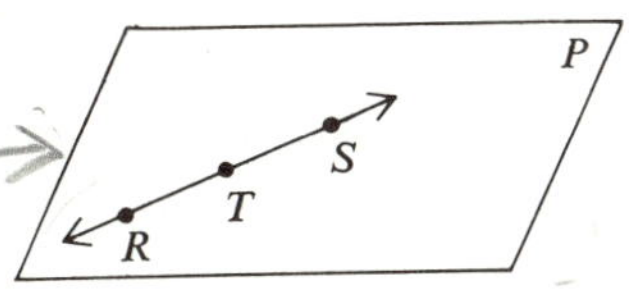

Horizontal Plane P

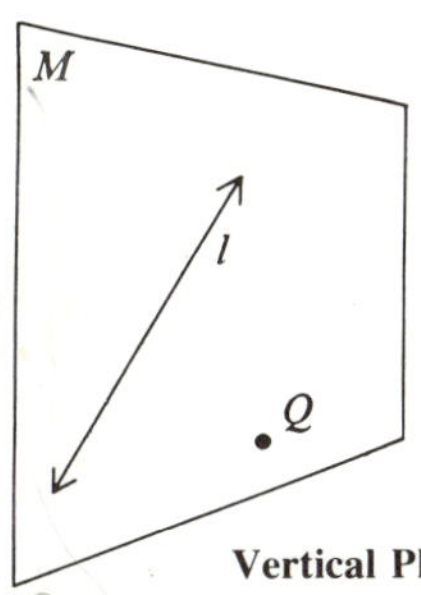

Vertical Plane M

It is customary to refer to points with capital letters and to represent them with dots. A line is named by referring to two points of the line, for example, points R and S of $\overleftrightarrow{RS}$. The symbol $\overleftrightarrow{RS}$ is read "line RS." We may also use a lower case letter to name a line. In the vertical plane, we see line l. There are several ways of referring to a plane, but for the present, we shall name a plane with a single capital letter preceded by the word "plane." In the figures above, we see models representing a horizontal plane P and a vertical plane M. In the models, point T is on $\overleftrightarrow{RS}$, $\overleftrightarrow{RS}$ is contained in plane P, line l is contained in plane M, and point Q lies in plane M but not on l.

Most sciences have their sets of **undefined terms**. Those are the terms that are not defined, but whose meaning is generally agreed upon. Can you give a definition of addition? What is time? Using these terms, we can define other terms. Then more terms are defined using the undefined terms and the terms already defined. We can now, for example, use the undefined terms *point* and *line* to give a definition of *collinear points*.

D **Definition 1-11** A set of points is **collinear** if all the points lie in a line.

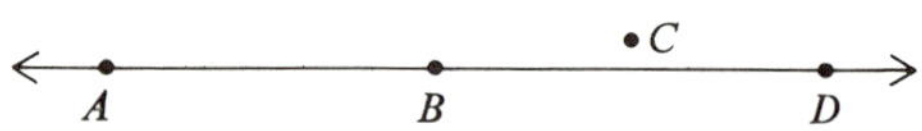

Points A, B, and D are collinear; they lie on $\overleftrightarrow{AD}$. Can we say they lie on $\overleftrightarrow{AB}$? How many lines can be drawn containing point A? How many lines can be drawn containing both points A and B? Are two points all that are needed to determine a line? "To determine a line" means to specify exactly one line. Points A, B, and C above do not all lie on the same line. Do they all lie in the same plane? In how many different planes can the three noncollinear points lie? When two points of a line lie in a plane, does the entire line lie in the plane?

D **Definition 1-12** A set of points is **coplanar** if all the points lie in the same plane.

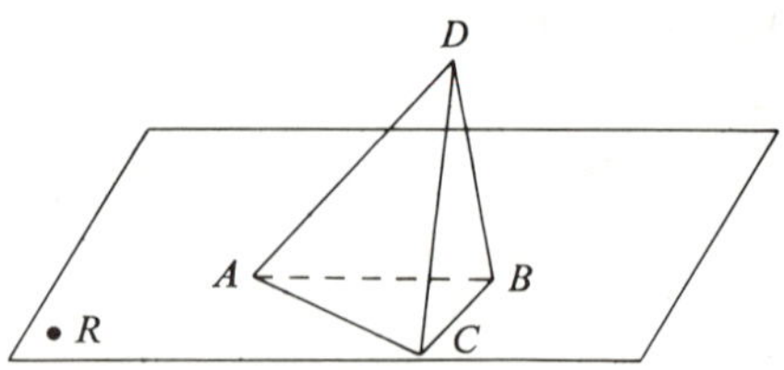

In the model at right, how many planes can you name? Name the coplanar points in plane R. Are points A, B, C, and D coplanar?

For the figure under Definition 1-11, describe the figure formed by the points A

and B and all the points between them. The subset you described is an example of a line segment.

Definition 1-13 The **line segment** AB is the set of points consisting of A, B, and all points between them. We denote line segment AB by $\overline{AB}$.

Class Exercises

1. Give an example of a point belonging to a line $\overleftrightarrow{PQ}$.
2. Describe the relationship between a line and the plane in which the line lies.
3. Does the relationship described in Exercise 2 apply to the relationship between a line segment and the line of which it is a part?
4. Are any two points collinear?
5. Are any four points coplanar?
6. Are any three points coplanar?
7. Are two intersecting lines coplanar?

In our discussion of the distance between two points, we said that AB represents the distance between A and B. This distance is also the length or measure of $\overline{AB}$.

Definition 1-14 AB is the **measure** of the distance between A and B. We denote the measure of $\overline{AB}$ by AB.

The point that divides a segment into two segments of equal measure is the midpoint of the segment. In the figure at the right, C is the midpoint of $\overline{AB}$. How many midpoints can a segment have?

Definition 1-15 Point C is the **midpoint** of $\overline{AB}$ if C lies between A and B, and $AC = CB$.

Segments with equal measures are said to be congruent segments. In the figure above, $\overline{AC}$ is congruent to $\overline{CB}$.

Definition 1-16 **Congruent segments** are segments with equal measures. To indicate that $\overline{AC}$ is congruent to $\overline{CD}$, we write $\overline{AC} \cong \overline{CD}$.

Since $\overline{AB} \cong \overline{CD}$ and $AB = CD$ are equivalent statements, we may replace one with the other whenever convenient. Remember, however, that AB and CD are the measures of segments and therefore are numbers. $\overline{AB}$ and $\overline{CD}$ are segments. The numbers AB and CD can be used in computation just as any other numbers are used in computation.

Example Find point C between A and B below such that $\overline{AC} \cong \overline{CB}$.

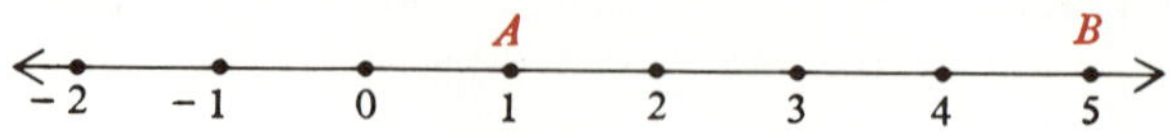

Solution If C is to be between A and B, C must be on $\overline{AB}$ such that $AC + CB = AB$. We want point C such that $AC = CB$. Thus,

$$2(AC) = AB$$

Therefore $$AC = \tfrac{1}{2}(AB)$$

The coordinates of A and B are 1 and 5. Thus,

$$AB = |5 - 1|$$
$$AB = 4$$

Substituting 4 for AB, we have

$$AC = \tfrac{1}{2}(4)$$
$$AC = 2$$

This means that C lies 2 units from A in the direction of B. Why? The coordinate of C is 3.

Do you think that a line can have a midpoint? Since any line is composed of an infinite number of points, no matter which point of the line we choose, the same number of points of the line are on each side of the chosen point.

P **Postulate 1-3 The Line Separation Postulate** Any point P of a line separates the line into two distinct sets of points, one set on each side of P.

D **Definition 1-17** The sets of points described in the line separation postulate are **half-lines**, with P as the endpoint of each half-line. We denote the half-line from P through B by $\overset{\circ}{\overrightarrow{PB}}$

In the figure above, the other half-line formed by P is $\overset{\circ}{\overrightarrow{PA}}$. If we consider the union of the half-line and its endpoint, we have a ray. Thus, the union of half-line PA and point P above is ray PA.

D **Definition 1-18** A **ray** is the union of a half-line and its endpoint. We denote the ray PQ by $\overrightarrow{PQ}$.

The endpoint of a ray is often referred to as the **vertex** of the ray. Choose any point of a line. The half-lines formed can each be considered with the endpoint. Thus, two rays are formed. The rays are called **opposite rays**. Opposite rays form a line. What other figures can rays form?

EXERCISES

A

Match each symbol in the left column with the proper name from the right column:

1. AB — *a*. half-line AB
2. $\overline{AB}$ — *b*. line AB
3. $\overset{\circ\!\rightarrow}{AB}$ — *c*. segment AB
4. $\overrightarrow{AB}$ — *d*. measure of segment AB
5. $\overleftrightarrow{AB}$ — *e*. ray AB

Using the accompanying three-dimensional model, determine whether each of the following sets is (*a*) collinear and/or (*b*) coplanar:

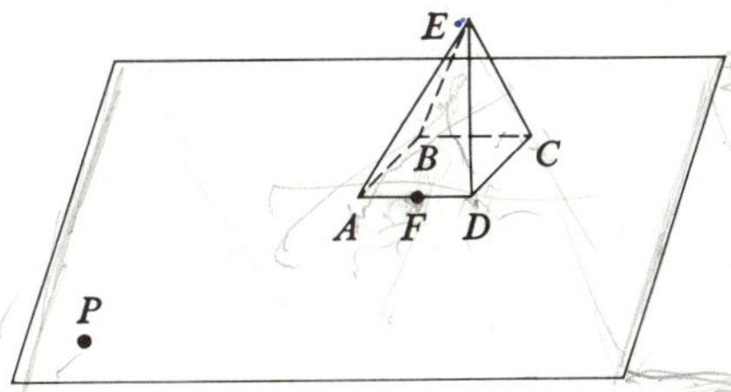

6. $\{A, F, D\}$
7. $\{P, A, F\}$
8. $\{A, B, C, E\}$
9. $\{A, F, D, E\}$

Describe each of the following sets, using the illustration above:

10. $\overline{AF} \cap$ plane P
11. $\overrightarrow{AD} \cap \overrightarrow{AB}$
12. $\overline{AF} \cup \overline{FD}$
13. $E \cap \overleftrightarrow{AD}$
14. $B \cup \overset{\circ\!\rightarrow}{BE}$
15. $\overset{\circ\!\rightarrow}{DE} \cap \overrightarrow{DC}$

B

16. How many lines are determined by two distinct points?
17. Is it proper to talk about the length of $\overleftrightarrow{AB}$? of $\overrightarrow{AB}$? of AB?

Indicate whether each of the following is true or false:

18. If points A and B lie in a plane, then all points of $\overleftrightarrow{AB}$ lie in the plane.
19. If $\overleftrightarrow{RA}$ and $\overleftrightarrow{RB}$ lie in one plane, then R, A, and B are collinear.
20. Three noncollinear points are also noncoplanar.
21. Rays with common vertices are opposite rays.
22. If $AC = CB$, then A, B, and C are collinear.
23. The measure PQ is a real number.

C

24. A point of a line separates the line into two half-lines. Extend this idea to describe the separation of a plane by a line.
25. Name a geometric figure that is formed by three noncollinear points. Into how many parts does this figure divide the plane in which it lies?
26. Name eight geometric figures formed by a pair of intersecting lines.

1-4 ANGLES

If two rays share a common endpoint, must they be opposite rays? What other kind of figure can they form? Here are some examples:

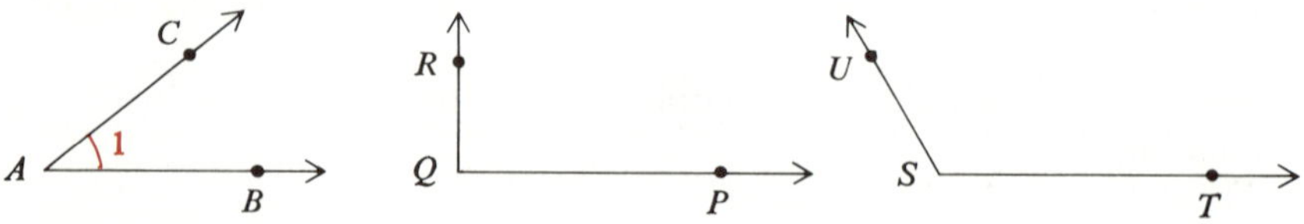

We recognize these figures as angles. From the figures, we see that we can define an angle in terms of the rays that form the angle.

D **Definition 1-19** An **angle** is the union of two noncollinear rays with a common endpoint. The common endpoint of the rays is the vertex of the angle. The two rays are the sides of the angle.

In the figure at left above, we can describe the angle formed by rays $\overrightarrow{AB}$ and $\overrightarrow{AC}$ as $\angle BAC$, $\angle CAB$, or $\angle 1$. We use the symbol $\angle$ for *angle*. When we use three letters to name an angle, we place the vertex letter in the middle position. If there is no danger of confusion, we can denote the angle by using the name of the vertex. Thus, $\angle CAB$ can also be named $\angle A$.

How many angles in the figure on the right have N as their vertex? Name $\angle 1$ using three letters.

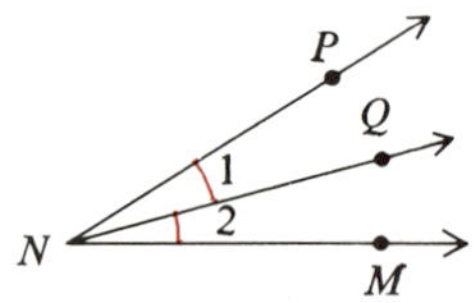

Any line in a plane separates the plane into two *half-planes*. If the half-plane includes the separating line, or edge of the half-plane, we say this is a closed half-plane.

P **Postulate 1-4 The Plane Separation Postulate** Any line m of plane P separates the points of P that are not in m into two regions such that:

1. for any two points A and B of a region, $\overline{AB}$ lies entirely in that region;
2. if A is in one region and B is in the other region, then $\overline{AB}$ intersects m.

D **Definition 1-20** The regions described in the Plane Separation Postulate are **half-planes**, with m as the edge of each half-plane.

In the model at right, $\overleftrightarrow{MN}$ is the edge of the half-plane containing P and the opposite half-plane containing R. Describe the half-planes formed by $\overleftrightarrow{RP}$.

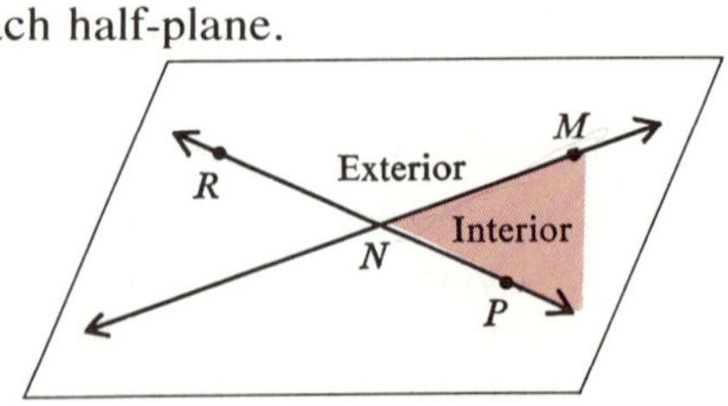

The intersection of the half-plane of $\overleftrightarrow{MN}$ containing P and the half-plane of $\overleftrightarrow{PN}$ containing M is the interior of $\angle MNP$. All other points of the plane lie either in the exterior of $\angle MNP$ or on $\angle MNP$.

Example 1 Is the vertex of an angle in the interior of the angle?

Solution The vertex of the angle lies on both of the sides of the angle. Each side lies on an edge of the half-planes that form the interior of an angle. Thus, the vertex is not in the interior of the angle.

Angle Measure

When we discussed lines and segments, we used a ruler as our measuring device. Can we use the ruler to measure angles? Angles differ in the amount of separation between their rays. The unit we use to measure this separation is the degree. The measuring device we use to determine the number of degrees in an angle is the protractor. The protractor uses a semicircular scale with units from 0 to 180. Usually the scale is marked in both directions.

P **Postulate 1-5 The Angle Measure Postulate** To every angle there corresponds a real number between, but not including, 0 and 180.

D **Definition 1-21** The **measure of an angle** is the number given in the Angle Measure Postulate. We denote the measure of $\angle A$ by $m\angle A$.

Example 2 Find the measure of $\angle ABC$.

Solution Place your protractor over $\overrightarrow{BA}$, with the center mark of the protractor over the vertex B and the straight line along $\overrightarrow{BA}$. On the semicircular scale, read the number under which $\overrightarrow{BC}$ passes. We see that there are 40° ("40 degrees") in $\angle ABC$. The measure is 40. We write $m\angle ABC = 40$.

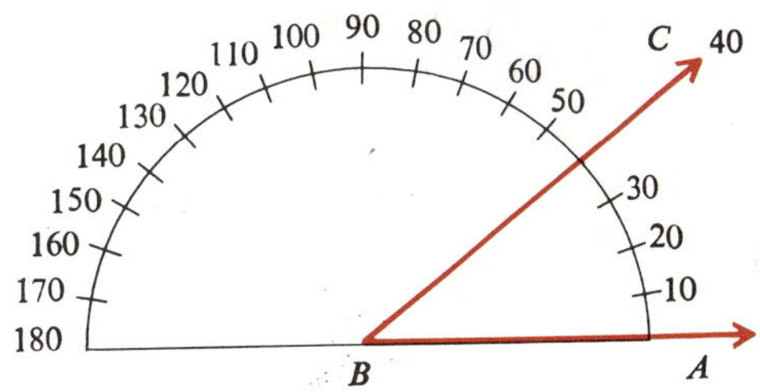

Example 3 Find the measure of each angle having O as its vertex.

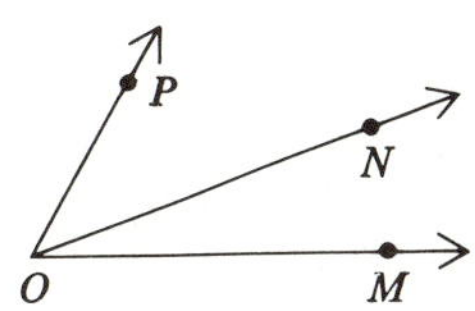

Solution There are three angles with vertex O. They are $\angle MON$, $\angle NOP$, and $\angle MOP$. Use your protractor to check each of the following:

$m\angle MON = 20$ $\quad$ $m\angle NOP = 40$ $\quad$ $m\angle MOP = 60$

In Example 3, point N and $\overrightarrow{ON}$ lie in the interior of $\angle MOP$. We see that $m\angle MOP = m\angle MON + m\angle NOP$, that is, $60 = 20 + 40$. This suggests the following definition for angle addition:

D **Definition 1-22** If point N is in the interior of $\angle MOP$, then $m\angle MOP = m\angle MON + m\angle NOP$.

Class Exercises

Determine which of the following are true:

1. The intersection of two rays is a point.
2. A ray separates a plane into two half-planes.
3. A half-plane does not include the edge.
4. The sides of $\angle ABC$ are $\overrightarrow{AB}$ and $\overrightarrow{BC}$.
5. Name all of the angles in the model having O as vertex.
6. Describe the relationship between $\angle AOB$ and $\angle 1$.
7. Name another angle that shares the common side $\overrightarrow{OB}$ with $\angle BOC$.
8. Which points of the set $\{A,B,C,D,E,O\}$ lie in the interior of $\angle AOC$?
9. Describe the closed half-plane containing the points B and C.
10. Does $\overrightarrow{OB}$ lie in the interior of $\angle AOC$?

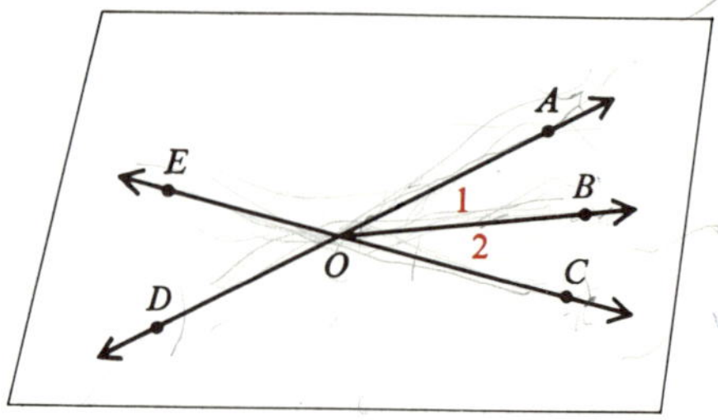

Use your protractor to find the measure of each of the angles of the following triangle:

11. $m\angle XYZ =$ ____
12. $m\angle YXZ =$ ____
13. $m\angle XZY =$ ____

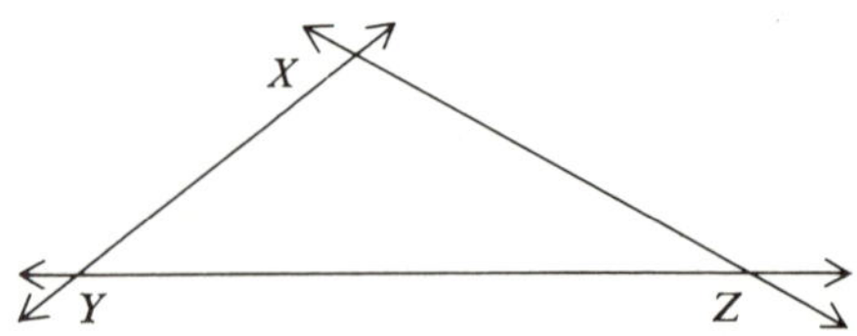

Angle Relationships

Consider the pair of angles PQR and RQS. The angles share the same vertex and share the common side $\overrightarrow{QR}$. Now consider the pair of angles PQS and PQR. These two pairs are alike in that the angles in each have a common

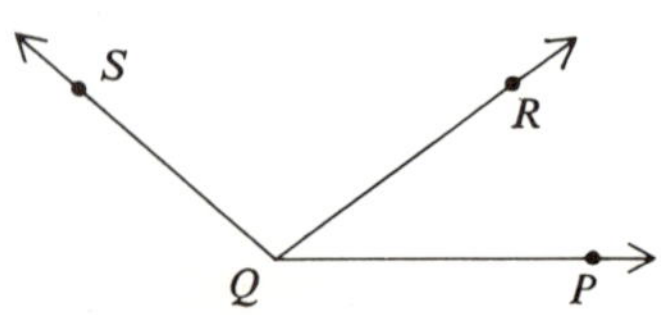

vertex and a common side, but they also differ in one respect. Can you tell how they differ? The first pair shows an example of adjacent angles.

(D) **Definition 1-23** **Adjacent angles** are two angles with a common vertex and a common side, but no common interior points.

We classify angles by their measure. The three types of angles we have seen so far are acute, obtuse, and right angles.

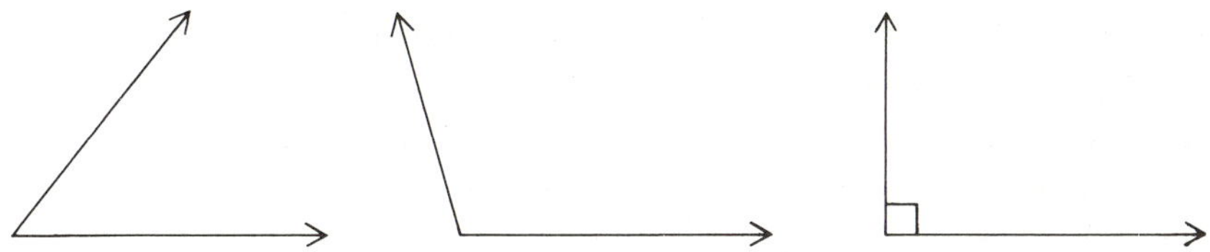

Acute Angle **Obtuse Angle** **Right Angle**

We use the small square ∟ at the vertex of an angle to indicate that it is a right angle.

(D) **Definition 1-24** An angle whose measure is between 0 and 90 is called an **acute angle**.
An angle whose measure is between 90 and 180 is called an **obtuse angle**.
An angle whose measure is 90 is called a **right angle**.

(D) **Definition 1-25** If when $\overleftrightarrow{AB}$ and $\overleftrightarrow{AC}$ intersect, they form adjacent angles of equal measure then the lines are **perpendicular**. To indicate that $\overleftrightarrow{AB}$ is perpendicular to $\overleftrightarrow{AC}$ we write $\overleftrightarrow{AB} \perp \overleftrightarrow{AC}$.

(D) **Definition 1-26** If the two noncommon sides of adjacent angles are collinear—that is, form opposite rays—we say that the angles are a **linear pair**.

In the figure below, the adjacent angles, $\angle 1$ and $\angle 2$, form a linear pair. Traditionally, we refer to a linear pair as forming a **straight angle**.

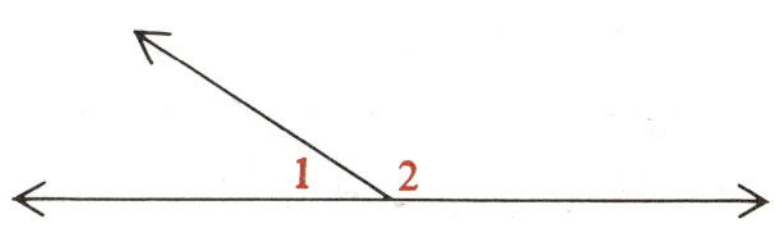

(D) **Definition 1-27** If the sum of the measures of two angles is 90, the angles are **complementary angles**. Each angle is the complement of the other.

Example 4 If $m\angle B = 10$, find the measure of the angle complementary to $\angle B$.

Solution Let the complement of $\angle B$ be labeled $\angle X$. Since the angles are complementary, $m\angle B + m\angle X = 90$. Substituting 10 for $m\angle B$, we have

$$10 + m\angle X = 90$$

Thus,

$$m\angle X = 80$$

D **Definition 1-28** If the sum of the measures of two angles is 180, the angles are **supplementary angles**. Each angle is the supplement of the other.

The definition of supplementary angles and our notion of a linear pair lead to the next postulate:

P **Postulate 1-6 The Supplementary Angles Postulate** Two angles that form a linear pair are supplementary.

Not every pair of supplementary angles forms a linear pair; the angles must also be adjacent.

Example 5 If $m\angle 1 = 50$, find $m\angle 2$ so that $\angle 1$ and $\angle 2$ are a linear pair.

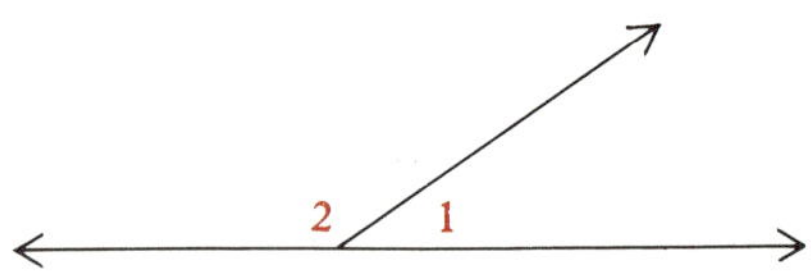

Solution The angles of a linear pair are supplementary, that is, their measures total 180. Thus, $m\angle 1 + m\angle 2 = 180$. Since $m\angle 1 = 50$ we write

$$50 + m\angle 2 = 180$$

or

$$m\angle 2 = 130$$

When a ray shares its vertex with an angle and lies in the interior of the angle so that two angles with equal measure are formed, the ray bisects the angle.

D **Definition 1-29** If point P lies in the interior of $\angle BAC$, such that $m\angle BAP = m\angle PAC$, then $\overrightarrow{AP}$ is the **bisector** of $\angle BAC$.

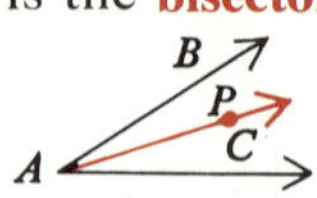

What we said about segments with equal measures, we can say about angles with equal measures:

D

Definition 1-30 **Congruent angles** are angles of equal measure. To indicate that $\angle ABC$ is congruent to $\angle DEF$, we write $\angle ABC \cong \angle DEF$.

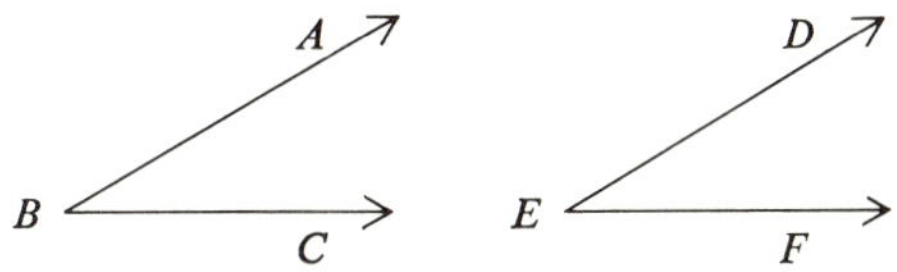

Since $m\angle ABC = m\angle DEF$ and $\angle ABC \cong \angle DEF$ are equivalent statements, we can replace one with the other whenever it is convenient. Remember, however, that $m\angle ABC$ and $m\angle DEF$ are the measures of angles and therefore are numbers. $\angle ABC$ and $\angle DEF$ are angles. The numbers $m\angle ABC$ and $m\angle DEF$ can be used in computation just like any other numbers.

Example 6 Classify the angles formed by a bisector of a straight angle as acute, right, or obtuse.

Solution We said earlier that we refer to a linear pair as forming a straight angle. Since a linear pair is formed by supplementary angles, the bisector of the straight angle also forms supplementary angles. If we label the angles formed by the bisector as $\angle 1$ and $\angle 2$, we can write

$$m\angle 1 + m\angle 2 = 180$$

These angles, however, are congruent. Thus, we can write

$$2(m\angle 1) = 180$$

or

$$m\angle 1 = 90$$

This means that the two angles formed are right angles.

EXERCISES

A

Use the model at right for Exercises 1–10. $\angle POS$ is a right angle.

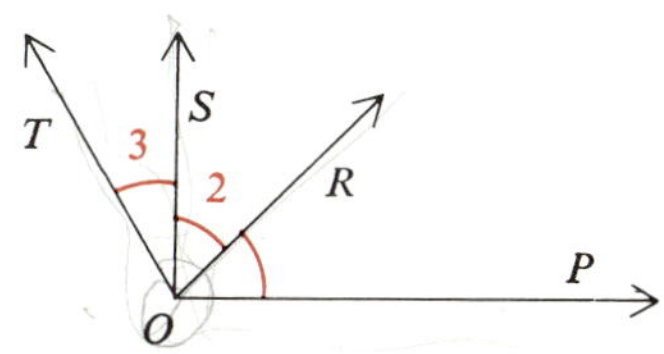

Classify each of the following as acute, obtuse, or right:

1. $\angle POR$ 2. $\angle POS$ 3. $\angle POT$
4. $\angle ROS$ 5. $\angle TOP$ 6. $\angle SOT$
7. Name the two angles for which $\overrightarrow{OS}$ is in the interior.
8. $\overrightarrow{OS}$ is in the exterior of what angle?
9. Name an angle complementary to $\angle ROS$.
10. If $m\angle ROS = \frac{1}{2}m\angle POR$ and $m\angle TOS = \frac{1}{2}m\angle ROS$, find $m\angle TOP$.

Name the degree measure of each of the following:

11. $\frac{2}{3}$ of a right angle
12. $1\frac{2}{3}$ of a right angle
13. two right angles
14. $\frac{1}{2}$ of a right angle
15. a pair of complementary angles
16. a linear pair

Using your protractor, determine the measure of each of the following angles:

17. 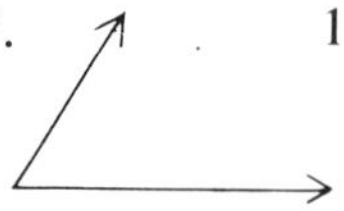

18. 

19. 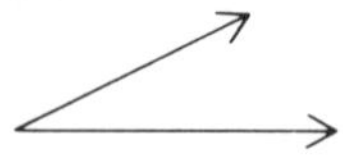

20. 

Given the following angle degree measures, classify each angle:

21. 38 22. 89 23. 100 24. 152 25. 90 26. 170

Using your protractor, form an angle for each of the following:

27. 54° 28. 72° 29. 85° 30. 142° 31. 115° 32. 38°

Determine the measure of the complement of each of the following angles:

33. 55° 34. 72° 35. 15°

Determine the measure of the supplement of each of the following angles:

36. 55° 37. 172° 38. 8° 39. 90°

B

40. If $\angle ABD$ and $\angle ABC$ form a linear pair and $m\angle ABD = 75$, find $m\angle ABC$.

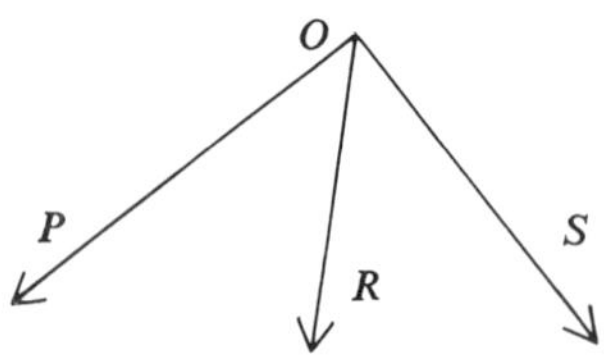

Exercise 41

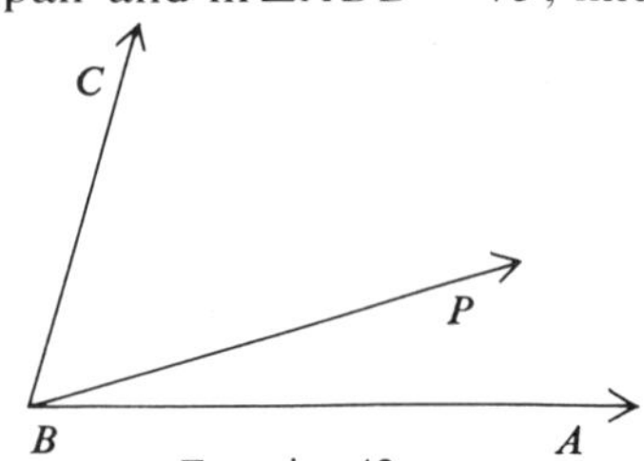

Exercise 42

41. If $\angle POR$ and $\angle ROS$ are complementary angles and $\overrightarrow{OR}$ bisects $\angle POS$, what is $m\angle POR$?
42. If P is in the interior of $\angle ABC$, if $m\angle ABC = 75$, and if $m\angle ABP = 17$, find $m\angle PBC$.

Using your protractor, determine the measure of each angle:

43. 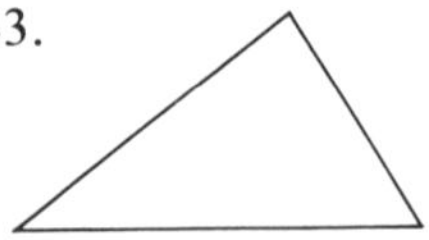

44. 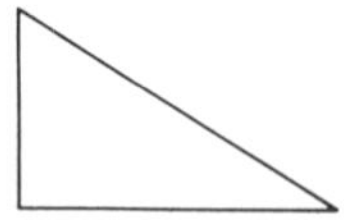

45. 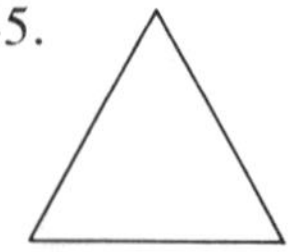

Classify the angle supplementary to each of the following angles:

46. a right angle 47. an obtuse angle 48. an acute angle

Describe the angles resulting from bisecting each of the following:

49. an obtuse angle 50. a right angle 51. an acute angle

C

52. Two angles are complementary and one is twice the measure of the other. Find the measures of these angles.
53. Two angles are complementary and the measure of one angle is 36 less than the other. Find the measures of the two angles.
54. The supplement of an angle exceeds five times the measure of the angle by 6. Find the measures of the supplementary angles.
55. Two congruent angles form a linear pair. Find the measures of the angles.
56. The supplement of an angle exceeds five times the complement of the angle by 10. Find the measure of the angle.

A LOOK AT THE PAST

Euclid's Elements

The first geometric concepts evolved as ancient civilizations attempted to explain the world in which they lived. In the records of the Egyptian and Babylonian civilizations, we find many references to geometric ideas. The Greeks, however, were the first people to make an extensive and well-organized study of the properties of geometry.

The geometry that we are studying was organized by the Greek mathematician Euclid (about 300 B.C.). His thoughts were set down in *The Elements*, a series of 13 books dealing with plane and solid geometry. In *The Elements*, Euclid included conclusions that had been discovered and developed by earlier Greek schools, such as the Pythagorean School and the Academy of Plato, as well as by individual Greek mathematicians such as Thales. Much of the content had been refined over a period of several hundred years before Euclid. Euclid's major contribution is the simplicity of his organization. He arranged all the concepts in a logical order, showing how each one could be deduced from a basic set of definitions and assumptions.

1-5 POLYGONS

We have briefly discussed the line and the angle. In this section we shall informally consider other plane figures called **polygons**. Polygons are figures formed by joining segments at their endpoints, if the segments do not intersect at any other points. The segments become the sides of the polygon. In Chapter 6 we shall formally define *polygon*.

We classify polygons according to the number of sides. An n-sided polygon is referred to as an **n-gon**. The more frequently used polygons, however, have special names. You probably learned these names for polygons of 3, 4, 5, 6, 7, 8, 9, 10, and 12 sides. Recall these names now.

What is the least number of sides a polygon can have? It is this polygon that we shall now discuss.

D **Definition 1-31** A **triangle** is a three-sided polygon.

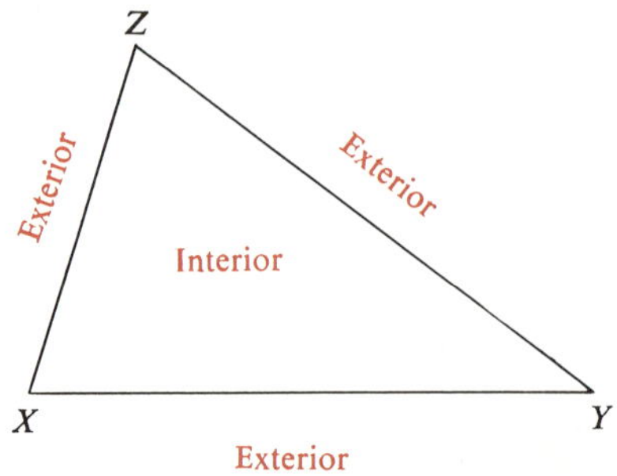

A triangle has three sides and three vertices. In the figure at the right, the three noncollinear points X, Y, and Z form $\overline{XY}$, $\overline{XZ}$, and $\overline{YZ}$. The union of these segments forms triangle XYZ. We denote triangle XYZ by $\triangle XYZ$. $\overline{XY}$, $\overline{XZ}$, and $\overline{YZ}$ are the sides of the triangle. The points X, Y, and Z are the vertices. The angles of the triangle are $\angle X$, $\angle Y$, and $\angle Z$.

The triangle separates the remaining points of the plane into two distinct sets of points: the set of points *inside* the triangle and the set of points *outside* the triangle. The set of points inside forms the **interior** of the triangle, and the set of points outside forms the **exterior** of the triangle. The interior of a triangle is the intersection of the interior of any two angles of the triangle.

Example 1 Name all the triangles in the figure at right.

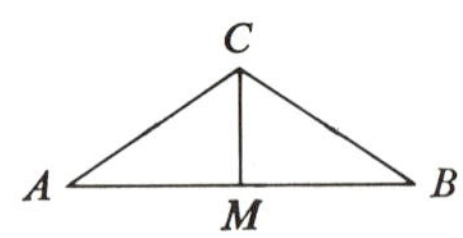

Solution Any three noncollinear points determine a triangle. We select any two and use all the possible third points. If we pick A and C we form $\triangle ACM$ and $\triangle ACB$. If we pick C and B we have $\triangle CBM$. We do not list $\triangle CBA$. If we select any other two points, we will repeat triangles that we have named. Why?

If point M in Example 1 is the midpoint of $\overline{AB}$, then $\overline{CM}$ is called a **median** of $\triangle ABC$. We shall discuss the median in greater detail in a later chapter. For now, it is sufficient to say that a median is the segment connecting a vertex of a triangle to the midpoint of the opposite side. In the figure above, a median is the segment from the vertex C to point M of $\overline{AB}$. If D is the midpoint of $\overline{CB}$ then a second median of $\triangle ABC$ is $\overline{AD}$ from the vertex A to $\overline{CB}$. Assume that N is the midpoint of $\overline{AC}$; name the third median of $\triangle ABC$. Since there is only one midpoint for each side, how many medians do you think a triangle has?

Classification of Triangles

If all three measures of the sides of a triangle are different, the triangle is **scalene.** If only two sides are of equal measure the triangle is **isosceles.** If all three sides are of equal measure, the triangle is **equilateral.**

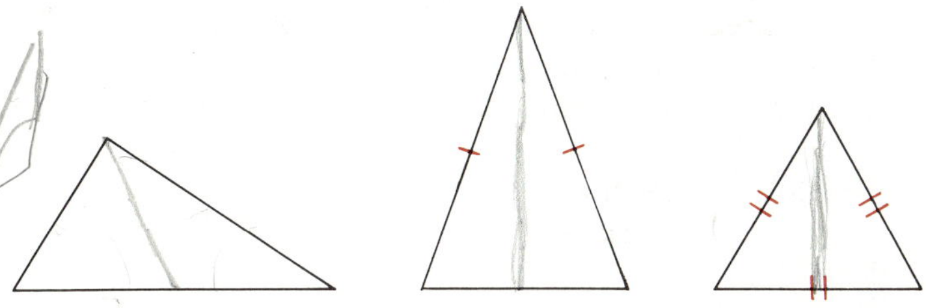

Scalene Isosceles Equilateral

Slash marks on the sides in the figures indicate sides that are equal in measure.

Example 2 In the model at the right, classify the triangles according to their sides. Congruent segments are marked.

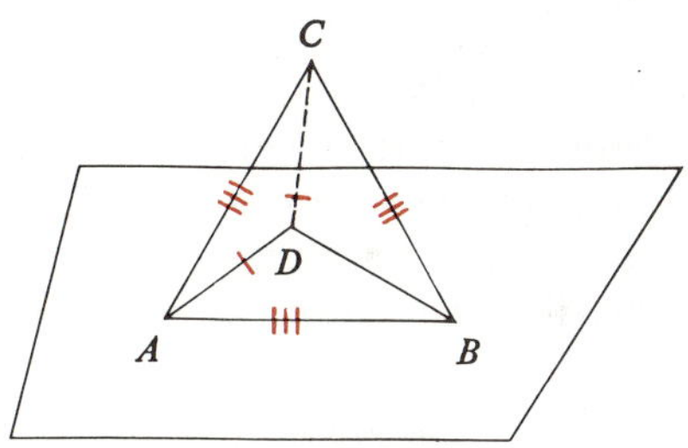

Solution
There are four triangles formed: $\triangle ABC$, $\triangle ADC$, $\triangle ADB$, and $\triangle CDB$.

$AB = BC = CA$	$\triangle ABC$ is equilateral.
$AD = DC \neq AC$	$\triangle ADC$ is isosceles.
$AD \neq DB \neq BA$	$\triangle ADB$ is scalene.
$CD \neq DB \neq BC$	$\triangle CDB$ is scalene.

Triangles can also be classified by the types of angles they contain.

D **Definition 1-32** An **acute triangle** is a triangle with three acute angles.
An **obtuse triangle** is a triangle with one obtuse angle.
A **right triangle** is a triangle with one right angle.

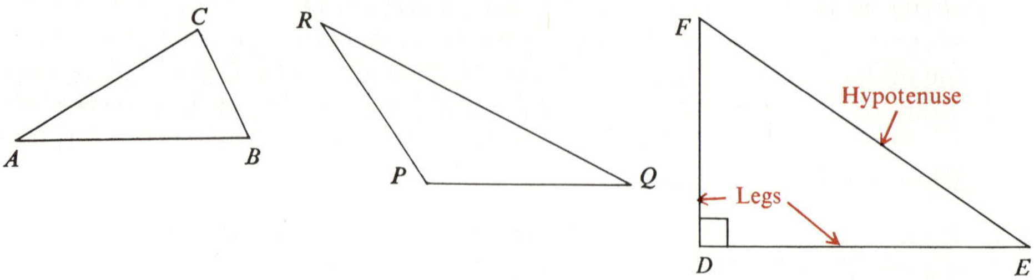

In the figures above, $\triangle ABC$ is an acute triangle, $\triangle PQR$ is an obtuse triangle, and $\triangle FDE$ is a right triangle. In $\triangle FDE$, the sides of the right angle, $\overline{DF}$ and $\overline{DE}$, are the **legs** of the triangle. The side opposite the right angle, $\overline{FE}$, is the **hypotenuse** of the right triangle.

If the three angles of a triangle are the same measure, the triangle is **equiangular**.

Class Exercises

Determine which of the following are polygons and which are not:

1. 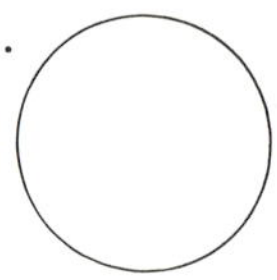
2. 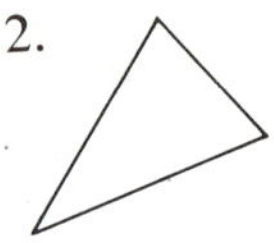
3. 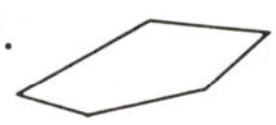
4. 
5. 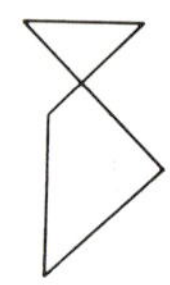

6. Name the triangles in the model at right.
7. Classify the triangles ADB, AOD, and AOB in Exercise 6, first according to their sides and then according to their angles.

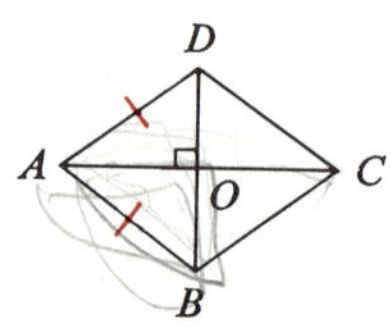

8. If A, B, and C are the midpoints of the sides of $\triangle PQR$, name the three medians.

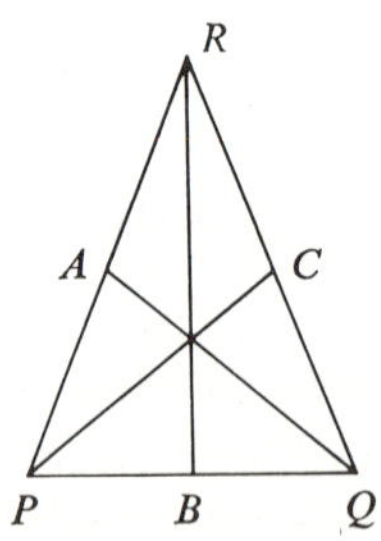

9. In $\triangle ABC$, K, M, and N are the midpoints of the sides. The segments $\overline{MN}$, $\overline{KN}$, and $\overline{MK}$ form four smaller triangles. They also help form six quadrilaterals, for example $AKNM$. Name the other five quadrilaterals.

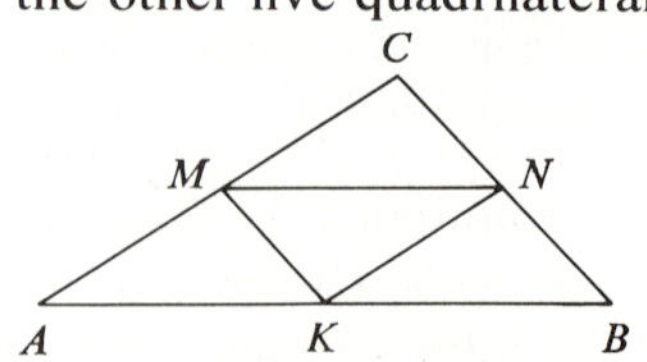

Quadrilaterals

The triangle is the polygon with the smallest number of sides. We will spend considerable time this year studying the triangle in more detail. We are also going to examine closely other polygons, particularly the four-sided polygons.

D

Definition 1-33 A **quadrilateral** is a four-sided polygon.

The quadrilateral has four sides and four vertices. The endpoints of a side are **consecutive vertices.** Vertices that are not consecutive in a quadrilateral are **opposite vertices.** Sides with a common endpoint are **adjacent sides.** In a quadrilateral, sides that are not adjacent are **opposite sides.**

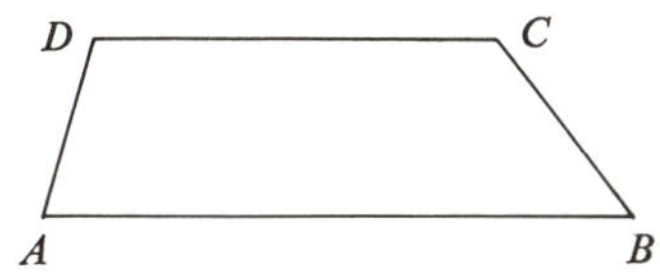

In naming a quadrilateral, we list consecutive vertices in order. Thus, for example, when we speak of quadrilateral $ABCD$, we know that vertices A and B, B and C, C and D, as well as A and D are all pairs of consecutive vertices. Which sides of quadrilateral $ABCD$ are adjacent to $\overline{AD}$? Which side is opposite $\overline{AB}$?

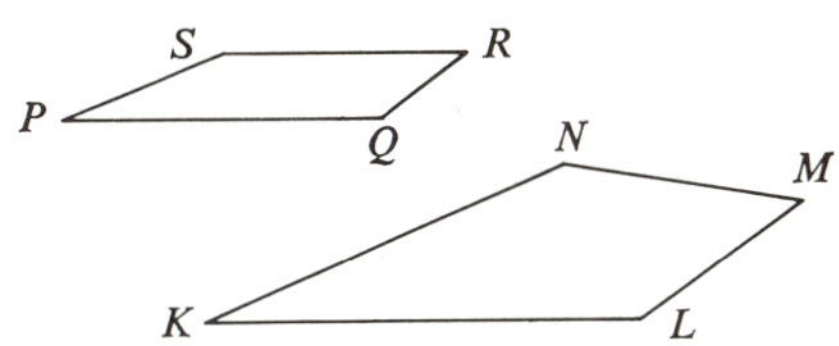

Example 3 If the vertices of quadrilaterals $PQRS$ and $KLMN$ are made to correspond, P with K, Q with L, R with M, and S with N, list the corresponding sides and angles.

Solution Since consecutive vertices are the endpoints of the sides, we can list the corresponding sides as $\overline{PS}$ and $\overline{KN}$; $\overline{SR}$ and $\overline{NM}$; $\overline{RQ}$ and $\overline{ML}$; and $\overline{QP}$ and $\overline{LK}$. Corresponding angles will have corresponding adjacent sides. Thus, we have the following corresponding angles: $\angle PSR$ and $\angle KNM$; $\angle SRQ$ and $\angle NML$; $\angle RQP$ and $\angle MLK$; and $\angle QPS$ and $\angle LKN$.

The segment joining a pair of nonconsecutive vertices of a polygon is called a *diagonal* of the polygon. In quadrilateral $KLMN$ of Example 3, $\overline{KM}$ is a diagonal. Can you name another diagonal of quadrilateral $KLMN$? Name the diagonals of quadrilateral $PQRS$ of Example 3. Assume that the correspondence given in the example still applies; name the corresponding diagonals of the two quadrilaterals. How many diagonals does a triangle have?

Example 4 Name the diagonals of pentagon *ABCDE*.

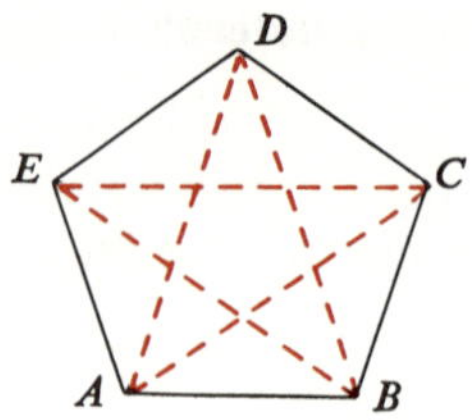

Solution We must not repeat segments. We consider the diagonal $\overline{AC}$ the same as the diagonal $\overline{CA}$. To organize the listing, pick a vertex and name the segments from this vertex to nonadjacent vertices.

Vertex A: $\overline{AC}$ and $\overline{AD}$ Vertex B: $\overline{BD}$ and $\overline{BE}$
Vertex C: $\overline{CE}$ and $\overline{CA}$; but we already have $\overline{AC}$.
Vertex D: $\overline{DA}$ and $\overline{DB}$; but we already have both.
Vertex E: $\overline{EB}$ and $\overline{EC}$; again, we already have both.
Thus, we have five diagonals: $\overline{AC}$, $\overline{AD}$, $\overline{BD}$, $\overline{BE}$, and $\overline{CE}$.

EXERCISES

A Classify each of the following polygons:

1. 
2. 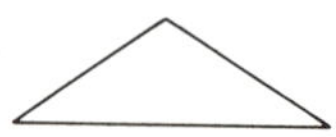
3. 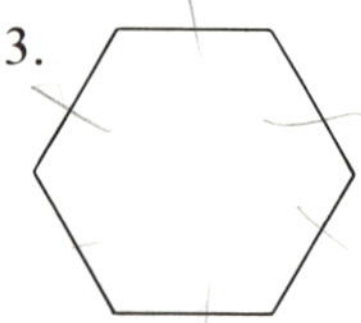
4. 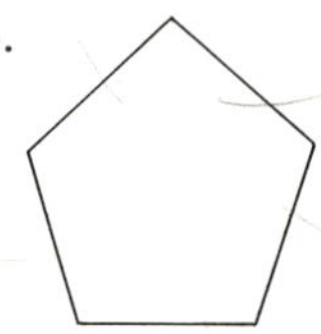

Complete each of the following:

5. Polygons are classified according to the number of ________.
6. A triangle is a(n) ________ polygon.
7. An n-sided polygon is called a(n) ________.
8. A median of a triangle is a segment connecting a ________ of a triangle and the ________ of the opposite side.
9. ________ triangles have one obtuse angle.
10. An isosceles triangle has ________ congruent sides.
11. The side opposite the right angle in a right triangle is the ________.
12. Every ________ triangle has three congruent angles.
13. In polygon *ABCD*, ________ is the vertex opposite $\angle B$.
14. If $\overline{PQ}$ is a diagonal of a polygon, P and Q are ________ vertices.
15. If the set of diagonals of a polygon is $\varnothing$, the polygon is a(n) ________.

Use the model at right for Exercises 16–19.

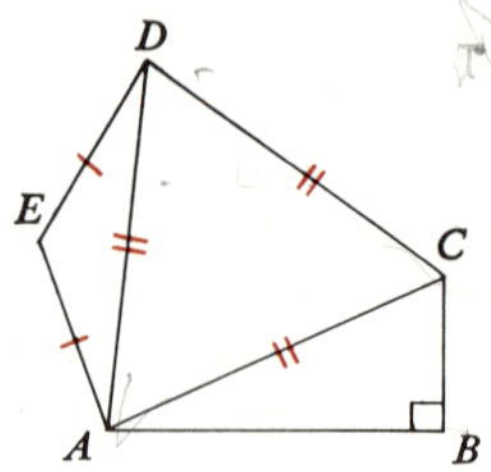

16. Name the triangles in polygon *ABCDE*.
17. Classify each of the triangles in polygon *ABCDE* according to its sides.
18. Classify each of the triangles formed in polygon *ABCDE* according to its angles.
19. If $\triangle ADC$ is equiangular, name the congruent angles.

Give another name to each of the following:

20. a 5-gon 21. a 4-gon 22. a 10-gon 23. a 12-gon
24. a nonagon 25. a heptagon 26. a triangle 27. a hexagon

Name all the diagonals in each of the following polygons:

28. 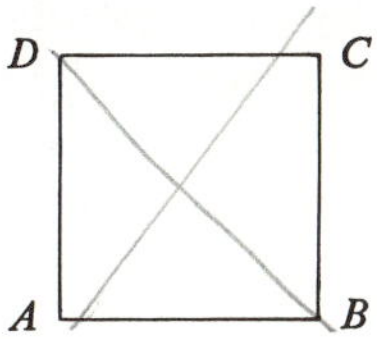

29. 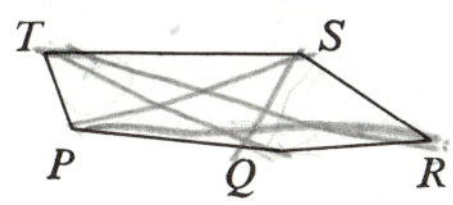

30. 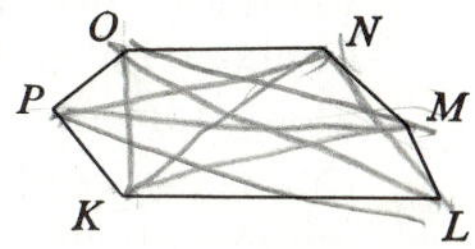

B Using your protractor, find the measure of the angles in each of the following:

31. 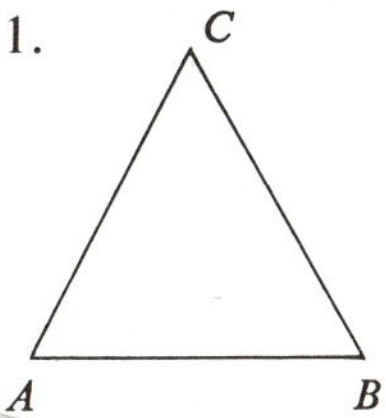

32. 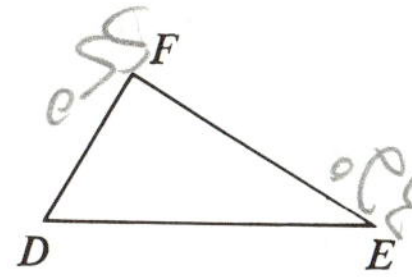

33. 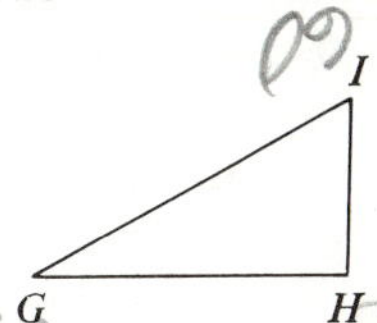

34. Using the measure of each of the angles of the triangles above, predict the sum of the measures of the angles of any triangle.

Using your protractor, find the measure of the angles in each of the following:

35. 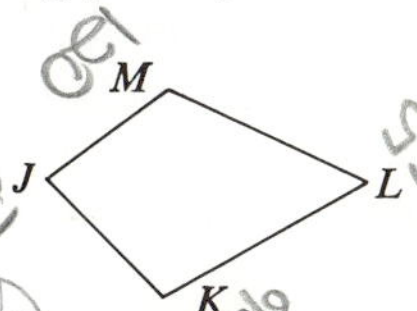

36. 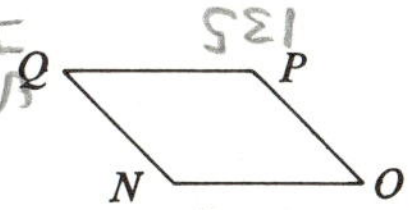

37. 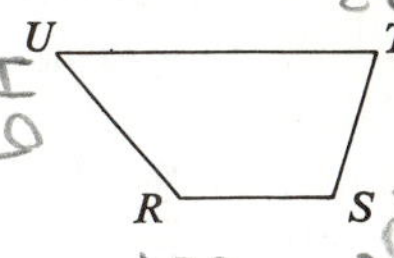

38. Using the measures found for each quadrilateral above, predict the sum of the measures of the angles of any quadrilateral.

1-6 LINES AND PLANES IN SPACE

In Section 1-3, we defined coplanar points as points lying in the same plane. It is obvious that all the points of a line are coplanar. However, since there is an infinite number of planes containing a given line, the line lies not just in one, but in each of those planes. What is the smallest number of points necessary to define one and only one plane? The next question should help you to answer the last question. What is the least number of points that can be noncollinear? To point out one and only one figure satisfying a given condition is **to determine** that figure. Any two distinct points determine a line segment; they are collinear. Three points can be noncollinear. If they are noncollinear, we can picture a triangle with these points as vertices. Since all points of the triangle lie in a plane, we can say that three noncollinear points determine a plane.

D **Definition 1-34** **Space** is the set of all points.

All points are space. However, if we are discussing only those points that are coplanar, we do not consider the points of space that do not lie in the plane. Our attention is only drawn to a spatial model when we have at least one point that is noncoplanar with the other points. If we extend the ideas of the paragraph above we can speak of any four noncoplanar points as determining space.

Much of what we discussed earlier can be extended to ideas about space. A point separates the line into two sets of points called half-lines. The separating point is the endpoint of each half-line. In the model, P separates $\overleftrightarrow{PA}$ into two half-lines. The line in turn separates the plane into two half-planes. One half-plane contains B and the other contains C. The line is the edge of each half-plane. The plane separates space into two regions. Name a point above the plane and a point below the plane.

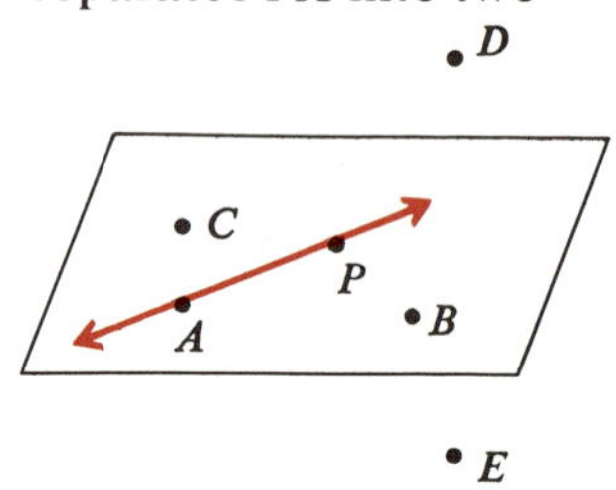

P **Postulate 1-7** **The Space Separation Postulate** Any plane P separates the points in space that are not in plane P into two distinct regions such that:
1. for any two points A and B of a region, $\overline{AB}$ lies entirely in that region;
2. if A is in one region and B is in the other, then $\overline{AB}$ intersects plane P.

D **Definition 1-35** The two regions described in the Space Separation Postulate are **half-spaces**, with plane P as the **face** of each half-space.

Consider two lines in a plane; call them $\overleftrightarrow{AB}$ and $\overleftrightarrow{MN}$. One of three conditions can occur with these two lines:

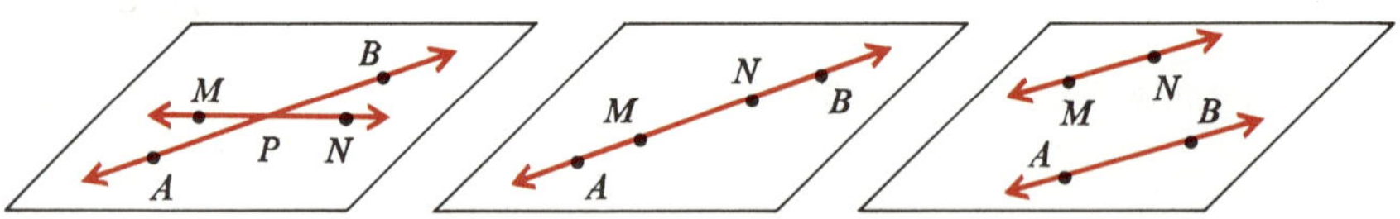

If $\overleftrightarrow{AB}$ and $\overleftrightarrow{MN}$ intersect, they intersect in a single point. When $\overleftrightarrow{AB}$ and $\overleftrightarrow{MN}$ have more than one point in common, they coincide. If $\overleftrightarrow{AB}$ and $\overleftrightarrow{MN}$ have no point in common, the lines are said to be parallel. Let us try to extend this to lines and planes. If a line m and plane P intersect, what is their intersection? If a line and a plane have more than one point in common, does the line lie completely in the plane? When the line and the plane have no points in common, would you say that the line is parallel to the plane?

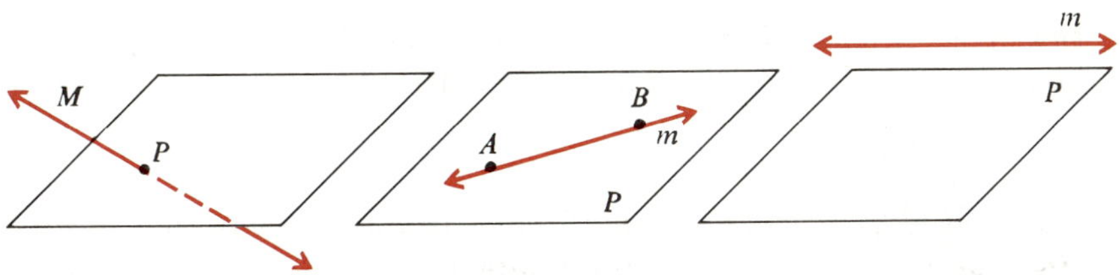

Before extending this last discussion to the relationship between two planes, let us pause to examine lines that determine a plane. We saw earlier that one line is contained not by a single plane, but by an infinite number of planes. If we consider two lines, will a plane be determined? Earlier we said that three noncollinear points are enough to determine a plane. Can you locate three noncollinear points on a pair of intersecting lines? on a pair of parallel lines? Would you say that intersecting lines determine a plane? Will a pair of parallel lines determine a plane? Can we say that any two lines determine a plane? Be careful.

In the models below, lines *l* and *m* are two lines that have no points in common but do not determine a plane. The lines are called **skew lines**.

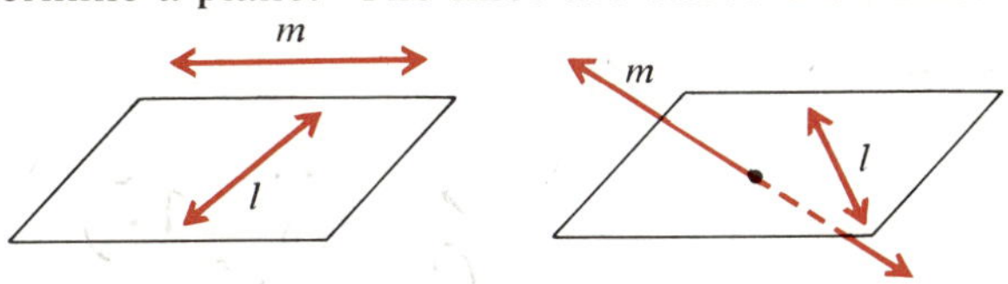

We shall now return to the relationship between planes. Study planes *P* and *M*.

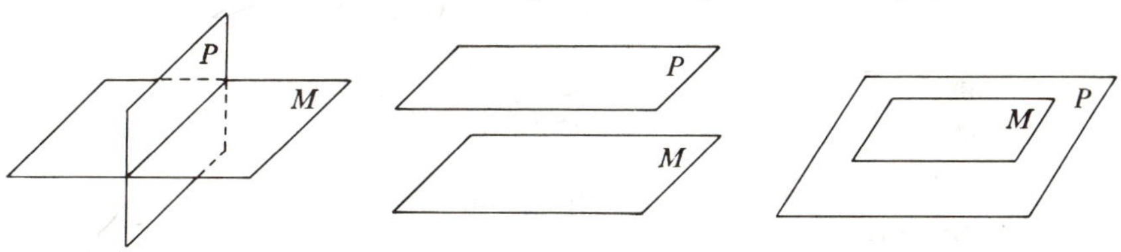

Intersecting **Parallel** **Coinciding**

When planes *P* and *M* intersect, their intersection is a line. If the planes intersect in more than a line, they coincide. If the planes have no points in common, they are said to be parallel.

Examples of intersecting planes are seen everywhere. The walls, floor, and ceiling of your classroom are obvious examples. The walls intersect in a line. The line of intersection lies in each of the planes. The floor and ceiling are parallel planes. We can even present a model for the intersection of three planes. Can you describe this intersection? Each intersection of a pair of the planes is a line. Are the three lines of intersection distinct? What is the intersection of distinct lines?

In a plane, when two lines intersect, angles are formed by their rays. When two planes intersect, their closed half-planes also form an angle. We call such an angle a dihedral angle.

D **Definition 1-36** A **dihedral angle** is an angle formed by the union of a line and the two noncoplanar half-planes that share the line as edge. The half-planes are the faces of the dihedral angle.

We name dihedral angles by referring to one point in one half-plane, the endpoints of the line segment, and finally, one point in the other half-plane. After studying the illustrations below, give three examples of dihedral angles.

Dihedral Angles

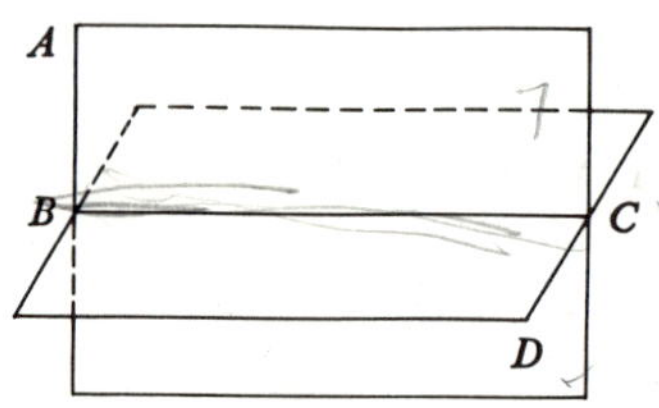

$\angle A\text{-}BC\text{-}D$
or
$\angle D\text{-}BC\text{-}A$

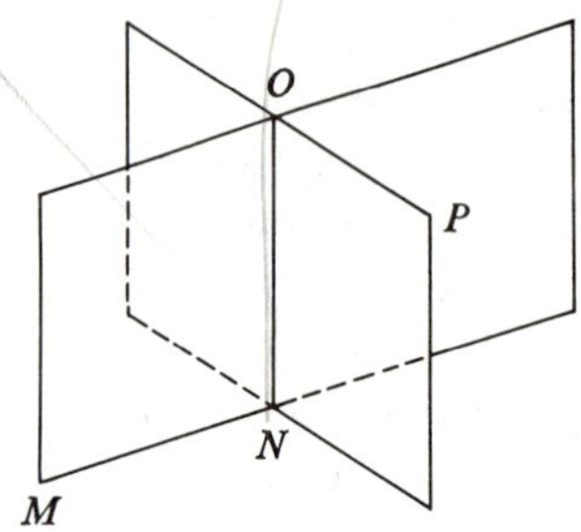

$\angle M\text{-}NO\text{-}P$
or
$\angle P\text{-}NO\text{-}M$

EXERCISES

A Draw a model representing each of the following:

1. horizontal plane
2. vertical plane
3. dihedral angle
4. plane passing through $\overleftrightarrow{AB}$
5. skew lines
6. parallel planes

Indicate which of the following are true:

7. A set of three points is collinear.
8. The intersection of two planes is at least a line.
9. Three noncollinear points determine space.
10. Two distinct lines have at most one point in common.
11. Four distinct points are always noncollinear.

Complete each of the following:

12. A plane is a subset of ________.
13. The intersection of a half-plane and its edge is ________.
14. Three noncollinear points determine a(n) ________.
15. If a point is not in or below a plane it is ________ the plane.
16. A plane separates space into two ________.
17. If two points of a line lie in a plane, then ________ points of the line lie in the plane.
18. If three distinct planes intersect, with distinct lines of intersection for each pair, the intersection of the planes is a(n) ________.

B Draw a model to represent each of the following:

19. a line intersecting a plane
20. a line parallel to a plane
21. four noncoplanar points
22. the plane formed by intersecting lines
23. three intersecting planes
24. the plane formed by parallel lines

Indicate which of the following are true and which are false:

25. Three parallel lines must be coplanar.
26. The intersection of two skew lines is a point.
27. Every triangle is a plane figure.
28. Any geometric figure with four vertices is a plane figure.
29. Name all the dihedral angles in the figure at right.
30. Name the plane angles in the plane determined by C, B, and G in the figure at right.

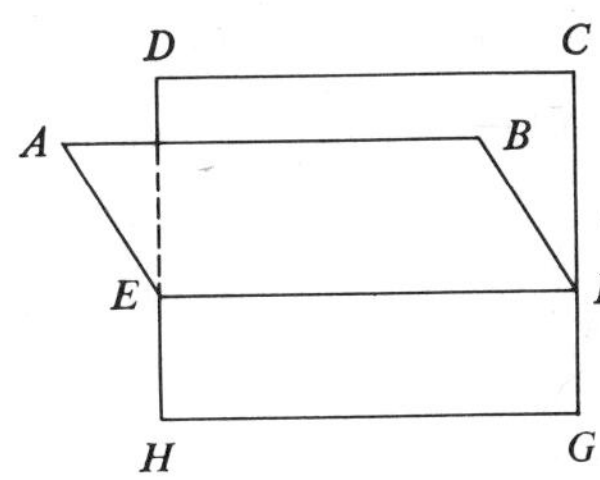

C

31. Extending your understanding of adjacent plane angles, name the requirements for adjacent dihedral angles.
32. Using Exercise 31, name adjacent dihedral angles in the model above.

HOW IS YOUR MATHEMATICAL VOCABULARY?

The key words and phrases introduced in this chapter are listed below. How many do you know and understand?

absolute value (1-2)	addition property (1-2)	angle (1-4)
acute angle (1-4)	adjacent angle (1-4)	Angle Measure Postulate (1-4)
acute triangle (1-5)	adjacent sides (1-5)	

between (1-2)
bisect (1-4)
bisector (1-4)
collinear points (1-3)
complementary angle (1-4)
congruent angles (1-4)
congruent segments (1-3)
consecutive vertices (1-5)
coordinate (1-2)
coplanar (1-3)
determine (1-6)
diagonal (1-5)
dichotomy property (1-2)
dihedral angle (1-6)
disjoint sets (1-1)
Distance Postulate (1-2)
edge (1-4)
element (1-1)
empty set (1-1)
equality properties (1-2)
equal sets (1-1)
equiangular triangle (1-5)
equilateral triangle (1-5)
equivalent sets (1-1)
exterior (1-5)
face (1-6)
graph (1-2)
half-line (1-3)
half-plane (1-4)
half-space (1-6)
hypotenuse (1-5)
infinite set (1-1)
interior (1-5)
intersection (1-1)
irrational numbers (1-2)
isosceles triangle (1-5)
legs (1-5)
linear pair (1-4)
line segment (1-3)
Line Separation Postulate (1-3)
measure (1-3)
median (1-5)
midpoint (1-3)
multiplication property (1-2)
n-gon (1-5)
obtuse angle (1-4)
obtuse triangle (1-5)
opposite rays (1-3)
opposite sides (1-5)
opposite vertices (1-5)
order properties (1-2)
origin (1-2)
perpendicular lines (1-4)
Plane Separation Postulate (1-4)
polygon (1-5)
proper subset (1-1)
quadrilateral (1-5)
ray (1-3)
real numbers (1-2)
reflexive property (1-2)
right angle (1-4)
right triangle (1-5)
Ruler Postulate (1-2)
scalene triangle (1-5)
set (1-1)
set-builder notation (1-2)
skew lines (1-6)
Space Separation Postulate (1-6)
straight angle (1-4)
subset (1-1)
supplementary angle (1-4)
Supplementary Angles Postulate (1-4)
symmetric property (1-2)
transitive property (1-2)
triangle (1-5)
trichotomy property (1-2)
undefined terms (1-3)
union (1-1)
vertex (1-3)
well-defined set (1-1)

REVIEW EXERCISES

1-1 Sets

Let U = natural numbers less than 15, $A = \{1, 3, 5, 7, 9, 11, 13\}$, and $B = \{2, 4, 6, 8, 10, 12, 14\}$. List the members of each of the following sets:

1. $A \cap B$ 2. $A \cap U$ 3. $B \cap U$ 4. $A \cup B$
5. $A \cup U$ 6. $B \cup U$ 7. $A \cap (B \cup U)$ 8. $(A \cap B) \cup (A \cap U)$
9. List all possible subsets of $\{1, 4, 6, 8\}$.
10. Describe the possible intersections of the sets of points of two lines in the same plane.

1-2 Real numbers

Using set-builder notation, describe each of the following:

11. pairs of numbers whose sum is 9 12. the set of rational numbers
13. the integers greater than 5 and less than 22
14. the numbers whose square is 25

Solve each of the following equations:

15. $3x + 2 \le 5 - 2x$ 16. $|x| \ge 7$ 17. $3|x - 1| = 15$
18. $|2x| < 4$ 19. $2|x + 2| - 7 \le 0$ 20. $2|x| = |x|$
21. Order -2.7, $3\frac{1}{7}$, 0, $\sqrt{2}$, $-\sqrt{9}$ as they appear on the number line.
22. If M, N, and L lie on a line such that $MN = 15$ and $ML = 23$, describe N's relative position to M and L.

1-3 Basic terms

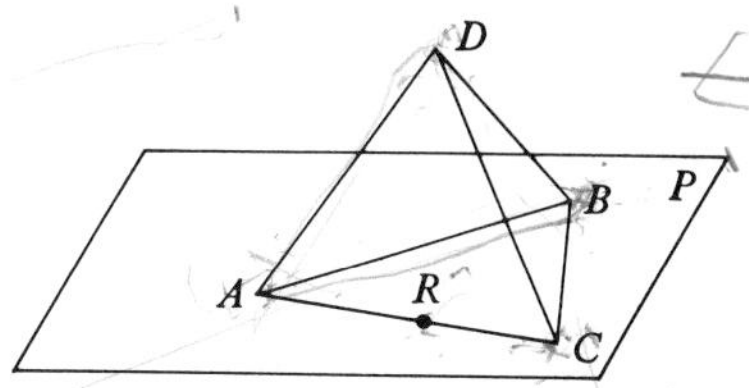

23. Describe the intersection of $\overrightarrow{AB}$ and $\overrightarrow{BA}$.
24. Using the figure at right, name four sets of coplanar points.
25. In the figure at right, what three points are collinear?

Using the figure above, describe the following sets:

26. $\overline{AD} \cap$ plane P 27. $A \cup \overleftrightarrow{AB}$ 28. $D \cap$ plane P
29. $R \cap \overrightarrow{AC}$ 30. $\overline{AC} \cap$ plane P 31. $\overline{AB} \cup$ plane P

1-4 Angles

Use your protractor to determine the measure of each of the following angles:

32. 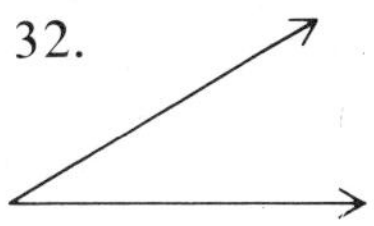

33. 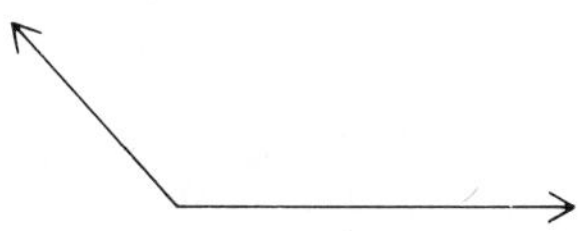

34. 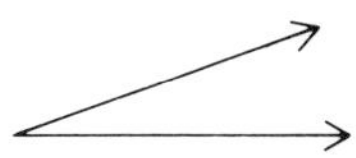

Use the figure at right to answer Exercises 35-40. Classify each of the following angles as acute, obtuse, or right. $\angle PQT$ and $\angle TQR$ are a linear pair.

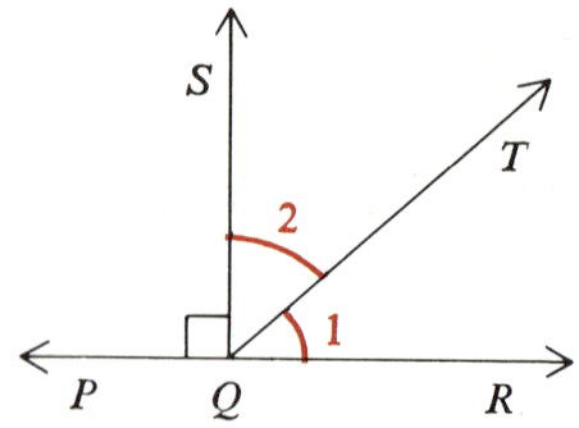

35. $\angle PQT$ 36. $\angle PQS$
37. $\angle SQT$ 38. $\angle TQR$
39. If $\angle RQT = 30$, then $m\angle TQS =$ ________.
40. If $\angle RQT$ and $\angle TQS$ are congruent, then $m\angle RQT =$ ________.

Form an angle for each of the following:
41. 175° 42. 80° 43. 15° 44. 30°

Determine the measure of the complement of each of the following angles:
45. 26° 46. 9° 47. 60° 48. 45°

Determine the measure of the supplement of each of the following angles:
49. 120° 50. 135° 51. 90° 52. 45°

1-5 Polygons

Supply another name for each of the following:
53. a 7-gon 54. a 9-gon 55. a pentagon 56. a hexagon

List the diagonals for each of the following polygons:

57.
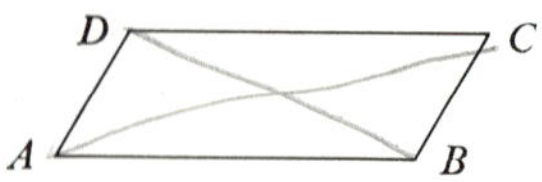

58.
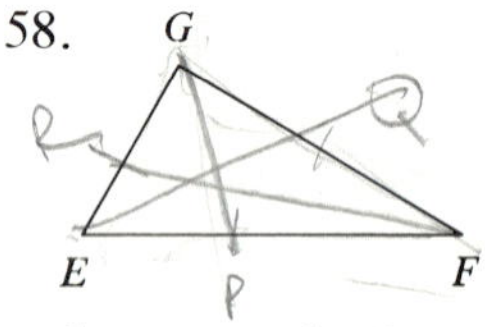

59. If $\angle G$ in Exercise 58 is a right angle, name the legs and hypotenuse of $\triangle EGF$.
60. If P, Q, and R are the midpoints of $\overline{EF}$, $\overline{FG}$, and $\overline{GE}$, respectively, name the medians of $\triangle EFG$.

1-6 Lines and planes in space

Indicate which of the following are true and which are false:
61. Four distinct points determine a plane.
62. Two planes intersect in a line.
63. Skew lines determine a plane.
64. Any three planes intersect in a point.
65. Extending the notion of adjacent angles, name the adjacent dihedral angles in the figure at right.

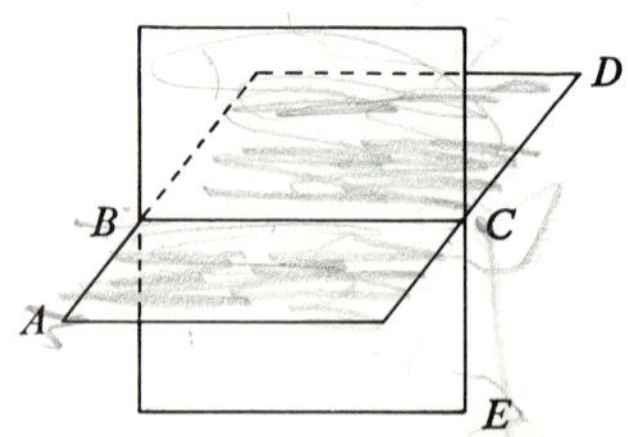

CHAPTER TEST

1. Describe the intersection of $\overline{AG}$ and $\overline{AD}$ if the segments are distinct.
2. Describe the union of $\overrightarrow{DE}$ and $\overrightarrow{DH}$ if D, E, and H are noncollinear.
3. Describe the union of $\overline{AB}$, $\overline{BF}$, and $\overline{FA}$ when the points A, B, and F are noncollinear.

Evaluate each of the following:

4. $|5-7|$ 5. $||5|-|7||$ 6. $|x|$

Solve each of the following:

7. $x \geq |-2|$ 8. $2|x| = |x-1|$

Translate each of the following symbols:

9. AB 10. $\overline{AB}$ 11. $\overleftrightarrow{AB}$ 12. $\overrightarrow{AB}$

Indicate which of the following are true and which are false. Make the necessary changes in the false statements to make them true.

13. Sets with the same number of elements are equal sets.
14. For all $x \geq 0$, $|x| = x$.
15. To every point of the line, there corresponds a positive integer.
16. Three or more points are collinear points.
17. If two lines form a right angle, the lines are perpendicular.
18. Any two points are collinear.

19. If $m\angle A = 2x$, and $m\angle B = 3x$, and $\angle A$ is the complement of $\angle B$, find x.
20. If in Exercise 19 $\angle A$ and $\angle B$ are supplementary, find x.
21. What is the measure of congruent angles that form a linear pair?
22. If two angles are complementary, both angles are ________ angles.
23. If two angles are supplementary, and one angle is acute, the other is ________.
24. One angle of a pair of supplementary angles is always acute. True or false?

SUGGESTED RESEARCH

1. We usually attribute the origin of geometry to the Greeks, yet the Egyptians and Babylonians helped to establish many of its basic concepts. Check the library for such books as the *Thirty-first Yearbook of the National Council of Teachers of Mathematics* or *History of Mathematics*, by D. E. Smith (New York: Dover Publications, 1951). Prepare a report about the early history of geometry.

2. The Greeks thought of an angle as a broken line. Referring to a dictionary or mathematical history books, compile a set of definitions of *angle*. Compare these to the definition we are using.

3. Using the books suggested above, prepare a paper on Euclid.

MATHEMATICAL EXCURSION

Finite Geometry

In Section 1-3 we agreed that *point*, *line*, and *plane* would be undefined terms. Because of this, let us assume that we can interpret them as we wish, as long as our interpretations are consistent with the assumptions we make.
Suppose we assume the following six postulates:

A_1 A line is the set of at least two points.

A_2 Two distinct points are elements of one and only one line.

A_3 A plane is a set of at least three points that are not all elements of the same line.

A_4 Three points not elements of one line belong to one and only one plane.

A_5 If a plane contains two points of a line, it contains all points of the line.

A_6 If one point is an element of the intersection of two planes, then at least one other point is also an element.

Can we construct a model that will satisfy these six postulates? Consider noncollinear points A, B, and C. If $\{A, B\}$, $\{B, C\}$, and $\{A, C\}$ are lines and $\{A, B, C\}$ is a plane, all the postulates are satisfied. The symbol $\{A, B\}$ is read "line AB." The symbol $\{A, B, C\}$ is read "plane ABC."

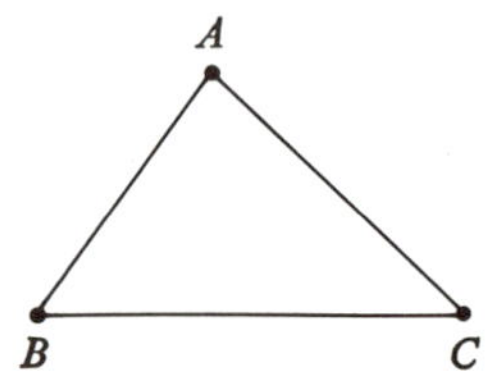

The figure above is a model showing the relationship of the three points and the three lines. If we compare the figure with the postulates, we can see that all the postulates are satisfied by the three points and the three lines.

A_1 is satisfied since each line contains two points. A_2 is satisfied since each pair of points belongs to one and only one line. A_3 is satisfied since all three points are not in one line. A_4 is satisfied since there is exactly one plane. A_5 is satisfied because each line has only two points. A_6 is satisfied since there is only one plane.

In this discussion, *point* and *line* remained undefined. The figure is just a visual aid to the system. The streaks of ink connecting A with B, B with C, and C with A are not lines, but are used to indicate the sets of undefined points that are called lines. The dots are not points, but are merely used to represent the undefined points.

Before we attempt a physical example of the model, we shall prove some theorems.

T **Theorem 1** If there are exactly three points in each plane, there are exactly two points in each line.

PROOF OUTLINE By Postulate A_1, we know that there are at least two points in each line. We are given the information that there are exactly three points in each plane. By Postulate A_3, a plane is a set of at least three noncollinear points.

We can conclude from these statements that a line can contain at most two points *that belong to a given plane*. But, by Postulate A_5, we know that if a plane contains two points of a line, it contains all points of the line. Therefore, we can conclude that if a plane contains exactly three points, there are exactly two points in each line.

T **Theorem 2** If there are exactly three points in each plane, there are exactly three lines in each plane.

PROOF OUTLINE From Theorem 1 we know that each line contains exactly two points. Then the three points of the plane, A, B, and C, determine at least three lines, $\{A, B\}$, $\{A, C\}$, and $\{B, C\}$. If there was a fourth line in the plane, it would have to contain exactly two of the three points A, B, and C. But this would contradict Postulate A_2, which states that two distinct points can be elements of only one line. Therefore, if there are exactly three points in a plane, there are also exactly three lines in the plane.

T **Theorem 3** If there are exactly three points in a plane, then two lines have exactly one point in common.

PROOF OUTLINE We saw in Theorem 2 that if a plane has exactly three points, A, B, and C, it has exactly three lines, $\{A, B\}$, $\{A, C\}$, and $\{B, C\}$. Since $\{A, B\} \cap \{A, C\} = \{A\}$, $\{A, B\} \cap \{B, C\} = \{B\}$, and $\{A, C\} \cap \{B, C\} = \{C\}$ we can conclude that two lines have exactly one point in common.

As an example of this system, consider a particular machine that requires a crew of two men to operate it. The machine is to be operated for three 8-hour shifts. No man is permitted to work more than two shifts per day. Brown, Jones, and Smith are the only available operators. If *point* is used to represent an operator and *line* to represent a crew, we can form an example of the situation we presented.

We have points: Brown, Jones, Smith
lines: {Brown, Jones}, {Brown, Smith}, {Jones, Smith}
plane: {Brown, Jones, Smith}

We leave it as an exercise to show that this model satisfies each of the six postulates.

Can we have a set of four points that satisfies the six postulates? Suppose we have line $\{A, B, C\}$ and a point D not in $\{A, B, C\}$. If at least three points are in a line, we know that the line determined by A and D is different from $\{A, B, C\}$. We also know that it must contain a third point. Therefore, there are at least five points. This contradicts the assumption that there are exactly four points.

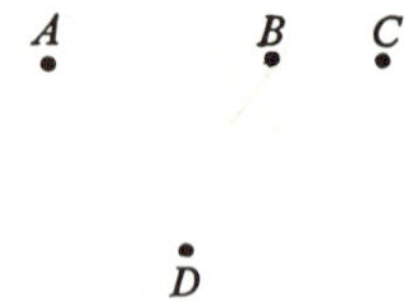

Suppose we consider line $\{A, B\}$ and point C not in $\{A, B\}$. We then have plane $\{A, B, C\}$ containing lines $\{A, B\}$, $\{A, C\}$, and $\{B, C\}$. If we now have a fourth point D not in plane $\{A, B, C\}$, we have three more planes: $\{A, B, D\}$, $\{A, C, D\}$, and $\{B, C, D\}$.

In the figure at right, can you prove that there are six lines? Is there a line containing D that has no point in common with $\{A, B\}$? Is there a line containing B that contains no point of line $\{A, C\}$? Through a point not in a given line there is always exactly one other line that does not contain any point of the first line.

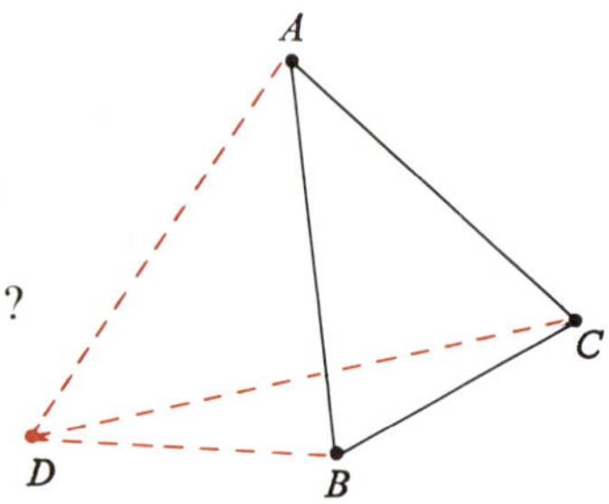

Suppose we have four points in one plane, no three of which are collinear, as in the figure at the right. Can you prove there are actually six lines rather than the four shown? Can you show that any point not contained in a given line is contained in a separate line that has no point in common with the first line? For example, A is not in line $\{B, C\}$, and $\{A, D\}$ is a line through A that does not contain any point of $\{B, C\}$.

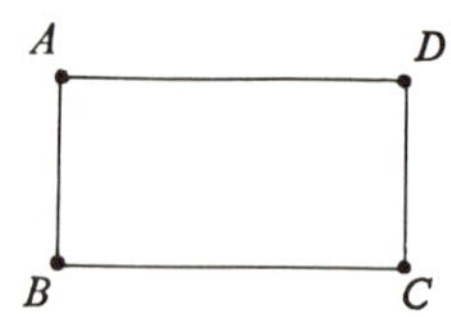

EXERCISES

1. Show that the Brown, Jones, Smith model given on page 41 satisfies Postulates A_1 through A_6.

If there are exactly four points, not all in the same plane, prove the following, using Postulates A_1 through A_6:

2. There are exactly two points in each line.
3. There are four planes.
4. There are six lines.
5. Through any point not in a line, there is one line that has no point in common with the first line.
6. In Exercise 5, show that the two lines involved do not lie in one plane.

The other side of the moon was never visible to men on Earth. Do you think we could have proved that the other side was round before we began our space-exploration program? If you think that a photograph would have supplied a good proof, consider this optical illusion.

2 Methods of Proof

The study of any subject is based on statements. They are the wires along which ideas are transmitted. Some statements are by their nature unacceptable, so we must examine the development and uses of our statements very carefully. We can develop statements by several methods. The method we use often determines whether or not a statement is acceptable. In geometry, we develop statements by logical reasoning based on definitions, postulates, or previously proved theorems.

2-1 INTUITION AND INDUCTION

Two unacceptable methods of arriving at geometric statements are intuition and induction. Although the following definition is greatly simplified, it will be sufficient for our purposes.

D **Definition 2-1** **Intuition** is the process of making generalizations based on insight.

An intuition is a prediction made on a hunch. "Most new cars this year will be green" is an example of intuition if there is no better reason for saying it than feeling that way. We sometimes make intuitive statements on the basis of quick examination. This can often lead to difficulties, as in the case of optical illusions. Is the figure on the left a plane figure, or can you see a box? In each of the two middle figures, which segment appears to be longer, $\overline{AB}$ or $\overline{CD}$?

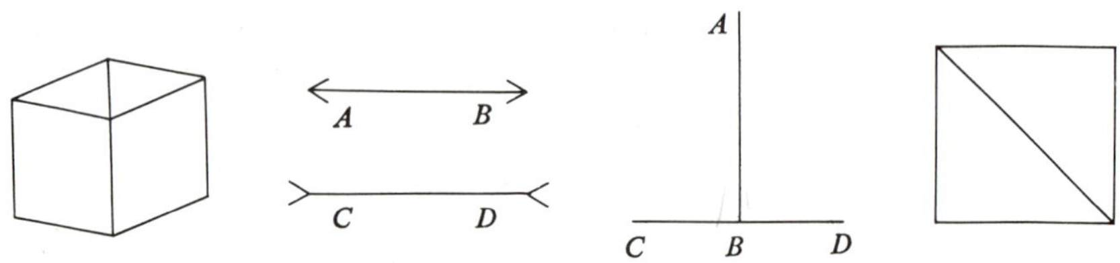

Look at the figure on the right. Does the diagonal of a square separate the square into two equal parts? A unit of measure to determine equal parts has not been discussed; yet some people may be willing to make a statement about the parts of a square without this knowledge of measure. Their statement will be intuitive—based only on a hunch.

Class Exercises

1. Your brother's new friend wants to sell you his old motor bike. You decide not to buy the bike because you do not like his friend's looks. You think that he looks as if he would cheat you. Is this intuitive thinking?
2. Just by looking at the figure at right, would you say that $BF = FD$?
3. After observing the trains that stop at your station for two months, you tell a friend who is coming to visit you when his train will arrive. Is your stated arrival time an intuition?

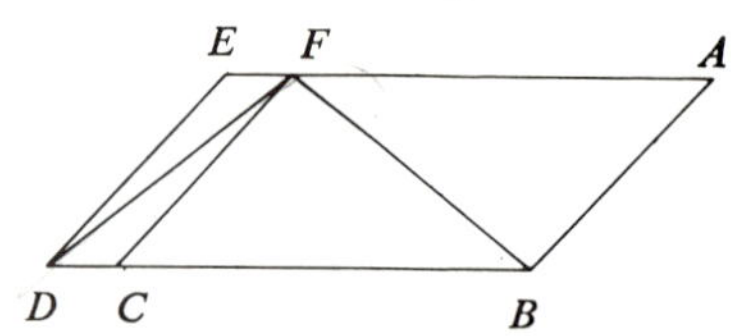

Complete each of the following with the number that makes the statement true:

4. $1 + 2 = \frac{2(\ \ )}{2}$
5. $1 + 2 + 3 = \frac{3(\ \ )}{2}$
6. $1 + 2 + 3 + 4 = \frac{4(\ \ )}{2}$

7. $1+2+3+4+5=\frac{5(\quad)}{2}$ 8. $1+2+3+4+5+6=\frac{6(\quad)}{2}$

9. $1+2+3+4+5+6+\cdots+n=\frac{n(\quad)}{2}$

10. Having completed Exercises 4–9, would you call your answer to Exercise 9 an intuition?

Induction

Induction is also a guessing process, but differs from intuition in one important aspect. Induction is based on observations and reasonable conjectures.

D **Definition 2-2** **Induction** is the process of drawing conclusions or generalizations from several known particular cases.

The resulting conclusion is called an induction. Scientists often use this method. From making many observations, they draw their inductions. Many mathematical facts were originally inductions and were later proved. We use this process ourselves in algebra, as in Class Exercise 10 above.

Example 1 Predict the twelfth term of the set of numbers whose pattern for the first five terms is 2, 4, 6, 8, 10, ...

Solution

First term: 2 is twice 1.

Second term: 4 is twice 2.

Third term: 6 is twice 3.

It appears that each term is twice the number of the term. Thus, by induction, the twelfth term is 24.

Example 2 Make an induction about the sum of the first n odd integers.

Solution To form the induction, add progressively larger sets of consecutive odd integers, always starting with the integer 1. This may establish a pattern.

First two: $1+3=4$, or 2^2

First three: $1+3+5=9$, or 3^2

First four: $1+3+5+7=16$, or 4^2

Induction: The sum of the first n odd integers is n^2. The last integer is twice n minus 1. Thus,

$$1+3+5+7+\cdots+(2n-1)=n^2$$

Example 3 How many triangles are formed by all the diagonals that can be drawn through one vertex of a 12-gon?

Solution Make several models to form a reasonable induction.

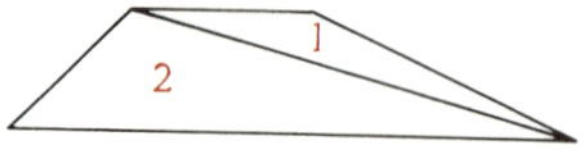

4-gon: two triangles

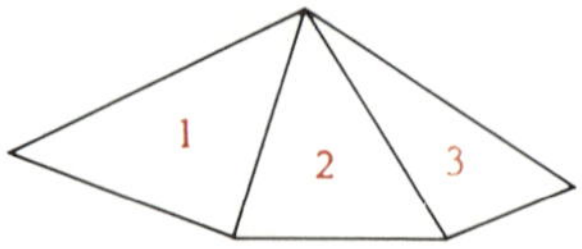

5-gon: three triangles

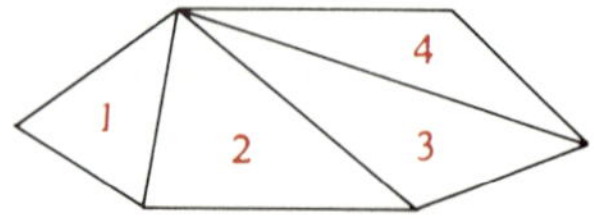

6-gon: four triangles

In each case, there seem to be two fewer triangles than sides. The induction for a 12-gon is that 10 triangles would be formed.

While inductions suggest many facts, they do not form proved statements. When there is an endless number of cases, each case cannot be checked. This can cause difficulty, since not all inductions are true. To add certainty to our study, we need a form of reasoning in which statements can be proved without considering each case. This form of reasoning is called **deductive reasoning.** Deductive reasoning gives us a process for proving statements, many of which were originally inductions. We shall delay studying deductive reasoning for a few sections in order to gain a better background.

EXERCISES

A Classify each of the following as an intuition or an induction:

1. Your best friend asks you to go out with a blind date and you accept. After seeing your date the first time, you know that you will not have a good time.
2. After eating at Lou's Diner for a week, you decide that the lunches there are not very good. You decide to eat somewhere else.
3. The measure of two sides of a triangle are equal. You can tell by looking.
4. "The chemist said that we will get copper sulfate when we do this, because he always gets copper sulfate when he does the same thing."

Just by looking, decide if l_1 and l_2 are straight or bent in each of the following:

5.

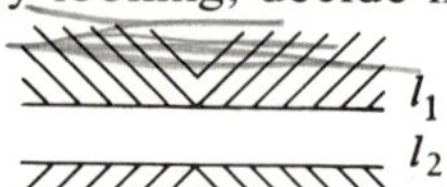

6.

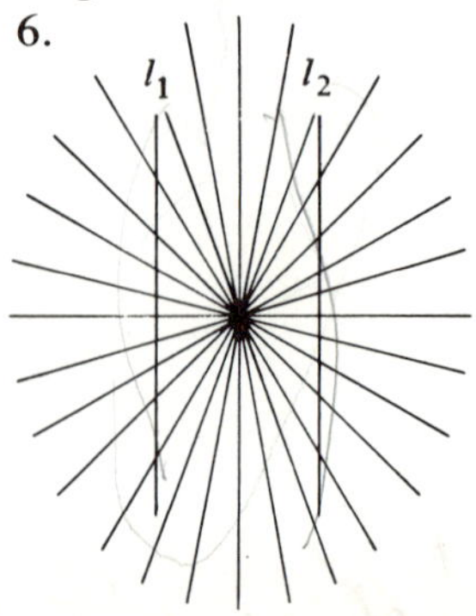

7.

8.

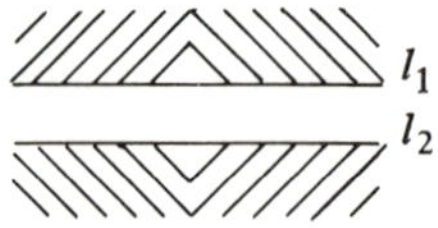

9. Without measuring the circles below, decide if circles A and B are the same size. To check, trace circle A and see if it can be placed over circle B.

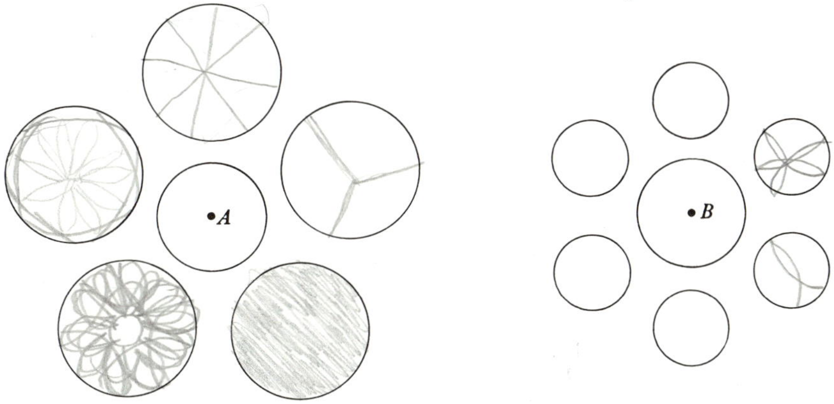

10. Without tracing or measuring, what conjecture can you make about the five circles surrounding circle A above? Is this an intuition or an induction?
11. Repeat Exercise 10, using the configuration around circle B of Exercise 9.
12. Draw four triangles of different shapes. Find the measures of the angles of each triangle. For each triangle, find the sum of the angle measures. What induction can you make about the sum of the measures of three angles of any triangle?
13. Repeat Exercise 12, using four pentagons of different shapes.

Supply the tenth term in each of the following:

14. $1, 5, 9, 13, \ldots$
15. $-12, -9, -6, -3, \ldots$
16. $\frac{1}{2}, \frac{1}{3}, \frac{1}{4}, \frac{1}{5}, \ldots$
17. $\frac{2}{3}, \frac{4}{5}, \frac{6}{7}, \frac{8}{9}, \ldots$
18. $9, 6, 4, \frac{8}{3}, \ldots$
19. $12, 18, 27, \frac{81}{2}, \ldots$

Make an induction for each of the following groups of observations:

20. $2 + 4 = 2(3)$
$2 + 4 + 6 = 3(4)$
$2 + 4 + 6 + 8 = 4(5)$

21. $5 + 10 = (\frac{5}{2})(2)(3)$
$5 + 10 + 15 = (\frac{5}{2})(3)(4)$
$5 + 10 + 15 + 20 = (\frac{5}{2})(4)(5)$

22. $1(2) = \frac{1(2)(3)}{3}$
$1(2) + (2)(3) = \frac{(2)(3)(4)}{3}$
$1(2) + (2)(3) + (3)(4) = \frac{3(4)(5)}{3}$

23. $3 + 5 = 2(4)$
$3 + 5 + 7 = 3(5)$
$3 + 5 + 7 + 9 = 4(6)$

B Use your protractor to measure each of the angles in the following figures:

24.

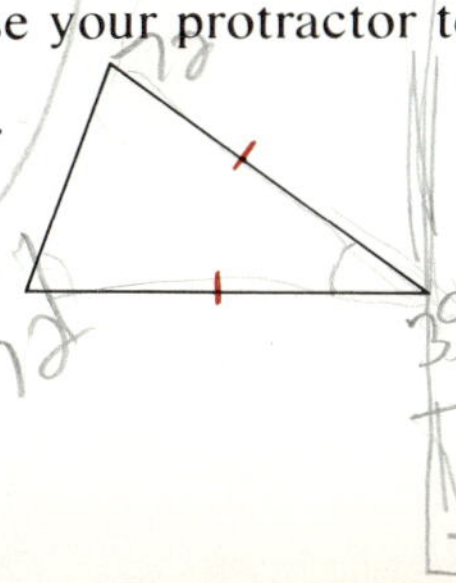

25.

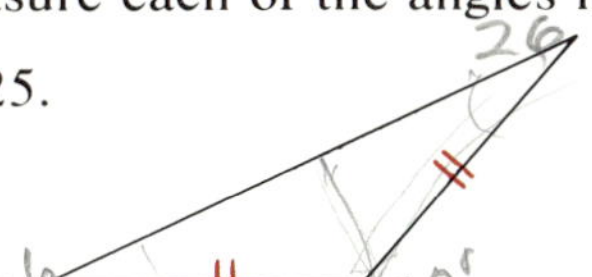

26.

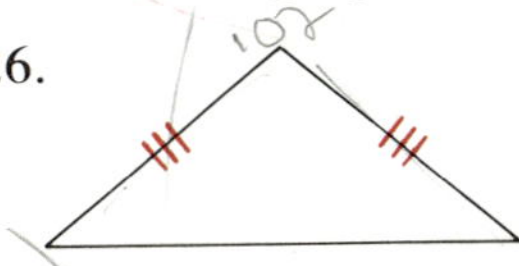

27. What induction can you form about the measures of the angles opposite the congruent sides of an isosceles triangle?
28. Draw three different quadrilaterals and measure their angles. What induction can you form about the sum of the measures of the angles of a quadrilateral?
29. Draw three pairs of intersecting lines and measure the nonadjacent angles. What induction can you form about the nonadjacent angles formed by intersecting lines?

2-2 A MATHEMATICAL SYSTEM

We attribute the usual high school geometry course to Euclid. Euclid was a Greek who lived about 300 B.C. We refer to "Euclidean" geometry even though we know that Euclid did not originate it. The Egyptians, Babylonians, and other Greeks had been developing geometric concepts for hundreds of years before Euclid was born. Euclid's contribution to geometry was his collection, organization, and presentation of these concepts as a **logical system** of mathematics. This course of geometry is not exactly the same as Euclid's, but it too is a logical mathematical system. It is not, however, the only mathematical system we study. Algebra is also a logical system of mathematics.

A logical system of mathematics has four basic features. There are *undefined terms*, *definitions*, *assumptions* (postulates), and *theorems*. Theorems are the conclusions derived logically from the assumptions and definitions. It is necessary that these features be present before we can say that what we are studying is a system. Some of these features we have already introduced. We shall study them further as we examine all the parts of a logical system.

Undefined Terms

Not all terms or expressions can be defined. Attempting to define everything leads to circular definitions. The statement "a point is the intersection of two lines, while a line is an infinite set of points" is an example of a circular definition. We have defined *point* in terms of lines and *line* in terms of points. We avoid this by leaving some words undefined. The **undefined terms** we shall use—*point*, *line* (meaning a straight line), and *plane*—were introduced in the last chapter. These words, together with common English terms, enable us to give meaningful definitions to other words.

Example 1 In Chapter 1, we gave the following definition for collinear points: "A set of points is collinear if all the points lie in a line." Describe the words used in the definition as common English, undefined, or previously defined.

Solution *Collinear* is the term we are defining. The common English terms, as they appear in the sentence, are *a*, *set*, *of*, *is*, *if*, *all*, *the*, *lie*, and *in*. The remaining two words, *points* and *line*, are the undefined terms.

Defined Terms

Besides the wording of the definition, we can also discuss the restrictiveness of the definition. If a definition states every fact applicable to the word being defined, it is ineffective. We might not be able to understand it, much less determine if a particular item satisfies the definition. On the other hand, if a definition is not restrictive enough, then many particular items may be improperly classified. We can summarize the requirements for a good definition in the following way:

Summary. Characteristics of a Good Definition

A good definition must:

1. name the term being defined;
2. use only known terms or accepted undefined terms;
3. place the term into the smallest set to which it belongs;
4. state the characteristics of the defined term which distinguish it from the other members of the set;
5. contain the least possible amount of information; and
6. always be reversible.

The last characteristic means that if the subject and predicate are interchanged, an equally true statement will result.

Example 2 Describe the reversibility of this definition: "Congruent segments are segments with equal measures."

Solution The subject of the definition is "congruent segments." The predicate is "segments with equal measures." If we interchange the position of the subject and the predicate, we have the statement, "Segments with equal measures are congruent segments." Since both statements describe the same situation, they are equally true statements.

Class Exercises

1. What reason is there for having undefined terms in a system?
2. Define an equilateral triangle.
3. List the undefined terms used in your definition in Exercise 2.

4. What previously defined terms did you use in your definition in Exercise 2?
5. Suppose you define a line as length without breadth. What terms are you using as undefined?
6. Criticize the statement "A line segment is a small part of a line."

Using our three undefined terms, and assuming that we know the meaning of *measurement*, *angle*, and *polygon*, define each of the following figures:

7. isosceles triangle
8. square
9. right triangle
10. scalene triangle

Postulates

In the preceding discussion on definitions, we started with a set of terms accepted without definition. We needed these undefined words because we cannot define everything.

The same is true with statements. Some, we shall prove. Others are **assumptions** that we must accept without proof. Those we accept are called postulates.

D

Definition 2-3 **Postulates** are statements accepted without proof.

Postulates, like definitions, were introduced in the first chapter. Can you name any of these postulates? Postulates make up the third feature of a logical system. What are the first two features we mentioned? Just as the undefined terms serve as a base from which we expanded our set of terms, the postulates serve as a base from which other statements or conclusions are logically reasoned. In certain instances, we can combine the postulates in a logical way to arrive at new conclusions, or to prove some of our statements. The proved statements are called **theorems**. Theorems are the fourth feature of a logical mathematical system.

EXERCISES

A

1. Use set notation to describe the relationship between definitions and undefined terms.
2. Use set notation to describe the relationship between all geometric statements of fact and the postulates of geometry.
3. Give a reason for including postulates as part of a logical system.

Criticize each of the following definitions:

4. A triangle is a three-sided polygon with three angles.
5. A half-line is either half of a line.
6. A straight line segment is the set of points that lie evenly between any two points.
7. A plane is a set of flat points.

8. Space is the set of all points.
9. A square is a four-sided polygon.
10. A plane is a set of coplanar points.
11. A median of a triangle is a segment of a triangle.

Determine which of these statements are true and which are false:

12. Any acceptable definition contains at least one undefined term.
13. "A set is a definite collection of objects" is an acceptable definition.
14. The statements of fact in geometry come from inductive reasoning.
15. Deductions are statements assumed to be true.
16. All properties of a geometric figure are listed in the definition of the figure.
17. Deduction is only one form of thinking.
18. Postulates could be omitted from a logical system.
19. There are no assumptions in algebra.
20. Deductions are established from postulates and other proved statements.

B

21. Can postulates be proved? Explain your answer.
22. State a postulate used for the equality of real numbers.
23. Name one assumption already used for order among the real numbers.
24. Name a second postulate used for the equality of real numbers.
25. Name a second assumption used for order among the real numbers.
26. Recall our definition of space. Could we have listed space as an undefined term? Explain your answer.
27. If we assume that the statements "All triangles have three sides" and "ABC is a triangle" are true, what conclusion can we draw?
28. Angles that form linear pairs are supplementary angles. $\angle A$ and $\angle D$ form a linear pair. What conclusion can you draw from these two statements?

C

Complete each of the following:

29. $\angle A$ and $\angle C$ are supplementary angles; $m\angle A = 40$. Thus, $m\angle C =$ ____.
30. If $\overline{AB} \cong \overline{AC}$ and $\overline{AC} \cong \overline{BD}$, then $\overline{AB} \cong$ ____.
31. $\triangle ABC$ is an isosceles triangle. $AB \neq BC$ and $AB \neq AC$. Thus, we can say that BC equals ____.
32. If in $\triangle ABC$, $\angle A$ is a right angle, then $\triangle ABC$ is a(n) ______ triangle.

2-3 POSTULATES: ALGEBRAIC AND GEOMETRIC

Algebraic Postulates

How many of the assumptions or postulates of algebra can you name? In algebra, postulates are often called **properties**. The properties of equality and order were presented in Chapter 1. Can you state the transitive property of equality?

The transitive property of equality is often used as a substitution postulate, but a substitution postulate must be broad enough to be used with any relationship. The substitution postulate we shall use is stated as follows:

Postulate 2-1 The Substitution Postulate If $a = b$, then either a or b may be replaced by the other in any statement.

In algebra, much time was spent on a detailed examination of the application of properties to various sets of numbers. The properties presented in Chapter 1 are only a few of the properties studied in algebra. An equally important set of properties for the real numbers is the set of **field properties**, which should be familiar to you by now. They are, however, listed for reference on page 605.

Geometric Postulates

We learned seven basic postulates of geometry in Chapter 1. If you do not remember what they are, look back in Chapter 1 and review them.

We can present the postulates in different ways. Grouping them by concepts, however, makes them more meaningful.

The first postulates, the Distance and Ruler Postulates, aid us in making and finding measures. They act as a bridge between algebra and geometry.

Example 1 Use the coordinates of A and B to find the distance between them.

Solution We find the absolute value of the difference of the coordinates of the points A and B.

$$\begin{aligned} AB &= |-25 - 12.5| \\ &= |-37.5| \\ &= 37.5 \end{aligned}$$

Thus, the distance between A and B is 37.5 units.

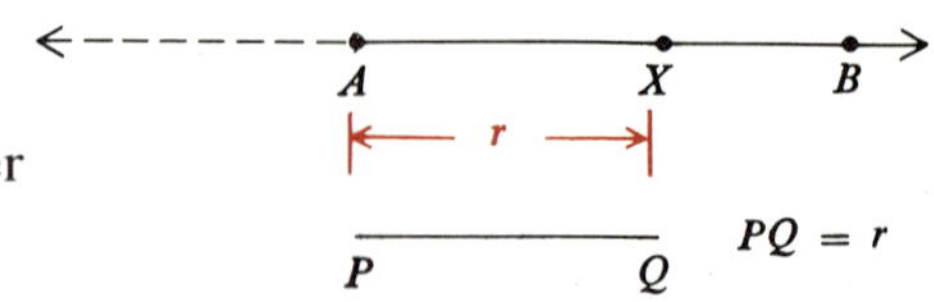

The correspondence between real numbers and points of a line can be made in an infinite number of ways. Everyone in the class can use a different point to correspond to zero. Can we assign zero to the endpoint of a ray and the positive real numbers to the rest of the points of the ray? If so, how many points of the ray $\overrightarrow{AB}$ are there such that the measure of the segment from A to the point X is equal to the measure of a given segment, say $\overline{PQ}$? Remember that distance is a unique number. If $PQ = r$, then the distance from A to the point X must also be r. This leads to the next postulate.

P **Postulate 2-2 The Point Uniqueness Postulate** If n is any positive number, then there is exactly one point N of $\overrightarrow{PQ}$ such that $PN = n$.

If $XY = n$, then the Point Uniqueness Postulate states that there is one and only one point R of $\overrightarrow{PQ}$ such that $PR = XY$.

Points and Lines Postulates

Postulates 2-1 and 2-2 lead into a consideration of points, lines, and line segments. The next set of postulates will state formally many ideas about points and lines.

P **Postulate 2-3 The Line Postulate** Any two distinct points determine exactly one line that contains both points.

From the postulates in the last group, we know that there is a distance between any two distinct points. How many points of the line do you think there are between any two points of the line?

P **Postulate 2-4 The Point Betweenness Postulate** Between any two points, there is at least one point of the line determined by the two points. That is, if P is between A and B, then $AP + PB = AB$.

What unique point have we already studied that is an instance of the Point Betweenness Postulate? Do you recall the Line Separation Postulate?

Example 2 Point B is between points P and Q of the line determined by P and Q. We indicate this by $\overleftrightarrow{PBQ}$. If $PB = 4$ and $BQ = 17$, find PQ.

Solution Since P, B, and Q are collinear, and B is between P and Q, we can write $PB + BQ = PQ$. By the Substitution Postulate, we can write $4 + 17 = PQ$. Thus, $PQ = 21$.

Planes and Space Postulates

If points P, Q, and R are not collinear, we say that $\triangle PQR$ is determined. As we said earlier, the triangle is a plane figure. Is the triangle the n-gon with the least number of sides that

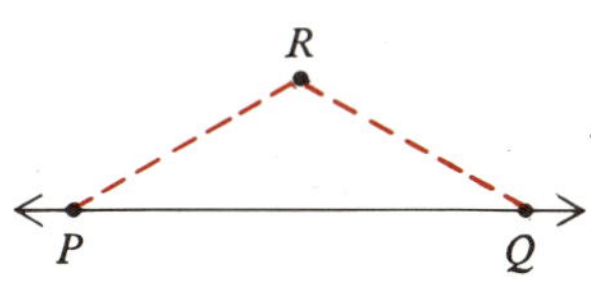

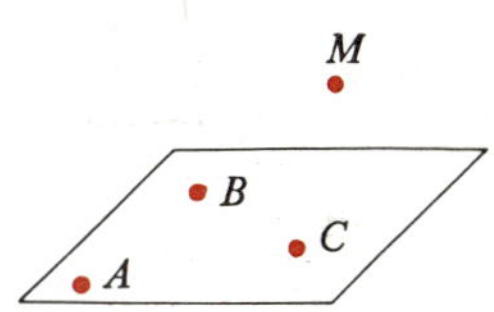

determine a plane? When do we know we are discussing points in space and not just points in a plane? If we discuss the noncoplanar points A, B, C, and M, can we say that we are discussing points in space? Our past experience with instances similar to the models above leads us to state the following postulates:

P **Postulate 2-5 The Plane Postulate** Any three noncollinear points determine exactly one plane that contains the three points.

You have seen the situation described by this postulate put to use in everyday life. Have you ever seen a three-legged table or a tripod wobble? What do you think keeps them from wobbling?

Do three collinear points lie in a plane? exactly one plane? How many planes do you think pass through a line? Could we say that the line lies in each of the planes? Where have we seen two distinct planes intersect in a line? Look around your classroom. Do you see any model for intersecting planes? If two distinct planes intersect, do you think they could intersect in anything other than a line? If so, what? In the model at left below, l is the intersection of planes P and Q. Name the intersection of planes R and S.

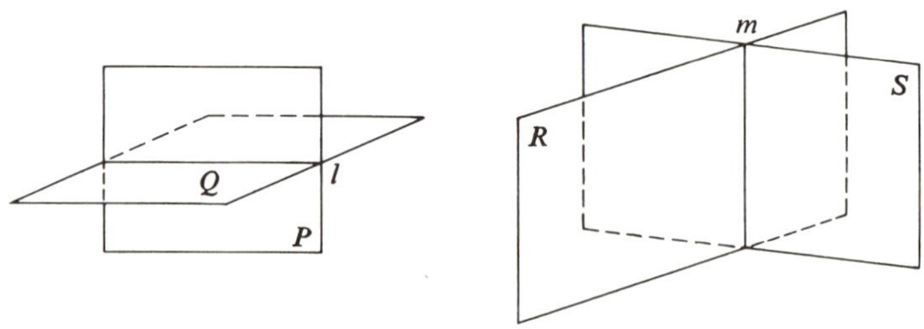

P **Postulate 2-6 The Plane Intersection Postulate** The intersection of two distinct planes is a line.

Example 3 If points A and B lie in the intersection of distinct planes M and N, describe the intersection of the planes M and N.

Solution The intersection of two distinct planes is a line. Since A and B lie in this line, we can say that the intersection is $\overleftrightarrow{AB}$.

P **Postulate 2-7 The Points-in-a-Plane Postulate** If two distinct points of a line lie in a plane, then the line lies in that plane.

Two points determine a line. Similarly, two intersecting lines determine a plane. When two lines intersect, their intersection is a point. When two planes intersect, their intersection is a line. At this time it would be a good idea to review Postulate 1-4, the Plane Separation Postulate.

If a line does not lie in a plane, but intersects the plane, how many points are in the intersection? In the figure P is the point of intersection of $\overleftrightarrow{AB}$ and the plane determined by points X, P, and Z.

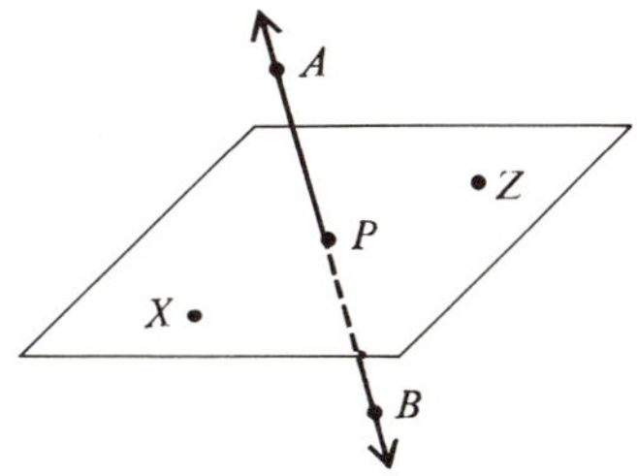

What do we have when we find four noncoplanar points? What is determined in the figure above by points X, P, Z, and A? Do the following postulate and Postulate 1-7, the Space Separation Postulate, agree with your observations and experiences?

P **Postulate 2-8 The Space Postulate** Space contains at least four noncoplanar points.

Angle Postulates

The set of postulates for angles is the last group we shall discuss at this time. Do you recall the Angle Measure Postulate, Postulate 1-5?

If we have a ray, say $\overrightarrow{PQ}$, and a real number n such that n is between 0 and 180 (that is, $0 < n < 180$), then we can form an angle with measure n and side $\overrightarrow{PQ}$. We can do this using our protractor. This leads to the following postulate:

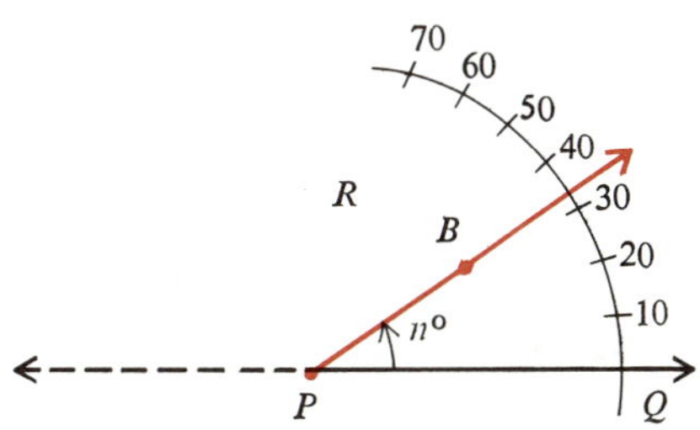

P **Postulate 2-9 The Angle Uniqueness Postulate** Given $\overrightarrow{PQ}$ on the edge of half-plane R: For any real number n, where $0 < n < 180$, there is one and only one ray $\overrightarrow{PB}$, where B is in R, such that $m\angle QPB = n$.

If $m\angle BPQ = n$ and $m\angle APB = r$, what is $m\angle APQ$? The following postulate gives us a way to find this measure:

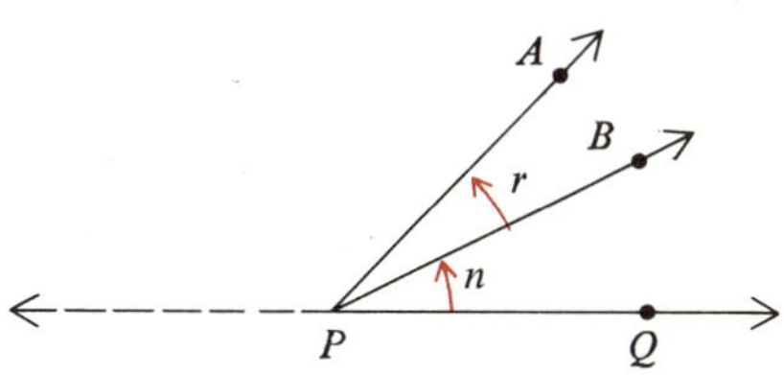

P **Postulate 2-10 The Angle Sum Postulate** If B is in the interior of $\angle APQ$, then $m\angle APQ = m\angle APB + m\angle BPQ$.

Example 4 If in the figure above $m\angle BPQ = 12$ and $m\angle APB = 37$, find $m\angle APQ$.

Solution By the Angle Sum Postulate and the given measures, we may write

$$m\angle APQ = 37 + 12$$

or

$$m\angle APQ = 49.$$

Example 5 If $m\angle ABC = 112$ and $\angle ABD$ is a right angle, find $m\angle DBC$.

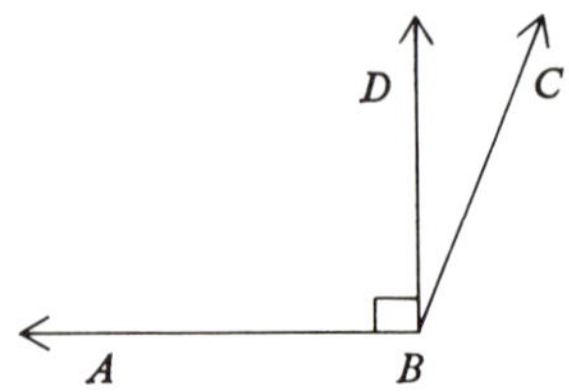

Solution By the Angle Sum Postulate, we have $m\angle ABC = m\angle ABD + m\angle DBC$. But since $\angle ABD$ is a right angle, $m\angle ABD = 90$. Thus we can write

$$112 = 90 + m\angle DBC$$

or

$$22 = m\angle DBC.$$

This last example leads to the next postulate. (Refer to the figure above.)

P **Postulate 2-11 The Angle Difference Postulate** If A is in the exterior of $\angle DBC$ and in the same half-plane, for edge $\overleftrightarrow{BC}$, as D, then $m\angle ABD = m\angle ABC - m\angle DBC$.

Postulate 1-6, the Supplementary Angles Postulate, is the last postulate that you need to review from Chapter 1. This concludes our discussion of postulates in Chapter 2. New postulates will be added as they are needed.

Example 6 For the linear pair $\angle ABC$ and $\angle CBD$, we know that $m\angle ABC = 150$. What is $m\angle CBD$?

Solution Since $\angle ABC$ and $\angle CBD$ are a linear pair, the angles are supplementary. The sum of their measures is 180. Thus, $m\angle CBD + 150 = 180$, or $m\angle CBD = 30$.

See how well you understand the postulates you have studied by answering the following exercises. Be sure you understand these postulates before moving on. They are essential to the study of geometry.

EXERCISES

A State the property illustrated by each of the following:

1. If $3x = 24$, then $x = 8$.

2. If $AB = CD$ and $CD = EF$, then $AB = EF$.
3. If $5x > -y$, then $-15x < 3y$.
4. $3 + (\sqrt{2} + x) = (x + \sqrt{2}) + 3$
5. $x = 0$, $x < 0$, or $x > 0$
6. $\sqrt{3}\,(\sqrt{2} + \sqrt{5})$ is a real number.
7. $AB = \frac{1}{2} AC$; thus $2(AB) = AC$
8. If $m\angle ABC = m\angle XYZ$, then $m\angle XYZ = m\angle ABC$.
9. $5(AB + CD) = 5(AB) + 5(CD)$
10. If $m\angle ABC = 90$ and $90 = m\angle PQR$, then $m\angle ABC = m\angle PQR$.

State the property, postulate, or definition that makes each of the following true:

11. If $\angle MPQ$ is a right angle, then $m\angle MPQ = 90$.
12. If $\angle MPR$ and $\angle QPR$ are complementary, then $m\angle MPR + \angle QPR = 90$.
13. If $m\angle MPR + m\angle QPR = 90$ and $m\angle MPR = 15$, then $15 + m\angle QPR = 90$.
14. If $15 + m\angle QPR = 90$, then $m\angle QPR = 75$.

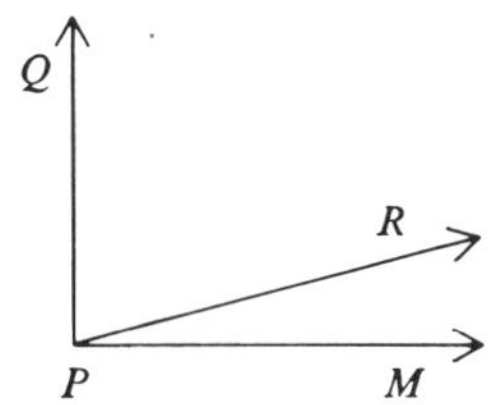

Exercises 11–14

15. If $\angle A$ and $\angle B$ are a linear pair, then $\angle A$ and $\angle B$ are supplementary angles.
16. If $m\angle A = m\angle B$ and $m\angle A + m\angle B = 180$, then $2(m\angle A) = 180$.
17. If $2(m\angle A) = 180$, then $m\angle A = 90$.
18. If $m\angle A = 90$, then $\angle A$ is a right angle.

Complete each of the following statements:

19. To every real number there ________ exactly one point of the line.
20. Space contains at least four ________ points.
21. To every angle there corresponds a real number between ________ and ________.
22. ________ noncollinear points determine a triangle.
23. Two distinct planes intersect in a(n) ________.
24. Two intersecting lines determine a(n) ________.
25. Two angles forming a linear pair are ________.
26. If $m\angle X + m\angle Y = m\angle Z$ and if $m\angle X = 10$, and $m\angle Y = 35$, then $m\angle Z =$ ____.
27. If $\angle A$ and $\angle B$ are a linear pair such that $m\angle A = 110$, then $m\angle B =$ ____.

B State the postulate or properties of which each of the following is an instance:

28. The measure of any angle is equal to itself.
29. If $m\angle ABC = 30$ and $m\angle ABD = 60$, $m\angle ABC = \frac{1}{2}(m\angle ABD)$.
30. If $m\angle CBD = 12$ and $m\angle CBA = 13$, then $m\angle ABD = 25$.

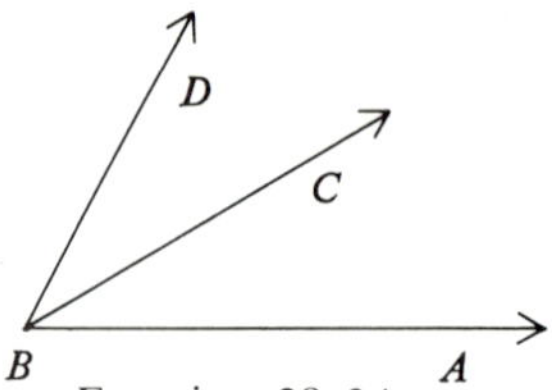

Exercises 28–34

31. If $m\angle ABD = 30$ and $m\angle CBA = 13$, then $m\angle CBD = 17$.
32. $\overrightarrow{BC}$ is the only ray in the half-plane with D such that $m\angle ABC = 30$.
33. Given $\overleftrightarrow{EBA}$ and C not on $\overleftrightarrow{EBA}$, $m\angle EBC + m\angle CBA = 180$.
34. If $\overleftrightarrow{AB} \cap$ plane P contains points A and B, then $\overleftrightarrow{AB}$ lies in plane P.

Classify each of the following as either true or false:

35. "Every line segment has a midpoint" is an example of the Point Betweenness Postulate.
36. There is only one point P on $\overrightarrow{AR}$ such that $\overline{AP} \cong \overline{BD}$, where $\overline{BD}$ is a given segment.
37. Any pair of supplementary angles forms a linear pair.
38. There are only two planes that can contain a given line.

C What conclusion can be drawn from each of the following? Each pair of statements given is considered true.

39. $\overleftrightarrow{AB}$ and $\overleftrightarrow{CD}$ are lines. $\angle AOC$ and $\angle COB$ are a linear pair. Thus, __________. Why?
40. $\angle COB$ and $\angle BOD$ are supplementary. $\angle BOD$ and $\angle DOA$ are supplementary. Therefore, __________. Why?
41. $\angle MYN$ and $\angle XYZ$ are right angles; $m\angle 1 + m\angle 2 = 90$; and $m\angle 2 + m\angle 3 = 90$. Thus, $m\angle 1 =$ __________.

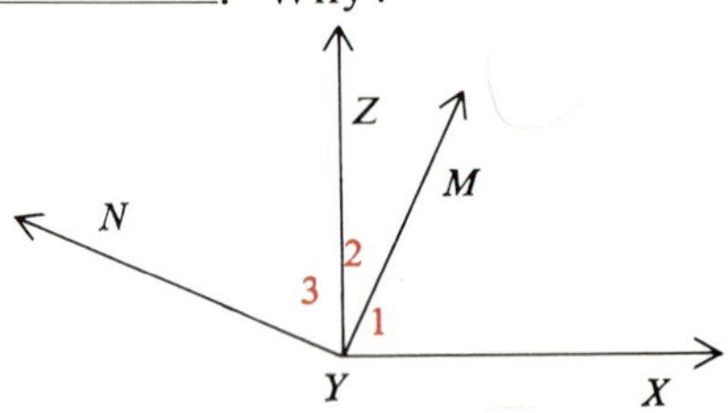

Exercise 41

2-4 LOGIC

Consider this set of statements:

All A's are B's.
This object is an A.

What can you conclude about the object mentioned in the second sentence?

You should have concluded, "Therefore, this object is a B." *Therefore* indicates that the statement it introduces follows from the given statements.

All A's are B's.
This object is an A.
$\therefore$ This object is a B.

The three dot arrangement $\therefore$ is the symbol for *therefore*. This set of three statements is an example of the use of deductive reasoning in a logical argument. When you reason from accepted statements to a conclusion or new fact, you are using deductive reasoning. The conclusion is a **deduction.** Care must be taken when reasoning is used. For example, consider these statements:

Serpents are egg eaters.
You are an egg eater.

What is your conclusion? Did you conclude, "You are a serpent"? This is not a deduction that can be drawn from these sentences. We did not say that serpents are the only egg eaters.

In geometry, we are concerned with ideas that are expressed as statements. We wish to draw deductions from the postulates and definitions. We see from the discussion above that we cannot guess what to deduce. We must master logical thinking to arrive at our deductions. Before we can learn deductive reasoning we must start with a study of statements.

Statements are usually declarative sentences containing a subject and predicate. They declare some fact, such as "It is cold." Every statement has its **negation**, the denial of the statement—for example, "It is not true that it is cold." This negation can be shortened to "It is not cold."

Example 1 Write the negation of the statement "$\triangle ABC$ is isosceles."

Solution The negation must deny the statement. Thus, the negation is "$\triangle ABC$ is not isosceles." We could also state the negation, "It is false that $\triangle ABC$ is isosceles."

To simplify the study of logic, we use letters to represent the statements. We can let the letter p represent the statement "$\angle PQR$ is a right angle." Thus,

p: $\angle PQR$ is a right angle.

The negation of p is represented by $\sim p$. Thus,

$\sim p$: $\angle PQR$ is not a right angle.

Both p and $\sim p$ are statements. The above example shows that a statement may be either true or false.

Compound Statements

Compound statements are formed by combining two or more statements with the connectives *and*, *or*, or *if . . . then*. Using the connectives respectively, we form **conjunctions**, **disjunctions**, and **implications**.

The first compound statement we shall study is the conjunction. Its connective is *and*. The conjunction of statements p and q is written $p \wedge q$, which is read "p and q."

Example 2 Write the conjunction of p and q, given:

$$p\text{: } m\angle MON = 130 \qquad q\text{: } m\angle NOP = 50$$

Solution A conjunction connects statements with *and*.

$$p \wedge q\text{: } m\angle MON = 130 \text{ and } m\angle NOP = 50$$

When we use the connective *or* with the statements p and q, we form a disjunction. The disjunction "p or q" is written $p \vee q$. The disjunction of the p and q statements from the example above is:

$$p \vee q\text{: } m\angle MON = 130 \text{ or } m\angle NOP = 50$$

So far, we have formed the following:

p: a statement

$\sim p$: the negation of p

$p \wedge q$: statement p and statement q

$p \vee q$: statement p or statement q

Example 3 Translate the statement below into symbols.

$\angle ABC$ and $\angle CBD$ form a linear pair and $m\angle ABC = 90$.

Solution Let p represent "$\angle ABC$ and $\angle CBD$ form a linear pair," and q represent "$m\angle ABC = 90$." In symbols, we have:

p: $\angle ABC$ and $\angle CBD$ form a linear pair. $\qquad q$: $m\angle ABC = 90$

The statement "$\angle ABC$ and $\angle CBD$ form a linear pair and $m\angle ABC = 90$" is a conjunction, which can now be represented thus: $p \wedge q$.

Example 4 Translate $\sim p \vee \sim q$ into English.

p: It is cold. $\qquad q$: It is snowing.

Solution $\sim p$: It is not cold. $\qquad \sim q$: It is not snowing.

$\sim p \vee \sim q$: It is not cold or it is not snowing.

Example 5 Translate the statement $\sim(r \vee s)$.

$$r\text{: } \triangle PQR \text{ is isosceles.} \qquad s\text{: } m\angle P = 30$$

Solution $(r \vee s)$: $\triangle PQR$ is isosceles or $m\angle P = 30$. We can read the negation sign, "It is not true that."

$$\sim(r \vee s)\text{: It is not true that } \triangle PQR \text{ is isosceles or } m\angle P = 30.$$

Can you write the statement from Example 5 in a different way? Is this the same as $\sim r \wedge \sim s$? Try translating this, letting r and s represent the two sentences from Example 5.

So far in our discussion of statements, we have not considered the truth or falsity of the statements. All statements must obey two basic laws of logic, presented below.

The Law of the Excluded Middle Any statement is either true or it is false.

The Law of Contradiction A statement cannot be both true and false.

While these laws are easily understood, their simplicity should not minimize their usefulness. If p is true, what can we say about $\sim p$? Can $\sim p$ ever be true? Can we talk about the truth of the conjunctions and disjunctions? This depends on how we define the use of the connectives.

The connective *and* and the set operation intersection are similar. The intersection ($\cap$) of two sets is the set of elements common to both sets. A given element a satisfies one of four possible conditions:

a is in A and a is in B — a is not in A but a is in B
a is in A but not in B — a is not in A and a is not in B

For a to be in the set $A \cap B$, a must be in both sets A and B, thus satisfying the first condition above.

Now consider the conjunction $p \wedge q$. As with intersection, four different possibilities arise:

p is true, q is true — p is false, q is true
p is true, q is false — p is false, q is false

We limit the truth of statements connected by *and* to the case when both statements are true. Thus, only the first case above is true.

D **Definition 2-4** The statement $p \wedge q$ is true if and only if p and q are both true.

Do you see why we said that set intersection and the connective *and* are similar? Even the symbols $\cap$ and $\wedge$ have a certain similarity.

We use a **truth table** to determine the truth of compound statements. To construct a truth table, we list the letters for the statements we are considering. In columns under the letters, we list all the possible combinations of true and false values for both letters. For example, for the statements p and q, we have the table at right.

p	q
T	T
T	F
F	T
F	F

To set up a truth table for a conjunction, use the headings p, q, and $p \wedge q$. Enter the four possible instances mentioned above. Apply Definition 2-4 and keeping in mind the Law of the Excluded Middle, enter the truth values of $p \wedge q$.

p	q	$p \wedge q$
T	T	T
T	F	F
F	T	F
F	F	F

Using the Law of Contradiction, we can form the truth table for $\sim r$, as shown at right.

r	$\sim r$
T	F
F	T

We shall now consider a second compound statement, the disjunction. The connective *or* and the set operation union are similar. The union of two sets contains those elements that are elements of at least one of the sets. As stated before, a given element satisfies one of the following cases:

a is in A and a is in B
a is in A but a is not in B
a is not in A but a is in B
a is not in A and a is not in B

For a to be in the set $A \cup B$, a must be in either set A or set B, or both. Thus, the first three cases satisfy the condition for a union.

Consider the disjunction "p or q" ($p \vee q$). As with conjunction, four possibilities arise, as listed in the truth table at right. The disjunction $p \vee q$ is true if at least one of the statements is true. This means that $p \vee q$ is true if p is true, if q is true, or if they are both true. When is $p \vee q$ not true?

p	q
T	T
T	F
F	T
F	F

Do you agree that there is a similarity between the disjunction *or* and the set operation of union? Do you think that the symbols $\cup$ and $\vee$ are also similar?

D **Definition 2-5** The statement $p \vee q$ is true whenever p is true or q is true or both are true.

Example 6 Construct the truth table for $p \vee q$.

Solution Use headings p, q, and $p \vee q$. Enter the possible pairings of p and q. Apply Definition 2-5 and enter the truth values of $p \vee q$.

p	q	$p \vee q$
T	T	T
T	F	T
F	T	T
F	F	F

Example 7 Compare the truth tables of $p \wedge q$ and $q \wedge p$.

Solution

p	q	$p \wedge q$	$q \wedge p$
T	T	T	T
T	F	F	F
F	T	F	F
F	F	F	F

Note in the last example that the two truth tables are the same. When two statements have the same truth table, we say they are **equivalent statements**.

Example 8 Construct the truth table for $r \vee \sim r$

Solution

r	$\sim r$	$r \vee \sim r$
T	F	T
F	T	T

When the truth values of a statement are all true, the statement is a **tautology**. Thus, $r \vee \sim r$ is a tautology. How else can we justify that $r \vee \sim r$ is a tautology?

Class Exercises

Can you state a possible conclusion for each of the following sets of statements? If so, what is the conclusion?

1. All people who study pass geometry.
   Al studies.
2. All dogs have four legs.
   Tabby has four legs.

Let r represent "I am taking you to the dance" and q represent "I like movies" for Exercises 3–13. Translate each of the following into sentences:

3. $r \wedge \sim q$
4. $\sim q \vee \sim r$
5. $\sim r \wedge \sim q$
6. $\sim(r \vee q)$
7. $\sim(\sim r)$
8. $\sim r \vee q$
9. Which of the statements in Exercises 3–8 are equivalent?

Translate each of the following into symbolic form:

10. I do not like movies.
11. I am taking you to the dance and I like movies.
12. It is false that I am taking you to the dance and I do not like movies.
13. I am not taking you to the dance and I like movies.

Construct truth tables for each of the following:

14. $p \wedge \sim q$
15. $\sim p \vee \sim q$
16. $\sim(\sim p \wedge q)$
17. Which of the statements in Exercises 14–16 are tautologies?

The last compound statement we shall consider is *implication*. Its connective consists of the two words *if* and *then*. The *if* precedes a statement called the

hypothesis. The *then* precedes a statement called the **conclusion**. The usual form for an implication is "if p, then q." In symbols, we write $p \rightarrow q$. We can also read this "p implies q." The implication is sometimes called a **conditional statement**. As with other compounds, we must consider the truth values of the implication. Consider the following implications:

p: You pass this test. q: I'll give you a C.
$p \rightarrow q$: If you pass this test, then I'll give you a C.

As we did with the other connectives, let us examine the various possibilities of truth values for p and q. Then we can construct a truth table for $p \rightarrow q$ using the headings p, q, and $p \rightarrow q$.

Case I p is true and q is true. You pass the test and I give you a C. I have kept my word; the implication held. $p \rightarrow q$ is true.

Case II p is true and q is false. You pass the test but I do not keep my word. I do not give you a C. The implication does not hold. $p \rightarrow q$ is false.

Case III p is false and q is true. You fail the test but on past grades I give you a C. I did not break my word. The implication is still true. $p \rightarrow q$ is true.

Case IV p is false and q is false. You fail the test and on past grades I do not give you a C. Again I do not break my word. The implication is still true. $p \rightarrow q$ is true.

	p	q	$p \rightarrow q$
Case I	T	T	T
Case II	T	F	F
Case III	F	T	T
Case IV	F	F	T

From the discussion above and the truth table we draw the following definition:

D **Definition 2-6** The statement $p \rightarrow q$ is true except when p is true and q is false.

How would you read $q \rightarrow p$? Are $p \rightarrow q$ and $q \rightarrow p$ equivalent? Using the above examples, what is the hypothesis of $q \rightarrow p$? What is the conclusion?

Example 9 Determine the truth value of $p \rightarrow q$ where p: $|x| = -5$ and q: $|5| = 5$.

Solution p is false since the absolute value is never negative; q is true. By definition, $p \rightarrow q$ is true. This is an instance of Case III.

Example 10 Construct a truth table for $(p \wedge q) \rightarrow p$.

Solution Use as headings p, q, $p \wedge q$, and $(p \wedge q) \rightarrow p$. Under the last heading, enter the truth values for $p \wedge q$ and for p, and apply the definition of implication. Enter the values for the implication under the arrow. $(p \wedge q) \rightarrow p$ is a tautology.

p	q	$p \wedge q$	$(p \wedge q) \rightarrow p$
T	T	T	T TT
T	F	F	F TT
F	T	F	F TF
F	F	F	F TF

The last column can become confusing with so many T's and F's. If we are careful, we can enter only the implication values under the arrow, by using the third and first column values to determine these implication values.

Where and when do we use the concepts of logic? We use them in deductive reasoning. We use them in making logical arguments.

D **Definition 2-7** Arguments are **valid** if acceptance of the hypothesis as true means that the conclusion must be true.

Usually the hypothesis is a compound sentence in the form of a disjunction or implication along with a true simple statement. By analyzing the truth table for the hypothesis, we can determine the truth of the conclusion. Study the following examples to see how this analysis works.

Example 11 Determine if the following argument is valid:
Hypothesis: If $m\angle A = 90$, then $\triangle ABC$ is a right triangle; and it is true that $m\angle A = 90$.
Conclusion: $\triangle ABC$ is a right triangle.

Solution p: $m\angle A = 90$
q: $\triangle ABC$ is a right triangle.
Hypothesis: If $(p \rightarrow q) \wedge p$
Conclusion: q
Study the table. Do you see that the only time that the hypothesis is true, q is also true. That is, the only time the implication $p \rightarrow q$ is true and p is true is where q is true. Thus the argument is valid.

p	q	$p \rightarrow q$	$(p \rightarrow q) \wedge p$
T	T	T	T
T	F	F	F
F	T	T	F
F	F	T	F

The hypothesis is often a conjunction of the form $(p \rightarrow q) \wedge p$. Each statement can be put on a separate line, with the connective *and* understood. The statements in Example 11 could have been set up this way. Although the connectives in Example 12 are different, the form follows this pattern.

Example 12 Determine if the following argument is valid:
Hypothesis: $x < 5$ or $x \geq 5$
$x \geq 5$
Conclusion: $x < 5$

Solution
p: $x < 5$ q: $x \geq 5$
Hypothesis: $(p \vee q) \wedge q$
Conclusion: p
Complete the table. When the hypothesis is true, the conclusion p is either true or false. The argument is *not* valid.

p	q	$p \vee q$	$(p \vee q) \wedge q$
T	T	T	T
T	F	T	F
F	T	T	T
F	F	F	F

We shall explore these ideas further in the next section, as we attempt to prove some geometric implications.

EXERCISES

A

Complete the following statements:

1. The conclusion in deductive reasoning is called a(n) ________.
2. Statements are usually ________ sentences.
3. Compound statements are formed by combining ________ or more statements by ________.
4. The ________ of p and q is written as $p \vee q$.
5. Every statement has as its ________ the denial of the statement.
6. The conjunction of p and q is indicated symbolically by ________.
7. "Any statement is either true or false" is the Law of ________.
8. The connective *or* and the set operation ________ are similar.
9. In $p \rightarrow q$, p is the ________ and q is the ________ of the implication.
10. Usually the hypothesis is a ________ statement.

State a valid conclusion, if possible, for each of the following hypotheses:

11. If Bill works, then he will get a raise.
    Bill works.
12. It is raining or it is snowing.
    It is not snowing.
13. The sides are congruent and the angles are congruent.
    The sides are congruent.
14. If it is not snowing, then we can go to the show.
    It is not snowing.
15. If you pass the test, then you will pass for the quarter.
    You do not pass the test.

For Exercises 16–28, let p: It is isosceles; q: They are congruent; and r: The measures are equal. Translate each of the following:

16. $\sim p \wedge q$ 17. $\sim q \vee \sim p$ 18. $p \to \sim q$
19. $p \wedge q \wedge r$ 20. $p \vee (q \wedge r)$ 21. $\sim(p \wedge q)$

B

22. It is not isosceles.
23. They are congruent and the measures are equal.
24. It is false that they are congruent and it is isosceles.
25. It is not isosceles or they are congruent and equal.
26. It is false that the measures are not equal.
27. If the measures are not equal, then they are not congruent and it is not isosceles.

Construct truth tables for each of the following:

28. $\sim p \wedge q$ 29. $p \vee \sim q$ 30. $\sim(\sim q)$
31. $\sim(p \wedge q)$ 32. $p \vee q \vee r$ 33. $(p \vee q) \vee \sim q$
34. $p \to \sim q$ 35. $p \to (q \wedge p)$ 36. $p \to \sim p$
37. $(p \wedge q) \to p$ 38. $(p \vee q) \to (q \vee p)$ 39. $(p \vee \sim q) \to \sim q$

C

Use truth tables to determine if the following are valid. Choose your own symbols for each statement.

40. Jack lives here or he lives somewhere else.
    Jack does not live here.
    ∴ Jack lives somewhere else.
41. Joe works in the office and Joe is lazy.
    Joe is lazy.
    ∴ Joe works in the office.
42. If $n \geq 0$, then n is nonnegative.
    $n = 0$
    ∴ n is nonnegative.
43. Betty will fix lunch or she will go downtown.
    Betty fixes lunch.
    ∴ Betty does not go downtown.
44. $2p < 7$ or $3p > 9$
    $3p > 9$
    ∴ $2p < 7$
45. If $\triangle ABC$ is isosceles, then $AB = AC$.
    $\triangle ABC$ is not isosceles.
    ∴ $AB \neq AC$

A LOOK AT THE PAST

A Writer of Logic

Alice's Adventures in Wonderland by Lewis Carroll is a study in logic. The Reverend Charles L. Dodgson, a mathematician, used Lewis Carroll as his pen name. While Dodgson was teaching at Oxford, he wrote "Alice's Adventures Under Ground," which he later expanded to *Alice's Adventures in Wonderland.* Most children read it as a fairy tale, but it is actually a study in logic.

The Alice stories are a science-fiction narrative about a girl who acts upon the accepted assumptions of her world during a visit to a world in which many of these assumptions do not hold.

Dodgson built into the story examples of accepted definitions, laws of logic, and logical argument. See if you can find the hypothesis and conclusion in the argument by which the pigeon tries to prove logically that Alice is not a girl, although she has just said that she is a girl. The pigeon says, "I've seen a good many little girls in my time, but never *one* with such a neck as that! No! You're a serpent . . . . "

Would you say that the pigeon postulated that no girl has such a long neck as Alice? This is just one instance of the many assumptions and implications used throughout the book. You might enjoy rereading the book to find other examples.

2-5 THEOREMS

The last part of a logical system is the set of statements that are logically proved from the postulates, definitions, and previously proved statements. This set of statements is more generally referred to as the set of theorems.

Theorems are usually in the form of implications: "If p, then q." To prove a theorem, we must show that the statement follows logically from definitions, postulates, or other theorems. The proof is a chain of implications. In a direct proof, we start with the hypothesis and show that this leads to the conclusion. The justifications for the statements in a proof are called **reasons**. After a theorem is proved, we may, in turn, use it as a reason to justify statements in the proofs of subsequent theorems.

The first few theorems follow almost directly from definitions and postulates. We shall state each theorem for future use and present a paragraph to justify it. To check your ability to justify our statements, see if you can cite reasons other than those that are given. Once we have presented these theorems we shall use them in the next section to help us examine formal proofs.

Reasoning from postulates for the plane, line, and points in a plane, we can derive the first theorem. The numbering 2-5.1 means that the theorem appears in Chapter 2, Section 5, and that it is the first theorem presented.

T **Theorem 2-5.1** If a point does not lie in a given line, then there is exactly one plane containing both the point and the line.

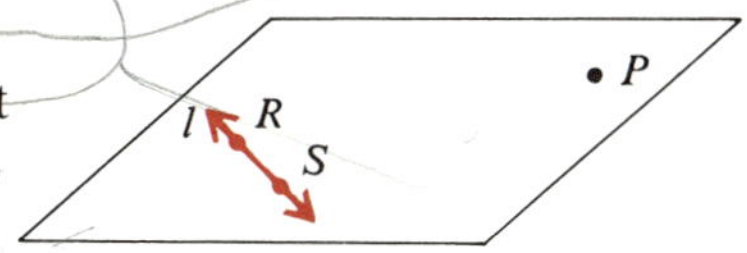

PROOF By the hypothesis, we are given a line *l* and a point *P* not in *l*. Since a line is determined by two distinct points, the line must contain at least two distinct points, *R* and *S*. Thus *R*, *S*, and *P* are three noncollinear points determining exactly one plane. If *R* and *S* lie in the plane, then we know that all points of *l* lie in the plane. Therefore, there is exactly one plane containing the point and the line.

Can you cite the exact postulates used in the proof of Theorem 2-5.1? The next theorem is based upon the Line Postulate.

T **Theorem 2-5.2** If two distinct lines intersect, then they intersect in exactly one point.

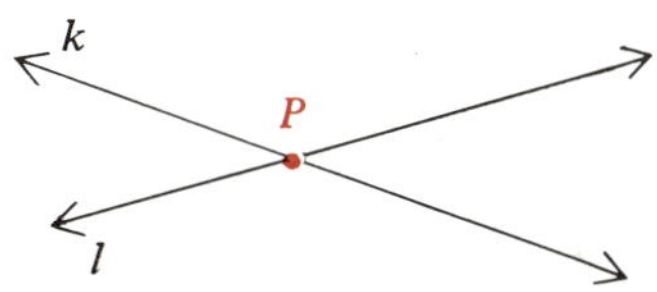

PROOF We are given two distinct lines *l* and *k* that intersect in point *P*. Suppose the lines also intersect in point *R*, a point distinct from point *P*. The Line Postulate, however, states that any two points determine exactly one line containing them both, not two distinct lines, as we are given. Thus, the lines cannot intersect in a second point.

We shall use the first two theorems to justify the next theorem.

T **Theorem 2-5.3** If two lines intersect, then there is exactly one plane containing them.

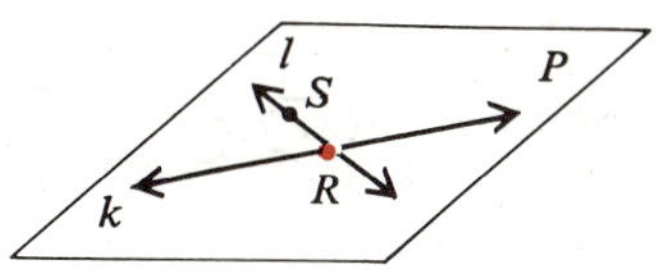

PROOF Lines *l* and *k* intersect in point *R*. Since any line contains at least two points, let *l* contain points *R* and *S*. We know that *S* does not lie in *k*: by Theorem 5-5.2, distinct lines intersect in only one point. We thus have *k*, and *S* not in *k*; by Theorem 2-5.1, there is exactly one plane containing both. Since *l* contains *S* and *R*, which lie in the same plane as *k*, *l* must lie in the same plane also.

Class Exercises

1. If M is between P and Q on $\overline{PQ}$, then $PM + MQ =$ ____.
2. If M is also the midpoint of $\overline{PQ}$, how are PM and MQ further related?
3. From Exercise 2, we conclude that $PM = k(PQ)$. What is the numerical value of k?
4. By the Distance Postulate, PQ is a unique positive number. From this what can we conclude about PM?
5. In Exercise 4, point M is the only midpoint of $\overline{PQ}$. True or false?

The Class Exercises suggest the following theorem:

Theorem 2-5.4 The Midpoint Uniqueness Theorem Any line segment has exactly one midpoint.

PROOF According to the Distance Postulate, to every pair of distinct points there corresponds a unique positive number. That is, to any segment $\overline{PQ}$ there corresponds its measure, PQ. Let $n = \frac{1}{2}(PQ)$. According to the Point Uniqueness Postulate, there corresponds *exactly one point M* on $\overrightarrow{PQ}$ such that $PM = n$. This point is the midpoint of PQ.

The next theorem follows directly from the definitions.

Theorem 2-5.5 All right angles are equal in measure.

PROOF Let $\angle PQR$ and $\angle MON$ be any two right angles. By the definition of a right angle, $m\angle PQR = 90$ and $m\angle MON = 90$. Since $\angle PQR$ and $\angle MON$ are any two right angles, we can say that all right angles are of equal measure.

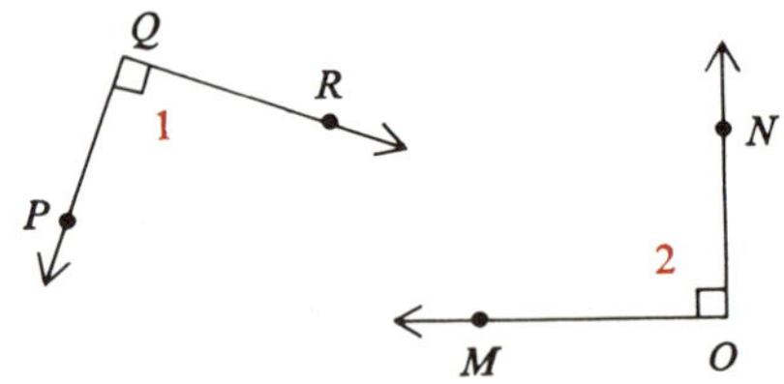

For the next theorem, remember that two angles are supplementary angles if the sum of their measures is 180.

Theorem 2-5.6 Two supplementary angles of equal measure are right angles.

PROOF In the figure for Theorem 2-5.5, let $\angle 1$ and $\angle 2$ be two supplementary angles such

that $m\angle 1 = m\angle 2$.	Reason: Hypothesis
$m\angle 1 + m\angle 2 = 180$	Reason: Definition of supplementary angles
$m\angle 1 + m\angle 1 = 180$	Reason: Substitution Postulate
$2m\angle 1 = 180$	Reason: Addition facts
$m\angle 1 = 90$	Reason: Multiplication property of equality
Since $m\angle 2 = m\angle 1 = 90$, $\angle 1$ and $\angle 2$ are right angles.	Reason: Definition of right angles

A

EXERCISES

State the postulate or definition that justifies each of the following:

1. A line is determined by two distinct points.
2. A line lies in a plane when two points of the line are in the plane.
3. R is the midpoint of $\overline{MN}$ when $MR + RN = MN$, and $MR = RN$.
4. ON is a unique number.
5. There is exactly one point P on $\overrightarrow{OR}$ such that $OP = n$, where n is a positive number.
6. If line l lies in a plane, then there must be a point in the plane not in l.

Indicate which of the following are true and which are false:

7. The proof of a theorem is composed of only definitions and postulates.
8. A direct proof starts with a hypothesis, which leads to a conclusion.
9. In the implication $p \rightarrow q$, p is the hypothesis.
10. Two points that lie in different planes are noncollinear.
11. If two lines intersect in two points, the lines are not distinct.

Complete each of the following:

12. Every line segment has ________ midpoint.
13. Two intersecting lines determine exactly one ________.
14. A line and a point not on that line determine exactly one ________.
15. Our theorems are usually in the form of ________.
16. Half of a positive number is a(n) ________ number.

B

To prove the statements in Exercises 17–25, you would have to determine the hypothesis and conclusion for each statement. Write the hypothesis for each of the following:

17. Any line segment has exactly one midpoint.
18. All right angles are equal in measure.
19. Two supplementary angles of equal measure are right angles.
20. Complements of congruent angles have the same measure.

21. Two perpendicular lines form right angles.
22. Intersecting lines that form right angles are perpendicular.
23. If two angles are complementary to the same angle, then the angles are congruent.
24. If two lines are perpendicular, then four right angles are formed.
25. If $\overrightarrow{OP}$ and $\overrightarrow{OR}$ are the noncommon sides of a linear pair, then $\overrightarrow{OP}$ and $\overrightarrow{OR}$ are opposite rays.

Write the conclusion for the statement from each of the exercises listed below.

26. Exercise 18 27. Exercise 20 28. Exercise 22 29. Exercise 24

For each of the exercises listed below, write the statement in the "if p, then q" form.

30. Exercise 17 31. Exercise 18 32. Exercise 19
33. Exercise 20 34. Exercise 21 35. Exercise 22

C

36. Write the hypothesis and conclusion for the statement, "Two angles that are complementary to the same angle are equal in measure."

Supply the reasons for the following statements, which support the conclusion from Exercise 40:

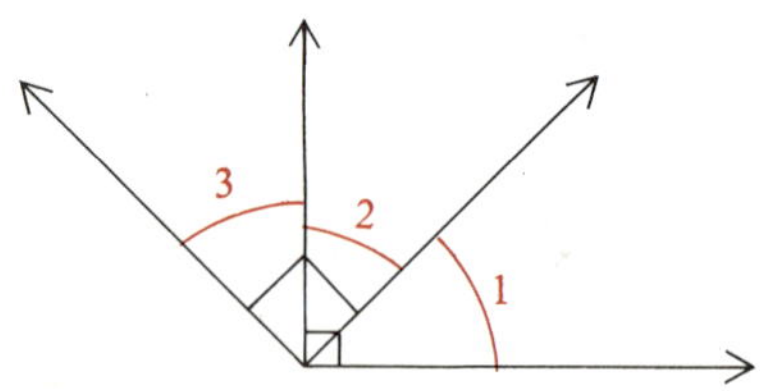

Exercises 36–40

STATEMENTS	REASONS
37. $\angle 1$ and $\angle 2$ are complementary; $\angle 2$ and $\angle 3$ are complementary.	
38. $m\angle 1 + m\angle 2 = 90$ and $m\angle 2 + m\angle 3 = 90$	
39. $m\angle 1 + m\angle 2 = m\angle 2 + m\angle 3$	
40. $m\angle 1 = m\angle 3$	

Write the hypothesis and conclusion for the following statement and, in the table below, give reasons for the statements supporting the conclusion.

STATEMENT Two angles that form a linear pair with the same angle have equal measures.

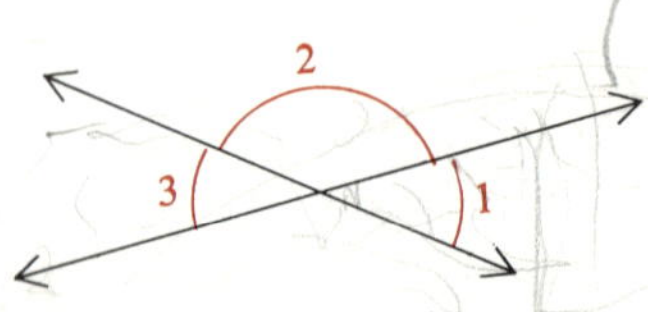

41. Hypothesis: ______

42. Conclusion: ______

STATEMENTS	REASONS
43. $\angle 1$ and $\angle 2$ form a linear pair; $\angle 2$ and $\angle 3$ form a linear pair.	
44. $\angle 1$ and $\angle 2$ are supplementary; $\angle 2$ and $\angle 3$ are supplementary.	
45. $m\angle 1 + m\angle 2 = 180$; $m\angle 2 + m\angle 3 = 180$.	
46. $m\angle 1 + m\angle 2 = m\angle 2 + m\angle 3$	
47. $m\angle 1 = m\angle 3$	

48. Prove that two angles complementary to congruent angles are congruent to each other.
49. Prove that supplements of the same angle are of equal measure.

SOMETHING TO THINK ABOUT

Other Systems

Geometry is not the only logical system in existence. We have noted that algebra is also a logical mathematical system. Not every logical system, however, is mathematical in nature. Consider the following set of postulates. The undefined terms are *club*, *committee*, and *person*.

Postulates 1 and 2 deal with relationships between committees:
P1 Any two committees have at least one person in common.
P2 Any two committees have not more than one person in common.

Postulates 3 and 4 relate people to committees:
P3 Every person in the club is on at least two committees.
P4 Every person in the club is on not more than two committees.

Postulate 5 tells us the number of committees in the club:
P5 The total number of committees in the club is four.

Logical conclusions can be drawn from these postulates. From the first two postulates we can conclude that any two committees have one and only one person in common. A similar conclusion can be obtained from the third and fourth postulates—every person in the club is on two and only two committees.

From P5 we know that there are exactly four committees. Label them L_1, L_2, L_3, and L_4. How many people are in the club? If we know the number of people in the club, we can determine how many people are on each committee. To answer these questions, draw a diagram of the system. Let a plane represent *club*, a line represent *committee*, and a point represent *person*. The intersection of two lines would represent a person on two committees. What does your diagram look like?

2-6 DIRECT PROOF

In the last section, we saw that **direct proof** starts with a hypothesis and leads to a conclusion. The theorems we discussed and presented were not proved formally. Rather, we used arguments supporting each theorem and giving the general outline of the proof. While not difficult, formal proofs might have caused some confusion as a beginning step. The proofs below provide a clear introduction to formal proof.

In this section, we shall set up our proofs according to an established form. Using this form, we begin by stating the theorem to be proved. Next we list the relationships given in the hypothesis and the relationship to be proved in the conclusion, usually illustrating the relationships given by a diagram. After this we are ready for the proof itself. Above separate columns, we write the headings *Statements* and *Reasons*. In the first column we list the logical statements of the proof. In the second we list the reasons for the statements.

The following is a model of the form just described:

T **Theorem 0-0.0** Statement of theorem

GIVEN Relationships from the hypothesis

PROVE Relationship to be proved

PROOF

STATEMENTS	REASONS
LIFE IS DULL	GEOMETRY CLASS

A proof using this structure is referred to as the **two-column proof.** We shall use the two-column format to prove the following theorem:

T

Theorem 2-6.1 Complements of angles of equal measure, or of the same angle, have the same measure.

GIVEN $\angle 1$ and $\angle 2$ are complementary angles; $\angle 3$ and $\angle 4$ are complementary angles; $m\angle 1 = m\angle 3$

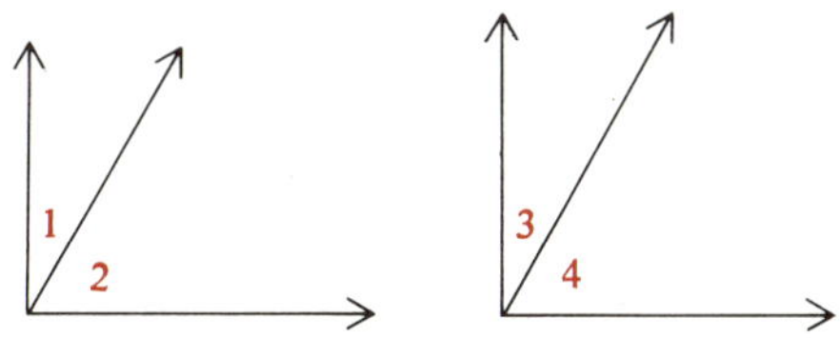

PROVE $m\angle 2 = m\angle 4$

PROOF

STATEMENTS	REASONS
1. $\angle 1$ and $\angle 2$ are complementary angles; $\angle 3$ and $\angle 4$ are complementary angles; $m\angle 1 = m\angle 3$	1. Given
2. $m\angle 1 + m\angle 2 = 90$; $m\angle 3 + m\angle 4 = 90$	2. Definition of complementary angles (1-27)
3. $m\angle 1 + m\angle 2 = m\angle 3 + m\angle 4$	3. Transitive property of equality
4. $m\angle 2 = m\angle 4$	4. Addition property of equality

Since we are adding the opposites of equal quantities to the equality in statement 3, the final reason stated is the addition property of equality. We have proved the theorem for angles of equal measures. By the reflexive property, the measure of any angle is equal to itself. Thus, the case of complements of the same angle is a special instance of the theorem.

The proof of the next theorem is almost identical to the proof above. This proof is left as an exercise for you to do.

T

Theorem 2-6.2 Supplements of angles of equal measure, or of the same angle, have the same measure.

Angles Formed by Intersecting Lines

When two lines intersect in a point, how many rays are formed? How many angles are formed? In the figure for Theorem 2-6.3, ∠1 and ∠3 are opposite each other. They are called *vertical* angles. ∠2 and ∠4 also form a pair of vertical angles.

D

Definition 2-8 If the sides of two angles form two pairs of opposite rays, then the angles are **vertical angles**.

In the figure below, how does the m∠1 compare to m∠3? Is m∠2 = m∠4? The answers to these questions lie in the conclusion of the next theorem.

T

Theorem 2-6.3 The Vertical Angle Theorem Vertical angles have the same measure.

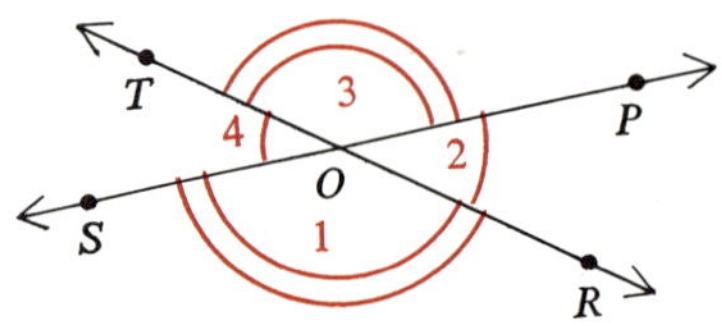

GIVEN ∠1 and ∠3 are vertical angles.

PROVE m∠1 = m∠3

PROOF

STATEMENTS	REASONS
1. ∠1 and ∠3 are vertical angles.	1. Given
2. $\overrightarrow{OP}$ and $\overrightarrow{OS}$ are opposite rays; $\overrightarrow{OR}$ and $\overrightarrow{OT}$ are opposite rays.	2. Definition of vertical angles. (2-8)
3. ∠1 and ∠2 form a linear pair; ∠3 and ∠2 form a linear pair.	3. Definition of linear pair. (1-26)
4. ∠1 and ∠2 are supplementary angles; ∠3 and ∠2 are supplementary angles.	4. Supplementary Angles Postulate. (1-6)
5. m∠1 = m∠3	5. Supplements of angles of equal measure, or of the same angle, have the same measure. (2-6.2)

Why is m∠2 = m∠4?

The next theorem follows directly from Theorem 2-6.3.

T **Theorem 2-6.4** If two intersecting lines form one right angle, then the lines form four right angles.

GIVEN Lines l and k intersect.
$\angle 1$ is a right angle.

PROVE $\angle 2$, $\angle 3$, and $\angle 4$ are right angles.

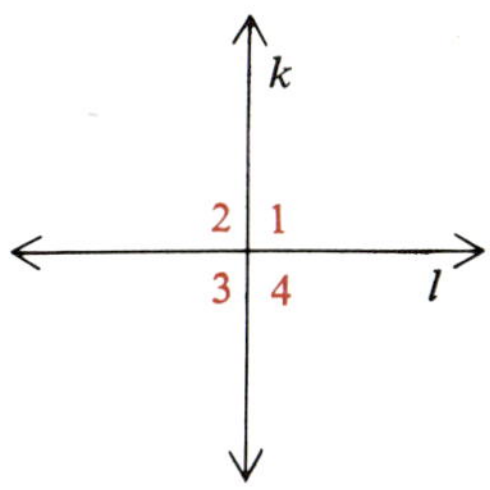

PROOF

STATEMENTS	REASONS
1. Lines l and k intersect; $\angle 1$ is a right angle.	1. Given.
2. $m\angle 1 = 90$	2. Definition of right angles. (1-24)
3. $m\angle 1 = m\angle 3$	3. Vertical angles have the same measure. (2-6.3)
4. $m\angle 3 = 90$	4. Substitution Postulate. (2-1)
5. $\angle 1$ and $\angle 2$ form a linear pair, as do $\angle 1$ and $\angle 4$.	5. Definition of linear pair. (1-26)
6. $\angle 1$ and $\angle 2$ are supplementary angles; $\angle 1$ and $\angle 4$ are supplementary angles.	6. Supplementary Angles Postulate. (1-6)
7. $m\angle 1 + m\angle 2 = 180$; $m\angle 1 + m\angle 4 = 180$	7. Definition of supplementary angles. (1-28)
8. $90 + m\angle 2 = 180$; $90 + m\angle 4 = 180$	8. Substitution Postulate. (2-1)
9. $m\angle 2 = 90$ and $m\angle 4 = 90$	9. Addition property of equality
10. $\angle 1$, $\angle 2$, $\angle 3$, and $\angle 4$ are right angles.	10. Definition of right angle. (1-24)

We can use Theorem 2-6.4 in the proof of the next theorem. Refer to the diagram for Theorem 2-6.4.

T **Theorem 2-6.5** If two intersecting lines are perpendicular, they form right angles.

GIVEN $l \perp k$

PROVE $\angle 1$, $\angle 2$, $\angle 3$, and $\angle 4$ are right angles.

PROOF

STATEMENTS	REASONS
1. $l \perp k$	1. Given.
2. $m\angle 1 = m\angle 2$	2. Definition of perpendicular lines. (1-25)
3. $\angle 1$ and $\angle 2$ form a linear pair.	3. Definition of a linear pair. (1-26)
4. $\angle 1$ and $\angle 2$ are supplementary angles.	4. Supplementary Angles Postulate. (1-6)
5. $m\angle 1 + m\angle 2 = 180$	5. Definition of supplementary angles. (1-28)
6. $2(m\angle 1) = 180$	6. Substitution Postulate (2-1) using statement 2.
7. $m\angle 1 = 90$	7. Multiplication property of equality
8. $m\angle 2 = 90$	8. Transitive property of equality
9. $m\angle 3 = 90$ and $m\angle 4 = 90$	9. Vertical angles have the same measure. (2-6.3)
10. $\angle 1$, $\angle 2$, $\angle 3$, and $\angle 4$ are right angles.	10. Definition of a right angle. (1-24)

Class Exercises

For the figure at the right, we are given that $\angle ABD$ and $\angle DBC$ form a linear pair. Use this figure to answer Exercises 1–10.

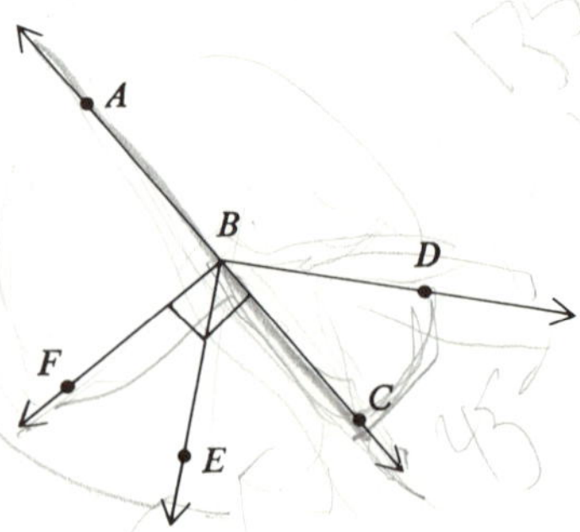

Name the supplementary angle of each of the following:

1. $\angle ABD$ 2. $\angle FBC$
3. $\angle EBC$
4. If $m\angle DBE = 90$, name two complementary angles.
5. If $m\angle EBC = 15$ and $\angle EBD$ is a right angle, find $m\angle DBC$.

6. If $\angle DBE$ and $\angle FBC$ are right angles, what relationship exists between $\angle DBC$ and $\angle FBE$?
7. If $m\angle ABD = m\angle ABE$, write a resulting equality.
8. Find the value of $m\angle ABE + m\angle EBC + m\angle CBD + m\angle DBA$.
9. If $\angle FBC$ is a right angle, find $m\angle ABF$.
10. If $m\angle ABF = m\angle FBC$, how are $\overleftrightarrow{FB}$ and $\overleftrightarrow{AC}$ related?

By interchanging the hypothesis and the conclusion of a theorem we form what is called the **converse** of the theorem. The converse is not, however, necessarily valid. We shall now prove the converse of Theorem 2-6.5.

Theorem 2-6.6 If two intersecting lines form a right angle, then they are perpendicular.

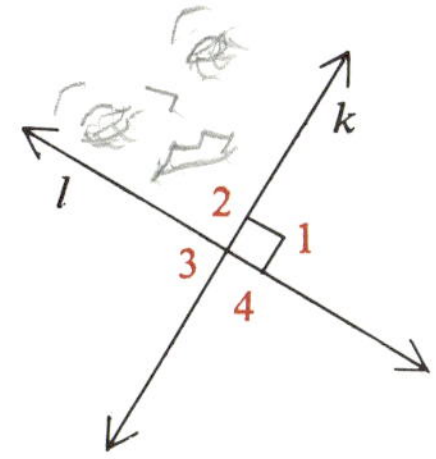

GIVEN l intersects k; $\angle 1$ is a right angle.

PROVE $l \perp k$

PROOF

STATEMENTS	REASONS
1. l intersects k; $\angle 1$ is a right angle.	1. Given
2. $m\angle 1 = 90$	2. Definition of right angle. (1-24)
3. $\angle 1$ and $\angle 2$ form a linear pair.	3. Definition of a linear pair. (1-26)
4. $\angle 1$ and $\angle 2$ are supplementary angles.	4. Supplementary Angles Postulate. (1-6)
5. $m\angle 1 + m\angle 2 = 180$	5. Definition of supplementary angles. (1-28)
6. $90 + m\angle 2 = 180$	6. Substitution Postulate. (2-1)
7. $m\angle 2 = 90$	7. Addition property of equality
8. $l \perp k$	8. Definition of perpendicular lines. (1-25)

EXERCISES

1. Prove Theorem 2-6.2.
2. Explain how the sixth reason of Theorem 2-6.5 is used.
3. Describe the relationship between Theorems 2-6.5 and 2-6.6.

Indicate which of the following are true and which are false:

4. The information labeled *Given* in a proof includes relationships from the conclusion.
5. Supplements of $\angle A$ are congruent.
6. The two-column format is used in formal proofs.
7. If $\angle A$ and $\angle B$ are a linear pair, $m\angle A + m\angle B = 180$.
8. Intersecting lines form two pairs of opposite rays.
9. Intersecting lines form only one pair of vertical angles.

Complete each of the following:

10. We list the ________ of the hypothesis as given.
11. The first three parts of a formal proof are the ________ of the theorem, the ________, and the ________.
12. If two lines intersect to form one right angle, then ________ right angles are formed.
13. Two lines intersecting to form adjacent angles of equal measure are ________.

If the lines $\overleftrightarrow{TOR}$, $\overleftrightarrow{QOS}$, and $\overleftrightarrow{PON}$ intersect at O, name another angle that forms vertical angles with each of the following:

14. $\angle 1$ 15. $\angle 3$
16. $\angle 5$ 17. $\angle POQ$
18. $\angle POR$ 19. $\angle QOR$

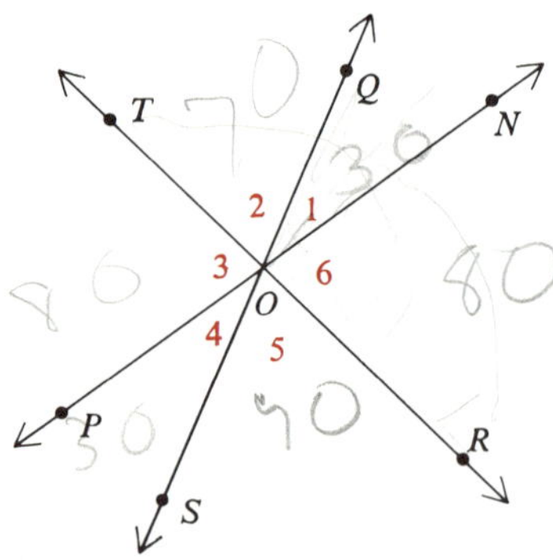

Exercises 14–25

If $m\angle 1 = 30$ and $m\angle 5 = 70$, find the measure of each of the following:

20. $\angle 2$ 21. $\angle 3$ 22. $\angle 4$ 23. $\angle 6$ 24. $\angle SON$ 25. $\angle POR$

B For Exercises 26–34 use the diagram at right below. In the diagram $\overleftrightarrow{FG}$ is perpendicular to the intersecting lines, $\overleftrightarrow{AC}$ and $\overleftrightarrow{DE}$, at point B.

26. Name two angles supplementary to $\angle EBC$.
27. If $\angle ABF$ is a right angle, name two angles supplementary to $\angle ABF$.
28. If $\angle CBD$ is a right angle, name the other right angles.

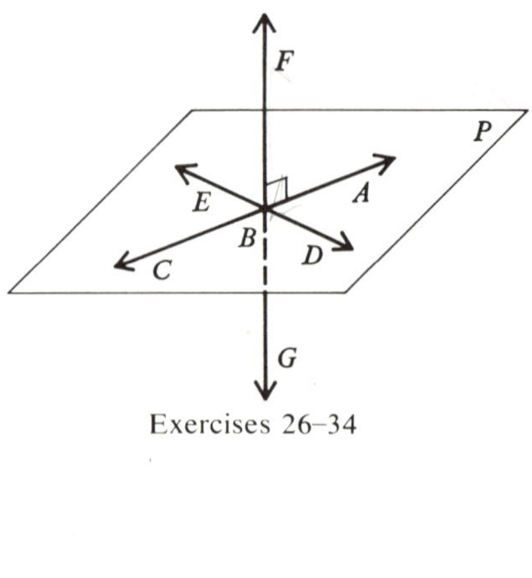

Exercises 26–34

If m$\angle ABE = x$, write the values of the following in terms of x:

29. $\angle EBC$ 30. $\angle CBD$ 31. $\angle ABD$
32. If m$\angle ABE = 2($m$\angle ABD)$, find m$\angle EBC$.
33. If $\overleftrightarrow{FG} \perp \overleftrightarrow{AC}$, name the angles with equal measures.
34. If $\overleftrightarrow{FG} \perp \overleftrightarrow{ED}$, name the angles with equal measures.

In Exercises 35–38 use the figure at right; m$\angle 2 =$ m$\angle 4$.

35. If m$\angle 1 = 120$, find m$\angle 3$.
36. If m$\angle 3 = 40$, find m$\angle 5$.
37. How are $\angle 3$ and $\angle 6$ related?
38. How are $\angle 1$ and $\angle 5$ related?

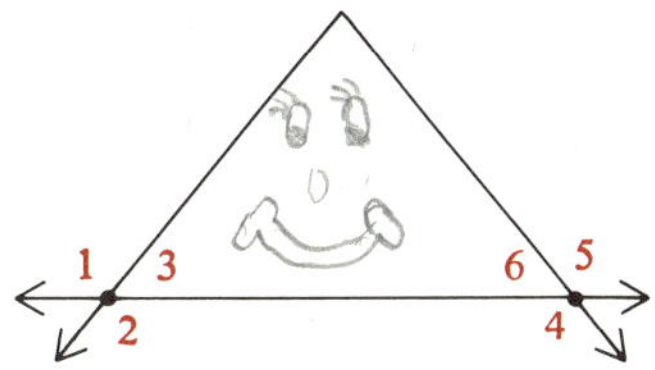

Exercises 35–38

HOW IS YOUR MATHEMATICAL VOCABULARY?

The following words and phrases were introduced in this chapter. How many do you know and understand?

- Angle Difference Postulate (2-3)
- Angle Sum Postulate (2-3)
- Angle Uniqueness Postulate (2-3)
- assumption (2-2)
- compound statements (2-4)
- conclusion (2-1)
- conditional statement (2-4)
- conjunction (2-4)
- converse (2-6)
- deduction (2-4)
- deductive reasoning (2-1)
- defined term (2-2)
- definition (2-2)
- direct proof (2-6)
- disjunction (2-4)
- equivalent statements (2-4)
- field properties (2-3)
- hypothesis (2-4)
- implication (2-4)
- induction (2-1)
- intuition (2-1)
- Law of Contradiction (2-4)
- Law of the Excluded Middle (2-4)
- Line Postulate (2-3)
- logical system (2-2)
- negation (2-4)
- Plane Postulate (2-3)
- Plane Intersection Postulate (2-3)
- Point Betweenness Postulate (2-3)
- Points-in-a-Plane Postulate (2-3)
- Point Uniqueness Postulate (2-3)
- postulate (2-2)
- property (2-3)
- reason (2-5)
- Space Postulate (2-3)
- Substitution Postulate (2-3)
- tautology (2-4)
- theorem (2-2)
- truth table (2-4)
- two-column proof (2-6)
- undefined terms (2-2)
- valid (2-4)
- vertical angle (2-6)

REVIEW EXERCISES

2-1 *Intuition and induction*

1. You meet your blind date and you know you are going to have a terrible time. Is this an intuition or an induction?
2. Define *induction* in your own words.

Supply the ninth term in each of the following sequences:

3. 81, 27, 9, 3, . . . 4. 1, 2, 4, 7, 11, . . . 5. 2, 3, 5, 8, . . .
6. Make the induction for the following:

$$1 + 2 = \tfrac{2}{2}(1 + 2);\ 1 + 2 + 3 = \tfrac{3}{2}(1 + 3);\ 1 + 2 + 3 + 4 = \tfrac{4}{2}(1 + 4)$$

2-2 *A mathematical system*

7. Name the four parts of a mathematical system.
8. Give an acceptable definition of complementary angles. Are any terms undefined? assumed to be defined already?
9. Criticize this definition: "A point is a position without length and breadth."

Complete each of the following:

10. Not all ________ can be defined.
11. Minimal ________ must be stated in an acceptable definition.
12. Not all ________ can be proved.
13. Assumptions are called ________.
14. The process of reasoning associated with a logical system is ________.
15. Deductions are formed from ________ and ________.

2-3 *Postulates: algebraic and geometric*

State the property that makes each of the following true:

16. If $3m\angle A = 84$, then $m\angle A = 28$.
17. If $AB = BC$ and $BC = CD$, then $AB = CD$.
18. $(\sqrt{3} + \sqrt{5})(\sqrt{3} - \sqrt{5})$ is a real number.
19. $\angle A$ and $\angle B$ are a linear pair; therefore $m\angle A + m\angle B = 180$.
20. If $\overleftrightarrow{AR} \cap$ plane P with A and R in plane P, then $\overleftrightarrow{AR}$ lies in plane P.
21. Q is the only point on $\overrightarrow{PM}$ such that $PQ = x$.
22. If $AB = x$ and $2(AB) + 3 = 21$, then $2x + 3 = 21$.

2-4 *Logic*

Given p: It is snowing; q: It is cloudy; and r: I am cold; translate each of the following:

23. $\sim r$ 24. $p \wedge q$ 25. $(\sim r \wedge p) \vee q$
26. $\sim(p \vee q)$ 27. $p \rightarrow r$ 28. $\sim(\sim p \rightarrow q)$

Write each of the following in symbols, using the statements given above:

29. It is snowing or it is not snowing.
30. If it is not cloudy, then I am not cold.
31. It is false that it is snowing and I am cold.
32. I am cold or it is snowing.

Construct the truth table for each of the following:

33. $\sim p \rightarrow q$ 34. $\sim (q \rightarrow p)$ 35. $(p \vee q) \rightarrow \sim p$

State a possible conclusion for each of the following:

36. $(p \rightarrow q) \wedge p$ 37. $(p \wedge q) \wedge q$

2-5 Theorems

Complete each of the following:

38. The intersection of two distinct lines is a(n) ________.
39. A plane is determined by ________ intersecting lines.
40. Any ________ has exactly one midpoint.
41. All right angles are ________.
42. If supplementary angles A and B are such that m $\angle A =$ m $\angle B$, then the supplementary angles are ________ angles.

Write the hypothesis and conclusion of each of the following:

43. Two perpendicular lines form right angles.
44. As linear pairs, m $\angle A +$ m $\angle B = 180$.
45. Opposite angles formed by intersecting lines are congruent.

2-6 Direct proof

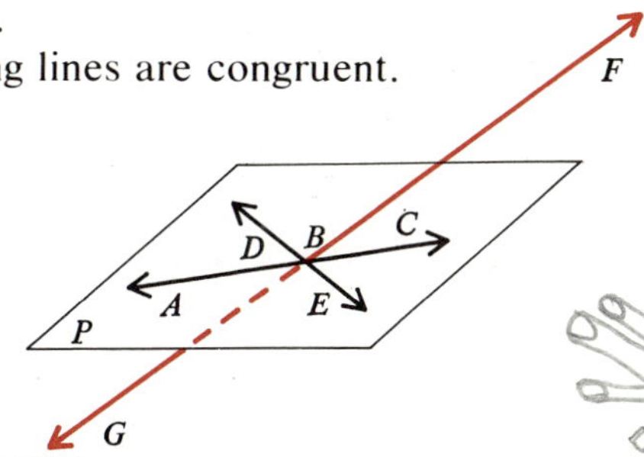

46. List the first three items of a two-column proof.
47. Name the two angles supplementary to $\angle ABF$ in the figure at right.
48. What angle is equal in measure to $\angle ABD$?
49. If m$\angle DBC = 90$, how are $\overleftrightarrow{AC}$ and $\overleftrightarrow{DE}$ related?

If m$\angle CBE = 30$ and m$\angle FBC = 70$, give the measure of each of the following:

50. $\angle ABD$ 51. $\angle CBG$ 52. $\angle ABF$
53. $\angle CBD$ 54. $\angle ABE$ 55. $\angle ABG$

CHAPTER TEST

1. Supply the ninth term of the sequence: 1, 1, 2, 3, 5, 8. . . .
2. Criticize the definition, "Perpendicular lines form four congruent angles."
3. Compare definitions and theorems and their relationships to assumed items.

4. State the distributive property.
5. Complete: If $\angle X$ and $\angle Y$ are a linear pair and m$\angle X = 10$, then m$\angle Y =$ ____.
6. Construct the truth table for $(p \vee q) \rightarrow (p \wedge \sim q)$.
7. Construct the truth table for $p \vee (q \wedge r)$ and $(p \vee q) \wedge (p \vee r)$
8. State a possible conclusion for these statements:
   I did not go to school or I did not go to the doctor. I did go to school.
9. Write the following statement in the "if p, then q" form: "Come over and you'll have to work."

Use the figure at right to answer Exercises 10–14.

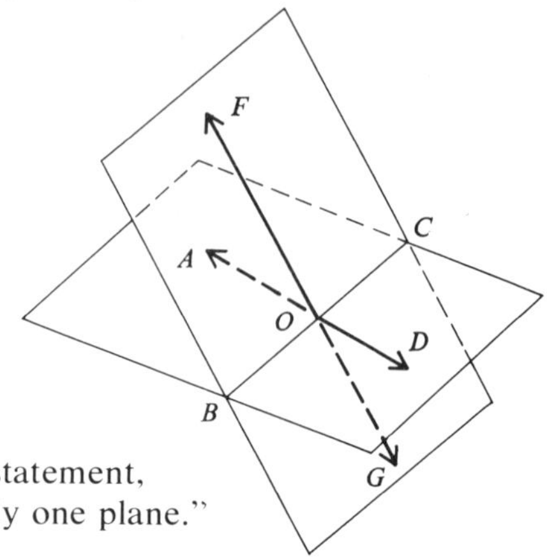

10. If m$\angle FOB = 90$, name four right angles.
11. If m$\angle FOB = 90$, what lines are perpendicular?
12. Name the angles supplementary to $\angle AOC$.
13. Name an angle with the same measure as $\angle AOC$.
14. Name the postulate that justifies the statement, "Points A, D, and C determine exactly one plane."
15. Use a truth table to decide if the reasoning in the following statements is valid. Choose your own symbols for the statements.
    $\triangle ABC$ is isosceles or $\triangle ABC$ is scalene.
    $\triangle ABC$ is not isosceles. $\therefore$ $\triangle ABC$ is scalene.
16. Use a two-column proof to prove Theorem 2-5.1: "If a point does not lie in a given line, then there is exactly one plane containing both the point and the line."

SUGGESTED RESEARCH

1. Lewis Carroll's *Alice's Adventures in Wonderland* contains many examples of deductive reasoning. Read enough of the book to write a report citing instances of this reasoning other than those mentioned in the chapter.
2. Prepare a slide or postcard show of the architecture in your community to demonstrate the uses of geometry.
3. Search the encyclopedias and prepare a paper on geometry in nature.
4. For further study of *proof*, read "Original Investigations or How to Attack an Exercise in Geometry," by Elisha S. Loomis (Columbus, Ohio: The Bonded Scale and Machine Co., 1965).

3 Congruent Triangles

In Chapter 1, we discussed segments of equal measure and angles of equal measure in terms of real numbers. At that time, we introduced the idea of congruence. At the end of Chapter 2, we examined several theorems about segments and angles of equal measure. In this chapter, we shall discuss segments and angles further as we explore congruent triangles.

3-1 CONGRUENCE: AN EQUIVALENCE RELATION

Geometric figures are congruent if they have exactly the same size and shape. Since all segments are the same shape, we need only compare lengths to check for congruence. Of the segments below, only $\overline{AB}$ and $\overline{CD}$ are the same length.

Definition 3-1 $\overline{AB}$ and $\overline{CD}$ are **congruent segments** if and only if they have the same measure. We denote their congruence by $\overline{AB} \cong \overline{CD}$.

Congruence is a relation between sets of points. In this case, the sets of points are the segments. The congruence relation $\overline{AB} \cong \overline{CD}$ is equivalent to the equality relation $AB = CD$. This means that either statement can replace the other at any time. Do you see why?

The phrase **if and only if** indicates the equivalence of an "if . . . then" statement and its converse. Thus, by the definition above, if segments are congruent, then their measures are automatically equal; and if segments have equal measures, then they are congruent.

Are all angles the same shape? the same size? When angles are the same size, are their shapes identical? If the angles at right could be placed one over the other so that the corresponding points of $\overrightarrow{DQ}$ and $\overrightarrow{CX}$ coincide and the corresponding points of $\overrightarrow{DP}$ and $\overrightarrow{CY}$ coincide, then the angles are congruent.

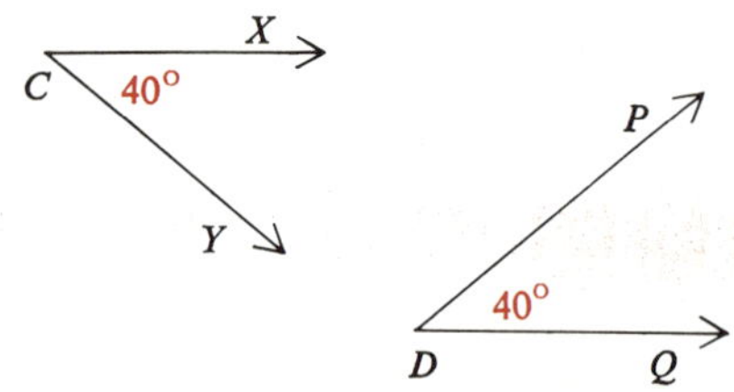

Definition 3-2 $\angle YCX$ and $\angle QDP$ are **congruent angles** if and only if they have the same measure. We denote their congruence by $\angle YCX \cong \angle QDP$.

The congruence relation $\angle YCX \cong \angle QDP$ is equivalent to the equality relation $m\angle YCX = m\angle QDP$. This means that either statement can replace the other.

These two definitions permit us to restate some of the theorems from the last chapter in terms of the congruence relation. Since we are merely restating them, we shall not prove them again. You may, however, wish to review the proofs in light of the definitions of congruence.

Theorem 3-1.1 All right angles are congruent.

Theorem 3-1.2 Two congruent supplementary angles are right angles.

Theorem 3-1.3 Complements of congruent angles, or of the same angle, are congruent.

Theorem 3-1.4 Supplements of congruent angles, or of the same angle, are congruent.

Theorem 3-1.5 Vertical angles are congruent.

Each of the above statements concerns the congruence of angles. The original statement of each theorem was in terms of the equality of the measures of the angles. Since measures are real numbers, equality relations have the reflexive, symmetric, and transitive properties of equality. But congruence statements and equality statements are interchangeable. Are these three properties also properties of congruence relations?

Relations which have these three properties are called **equivalence relations**. Is an equality relation an equivalence relation? Is congruence of segments an equivalence relation? Is congruence of angles? Proof of the following theorems shows that both congruence relations are equivalence relations.

Theorem 3-1.6 The Identity Theorem for Segments Every segment is congruent to itself.

GIVEN $\overline{AB}$

PROVE $\overline{AB} \cong \overline{AB}$

PROOF

STATEMENTS	REASONS
1. $\overline{AB}$	1. Given
2. For $\overline{AB}$, AB is a number.	2. Distance Postulate (1-1)
3. $AB = AB$	3. Reflexive property
4. $\overline{AB} \cong \overline{AB}$	4. Definition of congruent segments (1-16)

The proof of the next theorem is similar to the one above. It is left as an exercise for you to complete.

Theorem 3-1.7 The Identity Theorem for Angles Every angle is congruent to itself.

Theorem 3-1.6 and the results of the following Class Exercises will prove that segment congruence is an equivalence relation. This is the first part of Theorem 3-1.8. The second part, proving that angle congruence is an equivalence relation, is left as an exercise.

Theorem 3-1.8 Both segment congruence and angle congruence are equivalence relations.

Class Exercises

Complete each of the following statements:

1. If $AB = XY$, then $XY =$ ____, by the ________ property of equality.
2. If $XY = AB$, then $\overline{XY} \cong \overline{AB}$ by the definition of ________.
3. If $AB = XY$, then $\overline{AB} \cong$ ____, by the ________.
4. If $\overline{AB} \cong \overline{XY}$, then $\overline{XY} \cong \overline{AB}$, by the ________ property.
5. Thus, congruence of segments satisfies the ________ property.
6. If $\overline{AB} \cong \overline{XY}$ and $\overline{XY} \cong \overline{PQ}$, then $AB =$ ____ and ____ $= PQ$, by the definition of ________.
7. If $AB = XY$ and $XY = PQ$, then $AB = PQ$ by the ________ property of equality.
8. If $AB = PQ$, then $\overline{AB} \cong \overline{PQ}$ by the definition of ________.
9. If $\overline{AB} \cong \overline{XY}$ and $\overline{XY} \cong \overline{PQ}$, then $\overline{AB} \cong \overline{PQ}$ by the ________ property.
10. Thus, congruence of segments satisfies the ________ property.
11. The statement that every segment is congruent to itself (Theorem 3-1.6) satisfies the ________ property.
12. Exercises 1–11 can be used to prove that the congruence relation for ________ is a(n) ________ relation.

Summary. Number Equality and Angle and Segment Congruence Equivalence Relations

Property	Number Equality	Segment Congruence	Angle Congruence
Reflexive	$a = a$	$\overline{AB} \cong \overline{AB}$	$\angle ABC \cong \angle ABC$
Symmetric	If $a = b$, then $b = a$.	If $\overline{AB} \cong \overline{CD}$, then $\overline{CD} \cong \overline{AB}$.	If $\angle ABC \cong \angle PQR$, then $\angle PQR \cong \angle ABC$.
Transitive	If $a = b$ and $b = c$, then $a = c$.	If $\overline{AB} \cong \overline{CD}$ and $\overline{CD} \cong \overline{EF}$, then $\overline{AB} \cong \overline{EF}$.	If $\angle ABC \cong \angle PQR$ and $\angle PQR \cong \angle XYZ$, then $\angle ABC \cong \angle XYZ$.

Other Congruences

Segments may be congruent and angles may be congruent. But what about other geometric figures? Segments or angles are congruent when they have equal measures. When are triangles congruent? How can we tell if quadrilaterals are congruent? Earlier we discussed what would happen if we could overlay figures to get corresponding points to coincide. Will any one-to-one correspondence of the vertices assure a congruence?

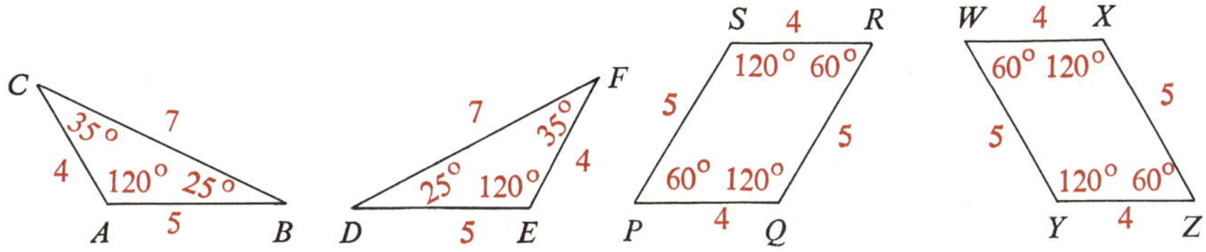

Study these correspondences between vertices of the paired figures above:

I	II	III	IV
$A \leftrightarrow E$	$A \leftrightarrow D$	$S \leftrightarrow W$	$S \leftrightarrow X$
$B \leftrightarrow D$	$B \leftrightarrow E$	$R \leftrightarrow X$	$R \leftrightarrow W$
$C \leftrightarrow F$	$C \leftrightarrow F$	$P \leftrightarrow Y$	$P \leftrightarrow Z$
		$Q \leftrightarrow Z$	$Q \leftrightarrow Y$

Example 1 Using correspondence I above, determine whether $\triangle ABC$ will coincide with $\triangle EDF$ if $\triangle ABC$ could be moved.

Solution Using correspondence I and the measures indicated in the diagram, we see that the sides that correspond have equal measures. That is,

$$AB = ED = 5 \qquad AC = EF = 4 \qquad BC = DF = 7$$

We also see that corresponding angles have equal measures. That is,

$$\text{m}\angle A = \text{m}\angle E = 120 \qquad \text{m}\angle B = \text{m}\angle D = 25 \qquad \text{m}\angle C = \text{m}\angle F = 35$$

Thus, if $\triangle ABC$ could be moved onto $\triangle DEF$, the corresponding vertices, sides, and angles would fit exactly. We say that $\triangle ABC$ could **coincide** with $\triangle EDF$.

Since the triangles could coincide by correspondence I, we say that correspondence I describes a congruence.

Example 2 Use correspondence IV and the appropriate measures to determine if quadrilateral $PQRS$ could coincide with quadrilateral $ZYXW$.

Solution We see that the measures of the sides and angles matched by this correspondence are equal. That is,

$$PQ = ZY = 4 \qquad \text{m}\angle P = \text{m}\angle Z = 60$$
$$QR = YW = 5 \qquad \text{m}\angle Q = \text{m}\angle Y = 120$$
$$RS = WX = 4 \qquad \text{m}\angle R = \text{m}\angle W = 60$$
$$SP = XZ = 5 \qquad \text{m}\angle S = \text{m}\angle X = 120$$

Thus, the quadrilaterals would coincide if we matched the vertices using this correspondence.

Since the quadrilaterals would coincide by correspondence IV, we say that correspondence IV describes a congruence.

Example 3 Determine if correspondence II above describes a congruence.

Solution If we use correspondence II and the appropriate measures, the vertices of $\angle B$ and $\angle E$ correspond; but $m\angle B = 25$ and $m\angle E = 120$. Thus, the angles could not coincide. Therefore, the triangles could not coincide. Correspondence II does not describe a congruence.

Any figure is congruent to itself. If each vertex is made to correspond with itself, the correspondence is a congruence. Thus the following correspondence is a congruence.

$$I \leftrightarrow I$$
$$F \leftrightarrow F$$
$$G \leftrightarrow G$$
$$H \leftrightarrow H$$

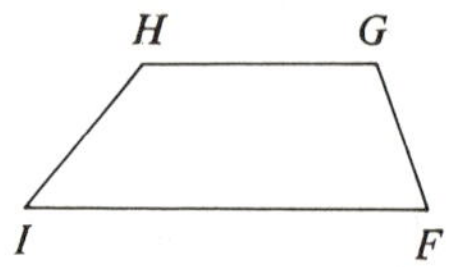

We can shorten the vertical correspondence notation to $IFGH \leftrightarrow IFGH$. This instance of the reflexive property is called an **identity congruence**.

EXERCISES

A Indicate whether or not the following pairs of figures appear to be congruent.

1. $\overline{AB}$ and $\overline{XY}$

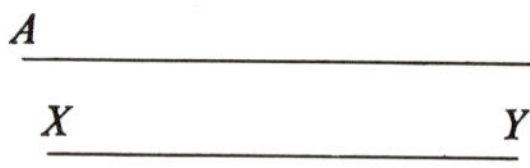

2. Circles ABC and ADE

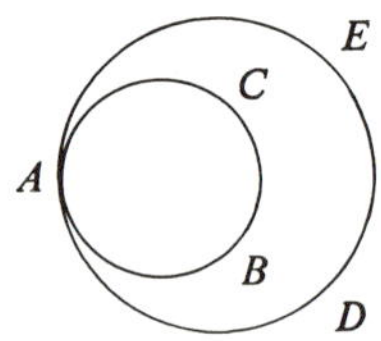

3. Triangles ABC and YXZ

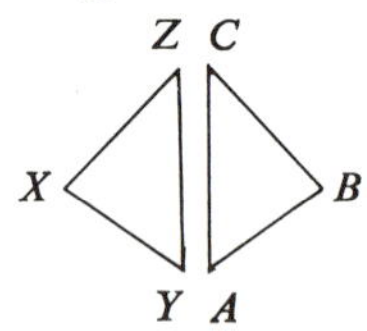

4. $\triangle JKN$ and $\triangle LKM$

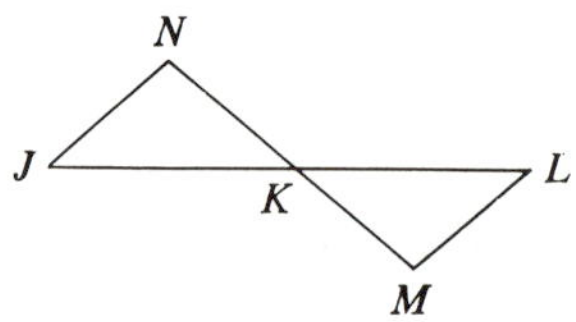

5. $\triangle ABM$ and $\triangle ACM$

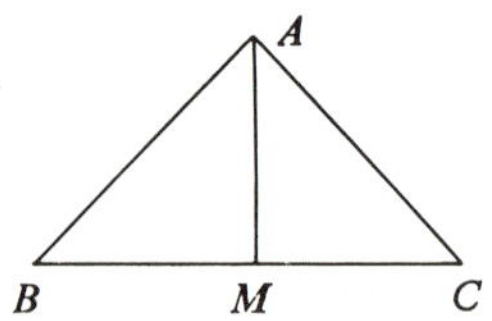

6. Figures $PQRS$ and $PXYT$

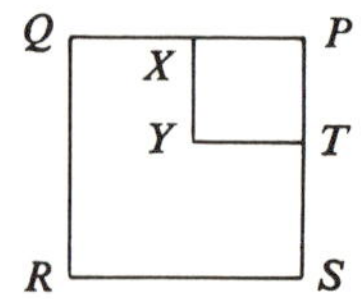

Write the correspondence for each pair of triangles below that appears to be congruent.

7.

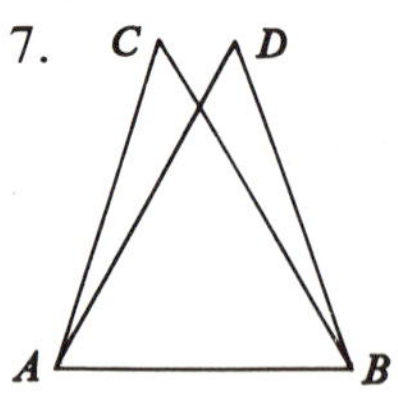

8.

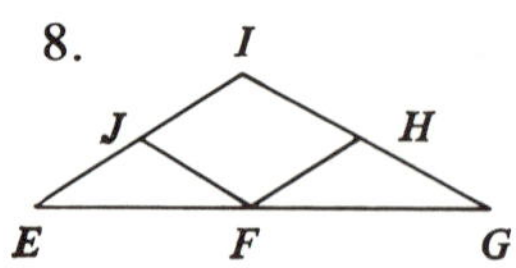

9.

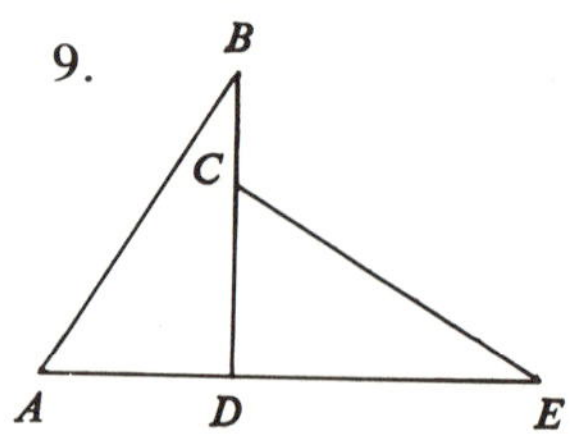

10. Show that correspondence III from the discussion preceding the exercises does not describe a congruence.
11. Prove Theorem 3-1.7.
12. Prove the statement, "Congruence of angles is an equivalence relation."

Indicate which of the following are true and which are false:
13. $\overline{XY} \cong \overline{PQ}$ is equivalent to $XY = PQ$.
14. All circles are congruent to each other.
15. "If and only if" indicates equivalence.
16. In the triangle correspondence $XYZ \leftrightarrow PQR$, vertex $Y \leftrightarrow$ vertex R.
17. $EFHJ \leftrightarrow JHGF$ is the identity congruence.

B What must we show before we say that each of the following is a congruent pair.
18. $\angle A$ and $\angle B$
19. $\overleftrightarrow{PQ}$ and $\overleftrightarrow{RS}$
20. $\triangle PQR$ and $\triangle XYZ$
21. $\overline{FG}$ and $\overline{HJ}$
22. $\triangle ABC$ and $\triangle ABC$
23. quadrilateral $ABCD$ and quadrilateral $PQRS$

Write the correspondence for each pair of triangles that appears congruent.

24.
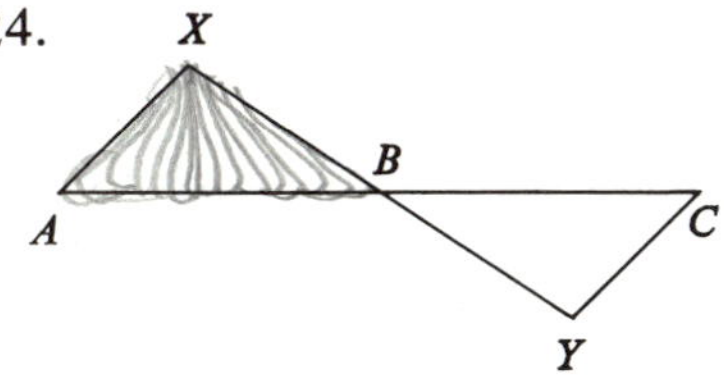

25.
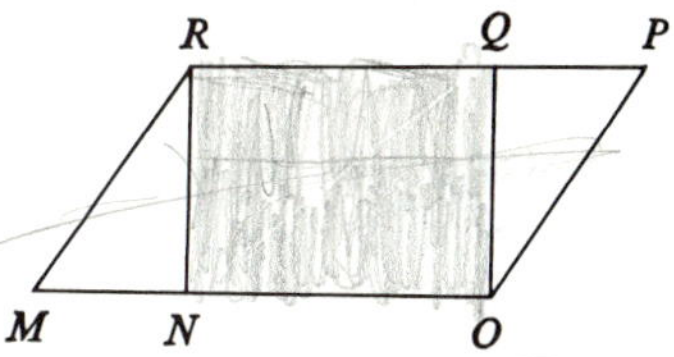

26.
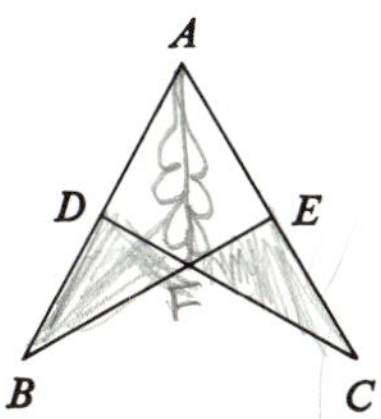

27.
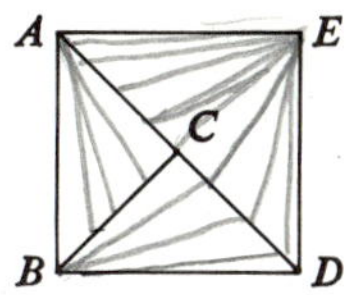

28.
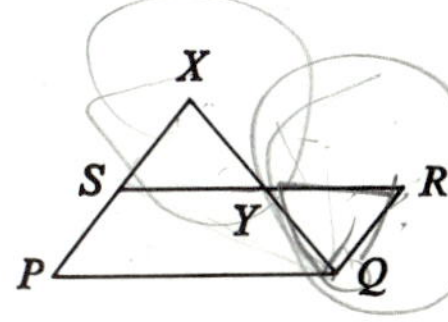

29. In Exercise 26, find a second set of triangles and write their correspondence. You will have to label another point.

C Complete each of the following using the congruence $ABM \leftrightarrow ACM$:

30. $\angle B \cong$ ____
31. $\angle BMA \cong$ ____
32. $\angle BAM \cong$ ____
33. $m\angle C =$ ____
34. $m\angle BAM =$ ____
35. $\overline{AM} \cong$ ____
36. $\overline{AB} \cong$ ____
37. $\overline{BM} \cong$ ____
38. $MC =$ ____
39. $AC =$ ____

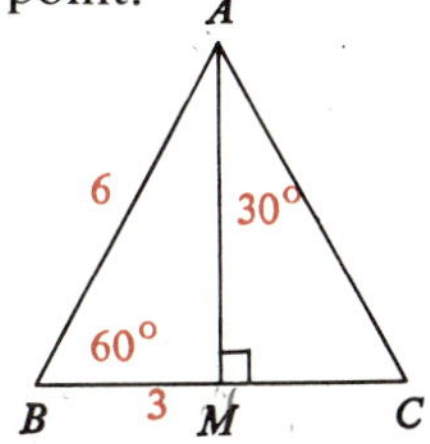

Exercises 30–39

40. If $PQRS \leftrightarrow ABCD$ is a congruence, draw quadilaterals $PQRS$ and $ABCD$, and show corresponding congruences by some type of marking.

3-2 CONGRUENT TRIANGLES

We have discussed one-to-one correspondences between the vertices of various geometric figures. We shall now examine *congruent* correspondences between triangles.

What do we need to know about two segments to establish congruence? Does equality of measure prove that two angles are congruent? To establish that triangles are congruent, we must be concerned with both shape and size. In our study, we shall refer to the sides and angles of a triangle as **the parts of the triangle**. How many parts does a triangle have?

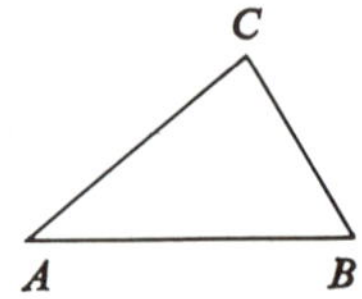

Consider the correspondence $ABC \leftrightarrow XYZ$ between the vertices of $\triangle ABC$ and $\triangle XYZ$. What matches besides the corresponding vertices? The matching angles are called **corresponding angles**. The matching sides are called **corresponding sides**.

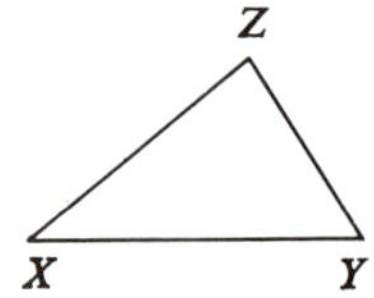

Corresponding Angles	Corresponding Sides
$\angle A \leftrightarrow \angle X$	$\overline{AB} \leftrightarrow \overline{XY}$
$\angle B \leftrightarrow \angle Y$	$\overline{BC} \leftrightarrow \overline{YZ}$
$\angle C \leftrightarrow \angle Z$	$\overline{AC} \leftrightarrow \overline{XZ}$

Since sides and angles, not vertices alone, can be made to correspond, we refer to $ABC \leftrightarrow XYZ$ as a correspondence between triangles. If all six of the corresponding parts of two triangles are congruent, we say the correspondence describes a congruence between the triangles.

D **Definition 3-3** If there is a correspondence $ABC \leftrightarrow XYZ$ such that the sides and angles of $\triangle ABC$ are congruent to the corresponding sides and angles of $\triangle XYZ$, then $ABC \leftrightarrow XYZ$ is a congruence, and the triangles are said to be **congruent triangles**. We denote their congruence by $\triangle ABC \cong \triangle XYZ$.

The short statement $\triangle PQR \cong \triangle FGH$ provides us with a considerable amount of information. In fact, it indicates six other congruences. What are these other congruences? What equivalent equations can we write?

Summary. Meaning of Triangle Congruence

If $\triangle PQR \cong \triangle FGH$,

then $\angle P \cong \angle F$ and $m\angle P = m\angle F$ $\quad$ $\overline{PQ} \cong \overline{FG}$ and $PQ = FG$

$\angle Q \cong \angle G$ and $m\angle Q = m\angle G$ $\quad$ $\overline{QR} \cong \overline{GH}$ and $QR = GH$

$\angle R \cong \angle H$ and $m\angle R = m\angle H$ $\quad$ $\overline{PR} \cong \overline{FH}$ and $PR = FH$

The congruences on the left above are equivalent to the corresponding equations on the right. Most of the time they can be used interchangeably.

However, we sometimes perform arithmetic operations with measures. Thus, for sides, we use $PQ = FG$ when lengths are necessary, and $\overline{PQ} \cong \overline{FG}$ when we are discussing the segments themselves. For angles, we use $\angle P \cong \angle F$, and for the measures of angles, we use $m\angle P = m\angle F$.

Class Exercises

1. If $CAB \leftrightarrow RST$ does not describe a congruence, but $ABC \leftrightarrow SRT$ describes a congruence, is $\triangle ABC \cong \triangle SRT$?

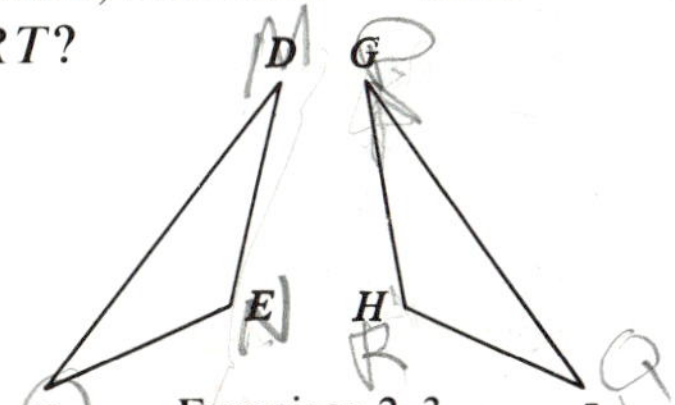

Exercises 2–3

For Exercises 2 and 3 consider $\triangle DEF \cong \triangle GHI$.

2. Name the corresponding angles.
3. Name the corresponding sides.

Assume $\triangle MNO \cong \triangle PQR$. Complete each of the following:

4. $\overline{MN} \cong$ ____ 5. $MO =$ ____ 6. $\angle M \cong$ ____ 7. $m\angle O =$ ____
8. What side of $\triangle PQR$ lies between $\angle P$ and $\angle R$?
9. What angle of $\triangle MNO$ is between $\overline{MN}$ and $\overline{NO}$?

We use many notations as shorthand in mathematics. As we said earlier, $\triangle ABC \cong \triangle DEF$ is a shorthand way of writing 12 relations. In illustrations, too, we use shorthand.

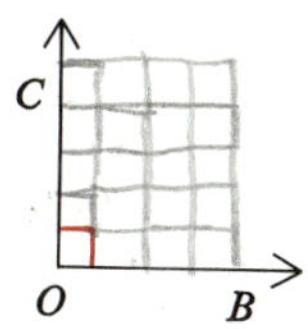

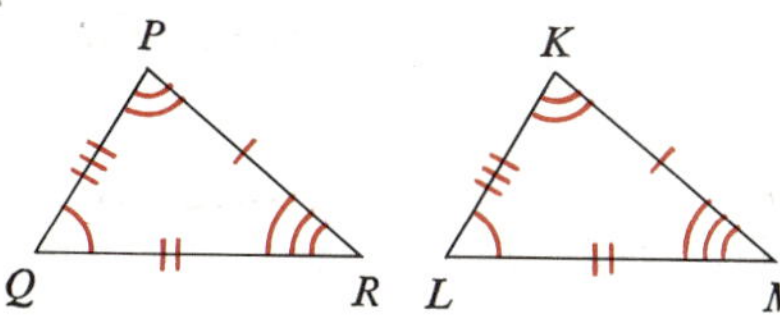

What can we tell about $\angle BOC$ from the figure on the left above? For $\triangle PQR \cong \triangle KLM$ on the right above, the markings indicate the six congruences.

Example 1 What congruences are indicated by the markings in the triangles at right?

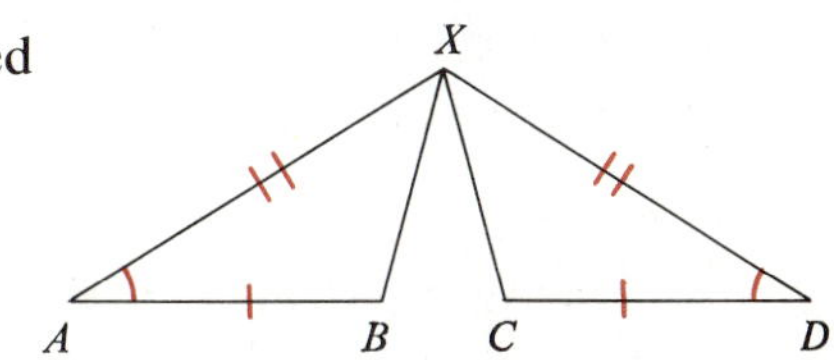

Solution The markings indicate that $\overline{AX} \cong \overline{DX}$, $\angle A \cong \angle D$, and $\overline{AB} \cong \overline{DC}$.

In the example, side $\overline{AB}$ has the vertices of $\angle A$ and $\angle B$ as endpoints. $\overline{AB}$ is said to be included between $\angle A$ and $\angle B$. What angles include $\overline{DC}$ in $\triangle DCX$? Angle A is said to be the angle included by sides $\overline{AB}$ and $\overline{AX}$. What angle do sides $\overline{CX}$ and $\overline{DX}$ include?

Definition 3-4 In a triangle, **an angle is included** by the sides of the triangle that lie in the sides of the angle.
In a triangle, **a side is included** by the angles whose vertices are endpoints of the segment.

Having defined triangle congruence above, we shall now examine its effects. Earlier we found that angle congruences and segment congruences are equivalence relations. The theorems proving these equivalence relations, as well as the definition of congruent triangles, are needed to justify various parts of the proof of the next theorem, which is left as an exercise for you.

Theorem 3-4.1 Triangle congruence is an equivalence relation.

EXERCISES

Make a drawing for each of the following. State and mark the correspondences.

1. $FGH \leftrightarrow UVW$ 2. $\triangle EFG \cong \triangle XYZ$ 3. $\triangle RST \cong \triangle MNO$

For each of the pairs of congruent triangles indicated below, list the six corresponding congruent parts.

4. $\triangle ABX$ and $\triangle CBY$

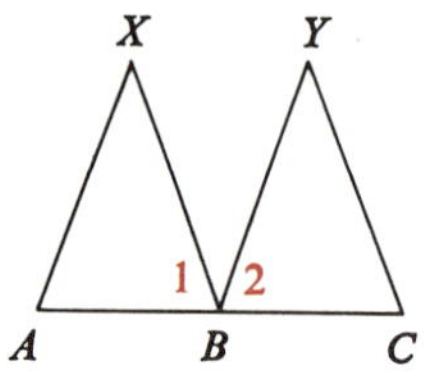

5. $\triangle POS$ and $\triangle QOR$

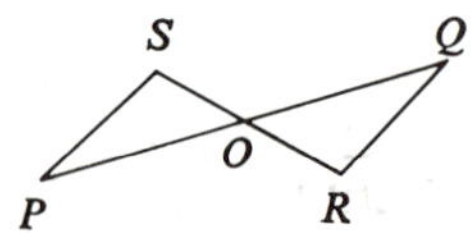

6. $\triangle AFD$ and $\triangle CFE$

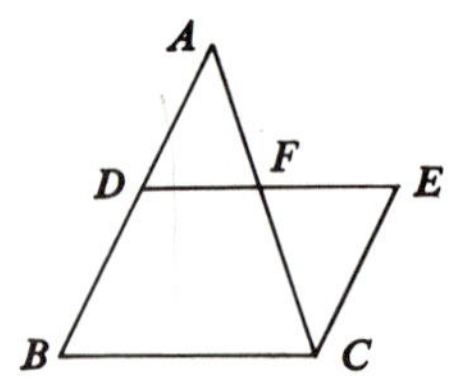

7. $\triangle PRX$ and $\triangle QSY$

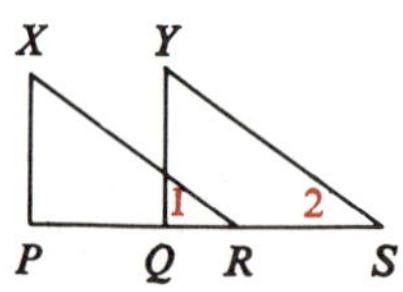

8. $\triangle AMP$ and $\triangle BNO$

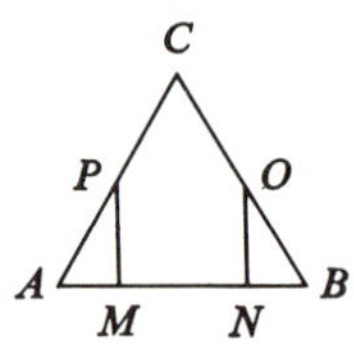

9. $\triangle MQP$ and $\triangle NPQ$

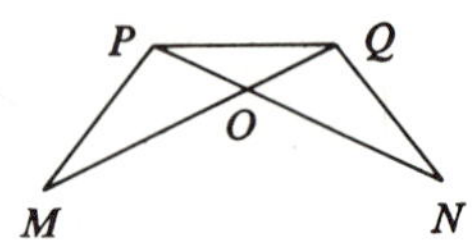

10. Given the correspondences $\overline{RS} \cong \overline{NM}$, $\angle R \cong \angle N$, $\overline{ST} \cong \overline{ML}$, $\angle S \cong \angle M$, $\overline{TR} \cong \overline{LN}$, $\angle T \cong \angle L$, write the triangular congruence.

Name the angles or sides included by each of the following:

11. $\overline{MQ}$ and $\overline{PQ}$ in $\triangle MQP$ 12. $\overline{RS}$ and $\overline{ST}$ in $\triangle RST$
13. $\angle A$ and $\angle C$ in $\triangle ACB$ 14. $\angle N$ and $\angle P$ in $\triangle NPQ$
15. Prove that for any $\triangle ABC$, $\triangle ABC \cong \triangle ABC$.

16. Prove that if $\triangle ABC \cong \triangle KLM$, then $\triangle KLM \cong \triangle ABC$.
17. Prove that if $\triangle ABC \cong \triangle KLM$ and $\triangle KLM \cong \triangle PQR$ then $\triangle ABC \cong \triangle PQR$. What theorem have you just proved?

B Given the following congruences, list the six pairs of corresponding parts:

18. $\triangle FGH \cong \triangle LMN$ 19. $\triangle FGE \cong \triangle XYZ$ 20. $\triangle STU \cong \triangle UTS$

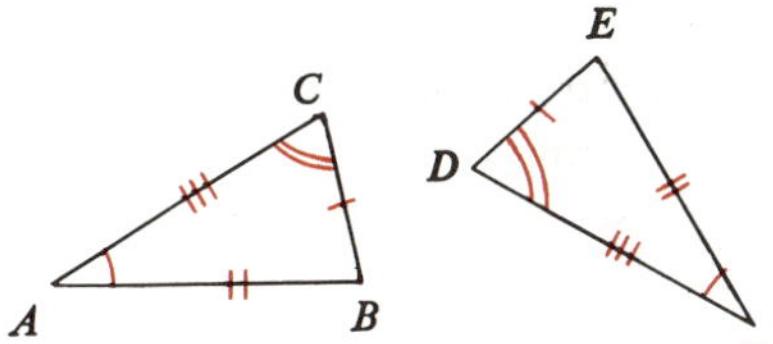

Exercise 21

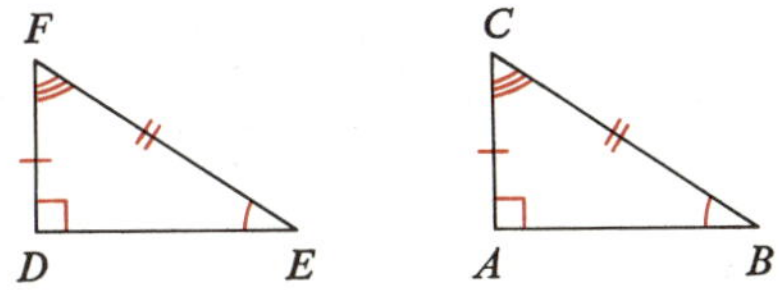

Exercise 22

21. If the markings in the figure at left above indicate congruences, what triangular congruence can you conclude?
22. If the markings in the figure at right above indicate congruences, what triangular congruence can you conclude?

C

23. Construct an angle of measure 36. On the sides of the angle, form segments of lengths 3 and 7. How many triangles are determined?
24. Repeat Exercise 23, with angle measure 110 and side lengths 6 and 8.
25. If $PQ = 5$, construct $\angle P$ and $\angle Q$ at the endpoints of $\overline{PQ}$ with common side $\overline{PQ}$, such that $m\angle P = 30$ and $m\angle Q = 40$. How many triangles are determined?
26. Repeat Exercise 25, using $PQ = 10$, $m\angle P = 50$, and $m\angle Q = 70$.
27. From the results of Exercises 23–26, is it always essential to know all six parts of a triangle before determining the triangle? Explain.

3-3 CONGRUENCE POSTULATES FOR TRIANGLES

Every triangle has six parts: three angles and three sides. How many of these parts must we know in order to determine the shape and size of a triangle?

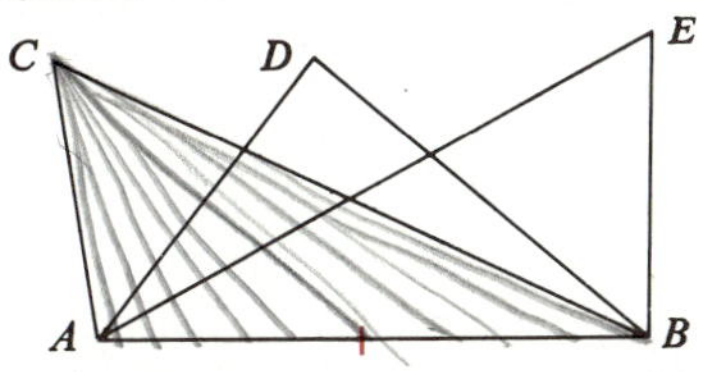

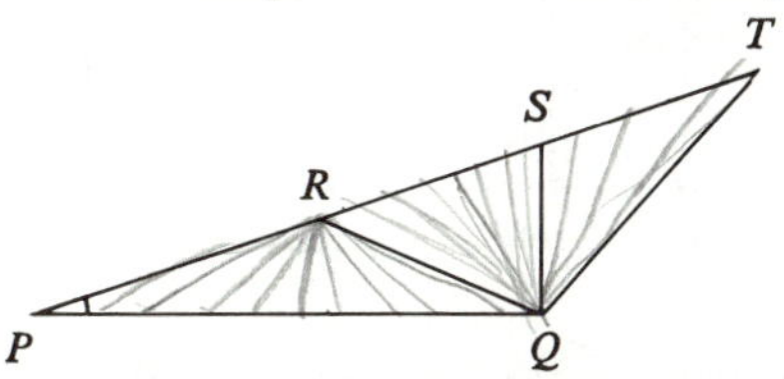

Many triangles share a common side. Name the triangles in the figure at left above with $\overline{AB}$ as a side. Many triangles can share a common side as well as a common angle. In the figure at right above, name three triangles that share $\angle P$ and $\overline{PQ}$. How many triangles can we determine if we know the measures of two sides and the included angle, that is, a side, an angle, and a side, in that order?

Example 1 If $AB = 5$, $BC = 7$, and $m\angle B = 40$, are any triangles determined?

Solution Using a protractor and ruler, construct a figure with the above measures. How many segments on $\overrightarrow{BA}$ have a measure equal to AB? How many segments on $\overrightarrow{BC}$ have a measure equal to BC? A and C are unique points. Since only one segment is determined by A and C, $\triangle ABC$ is the only triangle determined.

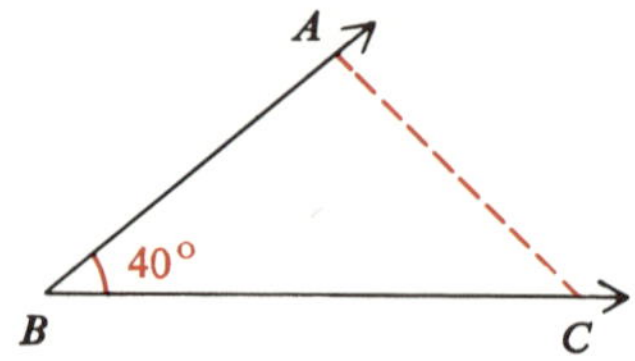

How many triangles, if any, can we determine if we know the measures of two angles and the length of the included side, that is, if we know the measures of an angle, a side, and an angle, in that order?

Example 2 If $m\angle A = 40$, $m\angle B = 60$, and $AB = 5$, how many triangles are determined?

Solution With a protractor and a ruler, we construct the figure at right using these measures. If we extend $\overrightarrow{AX}$ and $\overrightarrow{BY}$, and they intersect, in how many points will they intersect? One and only one triangle, $\triangle ABP$, is determined.

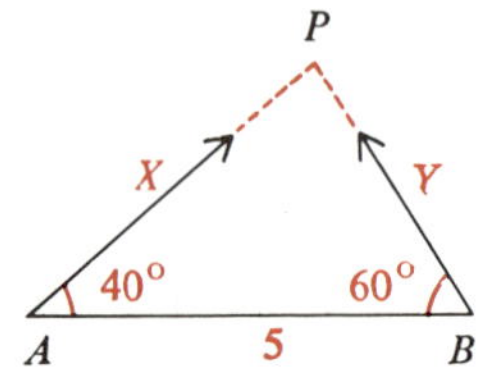

Will only one triangle be determined if the measures of the three sides are fixed?

Example 3 Three segments have measures 3, 4, and 5. Is only one triangle, if any, determined by these segments?

Solution Again, we construct a model with these measures. Can we form a triangle? Is only one triangle formed? If we change any angle in the model, will the measure of the sides be the same?

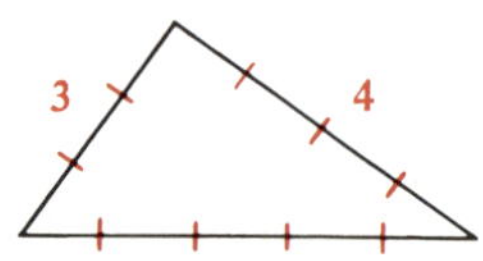

Do you think that knowing the measures of any three parts of a triangle allows us to determine a unique triangle?

Example 4 How many, if any, triangles are determined if $AB = 2$, $BC = 3$, and $m\angle C = 40$?

Solution Form $\angle QCP$ at the vertex of $\overrightarrow{CQ}$. On $\overrightarrow{CP}$, locate point B such that $CB = 3$. We know B is unique, because of the Point Uniqueness Postulate. If we put the zero point of a ruler at B and move the ruler along $\overrightarrow{CQ}$, we see that two points of $\overrightarrow{CQ}$, A and A', are found such that $BA = BA' = 2$. Thus, with the information given above, we find that two triangles are determined.

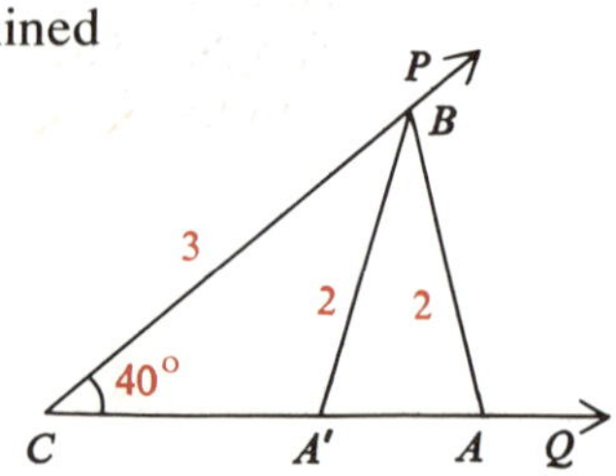

Suppose that a correspondence between two triangles is known such that three parts of one triangle are congruent to their corresponding parts of the second triangle. Can we say that the two triangles are congruent? After studying Example 4, you should see that the answer may be no. In the example, is $\triangle A'BC$ congruent to $\triangle ABC$? But what if the corresponding parts given are the parts known in Examples 1, 2, or 3? Would these triangles be congruent? In each instance the given parts determine exactly one triangle. We should agree that the triangles determined are congruent in each case. The following postulates formalize our discussion of these three instances.

D **Definition 3-5** By the **SAS correspondence**, two sides and the included angle of one triangle are congruent to the corresponding parts of a second triangle.

P **Postulate 3-1 The SAS Postulate** Any SAS correspondence is a congruence.

D **Definition 3-6** By the **ASA correspondence**, two angles and the included side of one triangle are congruent to the corresponding parts of a second triangle.

P **Postulate 3-2 The ASA Postulate** Any ASA correspondence is a congruence.

D **Definition 3-7** By the **SSS correspondence**, the three sides of one triangle are congruent to the corresponding sides of a second triangle.

P **Postulate 3-3 The SSS Postulate** Any SSS correspondence is a congruence.

The correspondence of the three angles of one triangle to congruent corresponding angles of a second triangle is not a congruence. The figure at right corresponds in this way, but the triangles have only the same shape, not the same size. The inner triangle is smaller than the outer triangle. The triangles are said to be **similar**.

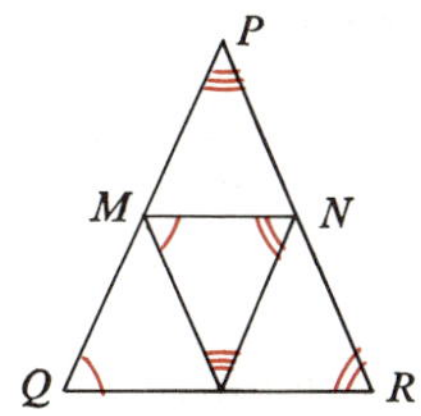

Example 5 If $\overline{AB} \cong \overline{PQ}$, $\overline{BC} \cong \overline{QR}$, and $\angle B \cong \angle Q$, show that $AC = PR$, $\angle A \cong \angle P$, and $\angle C \cong \angle R$.

Solution By the SAS postulate, $\triangle ABC \cong \triangle PQR$. By the definition of congruent triangles, the remaining corresponding parts are also congruent. That is, $\angle A \cong \angle P$, $\angle C \cong \angle R$, and $\overline{AC} \cong \overline{PR}$. Equivalent equations with measures also follow immediately from the definition. Thus, $AC = PR$.

Class Exercises

Assume that the markings in each pair of triangles indicate congruent correspondences. Determine which pairs of triangles are congruent and name the postulate that guarantees the congruence.

1.
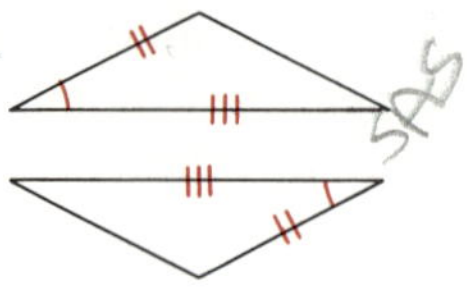

2.
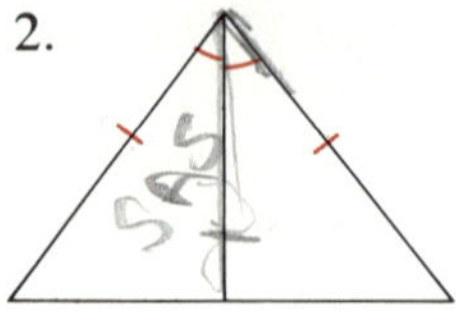

3.
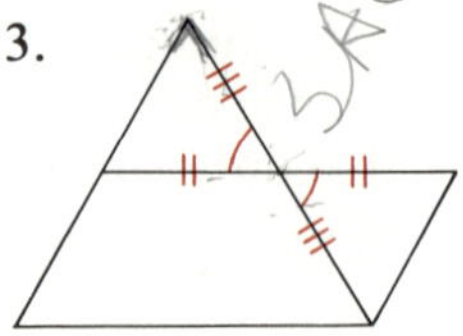

4.
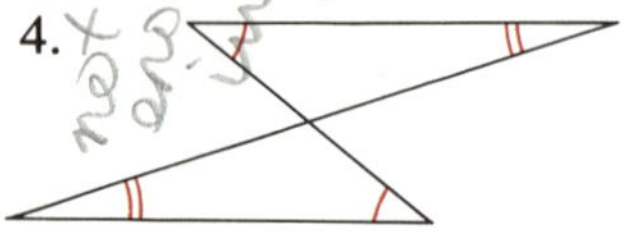

5.
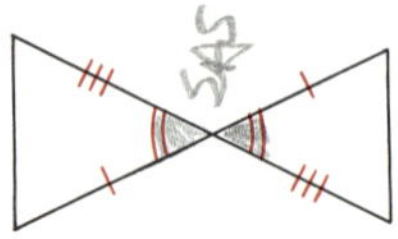

6.

Thus far, our proofs have followed almost directly from the definitions or postulates. The additional information that follows from proving that two triangles are congruent allows us to develop other proofs. These proofs will use the results of the triangle congruences.

Example 6

GIVEN $\overline{PM} \perp \overline{QR}$ and $QM = RM$

PROVE $\angle Q \cong \angle R$

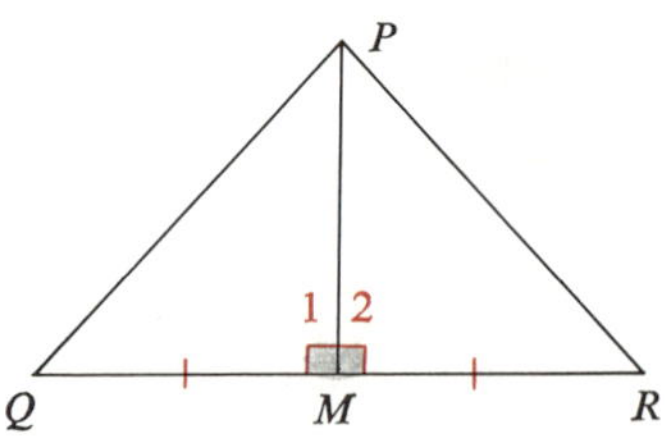

Solution In a proof, it is usually helpful to draw a figure and indicate the given parts with markings. We shall use the two-column proof format. We need to show $\angle Q \cong \angle R$. If $\triangle QMP$ and RMP are congruent, then the congruence of their corresponding angles follows immediately.

PROOF

STATEMENTS	REASONS
1. $QM = RM$; $\overline{PM} \perp \overline{QR}$	1. Given.
2. $\angle 1 \cong \angle 2$	2. Definition of perpendicular lines. (1-25)
3. $\overline{PM} \cong \overline{PM}$	3. Every segment is congruent to itself. (3-1.6)
4. $\triangle QMP \cong \triangle RMP$	4. The SAS Postulate. (3-1)
5. $\angle Q \cong \angle R$	5. Definition of congruent triangles. (3-3)

Since $QM = RM$ is equivalent to $\overline{QM} \cong \overline{RM}$, the statements are interchangeable.

Example 7 If $\angle 3 \cong \angle 4$ and $\overline{QM}$ bisects $\angle PQR$, prove that M is the midpoint of $\overline{PR}$.

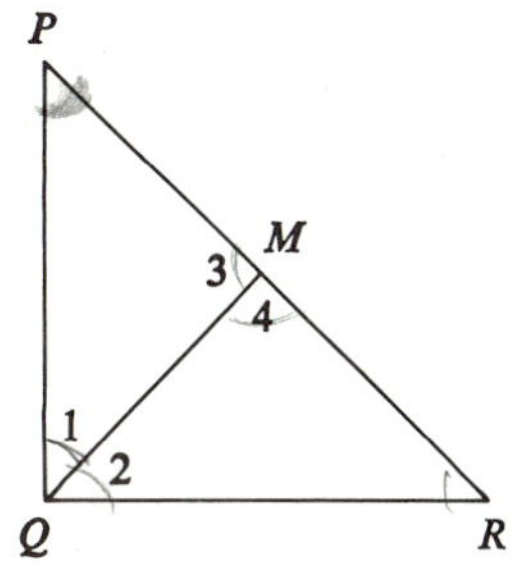

Solution

GIVEN $\overline{QM}$ bisects $\angle PQR$; $\angle 3 \cong \angle 4$

PROVE M is the midpoint of $\overline{PR}$.

PROOF

STATEMENTS	REASONS
1. $\angle 3 \cong \angle 4$; $\overline{QM}$ bisects $\angle PQR$.	1. Given.
2. $\angle 1 \cong \angle 2$	2. Definition of angle bisector. (1-29)
3. $\overline{QM} \cong \overline{QM}$	3. Every segment is congruent to itself. (3-16)
4. $\triangle PMQ \cong \triangle RMQ$	4. The ASA Postulate. (3-2)
5. $PM = RM$	5. Definition of congruent triangles. (3-3)
6. M is the midpoint of PR.	6. Definition of a midpoint. (1-15)

EXERCISES

A Determine if the following pairs of triangles are congruent, and, if so, the postulate which guarantees the congruence. The markings indicate congruences.

1.

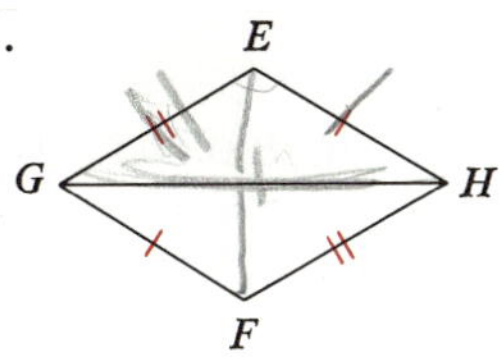

2.

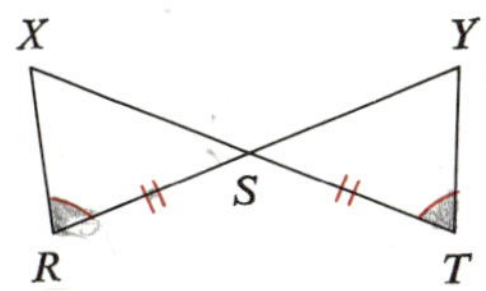

3.

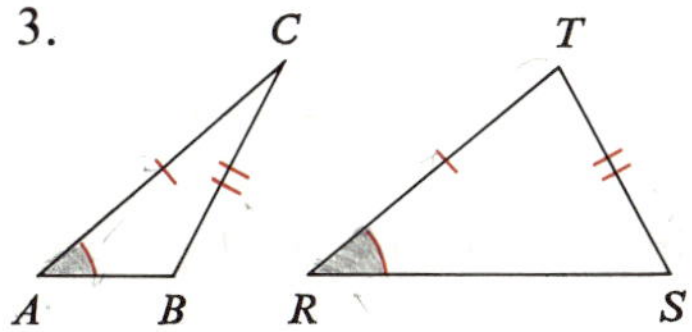

4.

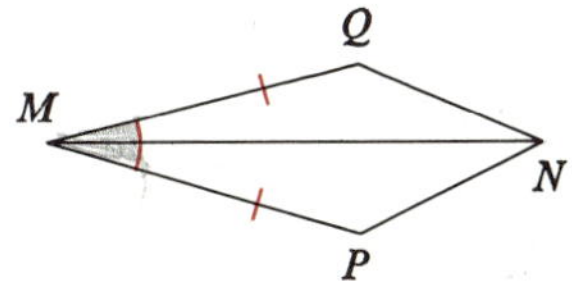

5.

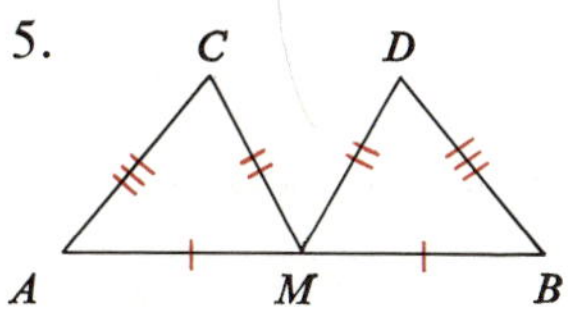

6.

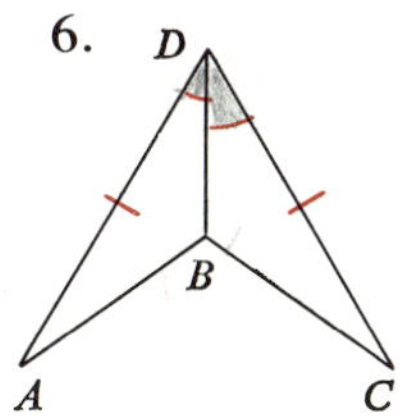

7.

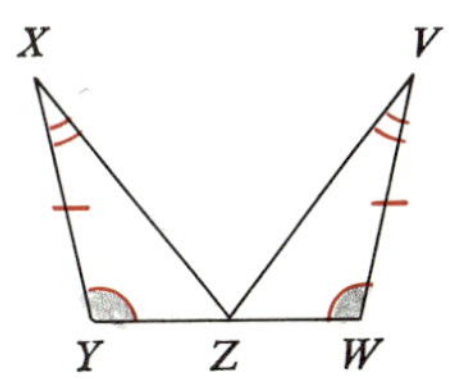

Determine which postulate of congruent triangles is appropriate to prove triangles congruent in each of the following:

8. 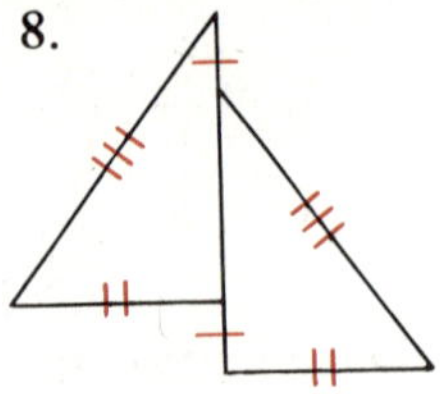

9. 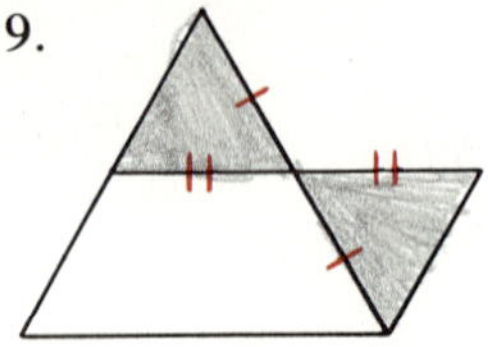

10. 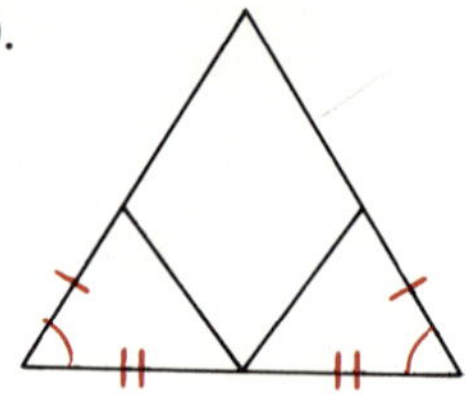

Determine which of the following prove congruence between $\triangle ABC$ and $\triangle DEF$. State which postulate was used.

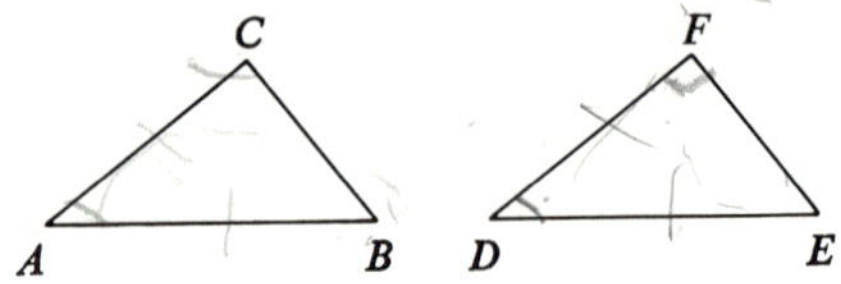

11. $AC = DF$, $\overline{BC} \cong \overline{EF}$, $\angle C \cong \angle F$

12. $\overline{BC} \cong \overline{EF}$, $\overline{AC} \cong \overline{DF}$, $\overline{BA} \cong \overline{ED}$

13. $BA = CA$, $ED = FD$, $m\angle A = m\angle D$

14. **GIVEN** R is the midpoint of $\overline{PT}$ and $\overline{QS}$.

PROVE $\angle P \cong \angle T$

PROOF

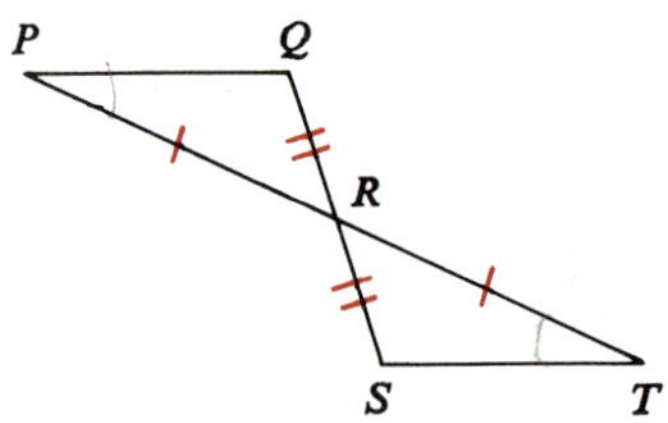

STATEMENTS	REASONS
1. R is the midpoint of $\overline{PT}$ and $\overline{QS}$.	1. Given
2. $PR = TR$ and $QR = SR$	2. Definition of midpoint (1-15)
3. $\angle PRQ \cong \angle TRS$	3. ________ Theorem
4. $\triangle PRQ \cong \triangle TRS$	4. ________ Postulate
5. $\angle P \cong \angle T$	5. Definition of ________

15. If $\angle 1 \cong \angle 2$ and $\angle 3 \cong \angle 4$, prove that $\angle H \cong \angle F$.

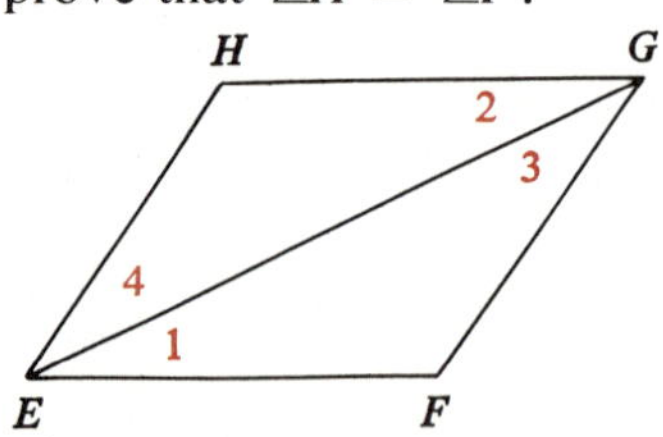

16. If $\angle AMP$ and $\angle BMC$ are right angles, $AM = CM$ and $\angle A \cong \angle C$, prove that $\overline{AP} \cong \overline{CB}$.

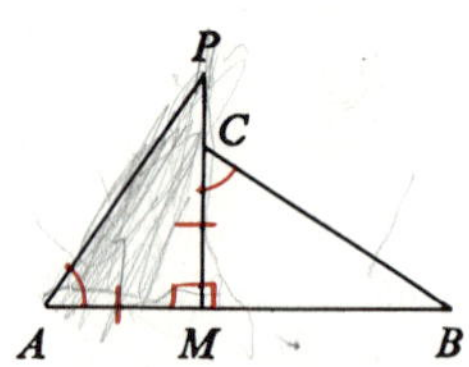

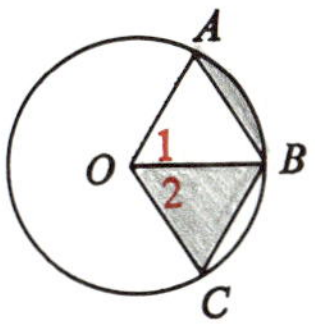

Exercise 17

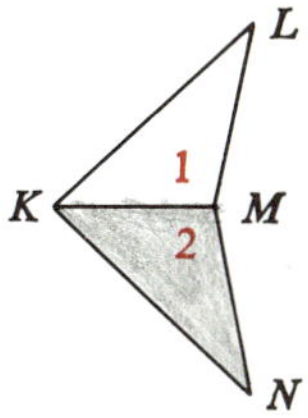

Exercise 18

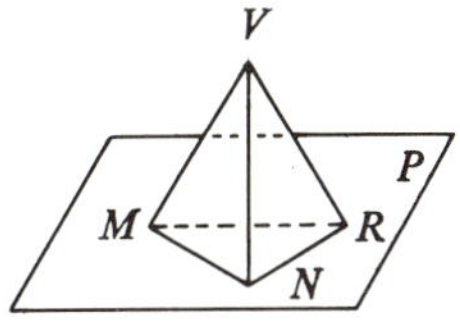

Exercise 19

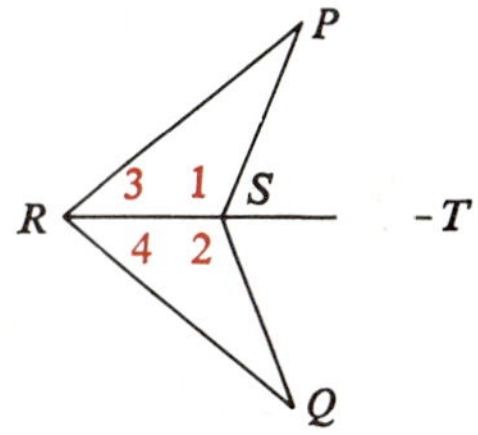

Exercise 20

17. If $OA = OB = OC$ and $\angle 1 \cong \angle 2$, prove that $AB = BC$.
18. If $KL = KN$ and $LM = NM$, prove $\angle 1 \cong \angle 2$.
19. If $\angle MVN \cong \angle RVN$ and $\angle MNV \cong \angle RNV$, then prove $MV = RV$.

B 20. If $\angle 3 \cong \angle 4$ and $PR = QR$, then prove $\angle PST \cong \angle QST$.

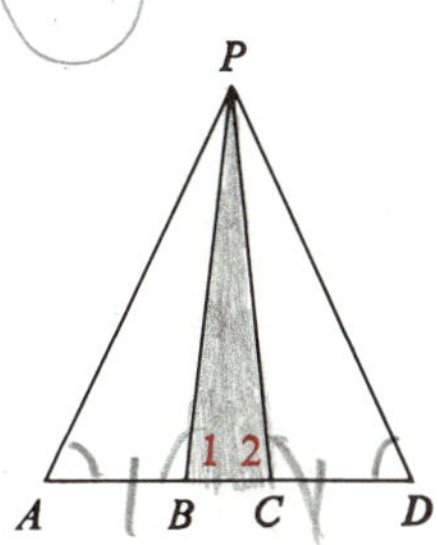

Exercise 21

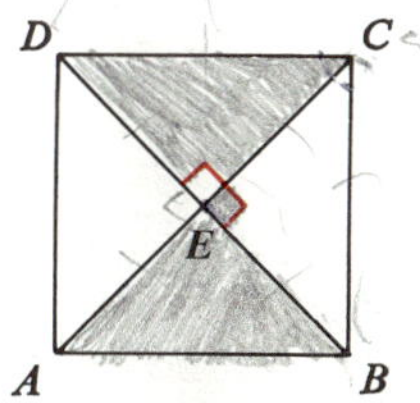

Exercise 22

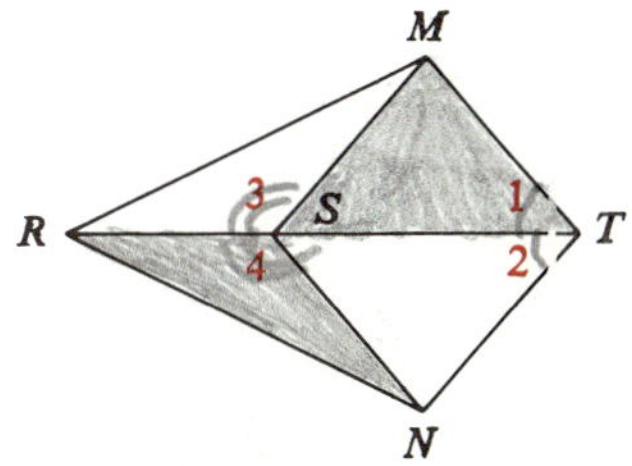

Exercise 23

21. If $\angle A \cong \angle D$, $AB = DC$, and $\angle 1 \cong \angle 2$, then prove $\triangle ACP \cong \triangle DBP$.
22. If $\overline{AC} \perp \overline{BD}$ and point E is the midpoint of $\overline{DB}$, then prove $DC = BC$.
23. Given $\angle 1 \cong \angle 2$ and $\angle 3 \cong \angle 4$ Prove $RM = RN$

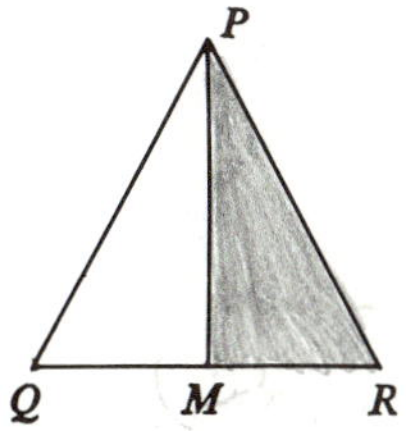

Exercises 24 and 27

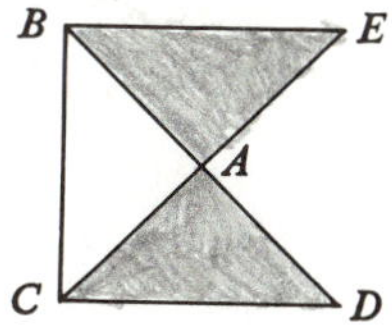

Exercise 25

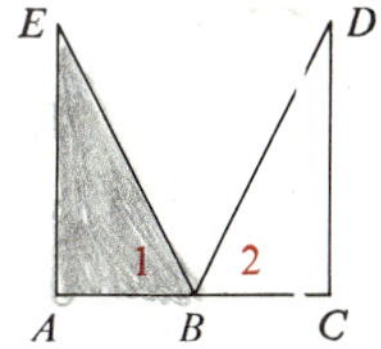

Exercise 26

24. If $\overline{PM}$ is the perpendicular bisector of $\overline{QR}$, prove that $\triangle PQR$ is isosceles.
25. Given $\overleftrightarrow{BAD}$ and $\overleftrightarrow{CAE}$ intersect at A such that $AB = AC$ and $AE = AD$.
    Prove $BE = CD$
26. Given $\overline{EA} \perp \overleftrightarrow{ABC}$, $\overline{DC} \perp \overleftrightarrow{ABC}$; B is the midpoint of $\overline{AC}$ and $\angle 1 \cong \angle 2$.
    Prove $AE = CD$
27. Using the hypothesis from Exercise 24, prove that $\angle Q \cong \angle R$.

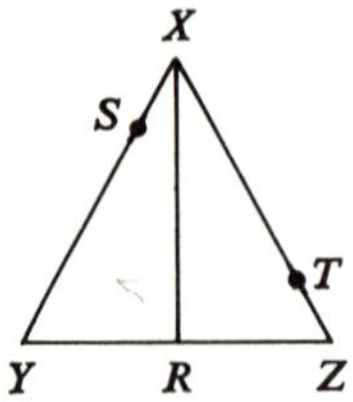

Exercise 28

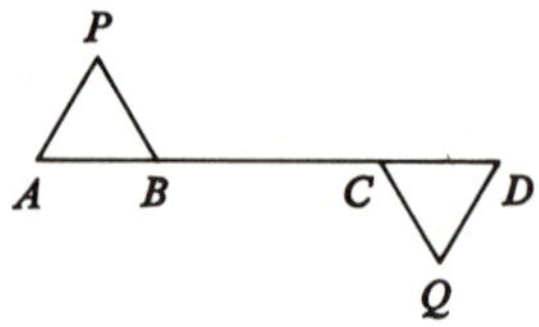

Exercise 29

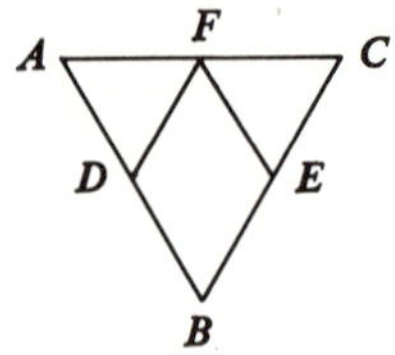

Exercise 30

28. **Given** $\overline{XR}$ is a median of $\triangle XYZ$; $XS = TZ$ and $SY = XT$
Prove $\angle Y \cong \angle Z$
29. **Given** $\overleftrightarrow{ABCD}$, $AP = DQ$, $AC = DB$; $BP = CQ$ **Prove** $\angle P \cong \angle Q$
30. D, E, and F are the midpoints of the sides of $\triangle ABC$. If $AB = BC$ and $\angle A \cong \angle C$, prove that $DF = EF$.

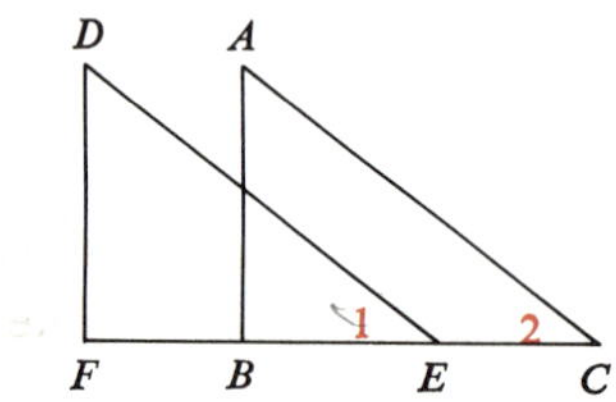

Exercise 31

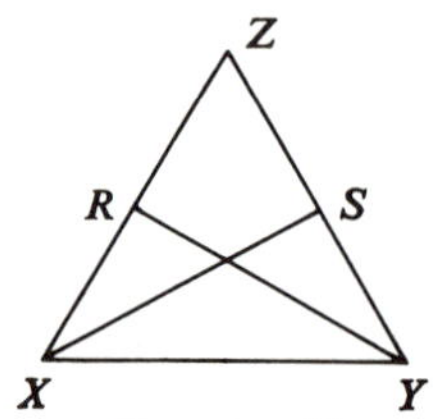

Exercise 32

C

31. **Given** $\overline{DF} \perp \overline{FE}$, $\overline{AB} \perp \overline{BC}$, $\angle 1 \cong \angle 2$, and $FB = EC$
Prove $\triangle DEF \cong \triangle ACB$
32. If $ZR = ZS$ and $ZX = ZY$, prove that $\triangle ZXS \cong \triangle ZYR$.

Use the figure and the information given below to answer Exercises 33–37. Do not use the ASA Postulate.

Given An ASA correspondence $PQR \leftrightarrow KLM$, such that $\angle P \cong \angle K$, $\angle Q \cong \angle L$, and $PQ = KL$

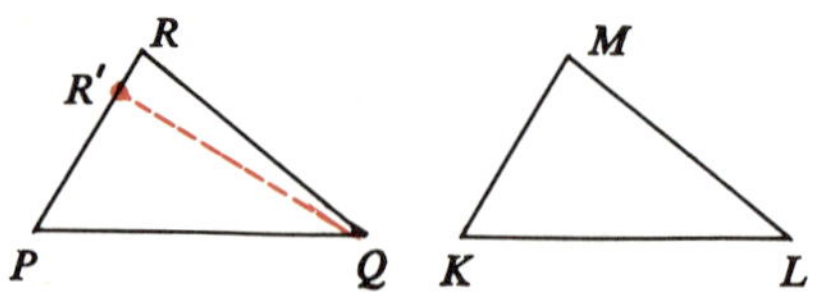

Exercises 33–37

33. If $\overrightarrow{PR}$ contains any point R' such that $PR' = KM$, prove that $\triangle PR'Q \cong \triangle KML$.
34. From the congruence of Exercise 33, we know that $\angle PQR'$ must be congruent to which angle of $\triangle KML$?
35. Provide a reason for the statement that $\overrightarrow{QR'}$ must be the same as $\overrightarrow{QR}$.
36. If two different lines intersect in at most one point, what point coincides with R'?
37. Using the information from Exercises 33–36, prove that $\triangle PQR \cong \triangle KLM$.

3-4 ISOSCELES AND EQUILATERAL TRIANGLES

In Chapter 1, we defined the angle bisector. Now we shall prove that there is only one bisector of any angle.

Theorem 3-4.1 Angle Bisector Theorem Every angle has exactly one bisector.

GIVEN $\angle BAC$

PROVE $\overrightarrow{AP}$ is the one and only bisector of $\angle BAC$.

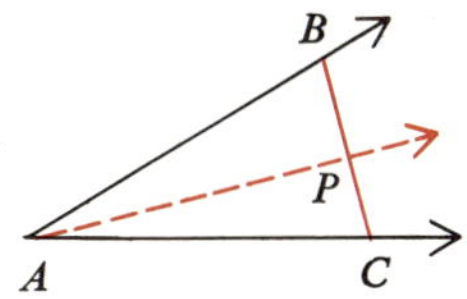

We must prove two things:
(1) There exists at least one bisector of $\angle BAC$.
(2) This bisector is the only bisector of $\angle BAC$.

PROOF of (1)

STATEMENTS	REASONS
1. Choose B on $\overrightarrow{AB}$ and C on $\overrightarrow{AC}$ so that $AB = AC$.	1. Point Uniqueness Postulate. (2-2)
2. Let P be the unique midpoint of $\overline{BC}$.	2. The Line Postulate (2-3); any line segment has exactly one midpoint. (2-5.4)
3. $BP = PC$	3. Definition of a midpoint. (1-15)
4. A and P determine $\overline{AP}$.	4. The Line Postulate. (2-3)
5. $\overline{AP} \cong \overline{AP}$	5. Every segment is congruent to itself. (3-1.6)
6. $\triangle APB \cong \triangle APC$	6. The SSS Postulate. (3-3)
7. $\angle PAB \cong \angle PAC$	7. Definition of congruent triangles. (3-3)
8. $\overrightarrow{AP}$ bisects $\angle BAC$.	8. Definition of angle bisector. (1-29)

PROOF of (2)

STATEMENTS	REASONS
1. $\overrightarrow{AP}$ bisects $\angle BAC$.	1. Just proved above.
2. $m\angle PAB = m\angle PAC = x$	2. Definition of congruent angles. (1-30)
3. P is in the interior of $\angle BAC$.	3. Definition of angle bisector. (1-29)
4. $m\angle BAC = m\angle PAB + m\angle PAC$	4. Definition of angle addition. (1-22)
5. $m\angle BAC = x + x = 2x$	5. The Substitution Postulate. (2-1)
6. $\frac{1}{2}m\angle BAC = x$	6. Multiplication property of equality.
7. $\overrightarrow{AP}$ is the only ray such that $m\angle PAC = x$	7. The Angle Uniqueness Postulate. (2-9)

D **Definition 3-8** An **angle bisector of a triangle** is a segment that lies on the ray bisector of an angle of the triangle and has its endpoints at the angle's vertex and at a point of the side opposite the angle.

In the figures below, which segment is an angle bisector of $\triangle BAC$? Assume that the figure markings indicate congruences.

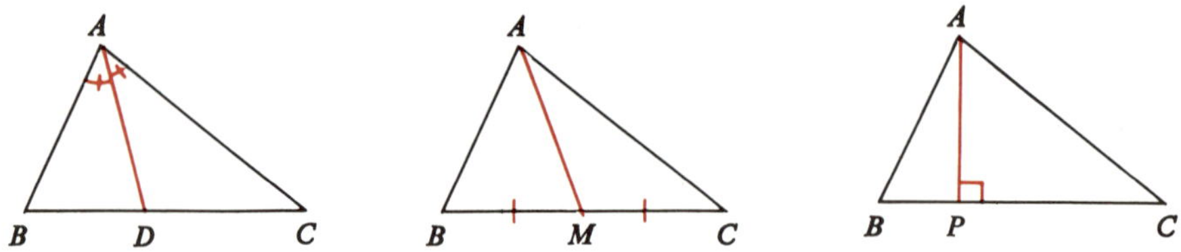

Judging by the markings, $\overrightarrow{AD}$ bisects $\angle BAC$. If you said that $\overline{AD}$ is the angle bisector of $\triangle BAC$, you are correct. Do you know what we call $\overline{AM}$ and $\overline{AP}$? The markings on the center figure show that M is the midpoint of $\overline{BC}$. Did you call $\overline{AM}$ a *median* of $\triangle ABC$? How many medians does any triangle contain?

D **Definition 3-9** A **median** of a triangle is a segment whose endpoints are a vertex of the triangle and the midpoint of the side opposite that vertex.

Since there is only one midpoint of any side, we can say that for each side there is a unique median. We first discussed medians in Chapter 1, but we have not discussed segments like $\overline{AP}$ before. $\overline{AP}$ is an *altitude* of $\triangle BAC$.

D **Definition 3-10** An **altitude** of a triangle is a perpendicular segment whose endpoints are a vertex of the triangle and a point of the line containing the opposite side.

Why do we say "to the line containing the opposite side" and not just "to the opposite side"? When all the angles of a triangle are acute angles, will the altitudes lie in the interior or exterior of the triangle? Draw an acute triangle, an obtuse triangle, and a right triangle. Where does the altitude from each vertex fall with respect to the interior of each triangle?

Class Exercises

1. If a right angle is bisected, find the measure of one of the two angles formed.
2. If K lies on the bisector of $\angle RLM$ such that $m\angle KLR = 22.5$, find $m\angle KLM$.

Complete each of the following:

3. Every triangle has ________ angle bisectors.
4. The endpoints of an angle bisector of a triangle are a(n) ________ and a point on the ________ side.
5. An altitude of a triangle is ________ to a side.
6. The ________ is a segment joining a vertex to the midpoint of the side opposite the vertex.
7. In a(n) ________ triangle some of the altitudes will lie in the exterior of the triangle.

Name the median, angle bisector, and altitude that are shown in each of the following triangles:

8.

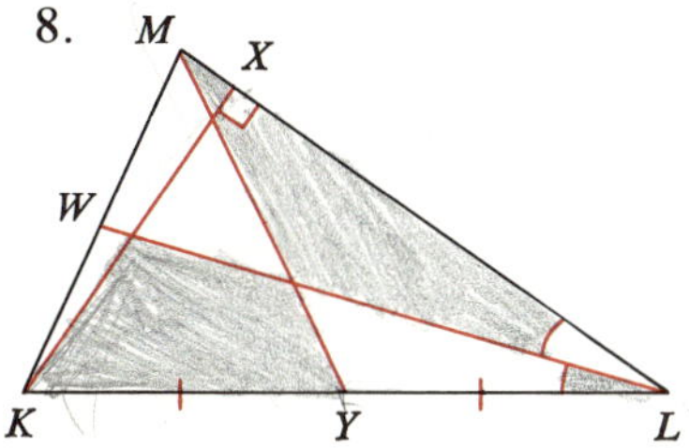

9.

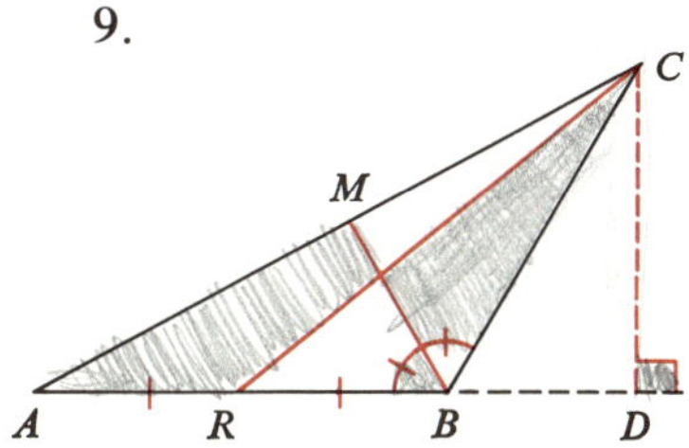

Special Triangles

We can classify angles as acute, right, or obtuse. Triangles, too, can be classified according to their angles.

all angles acute	*acute triangle*
one right angle	*right triangle*
one obtuse angle	*obtuse triangle*

What other possibility exists for the angles of a triangle?

D **Definition 3-11** A triangle is **equiangular** if all its angles are congruent.

Triangles can also be grouped according to sides.

D **Definition 3-12** A **scalene** triangle has no congruent sides. An **isosceles** triangle has two congruent sides. An **equilateral** triangle has all three sides congruent.

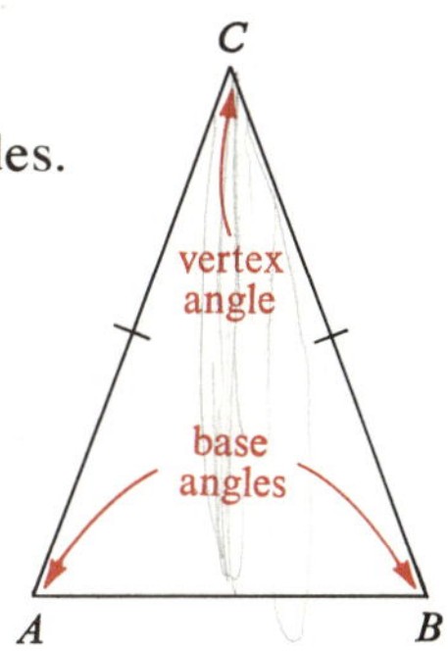

In the isosceles triangle at right, $\overline{AC} \cong \overline{BC}$. The third side, $\overline{AB}$, is called the **base**. The two angles that include the base, $\angle A$ and $\angle B$, are called **base angles**. The angle opposite the base of an isosceles triangle is called the **vertex angle**

We shall now state an important theorem about isosceles triangles.

T

Theorem 3-4.2 Isosceles Triangle Theorem The base angles of an isosceles triangle are congruent.

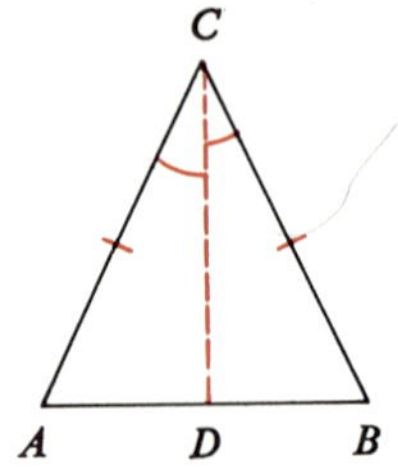

GIVEN In $\triangle ABC$, $\overline{AC} \cong \overline{BC}$.

PROVE $\angle A \cong \angle B$

We shall prove this theorem in two ways.

PROOF I

STATEMENTS	REASONS
1. In $\triangle ABC$, $\overline{AC} \cong \overline{BC}$	1. Given.
2. Let the bisector of $\angle C$ meet $\overline{AB}$ at D.	2. Definition of an angle bisector of a triangle. (3-8)
3. $\angle ACD \cong \angle BCD$	3. Definition of angle bisector. (1-29)
4. $\overline{CD} \cong \overline{CD}$	4. Every segment is congruent to itself. (3-1.6)
5. $\triangle ACD \cong \triangle BCD$	5. The SAS Postulate. (3-1)
6. $\angle A \cong \angle B$	6. Definition of congruent triangles. (3-3)

PROOF II Consider the correspondence $ACB \leftrightarrow BCA$ between $\triangle ABC$ and itself.

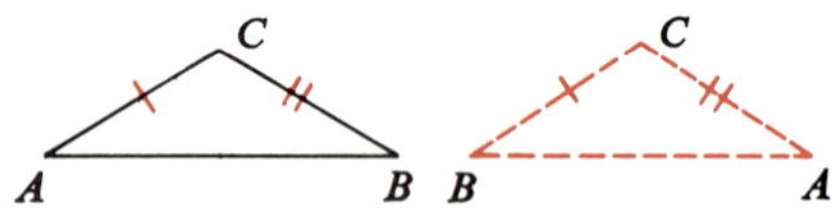

STATEMENTS	REASONS
1. $\overline{AC} \cong \overline{BC}$ and $\overline{BC} \cong \overline{AC}$	1. Given.
2. $\angle C \cong \angle C$	2. Every angle is congruent to itself. (3-1.7)
3. $\triangle ACB \cong \triangle BCA$	3. The SAS Postulate. (3-1)
4. $\angle A \cong \angle B$	4. Definition of congruent triangles. (3-3)

The next theorem follows almost immediately from the Isosceles Triangle Theorem. A theorem that follows directly from another theorem is called a **corollary** of that theorem. The *a* in the corollary number means that it is the first corollary following a specific theorem. In the proof of Corollary 3-4.2a, Theorem 3-4.2 is used twice. The proof is left as an exercise.

C

Corollary 3-4.2a Every equilateral triangle is equiangular.

The next theorem is the converse of Theorem 3-4.2. This theorem provides us with a method of establishing that a triangle is isosceles.

T

Theorem 3-4.3 In a triangle, if two angles are congruent, then the sides opposite these angles are congruent.

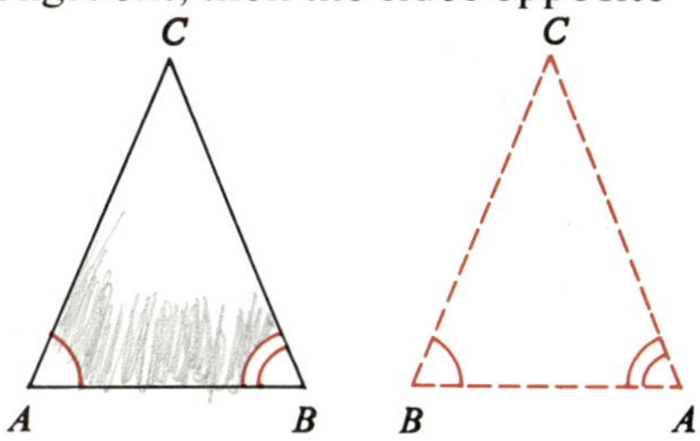

GIVEN In $\triangle ABC$, $\angle A \cong \angle B$

PROVE $\overline{AC} \cong \overline{BC}$

PROOF Consider $ABC \leftrightarrow BAC$ between the triangle and itself.

STATEMENTS	REASONS
1. $\angle A \cong \angle B$ and $\angle B \cong \angle A$	1. Given.
2. $\overline{AB} \cong \overline{BA}$	2. Every segment is congruent to itself. (3-1.6)
3. $\triangle ABC \cong \triangle BAC$	3. The ASA Postulate. (3-2)
4. $\overline{AC} \cong \overline{BC}$	4. Definition of congruent triangles. (3-3)

C

Corollary 3-4.3a Every equiangular triangle is equilateral.

The proof of the above corollary is left as an exercise.

EXERCISES

A

Complete the following. In $\triangle KLM$, $\overline{KL}$ lies on $\overleftrightarrow{KN}$ and $\overline{ML}$ lies on $\overleftrightarrow{MP}$.

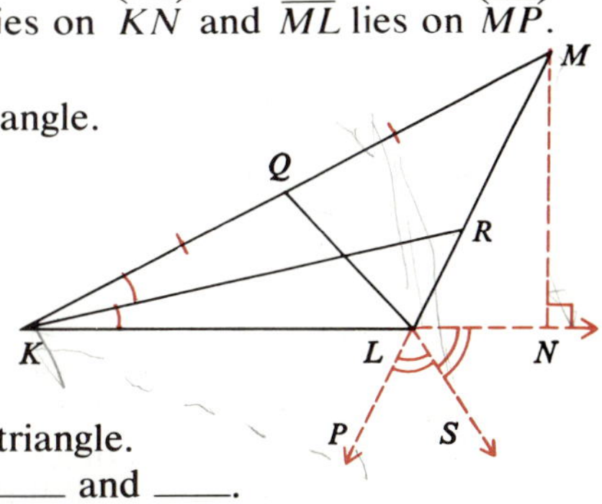

1. The altitude from vertex K will meet $\overleftrightarrow{MP}$ in the ________ of the triangle.
2. $\overline{QL}$ is a(n) ________ of $\triangle KLM$.
3. An angle bisector of $\triangle LKM$ is ____.
4. $\angle PLS \cong$ ____.
5. $\overrightarrow{LS}$ is the bisector of ____.
6. $\overline{MN}$ is the ________ from vertex M.
7. $\triangle KLM$ is a(n) ________ triangle.
8. If $KL = ML$, $\triangle KLM$ is a(n) ________ triangle.
9. If $KL = ML$, then the base angles are ____ and ____.

Indicate which of the following are true and which are false:

10. The base of an isosceles triangle is included by the base angles.
11. If $\angle A$ is the vertex angle of an equilateral triangle, any side may be considered the base.

12. If $\angle A$ is the vertex angle of equiangular $\triangle ABC$, the base angles are $\angle B$ and $\angle C$.
13. An angle bisector of a triangle is a ray.
14. At least one altitude of any triangle must lie in the interior of the triangle.
15. A corollary is not a theorem.
16. Prove Corollary 3-4.2a.
17. Prove Corollary 3-4.3a.
18. GIVEN $AB = AC$
PROVE $\angle 1 \cong \angle 2$

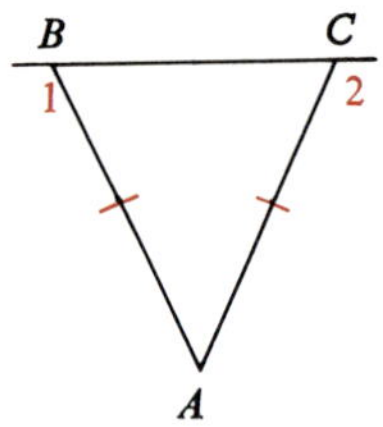

19. GIVEN $PQ = PR$ and $QM = RM$
PROVE $\overline{PM}$ is an angle bisector.

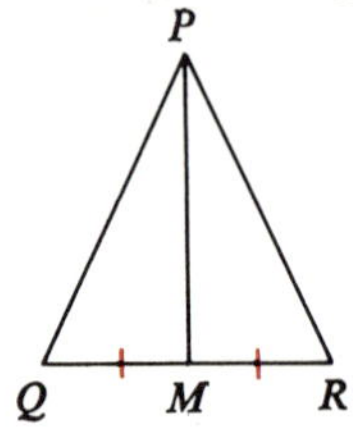

20. GIVEN $\overleftrightarrow{AKC}$, $\overleftrightarrow{CMB}$, and $\overleftrightarrow{ALB}$;
$KA = LB$,
$MB = LA$;
and $AC = BC$
PROVE $LM = KL$

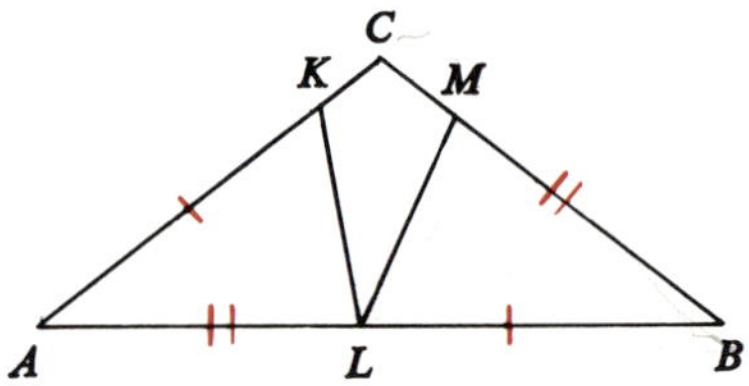

Exercise 20

21. GIVEN $\overleftrightarrow{DGHF}$; $DE = FE$;
$\angle 3 \cong \angle 4$
PROVE $\angle 1 \cong \angle 2$

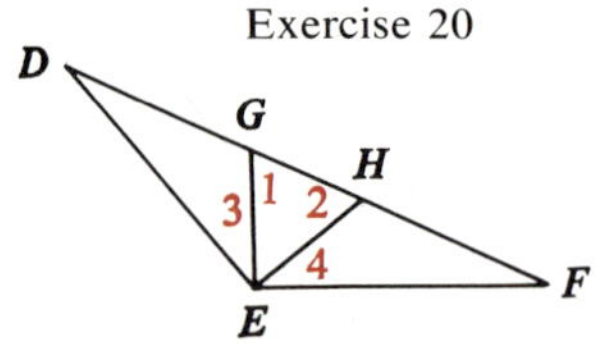

Exercise 21

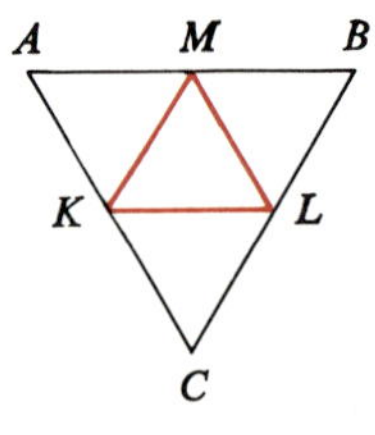

Exercise 22

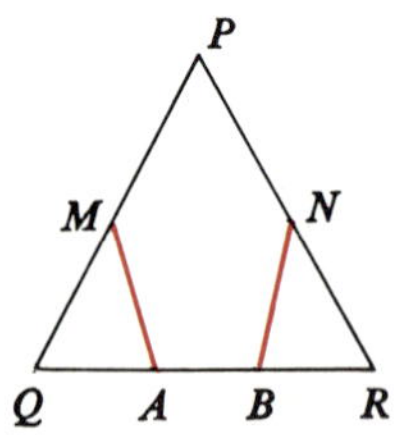

Exercise 23

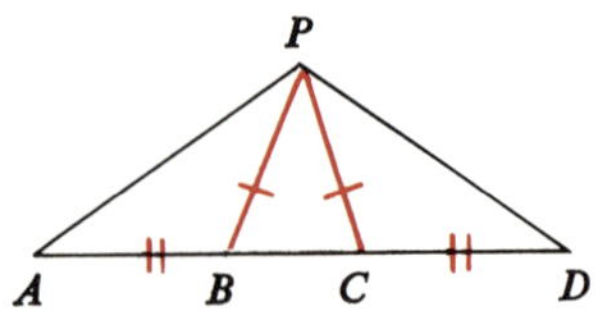

Exercise 24

B

22. If $\triangle ABC$ is equilateral, with M, K, and L as midpoints of $\overline{AB}$, $\overline{AC}$, and $\overline{BC}$, respectively, prove that $\triangle MKL$ is equilateral.
23. If in $\triangle PQR$, $PQ = PR$, M and N are the midpoints of $\overline{PQ}$ and $\overline{PR}$, respectively, and $QA = AB = BR$, prove that $MA = NB$.
24. GIVEN $AB = DC$ and $PB = PC$ PROVE $AP = DP$

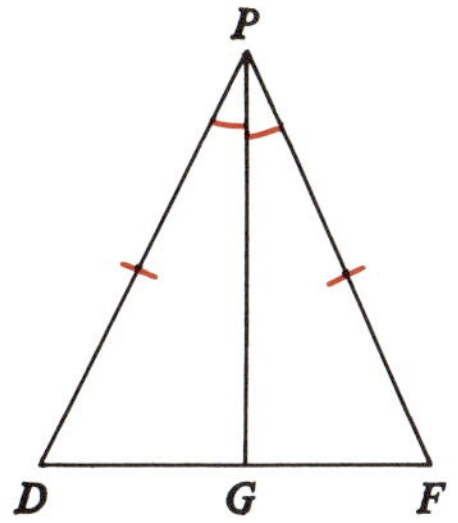

Exercise 25

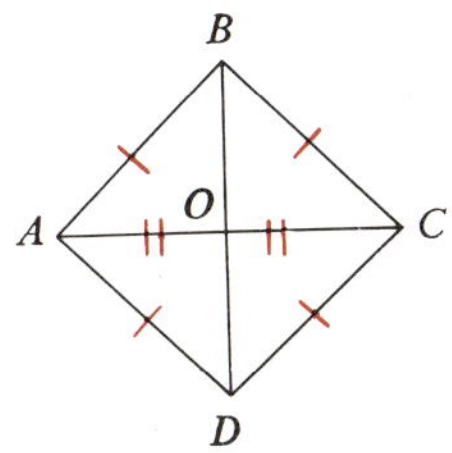

Exercise 26

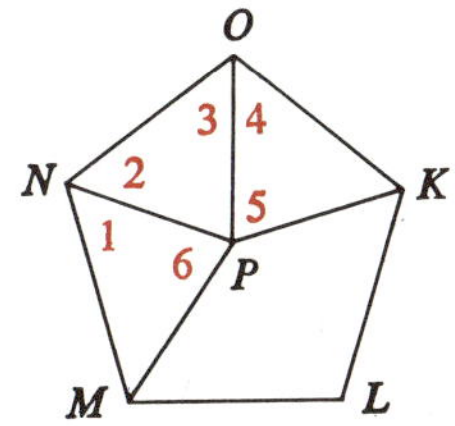

Exercise 27

25. **Given** $DP = FP$ and $\overline{PG}$ is the angle bisector of $\angle P$.
    Prove $\overline{PG}$ is an altitude of $\triangle DPF$, and $\overline{PG}$ is a median of $\triangle DPF$.
26. If $\overline{BD}$ intersects $\overline{AC}$ at O, $AB = BC = CD = DA$, and $AO = OC$, prove that $\overline{BD} \perp \overline{AC}$.
27. If $KL = LM = MN = NO = OK$, $\angle 1 \cong \angle 4$, and $\angle 2 \cong \angle 3$, prove that $\angle 5 \cong \angle 6$.

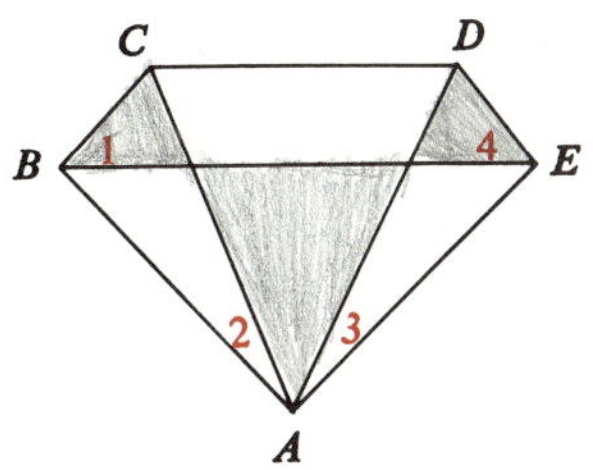

Exercise 28

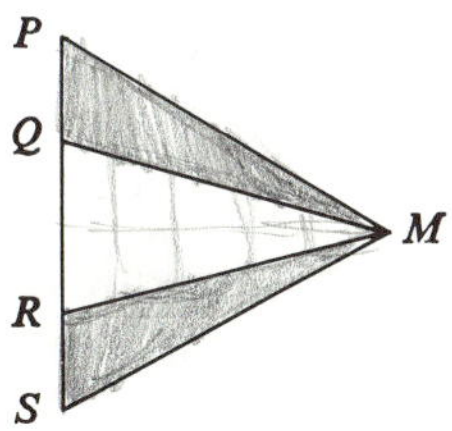

Exercise 29

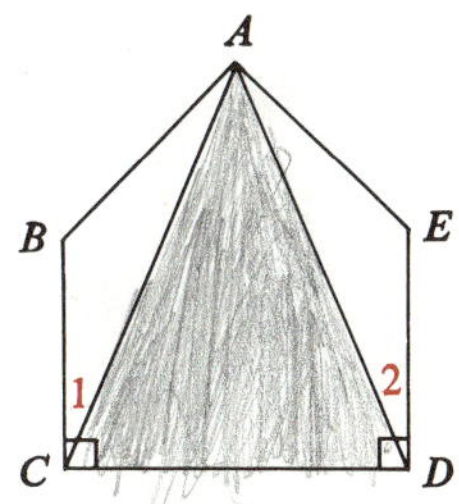

Exercise 30

28. **Given** $m\angle 2 = m\angle 3$, $AB = AE$, and $AC = AD$ **Prove** $m\angle 1 = m\angle 4$
29. If on $\overleftrightarrow{PQRS}$, $PQ = SR$, and $MQ = MR$, prove that $MP = MS$.
30. **Given** $\overline{BC} \perp \overline{CD}$, $\overline{ED} \perp \overline{CD}$, $BC = ED$, and $m\angle 1 = m\angle 2$
    Prove $\angle BAC \cong \angle EAD$

C

31. Prove that the bisector of the vertex angle of an isosceles triangle is perpendicular to the base.
32. Prove that the bisector of the vertex angle of an isosceles triangle bisects the base.
33. **Given** $\overline{QS}$ intersects $\overline{PR}$ at T such that $RQ = RS$ and $QT = ST$.
    Prove TP bisects $\angle SPQ$. (*Hint*: You may have to prove more than one pair of triangles congruent.)

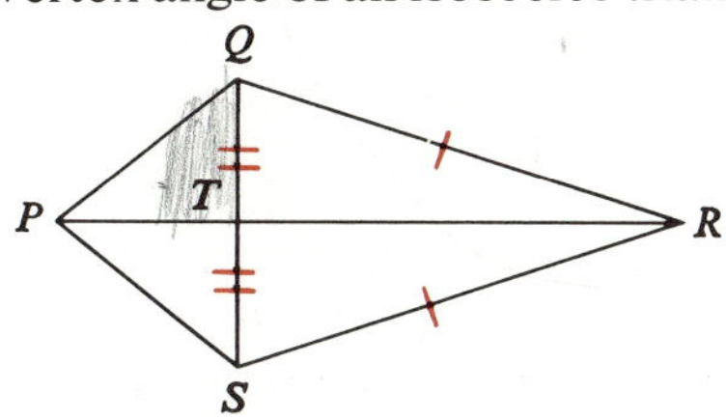

Exercise 33

34. If $\overline{CB} \perp \overline{AB}$, $\overline{CB} \perp \overline{DB}$, and $AB = DB$ prove $\angle ADC \cong \angle DAC$.

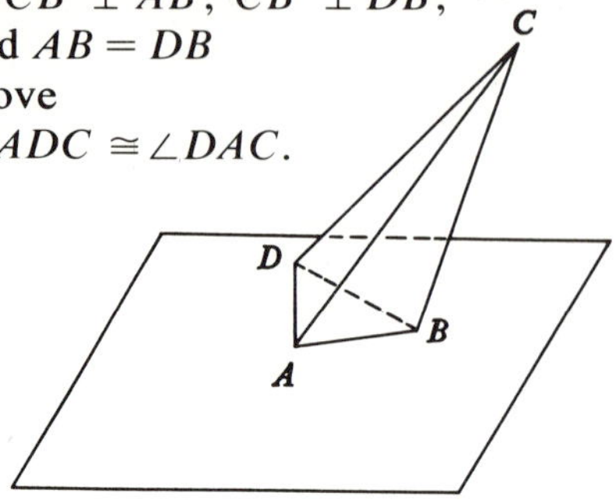

35. If $\triangle PQR$ and $\triangle PMN$ are each isosceles, prove that $MR = NQ$.

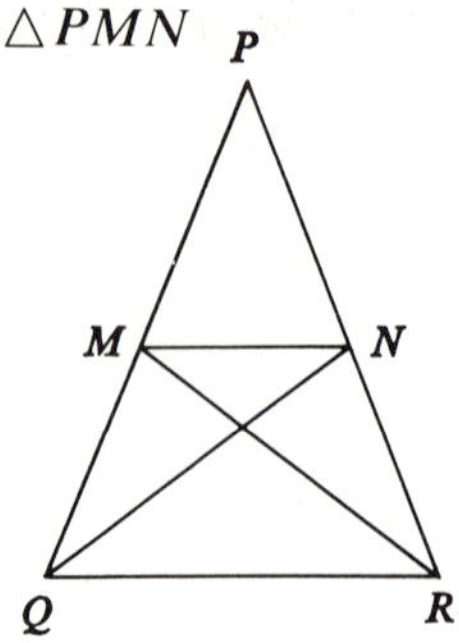

Given that the correspondence $PQR \leftrightarrow XYZ$ is such that $PQ = XY$, $QR = YZ$, and $PR = XZ$, answer the statements in Exercises 36–43 without using the ASA Postulate (3-2).

36. Let $\overrightarrow{PM}$ be the ray on the opposite side of $\overline{PR}$ from Q such that $\angle RPM \cong \angle X$ and $\overline{PN} \cong \overline{XY}$ Prove that $\triangle RPN \cong \triangle ZXY$.
37. From the congruence of Exercise 36, name the angle of $\triangle RPN$ congruent to $\angle Y$.

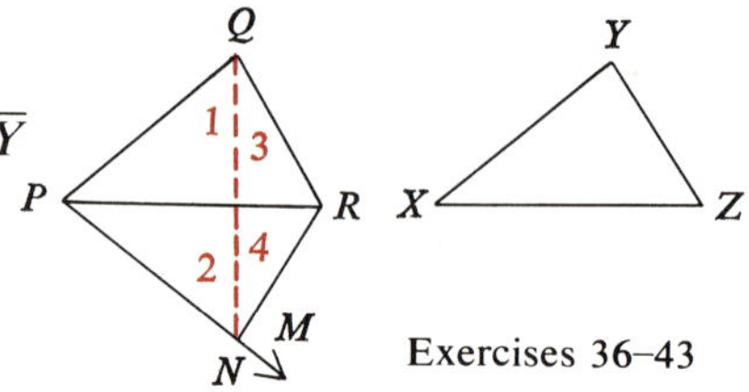

Exercises 36–43

38. If we join points Q and N, provide a reason why $PQ = PN$.
39. What angles in $\triangle PQN$ have equal measures?
40. Provide a reason for $RQ = RN$.
41. What angles in $\triangle RQN$ have equal measures?
42. Justify the statement: $\angle PQR \cong \angle Y$.
43. Prove: $\triangle PQR \cong \triangle XYZ$.

SOMETHING TO THINK ABOUT

A Dilemma

We have proved that if two angles of a triangle are congruent, the sides of the triangle opposite the angles are also congruent. Suppose, however, that before we proved the theorem our intuition had told us that the sides were not congruent. Let us see if our intuition would have led us into an awkward situation. Consider the following reasoning:

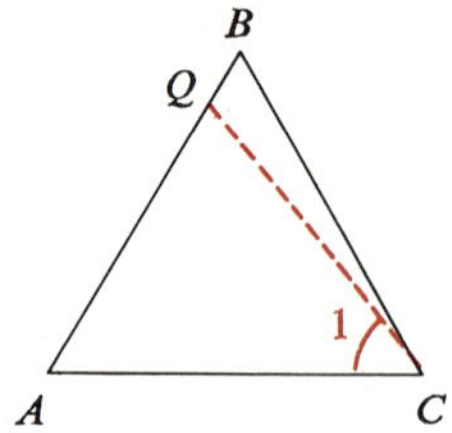

We assume that while $\angle BAC \cong \angle ACB$, the sides $\overline{BA}$ and $\overline{BC}$ are not congruent.
Since $\overline{BA}$ is not congruent to $\overline{BC}$, then $BA \neq BC$.
Therefore, $BA < BC$ or $BA > BC$. Why?
If $BA > BC$, by the Point Uniqueness Postulate there is a point Q on $\overline{BA}$ such that $AQ = BC$. Connect Q and C.

In $\triangle AQC$ and $\triangle CBA$, $QA = BC$, $\angle A \cong \angle ACB$, and $AC = CA$.
Thus, $\triangle AQC \cong \triangle CBA$ and $\angle 1 \cong \angle A$.
But $\angle A \cong \angle ACB$. Therefore, $\angle 1 \cong \angle ACB$.

But this is absurd! By applying the given facts to the drawing, we can see that $m\angle 1 \neq m\angle ACB$. Therefore, the two angles cannot be congruent. Are our steps logical? Can we support each step with a reason? If our steps are logical and we have good reasons, then what is causing us to conclude that angles of unequal measures are congruent?

3-5 OVERLAPPING TRIANGLES

Consider the figure for Example 1 below. How many different triangles do you see? Name them. Did you name $\triangle ABC$ and $\triangle DCB$? Sometimes such triangles are hard to pick out because they overlap, that is, they lie in the same half-plane and have a common side. Triangles are **overlapping** if they have some interior region in common and if one of the following is true:

1. A side of either triangle shares part of a line with a side of the other triangle;
2. An angle of either triangle shares a vertex with the other triangle.

The first type of overlapping triangles is shown in Example 1 and the second type is shown in Example 2.

Example 1

GIVEN $\overline{AB} \cong \overline{DC}$; $\overline{BD} \cong \overline{CA}$

PROVE $\angle ABC \cong \angle DCB$

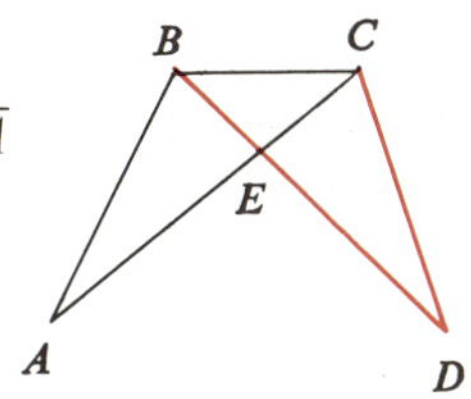

Solution

PROOF

STATEMENTS	REASONS
1. $\overline{AB} \cong \overline{DC}$ and $\overline{BD} \cong \overline{CA}$	1. Given.
2. $\overline{BC} \cong \overline{CB}$	2. Every segment is congruent to itself. (3-1.6)
3. $\triangle ABC \cong \triangle DCB$	3. The SSS Postulate. (3-3)
4. $\angle ABC \cong \angle DCB$	4. Definition of congruent triangles. (3-3)

As we work, we must be certain to make the proper correspondences. In the example, $\overline{BC} \leftrightarrow \overline{CB}$. The vertex B in $\triangle ABC$ has to correspond to the vertex C in $\triangle DCB$.

Example 2

GIVEN $\overline{AB} \cong \overline{AC}$ and $\overline{AQ} \cong \overline{AP}$

PROVE $\angle BQA \cong \angle CPA$

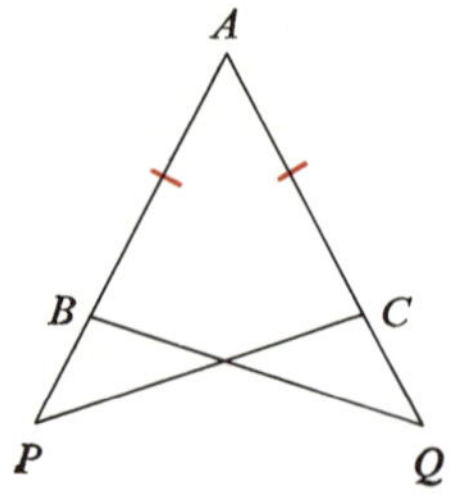

Solution

PROOF

STATEMENTS	REASONS
1. $\overline{AB} \cong \overline{AC}$ and $\overline{AQ} \cong \overline{AP}$	1. Given
2. $\angle A \cong \angle A$	2. Every angle is congruent to itself. (3-1.7)
3. $\triangle ABQ \cong \triangle ACP$	3. The SAS Postulate. (3-1)
4. $\angle BQA \cong \angle CPA$	4. Definition of congruent triangles. (3-3)

In the example above, we use the correspondence $ABQ \leftrightarrow ACP$. For this correspondence, are $\overline{AB} \leftrightarrow \overline{AC}$ and $\overline{PC} \leftrightarrow \overline{QB}$ proper correspondences?

Class Exercises

Using the points named in each of the following figures, identify overlapping triangles:

1. 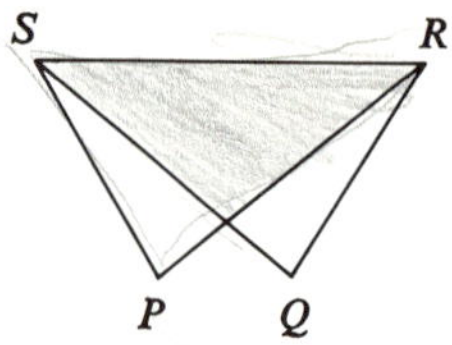

2. 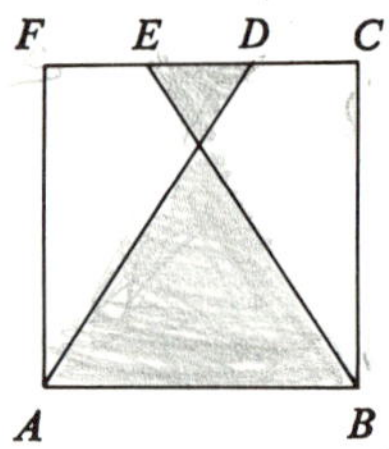

3. 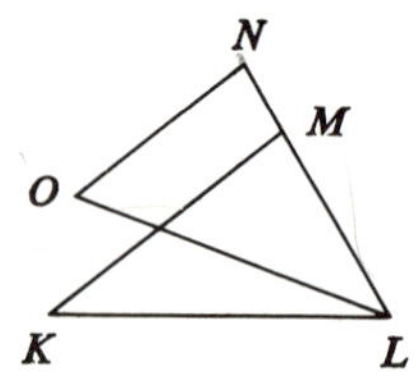

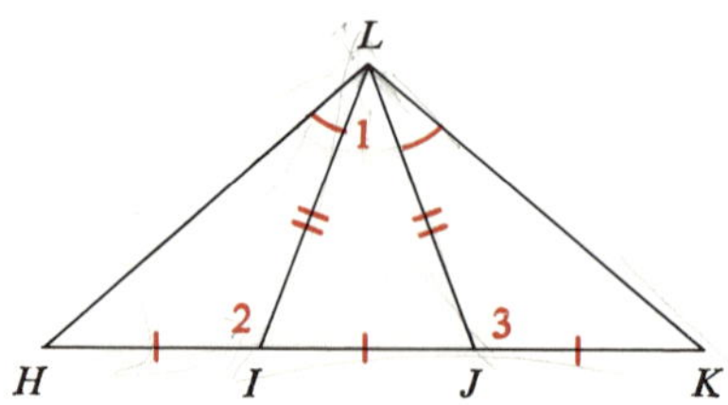

Exercises 4–12

4. $\triangle LHJ$ and $\triangle LKI$, in the figure at right, overlap. Name the shared segments and angles.
5. In $\triangle LHJ$, name the angle included by $\overline{HL}$ and $\overline{LJ}$.
6. Assuming that the figure markings indicate congruence, show that $\angle HLJ \cong \angle KLI$.
7. Give a reason why $HJ = KI$.
8. Name the postulate that proves that $\triangle LHJ \cong \triangle LKI$.
9. Using the congruence of Exercise 8, is $HL = KL$?
10. Is $\triangle HIL \cong \triangle KJL$? Give a reason for your answer.

11. Name the angle included by $\overline{HI}$ and $\overline{LI}$; by $\overline{KJ}$ and $\overline{LJ}$.
12. Give a reason why $\angle 2 \cong \angle 3$.

Auxiliary Lines and Segments

When we proved the Isosceles Triangle Theorem, we used a segment not given in the original hypothesis. We also added a segment in the proof of the Angle Bisector Theorem. The addition of these **auxiliary lines** is justified by postulates and other theorems. It is, however, difficult to know when to use these segments. Only experience will help you determine this. If the use of auxiliary figures in a plane is likely to aid a proof, use them. Only by trying will you get a feel for their use. But be careful—auxiliary line segments must be uniquely determined. For example, a segment that joins two known distinct points is acceptable. It is not acceptable just to add a line through a point if there is nothing else to determine the line. To show the inclusion of auxiliary lines, we usually draw dashed lines instead of solid ones. However, this is not the only use for dashed lines.

Example 3

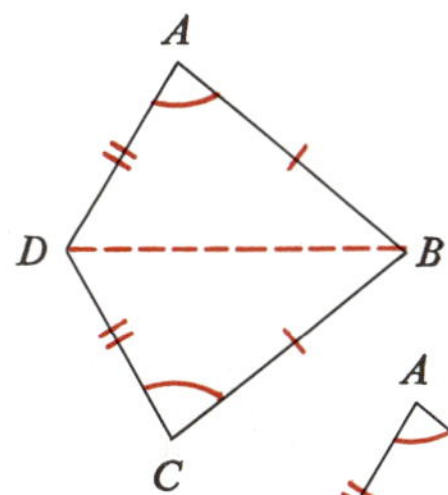

GIVEN $\overline{AD} \cong \overline{CD}$ and $\overline{AB} \cong \overline{CB}$

PROVE $\angle A \cong \angle C$

Solution $\angle A$ is included by $\overline{AD}$ and $\overline{AB}$; $\angle C$ is included by $\overline{CD}$ and $\overline{CB}$. Can you name a segment that gives us two triangles in which $\angle A \leftrightarrow \angle C$? If we consider $\overline{AC}$, we split the angles under consideration. $\overline{BD}$ seems to be the segment we need.

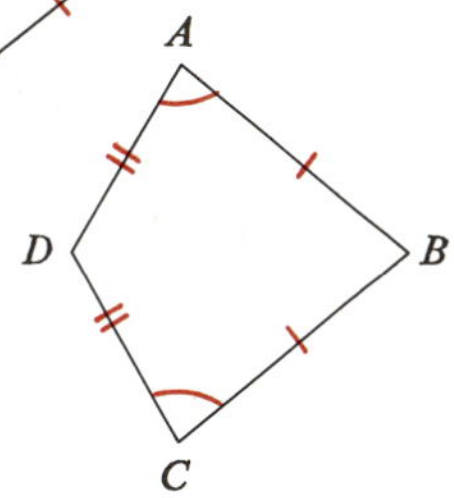

PROOF

STATEMENTS	REASONS
1. $\overline{AB} \cong \overline{CB}$ $\overline{AD} \cong \overline{CD}$	1. Given.
2. $\overline{DB} \cong \overline{DB}$	2. Every segment is congruent to itself. (3-1.6)
3. $\triangle ADB \cong \triangle CDB$	3. The SSS Postulate. (3-3)
4. $\angle A \cong \angle C$	4. Definition of congruent triangles. (3-3)

The use of auxiliary lines will also aid us in the next example.

Example 4 Prove that any point of the line perpendicular to a segment at its midpoint lies the same distance from one endpoint of the segment as it does from the other endpoint.

Solution

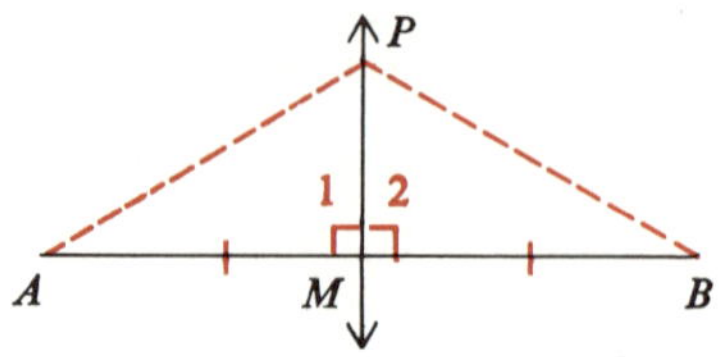

GIVEN P is any point on the perpendicular bisector of $\overline{AB}$; $\overline{PM} \perp \overline{AB}$; $AM = BM$

PROVE $PA = PB$

PROOF

STATEMENTS	REASONS
1. Let P be any point in $\overleftrightarrow{PM}$; $\overleftrightarrow{PM} \perp \overline{AB}$ and $AM = BM$	1. Given.
2. $\angle 1$ and $\angle 2$ are right angles.	2. If two intersecting lines are perpendicular, they form right angles. (2-6.5)
3. $\angle 1 \cong \angle 2$	3. All right angles are congruent. (3-1.1)
4. Draw auxiliary segments $\overline{PA}$ and $\overline{PB}$.	4. The Line Postulate. (2-3)
5. $\overline{PM} \cong \overline{PM}$	5. Any segment is congruent to itself. (3-1.6)
6. $\triangle PAM \cong \triangle PBM$	6. The SAS Postulate. (3-1)
7. $PA = PB$	7. Definition of congruent triangles. (3-3)

The next theorem will be proved as an exercise.

T **Theorem 3-5.1** The angle bisectors of the base angles of an isosceles triangle are congruent.

A further comment is needed before we suspend our discussion of the congruence relation. When developing a proof, we must pay close attention to what is given. We must be careful not to use the figure to assume facts that are not given. We cannot use figures to prove anything. We use the figure to determine, for example, whether points are or are not collinear, and the order, *but not* the congruence, of segments or angles. However, if markings are shown on the figure, we can assume that these indicate congruence. To be completely safe, it is a good idea to include all information in the hypothesis.

A further word of caution: We must not expect that we can always prove a statement directly from what is given. For example, we have seen that we may have to prove a pair of triangles congruent before we can prove that corresponding angles are congruent. We may even have to prove one pair of triangles congruent and use this information to prove that a second pair of triangles is congruent before we can prove that corresponding parts of the second pair are congruent.

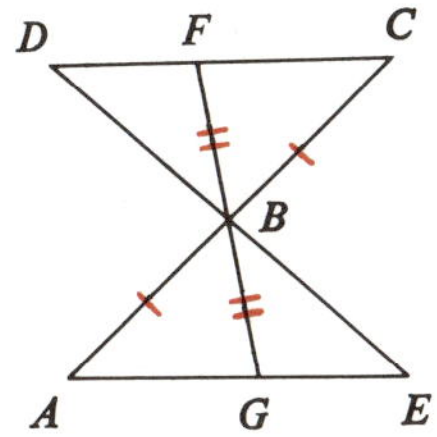

Example 5

GIVEN $\overleftrightarrow{ABC}$; $\overleftrightarrow{DBE}$; $\overleftrightarrow{FBG}$; $\overline{AB} \cong \overline{CB}$; $\overline{FB} \cong \overline{GB}$

PROVE $\overline{AE} \cong \overline{CD}$

Solution Vertical angles are congruent; thus, $\angle DBC \cong \angle EBA$. We are given $\overline{AB} \cong \overline{CB}$. Therefore, if $\angle A \cong \angle C$, triangles BCD and BAE are congruent by the ASA postulate. But we are not given that $\angle A \cong \angle C$. Can you see, however, that if we first prove $\triangle FBC \cong \triangle GBA$, then we have the last congruency statement we need? Once we know that $\triangle BCD \cong \triangle BAE$, $\overline{AE} \cong \overline{CD}$ follows.

PROOF

STATEMENTS	REASONS
1. $\overleftrightarrow{ABC}$, $\overleftrightarrow{DBE}$, $\overleftrightarrow{FBG}$, $\overline{AB} \cong \overline{CB}$, and $\overline{FB} \cong \overline{GB}$	1. Given.
2. $\angle FBC \cong \angle GBA$	2. Vertical angles are congruent. (3-1.5)
3. $\triangle FBC \cong \triangle GBA$	3. SAS Postulate. (3-1)
4. $\angle C \cong \angle A$	4. Definition of congruent triangles. (3-3)
5. $\angle DBC \cong \angle EBA$	5. Vertical angles are congruent. (3-1.5)
6. $\triangle BCD \cong \triangle BAE$	6. ASA Postulate. (3-2)
7. $\overline{AE} \cong \overline{CD}$	7. Definition of congruent triangles. (3-3)

EXERCISES

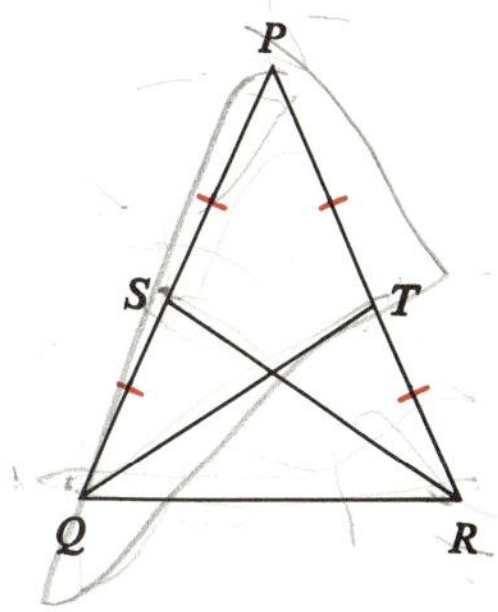

A Complete each of the following using the figure at right:

1. Two overlapping triangles are $\triangle QTR$ and ____.
2. A pair of overlapping triangles is $\triangle PTQ$ and ____.
3. The common angle in $\triangle PQT$ and $\triangle PRS$ is ____.
4. ____ is the side included by $\angle TQR$ and $\angle SRQ$.
5. From the figure markings, $\overline{QT}$ is a(n) ________ of $\triangle PQR$.

6. The markings indicate that $\triangle PQR$ is a(n) ______ triangle.

Indicate which of the following are true and which are false:

7. Congruence can be assumed by looking at a figure.
8. Auxiliary lines should be used whenever needed.
9. If two triangles share a side, they are overlapping triangles.
10. If overlapping triangles have a side on a common line, they have a common vertex.
11. If the opposite sides of a quadrilateral are congruent, a diagonal forms two congruent triangles.
12. Halves of equal quantities are equal.
13. Prove Theorem 3-5.1. (*Hint*: Use the multiplication property of equality as a reason for $\frac{1}{2}(m\angle A) = \frac{1}{2}(m\angle B)$, when $m\angle A = m\angle B$.
14. Prove that the median to the base of an isosceles triangle bisects the vertex angle.
15. Prove that the medians to the congruent sides of an isosceles triangle are congruent.
16. Prove that the three medians of any equilateral triangle are congruent.
17. GIVEN All sides of polygon $ABCDE$ are congruent; $AD = BE$

PROVE $\angle AED \cong \angle BAE$

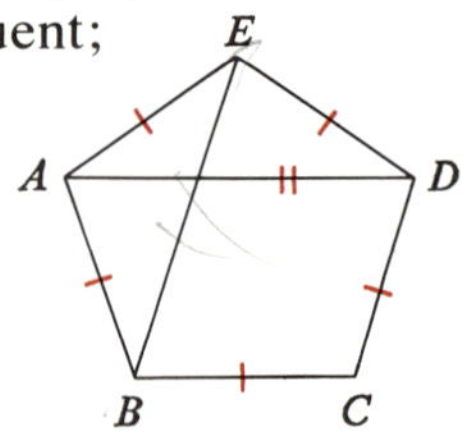

18. GIVEN $BD = CE$, $BE = CD$

PROVE $AB = AC$

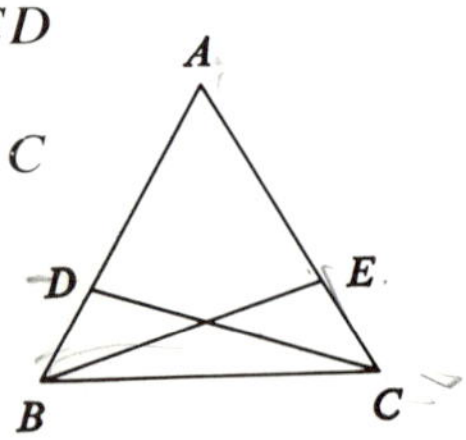

19. GIVEN $\angle 2 \cong \angle 3$, $\angle 1 \cong \angle 4$

PROVE $\overline{AC} \cong \overline{DB}$

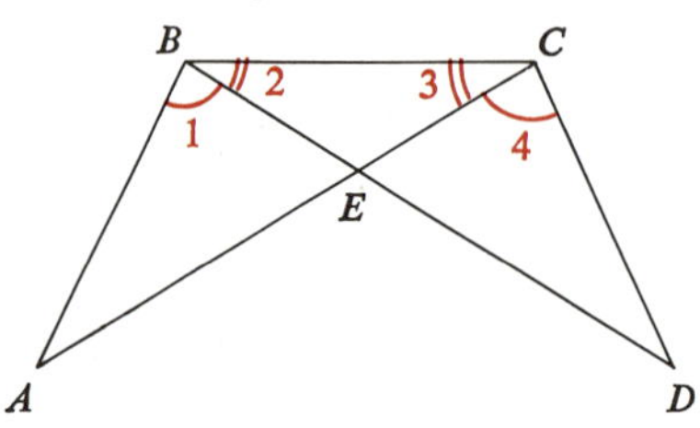

20. GIVEN $OR = OQ$, $\angle 1 \cong \angle 2$

PROVE $\overline{PR} \cong \overline{SQ}$

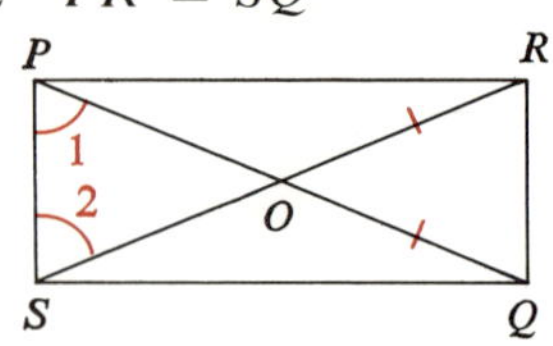

21. Prove that the angle bisector of the vertex angle of an isosceles triangle is perpendicular to the base.
22. Prove that any median of an equilateral triangle is perpendicular to its side.
23. GIVEN $\angle 1 \cong \angle 2$, $AB = DC$

PROVE $PA = PD$

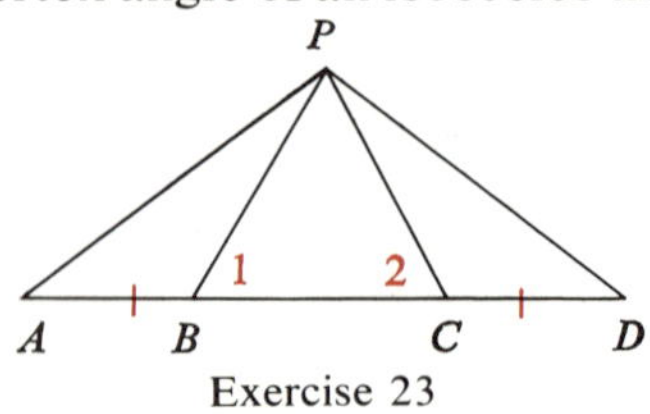

Exercise 23

24. **Given** $\overleftrightarrow{PQR}$; $\overline{SR} \cong \overline{TR}$;
$\angle 1 \cong \angle 2$
Prove $\angle PSQ \cong \angle PTQ$

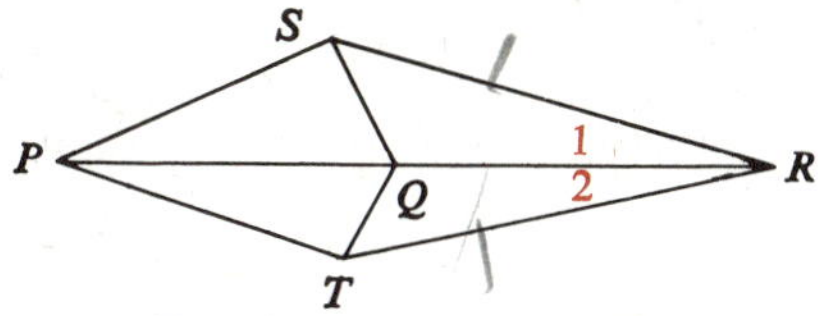

Exercises 24–26

25. **Given** $\overleftrightarrow{PQR}$; $\angle 1 \cong \angle 2$; $\angle SQR \cong \angle TQR$ **Prove** $\overline{PS} \cong \overline{PT}$
26. **Given** $\overline{PS} \cong \overline{PT}$; $\overline{RS} \cong \overline{RT}$ **Prove** $\angle PQS \cong \angle PQT$

B

27. Write the converse of the statement in Exercise 15.
28. Can you prove the statement you wrote for Exercise 27?
29. Prove in two ways that the medians of an equilateral triangle bisect the angles.

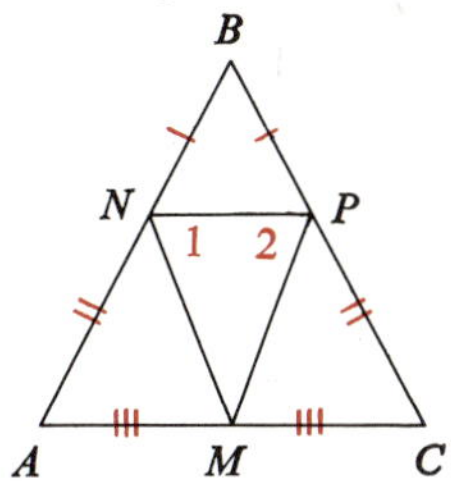

Exercise 30

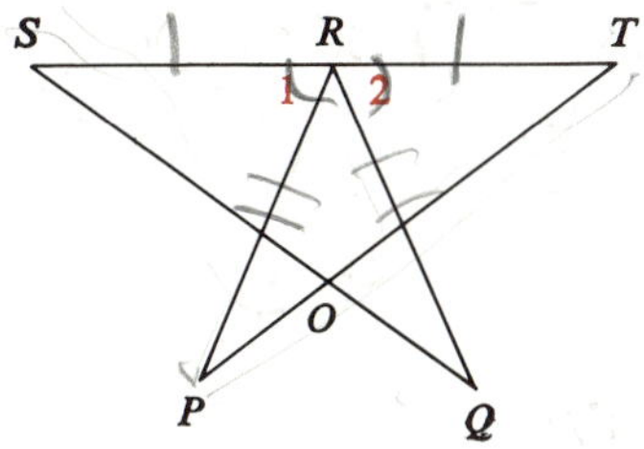

Exercise 31

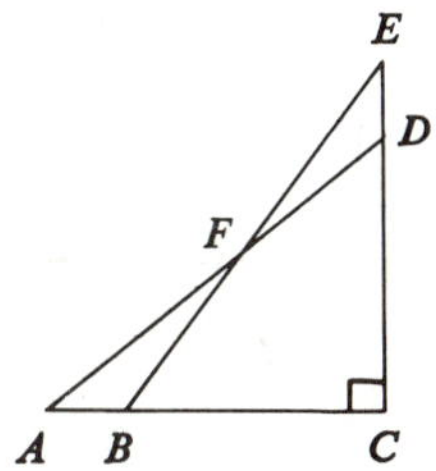

Exercise 32

30. **Given** $NB = PB$, $AN = CP$, and M is midpoint of $\overline{AC}$.
Prove $\angle 1 \cong \angle 2$

31. **Given** $\overline{PT}$ and $\overline{QS}$ intersect at O, $RS = RT$, and $\angle 1 \cong \angle 2$; $\overline{RP} \cong \overline{RQ}$
Prove $\overline{PT} \cong \overline{QS}$

32. **Given** $\overleftrightarrow{ABC} \perp \overleftrightarrow{EDC}$, $AB = ED$, $AC = EC$ **Prove** $\angle A \cong \angle E$

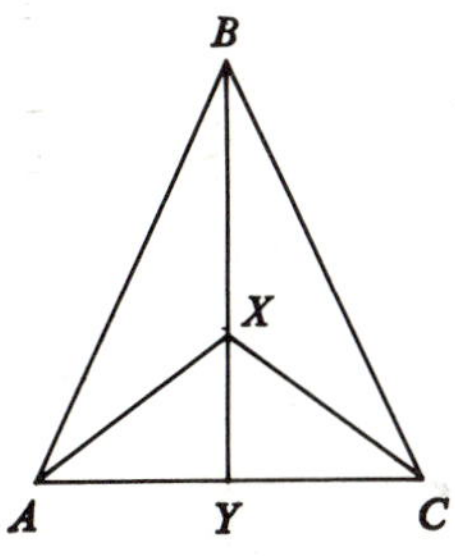

Exercise 33

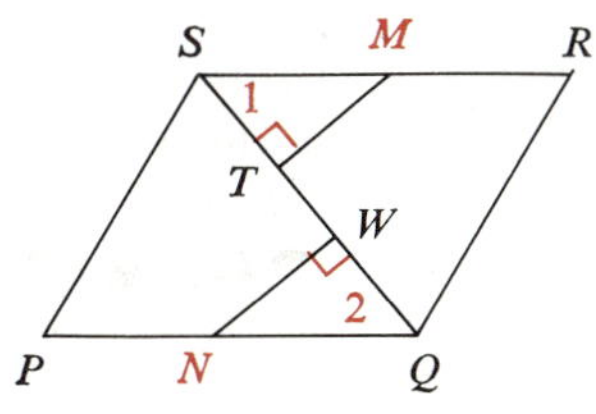

Exercise 34

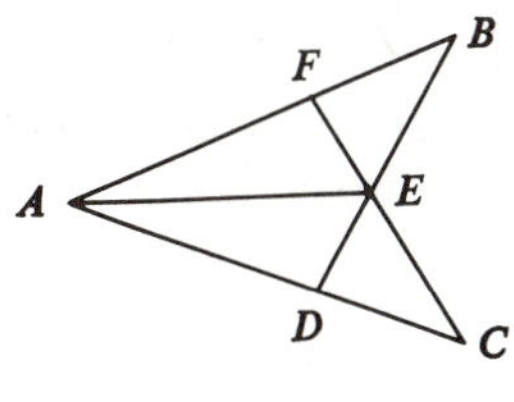

Exercise 35

33. **Given** $\overleftrightarrow{BXY}$; $\overline{BA} \cong \overline{BC}$, $\overline{XA} \cong \overline{XC}$ **Prove** $\overline{AY} \cong \overline{CY}$
34. **Given** $\overline{SM} \cong \overline{MR} \cong \overline{PN} \cong \overline{NQ}$; $\overline{SP} \cong \overline{QR}$; $\angle WNQ \cong \angle TMS$
Prove $\overline{SW} \cong \overline{QT}$
35. **Given** $\overline{AB} \cong \overline{AC}$; $\overline{BE} \cong \overline{CE}$ **Prove** $\overline{DE} \cong \overline{FE}$

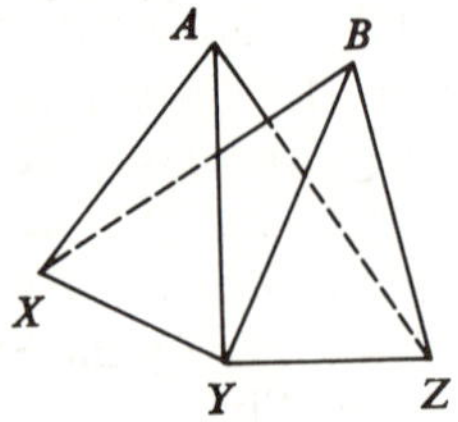

Exercise 36

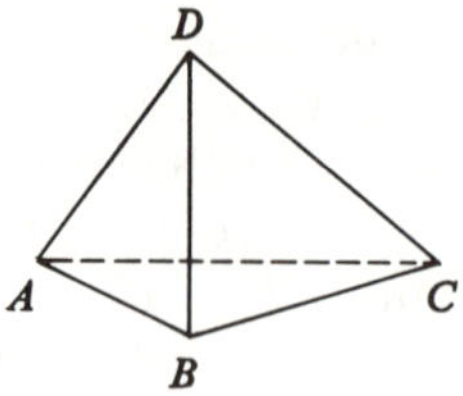

Exercise 37

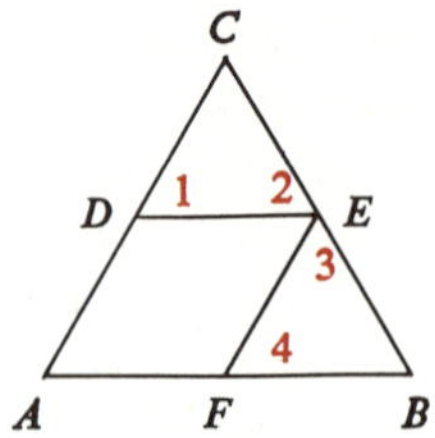

Exercise 38

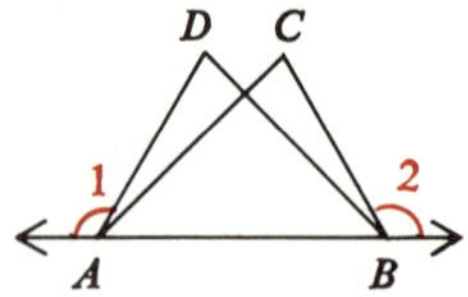

Exercise 39

C

36. GIVEN $ZB = XA$, $YZ = YX$
$YB = YA$
PROVE $XB = ZA$

37. GIVEN $\angle CAB \cong \angle CBA$
$\angle ACD \cong \angle BCD$
PROVE $\angle DAB \cong \angle DBA$

38. GIVEN $\triangle ABC$ is equiangular,
$\angle 1 \cong \angle 2$, and $\angle 3 \cong \angle 4$
PROVE $AD = BF$

39. GIVEN $\angle 1 \cong \angle 2$ and
$m\angle DAC = m\angle CBD$
PROVE $\angle ADB \cong \angle BCA$

40. Prove that if the diagonals of a quadrilateral are perpendicular to and bisect each other, the sides of the quadrilateral are congruent to each other.

A LOOK AT THE PAST

Pons Asinorum

In Book I of *The Elements*, Euclid presents a proof for the Isosceles Triangle Theorem. The original proof is credited to the Greek mathematician Thales (640–546 B.C.). Since Euclid had not yet discussed the construction of an angle bisector, he could not use it in his proof. The Euclidean proof used the Point-Uniqueness Postulate on the extensions of the congruent sides. The figure he used looked something like a bridge. This fact, and the difficulty of the proof, caused medieval scholars to refer to the figure as *Pons Asinorum*: The Bridge of Fools. It was this proof that seemed to separate the weak students from the better students during the Middle Ages.

Instead of using the two-column proof, we shall give Euclid's proof in paragraphic form. Make certain, however, that you are sure of the reason for each step.

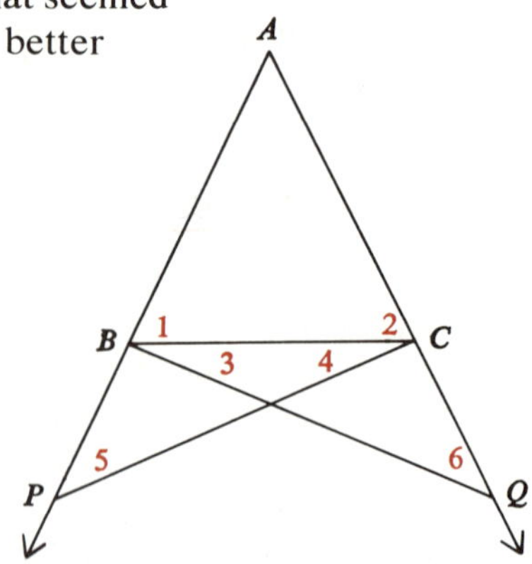

GIVEN In $\triangle ABC$, $AB = AC$ with $\overline{AB}$ and $\overline{AC}$ extended.

PROVE $\angle 1 \cong \angle 2$

Proof We know that $AB = AC$. Choose a point P on $\overrightarrow{AB}$ so that $\overleftrightarrow{ABP}$. Next choose a point Q on $\overrightarrow{AC}$ such that $\overleftrightarrow{ACQ}$ and $AQ = AP$. Using the proof in Example 2 of this section, the overlapping triangles, $\triangle ABQ$ and $\triangle ACP$, are congruent. As a result, $\angle 5 \cong \angle 6$, $\angle ABQ \cong \angle ACP$, and $PC = QB$. We know that $AP = AB + BP$ and $AQ = AC + CQ$. Thus $BP = AP - AB$, $CQ = AQ - AC$, and $BP = CQ$. By the SSS Postulate, $\triangle BPC \cong \triangle CQB$. Thus, $m\angle 3 = m\angle 4$. Since $m\angle ABQ = m\angle 1 + m\angle 3$ and $m\angle ACP = m\angle 2 + m\angle 4$, by the Angle Difference Postulate, $m\angle ABQ - m\angle 3 = m\angle 1$ and $m\angle ACP - m\angle 4 = m\angle 2$. Therefore, $m\angle 1 = m\angle 2$ or $\angle 1 \cong \angle 2$.

HOW IS YOUR MATHEMATICAL VOCABULARY?

The following words and phrases were introduced in this chapter. How many do you know and understand?

altitude of a triangle (3-4)
angle bisector of a triangle (3-4)
ASA Postulate (3-3)
auxiliary lines (3-5)
base (3-4)
base angles (3-4)
coincide (3-1)
congruent angles (3-1)
congruent segments (3-1)
congruent triangles (3-2)
corollary (3-4)
corresponding angles (3-2)
corresponding sides (3-2)
equiangular triangle (3-4)
equilateral triangle (3-4)
equivalence relation (3-1)
identity congruence (3-1)
"if and only if" (3-1)
included angle (3-2)
included side (3-2)
isosceles triangle (3-4)
median of a triangle (3-4)
overlapping triangles (3-5)
parts of a triangle (3-2)
SAS Postulate (3-3)
scalene triangle (3-4)
similar triangles (3-3)
SSS Postulate (3-3)
vertex angle of an isosceles triangle (3-4)

REVIEW EXERCISES

3-1 Congruence: an equivalence relation

Complete each of the following:

1. The correspondence $BHLM \leftrightarrow RTWX$ matches H with ____.
2. The statement "p if and only if q" means "if p then q," as well as "if ____ then ____."
3. The converse of "If the correspondence is a congruence, then it is an equivalence relation" is "If ____________, then ____________."

4. $XSW \leftrightarrow XSW$ is the ________ congruence.
5. $XW = SW$ is equivalent to ____ $\cong$ ____.
6. $\angle XMW \cong$ ____ is equivalent to ____ $= m\angle SMW$.
7. If $XMW \leftrightarrow SMW$ is a congruence, then $\angle X \cong$ ____.

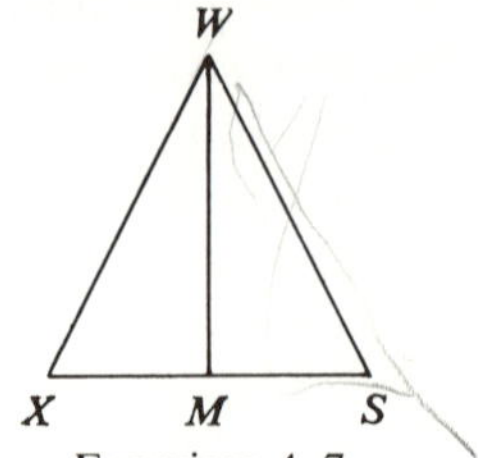

Exercises 4–7

3-2 Congruent triangles

8. Make a drawing of $FGHI \leftrightarrow KLMN$, using markings to indicate the congruence correspondence.

If $\triangle ABC \cong \triangle DEC$, list the corresponding angles and segments called for below.

9. $\overline{AB} \cong$ ____
10. ____ $\cong \overline{EC}$
11. $\overline{AC} \cong$ ____
12. $\angle ABC \cong$ ____
13. ____ $\cong \angle CDE$
14. $\angle ACB \cong$ ____
15. Name the sides that include $\angle X$.
16. Name the side included by $\angle ACB$ and $\angle CAB$.

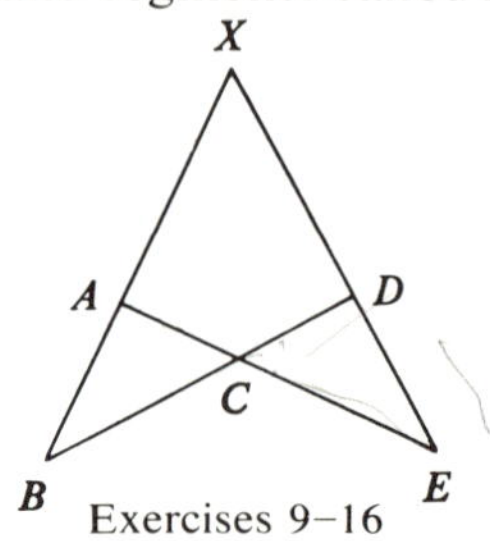

Exercises 9–16

17. If the sides of two triangles correspond such that $RS = ND$, $ST = DF$, and $TR = FN$, write the triangular congruence.

3-3 Congruence postulates for triangles

Determine which of the following always determines a triangle.

18. two sides and the included angle
19. three angles
20. an angle and two sides
21. three sides
22. two angles and the included side

Complete the following for each pair of triangles. The markings indicate congruences.

23. **Given** $\overline{FB} \perp$ ____; $\overline{GC} \perp$ ____;
____ $= BC =$ ____;
$m\angle A =$ ____
Prove $\triangle ABF \cong \triangle DCG$

24. **Given** $\overline{PC} \cong$ ____;
$\overline{AB} \cong$ ____;
$\angle C \cong$ ____
Prove $\triangle PAC \cong \triangle QDB$

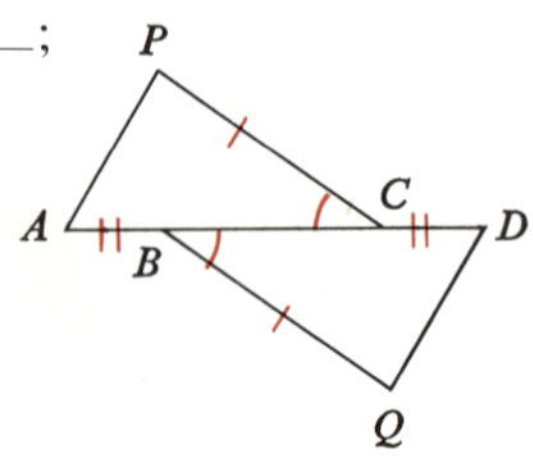

Exercise 24

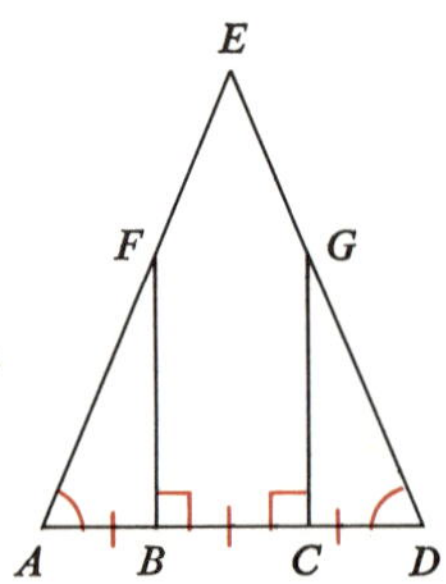

Exercise 23

3-4 Isosceles and equilateral triangles

25. If $\overleftrightarrow{ABCDE}$ and all sides are congruent in $\triangle BGC$, $\triangle GCF$, and $\triangle CFD$, prove that $\angle 1 \cong \angle 2$.

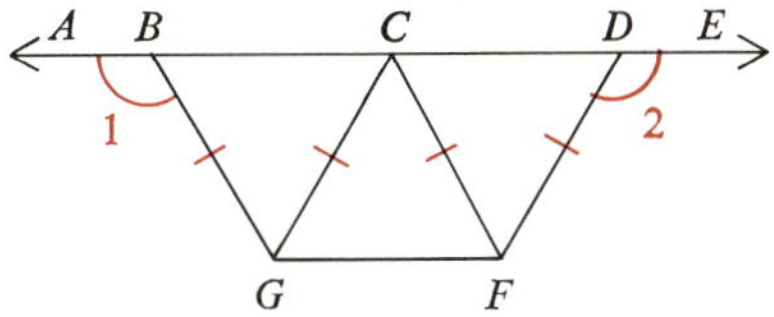

Complete each of the following. Use the figure at right for Exercises 26–30.

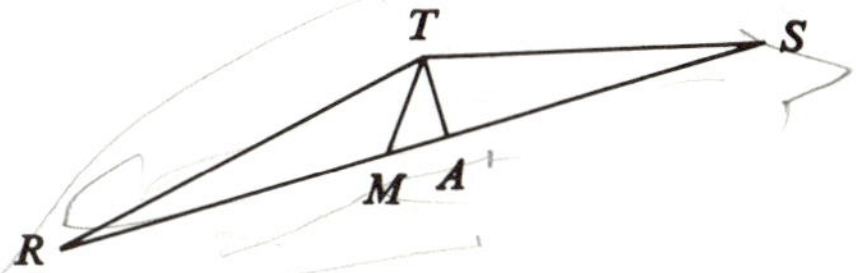

26. If $\overline{TA}$ is perpendicular to $\overleftrightarrow{RMAS}$, $\overline{TA}$ is the ________ from vertex T.
27. If $RM = SM$, $\overline{TM}$ is the ________ from vertex T.
28. If $\angle R \cong \angle S$, then sides ____ and ____ are congruent.
29. If $\triangle RST$ is equilateral, it is also ________.
30. The altitude from vertex R lies in the ________ of the triangle.
31. An angle bisector of a triangle lies in the ________ of the triangle.
32. A theorem that follows directly from another theorem is a(n) ________.
33. In a right triangle, the legs and two of the altitudes are the ________.
34. Justify the following by supplying the missing reasons for Statements 2–6.

Given $\triangle ABC$ is an isosceles triangle; $AB = AC$; M is the midpoint of $\overline{BC}$.

Prove The median is perpendicular to the base; that is, $\overline{AM} \perp \overline{BC}$.

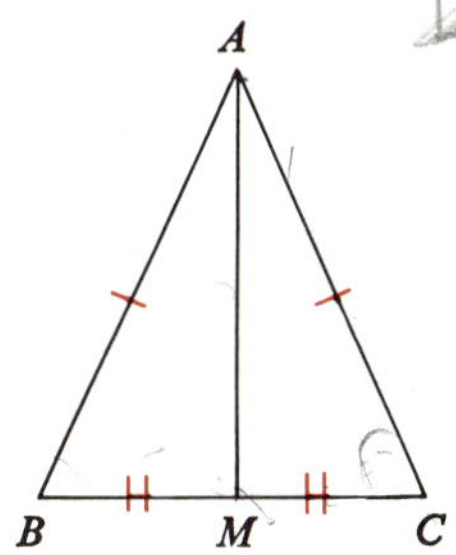

Proof

STATEMENTS	REASONS
1. $AB = AC$; M is the midpoint of $\overline{BC}$.	1. Given
2. $\angle B \cong \angle C$	2. ________
3. $BM = CM$	3. ________
4. $\triangle ABM \cong \triangle ACM$	4. ________
5. $m\angle BMA = m\angle CMA$	5. ________
6. $\overline{AM} \perp \overline{BC}$	6. ________

3-5 Overlapping triangles

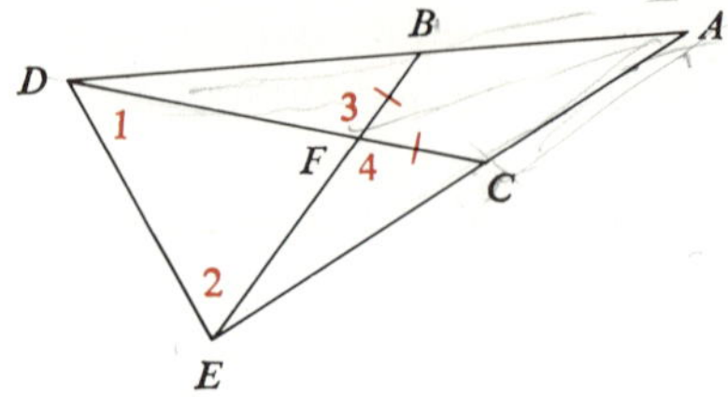

35. Write a correspondence for the overlapping triangles at right that have $\angle A$ in common.
36. Name the triangles that have $\overline{DE}$ as a common side.
37. If $AB = AC$ and $BF = CF$, what auxiliary segment would you draw to prove $\angle FBA \cong \angle FCA$?
38. If $DB = EC$ and $AD = AE$, what congruence postulate proves that $\triangle DCA \cong \triangle EBA$?

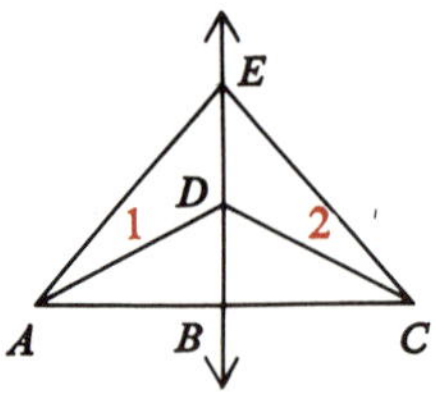

Exercise 39

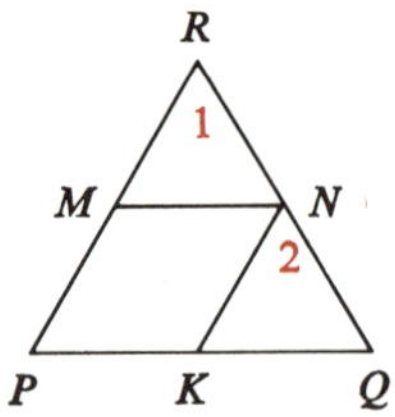

Exercise 40

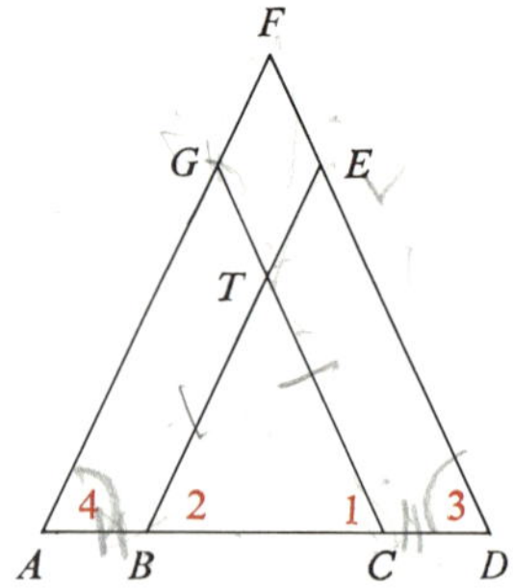

Exercise 41

39. If $\overline{EB} \perp \overleftrightarrow{ABC}$ at B, $\overline{AB} \cong \overline{BC}$, and $\angle 1 \cong \angle 2$, prove that $\triangle AEC$ is isosceles.
40. If M and N are the midpoints of $\overline{PR}$ and $\overline{QR}$, respectively, if $\angle 1 \cong \angle 2$, and if $\overline{MP} \cong \overline{NK}$, prove that $\angle RMN \cong \angle NKQ$.
41. $\overline{BE}$ and $\overline{GC}$ intersect at T such that $\overline{BT} \cong \overline{CT}$; $\angle 3 \cong \angle 4$; and B and C are on $\overline{AD}$ such that $AB = CD$. Prove $\overline{AG} \cong \overline{DE}$.

CHAPTER TEST

1. If the correspondence $ABP \leftrightarrow DCQ$ describes a congruence, complete the statement, "$\overline{AB} \cong \overline{DC}$ is equivalent to ________."
2. Defend the statement, "Perpendicularity is not an equivalence relation."
3. If the quadrilateral correspondence $ABCD \leftrightarrow CDAB$ describes a congruence, what is true about the sides of the quadrilaterals?
4. If, for two angles, $\overline{PQ} \cong \overline{MN}$ and $\overline{QR} \cong \overline{NL}$, is it true that $\angle PQR \cong \angle MNL$? Justify your answer.
5. If $\triangle ARX \cong \triangle BSZ$, name the corresponding parts of the two triangles.
6. If $\overline{PR}$ and $\overline{TS}$ bisect each other at Q, what other statement can we make to prove that $\triangle TPQ \cong \triangle SRQ$?

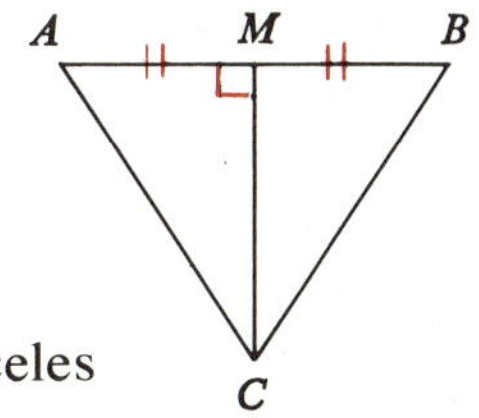

7. If the markings in $\triangle ABC$ indicate congruence, what postulate will prove that $\triangle AMC \cong \triangle BMC$?
8. Prove that the angle bisectors of the base angles of an isosceles triangle are congruent.
9. Prove that the median to the hypotenuse of an isosceles right triangle forms two additional right triangles.

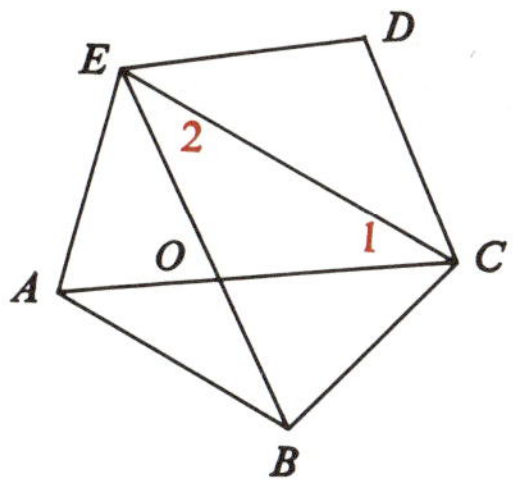

Exercises 10–11

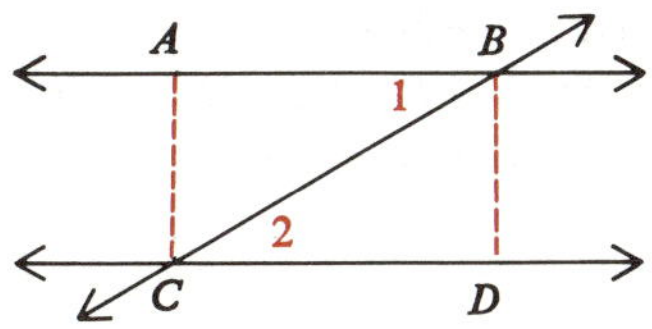

Exercise 12

10. Name the overlapping triangles that have $\overline{AB}$ as a common base.
11. GIVEN $AE = BC$; $AC = BE$; $\angle AEO \cong \angle BCO$ PROVE $EO = CO$
12. If $\overleftrightarrow{CB}$ intersects $\overleftrightarrow{AB}$ at B and $\overleftrightarrow{CD}$ at C such that $\overline{AC} \perp \overleftrightarrow{CD}$ and $\overline{DB} \perp \overleftrightarrow{AB}$ and $\angle 1 \cong \angle 2$, prove that $\overline{AC} \cong \overline{DB}$.

SUGGESTED RESEARCH

1. Every triangle has three special centers: the incenter, the circumcenter, and the centroid. These are formed by the intersection of segments we have discussed. Prepare a report describing the formation of each center and its special properties.

2. In designing the geodesic dome, Buckminster Fuller used equilateral triangles to line the dome. Write a report on the geodesic dome, and explain the significance of using the equilateral triangle.

3. State and prove the converse of Theorem 3-5.1. As a reference, you can use *Challenging Problems in Geometry 2,* by Alfred S. Posamentier and Charles T. Salkind (New York: Macmillan Co., 1970).

MATHEMATICAL EXCURSION

Equivalence Classes

In Chapter 3 we defined an equivalence relation as a relation which is reflexive, symmetric, and transitive. In this excursion we shall explore equivalence relations in further detail.

Consider a set S and an operation, the Cartesian product, on the set.

Definition The Cartesian product, $S \times S$, of a set S is the set of all ordered pairs obtained by pairing one element of S with another, but not necessarily different, element of S.

For the set $A = \{0, 1, 2\}$, $A \times A = \{(0, 0), (0, 1), (0, 2), (1, 0), (1, 1), (1, 2), (2, 0), (2, 1), (2, 2)\}$.

If we apply a rule to the elements of A, we can form a relation.

Definition For a and b, elements of set S, a is related to b if and only if $a < b$.

Applying the rule in the above definition, we form the relation $\{(0, 1), (0, 2), (1, 2)\}$ because $0 < 1$, $0 < 2$, and $1 < 2$. Which properties of the equivalence relation does this relation satisfy?

To satisfy the reflexive property, every element of S would have to be less than itself. Since we can show that none of the elements (0, 0), (1, 1), or (2, 2) of $A \times A$ are elements of the relation, $A \times A$ is not a reflexive relation.

For the symmetric property, whenever (a, b) is an element of the relation, (b, a) must also be an element of the relation. Since at least one of the elements (1, 0), (2, 0), and (2, 1) is not an element of the relation, $A \times A$ is not symmetric.

The transitive property requires that if (a, b) and (b, c) are elements of $A \times A$, then (a, c) must also be an element of $A \times A$. The only elements of $A \times A$ of the form (a, b), (b, c) are (0, 1) and (1, 2). If (a, c), in this instance (0, 2), is an element of $A \times A$, then $A \times A$ is transitive. Is $A \times A$ a transitive relation? Since the relation above does not satisfy the reflexive and symmetric properties, it is not an equivalence relation. In order to be an equivalence relation, *all three properties must be satisifed.*

Example 1 Consider the relation: the ordered pair (a, b) is an element of the relation if and only if $b = a^2$. Show that this relation is an equivalence relation for the set $T = \{0, 1\}$ but not for the set $S = \{0, 1, 2\}$. Over either T or S the relation is $\{(0, 0), (1, 1)\}$.

Solution

Set T	*Reflexive property*	yes	$0 = 0^2$, $1 = 1^2$
	Symmetric property	yes	(0, 0) and (1, 1) are symmetric
	Transitive property	yes	$(0, 0) \wedge (0, 0) \rightarrow (0, 0)$ $(1, 1) \wedge (1, 1) \rightarrow (1, 1)$
Set S	*Reflexive property*	no	$2 \neq 2^2$

Since at least one property is not true, the relation over the set S is not an equivalence relation.

Consider the following sets of integers called *equivalence classes.*

Name	Equivalence Class
[1]	$=\{1, 8, 15, 78, \ldots\}$
[2]	$=\{-5, 2, 9, 23, 79, \ldots\}$
[3]	$=\{10, 17, 24, 73, \ldots\}$
[4]	$=\{11, 25, 32, 81, \ldots\}$
[5]	$=\{-9, 5, 12, 19, 26, 47, \ldots\}$
[6]	$=\{13, 20, 27, 83, \ldots\}$
[7]	$=\{0, 7, 21, -28, \ldots\}$

What is the basis for determining into which set each element belongs? If you compare the differences between any two elements of the same set, you can discover the relation over the set of integers that yields equivalence classes.

Select an a and a b from any set and find their difference. This difference will always be a multiple of 7. Hence, the relation can be written $a - b = 7k$.

D

Definition For any integers a, b, and k, a is related to b if and only if $a - b = 7k$. We denote "a is related to b" as $a \mathscr{R} b$.
Is $a \mathscr{R} b$ an equivalence relation over the set of integers?

Reflexive For any integer a, $a - a = 0$ and $0 = 7 \cdot 0$.
Hence, for each integer a, $a \mathscr{R} a$.

Symmetric If $a - b = 7k$, then $b - a = -7k = 7(-k)$.
Hence, for any integers a and b, if $a \mathscr{R} b$, then $b \mathscr{R} a$.

Transitive If $a - b = 7k$ and $b - c = 7m$,
then $(a - b) + (b - c) = 7k + 7m$ and $(a - c) = 7(k + m)$.

Since $k + m$ is an integer if k and m are integers, we have: if $a \mathscr{R} b$ and $b \mathscr{R} c$, then $a \mathscr{R} c$.

The relation $a \mathscr{R} b$ if and only if $a - b = 7k$ partitions the integers into seven groups or equivalence classes. We have named these above. Each integer belongs to one and only one of these classes, and all integers having this relation to each other belong in the same class. That is, the difference between any integer and exactly one of the integers 1, 2, 3, 4, 5, 6, or 0 is a multiple of 7.

If you divide any element of class [6] by 7, what remainder do you get? Divide any element of class [4] by 7 and what remainder do you get? Check some of the elements of class [5]. If you repeat this activity with other classes, you will see that any two elements from the same class have the same remainder when they are divided by 7.

T

Theorem If a and b have the same remainder when they are divided by 7, then $a - b = 7k$. Conversely, if $a - b = 7k$, then a and b have the same remainder when divided by 7.

GIVEN $a = 7m + r$; m, r are integers, $0 \leq r < 7$
$b = 7n + r$; n, r are integers, $0 \leq r < 7$

PROVE $a - b = 7k$, k an integer

PROOF
$$\begin{aligned} a - b &= (7m + r) - (7n + r) \\ &= 7m + r - 7n - r \\ &= 7m - 7n \text{ or } 7(m - n) \end{aligned}$$

Since m and n are integers, $m - n$ is also an integer. Let $m - n = k$; then $a - b = 7k$.

In Chapter 2 we stated that since inverse and converse are equivalent statements, we can prove one in order to prove the other. We shall prove the inverse in order to prove the converse of the theorem. Therefore, we must prove that if a and b do not have the same remainder when they are divided by 7, then $(a - b) = 7k + r$, $0 < r < 7$.

GIVEN $a = 7m + r_1$, n and r_1 are integers, $0 \leq r_1 < 7$
$b = 7n + r_2$, n and r_2 are integers, $0 \leq r_2 < 7$, $r_2 \neq r_1$

PROVE $a - b = 7k + r$, $0 < r < 7$

PROOF
$$\begin{aligned} a - b &= (7m + r_1) - (7n + r_2) \\ &= 7m + r_1 - 7n - r_2 \\ &= 7m - 7n + r_1 - r_2 \text{ or } 7(m - n) + (r_1 - r_2) \end{aligned}$$

If $r_1 > r_2$, we have $a - b = 7k + r$, $r = r_1 - r_2$
If $r_1 < r_2$, we have $a - b = 7(k - 1) + 7 + r$, $r = r_1 - r_2$
$a - b = 7k_1 + (7 + r)$, $k_1 = (k - 1)$
$a - b = 7k_1 + r'$, $0 < r' = 7 + r < 7$

We can say that a and b are equivalent under the equivalence relation: $a \mathscr{R} b$ if and only if $a - b = 7k$, when they have equal remainders when divided by 7.

What happens if we add an element from class [4] to an element from class [5]? If we try 11 from class [4] and 12 from class [5], then $11 + 12 = 23$, which is an element of class [2]. Notice also that $4 + 5 = 9$, which is an element of class [2]. Show that for other choices from class [4] and class [5] the sum is a member of class [2].

Multiplication behaves in the same way. Using the same elements as above, $11 \times 12 = 132$, an element of class [6]. Other choices from classes [4] and [5] will also give a product that is in class [6]. For example, $4 \cdot 5 = 20$, which is an element of class [6].

Another example of equivalence classes is the equivalence relation over the Cartesian product $N \times N$, where N is the set of natural numbers.

D **Definition** $(a, b) \mathscr{R} (c, d)$ if and only if $a \cdot d = b \cdot c$ over the Cartesian product $N \times N$.

Is the relation defined above an equivalence relation?

Reflexive	yes	$(a, b) \mathscr{R} (a, b)$, because $a \cdot b = b \cdot a$,	
Symmetric	yes	If $(a, b) \mathscr{R} (c, d)$, then $a \cdot d = b \cdot c$. If $(c, d) \mathscr{R} (a, b)$, then $c \cdot b = d \cdot a$, but $a \cdot d = b \cdot c$ is equivalent to $c \cdot b = d \cdot a$.	
Transitive	yes	If $(a, b) \mathscr{R} (c, d)$, then $a \cdot d = b \cdot c$.	(1)
		If $(c, d) \mathscr{R} (e, f)$, then $c \cdot f = d \cdot e$.	(2)
		If $(a, b) \mathscr{R} (e, f)$, then $a \cdot f = b \cdot e$.	(3)

If we multiply statement (1) by statement (2) we have

$$a \cdot d \cdot c \cdot f = b \cdot c \cdot d \cdot e \tag{4}$$

and if we divide (4) by $c \cdot d = d \cdot c$ we have statement (3)

$$a \cdot f = b \cdot e$$

With this relation, we get infinitely many equivalence classes, such as

$$[\tfrac{2}{3}] = \{(2, 3), (4, 6), (6, 9), \ldots\}$$

$$[\tfrac{3}{5}] = \{(3, 5), (6, 10), (9, 15), \ldots\}$$

Each equivalence class is a positive rational number. We usually use the element whose members have no common factor as a representative member of each class. However, we can use any element of the class to represent the class.

EXERCISES

Determine whether each of the following is an equivalence relation.

1. Congruence, over the set of isosceles triangles.
2. Perpendicular to, over the set of lines in the plane.
3. Perpendicular to, over the set of lines in space.
4. In the same direction, over the set of lines in the plane.
5. $a \mathscr{R} b$ if and only if $a \cdot b = a$, over the set $S = \{0, 1\}$.
6. Verify that 17 and 38 are elements of the same class $(a - b) = 7k$.
7. Verify that 23 and 37 are elements of the same class $(a - b) = 7k$.
8. Use the numbers in Exercises 6 and 7 to show that the class in which the sum of two elements from two classes falls is independent of the elements.
9. Repeat Exercise 8 but replace addition with multiplication.
10. Generalize and prove Exercise 8.
11. Generalize and prove Exercise 9.

CUMULATIVE REVIEW

Chapters 1, 2, 3

These exercises are designed so that you can check your understanding of some of the principal concepts studied so far. The numbered items in the right column are the correct answers. To complete the review, first cover the answers, read and answer the questions, and check each response by uncovering the corresponding answer.

1. True or false: Two intersecting lines are coplanar.	1. true
2. The intersection of $\overrightarrow{PQ}$ and $\overrightarrow{QP}$ is ____.	2. $\overline{PQ}$
3. The angle complementary to an acute angle is ____.	3. acute
4. Polygons are classified according to the number of ____.	4. sides
5. Draw right $\triangle ABC$. If $\angle B$ is a right angle, then ____ is the hypotenuse.	5. $\overline{AC}$
6. In Exercise 5, the legs are ____ and ____.	6. $\overline{AB}$ and $\overline{BC}$
7. True or false: The union of two half-planes is a plane.	7. false
8. True or false: When two planes intersect in a line, four dihedral angles are formed.	8. true
9. Intuition or induction: You are in a courtroom. When the defendant comes in, you say to a friend, "You can tell that he is guilty."	9. intuition
10. List the four features of a mathematical system.	10. undefined terms, definitions, postulates, theorems
11. Acceptable or unacceptable definition: A circle is a set of points that circle a given point.	11. unacceptable
12. Any two points determine exactly one ____.	12. line
13. If $\angle P$ is a right angle, then $\frac{1}{2}(m\angle P) =$ ____.	13. 45
14. The assumption that makes Exercise 13 true is the ____ property of equality.	14. multiplication

For Exercises 15–18 use the following:
p: It is snowing. q: it is cold.

15. Translate: If it is cold, then it is snowing.
16. Translate: $\sim p \vee q$
17. State a possible conclusion for if $(\sim p \vee q) \wedge p$, then ____.
18. The truth values for $p \vee \sim p$ are ____.
19. True or false: Theorems are intuitively proved statements.
20. All right angles are ____ in measure.
21. If $\angle X$ and $\angle Y$ are supplementary, then $m\angle X =$ ____.
22. If $\triangle ABC \cong \triangle LMN$, $A \leftrightarrow L$, $B \leftrightarrow M$, and ____.
23. True or false: All squares are congruent.
24. The angle included by $\overline{DE}$ and $\overline{AE}$ in $\triangle ADE$ is ____.
25. The side included by $\angle P$ and $\angle R$ in $\triangle RXP$ is ____.
26. If $\triangle BAC \cong \triangle DEF$, $\angle B \cong \angle D$, $\angle A \cong \angle E$, ____, $\overline{BA} \cong \overline{DE}$, $\overline{AC} \cong \overline{EF}$, and ____.
27. True or false: From our postulates, $\triangle ABC \cong \triangle FED$ when $\overline{AB} \cong \overline{FE}$, $\overline{AC} \cong \overline{FD}$, and $\angle A \cong \angle F$.
28. If in $\triangle KLM$, $KL = ML$, then the base is ____ and the vertex angle is ____.
29. True or false: An endpoint of the median bisects a side of the triangle.
30. Auxiliary segments added to a diagram must be ____ determined.

15. $q \rightarrow p$
16. It is not snowing or it is cold.
17. q
18. all true
19. false
20. equal
21. $180 - m\angle Y$
22. $C \leftrightarrow N$
23. false
24. $\angle DEA$
25. $\overline{PR}$
26. $\angle C \cong \angle F$, $\overline{CB} \cong \overline{FD}$
27. true
28. $\overline{KM}$, $\angle L$
29. true
30. uniquely

4 More about Proof

Statements are not true just because we state them: we must prove that they are true. Until we do, they remain mere conjectures. Theorems are geometric statements that require proof. We have studied theorems derived from the basic definitions and postulates, and theorems based on observations of triangles. In this chapter we shall examine perpendicular lines and planes, but first we must continue our study of methods of proof.

4-1 FURTHER IMPLICATIONS

In Chapter 2, we studied compound statements of implication. Implications use *If . . . then* to connect a hypothesis and a conclusion, as in the following example:

If $\triangle ABC$ is equilateral [hypothesis], *then* $\triangle ABC$ is equiangular [conclusion].

Symbolically we denote this,

$\triangle ABC$ is equilateral $\rightarrow$ $\triangle ABC$ is equiangular.
p: $\triangle ABC$ is equilateral.
q: $\triangle ABC$ is equiangular.
$p \rightarrow q$

By definition, the only time $p \rightarrow q$ is not true is when p is true and q is false. The truth table at right shows the truth values for implication. We can see that in the only case where both $p \rightarrow q$ and p are true, q is also true. Thus, $((p \rightarrow q) \wedge p) \rightarrow q$. In this type of implication, there are two hypotheses. One is the statement $p \rightarrow q$ and the other is the statement p. If both are true, then the conclusion q is true. We usually write such an implication using the scheme at right:

p	q	$p \rightarrow q$
T	T	T
T	F	F
F	T	T
F	F	T

$$\begin{array}{l} p \rightarrow q \\ \underline{\quad p \qquad} \\ \therefore q \end{array}$$

For example, let $p \rightarrow q$ be the implication "If $\triangle ABC \cong \triangle PQR$, then $\angle A \cong \angle P$." If besides this implication we know that $\triangle ABC \cong \triangle PQR$, then we can conclude that $\angle A \cong \angle P$.

Hypothesis:	If $\triangle ABC \cong \triangle PQR$, then $\angle A \cong \angle P$.
Hypothesis:	$\triangle ABC \cong \triangle PQR$
Conclusion:	$\angle A \cong \angle P$

Most of our proofs have used the implication we just examined and the **transitive property of implication**. By the transitive property of implication, if $p \rightarrow q$ and $q \rightarrow r$, then $p \rightarrow r$. Notice that this statement is an implication containing three other implications. What is the hypothesis in the first implication? in the statement as a whole? Using truth tables, we can show that this statement of the transitive property of implication is true. As we said in Chapter 2, when this occurs we have a tautology.

Example 1 Construct a truth table showing that the statement of the transitive property of implication is a tautology.

Solution

p	q	r	$p \to q$	$q \to r$	$(p \to q) \wedge (q \to r)$	$p \to r$	$[(p \to q) \wedge (q \to r)] \to (p \to r)$
T	T	T	T	T	T	T	T
T	T	F	T	F	F	F	T
T	F	T	F	T	F	T	T
T	F	F	F	T	F	F	T
F	T	T	T	T	T	T	T
F	T	F	T	F	F	T	T
F	F	T	T	T	T	T	T
F	F	F	T	T	T	T	T

Example 2 Use the transitive property of implication to prove the following:

$$\text{In } \triangle ABC, \text{ if } \overline{AB} \cong \overline{AC}, \text{ then } m\angle B = m\angle C.$$

Solution Consider the following three implications:

1. In $\triangle ABC$, if $\overline{AB} \cong \overline{AC}$, then $\angle B \cong \angle C$.
2. If $\angle B \cong \angle C$, then $m\angle B = m\angle C$.
3. In $\triangle ABC$, if $\overline{AB} \cong \overline{AC}$, then $m\angle B = m\angle C$.

We know that the first implication is true because it is the Isosceles Triangle Theorem (3-4.2). We know also that the second implication is true because this is the definition of congruent angles (1-30). Thus, by the transitive property of implication, the third implication is true.

Example 3 Construct a truth table for the statement, $(p \to q) \wedge (q \to p)$.

Solution

p	q	$p \to q$	$q \to p$	$(p \to q) \wedge (q \to p)$
T	T	T	T	T
T	F	F	T	F
F	T	T	F	F
F	F	T	T	T

The statement in Example 3 is only true when p and q are equivalent, that is, when either both are true or both are false. In the last chapter, we said that equivalence is expressed by the phrase "if and only if." Using this phrase, restate the relation of p and q described in Example 3, in common English terms.

Class Exercises

1. The transitive property is true for implication. Are any other properties of the equivalence relation true for implication?
2. Complete the truth table below.

p	q	$p \rightarrow q$	$(p \rightarrow q) \wedge p$	$[(p \rightarrow q) \wedge p] \rightarrow q$

3. Is the implication at the top of the last column a tautology?
4. Using your knowledge of geometry, give an example of the statement $[(p \rightarrow q) \wedge p] \rightarrow q$.

Associated Implications

From any given implication, other basic implications can be formed. Consider the Isosceles Triangle Theorem: "If $\triangle ABC$ is isosceles, then the base angles of $\triangle ABC$ are congruent." In the last chapter, we formed the converse of this theorem by interchanging the hypothesis and the conclusion.

Definition 4-1 The **converse** of $p \rightarrow q$ is the implication $q \rightarrow p$ formed by interchanging the hypothesis p and the conclusion q.

Since we were able to prove the converse of the Isosceles Triangle Theorem, it is a theorem also. However, not all converses are valid theorems.

Example 4 Write the converse of this statement: "If two triangles are congruent, then they have the same shape."

Solution Let p and q represent the statements in the implication.

p: Two triangles are congruent. q: They have the same shape.

The implication is $p \rightarrow q$. The converse is $q \rightarrow p$ or, "If two triangles have the same shape, then they are congruent."

In the example, the given implication is always true because of our agreement about triangle congruence. The converse, however, is false.

Negation of the statements of an implication creates an additional implication. For example, if we negate statements p and q in the implication $p \rightarrow q$, we have $\sim p \rightarrow \sim q$. This is called the inverse of $p \rightarrow q$.

Definition 4-2 The **inverse** of $p \rightarrow q$ is the implication $\sim p \rightarrow \sim q$, formed by negating each statement of $p \rightarrow q$.

Example 5 Write the inverse of "In $\triangle ABC$, if $AB = AC$, then $\angle B \cong \angle C$."

Solution Representing the implication by $p \rightarrow q$, we have

$$p\text{: } AB = AC \qquad q\text{: } \angle B \cong \angle C$$

Therefore

$$\sim p\text{: } AB \neq AC \qquad \sim q\text{: } \angle B \ncong \angle C.$$

Thus the inverse of $p \rightarrow q$ is "In $\triangle ABC$, if $AB \neq AC$, then $\angle B \ncong \angle C$."

Likewise, substituting the negations of p and q in the converse of $p \rightarrow q$ produces a new implication called the contrapositive of $p \rightarrow q$.

D **Definition 4-3** The **contrapositive** of $p \rightarrow q$ is the implication $\sim q \rightarrow \sim p$, formed by negating the statements of the converse, $q \rightarrow p$.

Example 6 Write the contrapositive of "If two angles are right angles, then the two angles are congruent."

Solution Letting $p \rightarrow q$ represent the implication, we have

p: Two angles are right angles. q: The two angles are congruent.

Therefore,

$\sim p$: It is not true that the two angles are right angles.
$\sim q$: The two angles are not congruent.

Thus the contrapositive $\sim q \rightarrow \sim p$ is "If two angles are not congruent, then it is not true that the two angles are right angles."

We could show in a truth table that the converse is not equivalent to $p \rightarrow q$, nor is the inverse. This means that they cannot be substituted for $p \rightarrow q$. The truth values for the contrapositive are the same as those for $p \rightarrow q$. These two implications are equivalent. Thus, if one is true or false, the other is also true or false and one can be substituted for the other at any time. This is useful because sometimes it is easier to prove the contrapositive than to prove the original implication. The table below shows equivalent implications.

Summary. Equivalent Implications

	Positive Statements	Negative Statements	
Given	$p \rightarrow q$	Inverse	$\sim p \rightarrow \sim q$
Converse	$q \rightarrow p$	Contrapositive	$\sim q \rightarrow \sim p$

Diagonal implications are equivalent.

EXERCISES

A Complete each of the following:

1. $p \to q$ is true except when p is ________ and q is ________.
2. If $p \to q$ is true and p is true, then q is ________.
3. If $p \to q$ and $q \to r$, then ____.
4. The property in Exercise 3 is the ________ property of implication.
5. The converse of $r \to t$ is ____.
6. The inverse of ____ is $\sim x \to \sim y$.
7. The implication $\sim t \to \sim s$ is the ________ of $s \to t$.
8. The converse of a statement is ________ true. (always, sometimes, never)

Determine which of the following are true and which are false:

9. The converse of the inverse is the contrapositive.
10. The converse of "All right angles are congruent" is "All congruent angles are right angles."
11. The implication formed by negating the statements of $p \to q$ is the inverse of $p \to q$.
12. If the given implication is true, the contrapositive is true.

Write the converse, inverse, and contrapositive of each of the following:

13. If three points are noncollinear, then they determine exactly one plane.
14. A segment has exactly one midpoint.
15. If $\overline{AB} \cong \overline{AC}$ and $\overline{AC} \cong \overline{CD}$; then $\overline{AB} \cong \overline{CD}$.

B Write the converse of each of the following. Determine which statements are true and which are false.

16. If $\angle A$ and $\angle B$ are vertical angles, then $m\angle A = m\angle B$.
17. If the intersecting lines l_1 and l_2 form right angles, then $l_1 \perp l_2$.
18. If $PA = PB$, then P is on the perpendicular bisector of $\overline{AB}$.
19. If $\triangle ABC \cong \triangle PQR$, then $\angle A \cong \angle P$.

Supply a conclusion for each of the following:

20. If $l_1 \perp l_2$, then the vertical angles formed are right angles.
    $l_1 \perp l_2$
    Thus, ________________.
21. If $m\angle 1 = m\angle 2$ implies $AB = AC$, and $AB = AC$ implies $\triangle ABC$ is isosceles, then ________________.
22. If the corresponding sides of two triangles are congruent, the triangles are congruent.
    If two triangles are congruent, the corresponding angles are congruent.
    Thus, ________________.
23. If p then q.
    $\sim q$
    Thus, ____

C

The phrase $p \rightarrow q$ can also be read "p is sufficient for q." The phrase $q \rightarrow p$ can also be read "p is necessary for q." Let p: The two triangles are congruent, and q: Corresponding sides are congruent.

Write the following as implications:

24. A necessary condition for congruent triangles is congruent corresponding sides.
25. For triangles to be congruent, it is sufficient that they have congruent sides.
26. Congruence of sides is a necessary and sufficient condition for triangle congruence.

4-2 MORE ON PROOF

Knowing different methods of proof will provide us with more adequate means of proving various statements. One method is direct proof, which we discussed in Chapter 2 and used in Chapter 3. But direct proof is not the only method. In this section, we shall examine four other methods of proof: proof of existence, proof by enumeration, indirect proof, and proof by the contrapositive.

Proof of Existence

Some conjectures simply state that something exists—for example, statements about the existence of auxiliary lines and segments. We can prove that something exists if we can provide a justifiable example of it; that is, if we can provide a reason why our example fits the description of the thing said to exist.

Example 1 Prove the conjecture "Quadrilateral $ABCD$ has a diagonal."

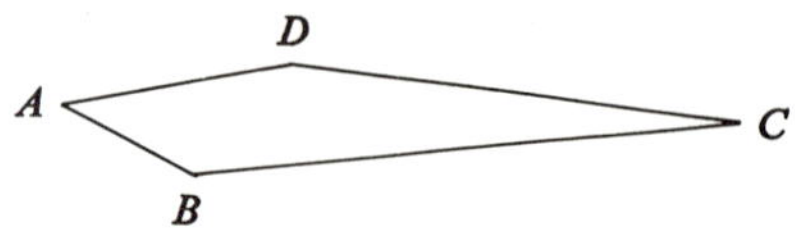

Solution From Chapter 1, we know that a diagonal is a segment joining nonconsecutive vertices of a polygon. To prove the conjecture above, we simply connect points A and C. We can justify this since the Line Postulate (2-3) states that any two points determine exactly one line. Thus, we have proved the conjecture.

Proof by Enumeration

This method of proof can be used only when a conjecture describes a finite number of instances. Such a conjecture is proved by showing that each instance is true.

Example 2 Using a proof by enumeration, prove that any triangle contains exactly three medians.

Solution A median is the segment connecting the vertex of a triangle to the midpoint of the opposite side. Any triangle ABC has exactly three vertices A, B, and C and exactly three sides $\overline{AB}$, $\overline{AC}$, and $\overline{BC}$. Each side has exactly one midpoint. We can label the midpoints of $\overline{BC}$, $\overline{AC}$, and $\overline{AB}$, M, N, and P, respectively. Thus, $\overline{AM}$, $\overline{BN}$, and $\overline{CP}$ are the medians of $\triangle ABC$. Therefore, we see that any triangle contains exactly three medians.

Usually we want conjectures to apply to an infinite number of cases. Thus, this method has limited value in mathematics. We must be careful not to think that we have proved a conjecture which applies to an infinite number of cases merely by showing that a limited number of cases are true.

Indirect Proof

This method of proof follows from the Law of the Excluded Middle, which states that a statement is either true or that it is false. From this we can conclude:

When q is true, then $\sim q$ is false.
When $\sim q$ is true, then q is false.

Sometimes when we are interested in proving that q is true, the use of direct proof is impractical. When this happens, we usually try using indirect proof.

Since the statement q is true if $\sim q$ is false, indirect proof attempts to prove that $\sim q$ is false. This is done by showing that assuming the opposite leads to a contradiction. The opposite of "$\sim q$ is false" is "q is false." This is also the opposite of "q is true."

The steps in an indirect proof of q are as follows:

1. Assume that q is false.
2. Draw logical conclusions until you come to a conclusion that contradicts a known fact.
3. Thus, the hypothesis and conclusion contradict each other. That is, we have either F $\rightarrow$ T or T $\rightarrow$ F. Since we cannot have T $\rightarrow$ F, we have F $\rightarrow$ T. Our hypothesis is false.
4. Thus, we reject the assumption in step 1.
5. Therefore, we have proved that q is true.

The proof of Theorem 4-2.1 illustrates indirect proof. When using indirect proof, we usually signal the assumption with either *Suppose* or *Let us assume*. The use of either of these introductions implies that what follows is questionable.

T **Theorem 4-2.1** If a plane and a line not in the plane intersect, the intersection is only one point.

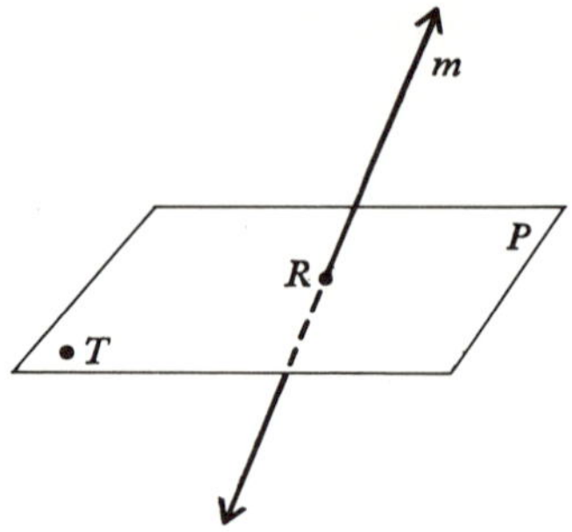

GIVEN Line m does not lie in plane P; line m intersects plane P in at least one point R.

PROVE R is the only point of intersection.

PROOF

STATEMENTS	REASONS
1. Line m does not lie in plane P; m intersects plane P in at least one point, R.	1. Given
2. Suppose m also intersects plane P at T, a point distinct from R.	2. Assumption
3. Line m lies in plane P.	3. Points-in-a-plane Postulate (2-7)

Statement 3, however, contradicts the given statement. Thus, the assumption we made in statement 2 is false, and the line and the plane intersect in only one point.

Proof by the Contrapositive

Proof by the contrapositive is a form of indirect proof. We deny q and reach the denial of p, but we know p is true. The process used in this method of proof is similar to that used in direct proof, but instead of starting with the hypothesis and proceeding directly to the conclusion, we attempt to prove the contrapositive of the original statement. As we saw at the end of the last section, a given implication and its contrapositive are equivalent. Thus, proving the contrapositive of an implication is equivalent to proving the original implication.

Example 3 Prove "In $\triangle ABC$, if $m\angle B \neq m\angle C$, then $AB \neq AC$."

Solution In the implication,

$$p:\ m\angle B \neq m\angle C \qquad q:\ AB \neq AC$$

Therefore

$$\sim p:\ m\angle B = m\angle C \qquad \sim q:\ AB = AC$$

The contrapositive of $p \to q$ is $\sim q \to \sim p$, that is, "In $\triangle ABC$, if $AB = AC$, then $m\angle B = m\angle C$." This is easily proved.

STATEMENTS	REASONS
1. In $\triangle ABC$, if $AB = AC$, then $\angle B \cong \angle C$.	1. The base angles of an isosceles triangle are congruent. (3-4.2)
2. If $\angle B \cong \angle C$, then $m\angle B = m\angle C$.	2. Definition of congruent angles (1-30)
3. In $\triangle ABC$, if $AB = AC$, then $m\angle B = m\angle C$.	3. Transitive property of implication
4. In $\triangle ABC$, if $m\angle B \neq m\angle C$, then $AB \neq AC$.	4. A statement and its contrapositive are equivalent.

Class Exercises

Complete each of the following:

1. Proof by enumeration proves a conjecture by showing that ________ is true.
2. Besides direct proof there are four other methods of proof: ________, ________, ________, and ________.
3. To prove existence, provide a justifiable ________.
4. The fact that p is true or p is false is the Law of ________.
5. The first step in an indirect proof is to assume that q is ________.
6. The contrapositive of an implication and the ________ are equivalent.
7. When $\sim p$ is true, p is ________.

Name a good method of proof to be used in proving each of the following:

8. Every quadrilateral has two diagonals.
9. If $m\angle AOC \neq m\angle BOD$, then $\angle AOC$ and $\angle BOD$ are not vertical angles.

Equally important as knowing how to prove that a conjecture is true is knowing how to prove that it is false. We shall discuss two methods of this type of proof: contradiction and counterexample.

Contradiction

If you try unsuccessfully to prove that a conjecture is true, it may be false. Based on this fact, proof by contradiction is somewhat similar to indirect proof. Both methods employ contradiction, but for different purposes. Indirect proof is used to prove that a statement is true; proof by contradiction is used to prove that a statement is false. The steps in proof by contradiction are given below. Compare these with the corresponding steps in Example 4.

1. Assume that the conjecture is true.
2. Draw logical conclusions until you come to a conclusion that contradicts some known fact.
3. Thus, the conclusion is false. That means that we have either T → F or F → F. Since we used logical reasoning, we cannot have T → F. Therefore, we have F → F. The hypothesis is false.
4. We reject the assumption in step 1 above.
5. Therefore, the conjecture is false.

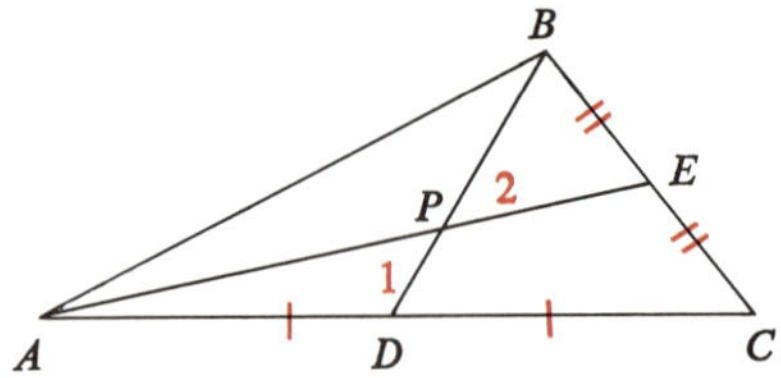

Example 4 Prove that the following conjecture is false: "Any two medians in a triangle bisect each other."

Solution Assume that the conjecture is true.

PROOF

STATEMENTS	REASONS
1. $AP = EP$ and $DP = BP$	1. Assumption.
2. $\angle 1 \cong \angle 2$	2. Vertical angles are congruent. (3-1.5)
3. $\triangle PDA \cong \triangle PEB$	3. SAS Postulate. (3-1)
4. $BE = AD$	4. Definition of congruent triangles. (3-3)
5. E and D are midpoints.	5. Definition of median of a triangle. (3-9)
6. $2(BE) = 2(AD)$	6. Definition of midpoint (1-15) and Substitution Postulate. (2-1)
7. $BC = AC$	7. Multiplication property of equality.
8. Since $\triangle ABC$ is any triangle, any triangle is isosceles.	8. Definition of isosceles triangle. (3-12)

But this is not true; we know that not all triangles are isosceles. Thus, we must reject the conjecture as false. This means that any two medians in a triangle do not necessarily bisect each other. Does this mean that they never bisect each other?

Counterexample

Most of our conjectures apply to an infinite number of instances. Each instance must be true before we can say that the conjecture is true. If we can find one instance that is not true, then the conjecture cannot be true.

Example 5 Prove that the following statement is false: "In a triangle, if two sides and the angle opposite one of the sides are congruent to the corresponding parts of a second triangle, then the triangles are congruent."

Solution By counterexample, we can prove that the conjecture is false. In the statement, we have

p: Two sides and the angle opposite one of these sides in a triangle are congruent to the corresponding parts of a second triangle.
q: The triangles are congruent.

In the isosceles $\triangle ABC$, $\overline{AB} \cong \overline{AC}$, with $\overline{AD}$ such that $\overline{BDC}$. Consider the correspondence $ABD \leftrightarrow ACD$. By the reflexive property, $\overline{AD} \cong \overline{AD}$. Angles B and C are congruent. Why? This is an instance of the conjecture, but since D is any point on $\overline{BC}$, $\triangle ABD \not\cong \triangle ADC$. Thus, by counterexample, we can say that the conjecture is false.

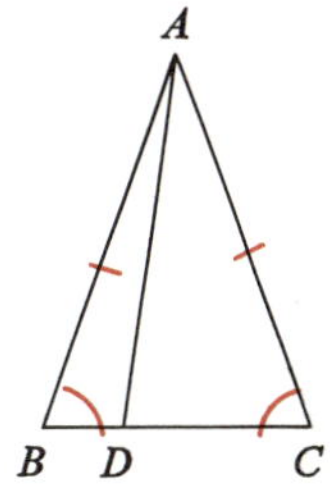

EXERCISES

A Complete each of the following:

1. Besides the method of direct proof, we have studied ________ other methods.
2. Proof by enumeration demands that there be a ________ number of cases.
3. Indirect proof follows from the Law of the ________.
4. ________ indicates that what follows is questionable.
5. When statement p is true, then ________ is false.
6. The equivalent of "$\sim p$ is true" is ________.
7. Proving the ________ is equivalent to proving $p \rightarrow q$.
8. The first step in proving by contradiction that a conjecture is false is to assume that the conjecture is ________.
9. Finding an untrue case proves that the conjecture is false by ________.

Assume you are using indirect proof to prove each of the following. Write the first step you would use in your proof.

10. Every angle has one bisector.
11. If $\angle A$ and $\angle B$ are right angles, then $\angle A \cong \angle B$.
12. Any segment has a midpoint.
13. There is only one altitude from a vertex to the opposite side.

14. Complete your indirect proof of Exercise 12.

Assume you are proving that each of the following is false by contradiction. Write the first step of each proof.

15. In an isosceles triangle, the three angles are congruent.
16. The altitude from vertex P of $\triangle PQR$ is also the bisector of $\angle P$.

17. Complete your proof in Exercise 15.

B

18. Prove that any triangle has at least one median.
19. Prove that if $\angle A$ is not congruent to $\angle B$, then $\triangle ABC$ is not equilateral.
20. Prove that the statement "All right triangles are congruent" is not true.
21. **Given** $\triangle PQR$ is scalene. **Prove** $\angle P \cong \angle Q$ is not true.
22. Prove that if no angles of $\triangle RST$ are congruent, $\triangle RST$ is not isosceles. Use indirect proof.
23. Prove Exercise 22 using the contrapositive.

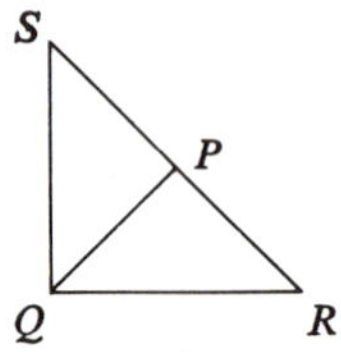

Exercise 25

C

24. Use the contrapositive method to prove that if a point is not the same distance from the endpoints of a segment, then the point is not on the line that bisects the segment.
25. Prove that if two lines are perpendicular, they form right angles. Use the contrapositive.
26. Prove that if the corresponding sides of two triangles have unequal measures, then the triangles are not congruent.

4-3 MATHEMATICAL METHODS

In the last section, we discussed how to prove conjectures true or false. But what mathematical methods were used to form these conjectures? We have already considered intuition and induction.

While intuition is not a reliable way of conjecturing, it does provide a motive for making further examinations. And while induction is a valuable way of forming conjectures, not every situation provides sufficient opportunities to make observations. There are, however, other methods we can use to form conjectures. Consider the following:

Example 1 Angles and triangles are both formed by sides; both have vertices. We know many properties of angles. From our knowledge of angles, can we infer anything about triangles?

Solution List some properties of angles, and make an item-by-item comparison between angles and triangles.

An angle—	A triangle—
1. has sides;	1. has sides;
2. has a vertex;	2. has vertices;
3. has an interior and an exterior;	3. has an interior and an exterior;
4. has angle measure;	4. Is there a triangle measure?
5. has angle addition;	5. If there is triangle measure, is there triangle addition?
6. may be acute, right, or obtuse;	6. may be acute, right, or obtuse;
7. has an angle bisector;	7. has an angle bisector of a triangle.
8. is congruent with another angle if their measures are equal.	8. Does congruence depend on measure?

We are able to list some triangle properties as a result of what we have already defined or proved about triangles, but we also have some questions: Can we represent triangles with a single measure? Is there a unique sum of the measures of the angles of a triangle? Does angle congruence necessitate triangle congruence?

Example 2 Can we say that the following statement is true?" "Two triangles are congruent if their corresponding angles are congruent."

Solution Use as a counterexample the figure at the right, where corresponding congruent angles are indicated.

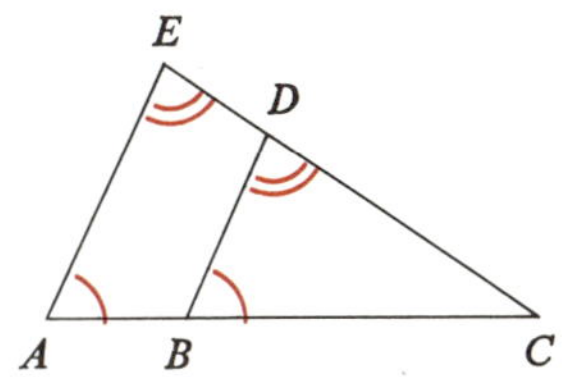

From the figure, we can see that corresponding sides are not congruent. Thus, the conjecture is not necessarily true.

We shall postpone discussion of the first two questions in Example 1 until later. At present, we wish to use the example only to show how to form conjectures by means of **analogy**.

An analogy exists between two things when they are alike in some way. By the process of analogy we infer additional areas in which they may be alike, thereby arriving at plausible conjectures which then have to be proved true or false. This type of conjecturing must be done carefully. Should we infer, for example, that you and Bill have the same singing potential because you have the same mathematics grade average? With proper care, however, the process of analogy offers a very good method of conjecturing.

Analogy and induction offer good examples of the uses of comparison. There is also value in *contrasting* the items under consideration.

Example 3 Describe all triangles by analyzing the six triangles below. The figure markings indicate corresponding congruences.

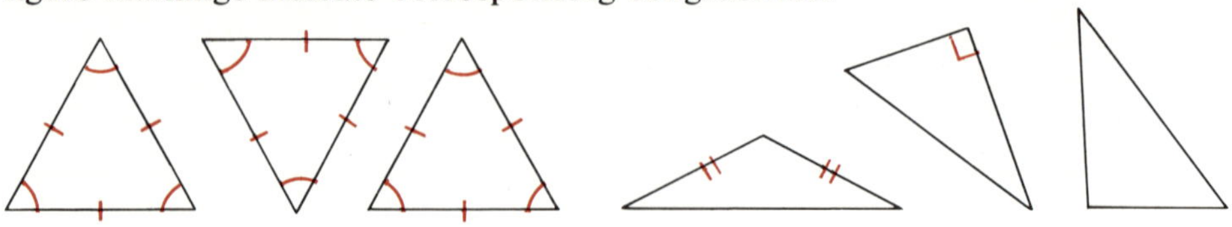

Solution If we had only three triangles on the left to compare, we might conclude that all triangles have three congruent sides and three congruent angles. When we contrast the three triangles on the left with those on the right, we see that congruent sides and congruent angles are not essential elements in forming triangles. Thus, by comparing and contrasting, we can conclude only that triangles are three-sided figures with three angles.

Class Exercises

1. List the methods that we have discussed for making plausible conjectures.
2. In what ways are triangles and quadrilaterals alike?
3. Would you say that the conjectures formed from the methods discussed are probable or necessary?

Use the following figures to answer the questions below.

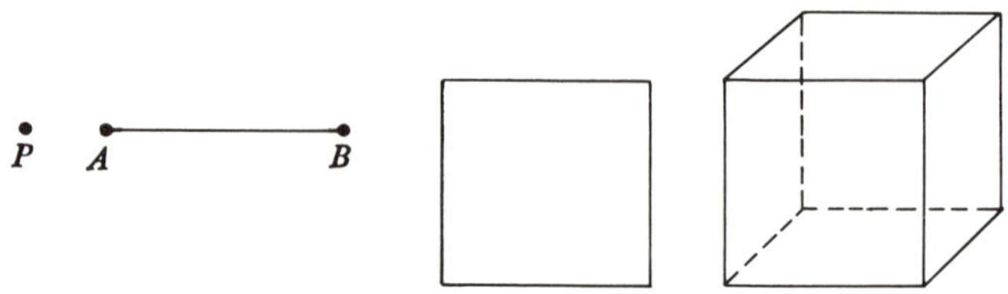

4. If we say a quadrilateral is two dimensional, how many dimensions do the other figures have?
5. If we say the segment is generated by a point, what can we say generates a cube?
6. Using the figures above as a model for comparison, name the figure missing below.

7. Planes, like lines, can intersect. Intersecting lines form angles. What can you conjecture about intersecting planes?

8. If the only three-dimensional figures we know about are the two above, what can you conjecture about the shapes of the plane regions that form them?
9. Include this figure with the two three-dimensional figures above and re-answer Exercise 8?

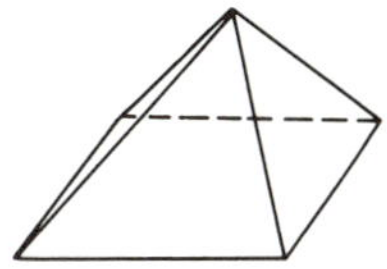

Constructing a picture or a model for figures under discussion helps us to see the relationship between what is given and what is to be proved. Models can also serve as a basis for conjectures, but we must be sure that the model represents a given figure exactly and is not more specific than demanded.

Example 4 Provide a model for a quadrilateral.

Solution

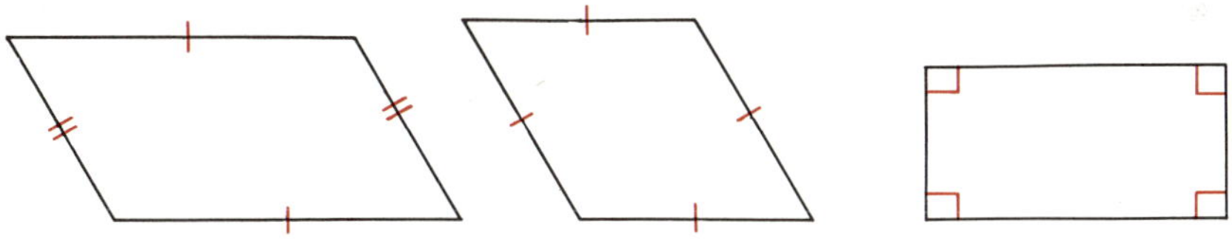

The first model is a quadrilateral, but it has opposite sides congruent. The second quadrilateral has all sides congruent. The third quadrilateral has four congruent angles. None of these figures is a good model for a quadrilateral because each gives us more properties than is required of a general quadrilateral.

In some cases, adding an auxiliary segment or line suggests additional conjectures. For example, consider the Isosceles Triangle Theorem. In the first proof in Section 3-4, we drew the bisector of the vertex angle. From looking at the model, we might conjecture that $AM = MC$ and that $\overline{BM} \perp \overline{AC}$. We cannot just state that these conjectures are true. If an accurate model suggests them, we must prove that they are true before we can accept them.

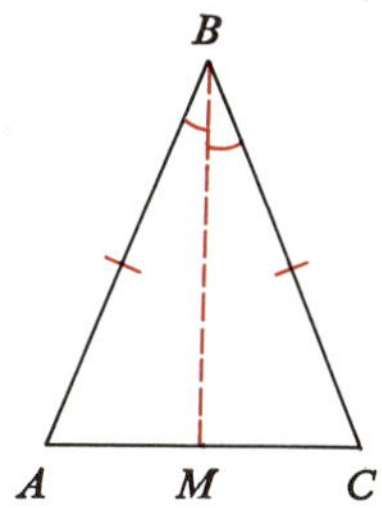

We usually introduce auxiliary lines or segments in a proof with the word *let*. In Example 5, we shall prove the ASA Postulate using the SAS Postulate and an auxiliary point.

Example 5

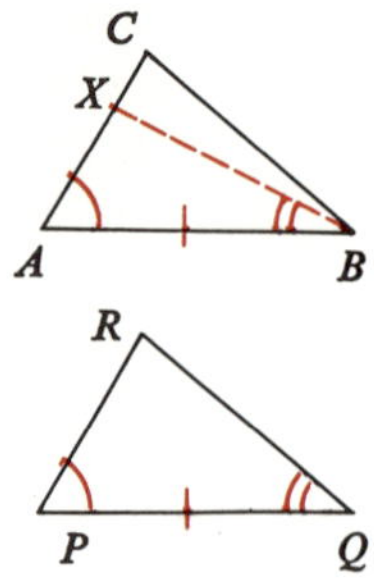

GIVEN The ASA correspondence:
$\angle A \cong \angle P, AB = PQ, \angle CBA \cong \angle RQP$

PROVE $\triangle ABC \cong \triangle PQR$

Solution

PROOF

STATEMENTS	REASONS
1. $\angle A \cong \angle P, AB = PQ,$ $\angle CBA \cong \angle RQP$	1. Given.
2. Let X be the point of $\overrightarrow{AC}$ such that $AX = PR$.	2. Point Uniqueness Postulate. (2-2)
3. $\triangle XAB \cong \triangle RPQ$	3. SAS Postulate. (3-1)
4. $\angle XBA \cong \angle RQP$	4. Definition of congruent triangles. (3-3)
5. $\angle XBA \cong \angle CBA$	5. Angle Uniqueness Postulate. (2-9)
6. X must be C.	6. If two distinct lines intersect, then they intersect in exactly one point. (2-5.2)
7. $\triangle ABC \cong \triangle PQR$	7. Substitution Postulate. (2-1)

EXERCISES

A

Complete each of the following:

1. ________ reasoning is based on observation of several known true cases.
2. By the process of ________ we infer additional areas of similarity.
3. ________ and ________ make use of comparison.
4. We should ________ as well as compare.
5. A model for a problem cannot indicate more ________ than is demanded by the problem.
6. ________ lines or segments can suggest additional conjectures.

Using the process of analogy and the following statements about triangles, write five conjectures for quadrilaterals:

7. An angle of a triangle is included by the sides of the triangle that lie in the sides of the angle.
8. A triangle is equiangular if all its angles are congruent.
9. A triangle is equilateral if all three sides are congruent.
10. Every equilateral triangle is equiangular.
11. Every equiangular triangle is equilateral.

12. Using counterexample, prove that this conjecture is false: "Every equilateral quadrilateral is equiangular."
13. Using counterexample, prove that this conjecture is false: "Every equiangular quadrilateral is equilateral."
14. Compare the three figures below. Based on this comparison, describe quadrilaterals.

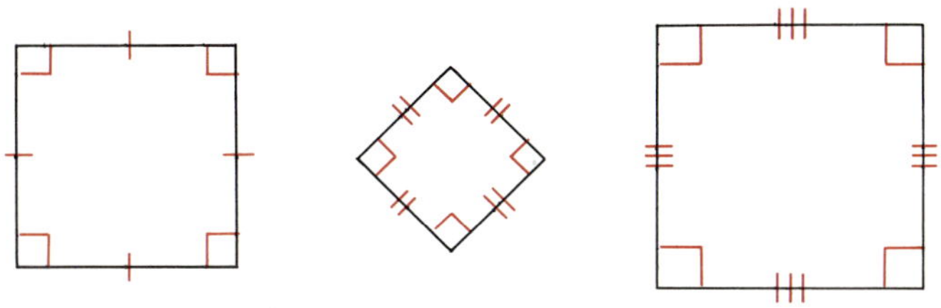

15. Using the figure at right and the figures above, give a better description of quadrilaterals.

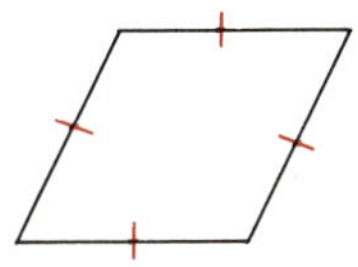

16. Using the figure at right and the figures in Exercises 14 and 15, give a still better description of quadrilaterals.

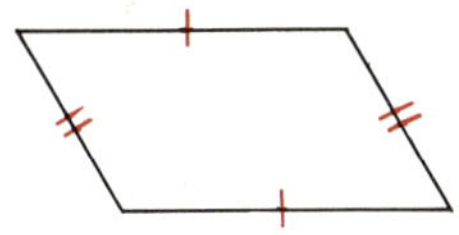

17. Does this figure restrict your description of a quadrilateral any further? If so, explain how.

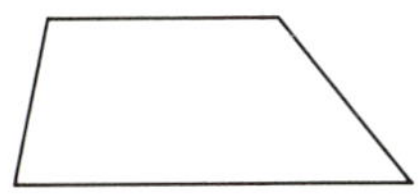

Using your ruler and protractor, draw a model for each of the following:

18. A quadrilateral such that $\angle P \cong \angle Q \cong \angle R$ and $m\angle S = 120$.
19. $\triangle PQR$ such that $\angle Q \cong \angle R$ and $\overline{QR}$ is extended to S such that $RS = QR$.
20. $\triangle KLM$ such that the bisector of $\angle LKM$ intersects $\overline{LM}$ at N.
21. The three altitudes in a scalene triangle.

B Draw a model for each of the following. Then prove each one true or false.

22. If in $\triangle PQR$ and $\triangle PQS$, $\angle RPQ \cong \angle SQP$, and $\angle RQP \cong \angle SPQ$, then $PR = QS$.
23. If in quadrilateral $ABCD$, $\overline{AC}$ bisects $\angle BAD$ and $AB = AD$, then $\angle D \cong \angle B$.
24. If in $\triangle ABC$, the bisectors of $\angle B$ and $\angle C$ intersect at X and $\overline{AB} \cong \overline{AC}$, then $\overline{XB} \cong \overline{XC}$.
25. If $LM = MN$ and $KL = KN$, then $\angle MLK \cong \angle MNK$.
26. In $\triangle ABC$, $AB = AC$, the bisector of $\angle B$ intersects $\overline{AC}$ at P, and the bisector of $\angle C$ intersects $\overline{AB}$ at Q. Prove that $\overline{AP} \cong \overline{AQ}$.
27. The segments connecting the midpoints of the sides of an equilateral triangle form an equilateral triangle.

A LOOK AT THE PAST

Halley's Comet

Edmund Halley (1656–1742) was an English astronomer. While checking the data on the 24 recorded comets, he noticed that 3 comets seemed to have traveled in identical paths. He had a hunch that this was no mere coincidence. The time between the appearances of the first two comets was 76 years; between the second and third, 75 years. After making a careful analysis of the observed orbits, Halley conjectured that instead of three comets, there was only one traveling in a fixed orbit, which it took 75 or 76 years to complete. He predicted the comet's return in 1759. In 1759, several years after his death, the comet appeared just as he had predicted. The next appearance of the comet, now named after Halley, will be in 1985 or 1986.

4-4 PERPENDICULARS IN A PLANE

In Chapter 1, Definition 1-25, we first introduced perpendicular lines. Definition 1-25 led to Theorem 2-6.5, and the converse of Theorem 2-6.5, Theorem 2-6.6, provided our first introduction to converses. Theorems 2-6.5 and 2-6.6 can be combined and restated to form the following theorem:

T **Theorem 4-4.1** Two lines are perpendicular if and only if they form right angles.

Now let us consider a point P on any line m. If we limit our discussion to a plane, how many lines can be perpendicular to m at P?

T **Theorem 4-4.2 Perpendicular Uniqueness Theorem** In a plane, there is one and only one line perpendicular to a given line through a given point on the line.

GIVEN Point P lies in line m in plane R.

PROVE (1) There is a line perpendicular to m at P.
(2) There is only one line perpendicular to m at P in plane R.

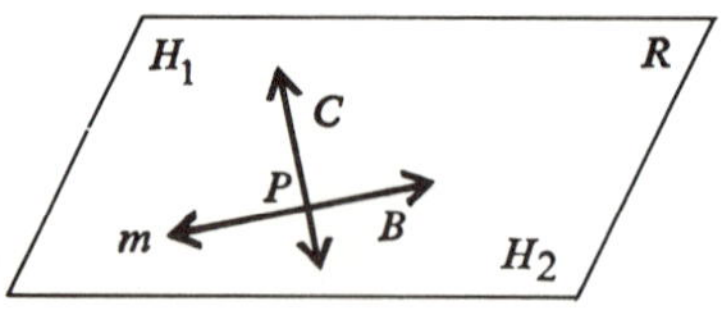

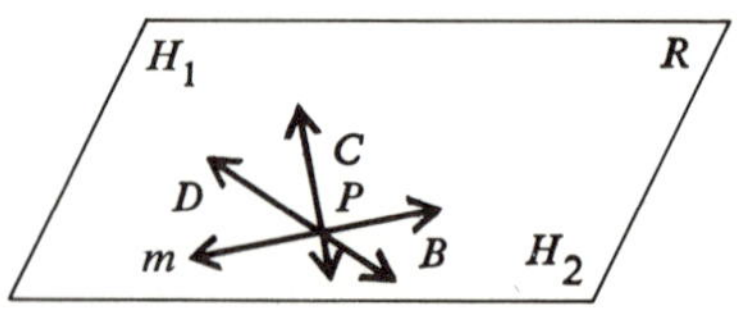

PROOF OF (1)

STATEMENTS	REASONS
1. Point P lies in line m in plane R.	1. Given.
2. Let m separate plane R into two half-planes H_1 and H_2.	2. Definition of half-planes. (1-20)
3. There is a ray, $\overrightarrow{PC}$, in H_1 such that $m\angle BPC = 90$.	3. Angle Uniqueness Postulate. (2-9)
4. $\overleftrightarrow{CP} \perp m$ at P.	4. Theorem 4-4.1.

PROOF OF (2)

STATEMENTS	REASONS
1. $\overleftrightarrow{CP} \perp$ m at P.	1. Proved above.
2. Suppose that $\overleftrightarrow{DP}$, distinct from $\overleftrightarrow{CP}$, is also perpendicular to m at P.	2. Assumption.
3. $\angle CPB$ and $\angle DPB$ are right angles.	3. Two lines are perpendicular if and only if they form right angles. (4-4.1)
4. Thus, there are two rays in half-plane H_1 such that $m\angle CPB = m\angle DPB = 90$.	4. Definition of right angle. (1-24)

Having more than one such ray contradicts the Angle Uniqueness Postulate (2-9). Thus, $\overleftrightarrow{CP}$ is the only line perpendicular to m at P.

In this theorem we stated that the lines must be in a plane. If we remove this restriction, do you see that there will be many lines perpendicular to m at P. Lines h, k, l, and n are all perpendicular to m at P. *Henceforth, unless we state otherwise, we shall restrict our discussion to a plane.*

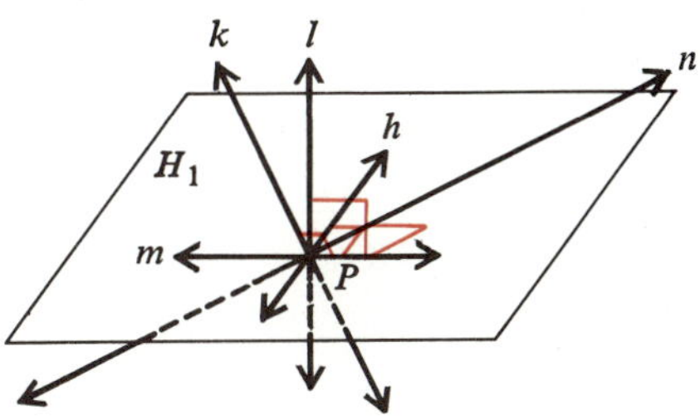

If lines m and n are perpendicular to each other at point P, then any subset of m, either a segment or a ray, containing P is also perpendicular to n. Thus, we can have a line perpendicular to a segment or to a ray.

Every segment has a unique midpoint. Every line perpendicular to a line or segment at a given point is also unique. A line perpendicular to a given segment at the midpoint of the given segment is a unique line called the perpendicular bisector of the given segment. In the figure at right, M is the unique midpoint of $\overline{AB}$. If $\overleftrightarrow{RM}$ is perpendicular to $\overline{AB}$ at M, then $\overleftrightarrow{RM}$ is the perpendicular bisector of $\overline{AB}$.

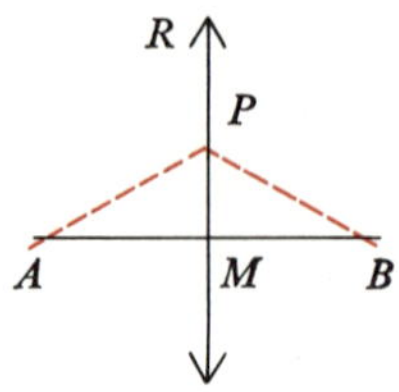

D **Definition 4-4** The **perpendicular bisector** of a segment is the line perpendicular to the segment at its midpoint.

T **Theorem 4-4.3 The Perpendicular Bisector Theorem** A line is the perpendicular bisector of a segment if and only if it is the set of all points in the plane equidistant from the endpoints of the segment.

Refer to the diagram above Definition 4-4 for the following proof:

GIVEN $\overleftrightarrow{RM}$ is the perpendicular bisector of $\overline{AB}$, with M the midpoint of $\overline{AB}$.

PROVE (1) If P is on the perpendicular bisector of $\overline{AB}$, then $PA = PB$.
(2) If $PA = PB$, then P is on the perpendicular bisector of $\overline{AB}$.

Implication (1) was proved in Example 4, Section 3-5.

PROOF OF (2)

STATEMENTS	REASONS
1. $\overleftrightarrow{RM}$ is the perpendicular bisector of $\overline{AB}$; $\overline{AM} \cong \overline{MB}$; $PA = PB$.	1. Given.
2. If P lies on $\overline{AB}$, then P and M are the same point.	2. Any line segment has exactly one midpoint. (2-5.4)
3. Since P is not on $\overline{AB}$, draw $\overline{PA}$, $\overline{PB}$, and $\overline{PM}$.	3. The Line Postulate. (2-3)
4. $\overline{PM} \cong \overline{PM}$	4. Every segment is congruent to itself. (3-1.6)
5. $AM = BM$	5. Definition of midpoint. (1-15)
6. $\triangle PAM \cong \triangle PBM$	6. The SSS Postulate. (3-3)
7. $\angle PMA \cong \angle PMB$	7. Definition of congruent triangles. (3-3)

(*continued*)

STATEMENTS	REASONS
8. $\overleftrightarrow{PM} \perp \overline{AB}$ at M	8. Definition of perpendicular lines. (1-25)
9. $\overleftrightarrow{PM}$ is the same as $\overleftrightarrow{RM}$; thus P is on $\overleftrightarrow{RM}$.	9. Theorem 4-4.2.

The following corollary follows directly from the Perpendicular Bisector Theorem. Its proof is left for you to do as an exercise.

C **Corollary 4-4.3a** If two distinct points are both equidistant from the endpoints of a segment, the two points determine the perpendicular bisector of the segment.

Class Exercises

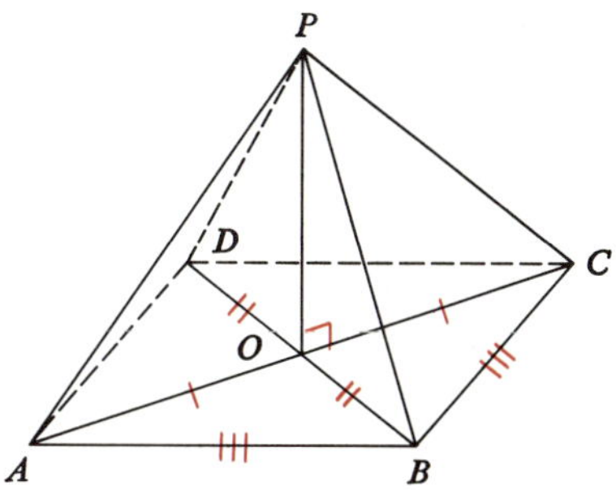

Use the figure at right to answer Exercises 1–3.

1. Name two points equidistant from B.
2. Name three points equidistant from A and C.
3. Name two planes containing $\overline{AC}$.

Complete each of the following:

4. If lines m and n form congruent adjacent angles, then m ________ n.
5. Line $m \perp n$ if and only if m and n form ________.
6. If we do not restrict our discussion to a plane, there are ________ number of perpendiculars to a line at a point.
7. If for $\overline{AB}$, $AM = MB$ when $m \perp \overline{AB}$ at M, then m is the ________ of $\overline{AB}$.

Starting with the figure for Theorem 4-4.4, construct a model satisfying each of the following:

8. Draw $\overrightarrow{RP}$ and find the measure of $\angle SRP$.
9. In half-plane H_2, use your protractor to form $\angle SRT$ such that $m\angle SRT = m\angle SRP$.
10. Locate the point Q on $\overrightarrow{RT}$ such that $RQ = RP$.
11. Draw the segment $\overline{PQ}$ and name the triangles other than $\triangle PRQ$ that are formed if $\overline{PQ}$ intersects $\overrightarrow{RS}$ at M.
12. If the triangles of Exercise 11 are congruent, what can we conclude about $\overleftrightarrow{RS}$ and $\overleftrightarrow{PQ}$?

By working through the last set of exercises, you have seen the essentials of the proof of the following theorem.

Theorem 4-4.4 If a point is not on a line, then there is a line through the point perpendicular to the given line.

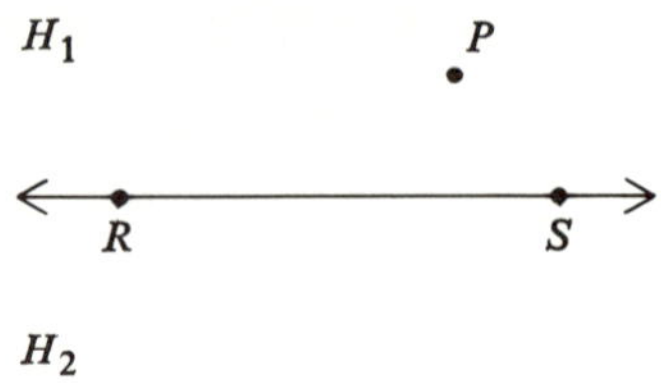

GIVEN Point P is not on $\overleftrightarrow{RS}$.

PROVE There is a line $\overleftrightarrow{PQ}$ such that $\overleftrightarrow{PQ} \perp \overleftrightarrow{RS}$.

PROOF

STATEMENTS	REASONS
1. Point P is not on $\overleftrightarrow{RS}$.	1. Given
2. $\overleftrightarrow{RS}$ separates the plane into half-planes H_1 and H_2.	2. Definition of half-planes. (1-20)
3. Let $\overrightarrow{RT}$ be ray in H_2 such that $m\angle SRT = m\angle SRP$.	3. The Angle Uniqueness Postulate. (2-9)
4. Let Q be the point on $\overrightarrow{RT}$ such that $RQ = RP$.	4. The Point Uniqueness Postulate. (2-2)
5. P and Q determine $\overline{PQ}$.	5. The Line Postulate. (2-3)
6. Let M be the intersection of $\overline{PQ}$ and $\overleftrightarrow{RS}$.	6. Definition of half-planes. (1-20)
7. $\overline{RM} \cong \overline{RM}$	7. Every segment is congruent to itself. (3-1.6)
8. $\triangle PRM \cong \triangle QRM$	8. SAS Postulate. (3-1)
9. $m\angle PMR = m\angle QMR$	9. Definition of congruent triangles. (3-3)
10. $\overleftrightarrow{PQ} \perp \overleftrightarrow{RS}$	10. Definition of perpendicular lines. (1-25)

Proving this theorem establishes that there is at least one perpendicular to the line from an **external point**, a point not on the line. The next theorem shows the uniqueness of this perpendicular.

Theorem 4-4.5 Through a point external to a line, there is at most one line perpendicular to the given line.

GIVEN P is external to line m

PROVE $\overleftrightarrow{PR}$ is the only perpendicular to m through P

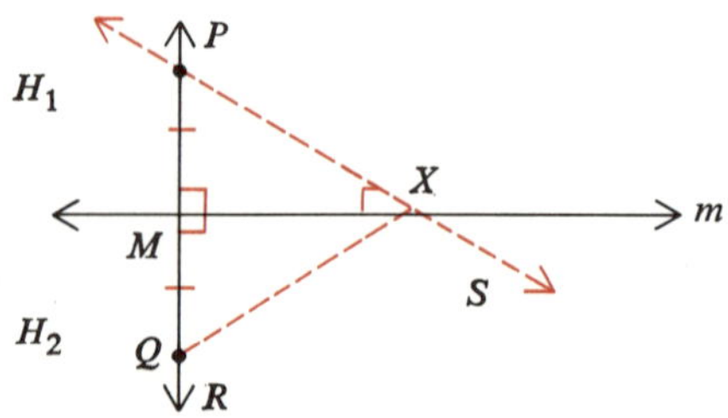

Proof

STATEMENTS	REASONS
1. P is external to line m.	1. Given.
2. $\overleftrightarrow{PR}$ is a line perpendicular to m.	2. Theorem 4-4.4
3. Suppose $\overleftrightarrow{PS}$, distinct from $\overleftrightarrow{PR}$, is also perpendicular to m.	3. Assumption.
4. $\overleftrightarrow{PR}$ intersects m at M; $\overleftrightarrow{PS}$ intersects m at X.	4. If two distinct lines intersect, then they intersect in exactly one point. (2-5.2)
5. Let Q be the point on $\overrightarrow{MR}$ such that $MQ = MP$.	5. Point Uniqueness Postulate. (2-2)
6. Q and X determine $\overline{QX}$.	6. Line Postulate. (2-3)
7. $\overline{MX} \cong \overline{MX}$	7. Every segment is congruent to itself. (3-1.6)
8. $\triangle PMX \cong \triangle QMX$	8. SAS Postulate. (3-1)
9. $\angle QXM \cong \angle PXM$	9. Definition of congruent triangles. (3-3)
10. Thus, $\overleftrightarrow{PS} \perp m$ at X; $\overleftrightarrow{QX} \perp m$ at X	10. Definition of perpendicular lines. (1-25)

But this contradicts the Perpendicular Uniqueness Theorem (4-4.2). This means that our assumption must be false. Thus, there is only one line perpendicular to a given line through an external point.

Theorems 4-4.4 and 4-4.5 taken together can be stated as follows:

T **Theorem 4-4.6** Through a point external to a line, there is one and only one line perpendicular to the given line.

In the model for Theorem 4-4.5, $\triangle PMX$ had two right angles. Do you see why this is not possible? The proof of the next theorem is left as an exercise.

C **Corollary 4-4.6a** No triangle has two right angles.

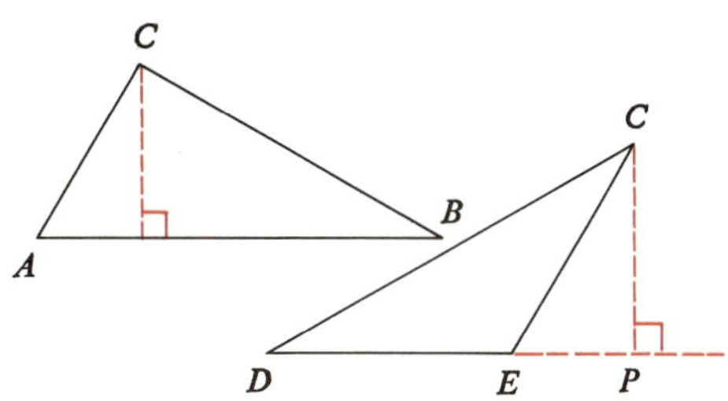

In a triangle, any vertex is an external point to the line containing the opposite side. By the theorems above, there must exist a unique perpendicular from a vertex to this opposite line. In Section 3-4, we defined the perpendicular segment from the vertex to the opposite side as the altitude. Each vertex has a unique altitude. How many altitudes does a triangle have?

Often when we speak of an altitude, we mean the *length* of the altitude. The way the word is used should tell us what is meant. In such a context, the altitude is the distance from a point (a vertex) to a line (the line containing the opposite side.)

D **Definition 4-5** The **distance to a line from an external point** is the measure of the perpendicular segment from the point to the line.

Example 1 Find the distance from C to the line not containing C in the figures below.

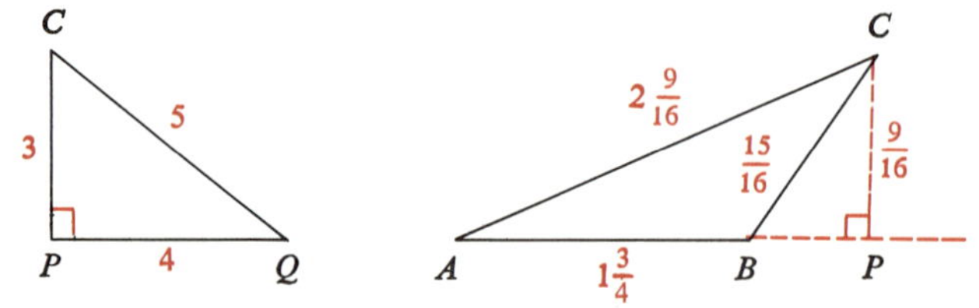

Solution In each case, $\overline{CP}$ is an altitude. Thus, it is a perpendicular segment from a point to a line. The measure of $\overline{CP}$ is the distance from point C to the opposite line. In the first figure, $CP = 3$; in the second figure, $CP = \frac{9}{16}$.

In the example, P is the point of intersection of $\overleftrightarrow{AB}$ and the perpendicular segment from an external point to the line. P is called the **foot** of the perpendicular segment. Suppose, in a physical model, we hold a flashlight aimed perpendicular to the line from above an external point. Do you see that the projection of this point onto the line is the foot of the perpendicular segment?

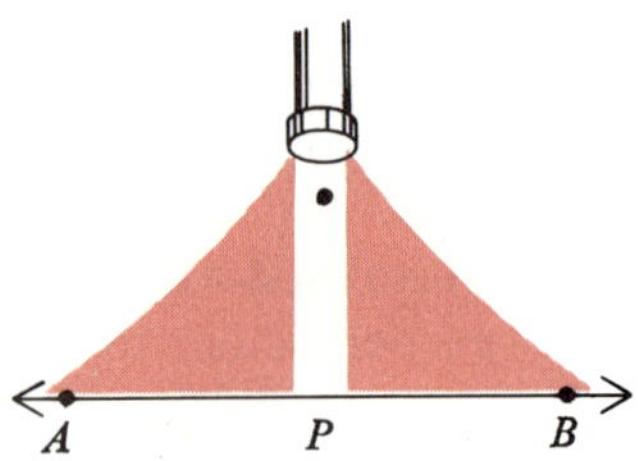

D **Definition 4-6** The **projection** of an external point onto a line in a plane is the foot of the perpendicular segment from the point to the line.

EXERCISES

A Complete each of the following:

1. If two lines are perpendicular, they form ________ angles.
2. ________ lines form right angles.
3. Every triangle has ________ altitudes.
4. There is a unique ________ from an external point to a given line.
5. The ________ is perpendicular to a segment at its midpoint.
6. A(n) ________ point of a plane is a point not in the plane.

7. No triangle can have ________ right angles.
8. The ________ from P to m is the length of the perpendicular segment from P to m.

9. Prove Corollary 4-4.3a.
10. Prove Corollary 4-4.6a.

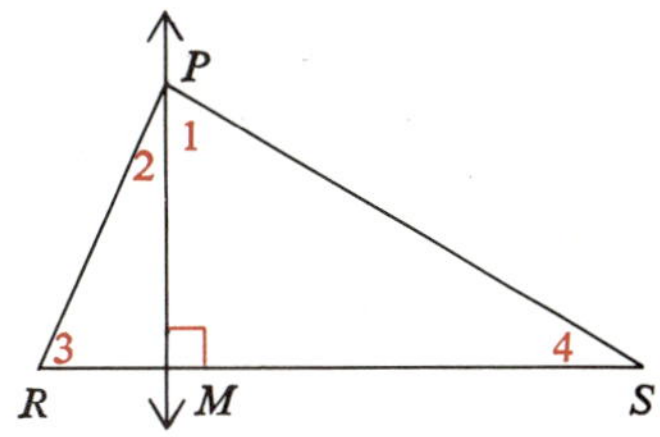

Exercises 11–16

In the figure at right, $\overleftrightarrow{PM} \perp \overline{RS}$.
11. What can we conclude about $\angle 1$?
12. What can we conclude about $\angle 3$?
13. What is the distance from P to $\overline{RS}$?
14. If $PR = PS$, is M the midpoint of $\overline{RS}$? Why?
15. Name the projection of S onto $\overleftrightarrow{PM}$.
16. If $\angle RPS$ is a right angle, $\overline{PS}$ is perpendicular to what line?

B

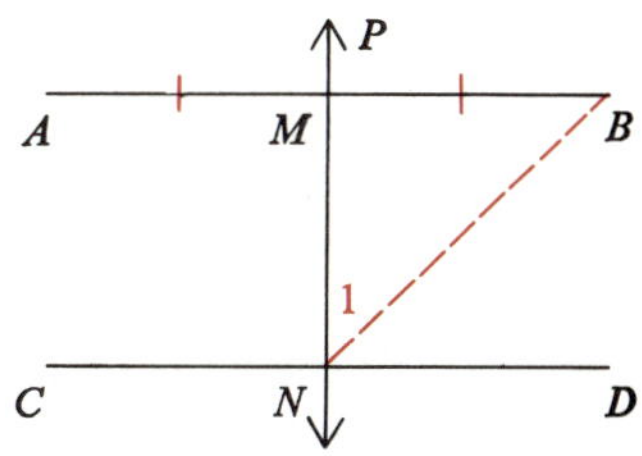

Exercises 17–22

In the figure at right, $AM = MB$.
17. If $\overleftrightarrow{PN} \perp \overleftrightarrow{AB}$, describe $\overleftrightarrow{PM}$.
18. If $\overleftrightarrow{PN} \perp \overleftrightarrow{CD}$, classify $\angle 1$.
19. If $\overleftrightarrow{PN} \perp \overleftrightarrow{AB}$ and $\overleftrightarrow{PN} \perp \overleftrightarrow{CD}$, at what point will the lines containing $\overleftrightarrow{AB}$ and $\overleftrightarrow{CD}$ meet?
20. Describe the lines $\overleftrightarrow{AB}$ and $\overleftrightarrow{CD}$ if the conditions in Exercise 19 are true.
21. In $\triangle MNB$, name the legs and hypotenuse.
22. If $CM = DM$ and $CN = DN$, what conclusion can we draw about $\overleftrightarrow{PN}$?

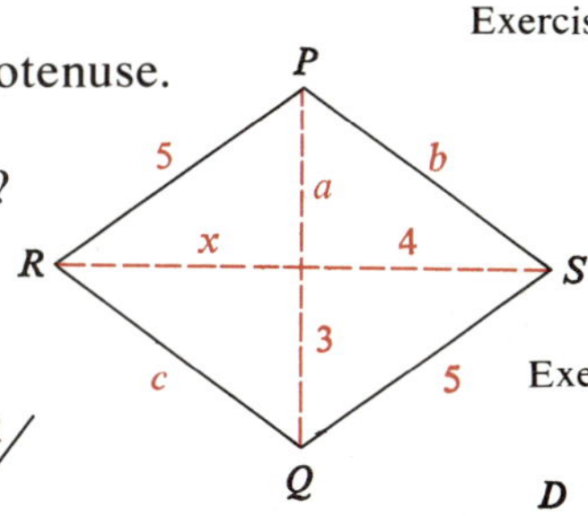

Exercises 23–26

In the figure $RQSP$, $\overline{PQ}$ and $\overline{RS}$ are perpendicular bisectors of each other.

23. $a =$ ____ 24. $b =$ ____
25. $x =$ ____ 26. $c =$ ____

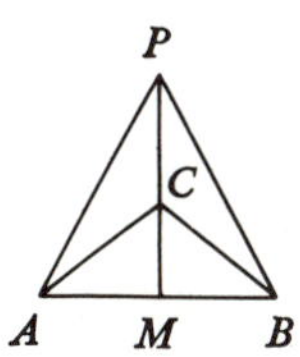

Exercise 27

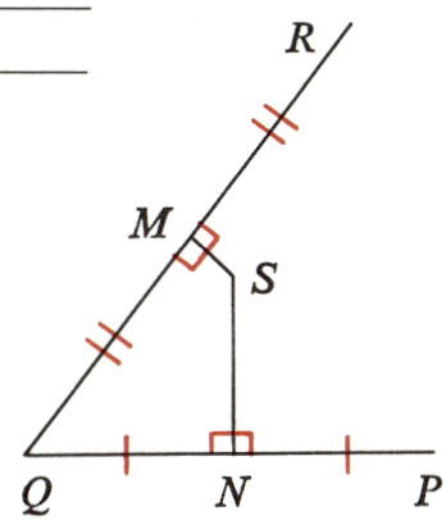

Exercise 28

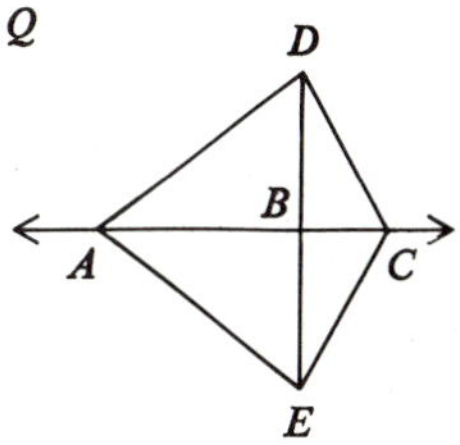

Exercise 29

C

27. Prove that if $\triangle AMP \cong \triangle BMP$, then $\triangle ABC$ is isosceles.
28. $\overline{SM}$ and $\overline{SN}$ are the perpendicular bisectors of $\overline{RQ}$ and $\overline{PQ}$, as shown. Prove that $SR = SP$.
29. If B is the midpoint of $\overline{DE}$ and $AD = AE$, use congruent triangles to prove that $DC = EC$.

30. Prove Exercise 29 without using congruent triangles.
31. Prove that if the diagonals of quadrilateral $ABCD$ are perpendicular bisectors of each other, the sides of the quadrilateral are congruent.
32. Prove that if the perpendicular bisector of the hypotenuse contains the vertex of the right angle in the triangle, then the triangle is isosceles.
33. Draw three triangles—one right, one acute, and one obtuse. Using your protractor, draw the three altitudes of each triangle.
34. If you extended the lines in which the altitudes lie, would the altitudes intersect?
35. What conjecture can you make about the three altitudes of any triangle?

4-5 PERPENDICULAR LINES AND PLANES

Consider the figure for Theorem 4-5.1. If we are not limited to one plane, there is an infinite number of lines perpendicular to a given line at a point of the line. Do you see that l_1, l_2 and l_3 all lie in one plane? Since m is perpendicular to each of these lines, we say that m is perpendicular to the plane containing these lines.

D

Definition 4-7 A **line and a plane are perpendicular** if and only if they intersect and all the lines in the plane which pass through the point of intersection are perpendicular to the given line.

In using the definition, how many lines perpendicular to m and lying in one plane are needed to prove that m is perpendicular to the plane? We know that the least number of lines needed to determine a plane is two. We use this fact to set minimum conditions for the perpendicularity of a line and a plane.

T

Theorem 4-5.1 If a line is perpendicular to each of two intersecting lines at their point of intersection, then the line is perpendicular to the plane determined by them.

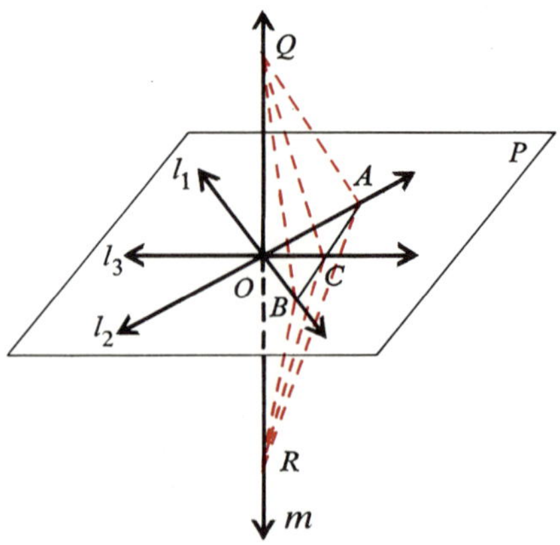

GIVEN Lines l_1 and l_2 determine plane P and intersect at O. $\overleftrightarrow{QR}$ not in plane P; $\overleftrightarrow{QR} \perp l_1$ and $\overleftrightarrow{QR} \perp l_2$

PROVE $\overleftrightarrow{QR}$ is perpendicular to any other line in plane P that passes through O, or $\overleftrightarrow{QR} \perp$ plane P.

PROOF

STATEMENTS	REASONS
1. l_1 and l_2 determine plane P and intersect at O; $\overleftrightarrow{QR} \perp l_1$ and $\overleftrightarrow{QR} \perp l_2$	1. Given.
2. Let Q be a point on $\overleftrightarrow{QR}$ in one half-space and R the point on $\overleftrightarrow{QR}$ in the opposite half-space such that $OQ = OR$.	2. Definition of half-spaces (1-35) and Point Uniqueness Postulate. (2-2)
3. Let l_3 be any line in plane P, containing O, but distinct from l_1 and l_2. If $\overline{AB}$ meets l_3, label the intersection C, where A lies in l_2 and B lies in l_1. l_1 and l_2 are both perpendicular bisectors of $\overline{QR}$.	3. Definition of perpendicular bisector. (4-4)
4. $QA = RA$; $QB = RB$	4. A line is the perpendicular bisector of a segment if and only if it is the set of all points in the plane equidistant from the endpoints of the segment. (4-4.3)
5. $\overline{AB} \cong \overline{AB}$	5. Every segment is congruent to itself. (3-1.6)
6. $\triangle QBA \cong \triangle RBA$	6. The SSS Postulate. (3-3)
7. $\angle QBA \cong \angle RBA$	7. Definition of congruent triangles. (3-3)
8. $\overline{BC} \cong \overline{BC}$	8. Same reason as step 5.
9. $\triangle QBC \cong \triangle RBC$	9. The SAS Postulate. (3-1)
10. $\overline{CQ} \cong \overline{CR}$	10. Definition of congruent triangles. (3-3)
11. $\overline{CO} \cong \overline{CO}$	11. Same reason as step 5.
12. $\triangle COQ \cong \triangle COR$	12. The SSS Postulate. (3-3)
13. $\angle COQ \cong \angle COR$	13. Definition of congruent triangles. (3-3)
14. l_3 is $\overleftrightarrow{CO}$ and $l_3 \perp \overleftrightarrow{QR}$	14. Definition of perpendicular lines. (1-25)
15. $\overleftrightarrow{QR} \perp$ plane P	15. Definition of a line perpendicular to a plane. (4-7)

Why does proving that $\overleftrightarrow{QR}$ is perpendicular to l_3 prove that it is perpendicular to all lines in plane P that pass through the point of intersection?

T **Theorem 4-5.2** A plane perpendicular to a line contains every line which is perpendicular to the line and which also intersects the line at the same point as the plane.

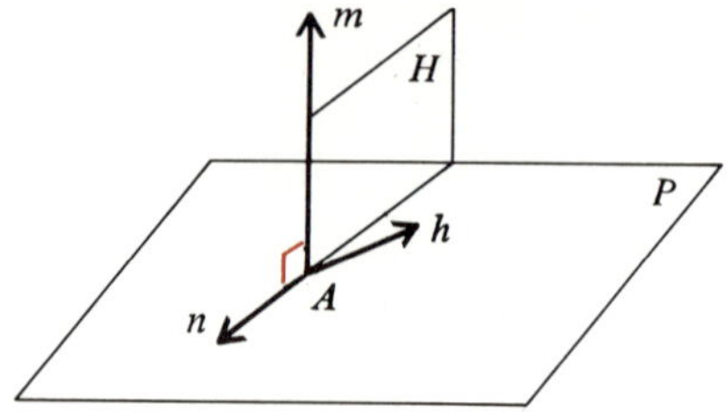

GIVEN $n \perp m$ at A;
plane $P \perp m$ at A.

PROVE Line n lies in plane P.

PROOF

STATEMENTS	REASONS
1. $n \perp m$ at A; plane $P \perp m$ at A.	1. Given.
2. Lines m and n determine plane H.	2. If two lines intersect, then there is exactly one plane containing them. (2-5.3)
3. Planes P and H intersect in a line; call it h.	3. Plane Intersection Postulate. (2-6)
4. $h \perp m$ at A	4. Definition of a line perpendicular to a plane. (4-7)
5. Line h must be the same as line n.	5. In a plane, there is one and only one line perpendicular to a given line through a given point of the line. (4-4.2)
6. Line n lies in plane P.	6. Substitution Postulate in step 3. (2-1)

Theorem 4-5.3 shows that when a line is perpendicular to a plane, the plane is in turn perpendicular to the line.

T **Theorem 4-5.3** Through a point in a given line, there passes one and only one plane perpendicular to the given line.

GIVEN Point A in line m

PROVE (1) There is at least one plane passing through A perpendicular to m.
(2) There is at most one plane passing through A perpendicular to m.

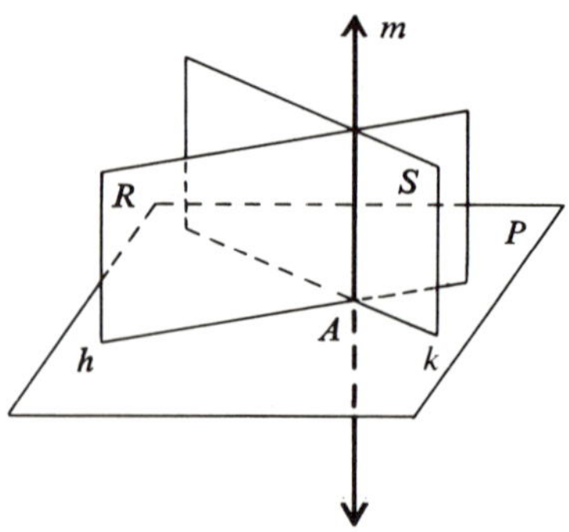

PROOF OF (1)

STATEMENTS	REASONS
1. Point A lies in line m.	1. Given.
2. Let S and R be two points not in m such that S, R, and A are noncollinear. S and m determine one plane; R and m determine a different plane.	2. If a point does not lie in a given line, then there is exactly one plane containing both the point and the line. (2-5.1)
3. Through point A, there is a line k in plane S such that $k \perp m$, and a line h in plane R such that $h \perp m$.	3. In a plane, there is one and only one line perpendicular to a given line through a given point of the line. (4-4.2)
4. Lines h and k determine plane P.	4. If two lines intersect , then there is exactly one plane containing them. (2-5.3).
5. Line $m \perp$ plane P at A.	5. Theorem 4-5.1.

PROOF OF (2)

STATEMENTS	REASONS
1. Plane $P \perp m$ at A.	1. Proved above.
2. Suppose that plane Q, distinct from plane P, is also perpendicular to m at A.	2. Assumption.
3. The intersection of two distinct planes is a unique line.	3. Plane Intersection Postulate. (2-6)
4. Both planes P and Q contain all lines perpendicular to m at the point of intersection of m at A.	4. Theorem 4-5.2.

Statement 3 contradicts statement 4. This means that our assumption is false. Thus, there is only one plane perpendicular to the line at a given point.

By interchanging the terms *line* and *plane* in the last theorem, we form Theorem 4-5.4.

Theorem 4-5.4 Through a point in a given plane there passes one and only one line perpendicular to the given plane.

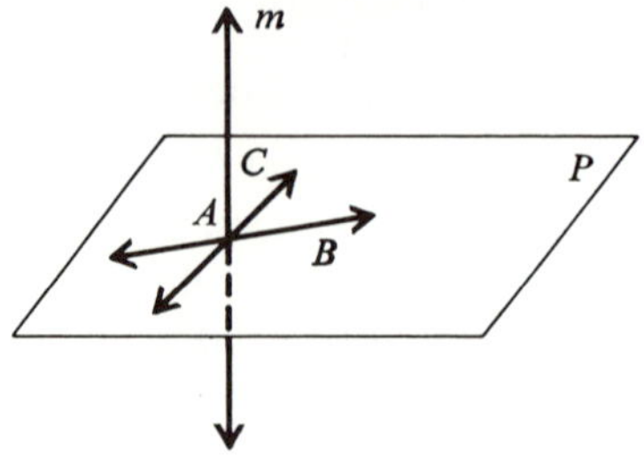

GIVEN Point A in plane P

PROVE (1) There is at least one line perpendicular to plane P at A.
(2) There is only one line perpendicular to plane P at A.

PROOF of (1)

STATEMENTS	REASONS
1. Point A is in plane P.	1. Given.
2. Let B and C be two other points in plane P such that A, B, and C are noncollinear.	2. The Plane Postulate. (2-5)
3. The points determine $\overleftrightarrow{AB}$ and $\overleftrightarrow{AC}$.	3. The Line Postulate. (2-3)
4. Let plane Q be the plane perpendicular to $\overleftrightarrow{AB}$ at A, with plane $R \perp \overleftrightarrow{AC}$ at A.	4. Theorem 4-5.3.
5. $\overleftrightarrow{AB}$ and $\overleftrightarrow{AC}$ are perpendicular to all lines of planes Q and R passing through A.	5. Definition of a line perpendicular to a plane. (4-7)
6. Plane Q intersects plane R in a line m through A such that $m \perp \overleftrightarrow{AB}$ and $m \perp \overleftrightarrow{AC}$.	6. Plane Intersection Postulate. (2-6)
7. $m \perp$ plane P.	7. Definition of a line perpendicular to a plane. (4-7)

PROOF OF (2)

STATEMENTS	REASONS
1. Suppose l_1 and l_2 are both perpendicular to plane P at A.	1. Assumption.
2. Lines l_1 and l_2 determine a plane Q.	2. If two lines intersect, then there is exactly one plane containing them. (2-5.3)
3. Planes Q and P intersect in line m.	3. Plane Intersection Postulate. (2-6)
4. $l_1 \perp m$ and $l_2 \perp m$.	4. Theorem 4-5.2.

Statement 4 contradicts the Perpendicular Uniqueness Theorem (4-4.2). Thus, our assumption is false. There is only one perpendicular to the given plane at A.

The two preceding theorems are incorporated in the more general theorems below.

T **Theorem 4-5.5** Through a given point there passes one and only one plane perpendicular to a given line.

T **Theorem 4-5.6** Through a given point there passes one and only one line perpendicular to a given plane.

Each theorem applies to two situations: the situation in which the given point lies *in* the given plane or line and the situation in which the point is *external* to the given plane or line. Because of the phrase "one and only one" in the theorems, there are two steps in the proof for each situation. Theorems 4-5.3 and 4-5.4 dealt with the cases in which a point is contained in the given plane or line. You will prove the external cases in the exercises.

The last theorem establishes a unique perpendicular from an external point to a given plane. This allows us to make the next definition.

D **Definition 4-8** The **distance to a plane from an external point** is the measure of the perpendicular segment from the point to the plane.

Class Exercises

Use the figure at right for the following exercises. In the figure, lines k, h, and l are in plane P and intersect at point M.

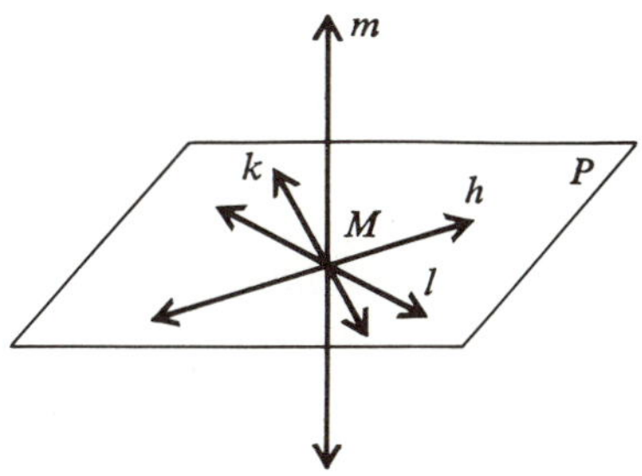

Exercises 1–6

1. If $h \perp l$, must l be perpendicular to k?
2. If $m \perp h$ and $m \perp k$, then $m \perp$ ____ and $m \perp$ ____.
3. Can m be perpendicular to h and k but not to l? Explain.
4. If $m \perp$ plane P at M, is l perpendicular to the plane determined by m and h?
5. If $m \perp$ plane P at M, is l perpendicular to the plane determined by m and k?
6. If $m \perp$ plane P at M, is l perpendicular to h?

Additional Properties of Perpendicularity

In a plane, every segment has a perpendicular bisector that is a line. In space, every segment has a perpendicular bisector that is a plane.

D **Definition 4-9** The **perpendicular bisecting plane** of a segment is the plane perpendicular to the segment at its midpoint.

The next theorem follows immediately from this definition. The exercises above supplied you with information that will help you prove the theorem as an exercise.

T **Theorem 4-5.7** The perpendicular bisecting plane of a segment is the set of all points equidistant from the endpoints of the segment.

We use Theorem 4-5.7 to prove the next theorem.

T **Theorem 4-5.8** Two lines perpendicular to the same plane are coplanar.

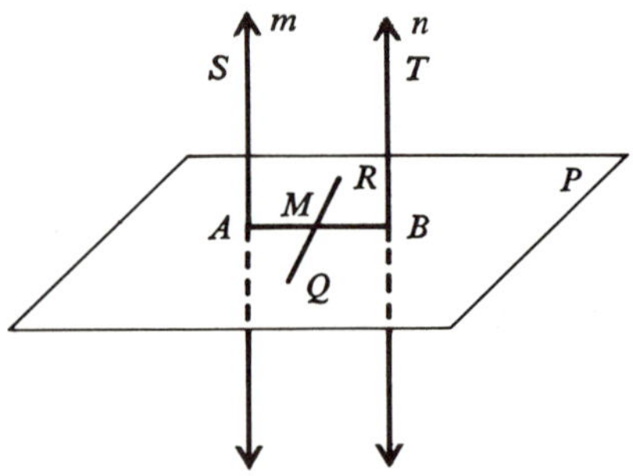

GIVEN Line $m \perp$ plane P at A.
Line $n \perp$ plane P at B.

PROVE Lines m and n are contained in one plane.

PROOF

STATEMENTS	REASONS
1. $m \perp$ plane P at A; $n \perp$ plane P at B	1. Given.
2. Let M be the midpoint of $\overline{AB}$, and in plane P, let $\overleftrightarrow{MR}$ be the perpendicular bisector of $\overline{AB}$.	2. Any line segment has exactly one midpoint (2-5.4). A line is the perpendicular bisector of a segment if and only if it is the set of all points in the plane equidistant from the endpoints of the segment. (4-4.3)
3. Let Q be the point of the opposite ray of $\overrightarrow{MR}$ such that $MQ = MR$.	3. The Point Uniqueness Postulate. (2-2)
4. $\overleftrightarrow{AB}$ is the perpendicular bisector of $\overline{QR}$.	4. Definition of perpendicular bisector. (4-4)

(continued)

STATEMENTS	REASONS
5. $AR = AQ$ and $BQ = BR$	5. A line is the perpendicular bisector of a segment if and only if it is the set of all points in the plane equidistant from the endpoints of the segment. (4-4.3).
6. Let S and T be any points on m and n distinct from A and B.	6. The Line Postulate. (2-3)
7. $\overline{AS} \cong \overline{AS}$ and $\overline{BT} \cong \overline{BT}$	7. Every segment is congruent to itself. (3-6.1)
8. $\triangle RAS \cong \triangle QAS$ and $\triangle RBT \cong \triangle QBT$	8. The SAS Postulate. (3-1)
9. $SR = SQ$ and $TR = TQ$	9. Definition of congruent triangles. (3-3)
10. Thus A and S lie in the perpendicular bisecting plane of $\overline{RQ}$, as do B and T.	10. Theorem 4-5.7.
11. Lines m and n lie in this plane.	11. The Line Postulate. (2-3)

From this theorem, we see that we have two planes perpendicular to each other.

D **Definition 4-10** **Two planes are perpendicular to each other** if and only if one plane contains a line perpendicular to the second plane.

You will need to use this definition when you prove the next theorem as an exercise.

T **Theorem 4-5.9** If a line is perpendicular to a plane, then every plane containing the line is perpendicular to the given plane.

What is the projection onto plane R of any point in m? Where do you think the projections of A and B lie in plane R? Is this true for all the points in $\overline{AB}$?

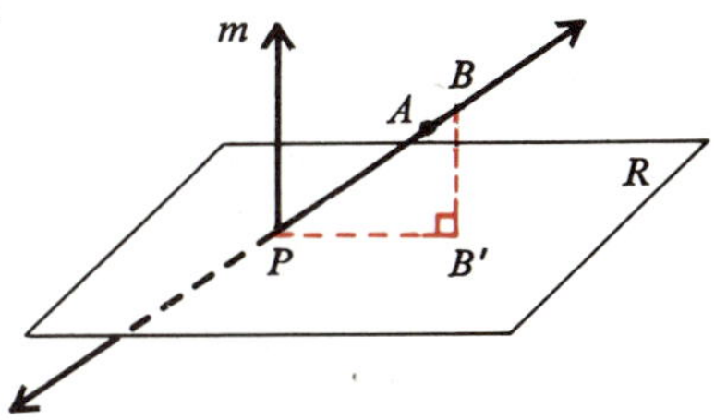

D **Definition 4-11** The **projection onto a plane of a segment** $\overline{AB}$ is the set of points in the plane which are the projections of the points of $\overline{AB}$.

Example How would you describe the projection of $\overline{PB}$ onto plane R in the figure above?

Solution P is its own projection. B' is the projection of B. Thus, we can say that $\overline{PB'}$ is the projection of $\overline{PB}$.

EXERCISES

A Complete each of the following:

1. If $m \perp$ plane P at M, m is ________ to all lines in P that pass through ____.
2. Every line perpendicular to m at M lies in one ________.
3. Through a given external point, there passes exactly one ________ perpendicular to a given line.
4. The distance to plane P from an external point H is the measure of the ________ from H to the plane.
5. Any plane containing line l is ________ to plane P if $l \perp$ plane P.
6. If n is the perpendicular bisector of $\overline{AB}$ in plane P, then the plane determined by n and l is the ________ of $\overline{AB}$.
7. The projection onto plane P of any point in l is ____.
8. If lines k and l are perpendicular to plane P, then l and k are ________.

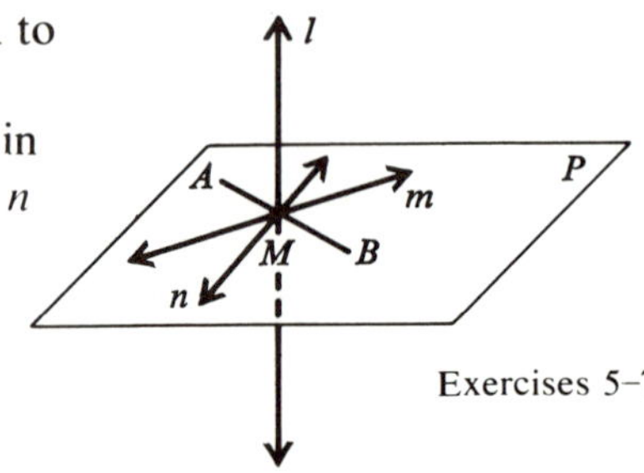

Exercises 5–7

Prove each of the following statements:

9. Through a point external to a line there passes at least one plane perpendicular to the line.
10. Through a point external to a line there passes only one plane perpendicular to the line.
11. Through a point external to a plane there passes at least one line perpendicular to the plane.
12. Through a point external to a plane there passes only one line perpendicular to the plane.
13. If two intersecting planes are perpendicular to a third plane, then their intersection is perpendicular to the third plane.
14. Two distinct lines perpendicular to the same plane never meet.
15. Prove Theorem 4-5.7.
16. Prove Theorem 4-5.9.
17. If $\overline{PM}$ is perpendicular to the plane containing circle M such that $MK = ML$, prove that P is equidistant from K and L.
18. S and T are points in plane P. R is an external point. If the projection of $\overline{RS}$ is congruent to the projection of $\overline{RT}$, prove $\overline{RS} \cong \overline{RT}$.

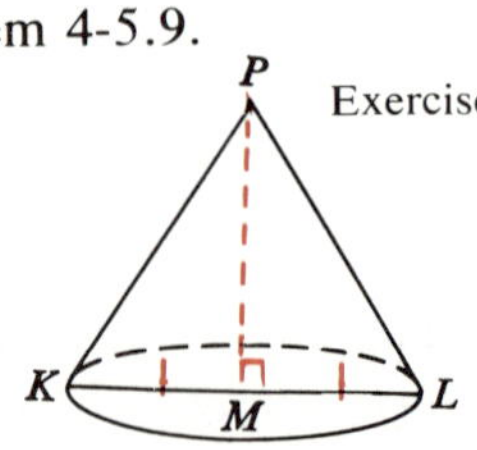

Exercise 17

B

In the figure at right, B, C, G, and F are the projections onto plane P of A, D, H, and E, respectively. All intersecting segments of the figure meet in right angles. Use this figure to answer the following:

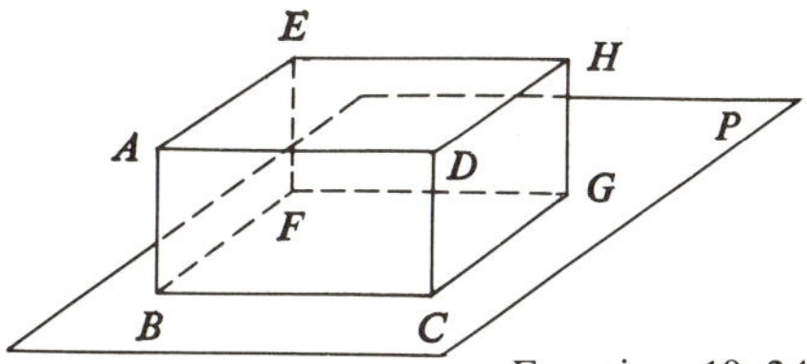

Exercises 19–24

19. Describe the projection of $\overline{EC}$ onto plane P.
20. Describe the projection of $\overline{EC}$ onto the plane through $EHGF$.
21. What is the measure of the projection onto plane P of D?
22. In $\triangle DCG$, $m\angle DCG = 90$ and $\overline{DG}$ is the hypotenuse. What is the projection of $\overline{DG}$ onto plane P?
23. What is the projection of $\overline{DG}$ onto the plane through $ABCD$?

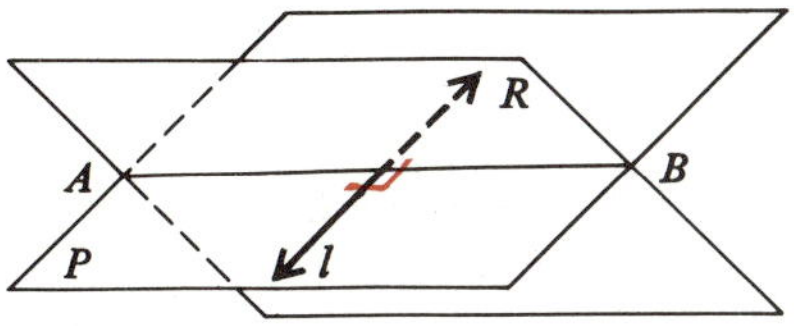

Exercise 24

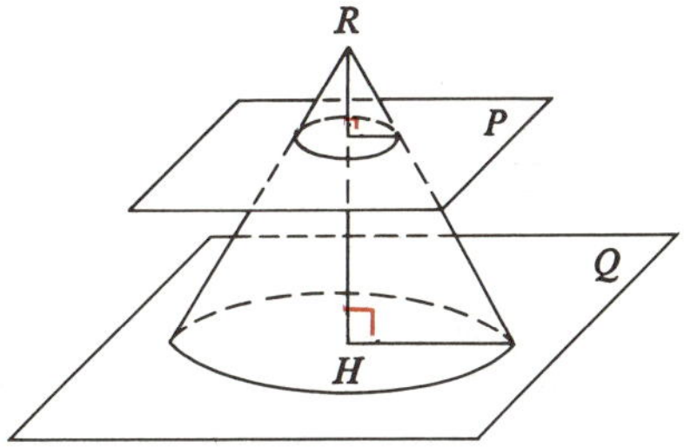

Exercise 25

24. Plane $P \perp$ plane R. If $l \perp \overline{AB}$, prove that $l \perp$ plane R.
25. If plane $P \perp \overline{RH}$ and plane $Q \perp \overline{RH}$, prove that they never intersect.

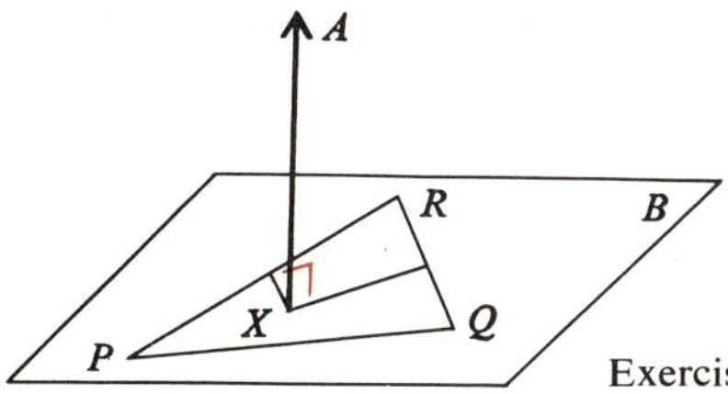

Exercise 26

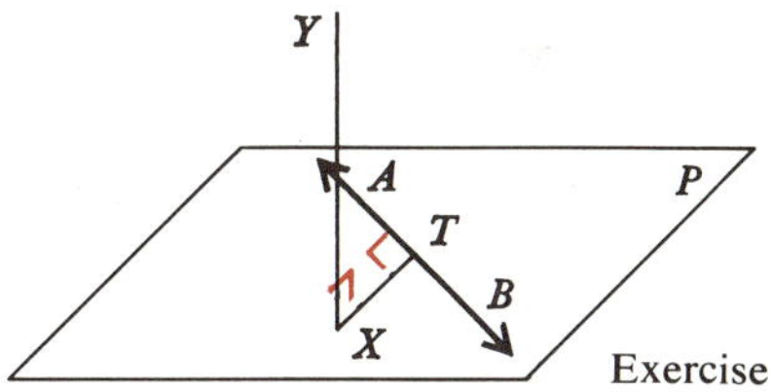

Exercise 27

C

26. $\triangle PQR$ lies in plane B. The perpendicular bisectors of $\overline{PR}$ and $\overline{RQ}$ meet at X. $\overline{AX}$ is perpendicular to plane B at X. Prove that $\overline{AP} \cong \overline{AQ} \cong \overline{AR}$.
27. $\overline{XY} \perp$ plane P and $\overline{XT} \perp \overleftrightarrow{AB}$. Prove that $\overline{YT} \perp \overleftrightarrow{AB}$.
28. If $\overline{RS} \perp \overline{AS}$ and $\overline{RS} \perp \overline{BS}$ with $\overline{AS}$ and $\overline{BS}$ in plane P such that $\angle ARS \cong \angle BRS$ prove that $\overline{RA} \cong \overline{RB}$.
29. If three lines are perpendicular to each other at their common point of intersection, prove that each line is perpendicular to the plane determined by the other two lines.

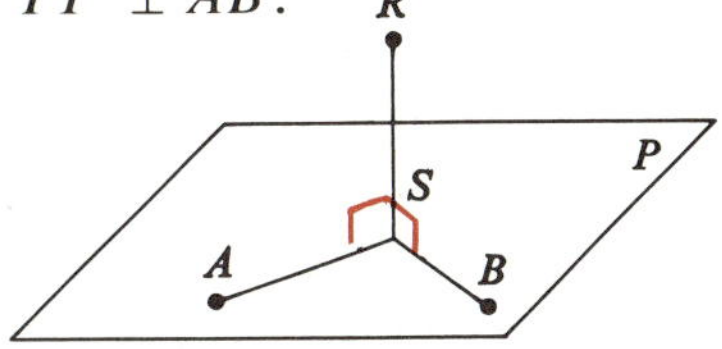

Exercise 28

SOMETHING TO THINK ABOUT

Circular Reasoning

We established undefined terms and postulates to avoid circular defining and reasoning. For example, we postulated three ways of proving triangles congruent. If we assume only the SAS Postulate, we can prove the ASA Postulate, see page 102. We can also prove the SSS Postulate.

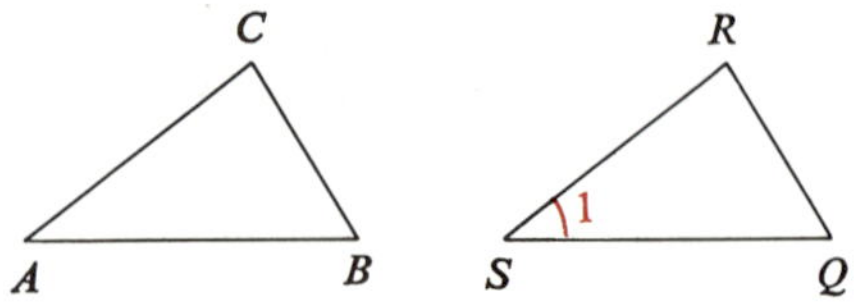

GIVEN $AB = SQ$, $BC = QR$, and $AC = SR$

PROVE $\triangle ABC \cong \triangle SQR$

PROOF Since $\overleftrightarrow{AB}$ separates the plane into half-planes P_1 and P_2, we can form in P_2 an angle congruent to $\angle 1$. Let M be the point on $\overrightarrow{AT}$ such that $AM = SR$. Thus $AM = AC$. Why? Since $AB = SQ$, $\triangle ABM \cong \triangle SQR$ by the SAS Postulate. Thus, $BM = QR = BC$.

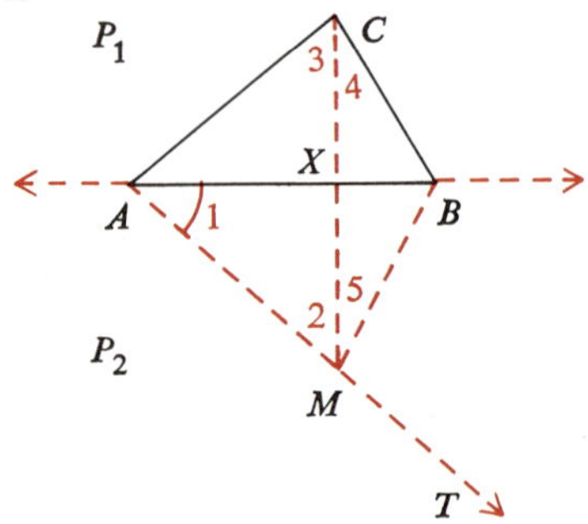

We know that $\overline{CM}$ intersects $\overline{AB}$ since C and M are in opposite half-planes. Let X be this point of intersection. $\triangle AMC$ and $\triangle CMB$ are both isosceles. Therefore, $m\angle 3 = m\angle 2$ and $m\angle 4 = m\angle 5$. By angle addition $\angle ACB \cong \angle AMB$. Using the SAS Postulate, we can conclude that $\triangle ABC \cong \triangle ABM$. The congruence relation for triangles is an equivalence relation; thus, $\triangle ABC \cong \triangle SRQ$.

Reexamine our first proof of the Isosceles Triangle Theorem (3-4.2) and our proof of the Angle Bisector Theorem (3-4.1). If we had not postulated SSS congruence and had to use what we gave as the alternate proof for the Isosceles Triangle Theorem, would we have had an example of circular reasoning?

4-6 DIHEDRAL ANGLES AND THEIR MEASURE

When two lines intersect, four angles are formed. Two pairs of angles are vertical angles. If the adjacent angles formed are right angles, then the lines are perpendicular. What analogy can you make for intersecting planes?

Example 1 Describe the intersection of two planes in terms of the angles formed.

Solution Two intersecting planes intersect in a line. Four dihedral angles are formed. Do you recall in Chapter 1 how we labeled them? The angles in the figure are $\angle A\text{-}BE\text{-}D$, $\angle A\text{-}BE\text{-}F$, $\angle C\text{-}BE\text{-}F$, and $\angle C\text{-}BE\text{-}D$. The pairs of vertical angles are $\angle A\text{-}BE\text{-}D$ and $\angle C\text{-}BE\text{-}F$, and $\angle A\text{-}BE\text{-}F$ and $\angle C\text{-}BE\text{-}D$. G is in the interior of $\angle C\text{-}BE\text{-}F$ while A is in the angle's exterior.

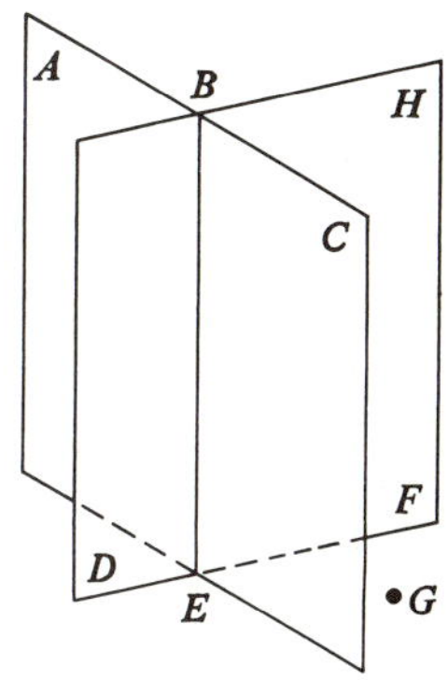

The figure also shows another angle, $\angle CBH$. Its vertex is in the edge $\overleftrightarrow{BE}$ of the dihedral angle, while its sides lie in the faces of the dihedral angle.

D **Definition 4-12** If a plane is perpendicular to the edge of a given dihedral angle, the intersection is called a **plane angle** of the dihedral angle.

If the measures of the plane angles $\angle ABC$ and $\angle EDF$ are equal, do you think all plane angles of a dihedral angle are congruent? The proof of the theorem for this idea uses properties of certain quadrilaterals. You can prove this theorem in Chapter 7 after you prove properties of quadrilaterals. However, we shall use it without proof in the meantime.

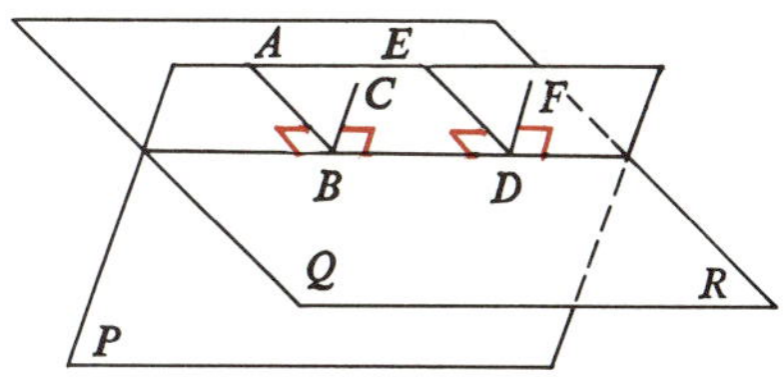

T **Theorem 4-6.1** All plane angles of a dihedral angle are congruent.

This theorem allows us to study other properties of a dihedral angle by studying any of its plane angles.

D **Definition 4-13** The **measure of a dihedral angle** is the same as the measure of any of its plane angles.

D **Definition 4-14** The **bisector of a dihedral angle** is the plane determined by the bisector of its plane angle and the **edge of the dihedral angle**.

A dihedral angle, like its plane angles, is classified as acute, right, or obtuse. Likewise, two dihedral angles are adjacent, complementary, supplementary, or congruent as their plane angles are adjacent, complementary, supplementary, or congruent.

The next two theorems follow from the above definitions:

T

Theorem 4-6.2 Two planes are perpendicular if they determine a right dihedral angle.

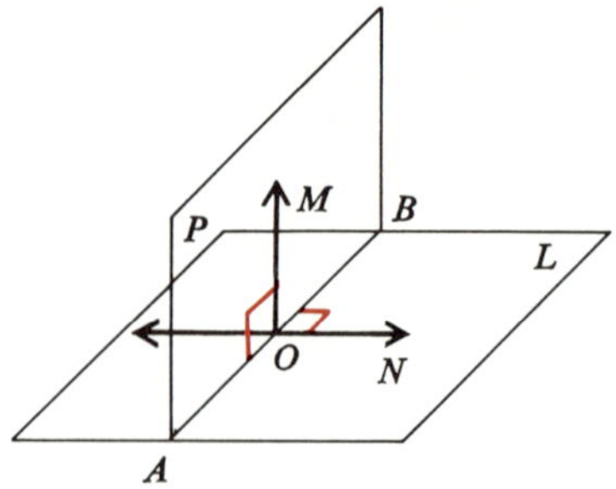

GIVEN Planes P and L intersect in $\overleftrightarrow{AB}$ such that $m\angle P\text{-}AB\text{-}L = 90$.

PROVE Plane $P \perp$ plane L

PROOF

STATEMENTS	REASONS
1. Planes P and L intersect in $\overleftrightarrow{AB}$ such that $m\angle P\text{-}AB\text{-}L = 90$.	1. Given.
2. Let $\angle MON$ be a plane angle of $\angle P\text{-}AB\text{-}L$ such that $\overleftrightarrow{MO} \perp \overleftrightarrow{AB}$.	2. Definition of plane angle. (4-12)
3. $m\angle MON = 90$	3. Definition of measure of dihedral angle. (4-13)
4. $\overleftrightarrow{MO} \perp \overleftrightarrow{ON}$	4. Two lines are perpendicular if and only if they form right angles. (4-4.1)
5. $\overleftrightarrow{MO} \perp$ plane L at O	5. If a line is perpendicular to each of two intersecting lines at their point of intersection, then the line is perpendicular to the plane determined by them. (4-5.1)
6. Plane $P \perp$ plane L	6. If a line is perpendicular to a plane, then every plane containing the line is perpendicular to the given plane. (4-5.9)

T

Theorem 4-6.3 If two planes are perpendicular, then in either plane any line perpendicular to their line of intersection is perpendicular to the second plane.

GIVEN Using the figure for the last theorem, plane $P \perp$ plane L at $\overleftrightarrow{AB}$ such that in plane P, $\overleftrightarrow{MO} \perp \overleftrightarrow{AB}$ at O.

PROVE $\overleftrightarrow{MO} \perp$ plane L at O.

PROOF Planes P and L are perpendicular at $\overleftrightarrow{AB}$ such that in plane P, $\overleftrightarrow{MO} \perp \overleftrightarrow{AB}$ at O. Since $m\angle P\text{-}AB\text{-}L = 90$, the measure of any plane angle is also 90. Let $\angle MON$ be one such plane angle through $\overleftrightarrow{MO}$. By definition, $\overleftrightarrow{MO} \perp \overleftrightarrow{ON}$. Since $\overleftrightarrow{MO}$ is perpendicular to two lines of plane L through O, $\overleftrightarrow{MO} \perp$ plane L.

EXERCISES

A Complete each of the following:

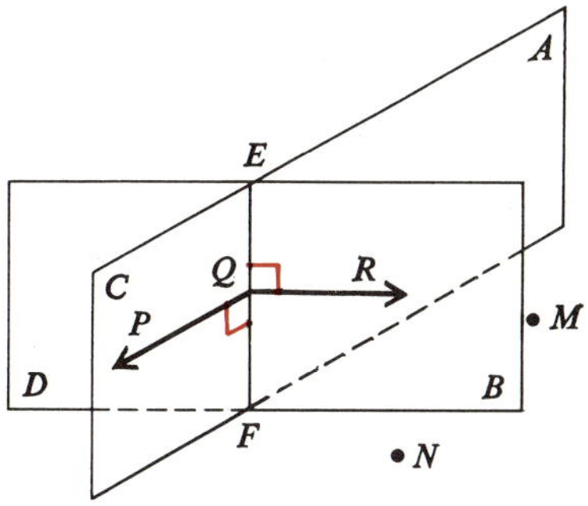

1. $\angle A\text{-}EF\text{-}B$ and ____ are vertical dihedral angles.
2. $m\angle PQR = m\angle$ ____ $\text{-}EF\text{-}$ ____.
3. N lies in the interior of $\angle$ ____.
4. If plane $A \perp$ plane B, line ____ $\perp$ plane A.
5. ____ is a supplementary dihedral angle to $\angle A\text{-}EF\text{-}B$.
6. The faces of $\angle C\text{-}EF\text{-}D$ are ____ and ____.

Define each of the following by analogy to definitions for plane angles:

7. vertical dihedral angles
8. supplementary dihedral angles
9. adjacent dihedral angles
10. right dihedral angle
11. acute dihedral angle
12. obtuse dihedral angle

B Prove each of the following:

13. If two planes intersect, the resultant vertical dihedral angles are congruent.
14. If two dihedral angles are supplementary and adjacent, their exterior faces are coplanar.
15. If two dihedral angles are complementary and adjacent, their exterior faces are perpendicular.

C We can classify angles by the number of their faces.

Dihedral: two faces

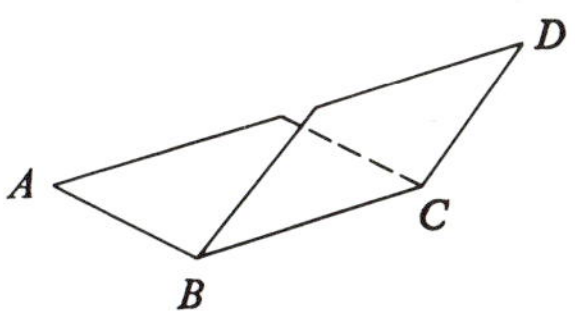

$\angle A\text{-}BC\text{-}D$

Trihedral: three faces

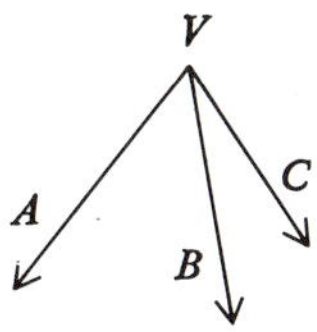

$\angle V\text{-}ABC$

Tetrahedral: four faces

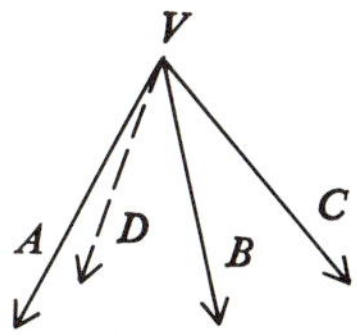

$\angle V\text{-}ABCD$

Copy the list below and fill in the names of the missing dihedral, trihedral, and tetrahedral angles for the figure at right.

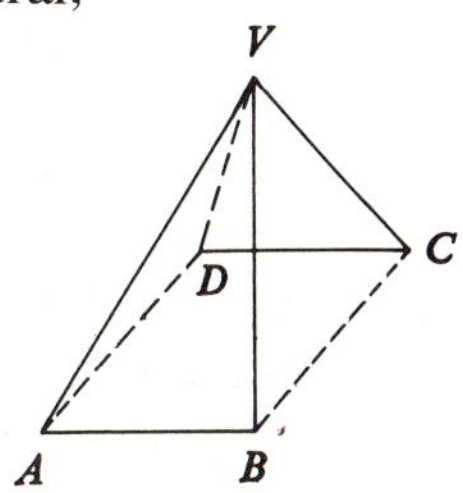

Dihedral	$\angle V\text{-}AB\text{-}C$	16. ______	17. ______	
	$\angle V\text{-}AD\text{-}C$	18. ______	19. ______	
	$\angle V\text{-}DC\text{-}B$			
	$\angle V\text{-}BC\text{-}D$			
Trihedral	$\angle A\text{-}BDV$	20. ____	21. ____	22. ____
Tetrahedral		23. ____		

HOW IS YOUR MATHEMATICAL VOCABULARY?

The following words and phrases were introduced in this chapter. How many do you know and understand?

analogy (4-3)
bisector of a dihedral angle (4-6)
contradiction (4-2)
contrapositive (4-1)
converse (4-1)
counterexample (4-2)
distance to a line from an external point (4-4)
distance to a plane from an external point (4-5)
edge of a dihedral angle (4-6)
enumeration (4-2)
equidistant (4-4)
external point (4-4)
foot of a perpendicular (4-4)
indirect proof (4-2)
inverse (4-1)
measure of a dihedral angle (4-6)
perpendicular bisecting plane (4-5)
perpendicular bisector (4-4)
perpendicular line and plane (4-5)
plane angle (4-6)
projection (4-4)
proof of existence (4-2)
proof by the contrapositive (4-2)
transitive property of implication (4-1)

REVIEW EXERCISES

4-1 Further implications

1. Show that $[(p \rightarrow q) \wedge p] \rightarrow q$ is valid.
2. Give an example of the transitive property of implication.

Write the converse, inverse, and contrapositive of each of the following:

3. Every segment has a unique midpoint.
4. If $\triangle ABC \cong \triangle XYZ$, then $\angle B \cong \angle Y$.

4-2 More on proof

Complete each of the following:

5. Enumeration requires a(n) ________ number of cases to be usable.
6. ________ follows from the Law of the Excluded Middle.
7. Proving the contrapositive is equivalent to proving the ________.
8. The first step in an indirect proof of a hypothesis is to assume the ________.

4-3 Mathematical methods

9. Describe the process of analogy.

Indicate the weaknesses of each of the following:

10. intuition
11. induction
12. process of analogy

Draw a model showing each of the following:

13. In $\triangle PQR$, $\overrightarrow{PQ}$ extended to S, $\overrightarrow{QR}$ to T, and $\overrightarrow{RP}$ to U.
14. The angle bisectors of scalene $\triangle KLM$.
15. The segments formed by the midpoints of the sides of right $\triangle ABC$.

4-4 Perpendiculars in a plane

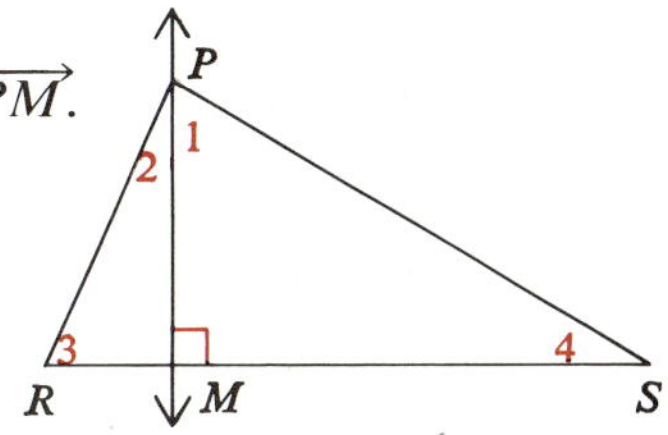

16. Explain why $\overline{PS}$ cannot be perpendicular to $\overleftrightarrow{PM}$.
17. If $RM = SM$, describe $\overleftrightarrow{PM}$.
18. Can $\overline{PS}$ be perpendicular to $\overline{PR}$? Explain.

Name the following projections:

19. P onto $\overline{RM}$
20. $\overline{RP}$ onto $\overline{RS}$
21. Name the foot of the perpendicular from R to $\overleftrightarrow{PM}$.
22. If two lines in a plane are perpendicular to line m, what conjecture can you make about the lines.

4-5 Perpendicular lines and planes

Complete each of the following:

23. If m is perpendicular at A to every line in plane E that passes through A, m is ________ to plane E at A.
24. Through a point external to a line, there passes a(n) ________ line perpendicular to the given line.
25. If $l \perp$ plane E at N, then every plane containing l is ________ to plane E.

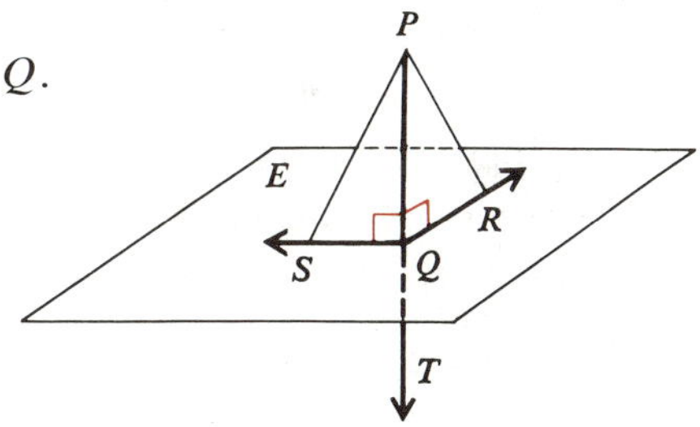

In the figure at right, $\overline{PQ} \perp$ plane E at Q.

26. Describe plane E if $PQ = TQ$.
27. If $PQ = TQ$, then $SP =$ ____.
28. Describe the projection of $\overline{PS}$ onto $\overline{PT}$.
29. Describe the projection of $\overline{PQ}$ onto $\overleftrightarrow{QR}$.

4-6 Dihedral angles and their measure

Complete each of the following:

30. A dihedral angle is formed by the union of a line and two noncoplanar ________ intersecting in the line.
31. The faces of the dihedral angle are the ________.
32. The intersection of the faces of the dihedral angle is the ________ of the angle.
33. The intersection of a dihedral angle and a plane perpendicular to the edge is ________.
34. A dihedral angle is measured by its ________.

CHAPTER TEST

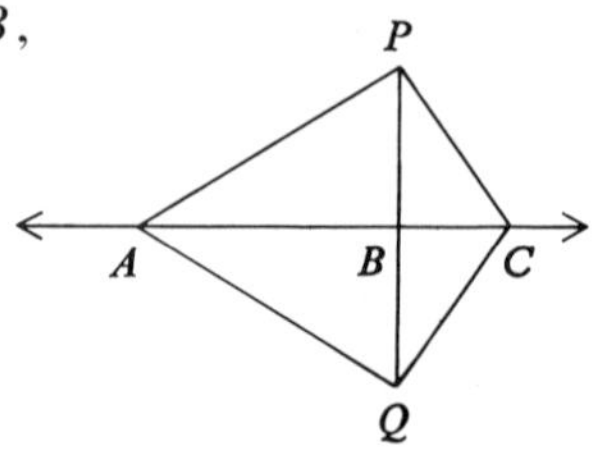

1. Write the contrapositive of the statement, "If $\angle A \cong \angle B$, then $BC = AC$."
2. Use an indirect proof to prove that if $\angle A \cong \angle B$, $\triangle ABC$ is not scalene.
3. Describe and evaluate the process of analogy.
4. Prove that if $\triangle PBC \cong \triangle QBC$, then $\angle APB \cong \angle AQB$.
5. If $\overline{PQ} \perp \overline{AC}$ in the figure, name the hypotenuse and legs in $\triangle ABP$.
6. Prove that if the diagonals of a quadrilateral are congruent and are the perpendicular bisectors of each other, then the quadrilateral has congruent sides and congruent angles.
7. Prove that if two planes are perpendicular to the same plane, then their intersection is also perpendicular to the plane.
8. Prove that vertical dihedral angles are congruent.
9. Prove that through a point external to a plane there passes one and only one line perpendicular to the plane.

SUGGESTED RESEARCH

1. Read about methods of proof in Carl B. Allendoerfer's article "Deductive Methods in Mathematics" in the Twenty-third Yearbook of the National Council of Teachers of Mathematics: *Insights into Modern Mathematics*. Prepare a paper for a class presentation.

2. Read "Proof," Chapter 4, of the Twenty-fourth Yearbook of the National Council of Teachers of Mathematics: *Growth of Mathematical Ideas*. Prepare a report on this article.

3. Read *Patterns of Plausible Inference*, Volume 2 of *Mathematics and Plausible Reasoning*, by George Polya (Princeton, N. J.: Princeton University Press, 1969) and prepare a report relating to the types of proofs discussed in this chapter.

4. Prepare a report on induction and analogy and how they are used in mathematics. Use *Induction and Analogy in Mathematics*, Volume 1 of *Mathematics and Plausible Reasoning*, by George Polya (Princeton, N. J.: Princeton University Press, 1954) for reference.

5 Geometric Inequalities

Up to this point, in studying pairs of angles and pairs of line segments we have been concerned with pairs with equal measure. In this chapter, we shall study conditions that enable us to conclude that two angles have different measures or that two segments have different lengths (measures): our concern will be with inequalities of measure.

5-1 PROPERTIES OF INEQUALITIES

An inequality between measures is essentially a comparison of numbers. We shall now review the basic properties of inequalities of numbers that we need for the study of geometric inequalities. After stating each property, we shall illustrate it by a geometric example.

Addition Property of Inequalities For all real numbers m, n, a, and b, if $m > n$ and $a \geq b$, then $m + a > n + b$.

Example 1 Point P is in the exterior of $\triangle ABC$, in the opposite half-plane of $\overleftrightarrow{BC}$ from A, so that $BP = CP$. Also, $m\angle ABC > m\angle ACB$. Prove that $m\angle ABP > m\angle ACP$.

Solution

GIVEN P is in the exterior of $\triangle ABC$, in the opposite half-plane of $\overleftrightarrow{BC}$ from A; $m\angle ABC > m\angle ACB$

PROVE $m\angle ABP > m\angle ACP$

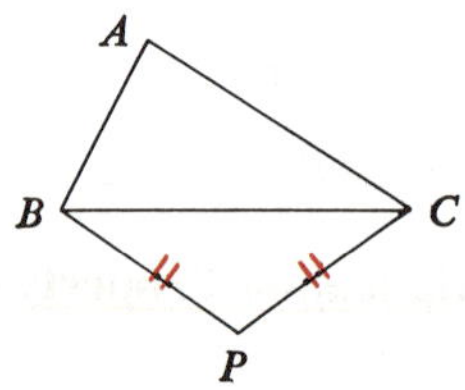

PROOF

STATEMENTS	REASONS
1. P is in the exterior of $\triangle ABC$, in the opposite half-plane of $\overleftrightarrow{BC}$ from A; $BP = CP$; $m\angle ABC > m\angle ACB$	1. Given.
2. $m\angle PBC = m\angle PCB$	2. The base angles of an isosceles triangle are congruent. (3-4.2)
3. $m\angle PBC + m\angle ABC > m\angle PCB + m\angle ACB$	3. Addition property of inequalities.
4. $m\angle ABP > m\angle ACP$	4. Angle Sum Postulate. (2-10)

Subtraction Property of Inequalities For all real numbers a, m, and n, if $m > n$, then (1) $m - a > n - a$ and (2) $a - m < a - n$.

Example 2 Point P is in the interior of $\triangle ABC$, so that $BP = CP$; $m\angle ABC > m\angle ACB$. Prove that $m\angle ABP > m\angle ACP$.

Solution

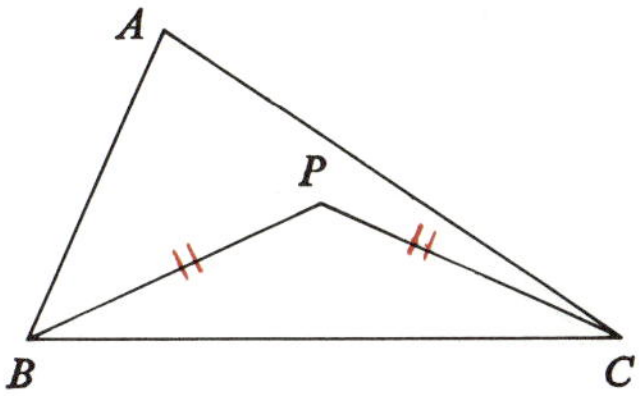

Given Point P is in the interior of $\triangle ABC$; $BP = CP$; $m\angle ABC > m\angle ACB$.

Prove $m\angle ABP > m\angle ACP$

Proof

STATEMENTS	REASONS
1. P is in the interior of $\triangle ABC$; $BP = CP$; $m\angle ABC > m\angle ACB$	1. Given.
2. $m\angle PBC = m\angle PCB$	2. The base angles of an isosceles triangle are congruent. (3-4.2)
3. $m\angle ABC - m\angle PBC >$ $m\angle ACB - m\angle PCB$	3. Subtraction property of inequalities.
4. $m\angle ABP > m\angle ACP$	4. Angle Sum Postulate. (3-10)

Multiplication Property of Inequalities For all real numbers a, m, and n, if $m > n$ and $a > 0$, then $am > an$.

Example 3 If D and E are the respective midpoints of sides $\overline{AB}$ and $\overline{AC}$ of $\triangle ABC$, and $AD < AE$, prove that $AB < AC$.

Solution

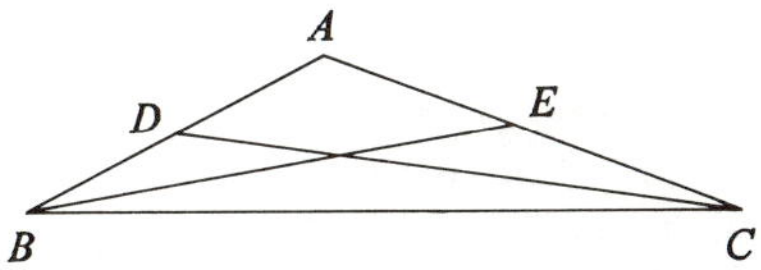

Given $\triangle ABC$ with D and E the respective midpoints of $\overline{AB}$ and $\overline{AC}$; $AD < AE$

Prove $AB < AC$

Proof

STATEMENTS	REASONS
1. $\triangle ABC$, with D and E the respective midpoints of $\overline{AB}$ and $\overline{AC}$; $AD < AE$	1. Given.
2. $2(AD) < 2(AE)$	2. Multiplication property of inequalities
3. $AB = 2(AD)$; $AC = 2(AE)$	3. Definition of a midpoint (1-15)
4. $AB < AC$	4. Substitution Postulate (2-1)

Division Property of Inequalities For all real numbers a, m, and n, if $m > n$ and $a > 0$, then (1) $\frac{m}{a} > \frac{n}{a}$ and (2) $\frac{a}{m} < \frac{a}{n}$.

Example 4

GIVEN $DB > AC$; $\overline{AC}$ and $\overline{BD}$ bisect each other at M.

PROVE $DM > AM$

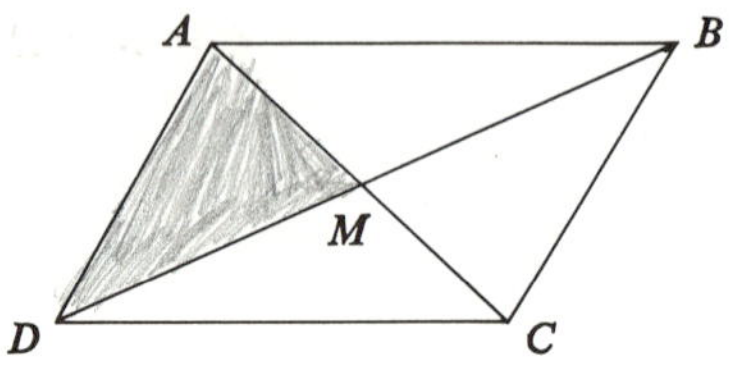

Solution

PROOF

STATEMENTS	REASONS
1. $DB > AC$; $\overline{AC}$ and $\overline{BD}$ bisect each other at M.	1. Given
2. $\frac{DB}{2} > \frac{AC}{2}$	2. Division property of inequalities
3. $DM = \frac{DB}{2}$; $AM = \frac{AC}{2}$	3. Definition of midpoint (1-15)
4. $DM > AM$	4. Substitution Postulate. (2-1)

Trichotomy Property For all real numbers m and n, either $m > n$, $m = n$, or $m < n$.

Example 5 Suppose we measured the sides of $\triangle ABC$. What relationships might exist between the lengths of $\overline{AB}$ and $\overline{BC}$?

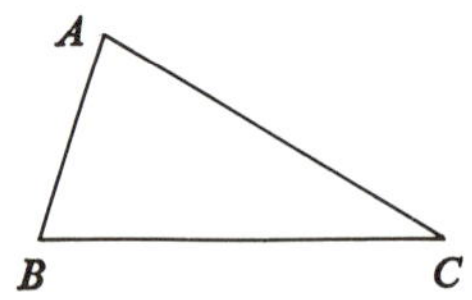

Solution By the trichotomy property we could have one and only one of the following:

$AB > BC$ $\quad$ $AB = BC$ $\quad$ or $\quad$ $AB < BC$

Transitive Property of Inequalities For all real numbers k, m, and n, if $k > m$ and $m > n$, then $k > n$.

Example 6 In $\triangle ABC$ above, if $AC > BC$ and $BC > AB$, compare the lengths of $\overline{AC}$ and $\overline{AB}$.

Solution By the transitive property of inequalities, we have $AC > AB$.

Class Exercises

State which property justifies each of the following:

1. If $3 > 2$, then $18 > 12$.
2. If $19 > 5$, then $17 > 3$.
3. If $x < 13$ and $y > 13$, then $y > x$.
4. If $AB > CD$, then $AB \neq CD$.
5. If $m\angle A > m\angle B$ and $m\angle C < m\angle B$, then $m\angle A > m\angle C$.
6. If $AB > PR$ and $CD > MN$, then $AB + CD > PR + MN$.
7. If $m\angle A \nless m\angle B$ and $m\angle A \neq m\angle B$, then $m\angle A > m\angle B$.
8. If $PQ > RS$, then $\frac{PQ}{2} > \frac{RS}{2}$.
9. If $m\angle A > m\angle B$, then $m\angle A - m\angle C > m\angle B - m\angle C$.
10. If $MN < GH$, then $2(MN) < 2(GH)$.

Now that you can apply the properties of inequalities to geometric relations, we shall prove our first theorem dealing with inequalities.

T **Theorem 5-1.1** For any numbers k, m, and n, if $n = m + k$, and $k > 0$, then $n > m$.

GIVEN $n = m + k$; $k > 0$

PROVE $n > m$

PROOF

STATEMENTS	REASONS
1. $n = m + k$; $k > 0$	1. Given
2. $n - m = k$	2. Subtraction property of equality
3. $n - m > 0$	3. Substitution Postulate (2-1)
4. $n - m + m > 0 + m$	4. Addition property of inequalities
5. $n > m$	5. Additive inverse and identity properties

As it is stated, Theorem 5-1.1 involves only numbers (measures). Restate the theorem using angles or line segments.

C **Corollary 5-1.1a** If P is a point of $\overline{AB}$ between A and B, then $AB > AP$ and $AB > BP$.

C **Corollary 5-1.1b** If P is a point in the interior of $\angle ABC$, then $m\angle ABC > m\angle ABP$ and $m\angle ABC > m\angle CBP$.

The proofs of these corollaries follow directly from Theorem 5-1.1. These proofs are left as exercises.

EXERCISES

A

State the property, theorem, or corollary which justifies each of the following:

1. If $AB = RS + PT$, then $AB > RS$.
2. If $MN > EF$, then $2MN > 2EF$.
3. If $PQ \nless NK$ and $PQ \neq NK$, then $PQ > NK$.
4. If $AB < CD$, then $\frac{AB}{3} < \frac{CD}{3}$.
5. If $m\angle A > m\angle B$ and $m\angle B > m\angle D$, then $m\angle A > m\angle D$.
6. If $m\angle P < m\angle R$, then $m\angle P + m\angle A < m\angle R + m\angle A$.
7. If $AB > PR$ and $CD = GH$, then $AB - CD > PR - GH$.
8. If $MN > RS$ and $PQ < RS$, then $MN > PQ$.

Supply each missing symbol. Justify your answers.

9. If $m\angle P > m\angle R$ and $m\angle S < m\angle R$, then $m\angle S$ ____ $m\angle P$.
10. If $AB > CD$ and $BP > DR$, then $AB + BP$ ____ $CD + DR$.
11. If $m\angle A \ngtr m\angle B$ and $m\angle A \neq m\angle B$, then $m\angle A$ ____ $m\angle B$.
12. If $HP < AP$, then $HP - NP$ ____ $AP - NP$.

Exercises 13–14

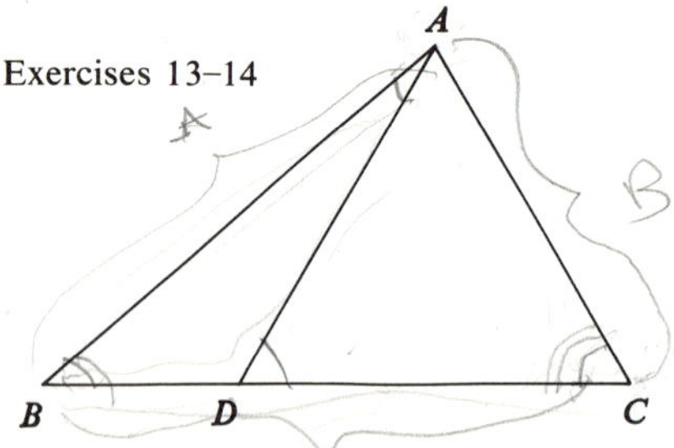

Exercises 15–16

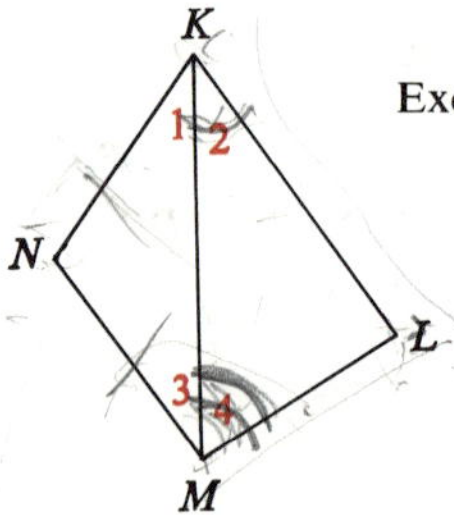

13. If $AB + AC > BC$ and $\angle ADC \cong \angle C$, then BC ____ $AB + AD$.
14. If $m\angle C > m\angle B$ and $m\angle BAD < m\angle B$, then $m\angle C$ ____ $m\angle BAD$.

B

15. If $KN = MN$ and $m\angle 2 < m\angle 4$, then $m\angle NML$ ____ $m\angle NKL$.
16. If $m\angle NKL = 2m\angle 2$, $m\angle NML = 2m\angle 4$, and $m\angle 4 > m\angle 2$, then $m\angle NKL$ ____ $m\angle NML$.
17. If $m\angle A > m\angle B$, what is the relationship between their complements?
18. If $m\angle A < m\angle B$, what is the relationship between their supplements?
19. Using the transitive property of inequalities, prove that any negative number is less than any positive number.
20. Prove that the supplement of an acute angle is an obtuse angle.
21. Prove Corollary 5-1.1a.
22. Prove Corollary 5-1.1b.

C

23. GIVEN Diagonals of quadrilateral $ABCD$ meet at E; $BE > EC$; $EC < DE$; E is the midpoint of $\overline{AC}$

    PROVE $BD > AC$

24. **Given** $\triangle ABC$ is isosceles, with $AB = AC$; $\overline{BDC}$; $m\angle DAC > m\angle C$
Prove $m\angle BAC > m\angle B$

25. **Given** $\angle 1 \cong \angle 2$; $\angle 3 \cong \angle 4$; $AB < AC$; $\overline{AEB}$; $\overline{AFC}$
Prove $EB < FC$

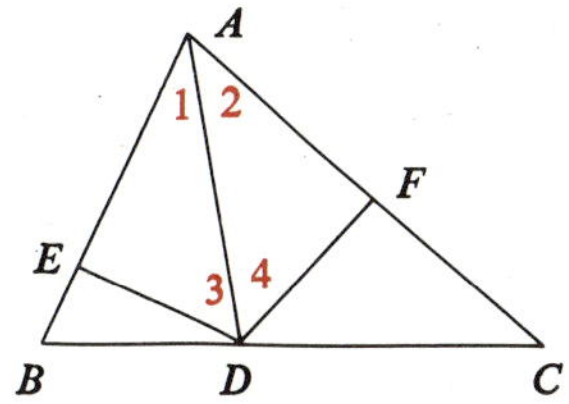

26. **Given** $\overline{AH}$ is the perpendicular bisector of $\overline{BD}$; $m\angle ADB > m\angle C$
Prove $m\angle C < m\angle B$

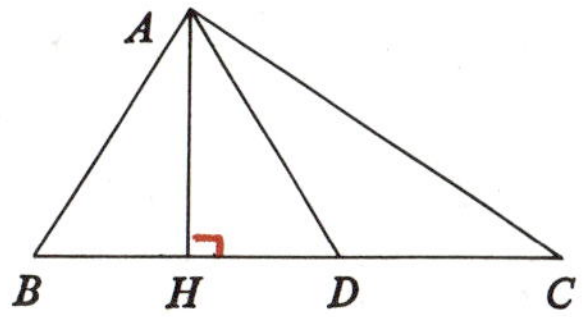

27. **Given** $\triangle ABC$; D is the midpoint of $\overline{AC}$ and $\overline{BP}$; P is in the interior of $\angle ACR$.
Prove $m\angle ACR > m\angle A$

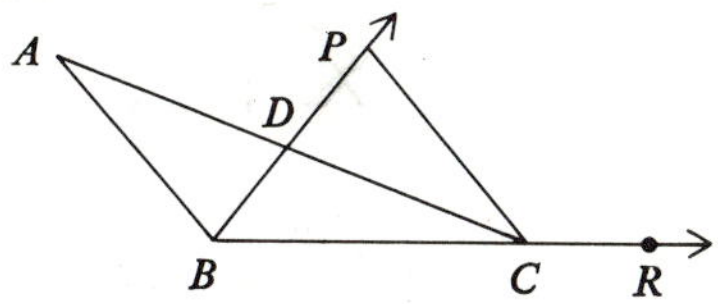

28. **Given** $\angle 1 \cong \angle 2$; $\angle 3 \cong \angle 4$; $\overline{GD} \cong \overline{HD}$; $EH = CG$; $\overline{BHD}$; $\overline{EGHC}$; $\overline{APGD}$
Prove $m\angle 5 < m\angle 6$

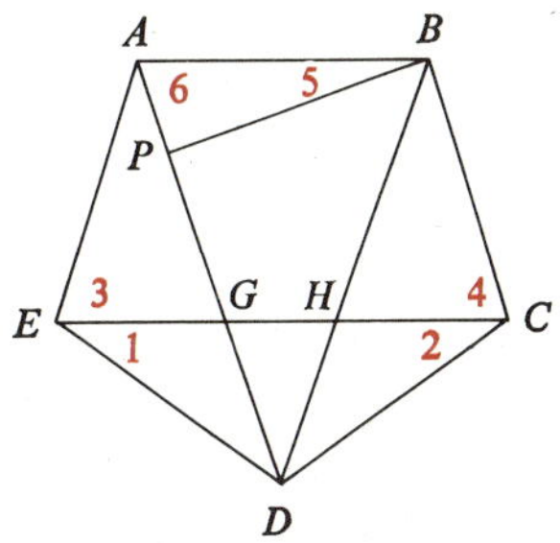

5-2 THE EXTERIOR-ANGLE THEOREM

The three angles of a triangle are sometimes referred to as the **interior angles of a triangle**. What do you think an exterior angle of a triangle is?

In each of the diagrams below, $\angle 1$ is called an exterior angle of $\triangle ABC$.

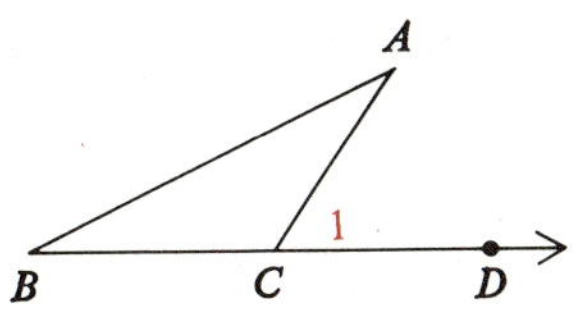

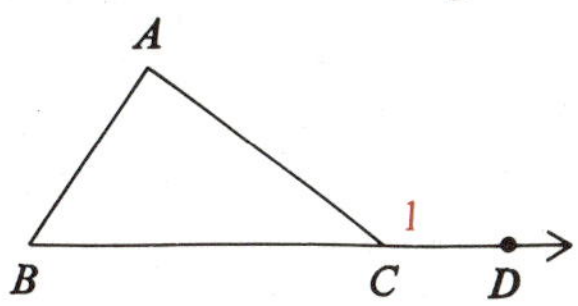

D **Definition 5-1** An **exterior angle of a triangle** is an angle that forms a linear pair with one of the interior angles of the triangle.

At each vertex of a triangle, there is a pair of exterior angles. In the diagram at left below, $\angle 1$ and $\angle 2$ form one such pair, $\angle 3$ and $\angle 4$ form another pair, and $\angle 5$ and $\angle 6$ form a third pair. What do you think is the relationship between the two exterior angles at each vertex of a triangle? Why?

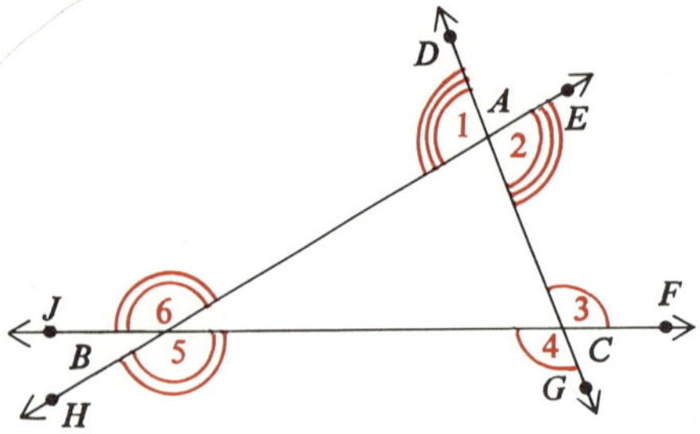

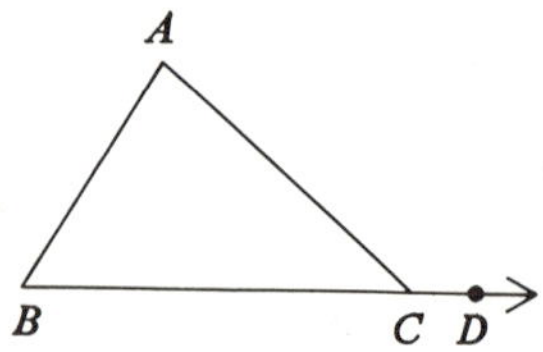

Before we are ready to state a theorem about the exterior angles of a triangle, we must define one more angle of a triangle. Consider $\angle ACD$ at right above, the exterior angle of $\triangle ABC$ at vertex C. In $\triangle ABC$, $\angle A$ and $\angle B$ are the remote interior angles or the nonadjacent angles of exterior $\angle ACD$.

Definition 5-2 In a triangle, the two interior angles which do not form a linear pair with an exterior angle are the **remote interior angles** of that exterior angle.

Class Exercises

Using this diagram, name the remote interior angles of each of the exterior angles listed below.

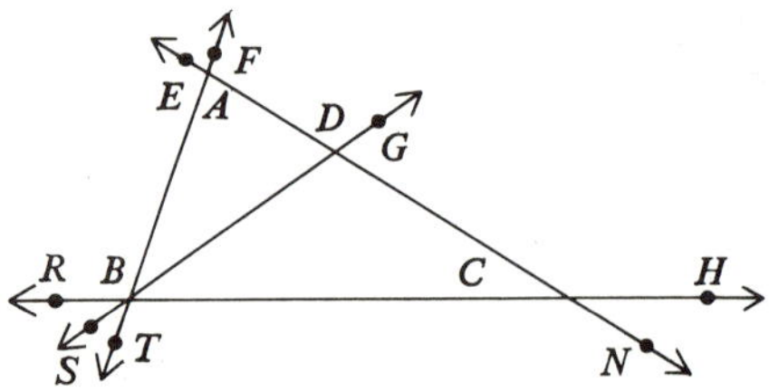

1. $\angle EAB$ of $\triangle ABC$
2. $\angle FAD$ of $\triangle ADB$
3. $\angle GDC$ of $\triangle BDC$
4. $\angle RBA$ of $\triangle ABC$
5. $\angle BCN$ of $\triangle BCD$
6. $\angle TBD$ of $\triangle ABD$
7. $\angle HCA$ of $\triangle ABC$
8. $\angle EAB$ of $\triangle ABD$
9. $\angle SBC$ of $\triangle BDC$

Theorem 5-2.1 The Exterior-Angle Theorem The measure of an exterior angle of a triangle is greater than the measure of either remote interior angle.

GIVEN $\triangle ABC$; $\overrightarrow{BCD}$

PROVE $m\angle ACD > m\angle A$ and
$m\angle ACD > m\angle ABC$

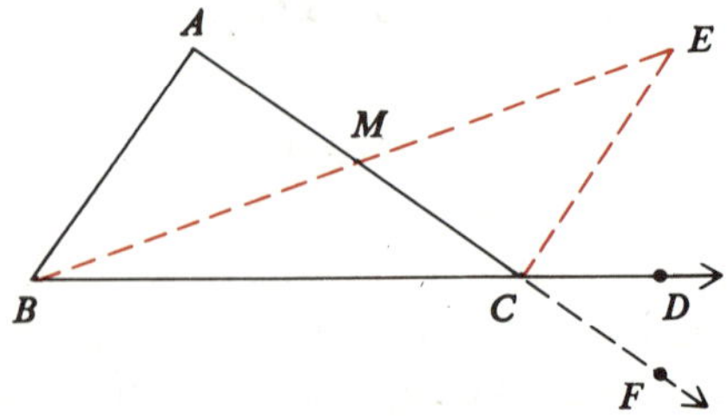

Proof

STATEMENTS	REASONS
1. $\triangle ABC$ and $\overrightarrow{BCD}$	1. Given.
2. Let M be the midpoint of $\overline{AC}$.	2. Any line segment has exactly one midpoint. (2-5.4)
3. $\overline{AM} \cong \overline{CM}$	3. Definition of a midpoint. (1-15)
4. Select the point E on $\overrightarrow{BM}$ so that $\overline{BM} \cong \overline{EM}$.	4. The Point Uniqueness Postulate. (2-2)
5. $\angle AMB \cong \angle CME$	5. Vertical angles are congruent. (3-1.5)
6. $\triangle AMB \cong \triangle CME$	6. The SAS Postulate. (3-1)
7. $m\angle A = m\angle MCE$	7. Definition of congruent triangles. (3-3)
8. $m\angle ACD = m\angle ACE + m\angle ECD$	8. Angle Sum Postulate. (2-10)
9. $m\angle ACD = m\angle A + m\angle ECD$	9. Substitution Postulate. (2-1)
10. $m\angle ACD > m\angle A$	10. For any numbers k, m, and n, if $n = m + k$, and $k > 0$, then $n > m$. (5-1.1)

By similar reasoning, using the median from A, we can conclude that $m\angle ACD > m\angle ABC$.

Class Exercises

In the diagram at right, $\triangle ABC$ is a right triangle with $m\angle ACD = 90$.

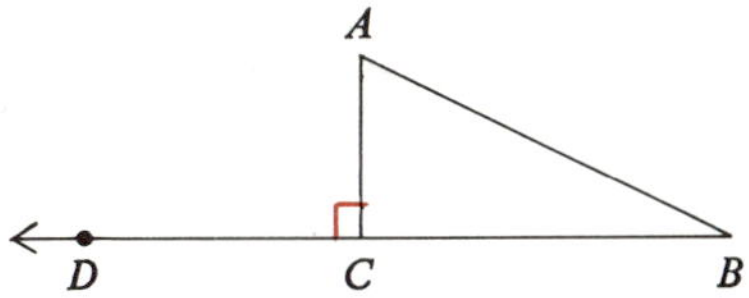

1. Name two angles whose measures are less than $m\angle ACD$.
2. State a theorem which justifies your answer to Exercise 1.
3. Why are $m\angle A$ and $m\angle B$ less than 90?
4. Define an acute angle.
5. What type of angles are $\angle A$ and $\angle B$? Why?

These exercises lead us to the following corollary:

C

Corollary 5-2.1a If a triangle has one right angle, then the other two angles must be acute.

If we had been able to use the Exterior-Angle Theorem (5-2.1) in Chapter 4, we would have been able to prove more easily that the perpendicular to a line from an external point is unique. The proof using Theorem 5-2.1 is as follows:

Suppose the perpendicular from point P to $\overleftrightarrow{AB}$ is *not* unique. We may then assume that there are two such perpendiculars. Let $\overleftrightarrow{PA} \perp \overleftrightarrow{AB}$ and $\overleftrightarrow{PB} \perp \overleftrightarrow{AB}$. Explain why $\angle x \cong \angle PAB$.

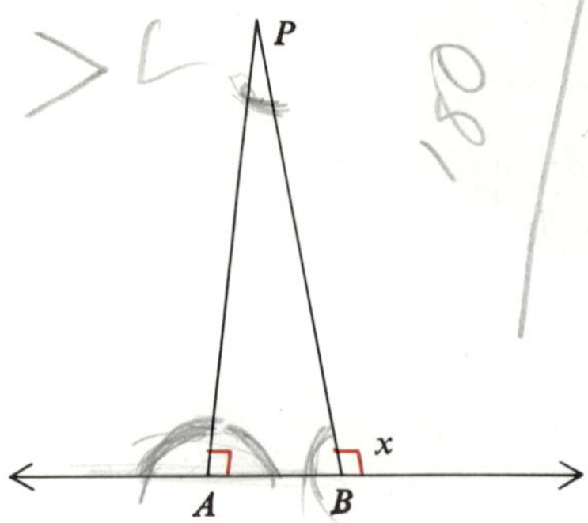

Is this possible? According to Theorem 5-2.1, what relationship exists between $m\angle x$, the exterior angle of $\triangle APB$, and $m\angle PAB$? What conclusion can you draw from this argument?

EXERCISES

A Refer to the diagram at right for Exercises 1-8.

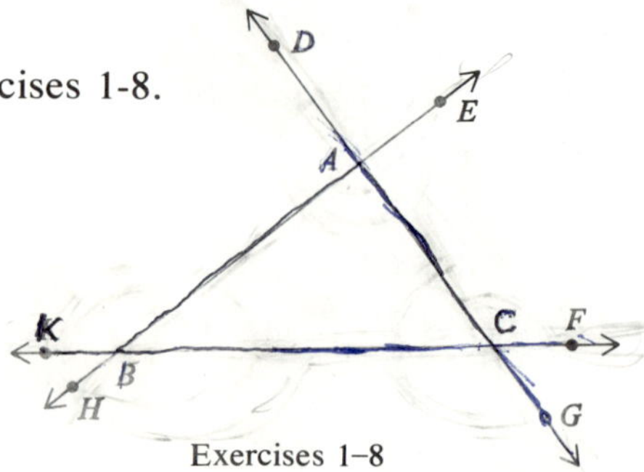

Exercises 1–8

1. What exterior angles of $\triangle ABC$ have as a remote interior angle $\angle BAC$?
2. Answer Exercise 1 for remote interior $\angle ABC$.
3. Name the remote interior angles of $\triangle ABC$ for exterior $\angle BCG$.
4. What is the relationship between $m\angle DAB$ and $m\angle ACB$?
5. What is the relationship between $m\angle ABK$ and $m\angle HBC$?
6. What is the relationship between $m\angle FCA$ and $m\angle ACB$?
7. By Theorem 5-2.1, $m\angle GCB$ ____ $m\angle$____ and $m\angle GCB$ ____ $m\angle$____.
8. By Theorem 5-2.1, $m\angle EAC$ ____ $m\angle$____ and $m\angle EAC$ ____ $m\angle$____.

Refer to the diagram at right for Exercises 9-15.

9. Which angles have measures less than $m\angle 7$?
10. Which angles have measures less than $m\angle 5$?
11. Which angles have measures greater than $m\angle 10$?
12. Which angles have measures greater than $m\angle 1$?
13. Name an angle which is supplementary to $\angle 6$.
14. Which angle is congruent to $\angle 16$?
15. Explain in two ways why $m\angle ABJ > m\angle 8$.

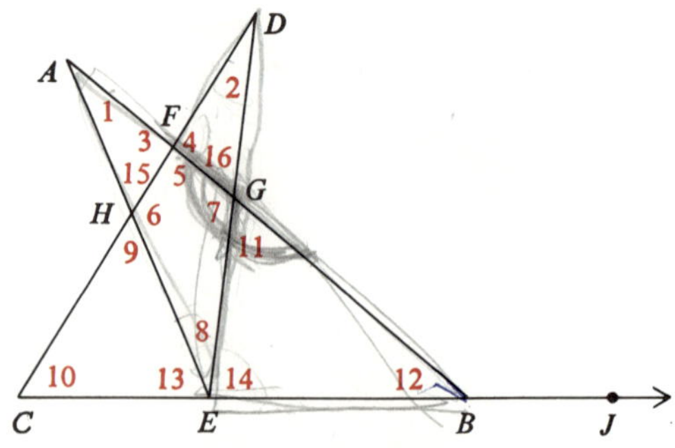

Exercises 9–15

B

16. Prove that if a triangle contains an obtuse angle, then the other angles are acute.

17. **Given** $\triangle ABC$; $\overline{ADC}$; $\overrightarrow{BCE}$
    Prove $m\angle ACE > m\angle ABD$

18. **Given** $\triangle ABC$; $\overline{BDC}$
    Prove $m\angle ADB \neq m\angle ACB$

19. **Given** $\triangle ABC$; $\overline{ADB}$; $m\angle ACD > m\angle ADC$
    Prove $m\angle ACD > m\angle B$

20. Prove that the sum of the measures of any two angles of a triangle is less than 180.

21. Prove that the base angles of any isosceles triangle are acute.

C

22. Prove that the sum of the measures of the angles of any triangle is less than 270.

23. Prove that the measure of an exterior angle of a quadrilateral is not always greater than the measure of one of its remote interior angles.

24. **Given** $\triangle ABC$; $\overline{ADC}$; $\overline{AD} \cong \overline{AB}$
    Prove $m\angle ABC > m\angle C$

25. **Given** $\overrightarrow{CD}$ bisects $\angle ACB$.
    Prove $m\angle 3 > m\angle 2$

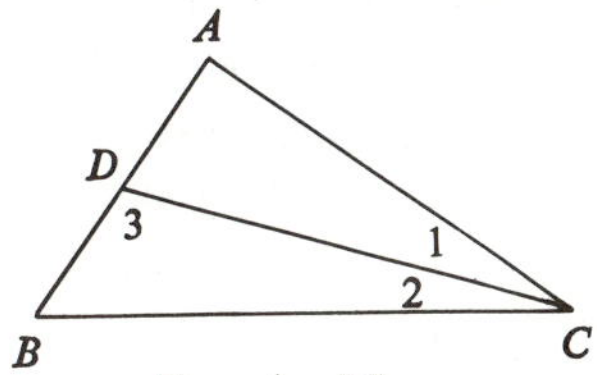

Exercise 25

26. **Given** Point D is in the interior of $\triangle ABC$.
    Prove $m\angle BDC > m\angle A$
    (*Hint:* Consider $\overrightarrow{BD}$.)

27. **Given** Point D is in the interior of $\triangle ABC$.
    Prove $m\angle BDA + m\angle ADC + m\angle BDC$ is greater than the sum of the measures of the angles of $\triangle ABC$.

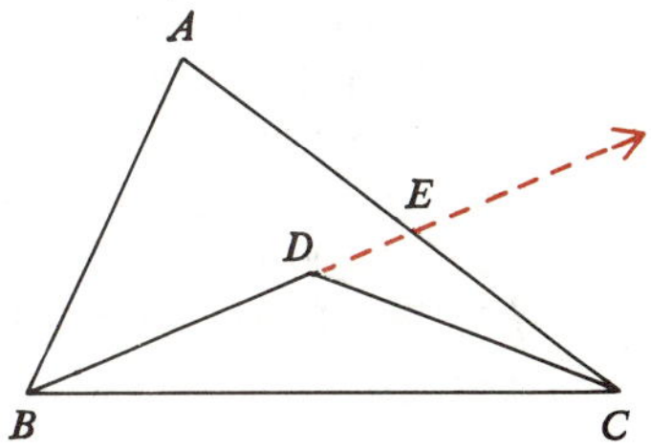

28. **Given** Quadrilateral $ABCD$; $\overrightarrow{ADF}$; $\overrightarrow{ABE}$
    Prove $m\angle EBC + m\angle FDC > \frac{1}{2}(m\angle A + m\angle C)$
    (*Hint:* Draw $\overline{AC}$.)

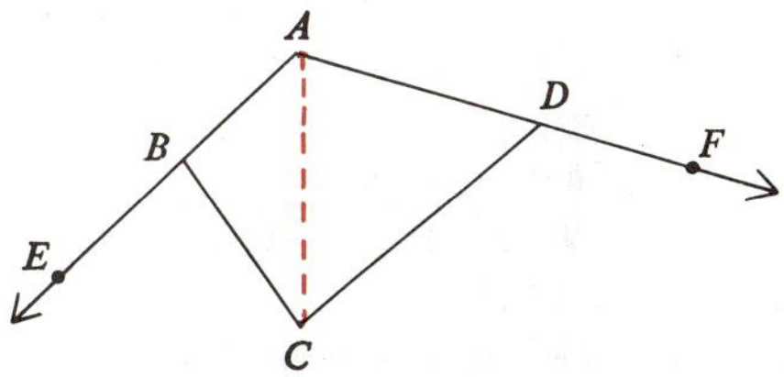

29. Using an indirect proof, prove that an altitude drawn from the vertex of an acute angle of an obtuse triangle is in the exterior of the triangle.

5-3 INEQUALITIES IN ONE TRIANGLE

In Chapter 3 we studied the theorem which states that if two sides of a triangle are congruent, then the angles opposite those sides are congruent. We also studied the converse of this theorem, which states that if two angles are congruent, then the sides opposite those angles are congruent. In this section, we shall consider theorems about triangles with two sides or angles which are not congruent.

In the diagrams below, $m\angle A > m\angle B$. By observation, how do you think the length of $\overline{BC}$ compares with the length of $\overline{AC}$?

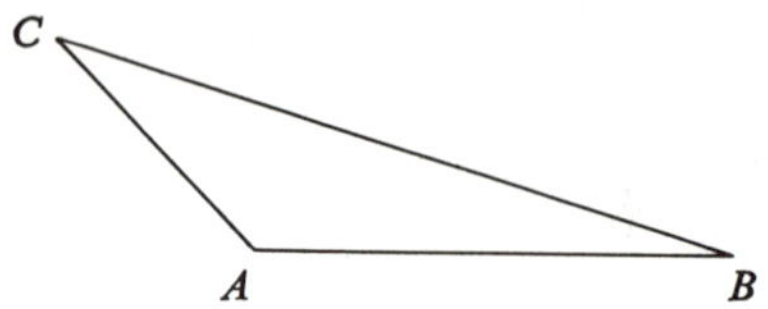

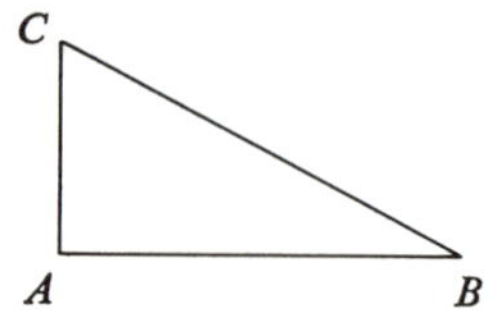

If $m\angle C < m\angle A$, which would you assume is greater, AB or BC? Can you make a reasonable conjecture regarding noncongruent sides and angles of a triangle?

T

Theorem 5-3.1 If two sides of a triangle are not congruent, then the angles opposite those sides are not congruent, the angle with the greater measure being opposite the longer side.

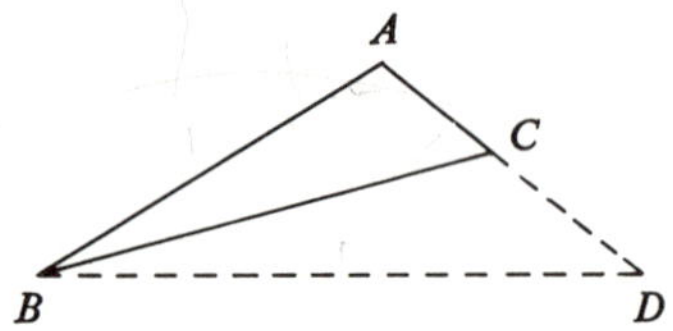

GIVEN $\triangle ABC$; $AB > AC$

PROVE $m\angle ACB > m\angle ABC$

PROOF

STATEMENTS	REASONS
1. In $\triangle ABC$, $AB > AC$.	1. Given.
2. Select D on $\overrightarrow{AC}$ so that $\overline{AD} \cong \overline{AB}$.	2. Point Uniqueness Postulate. (2-2)
3. $AD > AC$	3. Substitution Postulate. (2-1)
4. $\overline{ACD}$	4. Point Betweenness Postulate. (2-4)
5. $m\angle ABD = m\angle ABC + m\angle CBD$	5. Angle Sum Postulate. (2-10)
6. $m\angle ABD > m\angle ABC$	6. For any numbers k, m, and n, if $n = m + k$, and $k > 0$, then $n > m$. (5-1.1)

(continued)

STATEMENTS	REASONS
7. $\angle ABD \cong \angle D$	7. The base angles of an isosceles triangle are congruent. (3-4.2)
8. $m\angle D > m\angle ABC$	8. Substitution Postulate. (2-1)
9. $m\angle ACB > m\angle D$	9. The measure of an exterior angle of a triangle is greater than the measure of either remote interior angle. (5-2.1)
10. $m\angle ACB > m\angle ABC$	10. Transitive property of inequalities

In Theorem 5-3.1, we showed that an inequality between the measures of the sides of a triangle enables us to establish an inequality between the measures of the angles of the triangle. Let us now consider the converse.

T

Theorem 5-3.2 If two angles of a triangle are not congruent, then the sides opposite those angles are not congruent, the longer side being opposite the angle with the greater measure.

GIVEN $\triangle ABC$; $m\angle C > m\angle B$

PROVE $AB > AC$

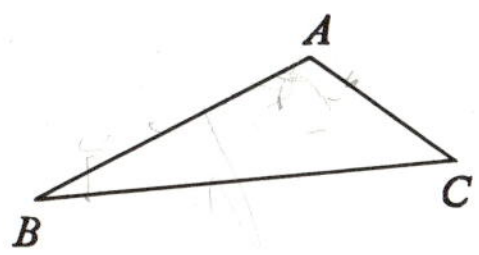

PROOF

STATEMENTS	REASONS
1. In $\triangle ABC$, $m\angle C > m\angle B$.	1. Given.
2. Assume AB is not greater than AC.	2. Assumption.
3. Either $AB = AC$, or $AB < AC$.	3. Trichotomy property.
4. If $AB = AC$, then $m\angle C = m\angle B$.	4. The base angles of an isosceles triangle are congruent. (3-4.2)
5. $AB \neq AC$	5. Contradiction of hypothesis.
6. If $AB < AC$, then $m\angle C < m\angle B$.	6. Theorem 5-3.1.
7. But $m\angle C \nless m\angle B$.	7. Contradiction of hypothesis.
8. Therefore $AB > AC$.	8. Since steps 4 and 6 both lead to contradictions, by the trichotomy property, this is the only possibility.

Example 1 Arrange the sides of $\triangle PQR$ in ascending order of length.

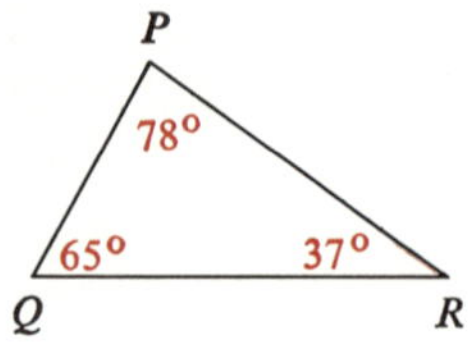

Solution According to Theorem 5-3.2, $PQ < PR$, since $m\angle R < m\angle Q$. Similarly, $PR < RQ$, since $m\angle Q < m\angle P$. Therefore, the order of the lengths of the sides is $PQ < PR < RQ$.

The following example will illustrate how the theorems presented in this section may be used in a proof:

Example 2

GIVEN $\triangle ABC$; $\overline{ADC}$; $BD > AB$

PROVE $BC > AB$

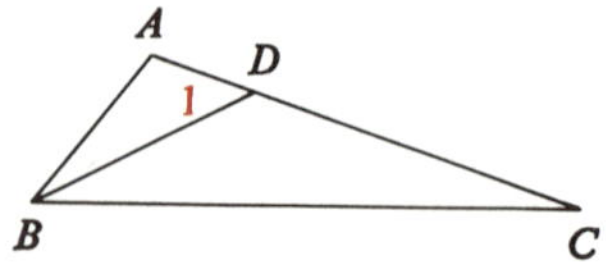

Solution

PROOF

STATEMENTS	REASONS
1. $\triangle ABC$; $\overline{ADC}$; $BD > AB$	1. Given.
2. $m\angle A > m\angle 1$	2. Theorem 5-3.1.
3. $m\angle 1 > m\angle C$	3. The measure of an exterior angle of a triangle is greater than the measure of either remote interior angle. (5-2.1)
4. $m\angle A > m\angle C$	4. Transitive property of inequalities.
5. $BC > AB$	5. Theorem 5-3.2.

EXERCISES

A In $\triangle ABC$, $AB = 5$, $BC = 11$, and $AC = 14$.

1. Which angle of $\triangle ABC$ has the greatest measure?
2. Which angle of $\triangle ABC$ has the least measure?

In $\triangle PRS$, $m\angle P = 30$, $m\angle R = 40$, and $m\angle S = 110$.

3. Which side of $\triangle PRS$ is the shortest?
4. Which side of $\triangle PRS$ is the longest?

In $\triangle DEF$, $DE > EF$ and $FD < EF$.

5. Which angle of $\triangle DEF$ has the greatest measure?
6. Which angle of $\triangle DEF$ has the least measure?

In $\triangle MNK$, $m\angle M < m\angle K$ and $m\angle M > m\angle N$.

7. Which is the longest side of $\triangle MNK$?
8. Which is the shortest side of $\triangle MNK$?

9. Which is the longest side of right $\triangle ASP$, with right $\angle S$?

10. Prove that the sum of the lengths of the altitudes of a triangle is less than the sum of the lengths of the sides.

11. If a median and an altitude are drawn from the same vertex of a scalene triangle, prove that the median is always longer than the altitude.

12. The measures of the angles in the diagram at right are indicated. Which line segment is the longest?

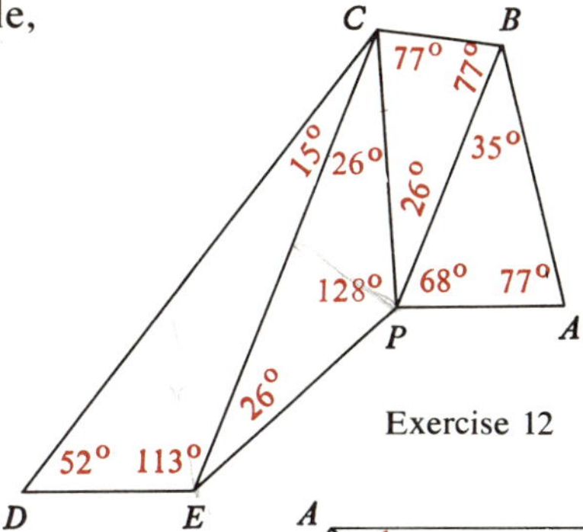

Exercise 12

13. **Given** Obtuse $\triangle ABC$; $\overrightarrow{BCD}$; $m\angle ACB > m\angle ACD$
Prove $BC < AB$

14. **Given** Quadrilateral $ABCD$; $m\angle 2 < m\angle 1$; $m\angle 4 < m\angle 3$
Prove $AC < BD$

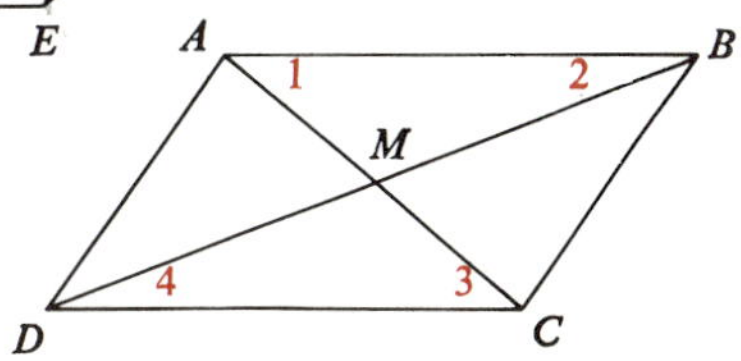

Exercise 14

B

15. **Given** $\triangle ABC$; $\overline{BDC}$; $\overline{AD}$ bisects $\angle BAC$
Prove $AB > BD$

16. **Given** $\triangle ABC$; $AB < AC$; $\overline{BP}$ bisects $\angle ABC$; $\overline{CP}$ bisects $\angle ACB$.
Prove $PC > PB$

17. **Given** Quadrilateral $ABCD$; $AD > AB$; $m\angle ABC = m\angle ADC$
Prove $DC < BC$ (*Hint*: Draw $\overline{BD}$.)

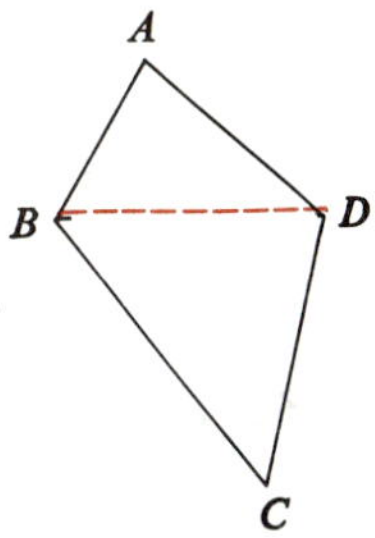

Exercise 17

18. **Given** $ABCD$ is a quadrilateral. $\overline{AD}$ is the shortest side, and $\overline{BC}$ is the longest side.
Prove $m\angle ABC < m\angle ADC$ (*Hint:* Draw $\overline{DB}$.)

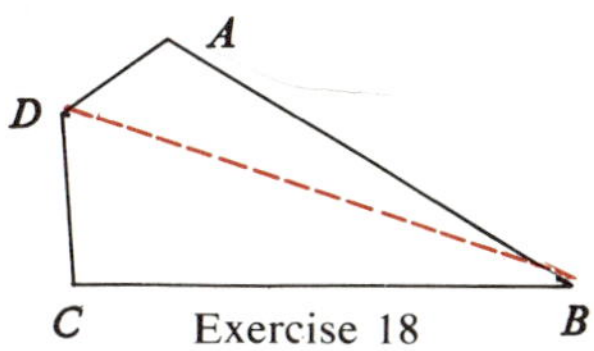

Exercise 18

19. **Given** Isosceles $\triangle ABC$; $AB = AC$; $\overline{ABE}$; $\overline{ACD}$; $CD < BE$
Prove $m\angle D > m\angle E$

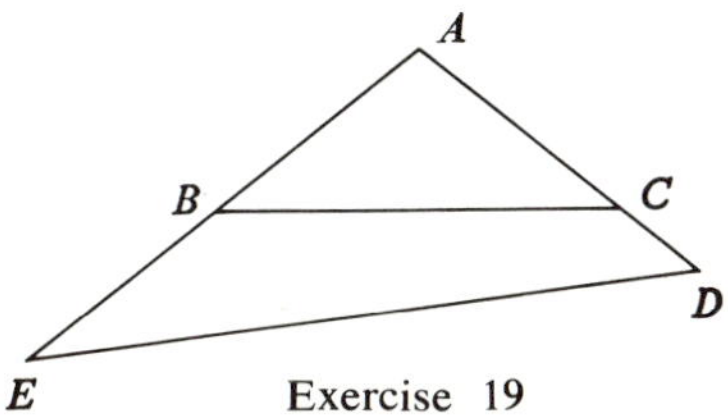

Exercise 19

20. GIVEN $\overline{AB} \cong \overline{DC}$;
$\overline{AD} \cong \overline{BC}$;
$AD < AB$
PROVE $\overline{AC}$ does not bisect $\angle BAD$.

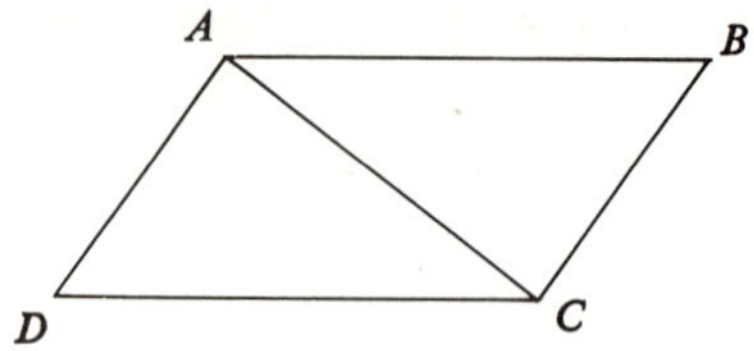

21. GIVEN $\triangle NFP$; $\overline{AFP}$;
$\overline{NF} \cong \overline{NP}$
PROVE $AN > NP$

22. GIVEN $\triangle NFP$; $\overline{FAP}$;
$\overline{NF} \cong \overline{NP}$
PROVE $AN < NP$

C

23. GIVEN Isosceles $\triangle ABC$; $AB = AC$
PROVE $2(AB) > BC$ (*Hint*: Draw the altitude from vertex A.)

24. Prove Theorem 5-3.1 by selecting point D on $\overline{AB}$ such that $\overline{AD} \cong \overline{AC}$.

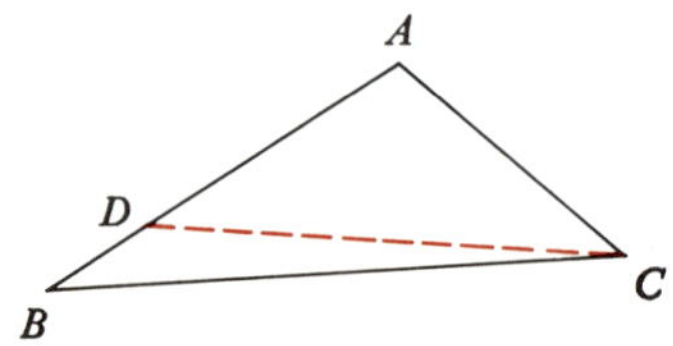

25. Prove Theorem 5-3.1 by selecting point E on $\overline{AB}$ such that $\overline{AE} \cong \overline{AC}$, and by selecting point D on $\overline{BC}$ such that $\overline{AD}$ bisects $\angle BAC$.

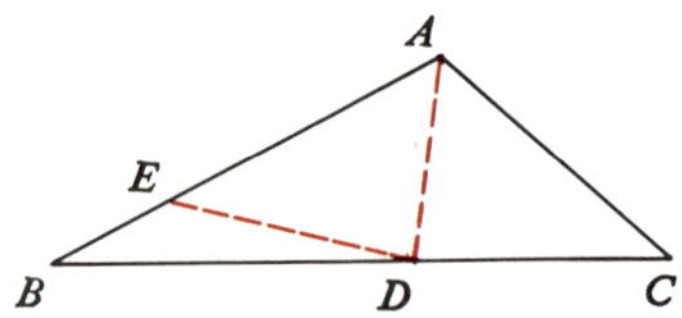

A LOOK AT THE PAST

The Origin of the Inequality Symbols

The symbols $<$ and $>$ first appeared in a book, *Artis Analyticae Praxis*, written by the English mathematician Thomas Harriot (1560–1621). The book was first published in 1631, ten years after the death of the author. It contained a great deal of symbolic algebra, including a section on the roots of equations, in which the now-familiar symbols $<$ and $>$ were introduced.

These symbols were not universally accepted, at first. In 1631, William Oughtred (1574–1660), another English mathematician, devised different inequality symbols, letting ⊏ represent "is greater than" and ⊐ represent "is less than." For a while, both Harriot's and Oughtred's symbols were being used. However, since Harriot's symbols were simpler and less confusing, they survived their competition.

5-4 TRIANGLE INEQUALITY

Past experience tells us that the shortest distance between two points is the measure of the straight line segment joining them. Suppose we wish to walk from point A to point B. Detouring to C, and then walking to B, will certainly require more walking than going from A to B along $\overline{AB}$. In symbols we write this as $AC + CB > AB$.

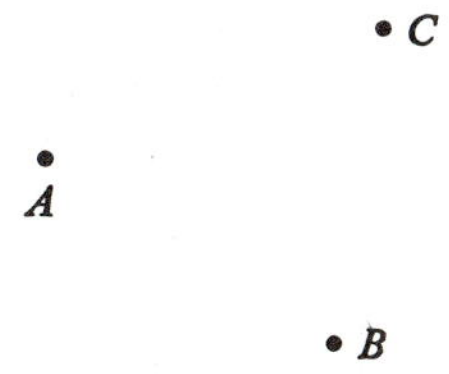

T **Theorem 5-4.1 The Triangle Inequality Theorem** The sum of the lengths of any two sides of a triangle is greater than the length of the third side.

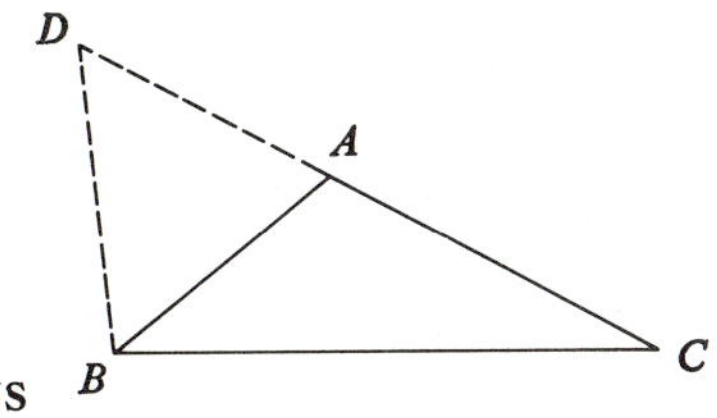

GIVEN $\triangle ABC$

PROVE $AB + AC > BC$

PROOF

STATEMENTS	REASONS
1. $\triangle ABC$	1. Given.
2. Select D on $\overrightarrow{CA}$, not between C and A, so that $AD = AB$.	2. Point Uniqueness Postulate. (2-2)
3. $DC = DA + AC$	3. Point Betweenness Postulate. (2-4)
4. $DC = AB + AC$	4. Substitution Postulate. (2-1)
5. $m\angle DBC = m\angle DBA + m\angle ABC$	5. Angle Sum Postulate. (2-10)
6. $m\angle DBC > m\angle DBA$	6. For any numbers k, m, and n, if $n = m + k$, and $k > 0$, then $n > m$. (5-1.1)
7. But $m\angle DBA = m\angle D$	7. The base angles of an isosceles triangle are congruent. (3-4.2)
8. Therefore $m\angle DBC > m\angle D$	8. Substitution Postulate. (2-1)
9. $DC > BC$	9. If two angles of a triangle are not congruent, then the sides opposite those angles are not congruent, the longer side being opposite the angle with the greater measure. (5-3.2)
10. $AB + AC > BC$	10. Substitution Postulate. (2-1)

This theorem also provides us with a method for determining whether a triangle can be formed by three given line segments. In order for a triangle to be formed from three line segments, the sum of the lengths of any two of the segments must be greater than the length of the third segment.

Example In $\triangle ABC$, $AB = 6$, $BC = 11$, and $AC = x$. What are the possible values of x?

Solution From Theorem 5-4.1 we derive the following inequalities:

(1) $AB + AC > BC$	(2) $AC + BC > AB$	(3) $AB + BC > AC$
$6 + x > 11$	$x + 11 > 6$	$6 + 11 > x$
$x > 5$	$x > -5$	$17 > x$

Combining these inequalities, we find $5 < x < 17$

After discussing the minimum distance between two points, we should discuss the minimum distance to a line from an external point.

T

Theorem 5-4.2 The shortest segment joining a line with an external point is the perpendicular segment from the point to the line.

GIVEN $\overleftrightarrow{AB}$ with the external point P; $\overline{PS} \perp \overleftrightarrow{AB}$

PROVE $\overline{PS}$ is the shortest segment from P to $\overleftrightarrow{AB}$.

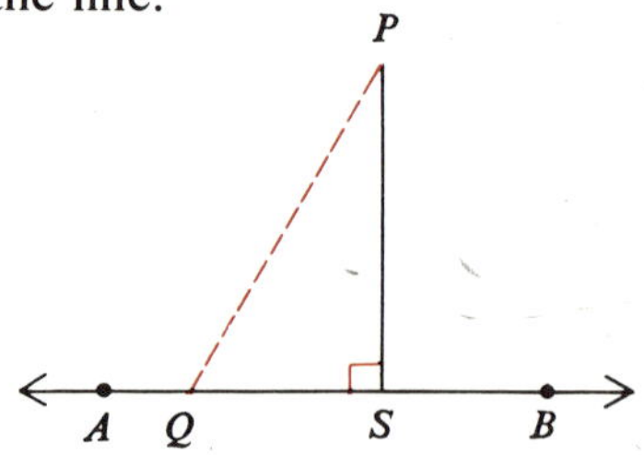

PROOF

STATEMENTS	REASONS
1. $\overleftrightarrow{AB}$ with external point P; $\overline{PS} \perp \overleftrightarrow{AB}$	1. Given.
2. $\angle PSA$ is a right angle.	2. Definition of perpendicular lines. (1-25)
3. Select any point Q on $\overleftrightarrow{AB}$.	3. Point Uniqueness Postulate. (2-2)
4. $\angle PQS$ is acute.	4. If a triangle has one right angle, then its other two angles must be acute. (5-2.1a)
5. $m\angle PQS < m\angle PSA$	5. Definition of an acute angle. (1-24)
6. $PS < PQ$	6. If two angles of a triangle are not congruent, then the sides opposite those angles are not congruent, the longer side being opposite the angle with the greater measure. (5-3.2)

Therefore PS is less than the measure of any other segment from P to $\overleftrightarrow{AB}$. Why? In Definition 4-5 we stated that the distance to a line from an external point is the measure of the perpendicular segment from the point to the line. We now see that this segment is shorter than any other from a given point to a line. Similarly, the distance to a plane from an external point, as stated in Definition 4-8, is the measure of the segment from the point and perpendicular to the plane. This, too, is the shortest distance from the point to the plane.

Class Exercises

Which of the following sets have elements that may represent the lengths of the sides of a triangle?

1. {5, 7, 7}   2. {6, 12, 6}   3. {3, 5, 9}   4. {11, 12, 13}

For each of the following, determine the distance from the given point to the given line:

5. From A to $\overline{BC}$.
6. From H to $\overline{AC}$.
7. From B to $\overline{AD}$.
8. From C to $\overline{BE}$.
9. From A to $\overline{BE}$.
10. From H to $\overline{BC}$.

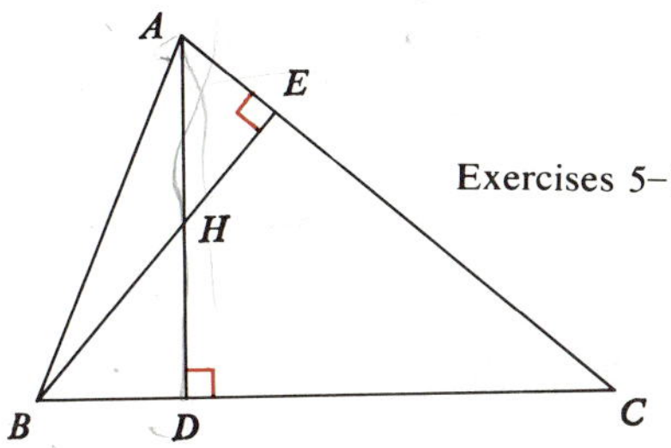

Exercises 5–10

EXERCISES

A

Complete the following inequalities:

1. $AB + AC >$ ____   2. $BG +$ ____ $> BD$
3. $GE +$ ____ $> EC$ 4. $AF + AC >$ ____
5. ____ $+ GC > AC$ 6. $AG +$ ____ $> AF$
7. $FG + BG >$ ____   8. ____ $+ DC > GC$
9. $AD +$ ____ $> AB$ 10. ____ $+ GC > DC$

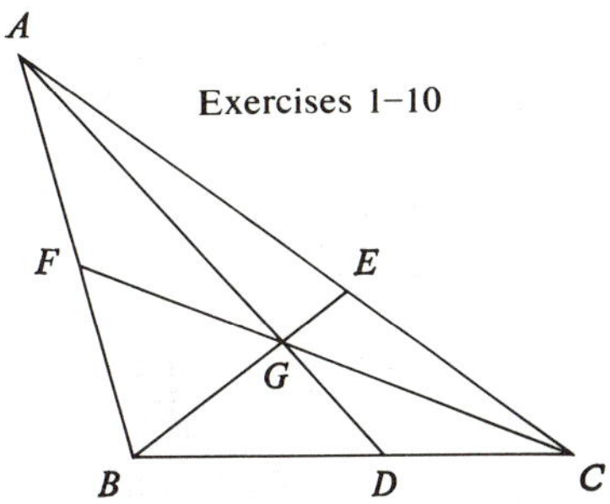

Exercises 1–10

Indicate whether the elements in each of the following sets may represent the lengths of the sides of a triangle:

11. {5, 6, 7}   12. {8, 5, 3}   13. {105, 94, 9}   14. {32, 58, 21}

The lengths of two sides of a triangle are given in each of the following. What values may the length of the third side have?

15. 5, 7   16. 3, 11   17. 1, 2   18. 7, 18

Complete the following inequalities:

19. $AE <$ ____ and $AE <$ ____.
20. $CD <$ ____ and $CD <$ ____.
21. $BF <$ ____ and $BF <$ ____.

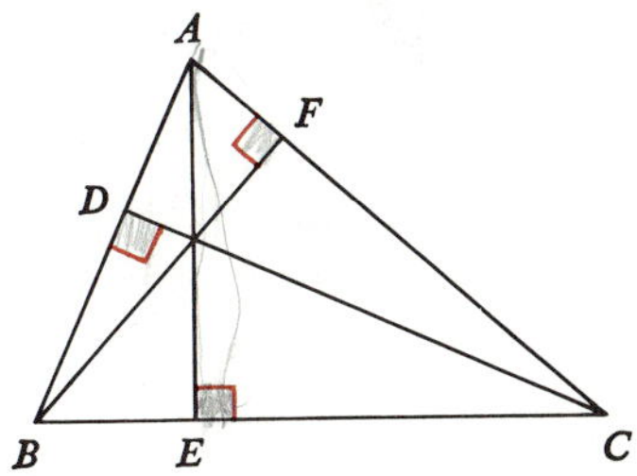

Exercises 19–21

22. Prove that the sum of the lengths of three sides of a quadrilateral is greater than the length of the fourth side.

B

23. GIVEN $\triangle ABC$; $AB = AC$
    PROVE $AB > \frac{1}{2}BC$

24. GIVEN $\triangle ABC$; $\overline{APC}$
    PROVE $PB < \frac{1}{2}(AB + BC + CA)$

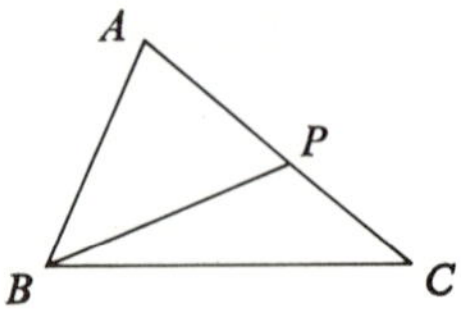

Exercise 24

25. GIVEN Quadrilateral $ABCD$
    PROVE $BD < \frac{1}{2}(AB + BC + CD + DA)$

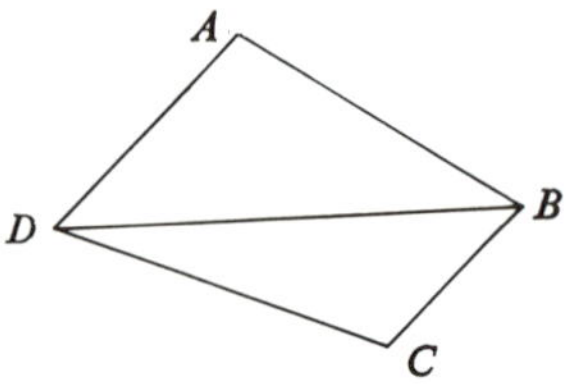

26. GIVEN Quadrilateral $ABCD$ and $\overline{APD}$
    PROVE $BP < \frac{1}{2}(AB + BC + CD + DA)$

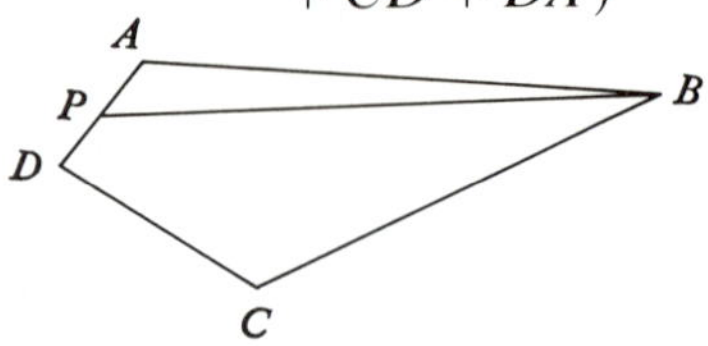

27. GIVEN $\triangle ADC$; $AB = BC$; $\overline{DBC}$
    PROVE $DC > AD$

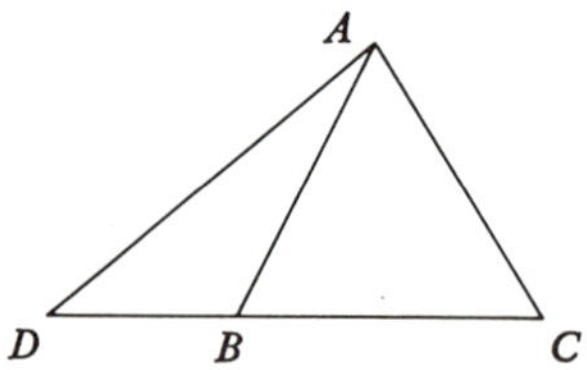

28. GIVEN $\triangle ABC$ with any points P and R such that $\overline{APB}$ and $\overline{ARC}$.
    PROVE $AB + AC > BP + PR + RC$

29. GIVEN $\overline{AB} \cong \overline{AD}$; $\overline{AE}$ bisects $\angle BAD$; $\overline{BEC}$
    PROVE $DC < BC$
    (*Hint*: Prove $BE = DE$)

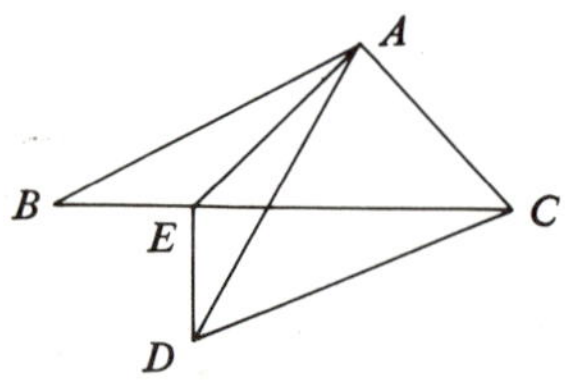

30. Prove that the difference between the lengths of any two sides of a triangle is less than the length of the third side.

C

31. GIVEN Median $\overline{AM}$ of $\triangle ABC$
    PROVE $AM < \frac{1}{2}(AB + AC)$
    (*Hint*: Consider $\overline{AMP}$ where $AM = MP$.)

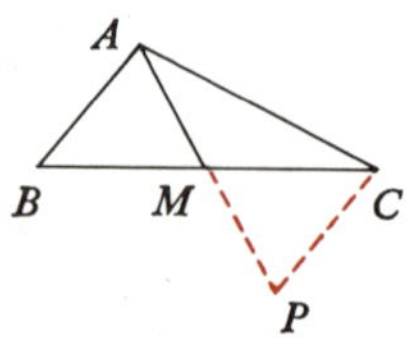

32. Prove that in any triangle, the sum of the lengths of the medians is less than the sum of the lengths of the three sides.

33. GIVEN Point P is an interior point of $\triangle ABC$.

PROVE $AB + AC > BP + PC$ (*Hint*: Consider $\overline{BPN}$ with N on $\overline{AC}$.)

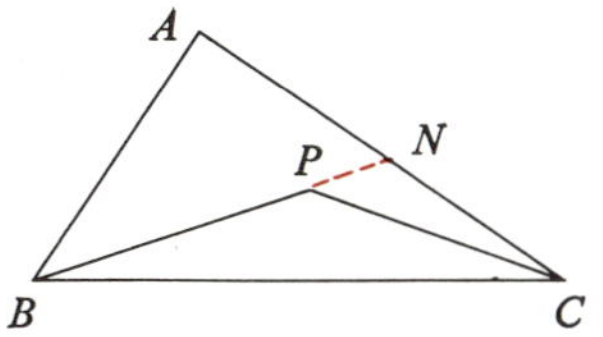

34. GIVEN Point P is in the interior of $\triangle ABC$.

PROVE $AP + PB + PC > \frac{1}{2}(AB + AC + BC)$

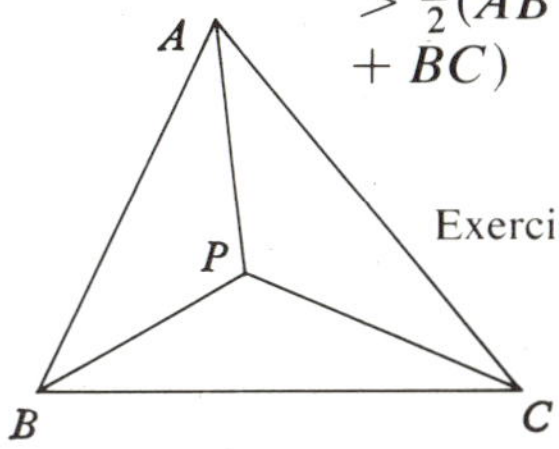

Exercises 34–35

35. GIVEN Point P is in the interior of $\triangle ABC$.

PROVE $AP + PB + PC < AB + AC + BC$ (*Hint*: Use the conclusion of Exercise 33.)

36. Prove that the shortest distance from an external point to a plane is the length of the segment perpendicular to the plane with one endpoint at the point and the other endpoint in the plane.

5-5 INEQUALITIES BETWEEN A PAIR OF TRIANGLES

So far in our discussion of geometric inequalities, we have studied theorems concerning inequalities of measure among the sides or angles of one triangle. Let us now compare the measures of sides and of angles in two triangles.

Suppose we form a triangle, fastening two wooden sticks of unequal length together at one end with a hinge (vertex B), and joining the remaining ends (vertices A and C) with a rubber band.

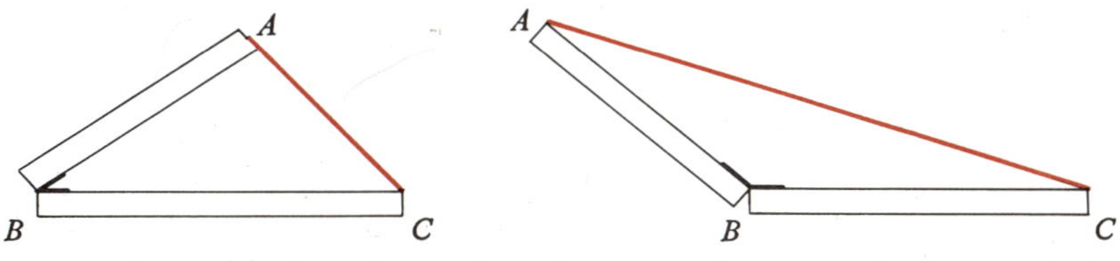

You will notice that when the angle at vertex B decreases in measure, the length of $\overline{AC}$ (the rubber band) also becomes smaller. Let us restate this observation in mathematical language.

T

Theorem 5-5.1 If two sides of a triangle are congruent respectively to two sides of a second triangle, and the measure of the included angle of the first triangle is greater than the measure of the included angle of the second triangle, then the measure of the third side of the first triangle is greater than the measure of the third side of the second triangle.

The complete proof of this theorem will be found on the next page.

Given $\triangle ABC$ and $\triangle DEF$; $AB = DE; BC = EF$; $m\angle B > m\angle E$

Prove $AC > DF$

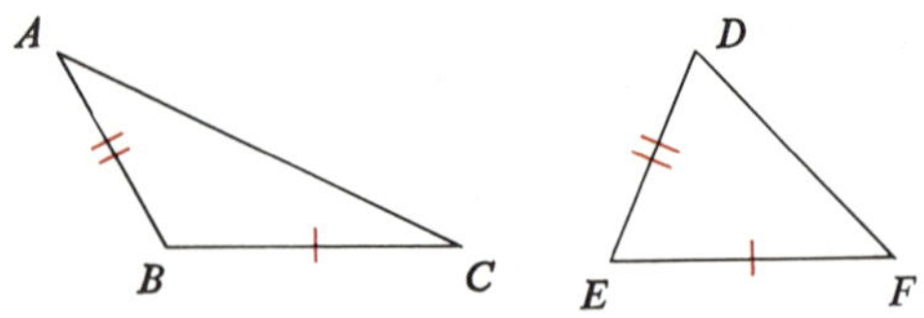

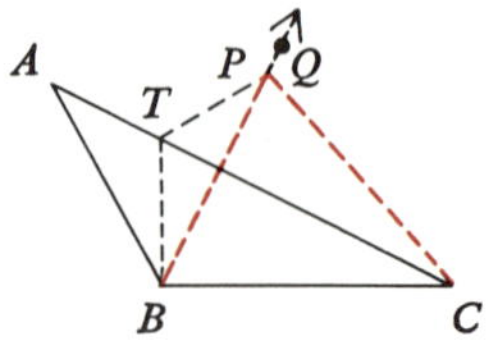

Proof

STATEMENTS	REASONS
1. $\triangle ABC$ and $\triangle DEF$; $AB = DE$; $BC = EF$; $m\angle B > m\angle E$	1. Given.
2. Select point Q on the same side of $\overleftrightarrow{BC}$ as A so that $\angle QBC \cong \angle E$.	2. Angle Uniqueness Postulate. (2-9)
3. Now consider point P on $\overleftrightarrow{BQ}$ so that $\overline{PB} \cong \overline{DE}$.	3. Point Uniqueness Postulate. (2-2)
4. $\triangle DEF \cong \triangle PBC$	4. SAS Postulate. (3-1)
5. Let T be the intersection of the bisector of $\angle ABP$ and $\overline{AC}$.	5. Point Uniqueness Postulate. (2-2)
6. $\angle ABT \cong \angle PBT$	6. Definition of an angle bisector. (1-29)
7. $AB = PB$	7. Transitive property of inequalities.
8. $\overline{TB} \cong \overline{TB}$	8. Every segment is congruent to itself. (3-1.6)
9. $\triangle ATB \cong \triangle PTB$	9. SAS Postulate. (3-1)
10. $AT = PT$	10. Definition of congruent triangles. (3-3)
11. $PT + TC > PC$	11. The sum of the lengths of any two sides of a triangle is greater than the length of the third side. (5-4.1)
12. $AT + TC > PC$	12. Substitution Postulate. (2-1)
13. $AC > PC$	13. Point Betweenness Postulate. (2-4)
14. $PC = DF$	14. Definition of congruent triangles. (3-3)
15. $AC > DF$	15. Substitution Postulate. (2-1)

Example 1

GIVEN $\overline{AM}$ is a median of $\triangle ABC$; $m\angle 2 > m\angle 1$

PROVE $AC > AB$

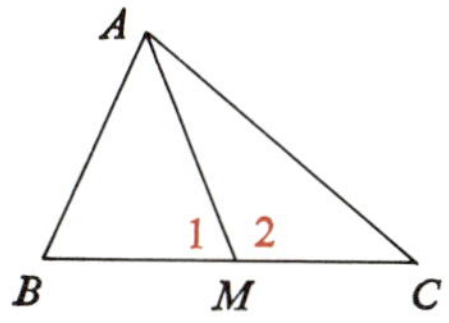

Solution

PROOF

STATEMENTS	REASONS
1. $\overline{AM}$ is a median of $\triangle ABC$; $m\angle 2 > m\angle 1$	1. Given.
2. M is the midpoint of $\overline{BC}$.	2. Definition of a median. (3-9)
3. $\overline{BM} \cong \overline{MC}$	3. Definition of a midpoint. (1-5)
4. $\overline{AM} \cong \overline{AM}$	4. Every segment is congruent to itself. (3-1.6)
5. $AC > AB$	5. Theorem 5-5.1.

Can you prove that $\angle 2$ is obtuse? It is the greater measure of two supplementary angles.

We shall now consider the converse of Theorem 5-5.1.

T **Theorem 5-5.2** If two sides of one triangle are congruent respectively to two sides of a second triangle, and the measure of the third side of the first triangle is greater than the measure of the third side of the second triangle, then the measure of the included angle of the first triangle is greater than the measure of the included angle of the second triangle.

GIVEN $\triangle ABC$ and $\triangle DEF$; $AB = DE$; $BC = EF$; $AC > DF$

PROVE $m\angle B > m\angle E$

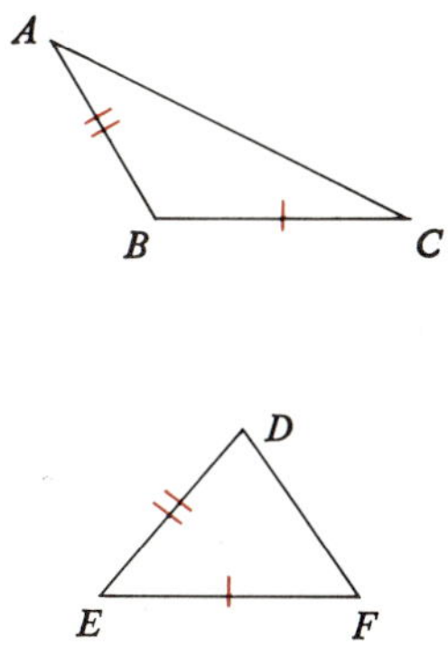

The proof of this theorem follows the same pattern as the proof of Theorem 5-3.2. We must show that $m\angle B < m\angle E$ and $m\angle B = m\angle E$ are impossible conclusions, thus leaving $m\angle B > m\angle E$ as the only possibility, by the trichotomy property. The detailed proof of this theorem is left as an exercise.

Example 2

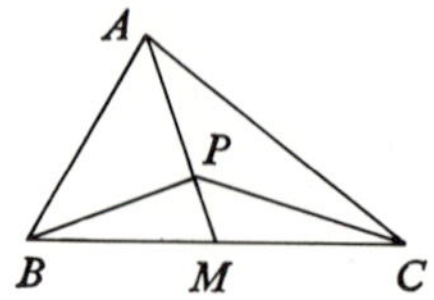

GIVEN $\overline{AM}$ is a median of $\triangle ABC$; $AC > AB$

PROVE $PC > PB$

Solution

PROOF

STATEMENTS	REASONS
1. $\overline{AM}$ is a median of $\triangle ABC$; $AC > AB$	1. Given.
2. M is the midpoint of $\overline{BC}$.	2. Definition of a median. (3-9)
3. $\overline{BM} \cong \overline{MC}$	3. Definition of a midpoint. (1-15)
4. $\overline{AM} \cong \overline{AM}$	4. Every segment is congruent to itself. (3-1.6)
5. $m\angle AMC > m\angle AMB$	5. Theorem 5-5.2.
6. $\overline{PM} \cong \overline{PM}$	6. Every segment is congruent to itself. (3-1.6)
7. $PC > PB$	7. Theorem 5-5.1.

Class Exercises

In the accompanying diagram, the measures of some of the angles are indicated and congruent segments are marked. Indicate the appropriate inequality symbol for each of the following:

1. DE ____ EC
2. AE ____ BC
3. EF ____ CE
4. BC ____ FE
5. EC ____ BC
6. ED ____ AE
7. DE ____ BC
8. EA ____ CE

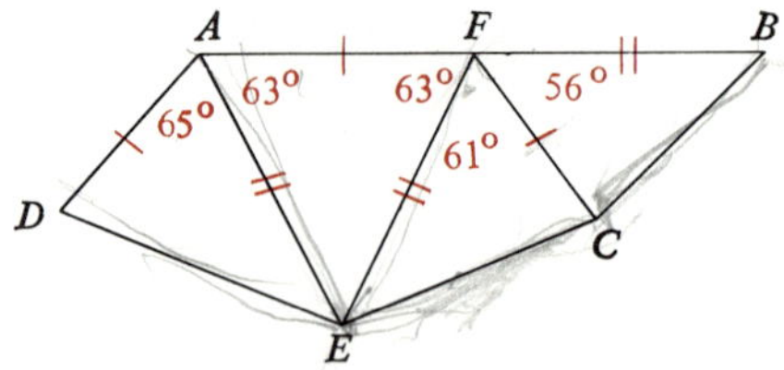

EXERCISES

A Indicate the appropriate inequality symbol for each of the following, referring to the diagrams:

1. $m\angle APB$ ____ $m\angle APD$
2. $m\angle BPC$ ____ $m\angle CPD$
3. $m\angle APD$ ____ $m\angle BPC$
4. $m\angle DPC$ ____ $m\angle APB$

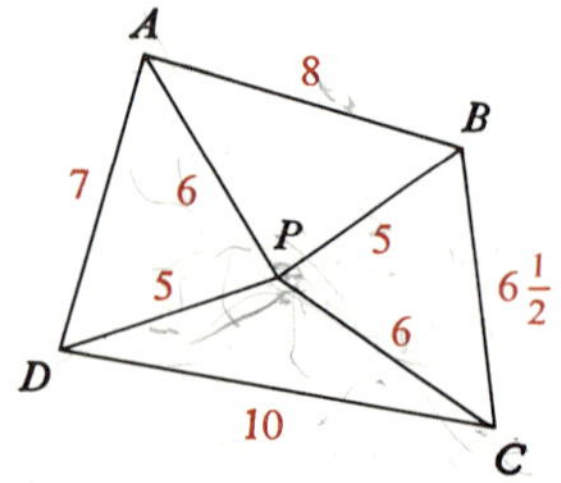

Exercises 1–4

5. PQ ____ TR 6. TS ____ TR
7. QT ____ TS 8. PQ ____ QT
9. PQ ____ TS

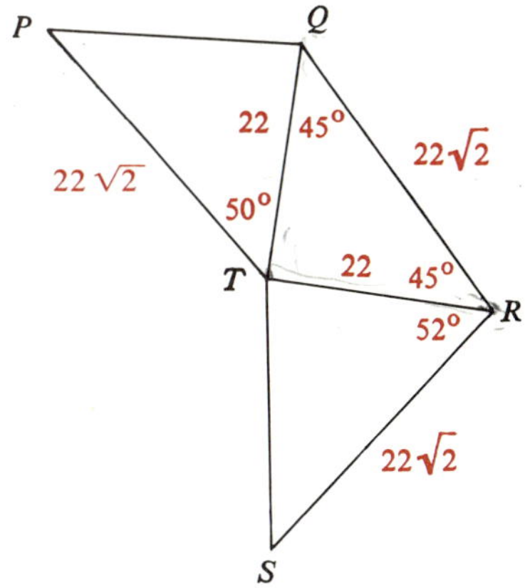

Exercises 5–9

10. **Given** $\overline{AM}$ is a median of $\triangle ABC$; $m\angle C > m\angle B$
Prove $\angle BMA$ is obtuse

B

11. **Given** $\overline{AM}$ is a median of $\triangle ABC$; $m\angle AMC < m\angle AMB$
Prove $m\angle C > m\angle B$
12. **Given** $\triangle PQR$; $\overline{PQ} \cong \overline{PR}$; $\overline{QTR}$; $m\angle QPT < m\angle TPR$
Prove $QT < RT$

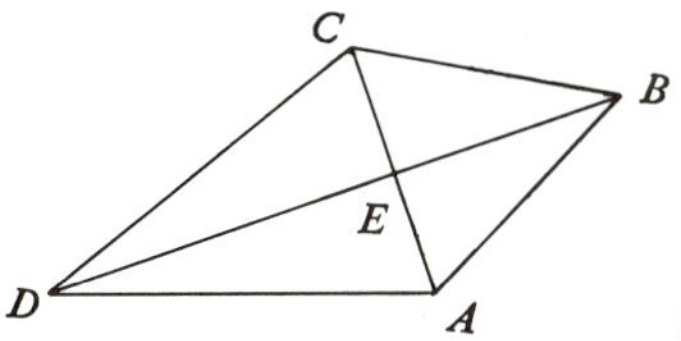

Exercise 13

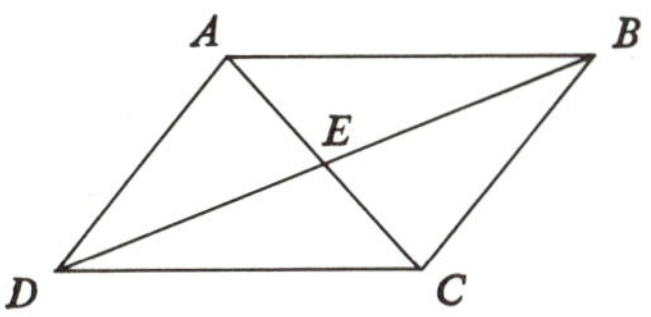

Exercise 14

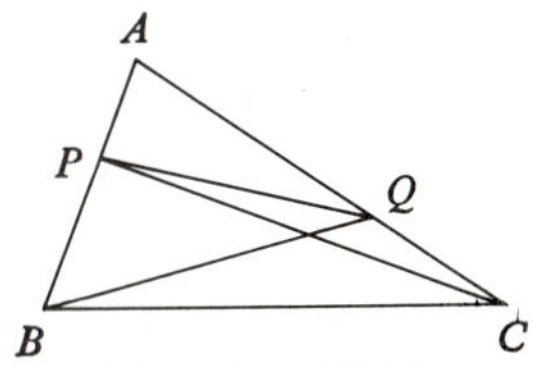

Exercises 15–16

13. **Given** Quadrilateral $ABCD$; $\overline{AB} \cong \overline{AC}$ **Prove** $BD > CD$
14. **Given** Diagonals of quadrilateral $ABCD$ bisect each other at E; $\angle ADC$ is supplementary to $\angle BCD$; $\angle BCD$ is obtuse.
Prove $AC < BD$
15. **Given** $\triangle ABC$; $AC > AB$; $\overline{APB}$; $\overline{AQC}$; $\overline{BP} \cong \overline{CQ}$
Prove $PC > QB$
16. **Given** $\triangle ABC$; $\overline{APB}$; $\overline{AQC}$; $\overline{BP} \cong \overline{CQ}$; $PC > QB$
Prove $AC > AB$

C

17. **Given** $\triangle PQR$; $\overline{PSR}$; $\overline{PQ} \cong \overline{SR}$ **Prove** $QR > PS$
18. By showing that both $m\angle B = m\angle E$ and $m\angle B < m\angle E$ lead to contradictions, write a detailed indirect proof of Theorem 5-5.2.

SOMETHING TO THINK ABOUT

Reflected Images

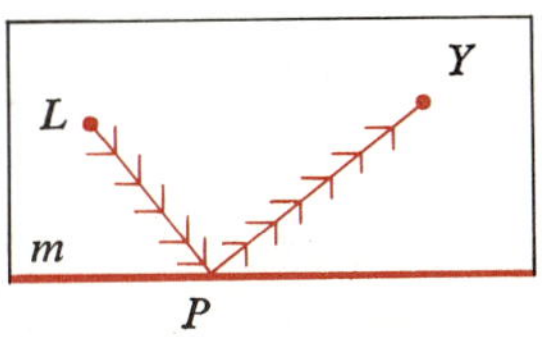

Suppose you (point Y in the diagram) are in a room, facing a wall covered with a mirror (line m). On the opposite side of the room is a table lamp (point L). How can we locate the point in the mirror (P) where you will see the reflection of the lamp? What do you think will be true about the length of the path in which the ray of light will travel from the lamp to the mirror and then to you? We shall make the assumption that this path will be the shortest path, that is $LP + PY$ will be less than any other path from the lamp to the mirror to you.

In order to find point P consider the perpendicular from point L to line m, meeting line m at H. Let R be the point of $\overleftrightarrow{LH}$ such that $\overline{LH} \cong \overline{HR}$. The point R is called the *reflected image* of point L through line m. The point of intersection of $\overline{YR}$ and line m will determine point P.

We must now prove that $LP + PY$ is less than any other path from L to line m to Y, for example $LQ + QY$.

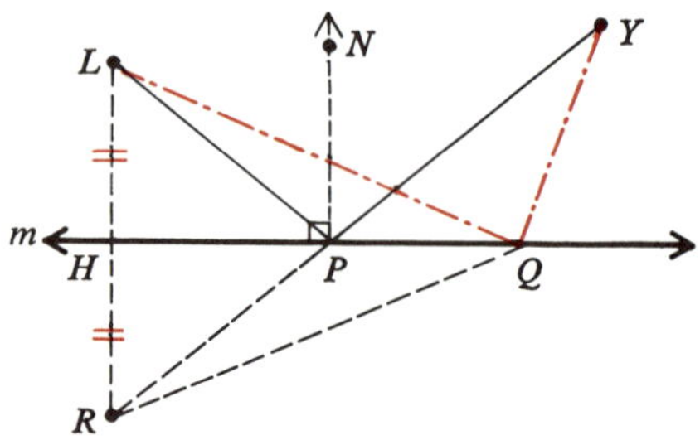

GIVEN Points L and Y lie in the same half-plane of line m; $\overline{LHR} \perp \overline{HPQ}$, where Q is any point of $\overleftrightarrow{HP}$; $\overline{LH} \cong \overline{HR}$; $\overline{RPY}$

PROVE $LP + PY < LQ + QY$

PROOF

STATEMENTS	REASONS
1. Points L and Y lie in the same half-plane of line m; $\overline{LHR} \perp \overline{HPQ}$, where Q is any point of $\overleftrightarrow{HP}$; $\overline{LH} \cong \overline{HR}$; $\overline{RPY}$	1. Given.
2. Line m is the perpendicular bisector of $\overline{LHR}$.	2. Definition of a perpendicular bisector. (4-4)
3. $LP = RP$; $LQ = RQ$	3. A line is the perpendicular bisector of a segment if and only if it is the set of all points in a plane equidistant from the endpoints of the segment. (4-4.3)
4. In $\triangle RQY$, $YR < RQ + QY$	4. The sum of the lengths of any two sides of a triangle is greater than the length of the third side. (5-4.1)
5. $YR = YP + PR$	5. Point Betweenness Postulate. (2-4)
6. $YP + PL < LQ + QY$	6. Substitution Postulate. (2-1)

So far we have found a way of locating point P which provides the shortest path from point L to line m to point Y.

Consider another property of point P. Why is $\angle HPR \cong \angle QPY$ and $\angle LPH \cong \angle QPY$? If $\overline{PN} \perp$ line m, then $\angle LPN$, the *angle of incidence*, is congruent to $\angle YPN$, the *angle of reflection*. Why?

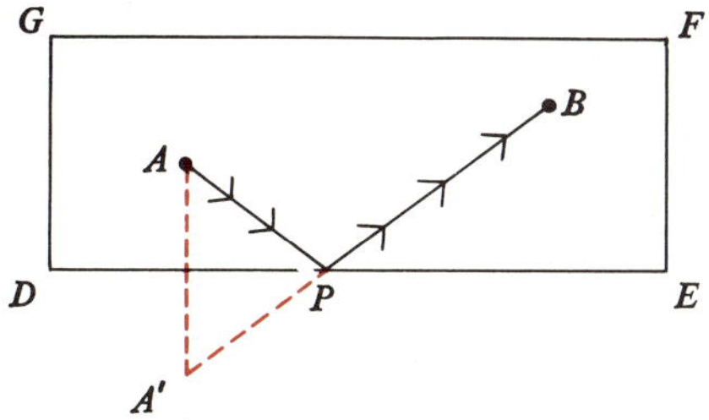

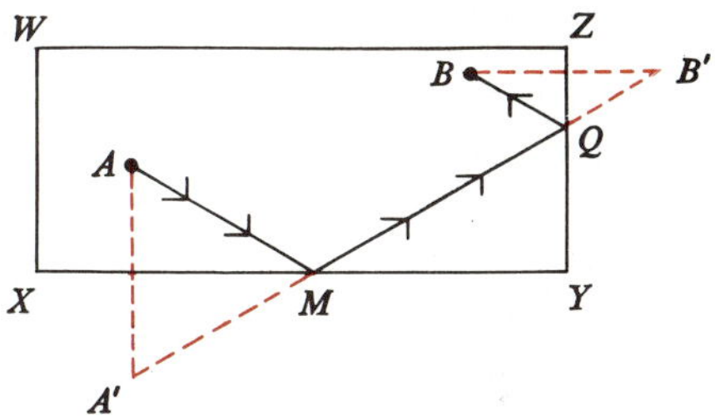

What we proved above has an application in billiards. If we wish ball A to strike the cushion $\overline{DE}$ so as to hit ball B, the point P must have the same properties discussed above. Suppose we place a mirror along cushion $\overline{DE}$ and look along $\overrightarrow{AP}$. What do you think we see in the mirror? You are correct if you said ball B. Why?

Explain how you would locate the point M of cushion $\overline{XY}$ which ball A would have to hit before hitting cushion $\overline{YZ}$ and ball B, in that order. (*Hint*: Find the reflected images of ball A in $\overline{XY}$ and ball B in $\overline{YZ}$. Treat the reflected images as you would treat balls A and B.)

HOW IS YOUR MATHEMATICAL VOCABULARY?

The key words and phrases introduced in this chapter are listed below. How many do you know and understand?

- addition property of inequalities (5-1)
- division property of inequalities (5-1)
- exterior angle of a triangle (5-2)
- interior angle of a triangle (5-2)
- multiplication property of inequalities (5-1)
- remote interior angles of a triangle (5-2)
- subtraction property of inequalities (5-1)
- transitive property of inequalities (5-1)
- trichotomy property of inequalities (5-1)

REVIEW EXERCISES

5-1 Properties of inequalities

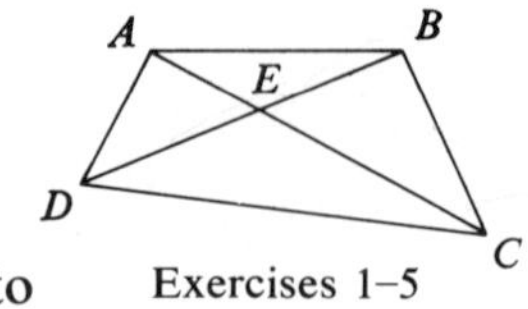

Exercises 1–5

In the following exercises, fill in the missing inequality symbol and then state which property, theorem, or corollary may be used to justify the completed conclusion, referring to the diagram above.

1. If $AC = AE + EC$, then AC ____ EC.
2. If $AB > AD$ and $DC > BC$, then $AB + DC$ ____ $AD + BC$.
3. If $AE = \frac{1}{2}BE$, $DE = \frac{1}{2}EC$, and $AE < DE$, then EC ____ BE.
4. If AD is not greater than BC and $AD \neq BC$, then AD ____ BC.
5. If $BE < EC$ and $AE < BE$, then EC ____ AE.
6. If $m\angle ABC = m\angle ABE + m\angle CBE$, then $m\angle CBE$ ____ $m\angle ABC$.
7. If the supplement of $\angle M$ is greater than the supplement of $\angle N$, what is the relationship between $\angle M$ and $\angle N$?
8. GIVEN Point A is not on $\overline{BCDEF}$. $\overline{AC} \cong \overline{AD}$; $\angle BAC \cong \angle EAD$

   PROVE $BD < CF$

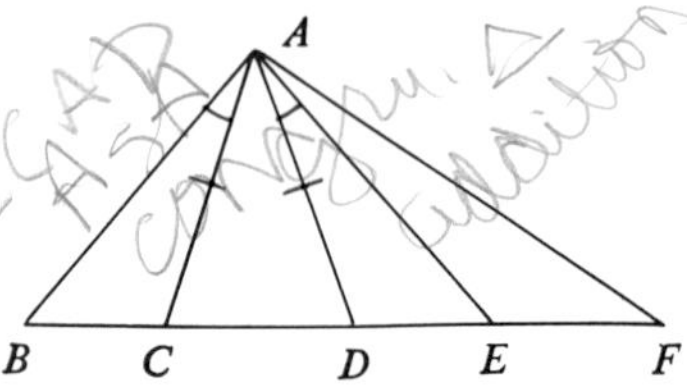

5-2 The exterior-angle theorem

In Exercises 12–15 you are given a triangle and one of its exterior angles. Tell which are the remote interior angles of the given exterior angle.

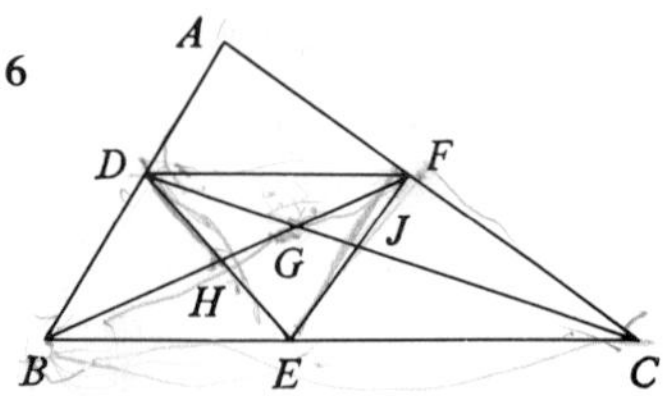

Exercises 9–16

9. $\angle ADE$ and $\triangle BDE$
10. $\angle DGF$ and $\triangle GFC$
11. $\angle ADC$ and $\triangle DBC$
12. $\angle CGB$ and $\triangle DGB$
13. Is $m\angle BHD > m\angle EDF$? Why?
14. Is $m\angle BEF < m\angle ACB$? Why?
15. Name 13 angles which have measures less than that of $\angle BGC$. Prove that each of your answers is correct.
16. Which angles have measures less than that of $\angle ADE$? Prove that each of your answers is correct.
17. GIVEN $\triangle ABE$; $\overline{BCDE}$

    PROVE $m\angle ACB > m\angle DAE$

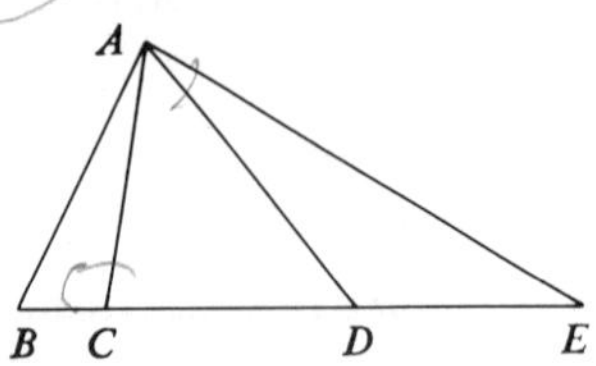

18. GIVEN $\triangle ABC$ and $\triangle NBQ$; $\overline{ANC}$; $\overline{BCQ}$; $\overline{BKN}$

    PROVE $m\angle Q < m\angle AKB$

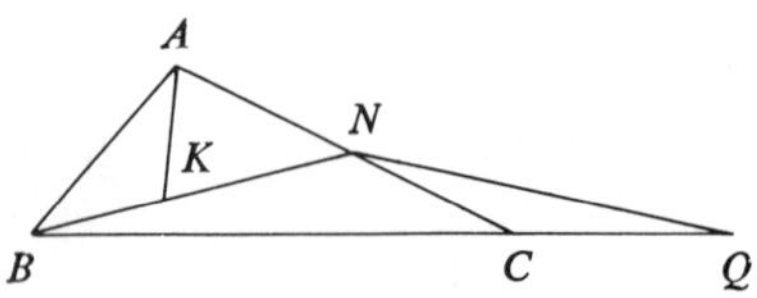

5-3 *Inequalities in one triangle*

19. List the angles of $\triangle PQR$ in ascending order of measure (smallest to largest), if $PQ = 7$, $QR = 8$, and $RP = 10$.
20. List the sides of $\triangle KMN$ in ascending order of length, if $m\angle K = 75$, $m\angle M = 35$, and $m\angle N = 70$.

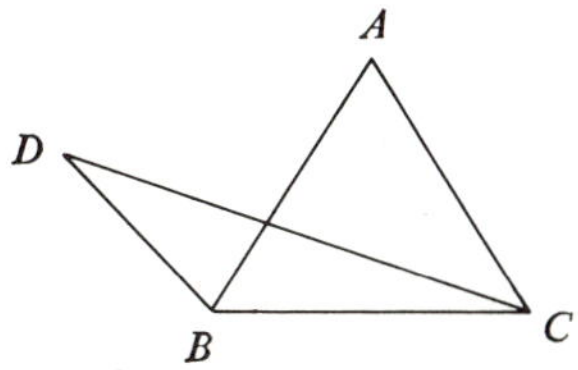

Exercise 21

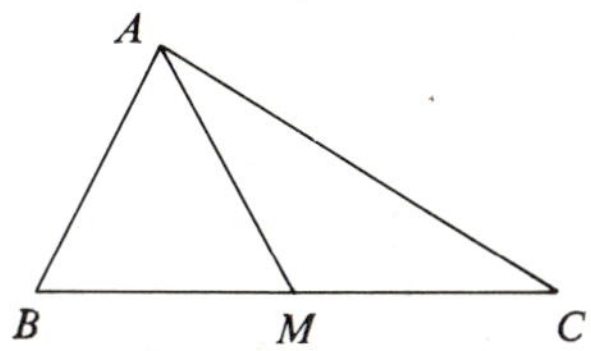

Exercises 22–23

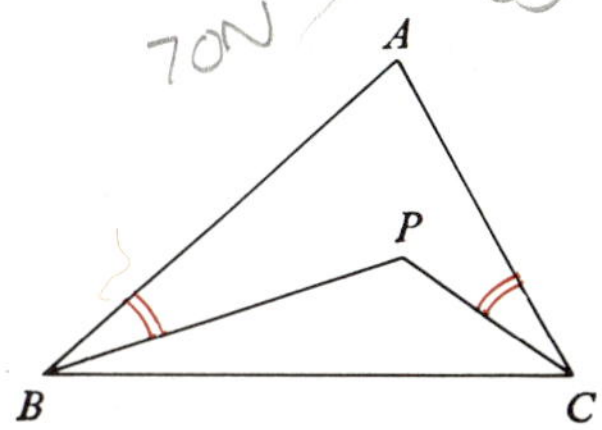

Exercise 24

21. GIVEN Isosceles $\triangle ABC$; $\overline{AB} \cong \overline{AC}$; D is any point in the interior of $\angle ACB$ exterior to $\triangle ABC$. PROVE $DC > DB$
22. GIVEN $\triangle ABC$; $\overline{BMC}$; $AC > AB$ PROVE $AM < AC$
23. GIVEN $\overline{AM}$ is a median of $\triangle ABC$; $AM > BM$ PROVE $m\angle B + m\angle C > m\angle BAC$
24. GIVEN Point P is in the interior of $\triangle ABC$; $m\angle ABP = m\angle ACP$; $BP > PC$ PROVE $AB > AC$

5-4 *Triangle inequality*

Complete each of the following inequalities by filling in the missing line segments:

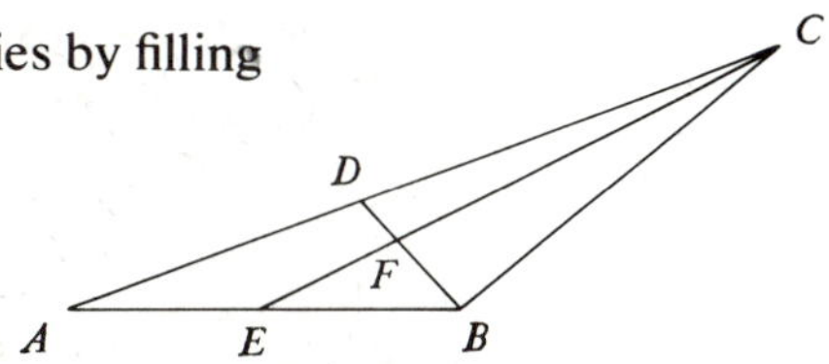

Exercises 25–28

25. $DF +$ ____ $> DC$
26. ____ $+ EB > CE$
27. $AD + DB >$ ____
28. $FB +$ ____ $> BC$

Which of the following sets have elements that may represent the lengths of the sides of a triangle?

29. {3, 6, 8} 30. {5, 12, 17} 31. {8, 10, 3}

Find the range of values of x in each of the following:

32. In $\triangle PQR$, $PQ = 5$, $RQ = 6$, and $RP = x$.
33. In $\triangle ABC$, $AB = 28$, $BC = 28$, and $AC = x$.
34. If point P is not on $\overline{ABCDE}$ and $m\angle PCE = 90$, then which is the shortest: $\overline{PA}$, $\overline{PB}$, $\overline{PC}$, $\overline{PD}$, or $\overline{PE}$? Why?

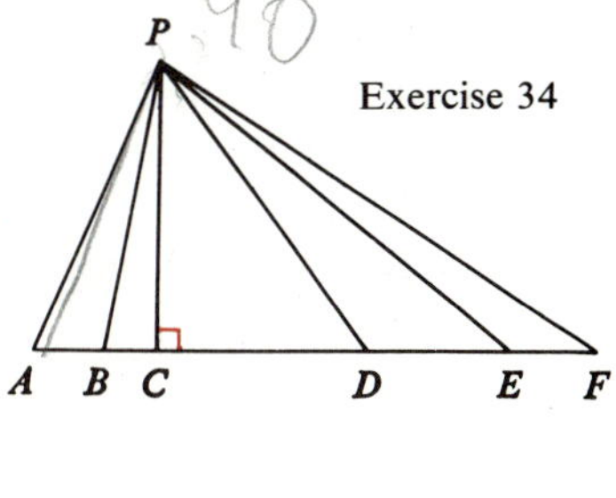

Exercise 34

35. Prove that the length of either of the congruent sides of an isosceles triangle is greater than half the length of the base.
36. Prove that the sum of the lengths of the segments joining an interior point of a quadrilateral with its vertices is greater than half the sum of the lengths of the sides (the perimeter) of the quadrilateral.
37. **Given** $\triangle ABC$; $\overline{BDC}$
Prove $2(AD) + BC > AB + AC$

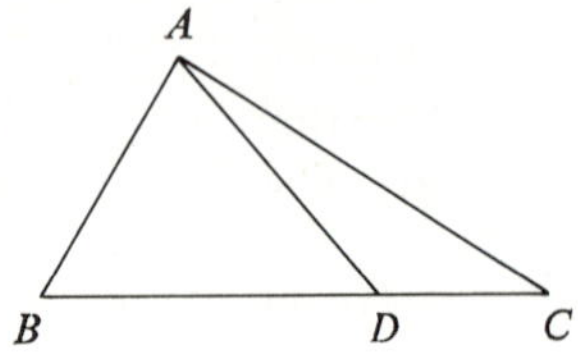

Exercise 37

5-5 *Inequalities between a pair of triangles*

Referring to the measures indicated in the diagram at right, supply the appropriate inequality symbol for each of the following:

38. $m\angle TPQ$ ____ $m\angle RPS$
39. $m\angle TPS$ ____ $m\angle SPR$
40. $m\angle RPQ$ ____ $m\angle SPT$
41. $m\angle SPR$ ____ $m\angle QPR$

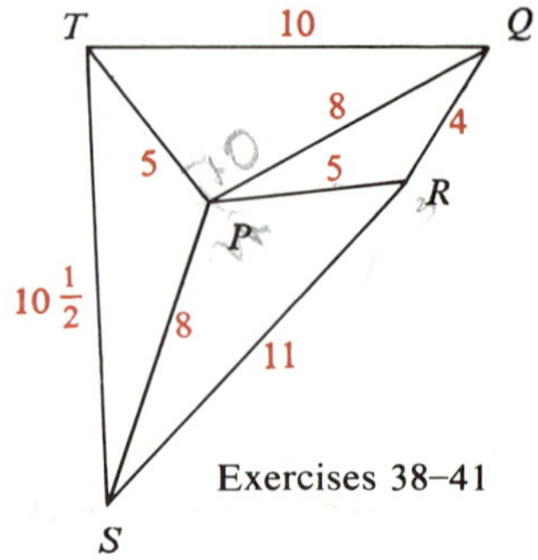

Exercises 38–41

42. **Given** $\overline{AB} \cong \overline{DC}$; $BD > AC$
Prove $m\angle DCB > m\angle ABC$

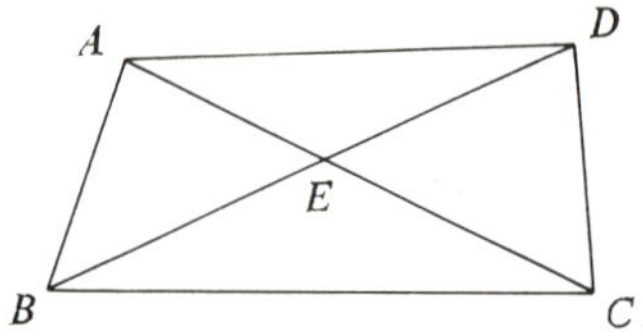

43. **Given** $\triangle ABC$; $\overline{ADB}$; $AD = CD = DB$; $m\angle BDC > m\angle ADC$
Prove $EC > AC$

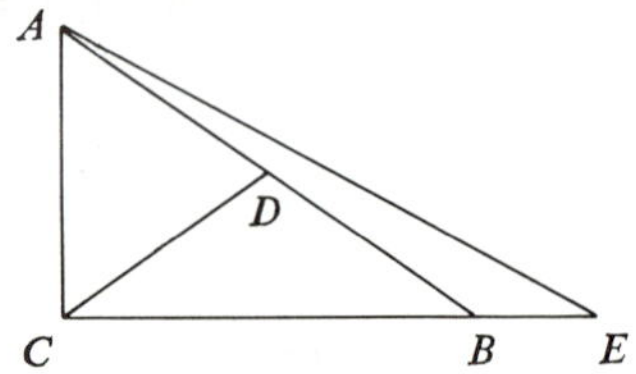

CHAPTER TEST

1. What is the least integral measure that the exterior angle at vertex R of $\triangle PRT$ may have if $m\angle P = 50$ and $m\angle T = 63$?
2. Name the largest and smallest angles of $\triangle KMN$, if $KM = 5$, $MN = 7$, and $KN = 10$.
3. Which is the longest segment in the diagram at right?
4. Which is the shortest segment?

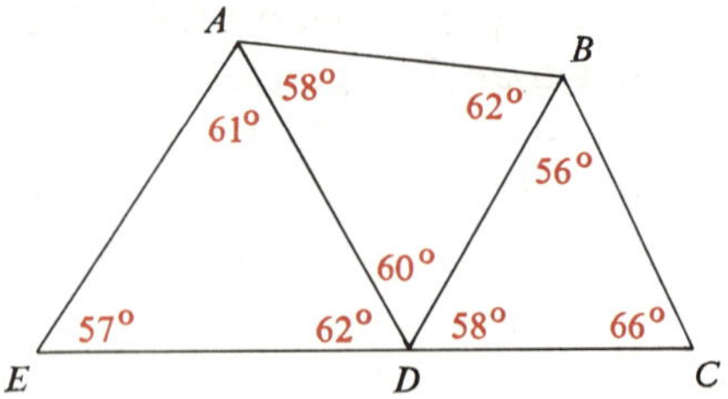

Exercises 3–4

Which of the following sets have elements that may represent the measures of the sides of a triangle?

5. $\{1, 2, 3\}$ 6. $\{5, 6, 7\}$ 7. $\{57, 80, 24\}$ 8. $\{43, 133, 89\}$

9. If the two congruent sides of an isosceles triangle have lengths of 17, then the length of the third side must be greater than ___ and less than ___.
10. $\overline{AM}$ is a median of $\triangle ABC$, and $\angle AMB$ is obtuse. State an inequality between two of the sides of $\triangle ABC$.
11. Explain how you would use the trichotomy property in an indirect proof on inequalities.
12. Explain why the sum of the lengths of the medians of a triangle is equal to or greater than the sum of the lengths of the altitudes of the triangle.
13. GIVEN Quadrilateral $ABCD$; $AD > DC$; $\overline{BA} \perp \overline{AD}$; $\overline{CD} \perp \overline{BC}$
    PROVE $BC > AB$

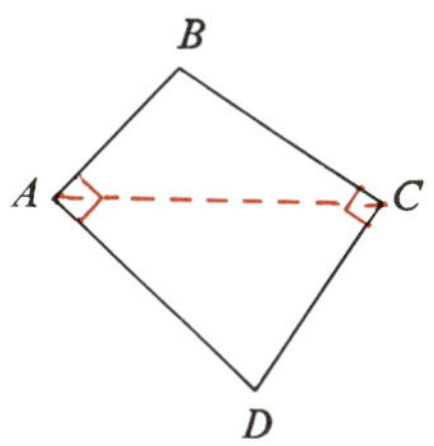

14. GIVEN Isosceles $\triangle ABC$; $\overline{AB} \cong \overline{AC}$; $\overline{APB}$; $\overline{AQC}$; $\overrightarrow{PQ}$ meets $\overrightarrow{BC}$ at D.
    PROVE $AP < AQ$

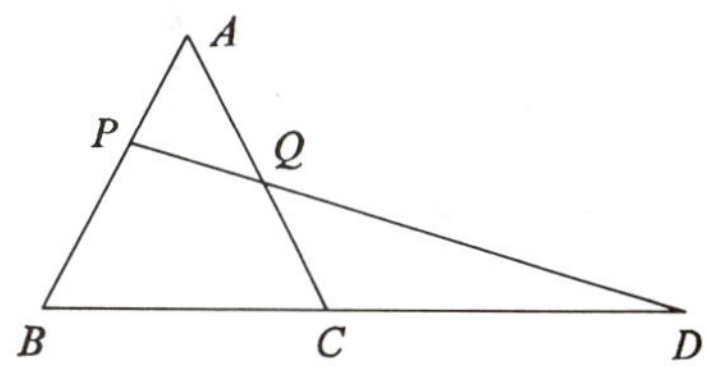

15. GIVEN Quadrilateral $ABCD$; $\overline{AB} \cong \overline{AD}$; $\overline{BC} \cong \overline{CD}$; $\overline{BPD}$; $m\angle BAP < m\angle PAD$
    PROVE $m\angle BCP < m\angle PCD$

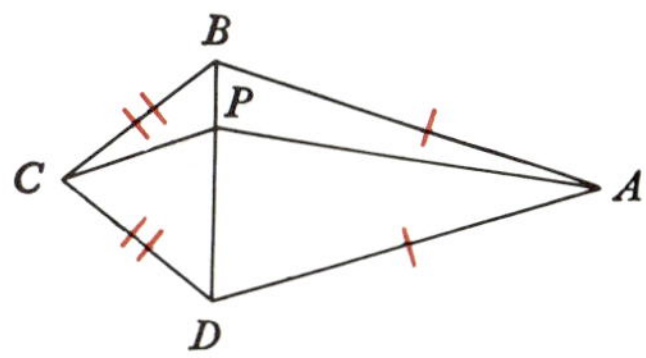

SUGGESTED RESEARCH

1. Prepare a report on isoperimetric theorems. An excellent source is *Geometric Inequalities*, by Nicholas D. Kazarinoff (New York: Random House, 1961).
2. Try to prove that the following statement is true: "If the length of one side of a triangle is greater than the length of another side, then the length of the angle bisector to the shorter side is longer than the angle bisector to the longer side." Is the statement true in all cases?
3. Choose a topic of interest to you from *An Introduction to Inequalities*, by E. F. Beckenback and R. Bellman (New York: Random House, 1961). Present a report on this topic to the rest of your class. You may wish to add algebraic inequalities to those discussed in this chapter.

MATHEMATICAL EXCURSION

Limits

Did you know that if you start walking from one side of a room you cannot walk to the other side? Before you reach the far side, you must reach the midpoint. Before you reach the midpoint, you must reach the midpoint of the segment between you and the first midpoint. And before you get there, you must reach the midpoint of that segment. Before any point is reached, other points before it must have been reached. Hence, *motion is impossible*. This is one of the paradoxes of Zeno of Elea, a Greek mathematician and philosopher who lived about 500 B.C.

Another of Zeno's paradoxes relates to an arrow. An *arrow is motionless throughout its flight*. At any instant the tip of the arrow occupies a specific position in space. While it is occupying this position, it is not moving. Therefore, at every instant of its flight, the arrow is motionless.

A third of Zeno's paradoxes is about the race between the tortoise and the hare. According to this paradox, it is impossible for the hare ever to catch the tortoise. For example, assume that the tortoise is 100 feet ahead of the hare. While the hare moves 10 feet, the tortoise moves 1 foot. Then, when the hare reaches the starting point of the tortoise, the tortoise will be 10 feet away. When the hare reaches that position of the tortoise, the tortoise will be 1 foot away. When the hare covers that foot, the tortoise will be $\frac{1}{10}$ foot away. Whenever the hare reaches a position the tortoise has been, the tortoise will have moved on. So the hare, although getting ever closer to the tortoise, will never catch it.

These paradoxes greatly confused the ancient Greek mathematicians and philosophers. Whether he was aware of it or not, Zeno had put his finger on one of the most serious weaknesses of the mathematics of ancient Greece. The Greeks did not know how to handle those types of situations which deal with *infinitesimals*. In fact, it was not until the eighteenth century that mathematicians dealt with them correctly.

The arrow paradox raises a question that is not involved in the others: What is motion? We can answer this by analyzing motion pictures. As you probably know, a motion picture is actually a sequence of still pictures shown in rapid succession, thus giving the illusion of movement. This concept of motion, being still in continuously changing positions, effectively voids the paradox.

Let us reconsider the trip across the room. We shall call the distance across the room 1 unit of length. Suppose you walk halfway across the room in half a minute, and at the end of the half-minute you continue walking half the remaining distance. If, in succeeding half-minutes, you walk half the distance that remains from the half-minute before, will you ever reach the opposite wall? Before many half-minutes have passed, it will become impossible to move without touching the far wall. But theoretically you will never arrive there. Essentially, this problem is the same as repeatedly bisecting a segment. We cannot successively bisect a segment until we ultimately reach a point—an indivisible unit. There are always other points between any two points. Thus, it is not possible to cross the room in the described manner.

Analyze the following situation to determine if there is any difference between it and the above situation. Walking at the same constant rate, you walk halfway across the room, then half the remaining distance, then half that remaining distance. If you continue, each time walking half the remaining distance, will you ever succeed in reaching the other side? In both instances, you travel $\frac{1}{2} + \frac{1}{4} + \frac{1}{8} + \frac{1}{16} + \cdots$ units of distance. In the first situation, the time required for the trip is $\frac{1}{2} + \frac{1}{2} + \frac{1}{2} + \frac{1}{2} + \frac{1}{2} + \cdots$ minutes. But in the second situation, the time required is $\frac{1}{2} + \frac{1}{4} + \frac{1}{8} + \frac{1}{16} + \cdots$ minutes. In the first instance, you never reach the other side because it would require an infinite amount of time, *even though the distance is finite*. In the second instance, you do reach the other side, in one minute, even though both distance and time are expressed as *a sum with infinitely many terms*.

The Greek mathematicians and philosophers were troubled by the hare and the tortoise paradox because they failed to consider time consumed in the same way as distance traveled.

The point of the above discussion is that the sum of infinitely many terms can be finite. For example, the sum $\frac{1}{2} + \frac{1}{4} + \frac{1}{8} + \cdots = 1$. But if there are infinitely many terms, we can never add them together, so how can we say that the sum is 1? Here, the term *sum* has a slightly different meaning than when we say the sum of 3 and 5 is 8.

A set of numbers ordered like the natural numbers, $a_1, a_2, a_3, \ldots, a_n$, is a sequence. A sequence can be finite or infinite. If it is infinite, then there must be, after a finite number of terms, a general rule of formation of terms. Sometimes the rule can be given as a *general term*. In the sequence

$$\frac{1}{2}, \frac{1}{4}, \frac{1}{8}, \frac{1}{16}, \ldots$$

the general term is $\frac{1}{2^n}$. If we wish to indicate that the sequence is infinite, we may write

$$\frac{1}{2}, \frac{1}{4}, \frac{1}{8}, \ldots, \frac{1}{2^n}, \ldots$$

In contrast to a sequence, a *series* is the indicated sum of the terms of a sequence. The series

$$\frac{1}{2} + \frac{1}{4} + \frac{1}{8} + \cdots + \frac{1}{2^n} + \cdots$$

is an *infinite series*, the nth term being $\frac{1}{2^n}$. Here the general term describes all terms. The first term of the series is $\frac{1}{2^1} = \frac{1}{2}$. In the series

$$\frac{1}{2} + \frac{1}{2} + \frac{1}{2} + \cdots + \frac{1}{2} + \cdots$$

the general term is $\frac{1}{2}$. A finite number of terms of this series can be found whose sum is greater than any number you care to choose. For example, consider any natural number N. The sum of the first $2N + 1$ terms is equal to $N + \frac{1}{2}$.

Intuitively, it seems certain that we can find a term of the sequence $\frac{1}{2}, \frac{1}{4}, \frac{1}{8}, \ldots, \frac{1}{2^n}, \ldots$ that is less than any number we care to name, and all the terms following it are also less than the number we named. But this does not guarantee that the series $\frac{1}{2} + \frac{1}{4} + \frac{1}{8} + \cdots + \frac{1}{2^n} + \cdots$ will not ultimately exceed any number we name. On the other hand, it seems reasonable that the series $\frac{1}{2} + \frac{1}{4} + \frac{1}{8} + \cdots + \frac{1}{2^n} + \cdots$ continues to get closer and closer to some finite number, without ever reaching it. The number we are getting closer and closer to is 1.

The sequence $\frac{1}{1}, \frac{1}{2}, \frac{1}{3}, \frac{1}{4}, \frac{1}{5}, \ldots, \frac{1}{n}, \ldots$ is getting closer and closer to zero. You might infer from this that $\frac{1}{1} + \frac{1}{2} + \frac{1}{3} + \cdots + \frac{1}{n} + \cdots$ would also be forever getting closer to some finite number. But your inference would be wrong. We can add enough terms of this series and get a sum greater than any number we care to name.

The ideas we have been discussing are somewhat vague. We shall now try to express them in a precise mathematical form. First, we should note that when we are concerned about an infinite series we are also concerned about an infinite sequence. For example, when we ask whether the series $\frac{1}{2} + \frac{1}{4} + \cdots + \frac{1}{2^n} + \cdots$ is approaching closer and closer to some finite number, we are really concerned with whether the sequence $\frac{1}{2}, \left(\frac{1}{2} + \frac{1}{4}\right), \left(\frac{1}{2} + \frac{1}{4} + \frac{1}{8}\right), \ldots, \left(\frac{1}{2} + \frac{1}{4} + \frac{1}{8} + \cdots + \frac{1}{2^n}\right), \ldots$ is approaching closer and closer to some finite number. We refer to this as *approaching a limit*.

We say that a sequence approaches a limit L if (1) a term of the sequence can be found that differs from L by less than any number we can name, and (2) all subsequent terms must differ from L by less than that amount.

To show how we can apply this definition, let us see whether the limit of the sequence $\frac{1}{2}, \frac{1}{4}, \frac{1}{8}, \cdots, \frac{1}{2^n}, \cdots$ is zero. Since the supposed limit is zero, we must show that we can find a term of the sequence that is less than any specified positive number. Suppose we wish to find a term $\frac{1}{2^n}$ that is less than any positive number M.

$$\frac{1}{2^n} < M$$

$$2^n > \frac{1}{M}$$

If we set $n \geq \frac{1}{M}$, then $2^n \geq 2^{\frac{1}{M}} > \frac{1}{M}$ provided that $\frac{1}{M} \geq 1$. If $\frac{1}{M} < 1$, then the first term of the sequence, $\frac{1}{2}$, is less than M because $\frac{1}{M} < 1$ implies that $M > 1$. Hence, $\frac{1}{2^n}$ is less than M provided that $n \geq \frac{1}{M}$. Each term after $\frac{1}{2^n}$ is less than $\frac{1}{M}$ because each such term is less than $\frac{1}{2^n}$.

D **Definition** The series $a_1 + a_2 + \cdots + a_n + \cdots$ has a sum if and only if the sequence $(a_1), (a_1 + a_2), (a_1 + a_2 + a_3), \cdots, (a_1 + a_2 + \cdots + a_n), \cdots$ has a limit. The limit of the sequence is the sum of the series.

The series $\frac{1}{2} + \frac{1}{4} + \frac{1}{8} + \cdots + \frac{1}{2^n} + \cdots$ has a sum if and only if $\left(\frac{1}{2}\right)$, $\left(\frac{1}{2} + \frac{1}{4}\right), \left(\frac{1}{2} + \frac{1}{4} + \frac{1}{8}\right), \cdots = \frac{1}{2}, \frac{3}{4}, \frac{7}{8}, \cdots$ has a limit.

The general term of this sequence is $\frac{2^n - 1}{2^n}$. We can show that the limit approached by the sequence is 1.

$$\left|1 - \frac{2^n - 1}{2^n}\right| < M$$

implies

$$\left|\frac{2^n - 2^n + 1}{2^n}\right| < M$$

or

$$\frac{1}{2^n} < M$$

This relationship between n and M is the same as the one we derived in showing that $\frac{1}{2}, \frac{1}{4}, \frac{1}{8}, \cdots, \frac{1}{2^n}, \cdots$ approaches the limit zero. Since $\frac{1}{2}, \frac{3}{4}, \frac{7}{8}, \cdots, \frac{2^n - 1}{2^n}, \cdots$ approaches the limit 1, the series from which it was derived, $\frac{1}{2} + \frac{1}{4} + \frac{1}{8} + \cdots + \frac{1}{2^n} + \cdots$ has the sum 1.

An infinite sequence $a_1, a_2, \cdots, a_n, \cdots$ that has a limit is said to be a *convergent* sequence. Otherwise, it is called a *divergent* sequence.

If the sequence $(a_1), (a_1 + a_2), (a_1 + a_2 + a_3), \cdots,$ $(a_1 + a_2 + a_3 + \cdots + a_n), \cdots$ converges, we say that the series

$$a_1 + a_2 + a_3 + a_4 + \cdots + a_n + \cdots$$

is convergent. A series that is not convergent is divergent.

The sequence $\frac{1}{1}, \frac{1}{2}, \frac{1}{3}, \cdots, \frac{1}{n}, \cdots$ is convergent because it has a limit of zero.

$$\left|0 - \frac{1}{n}\right| < M$$

$$\frac{1}{n} < M$$

$$n > \frac{1}{M}$$

If we choose n such that $n > \frac{1}{M}$ regardless of the M we choose, then we can show that $|L - a_n| < M$ for $L = 0$ and $a_n = \frac{1}{n}$.

Does the corresponding series $1 + \frac{1}{2} + \frac{1}{3} + \cdots + \frac{1}{n} + \cdots$ have a sum?

In advanced mathematics there is a theorem for determining whether a series converges or diverges. But even without this theorem we can determine that the series $1 + \frac{1}{2} + \frac{1}{3} + \cdots + \frac{1}{n} + \cdots$ diverges.

If we double the number of terms of the series, the sum is increased by $\frac{1}{2}$ or more. Going from one to two terms the sum is increased by $\frac{1}{2}$. Consider an increase from n to $2n$ terms.

$$1 + \frac{1}{2} + \frac{1}{3} + \cdots + \frac{1}{n} + \frac{1}{n+1} + \cdots + \frac{1}{2n}$$

The sum has increased

$$\frac{1}{n+1} + \frac{1}{n+2} + \cdots + \frac{1}{2n}$$

but

$$\frac{1}{n+1} > \frac{1}{2n}$$

$$\frac{1}{n+2} > \frac{1}{2n}$$

$$\cdots\cdots\cdots\cdots$$

$$\frac{1}{2n} = \frac{1}{2n}$$

$$\frac{1}{n+1} + \frac{1}{n+2} + \cdots + \frac{1}{2n} > \frac{1}{2n} + \frac{1}{2n} + \cdots + \frac{1}{2n}$$

$$\frac{1}{n+1} + \frac{1}{n+2} + \cdots + \frac{1}{2n} > \frac{n}{2n} = \frac{1}{2}$$

Since we can continue doubling the number of terms, we can make the sum exceed any number. Although the sequence

$1, \frac{1}{2}, \frac{1}{3}, \cdots, \frac{1}{n}, \cdots$ converges to zero, the corresponding series

$1 + \frac{1}{2} + \frac{1}{3} + \cdots + \frac{1}{n} + \cdots$ is divergent.

We have shown by the last example that showing that a series converges involves more than "it looks like it has a limit."

The concept of *limit* is important to many situations in elementary geometry. In Chapter 9 we shall define the circumference of a circle as "the limit of the perimeters of the inscribed regular polygons as the number of sides of the polygons increases without bound." Built into this definition is the assumption that the limit exists. Otherwise, we would have to prove that the limit exists.

We could use intuition to tell us that the limit exists, but, as we learned earlier, intuition is a dangerous tool. Using methods of higher mathematics, not only can we prove that the perimeter of the inscribed polygon approaches a limit as the number of sides increases, but we can also establish the value of the limit.

In Chapter 12, we shall define the area of a circle. The area will also be defined as a limit. In this definition, too, we assume that a limit exists.

The formula for finding the area of a rectangle is given in Chapter 12 as a postulate. Since the unit of area is defined as a square, we might wonder why it is necessary to postulate that the area of a rectangle is the product of its base and its altitude. We might wonder why we cannot prove that the formula is correct. If a unit can be found that will exactly divide the altitude into a units and the base into b units, we can prove that the rectangle can be subdivided into $a \cdot b$ units. The details of the proof are left as an exercise.

Suppose, however, that the base and the altitude are *incommensurable*; that is, there is no unit that will exactly divide both the base and the altitude. Now we can select a unit that exactly divides one of the sides. For example, the short side can be used as a unit. Suppose this unit divides the other side some whole number of times, giving a remainder which we call ϵ. We can then use ϵ as a unit. It is likely that ϵ will not divide either side exactly, but the remainders, α and β, will be less than ϵ on each side. Next, use the larger of α or β for a new unit. Obviously, if this process is continued without stopping, the unit will approach zero. But it will never become zero. For if it does, the base and the altitude are not incommensurable.

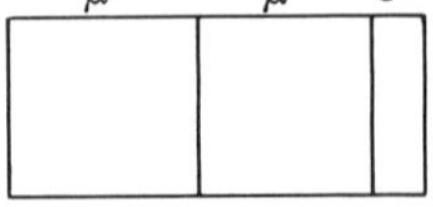

The altitude is divided into a units with remainder α, and the base is divided into b units with remainder β. The product is

$$(a + \alpha)(b + \beta) = ab + b\alpha + a\beta + \alpha\beta$$

As α and β both approach zero, then $b\alpha$ approaches zero. Similarly, $a\beta$ and $\alpha\beta$ approach zero. Then the sum $(b\alpha + a\beta + \alpha\beta)$ approaches zero and the area approaches $a \cdot b$. If, in the incommensurable case, we define the area as the value approached when the remainders approach zero, then the result is the same as in the commensurable case, namely area $= a \cdot b$.

EXERCISES

1. Consider the series $1 + \frac{1}{2} + \frac{1}{4} + \cdots$. Compare 2 minus the sum of the first n terms with the nth term. If it is agreed that we can find a term less than any assigned number, what can we say about the sum?
2. Prove that the sequence $\frac{1}{2}, \frac{2}{3}, \frac{3}{4}, \ldots, \frac{n}{n+1}, \ldots$ approaches the limit 1.
3. Does the sequence 2, 2, 2, $\cdots$ approach a limit? Justify your answer.
4. On the assumption that the circumference of a circle is 2π times the radius, use the limit process to show that the area is πr^2.
5. Prove the formula for the area of a rectangle in the commensurable case.

6 Parallelism

For nearly 2,100 years, mathematicians accepted Euclid's Parallel Postulate, although not without question. However, during the last two hundred years, geometries differing from Euclid's on this matter have been developed. These geometries have played an important role in recent scientific developments. In this chapter, we shall consider the existence of parallel lines and Euclid's Parallel Postulate. We shall also discuss methods of deriving additional information about parallel lines.

6-1 PARALLEL LINES

Railroad tracks and the lines on a football field are two familiar examples of parallel lines. From observing such examples, you may have various notions about parallel lines: that parallel lines do not intersect, that they are always the same distance apart, and that they have the same direction.

To the mathematician, these notions are informal ideas and do not form an acceptable definition. However, we shall build a definition of parallel lines based on these intuitive notions.

The first notion mentioned was that parallel lines do not intersect. However, we may have a pair of lines that do not intersect, yet are not parallel. Do you recall our discussion of skew lines in Chapter 1? In each figure below, l_1 and l_2 are skew lines.

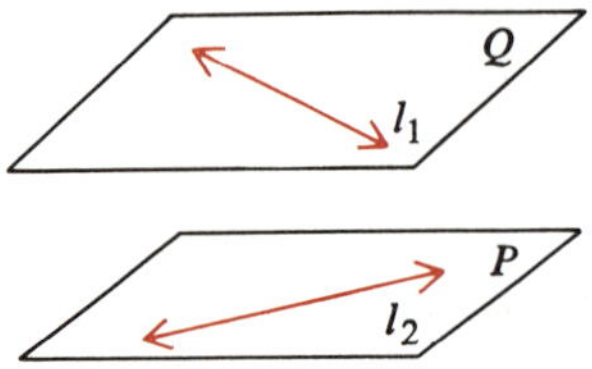

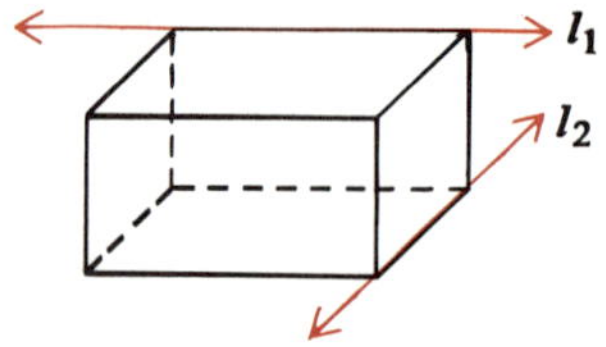

Are l_1 and l_2 coplanar?

We see from this that a definition of parallel lines must specify that lines are in the same plane. Furthermore, to satisfy the conditions of our second and third notions, the definition must specify that the lines do not intersect.

D **Definition 6-1** Two distinct lines are **parallel** if and only if they are coplanar and do not intersect.

By this definition, a line is not parallel to itself. We shall consider rays or line segments that are subsets of parallel lines to be parallel, provided they are not subsets of the same line. The symbol // is used to represent "parallel to." Thus, we may write

$$\overleftrightarrow{AB} \parallel \overleftrightarrow{CD} \qquad \overline{AB} \parallel \overrightarrow{CD} \qquad \overline{AB} \parallel \overline{CD} \qquad \overline{AB} \parallel \overleftrightarrow{CD}$$

Now that we have defined parallel lines, we must prove that parallel lines exist. First, however, we must prove Theorem 6-1.1.

T **Theorem 6-1.1** If two distinct lines in the same plane are perpendicular to the same line, then they are parallel.

Given $\overleftrightarrow{AB} \perp \overleftrightarrow{RQ}$; $\overleftrightarrow{CD} \perp \overleftrightarrow{RQ}$

Prove $\overleftrightarrow{AB} \parallel \overleftrightarrow{CD}$

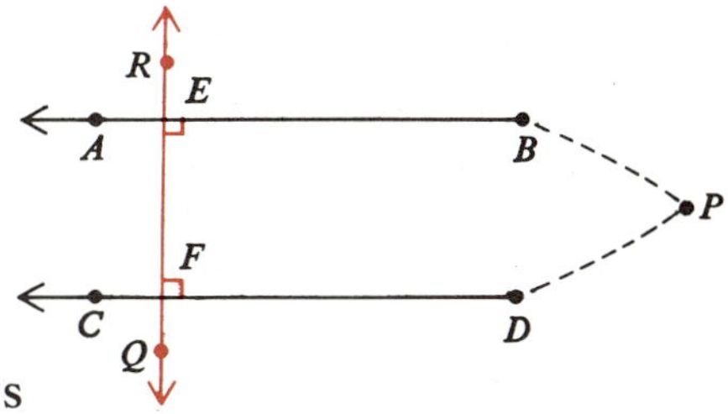

Proof

STATEMENTS	REASONS
1. $\overleftrightarrow{AB} \perp \overleftrightarrow{RQ}$ and $\overleftrightarrow{CD} \perp \overleftrightarrow{RQ}$	1. Given.
2. Assume that $\overleftrightarrow{AB}$ is not parallel to $\overleftrightarrow{CD}$, but that $\overleftrightarrow{AB}$ intersects $\overleftrightarrow{CD}$ in P.	2. Assumption.
3. $\triangle EPF$ is formed.	3. Definition of a triangle. (1-31)
4. $m\angle QFP > m\angle FEP$	4. The measure of an exterior angle of a triangle is greater than the measure of either remote interior angle. (5-2.1)
5. But $\angle QFP \cong \angle FEP$	5. All right angles are congruent. (3-1.1)
6. Therefore, it is false that $\overleftrightarrow{AB}$ intersects $\overleftrightarrow{CD}$ in P.	6. Step 4 contradicts step 5 which is known true.
7. $\overleftrightarrow{AB} \parallel \overleftrightarrow{CD}$	7. The contradiction of a false statement is true.

In order to prove the existence of parallel lines, we shall use Theorem 6-1.1 to prove the corollary below.

C

Corollary 6-1.1a Parallel lines exist in any given plane.

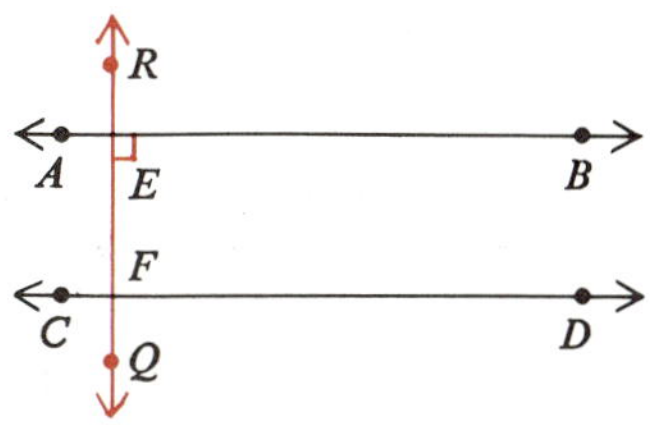

Proof Consider $\overleftrightarrow{AB} \perp \overleftrightarrow{RQ}$. To show that there exists a line parallel to $\overleftrightarrow{AB}$, we consider any point on $\overleftrightarrow{RQ}$, say F, and show that there exists a line through F parallel to $\overleftrightarrow{AB}$. We already know that in a plane, there is one and only one line perpendicular to a given line through a given point of the line. (Theorem 4-4.2) Therefore, there exists a line in the plane of $\overleftrightarrow{RQ}$ and $\overleftrightarrow{AB}$ through F perpendicular to $\overleftrightarrow{RQ}$. By Theorem 6-1.1, we proved that this perpendicular, $\overleftrightarrow{CD}$, is parallel to $\overleftrightarrow{AB}$; this then proves the existence of parallel lines.

It follows naturally that we should also want to prove the statement that through the given point, there exists only one line parallel to the given line. For more than 20 centuries, mathematicians have proved this statement in many ways, but never without implicit use of Euclid's Parallel Postulate. In the last two centuries, mathematicians realized that it is impossible to prove the statement without assuming a statement equivalent to the Parallel Postulate, and they showed that assumptions that contradict the Parallel Postulate lead to different but consistent conclusions. For further discussion of this point, see the Mathematical Excursion at the end of this chapter, and the discussion of non-Euclidean geometry on page 597. For our purposes, we shall assume that there is only one line parallel to a given line through a point not in the given line.

Postulate 6-1 The Parallel Postulate Through a given point not contained in a given line, there exists only one line parallel to the given line.

Example Restate the Parallel Postulate, using specific lines and a specific point.

Solution If F is not on $\overleftrightarrow{AB}$, then there is only one line, $\overleftrightarrow{CD}$, which contains F and is parallel to $\overleftrightarrow{AB}$. That is, no other line may contain F and be parallel to $\overleftrightarrow{AB}$.

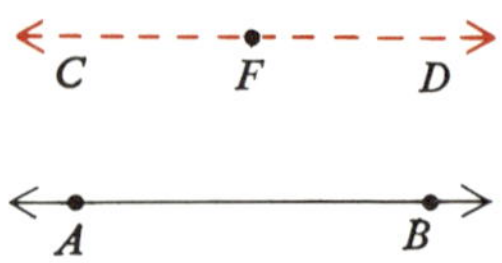

We shall now state three more corollaries to Theorem 6-1.1. The proofs of these corollaries will be left as exercises.

Corollary 6-1.1b In a plane, if a line is perpendicular to one of two parallel lines, then it is also perpendicular to the other line.

Corollary 6-1.1c If each of two lines is parallel to a third line, then they are parallel to each other.

Corollary 6-6.1d In a plane, if a line is perpendicular to one of two parallel lines and if another line is perpendicular to the second of the two parallel lines, then these two perpendicular lines are parallel to each other.

Transversals

In the figure at right, consider line t to be a transversal of the coplanar lines l_1 and l_2. Line k in the figure is *not* considered to be a transversal. How does the intersection of l_1 and l_2 with t differ from the intersection of l_1 and l_2 with k?

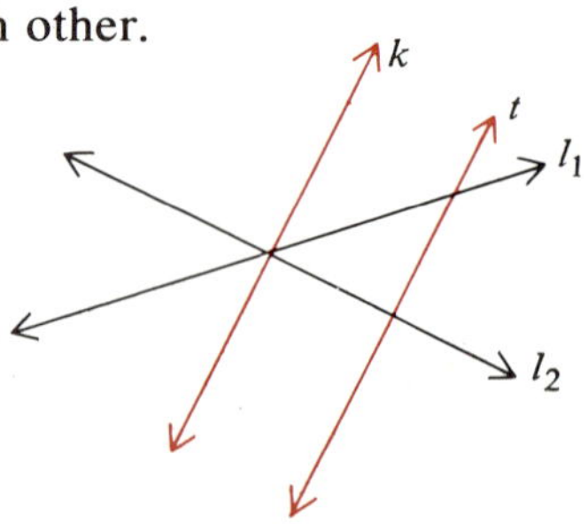

Definition 6-2 A line is a **transversal** of two or more coplanar lines if and only if it intersects each of these lines in different points.

How many angles are formed when two lines are intersected by a transversal? Several pairs of these angles are very useful to us as we study parallel lines.

These important pairs of angles are classified and illustrated below.

Alternate Interior Angles — $\angle 3$ and $\angle 6$; $\angle 4$ and $\angle 5$

Interior Angles on the Same Side of a Transversal — $\angle 3$ and $\angle 5$; $\angle 4$ and $\angle 6$

Corresponding Angles — $\angle 1$ and $\angle 5$; $\angle 2$ and $\angle 6$; $\angle 3$ and $\angle 7$; $\angle 4$ and $\angle 8$

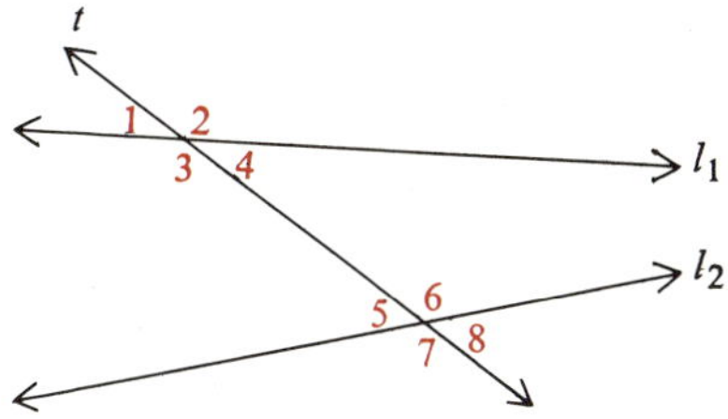

Class Exercises

Indicate whether or not each of the following statements is equivalent to our definition of parallel lines. Justify each answer.

1. Two lines that do not intersect are parallel.
2. Two lines are parallel if they lie in the same plane, do not intersect, and have a line that is perpendicular to each of them.
3. Two line segments that lie in the same plane and do not intersect are parallel.
4. Two lines are parallel if they lie in the same plane and do not intersect.
5. Two lines that share no points and lie in the same plane are parallel.

EXERCISES

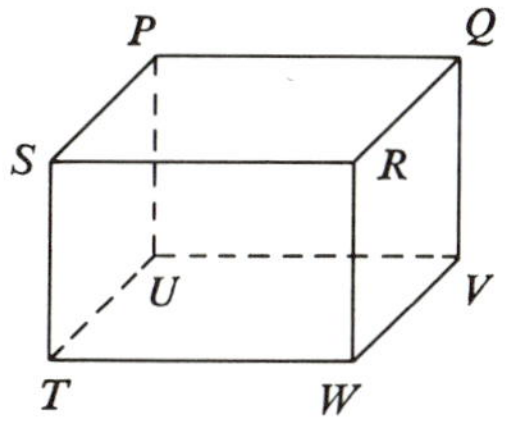

Exercises 1–6

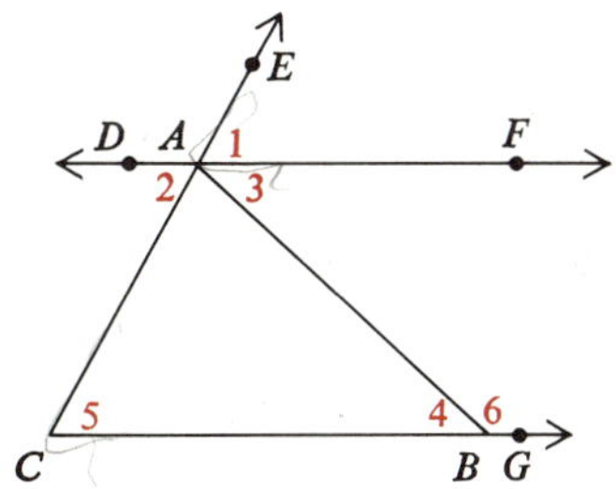

Exercises 7–12

A Refer to the figure at left above and complete Exercises 1–5 in at least two ways.

1. $\overline{PQ} \parallel$ ____ 2. $\overline{RW} \parallel$ ____
3. $\overline{VW} \parallel$ ____ 4. $\overline{ST} \parallel$ ____
5. $\overline{RQ} \parallel$ ____ 6. $\overline{UV} \parallel$ ____
7. In the figure at right above, is $\overleftrightarrow{AB} \parallel \overleftrightarrow{CB}$?
8. Why are two sides of a triangle never parallel?

Referring to the figure at right above, name each pair of angles given.

9. $\angle 1$ and $\angle 5$ 10. $\angle 3$ and $\angle 4$ 11. $\angle 2$ and $\angle 5$ 12. $\angle 3$ and $\angle 6$

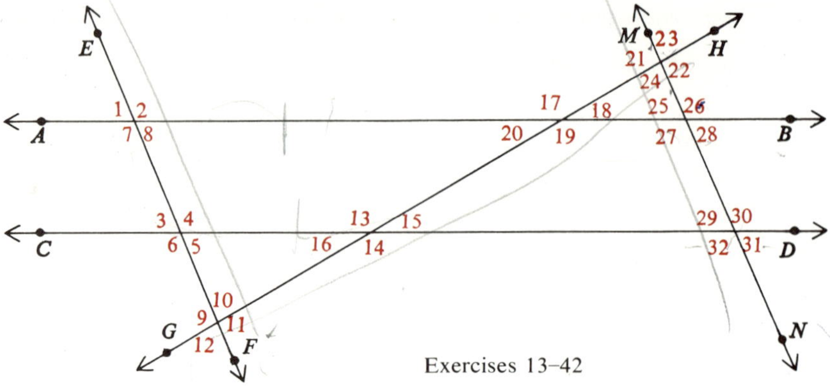

Exercises 13–42

B Identify two pairs of alternate interior angles formed by $\overleftrightarrow{AB}$, $\overleftrightarrow{CD}$, and each of the following transversals:

13. $\overleftrightarrow{EF}$ 14. $\overleftrightarrow{HG}$ 15. $\overleftrightarrow{MN}$

Identify four pairs of corresponding angles formed by $\overleftrightarrow{AB}$, $\overleftrightarrow{CD}$, and each of the following transversals:

16. $\overleftrightarrow{EF}$ 17. $\overleftrightarrow{HG}$ 18. $\overleftrightarrow{MN}$

Identify two pairs of interior angles on the same side of each of the following transversals of $\overleftrightarrow{AB}$ and $\overleftrightarrow{CD}$:

19. $\overleftrightarrow{EF}$ 20. $\overleftrightarrow{HG}$ 21. $\overleftrightarrow{MN}$

Identify two pairs of alternate interior angles formed by $\overleftrightarrow{MN}$, $\overleftrightarrow{EF}$, and each of the following transversals:

22. $\overleftrightarrow{AB}$ 23. $\overleftrightarrow{CD}$ 24. $\overleftrightarrow{HG}$

Identify four pairs of corresponding angles formed by $\overleftrightarrow{MN}$, $\overleftrightarrow{EF}$, and each of the following transversals:

25. $\overleftrightarrow{AB}$ 26. $\overleftrightarrow{CD}$ 27. $\overleftrightarrow{HG}$

Identify two pairs of interior angles on the same side of each of the following transversals of $\overleftrightarrow{MN}$ and $\overleftrightarrow{EF}$:

28. $\overleftrightarrow{AB}$ 29. $\overleftrightarrow{CD}$ 30. $\overleftrightarrow{HG}$

Identify the transversal for each of the following pairs of alternate interior angles:

31. $\angle 7$ and $\angle 4$ 32. $\angle 20$ and $\angle 15$ 33. $\angle 21$ and $\angle 11$
34. $\angle 32$ and $\angle 4$

Identify the transversal for each of the following pairs of corresponding angles:

35. $\angle 1$ and $\angle 3$ 36. $\angle 10$ and $\angle 23$ 37. $\angle 28$ and $\angle 31$
38. $\angle 12$ and $\angle 19$

Each line listed below is a transversal. Identify the angle which forms a pair of alternate interior angles with each angle given below, with respect to the given transversal.

39. $\angle 7, \overleftrightarrow{EF}$ 40. $\angle 29, \overleftrightarrow{MN}$ 41. $\angle 10, \overleftrightarrow{HG}$ 42. $\angle 15, \overleftrightarrow{CD}$

43. Prove that in a plane, if a line is perpendicular to one of two parallel lines, then it is also perpendicular to the other line. (Corollary 6-1.1b.)
44. Prove that in a plane, if each of two lines is parallel to a third line, then they are parallel to each other. (Corollary 6-1.1c.)
45. Prove that in a plane, if a line is perpendicular to one of two parallel lines and if another line is perpendicular to the second of the two parallel lines, then these two perpendiculars are parallel to each other. (Corollary 6-1.1d.)

C

46. Prove that in a plane, if a line is perpendicular to one of two nonparallel lines and if another line is perpendicular to the second of the nonparallel lines, then these perpendiculars are also nonparallel.

6-2 PROVING LINES PARALLEL

In the previous section, we introduced certain pairs of angles that are formed when two lines are intersected (cut) by a transversal: alternate interior angles, interior angles on the same side of a transversal, and corresponding angles. These pairs of angles can be used to prove lines parallel. Let us begin by examining the figures below.

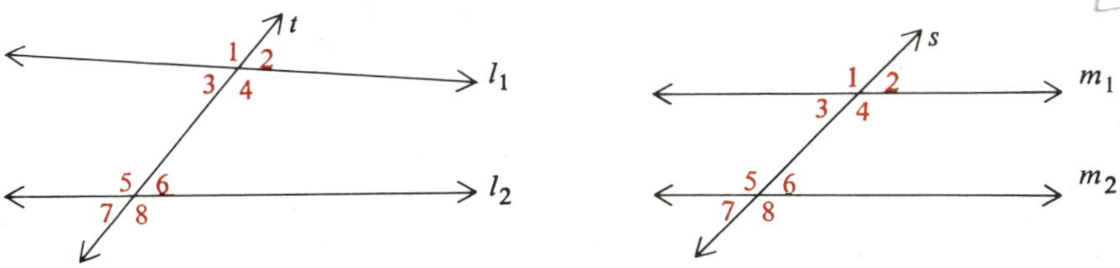

In the figure on the right, which pairs of angles appear congruent? Do the same pairs appear congruent in the figure on the left? Which pairs of angles appear supplementary in the figure on the right? Do the same pairs appear supplementary in the figure on the left? Which pair of lines appears to be parallel? What conclusion can you draw? This conclusion is stated by the following theorem and its corollaries:

T

Theorem 6-2.1 If two lines are cut by a transversal so that the alternate interior angles are congruent, then the lines are parallel.

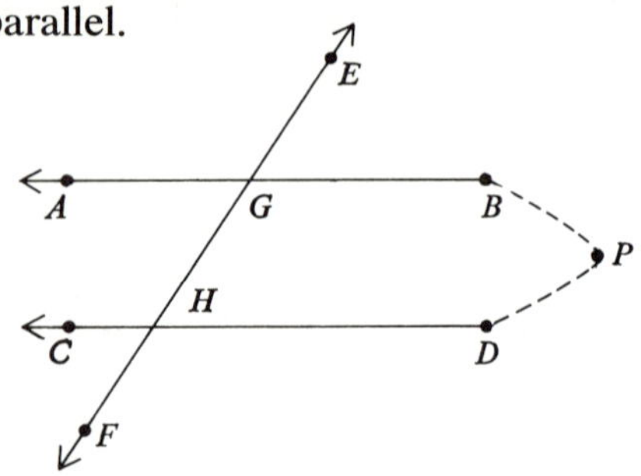

INDIRECT PROOF

GIVEN Transversal $\overleftrightarrow{EF}$ intersects $\overleftrightarrow{AB}$ and $\overleftrightarrow{CD}$ at points G and H, respectively; $\angle AGH \cong \angle DHG$.

PROVE $\overleftrightarrow{AB} \parallel \overleftrightarrow{CD}$

STATEMENTS	REASONS
1. Transversal $\overleftrightarrow{EF}$ intersects $\overleftrightarrow{AB}$ and $\overleftrightarrow{CD}$ at points G and H, respectively.	1. Given.
2. Assume $\overleftrightarrow{AB}$ is not parallel to $\overleftrightarrow{CD}$; therefore $\overleftrightarrow{AB}$ intersects $\overleftrightarrow{CD}$ at P.	2. Assumption.
3. $\triangle HGP$ is formed.	3. Definition of a triangle. (1-31)
4. $\angle AGH > \angle DHG$	4. The measure of an exterior angle of a triangle is greater than the measure of either remote interior angle. (5-2.1)
5. But $\angle AGH \cong \angle DHG$	5. Given.
6. Therefore "$\overleftrightarrow{AB}$ is not parallel to $\overleftrightarrow{CD}$" is false.	6. Step 4 contradicts step 5.
7. $\overleftrightarrow{AB} \parallel \overleftrightarrow{CD}$	7. If the negation of a statement is false, the statement is true.

Two useful corollaries follow directly from Theorem 6-2.1. The proofs are left as exercises.

C

Corollary 6-2.1a If two lines are cut by a transversal so that the corresponding angles are congruent, then the lines are parallel.

C

Corollary 6-2.1b If two lines are cut by a transversal so that the interior angles on the same side of the transversal are supplementary, then the lines are parallel.

In order to prove lines parallel, we may have to establish certain relationships among various angles formed by a transversal with the two lines. Many times it will be necessary to prove triangles congruent in order to establish these relationships.

Summary. Proving Two or More Lines Parallel

We now have five methods of proving that lines are parallel:

1. Prove a pair of alternate interior angles congruent.
2. Prove a pair of corresponding angles congruent.
3. Prove a pair of interior angles on the same side of the transversal supplementary.
4. Prove that the lines are parallel to the same line or to parallel lines.
5. Prove that the lines are coplanar and perpendicular to the same line or to parallel lines.

Example

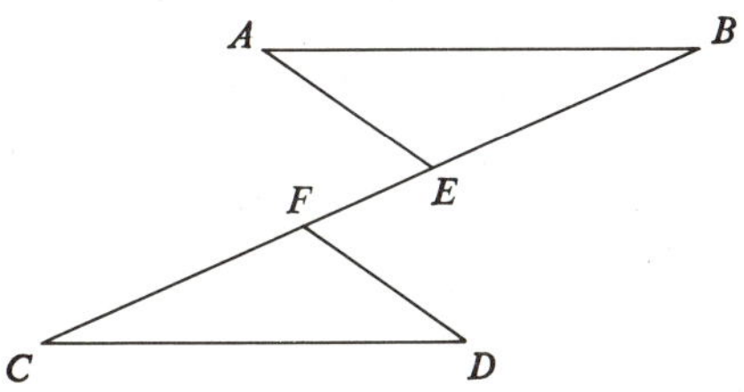

GIVEN $\overline{AE} \cong \overline{DF}$; $\overleftrightarrow{BEFC}$;
$CE = BF$;
$\angle AEF \cong \angle DFE$

PROVE $\overline{AB} \parallel \overline{CD}$

Solution

PROOF

STATEMENTS	REASONS
1. $\overline{AE} \cong \overline{DF}$; $\overleftrightarrow{BEFC}$; $CE = BF$; $\angle AEF \cong \angle DFE$	1. Given.
2. $\overline{FE} \cong \overline{FE}$	2. Every segment is congruent to itself. (3-1.6)
3. $CE - FE = BF - FE$	3. Addition property of equality
4. $CE - FE = CF$; $BF - FE = BE$	4. Point Betweenness Postulate. (2-4)
5. $CF = BE$	5. Substitution Postulate. (2-1)
6. $\angle AEB$ is supplementary to $\angle AEF$; $\angle CFD$ is supplementary to $\angle DFE$.	6. Supplementary Angles Postulate. (1-6)
7. $\angle AEB \cong \angle CFD$	7. Supplements of congruent angles, or of the same angle, are congruent. (3-1.4)
8. $\triangle ABE \cong \triangle DCF$	8. SAS Postulate. (3-1)
9. $\angle B \cong \angle C$	9. Definition of congruent triangles. (3-3)
10. $\overline{AB} \parallel \overline{CD}$	10. Theorem 6-2.1.

Class Exercises

In which of the following figures are $\overleftrightarrow{AB}$ and $\overleftrightarrow{CD}$ parallel? Why?

1. 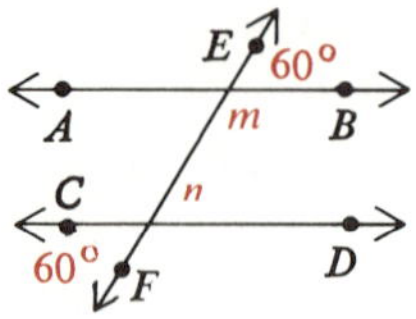

2. 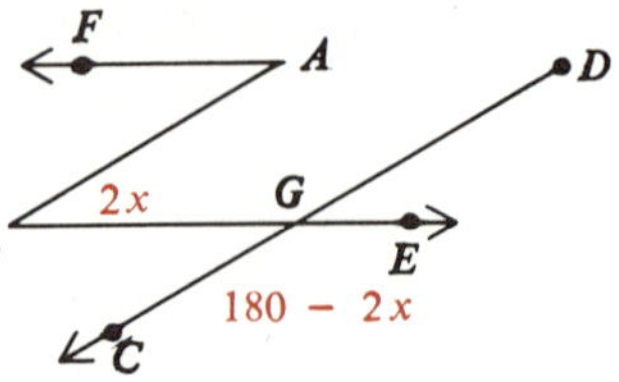

3. 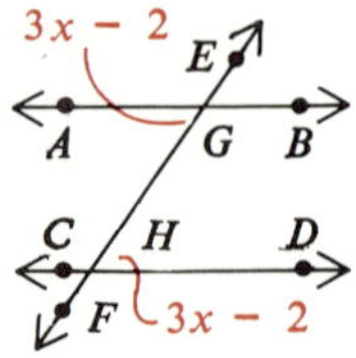

Which line segments are parallel in each of the following? Justify your answers. Congruent parts are marked.

4. 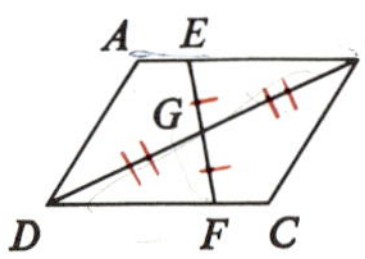

5. 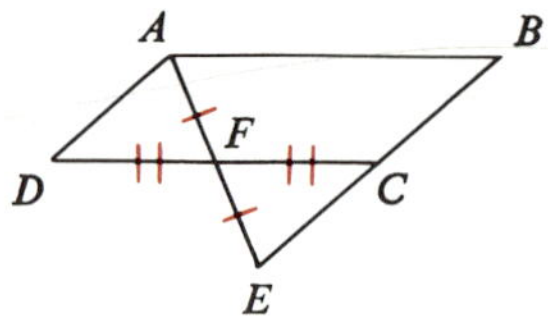

6. 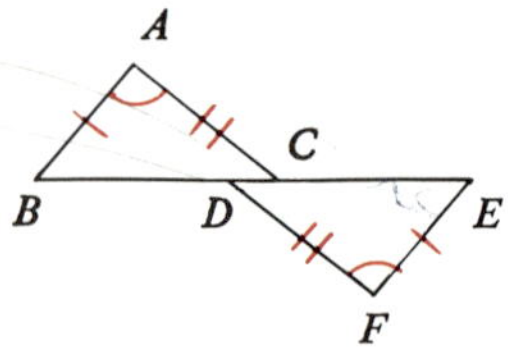

EXERCISES

A In the following, $x = y$. In Exercises 1–5, x and y represent angle measures. State which lines are parallel and justify your answers with a theorem or corollary.

1. 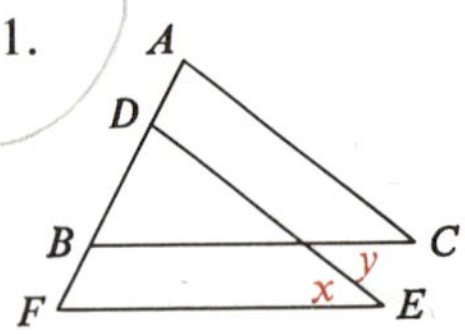

2. 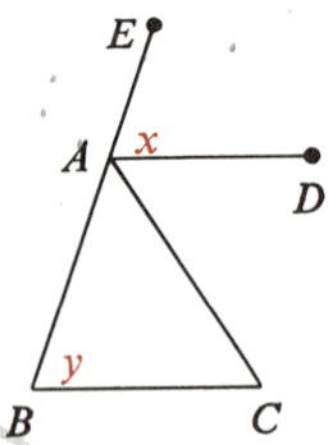

3. 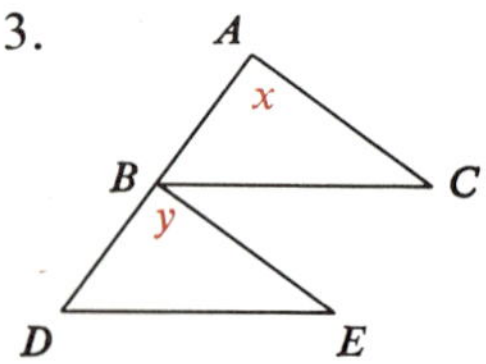

4. 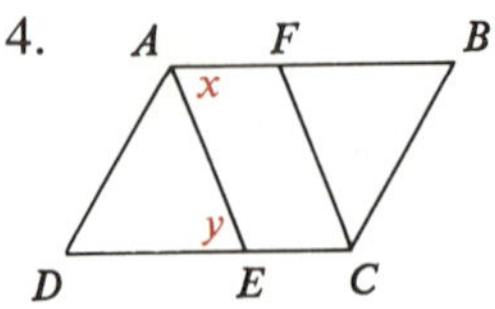

5. 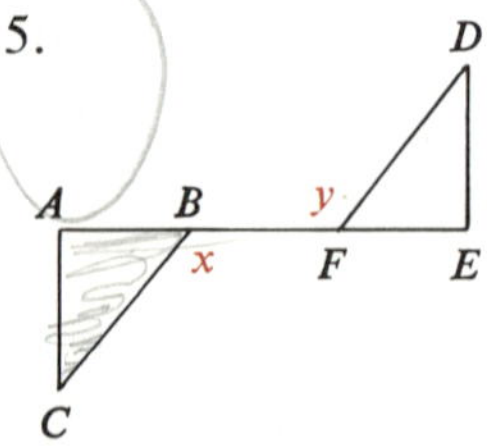

6. 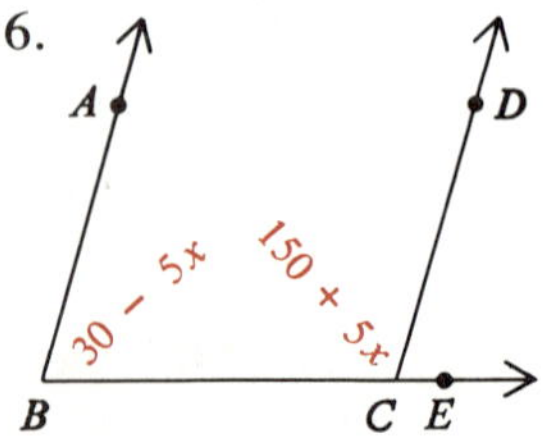

7. **Given** $\overline{BE} \cong \overline{BC}$;
$\angle ABC \cong \angle BEC$;
$\angle DCF \cong \angle BCE$
Prove $\overrightarrow{BA} \parallel \overrightarrow{CD}$

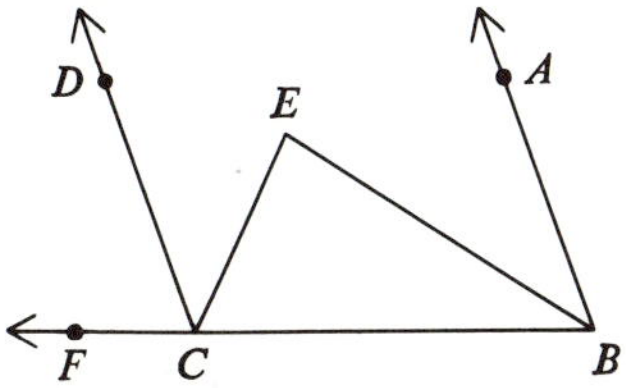

8. **Given** $\angle ABC \cong \angle FDE$;
$\angle CBD \cong \angle EDB$
Prove $\overleftrightarrow{CB} \parallel \overleftrightarrow{DE}$; $\overleftrightarrow{AB} \parallel \overleftrightarrow{DF}$

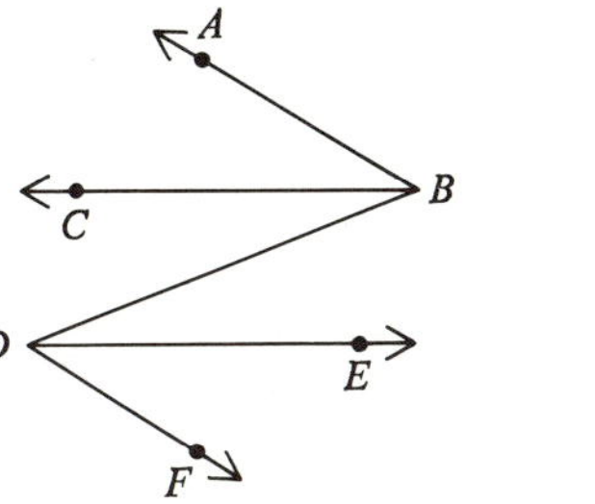

9. Prove Corollary 6-2.1a.

10. Prove Corollary 6-2.1b.

B

11. **Given** $m\angle ACB = 63$;
$m\angle BAC = 27$;
$\overline{AB}$ bisects $\angle DAC$;
$\overline{BC}$ bisects $\angle ACE$.
Prove $\overrightarrow{AD} \parallel \overrightarrow{CE}$

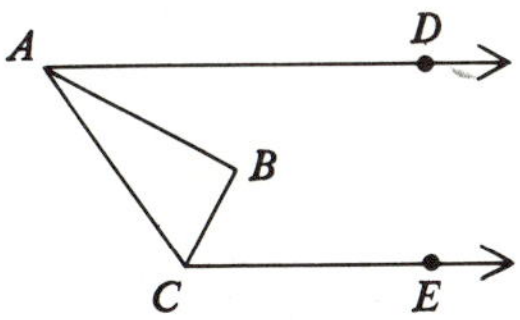

12. **Given** $\overline{AB} \perp \overline{BQ}$; $\overline{QR} \perp \overline{BQ}$;
$\overline{BPCQ}$; $\overline{BP} \cong \overline{QC}$;
$\overline{AB} \cong \overline{RQ}$
Prove $\overline{AC} \parallel \overline{PR}$

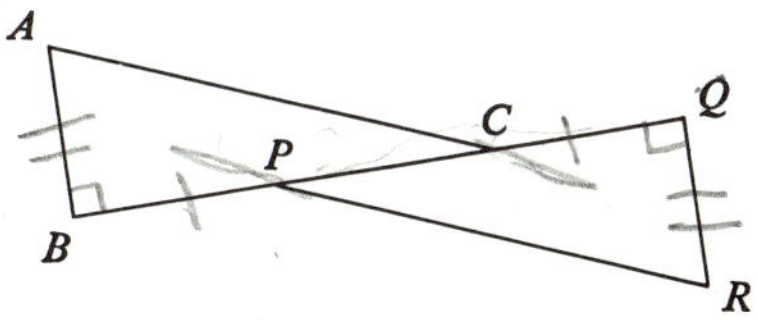

13. **Given** $\angle ADE \cong \angle BCF$;
$\overline{ADCF}$; $AD = FC$;
$\overline{DE} \cong \overline{CB}$
Prove $\overline{AB} \parallel \overline{EF}$

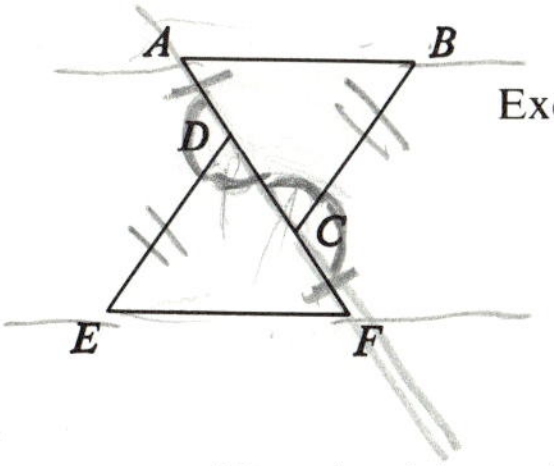

Exercise 13

14. **Given** $\overline{AD}$ and $\overline{BE}$ bisect each other at M.
Prove $\overline{ABC} \parallel \overline{DE}$

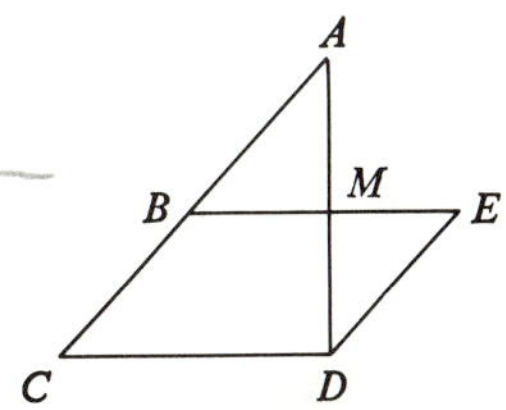

Exercise 14

C

15. Prove that if both pairs of opposite sides of a quadrilateral are congruent, then they are also parallel.

16. **Given** $\overline{PH} \perp \overline{AB}$; $\overline{AH} \cong \overline{BH}$;
$\overline{AB}$ bisects $\angle PAC$.
Prove $\overline{PB} \parallel \overline{AC}$

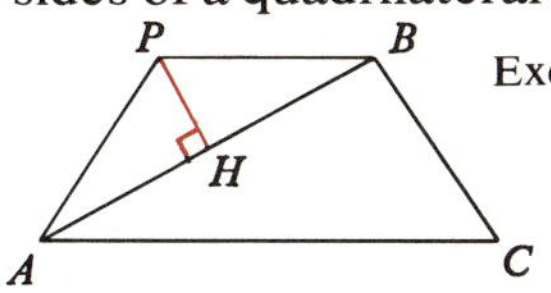

Exercise 17

17. In $\triangle ABC$, the bisector of $\angle A$ meets $\overline{BC}$ at D. The perpendicular bisector of $\overline{AD}$ meets $\overline{AB}$ at E. Prove that $\overline{DE} \parallel \overline{AC}$.

18. $\overline{CD}$ is the median of $\triangle ABC$. E is the midpoint of $\overline{AC}$. F is on $\overline{DE}$, so that $DE = EF$. Prove $\overline{AF} \parallel \overline{DC}$, $\overline{CF} \parallel \overline{AB}$, and $\overline{CF} \cong \overline{DB}$.

19. GIVEN $\overline{AD} \cong \overline{BC}$; $\overline{AEB}$ and $\overline{DFC}$; $AE = FC$; $EB = DF$; $\angle A \cong \angle C$

PROVE $\overline{DE} \parallel \overline{BF}$ (*Hint*: Consider transversal $\overline{EF}$ or $\overline{BD}$.)

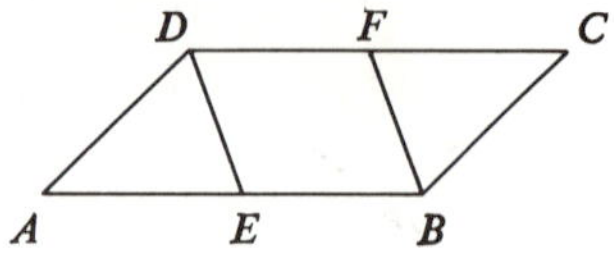

Exercise 19

6-3 PROPERTIES OF ANGLES DETERMINED BY PARALLEL LINES AND TRANSVERSALS

Now that we have established methods of proving lines parallel, let us examine some relationships between the angles formed when parallel lines are cut by a transversal. Can you anticipate the results? Consider the converse of each theorem from the previous section.

T **Theorem 6-3.1** If two parallel lines are cut by a transversal, then the alternate interior angles are congruent.

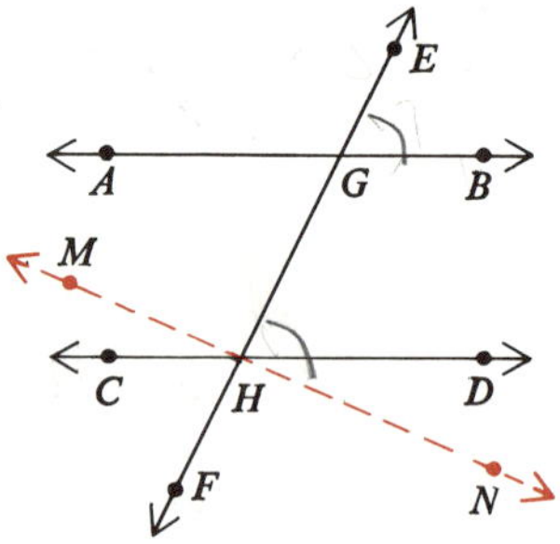

GIVEN $\overleftrightarrow{AB} \parallel \overleftrightarrow{CD}$; transversal $\overleftrightarrow{EF}$ intersects $\overleftrightarrow{AB}$ and $\overleftrightarrow{CD}$ at points G and H, respectively.

PROVE $\angle AGH \cong \angle DHG$

PROOF We are given that $\overleftrightarrow{AB} \parallel \overleftrightarrow{CD}$ and that transversal $\overleftrightarrow{EF}$ intersects $\overleftrightarrow{AB}$ and $\overleftrightarrow{CD}$ at points G and H, respectively. Assume that $\angle AGH \not\cong \angle DHG$. Then we can draw $\overleftrightarrow{MHN}$ distinct from $\overleftrightarrow{CHD}$ so that $\angle AGH \cong \angle NHG$. Since $\angle AGH$ and $\angle NHG$ are congruent alternate interior angles, $\overleftrightarrow{MHN} \parallel \overleftrightarrow{AB}$. But, we were given that $\overleftrightarrow{AB} \parallel \overleftrightarrow{CHD}$. The fact that two distinct lines, $\overleftrightarrow{MHN}$ and $\overleftrightarrow{CHD}$, are parallel to AB through the same point contradicts the Parallel Postulate (6-1). Thus, our original assumption leads to a contradiction. Consequently, the assumption must have been false and $\angle AGH \cong \angle DHG$.

Class Exercises

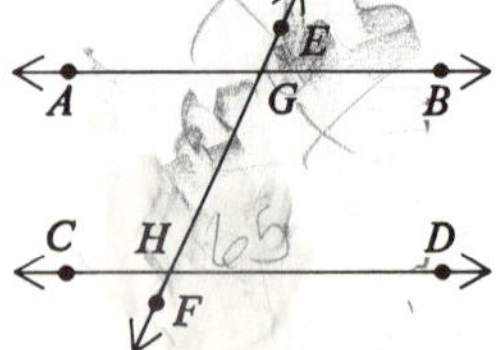

In the diagram at right $\overleftrightarrow{AB} \parallel \overleftrightarrow{CD}$, with $\overleftrightarrow{EF}$ as a transversal. If $m\angle AGH = 65$, answer the following:

1. $m\angle DHG =$ ____
2. $m\angle EGB =$ ____

3. What is the relation between m$\angle DHG$ and m$\angle EGB$?
4. What is this pair of angles called?
5. m$\angle CHG$ = ____.
6. What is the relation between m$\angle AGH$ and m$\angle CHG$?
7. What is this pair of angles called?
8. Give a conclusion based upon your answers to the preceding exercises.

These Class Exercises suggest the following corollaries to Theorem 6-3.1:

C **Corollary 6-3.1a** If two parallel lines are cut by a transversal, then the corresponding angles are congruent.

C **Corollary 6-3.1b** If two parallel lines are cut by a transversal, then the interior angles on the same side of the transversal are supplementary.

The proofs of these corollaries are left as exercises.

Example 1 In the figure at right, $\overrightarrow{BA} \parallel \overrightarrow{DC} \parallel \overrightarrow{EF}$; m$\angle BED = 25$; and m$\angle CDE = 135$. Find m$\angle ABE$.

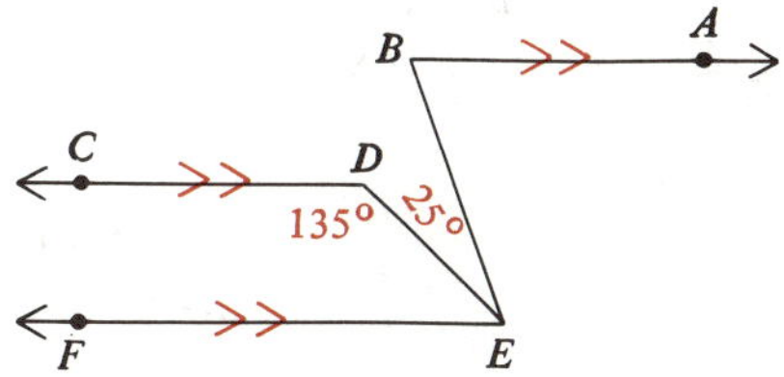

Solution By Corollary 6-3.1b,
m$\angle DEF = 180 -$ m$\angle CDE = 180 - 135 = 45$. Therefore m$\angle BEF = 25 + 45 = 70$. Since by Theorem 6-3.1, m$\angle ABE =$ m$\angle BEF$, m$\angle ABE = 70$.

Example 2 In the figure at right, $\overrightarrow{AB} \parallel \overrightarrow{CD}$, m$\angle BAP = 130$, and m$\angle APC = 70$. Find m$\angle DCP$.

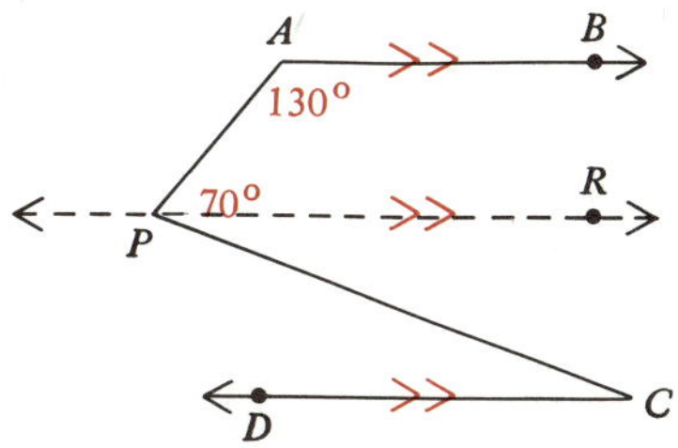

Solution Since there is no segment that is a transversal of the parallel rays, we must either introduce one or draw a line parallel to the given parallel rays. By drawing $\overrightarrow{PR} \parallel \overrightarrow{AB}$ we now have $\overleftrightarrow{AP}$ a transversal. Why is $\overleftrightarrow{PR}$ unique? It then follows that m$\angle APR = 180 -$ m$\angle BAP = 180 - 130 = 50$. Therefore, m$\angle CPR = 70 - 50 = 20$. By Theorem 6-3.1, m$\angle CPR =$ m$\angle DCP$; therefore, m$\angle DCP = 20$.

EXERCISES

A Find the value of x in each of the following.

1. 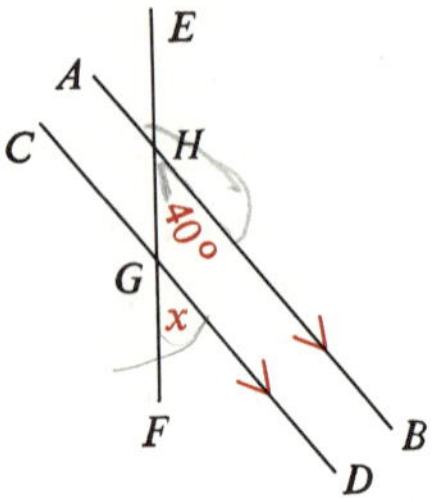

2. 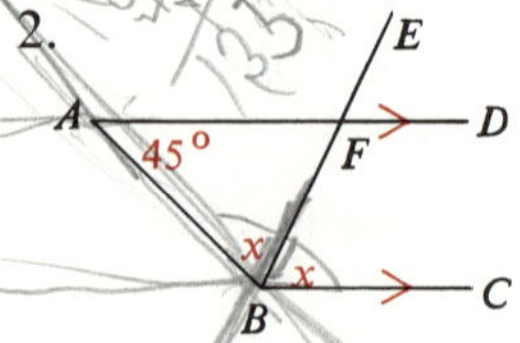

3. 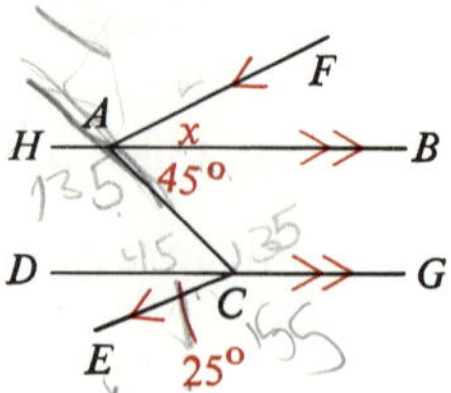

4. 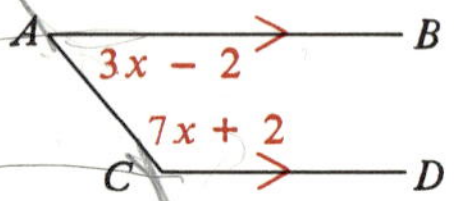

5. 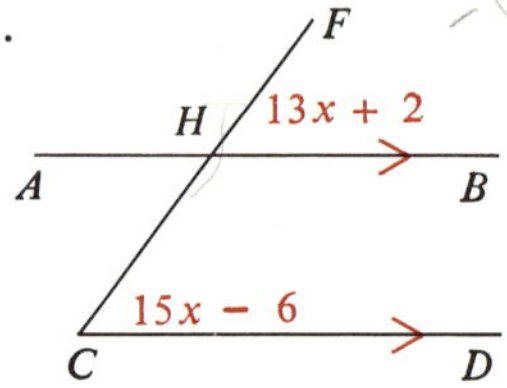

6. 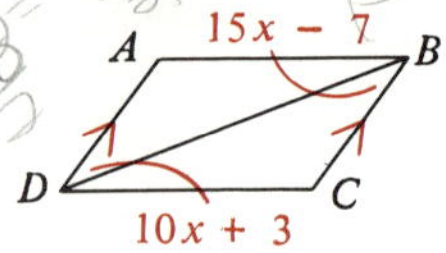

7. GIVEN $\overline{AB} \parallel \overline{CD}$; $\overline{AD} \parallel \overline{BC}$
PROVE $\angle B \cong \angle D$

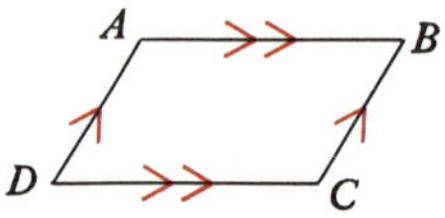

8. GIVEN $\overrightarrow{AB} \parallel \overline{CD}$; $\overline{AC} \parallel \overrightarrow{DE}$
PROVE $\angle A \cong \angle D$

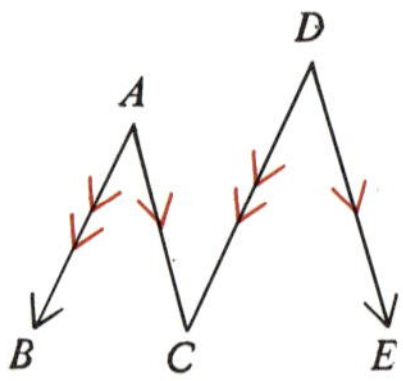

9. GIVEN $\overline{AB} \parallel \overline{CD}$; $\overline{AB} \cong \overline{CD}$; $\overline{AEC}$; $\overline{BED}$
PROVE $\overline{AC}$ and $\overline{BD}$ bisect each other.

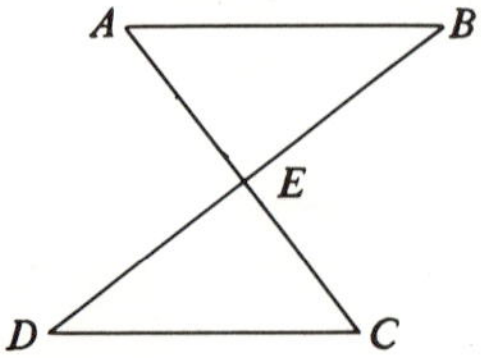

10. GIVEN $\overline{AB} \parallel \overline{CD}$; $\overline{AF} \parallel \overline{ED}$; $EB = FC$; $\overline{BEFC}$
PROVE $\overline{AF} \cong \overline{DE}$

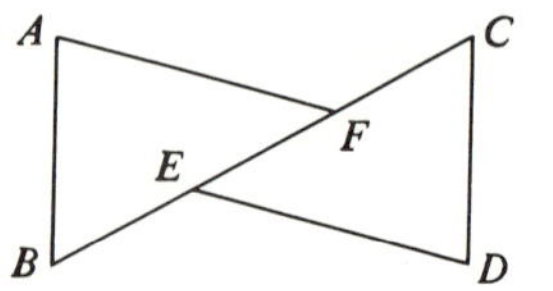

11. GIVEN $\overleftrightarrow{AB} \parallel \overrightarrow{EC}$; $\overrightarrow{PE}$ bisects $\angle GEC$; $\overline{AFPB}$
PROVE $\triangle PFE$ is isosceles

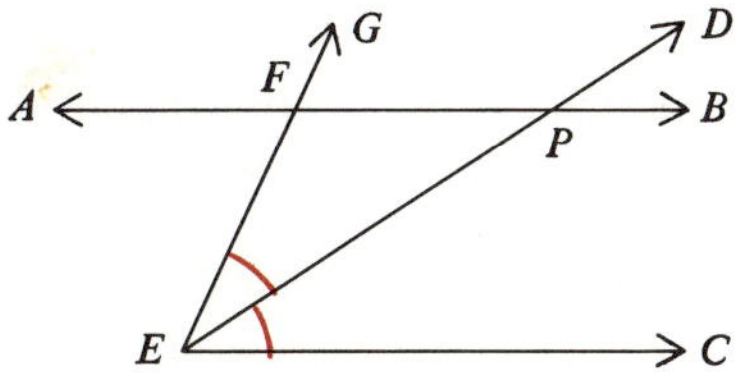

12. GIVEN $\overrightarrow{AD} \parallel \overline{BC}$; $\triangle ABC$ is isosceles; $\overline{AB} \cong \overline{AC}$; $\overrightarrow{BAF}$
PROVE $\overrightarrow{AD}$ bisects $\angle FAC$

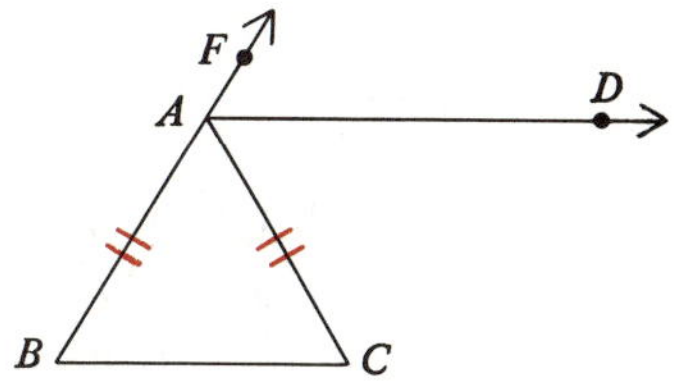

13. GIVEN Quadrilateral $ABCD$; $\overline{AB} \parallel \overline{DC}$; $\overline{AD} \parallel \overline{BC}$
PROVE $\overline{AC}$ and $\overline{BD}$ bisect each other.

14. GIVEN Quadrilateral $ABCD$; $\overline{AB} \parallel \overline{CD}$; $\overline{AD} \parallel \overline{BC}$
PROVE $\triangle ABD \cong \triangle CDB$

15. Prove Corollary 6-3.1a

16. Prove Corollary 6-3.1b.

B Find the values of x and y in the following diagrams. Parallel lines are marked.

17.

18.

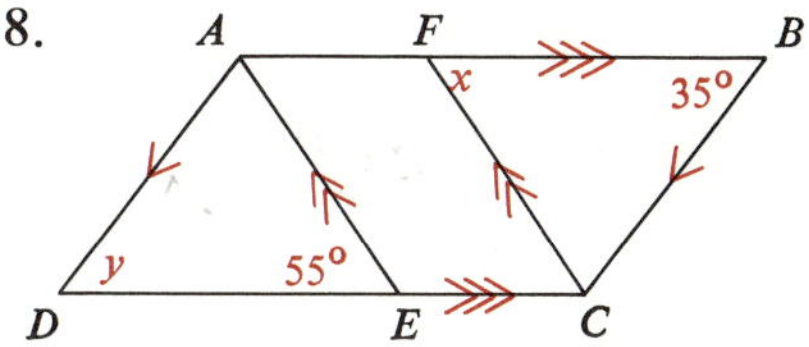

19.

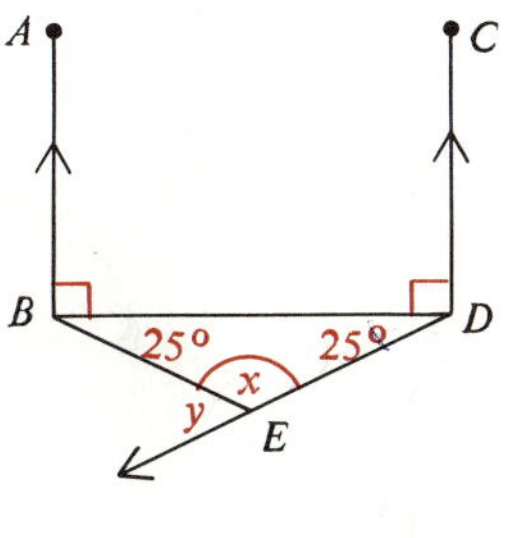

(*Hint*: Draw a line through E, parallel to $\overleftrightarrow{AB}$.)

20.

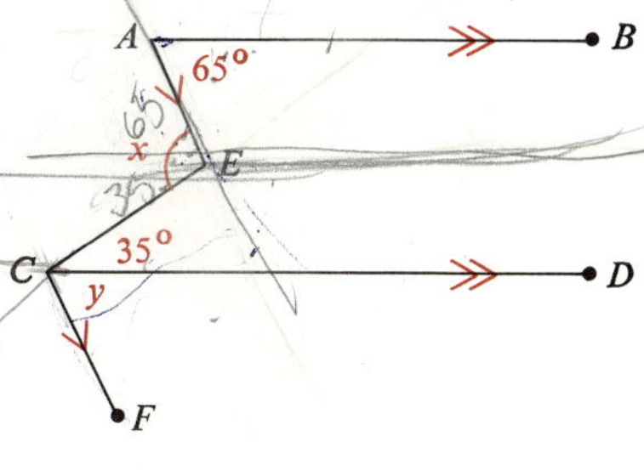

(*Hint*: Draw a line through E, parallel to $\overleftrightarrow{AB}$.)

C

21. Prove that if the line which bisects an exterior angle of a triangle is parallel to the opposite side, then the triangle is isosceles.
22. $\overleftrightarrow{AB}$ intersects $\overleftrightarrow{CD}$ at P. Through any point R of $\overrightarrow{PC}$, a line exists parallel to $\overleftrightarrow{AB}$ which intersects the bisectors of $\angle APC$ and $\angle BPC$ at points M and N, respectively. Prove that $\overline{MR} \cong \overline{NR}$.
23. Point P is any point of base $\overline{BPC}$ of isosceles $\triangle ABC$ other than the midpoint. Through P a perpendicular is drawn meeting $\overleftrightarrow{AB}$ at M and $\overleftrightarrow{CA}$ at N. Prove that $\triangle AMN$ is isosceles.
24. Side $\overline{BC}$ is the base of isosceles $\triangle ABC$. Point D is on $\overleftrightarrow{AC}$ not between A and C, while point E is on $\overline{AB}$ so that $\overline{CD} \cong \overline{BE}$. Prove that $\overline{BC}$ bisects $\overline{DE}$. (*Hint*: Draw a line through E, parallel to $\overleftrightarrow{AD}$.) Will the conclusion hold true if E is on $\overleftrightarrow{AB}$, not between A and B?

A LOOK AT THE PAST

How Eratosthenes Measured the Earth

One of the earliest measurements of the circumference of the earth was made by the Greek mathematician Eratosthenes, about 230 B.C. His measurement was remarkably accurate, being less than 2 percent in error. To make this measurement, Eratosthenes used what we call Theorem 6-3.1.

As librarian of Alexandria, Eratosthenes had access to records of calendar events. He discovered that at noon on a certain day of the year, in a town on the Nile called Syene (now called Aswan), the sun was directly overhead. As a result the bottom of a deep well was entirely lit and a vertical pole, being parallel to the rays hitting it, cast no shadow.

At the same time, however, a vertical pole in the city of Alexandria did cast a shadow. When that day arrived again, Eratosthenes measured the angle ($\angle 1$ in the figure below) formed by such a pole and the ray of light from the sun going past the top of the pole to the far end of the shadow. He found it to be about 7° 12′, or $\frac{1}{50}$ of 360°.

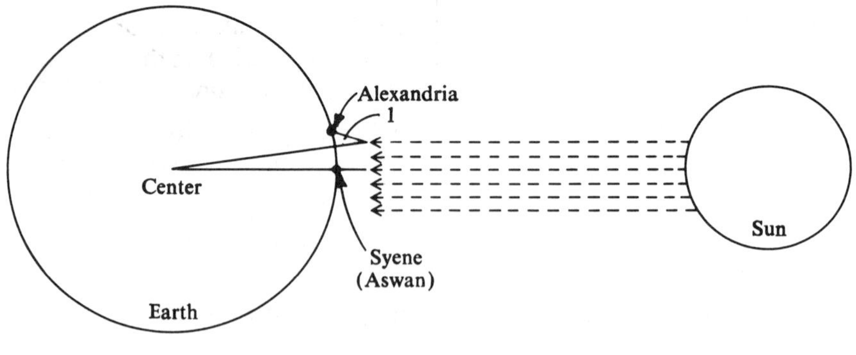

Assuming the rays of the sun to be parallel, he knew that by Theorem 6-3.1, the angle at the center of the earth must be congruent to $\angle 1$, and hence must also measure approximately $\frac{1}{50}$ of 360°. Since Syene and Alexandria were almost on the same meridian, Syene must be located on the radius of the circle which was parallel to the rays of the sun. Eratosthenes thus deduced that the distance between Syene and Alexandria was $\frac{1}{50}$ of the circumference of the earth. The distance from Syene to Alexandria was believed to be about 5,000 Greek *stadia*. A *stadium* was a unit of measurement equal to the length of an Olympic or Egyptian stadium. Therefore Eratosthenes concluded that the circumference of the earth was about 250,000 Greek stadia, or about 24,660 miles. This is very close to modern calculations.

6-4 ANGLE SUM OF A TRIANGLE

One of the most familiar facts from elementary mathematics is that the sum of the measures of the angles of a triangle is 180. Before we prove this, it will be helpful to prove a theorem about the measure of an exterior angle of a triangle.

T **Theorem 6-4.1** The measure of the exterior angle of a triangle is equal to the sum of the measures of the two remote interior angles.

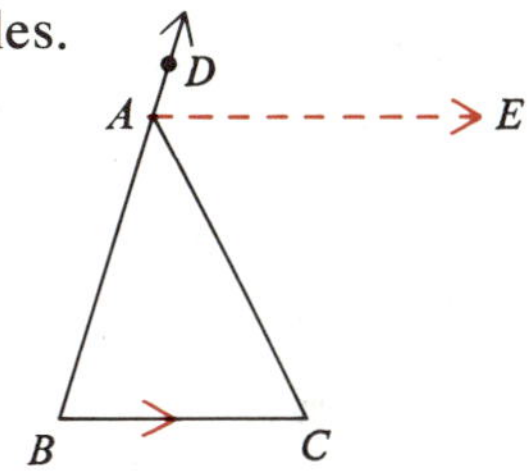

GIVEN $\triangle ABC$; $\overrightarrow{BAD}$

PROVE $m\angle DAC = m\angle B + m\angle C$

PROOF

STATEMENTS	REASONS
1. $\triangle ABC$; $\overrightarrow{BAD}$	1. Given.
2. Through A, draw $\overleftrightarrow{AE} \parallel \overline{BC}$.	2. The Parallel Postulate. (6-1)
3. $\angle DAE \cong \angle B$	3. If two parallel lines are cut by a transversal, then the corresponding angles are congruent. (6-3.1a)
4. $\angle EAC \cong \angle C$	4. If two parallel lines are cut by a transversal, then the alternate interior angles are congruent. (6-3.1)
5. $m\angle DAC = m\angle DAE + m\angle EAC$	5. Angle Sum Postulate. (2-10)
6. $m\angle DAC = m\angle B + m\angle C$	6. Substitution Postulate. (2-1)

Example 1 Find m$\angle A$.

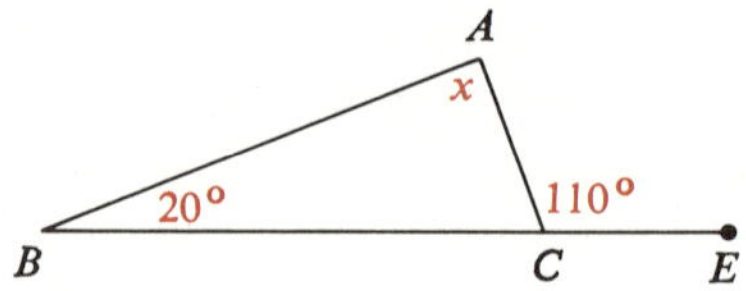

Solution Let $x = m\angle A$. Theorem 6-4.1 suggests the equation $110 = 20 + x$. Therefore, $x = 90$ and $m\angle A = 90$.

Example 2 Find m$\angle B$.

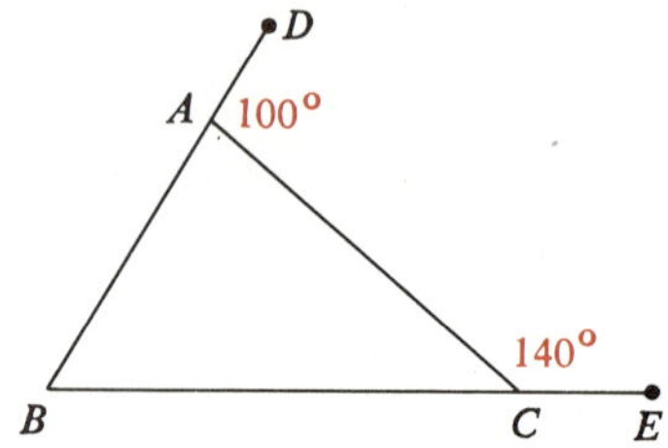

Solution Let $x = m\angle B$. Since $\angle DAC$ is supplementary to $\angle BAC$, $m\angle BAC = 80$. Theorem 6-4.1 suggests the equation $140 = 80 + x$. Therefore, $x = 60$ and $m\angle B = 60$.

Example 3 Find the numerical measure of each of the angles of $\triangle ABC$ if $m\angle A = 19x - 15$, $m\angle C = 9x + 25$, and $m\angle ABD = 26x + 20$.

Solution From Theorem 6-4.1 we obtain the following equation:

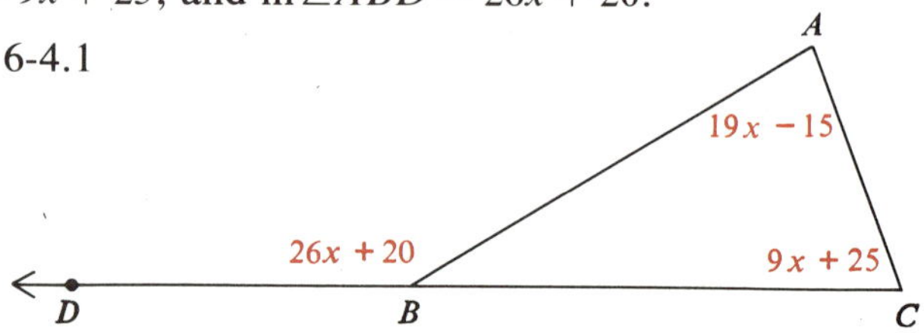

$$26x + 20 = (19x - 15) + (9x + 25)$$
$$26x + 20 = 28x + 10$$
$$x = 5$$

Therefore,

$$m\angle A = 19x - 15 = 19(5) - 15 = 80$$
$$m\angle C = 9x + 25 = 9(5) + 25 = 70$$
$$m\angle ABC = 180 - m\angle ABD = 180 - 150 = 30$$

Using Theorem 6-4.1 we can easily prove the theorem about the sum of the measures of the angles of any triangle:

T **Theorem 6-4.2** The sum of the measures of the angles of any triangle is 180.

GIVEN $\triangle ABC$; $\overrightarrow{BAD}$

PROVE $m\angle B + m\angle C + m\angle BAC = 180$

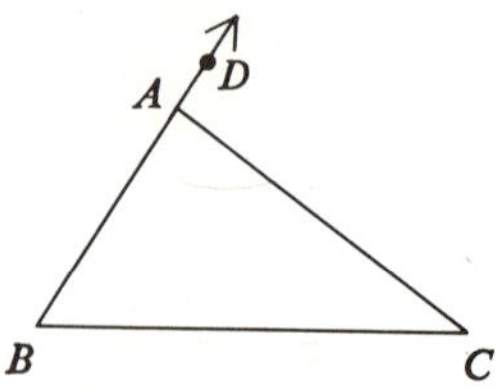

Proof

STATEMENTS	REASONS
1. $\triangle ABC$; $\overrightarrow{BAD}$	1. Given
2. $m\angle DAC + m\angle BAC = 180$	2. Definition of a linear pair (1-26)
3. $m\angle DAC = m\angle B + m\angle C$	3. Theorem 6-4.1
4. $m\angle B + m\angle C + m\angle BAC = 180$	4. Substitution Postulate (2-1)

This theorem proves that the sum of the measures of the angles of any triangle is a constant, thus leading to the following corollary:

C **Corollary 6-4.2a** If two angles of one triangle are congruent to two angles of another triangle, then the remaining angle of the first triangle is congruent to the remaining angle of the second triangle.

C **Corollary 6-4.2b** The acute angles of a right triangle are complementary.

The proofs of these corollaries are left as exercises.

Example 4 In isosceles $\triangle ABC$, $\overline{AB} \cong \overline{AC}$. Find $m\angle B$ where $m\angle A = 30$.

Solution If we let $m\angle B = x$, then $m\angle C = x$. Why? By using Theorem 6-4.2, we get the equation $x + x + 30 = 180$. Therefore, $2x + 30 = 180$ and $x = 75$. Hence $m\angle B = 75$.

Class Exercises

In $\triangle ABC$, find $m\angle A$, given the following information:

1. $m\angle B = 30$ and $m\angle C = 50$
2. $m\angle B + m\angle C = 70$
3. $m\angle B = 52$ and $\angle A \cong \angle C$
4. $m\angle C = 48$ and $m\angle B = 2(m\angle A)$
5. $\angle C$ is a right angle, and $m\angle B = 20$.

Find x and y in each of the following:

6.
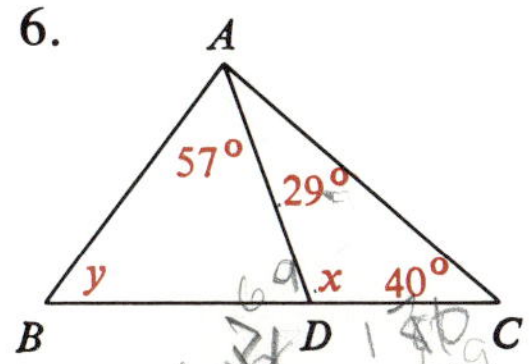

7.
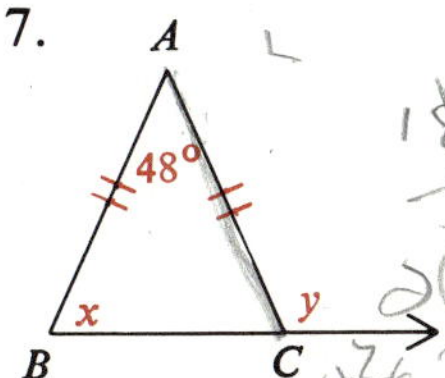

8.
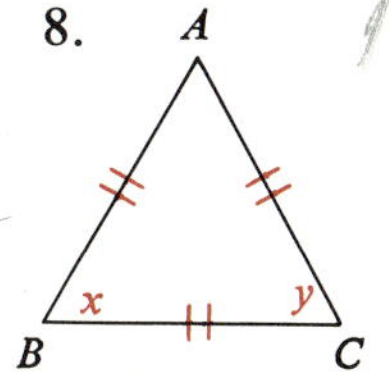

EXERCISES

A

In isosceles $\triangle ABC$, $\overline{AB} \cong \overline{BC}$. Find m$\angle A$ if m$\angle B$ is:

1. 60 2. 58 3. 112 4. 90 5. 155
6. 73 7. $2x$ 8. $5x$ 9. $4x - 10$ 10. $7x + 3$

In isosceles $\triangle PQR$, $\overline{PQ} \cong \overline{QR}$. Find the measure of $\angle Q$ if m$\angle P$ is:

11. 60 12. 25 13. 45 14. 1 15. $89\frac{1}{2}$
16. x 17. $3x$ 18. $180 - 2x$ 19. $5x - 3$ 20. $7x - 180$

21. Find the measures of the acute angles of a right triangle if one has twice the measure of the other.

Find the measure of the exterior angle at vertex B of isosceles $\triangle ABC$ where $\overline{AB} \cong \overline{AC}$, if m$\angle A$ is:

22. 20 23. 60 24. 90 25. 115 26. $180 - 8x$

Find the measures of $\angle y$ and/or $\angle x$ in each of the following figures.

27.
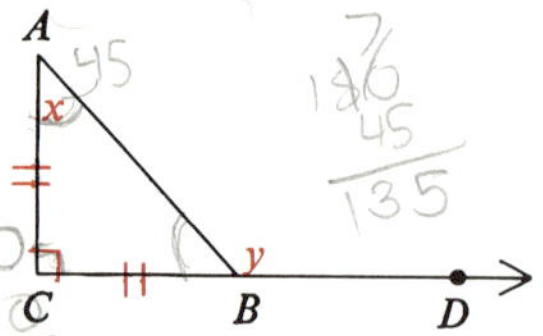

28.
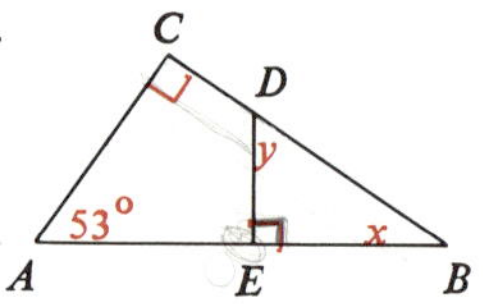

29.
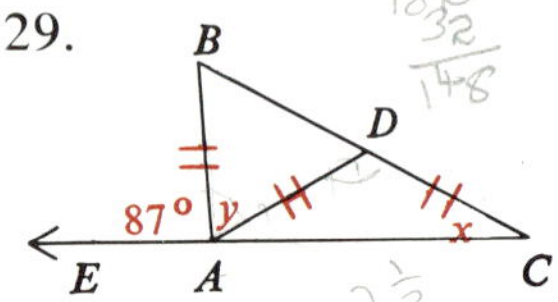

30. Prove Corollary 6-4.2a
31. Prove Corollary 6-4.2b
32. Prove that the measure of the obtuse angle formed by the bisectors of the base angles of an isosceles triangle equals the measure of one of the exterior angles at the base.
33. Prove that if the sum of the measures of two angles of a triangle equals the measure of the third angle, then the triangle is a right triangle.
34. GIVEN Right $\triangle ABC$ with m$\angle ACB = 90$; $\overline{CD} \perp \overline{AB}$; $\overline{ADB}$
    PROVE $\angle DCB \cong \angle A$

35. GIVEN $\overline{AE} \parallel \overline{BF}$; $\overline{AC} \cong \overline{AP}$; $\overline{BP} \cong \overline{BD}$; $\overline{APB}$; $\overrightarrow{ACE}$; $\overrightarrow{BDF}$
    PROVE m$\angle CPD = 90$

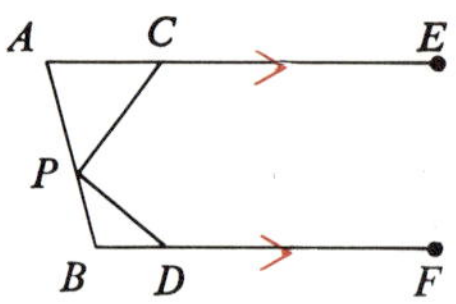

36. GIVEN $\overline{AD} \perp \overline{AEB}$; $\overline{BC} \perp \overline{AEB}$; $\overline{DA} \cong \overline{EB}$; $\overline{AE} \cong \overline{BC}$; $\overline{AEB}$
    PROVE $\triangle DEC$ is an isosceles right triangle.

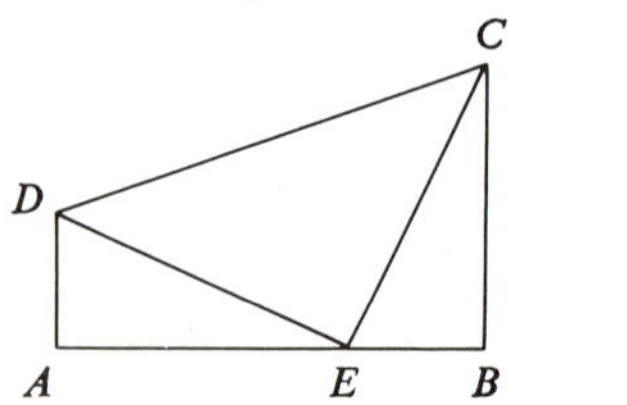

B Different sets of measures of the angles of $\triangle KNM$ are given below. Find the numerical measures of the angles and indicate what type of triangle they form in each case.

37. $m\angle K = x + 10$; $m\angle N = 2x$; $m\angle M = 2x - 30$
38. $m\angle K = 10x$; $m\angle N = 3x + 18$; $m\angle M = 4x + 9$
39. $m\angle K = 5x$; $m\angle N = 2x + 20$; $m\angle M = 8x + 10$
40. $m\angle K = \frac{5}{4}x$; $m\angle N = 2(2x - 3)$; $m\angle M = 2(x + 6)$

In $\triangle ABC$, $m\angle A = 50$ and $m\angle B = 60$.

41. Find the measure of the acute angle formed by the bisectors of $\angle A$ and $\angle B$.
42. Find the measure of the acute angle formed by the altitudes on sides $\overline{AC}$ and $\overline{BC}$.

Find the values of x and y in each of the following figures:

43.

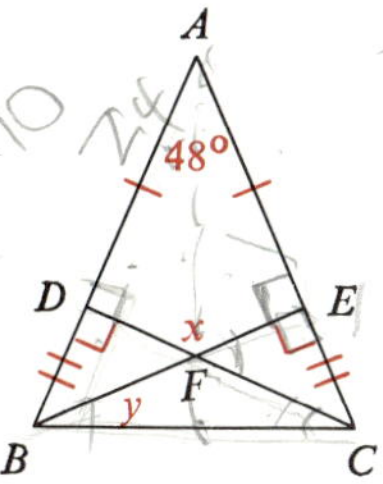

44.

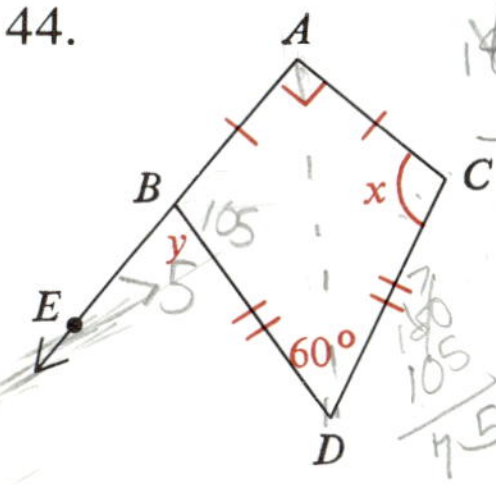

45.

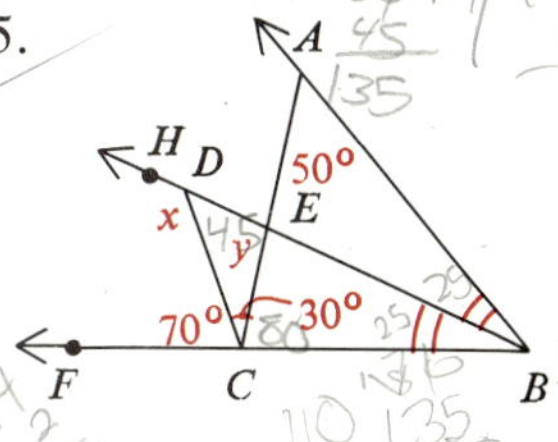

46. **Given** $\overline{AE}$ bisects $\angle BAC$; $\overline{AD} \perp \overline{BC}$; $\overline{BDEC}$; $m\angle B = 32$; $m\angle C = 25$
Find $m\angle DAE$

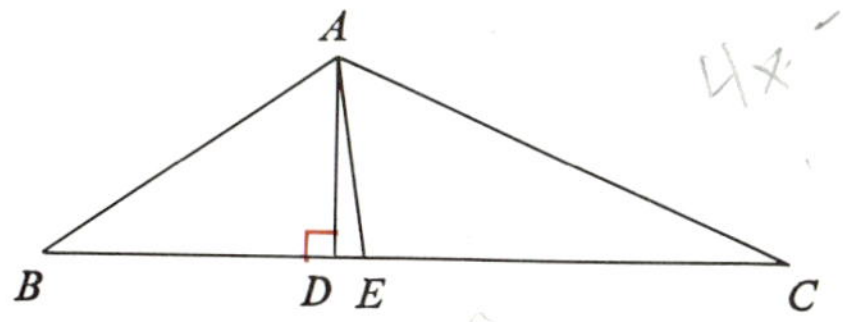

47. **Given** $\overline{AB} \cong \overline{AC}$; $\overline{CD} \cong \overline{CE}$; $m\angle AFD = 100$; $\overline{BC}$ meets $\overline{DF}$ at E.
Find $m\angle ADE$

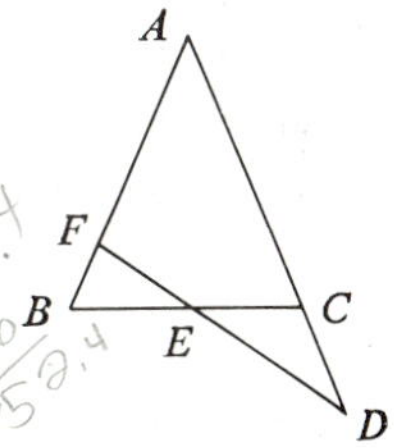

48. **Given** $\overline{AC} \cong \overline{BC}$; $AB = BE = ED = DC$; $\overline{AEC}$; $\overline{BDC}$
Find $m\angle C$

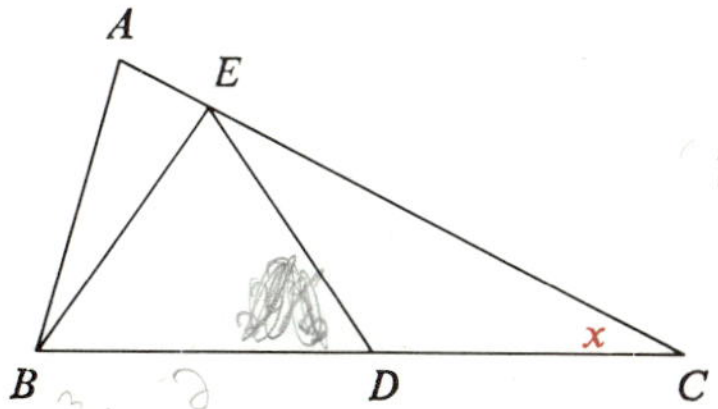

49. If $\overline{AB} \cong \overline{AC}$, $\overline{DC} \cong \overline{DE}$, and $\overline{DE} \parallel \overline{AB}$, what is the measure of $\angle BCE$? Prove your answer.

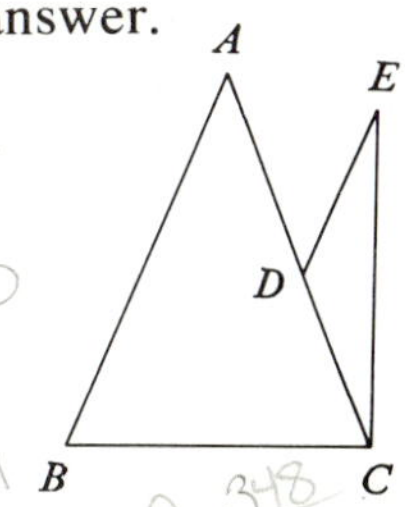

C Theorem 6-4.2 may be proved in a variety of ways. Prove this theorem using each of the following diagrams:

50.

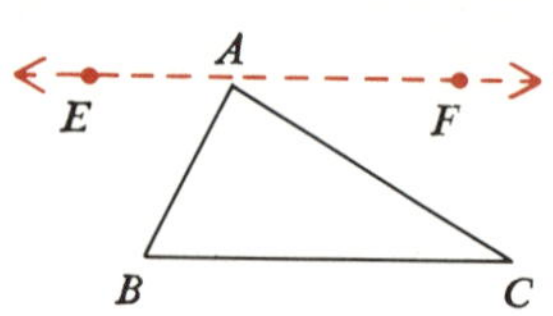

51.

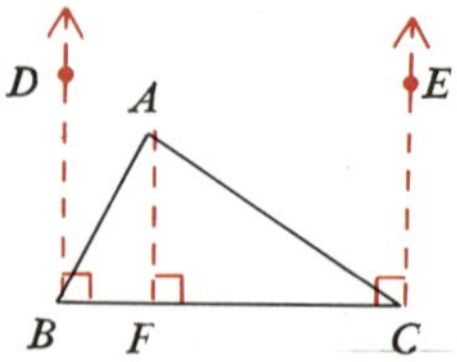

52.

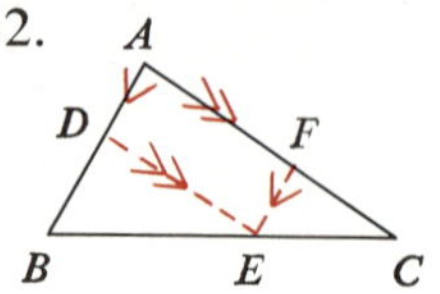

53. GIVEN $\overline{DB}$ bisects $\angle ABC$; $\overline{DC}$ bisects $\angle ACF$; $\overrightarrow{BCF}$; $\overline{AC}$ meets $\overline{BD}$ at E

PROVE $m\angle D = \frac{1}{2} m\angle BAC$

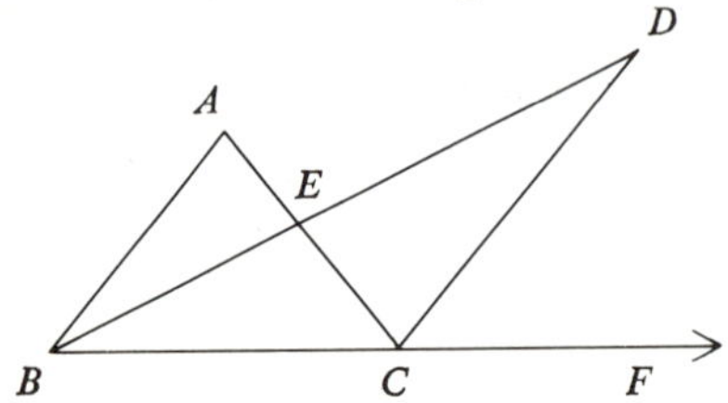

54. GIVEN Any $\triangle ABC$ and equilateral $\triangle$s ABD, ACE and BCF; $\overline{AF}$, $\overline{BE}$, and $\overline{CD}$ meet at P.

PROVE $\overline{AF} \cong \overline{BE} \cong \overline{CD}$

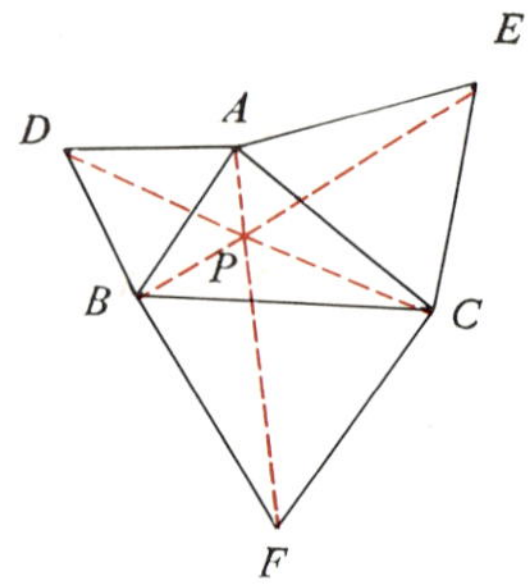

55. Externally on side $\overline{PQ}$ of equilateral $\triangle PQR$, another equilateral triangle, $\triangle QMN$, is constructed, with N on $\overline{PQ}$, between P and Q. Prove that $\overline{PM} \cong \overline{RN}$.
56. Prove that if median $\overline{AM}$ of $\triangle ABC$ is such that $\overline{AM} \cong \overline{BM}$, then $\triangle ABC$ is a right triangle.
57. GIVEN Isosceles $\triangle ABC$; $\overline{AB} \cong \overline{AC}$; $\overline{AKB}$; $\overline{AMC}$; $\overline{BNC}$; $\triangle KMN$ is equilateral.

    PROVE $m\angle MNC = \frac{1}{2}(m\angle BKN + m\angle AMK)$

6-5 POLYGONS

Polygons were discussed informally in Section 1-5. We shall state now the formal definition of a polygon.

D **Definition 6-3** Let $\{P_1, P_2, \ldots, P_n\}$ be a set of n distinct points in a plane, where $n \geq 3$. Let the n segments, $\overline{P_1P_2}$, $\overline{P_2P_3}$, . . . , $\overline{P_{n-1}P_n}$, $\overline{P_nP_1}$, have the following properties:

1. No two segments intersect except at their endpoints.
2. No two segments with a common endpoint are collinear.

The union of such segments is called a **polygon.**

The **vertices** of the polygon at right are the given points $P_1, P_2, P_3, \ldots, P_{n-1}, P_n$.
The **sides** of the polygon are the segments $\overline{P_1P_2}, \overline{P_2P_3}, \overline{P_3P_4}, \ldots, \overline{P_{n-1}P_n}, \overline{P_nP_1}$.
The **angles** of the polygon are $\angle P_nP_1P_2$, $\angle P_1P_2P_3$, $\angle P_2P_3P_4, \ldots,$ $\angle P_{n-1}P_nP_1$.

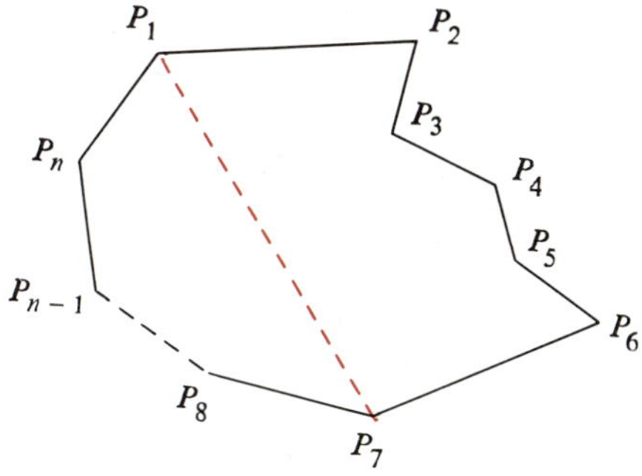

The **consecutive vertices of a polygon** are the endpoints of a side of the polygon. For example, P_1 and P_2 are consecutive vertices of the polygon above.

The **consecutive angles of a polygon** are angles of a polygon at consecutive vertices. For example, $\angle P_1P_2P_3$ and $\angle P_2P_3P_4$ are consecutive angles.

A **diagonal of a polygon** is a line segment joining any two nonconsecutive vertices. A diagonal of the above polygon is $\overline{P_1P_7}$.

Class Exercises

State why each of the following is or is not a polygon:

1. 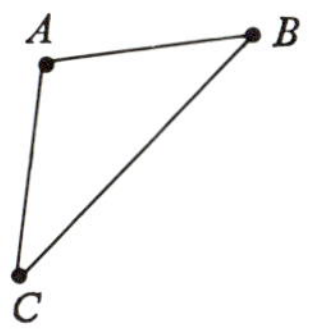

2. 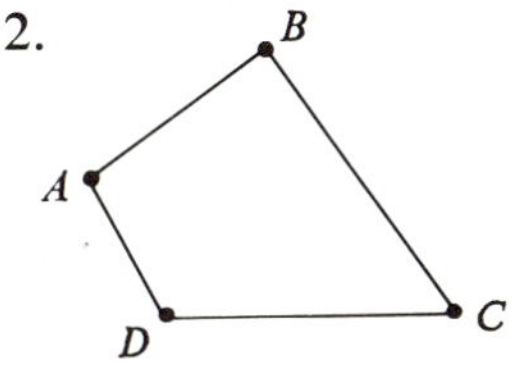

3. 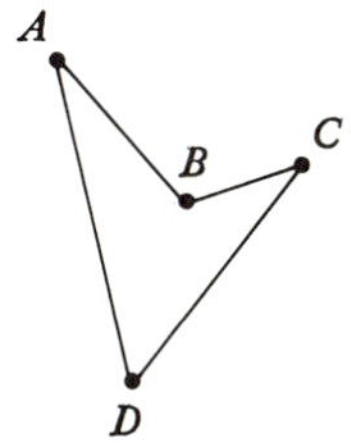

What differences can you find between the polygons for Exercises 2 and 3? The interior of the first polygon is contained within the interiors of all its angles. Is this true of the polygon in Exercise 3? This suggests the formal definition of a convex polygon.

D **Definition 6-4** A **convex polygon** is a polygon whose points all lie on one side of each line containing a side of the polygon.

In order to decide whether a certain polygon is convex or concave (nonconvex), we can consider the lines containing the sides of the polygon. If any line that contains a side of the polygon is so situated that points of the polygon lie on both sides of the line, then the polygon is concave. The red line in each of the polygons below should help you decide which of them are convex and which are concave.

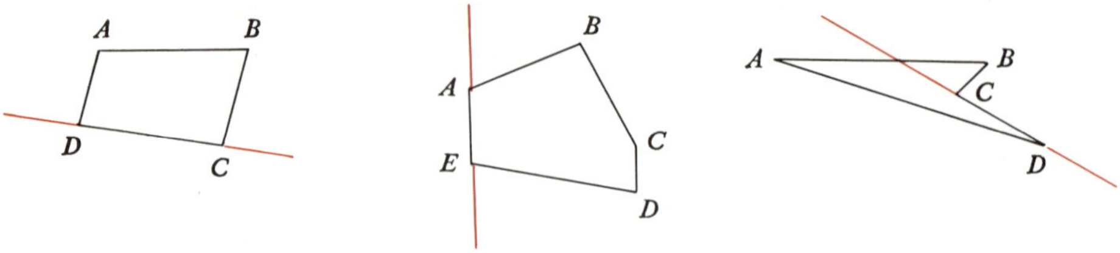

Hereafter, when we use the word *polygon*, we shall be referring to convex polygons.

Theorem 6-5.1 The sum of the measures of the interior angles of a convex polygon of n sides equals $(n-2)180$.

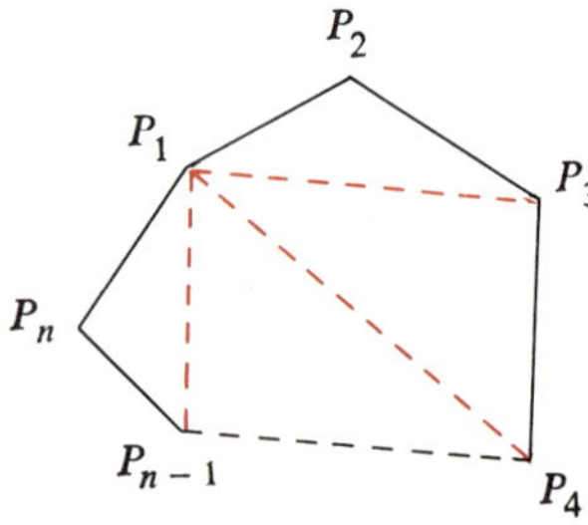

GIVEN Polygon $P_1P_2P_3P_4 \cdots P_n$ is any convex polygon.

PROVE $m\angle P_1 + m\angle P_2 + m\angle P_3 + m\angle P_4 + \cdots + m\angle P_n = (n-2)180$

The following is not a proof but the outline of a proof. There are $n-2$ triangles formed by the sides of a polygon and the diagonals from any one vertex. Two of the triangles thus formed contain two sides of the polygon; all the rest contain just one side of the polygon. The sum of the measures of the interior angles of the polygon equals the sum of the measures of the angles of the $n-2$ triangles formed. Thus the sum of the measures of the interior angles of the polygon, $m\angle P_1 + m\angle P_2 + m\angle P_3 + m\angle P_4 + \cdots + m\angle P_n$, equals $(n-2)180$.

Example 1 Find the sum of the measures of the interior angles of a decagon (10-sided polygon).

Solution From Theorem 6-5.1, we find that the sum of the measures of the interior angles of a decagon is $(10-2)180 = 1{,}440$.

Using Theorem 6-5.1, we are able to establish a theorem regarding the sum of the measures of the **exterior angles** of any polygon. We shall form the exterior angles by extending the sides of a polygon in the same order. The sum of the measures of the exterior angles will always be 360, regardless of how many sides the polygon has.

Theorem 6-5.2 The sum of the measures of the exterior angles of any convex polygon is 360.

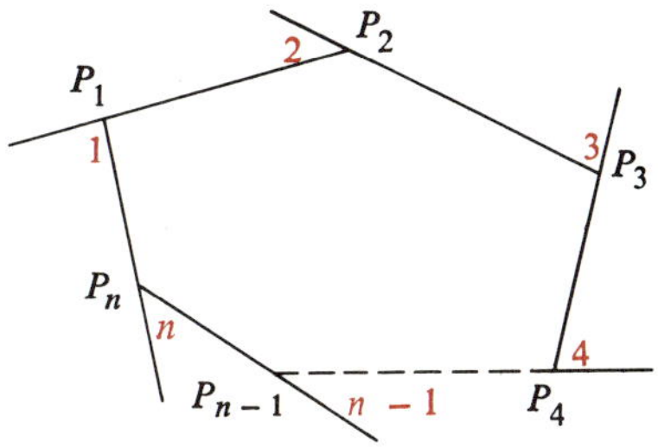

GIVEN Polygon $P_1P_2P_3P_4 \ldots P_{n-1}P_n$ is any convex polygon. $\angle 1, \angle 2, \angle 3, \ldots, \angle n-1, \angle n$ are the exterior angles.

PROVE $m\angle 1 + m\angle 2 + m\angle 3 + m\angle 4 + \cdots + m\angle n-1 + m\angle n = 360.$

The following outline should show you how to prove the theorem. At each vertex of a polygon, the sum of the measures of the interior angle and the exterior angle is 180. Since there are n vertices in the polygon above, the sum of the measures of all the interior angles and all the exterior angles is $180n$. However, the sum of the measures of the interior angles of the n-gon is $(n-2)180 = 180n - 360$. Hence, the sum of the measures of the exterior angles of the n-gon is $180n - (180n - 360) = 360$.

Regular Polygons

Now that we are reasonably familiar with polygons in general, let us consider a special type of polygon, the regular polygon.

Definition 6-5 A **regular polygon** is a polygon with all angles congruent and all sides congruent.

Corollary 6-5.1a The measure of each interior angle of a regular polygon of n sides is $\frac{(n-2)180}{n}$.

Corollary 6-5.2a The measure of each exterior angle of a regular polygon of n sides is $\frac{360}{n}$.

Each interior angle of a regular polygon is supplementary to an exterior angle. This suggests the following corollary:

Corollary 6-5.2b The measure of each interior angle of a regular polygon of n sides is $180 - \frac{360}{n}$.

The proofs of these corollaries are left as exercises.

Example 2 What is the measure of an exterior angle of a regular octagon (a polygon with eight sides)?

Solution Using Corollary 6-5.2a, we find that the measure of the exterior angle is $\frac{360}{8} = 45$.

Example 3 Find the measure of an interior angle of a regular dodecagon (a polygon with 12 sides).

Solution We may solve this problem by using either Corollary 6-5.1a or Corollary 6-5.2b. Using Corollary 6-5.1a, we get $\frac{(12-2)180}{12} = \frac{1800}{12} = 150$.

Using Corollary 6-5.2b, we get $180 - \frac{360}{12} = 180 - 30 = 150$.

EXERCISES

A State whether or not each of the following figures is a polygon. Indicate which of the polygons are convex. If a figure is not a polygon, give a reason why. Only those points labeled with letters are to be considered vertices.

1.

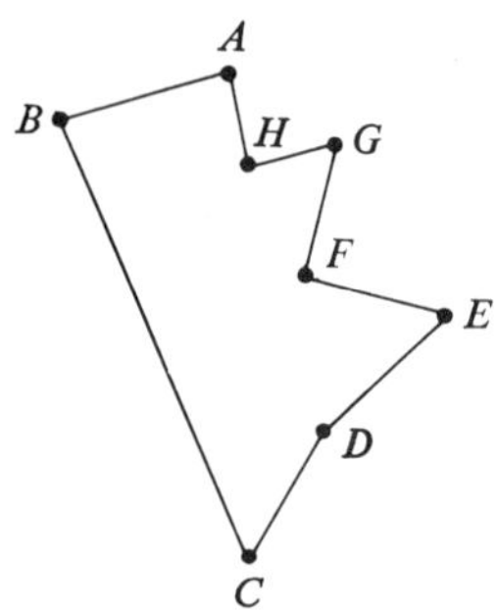

2.

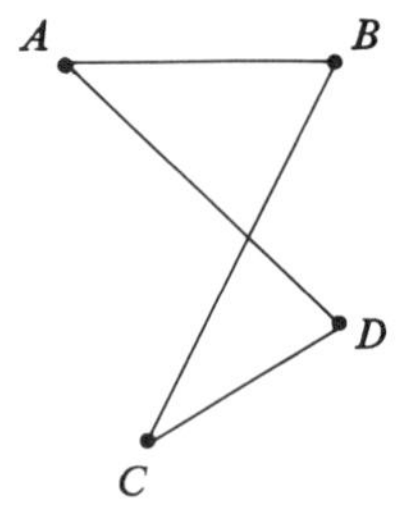

3.

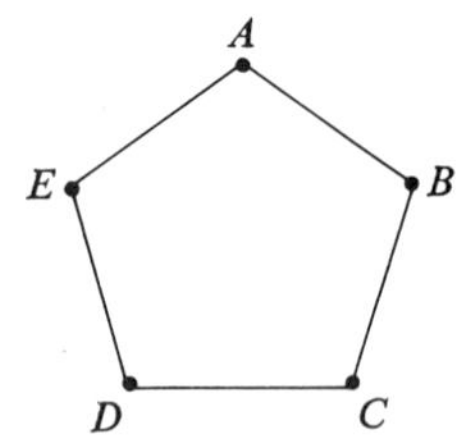

4.

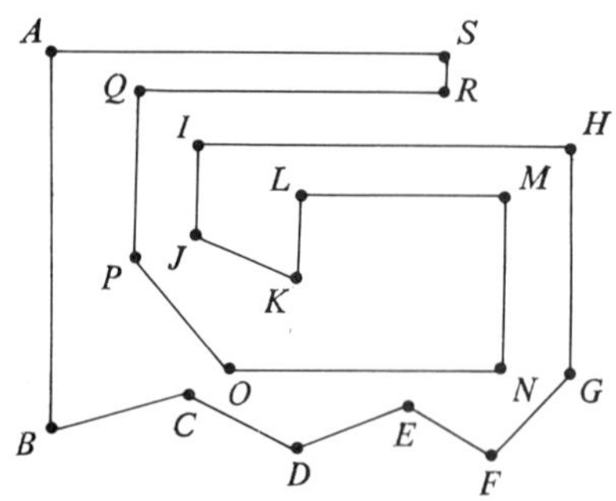

5.

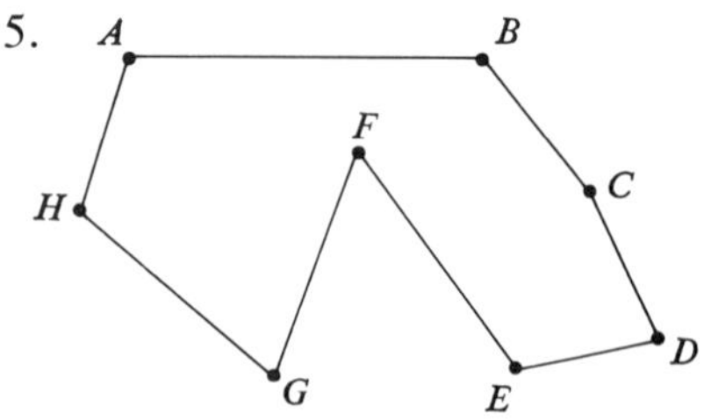

6. Copy the pentagon in Exercise 3. Insert and name all of its diagonals. How many diagonals are there?

7. What type of triangle is a regular 3-gon?
8. What type of quadrilateral is a regular 4-gon?
9. Is an equilateral quadrilateral always a regular polygon? Why?
10. Is an equiangular quadrilateral always a regular polygon? Why?

Find the sum of the measures of the interior angles of a polygon with each of the following number of sides:

11. 3 sides
12. 5 sides
13. 6 sides
14. 7 sides
15. 8 vertices
16. 10 vertices
17. 12 vertices
18. 20 vertices

Find the measure of an interior angle of each of the following regular polygons in two ways:

19. quadrilateral
20. pentagon
21. hexagon
22. octagon
23. nonagon
24. dodecagon

Find the measure of an exterior angle of a regular n-gon if n equals the following:

25. $n = 4$
26. $n = 18$
27. $n = 36$
28. $n = 360$

29. Prove Corollary 6-5.1a.
30. Prove Corollary 6-5.2a.
31. Prove Corollary 6-5.2b.

B

Find the number of sides of a regular polygon, if the measure of one of its interior angles equals:

32. the measure of the adjacent exterior angle.
33. one-half the measure of the adjacent exterior angle.
34. twice the measure of the adjacent exterior angle.
35. three times the measure of the adjacent exterior angle.
36. $1\frac{1}{2}$ times the measure of the adjacent exterior angle.
37. four times the measure of the adjacent exterior angle.

38. Prove that if the measure of each interior angle of a regular polygon is m times the measure of its adjacent exterior angle, then the polygon has $2(m + 1)$ sides.

Find the number of sides of a polygon if the sum of the measures of its interior angles is:

39. 540
40. 720
41. 1,800
42. 2,700

43. If the sum of the measures of six of the interior angles of a polygon with seven sides is 755, then what is the measure of the remaining angle?
44. The ratio of the measures of the interior angles of a quadrilateral is 3:4:5:6. Find the measures of these angles.

C

45. What is the smallest measure that an interior angle of a regular polygon can have?

46. As the number of sides of a polygon increases, what change is there in the sum of the measures of the interior angles of the polygon?
47. As the number of sides of a polygon increases, what change is there in the sum of the measures of the exterior angles of the polygon?
48. What is the smallest measure an obtuse interior angle of a regular polygon can have?
49. Can the measure of an exterior angle of a regular polygon be 50? Why?
50. Can the measure of an exterior angle of a regular polygon be 40? Why?

6-6 MORE TRIANGLE CONGRUENCE THEOREMS

In Chapter 3 we established three methods of proving triangles congruent, by proving an ASA, SAS, or SSS correspondence. In this section, we shall develop two more methods of proving triangles congruent.

T

Theorem 6-6.1 AAS Congruence Theorem If two angles and a nonincluded side of one triangle are congruent to the two corresponding angles and the nonincluded side of another triangle, then the two triangles are congruent.

GIVEN $\triangle ABC$ and $\triangle DEF$;
$\angle A \cong \angle D$;
$\angle B \cong \angle E$; $\overline{BC} \cong \overline{EF}$

PROVE $\triangle ABC \cong \triangle DEF$

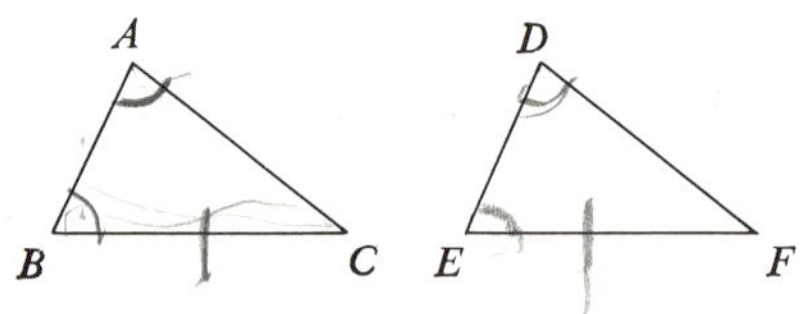

PROOF

STATEMENTS	REASONS
1. $\triangle ABC$; $\triangle DEF$; $\angle A \cong \angle D$; $\angle B \cong \angle E$; $\overline{BC} \cong \overline{EF}$	1. Given.
2. $\angle C \cong \angle F$	2. If two angles of one triangle are congruent to two angles of another triangle, then the remaining angle of the first triangle is congruent to the remaining angle of the second triangle. (6-4.2a)
3. $\triangle ABC \cong \triangle DEF$	3. ASA Postulate. (3-2)

Thus far, we have not mentioned the angle-angle-angle and side-side-angle correspondences. Suppose we assume that these correspondences cannot be used to prove triangles congruent. How can we verify this assumption? If we can find a pair of noncongruent triangles with congruent corresponding angles or a congruent side-side-angle correspondence, will this prove our assumption?

T **Theorem 6-6.2 HL Congruence Theorem** If the hypotenuse and one leg of a right triangle are congruent to the corresponding hypotenuse and leg of another right triangle, then the two triangles are congruent.

GIVEN Right $\triangle ABC$ and right $\triangle DEF$; $m\angle C = m\angle F = 90$; $\overline{AB} \cong \overline{DE}$; $\overline{AC} \cong \overline{DF}$

PROVE $\triangle ABC \cong \triangle DEF$

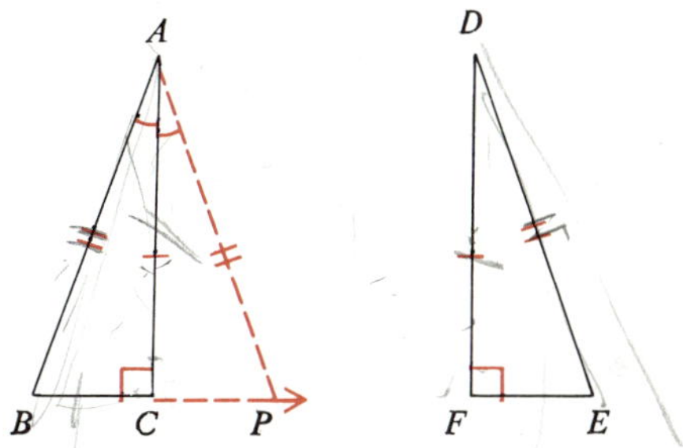

PROOF

STATEMENTS	REASONS
1. Right $\triangle ABC$ and right $\triangle DEF$; $m\angle C = m\angle F = 90$; $\overline{DE} \cong \overline{AB}$; $\overline{AC} \cong \overline{DF}$	1. Given.
2. Let P be the point of $\overrightarrow{BC}$ such that $\overline{CP} \cong \overline{FE}$.	2. Point Uniqueness Postulate. (2-2)
3. $m\angle ACP = 90$	3. If two intersecting lines form one right angle, then the lines form four right angles. (2-6.4)
4. $\angle ACP \cong \angle DFE$	4. All right angles are congruent. (3-1.1)
5. $\triangle APC \cong \triangle DEF$	5. SAS Postulate. (3-1)
6. $\overline{AP} \cong \overline{DE}$	6. Definition of congruent triangles. (3-3)
7. $\overline{AP} \cong \overline{AB}$	7. Transitive property.
8. $\angle B \cong \angle P$	8. The base angles of an isosceles triangle are congruent. (3-4.2)
9. $\triangle ABC \cong \triangle APC$	9. Theorem 6-6.1.
10. $\triangle ABC \cong \triangle DEF$	10. Transitive property.

Summary. Proving Triangles Congruent

We now have five methods of proving triangles congruent:

1. Prove two sides and the included angle of one triangle congruent to the corresponding parts of the other triangle. (SAS)
2. Prove two angles and the included side of one triangle congruent to the corresponding parts of the other triangle. (ASA)
3. Prove three sides of one triangle congruent to the corresponding sides of the other triangle. (SSS)
4. Prove two angles and a nonincluded side of one triangle congruent to the corresponding parts of the other triangle. (AAS)
5. Prove the hypotenuse and leg of one right triangle congruent to the corresponding parts of the other right triangle. (HL)

Class Exercises

The congruent sides and angles in each of the following diagrams are marked. Name each pair of congruent triangles and the reason why they are congruent. Indicate which, if any, of the diagrams contain no congruent triangles.

1.
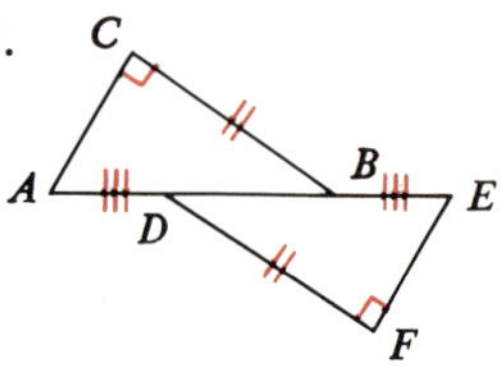

2.
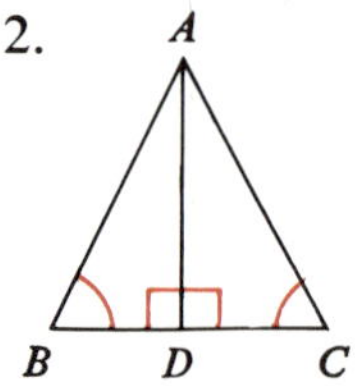

3.
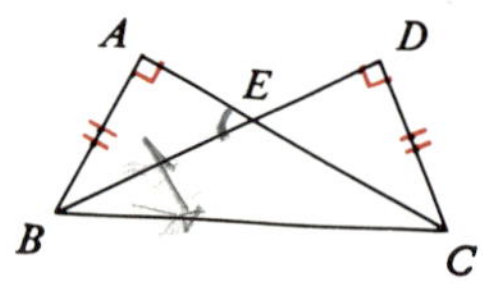

4.
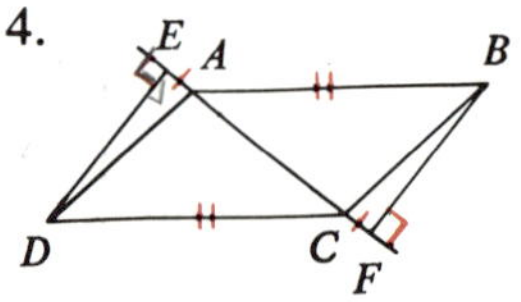

5.
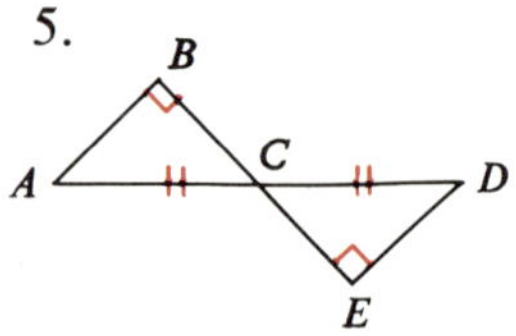

6.
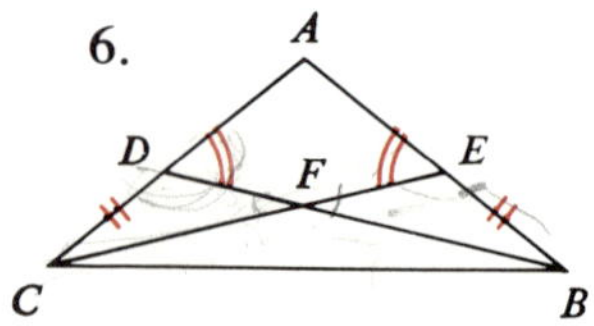

EXERCISES

A

1. **Given** $\overline{DB}$ bisects $\angle ADC$; $\angle EAB \cong \angle FCB$; $\overrightarrow{DCF}$; $\overrightarrow{DAE}$

   Prove $\triangle ABD \cong \triangle CBD$

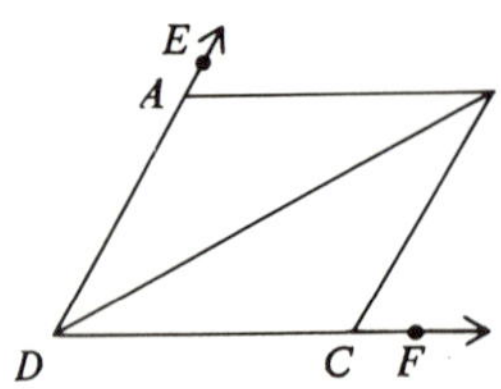

2. **Given** $\overline{AB} \perp \overline{BC}$; $\overline{DC} \perp \overline{BC}$; $\overline{BC}$ bisects $\overline{AD}$ at E.

   Prove $\triangle ABE \cong \triangle DCE$

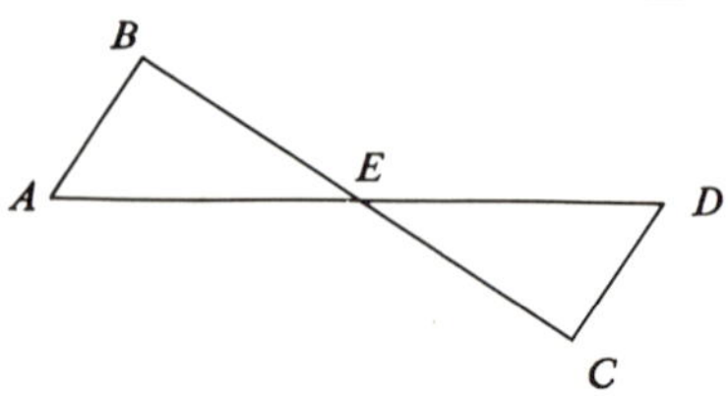

3. **Given** $\overline{FE} \perp \overline{AB}$ at E; $\overline{FD} \perp \overline{BC}$ at D; $\overline{FE} \cong \overline{FD}$
Prove $\overline{FB}$ bisects $\angle ABC$.

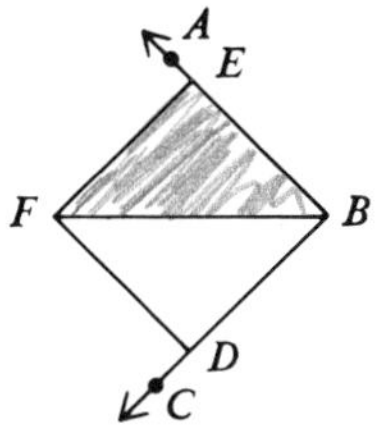

4. **Given** $\overline{AD} \parallel \overline{BC}$; $DF = EB$, $\overline{DFEB}$; $\angle DAE = \angle BCF$
Prove $\triangle ADE \cong \triangle CBF$

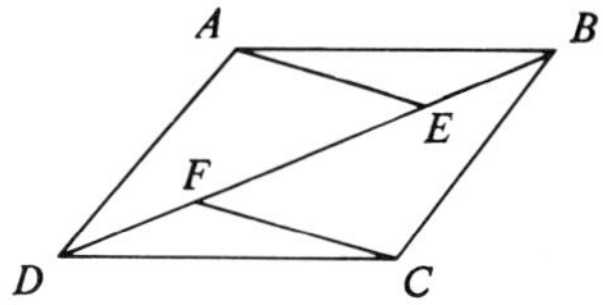

5. **Given** Quadrilateral $ABCD$; $\overline{AD} \perp \overline{AB}$; $\overline{BC} \perp \overline{AB}$; $\overline{AC} \cong \overline{BD}$
Prove $\overline{AD} \cong \overline{BC}$

6. **Given** $\triangle ABC$; $\overline{AB} \cong \overline{AC}$; $\angle BDC \cong \angle CEB$; $\overline{ADB}$; $\overline{AEC}$
Prove $\overline{CD} \cong \overline{BE}$ (in two ways)

B

7. **Given** $\overline{AB} \cong \overline{AC}$; $\overline{GD} \perp \overline{AB}$ at D; $\overline{FE} \perp \overline{AC}$ at E; $FB = CG$
Prove $\angle F \cong \angle G$

8. **Given** $\overline{DB} \cong \overline{EC}$; $\overline{GD} \perp \overline{AB}$ at D; $\overline{FE} \perp \overline{AC}$ at E; $FB = CG$; $\overline{FBCG}$
Prove $\angle F \cong \angle G$

9. **Given** $\overline{DB} \cong \overline{EC}$; $\overline{GD} \perp \overline{AB}$ at D; $\overline{FE} \perp \overline{AC}$ at E; $FB = CG$; $\overline{FBCG}$
Prove $\triangle ABC$ is isosceles.

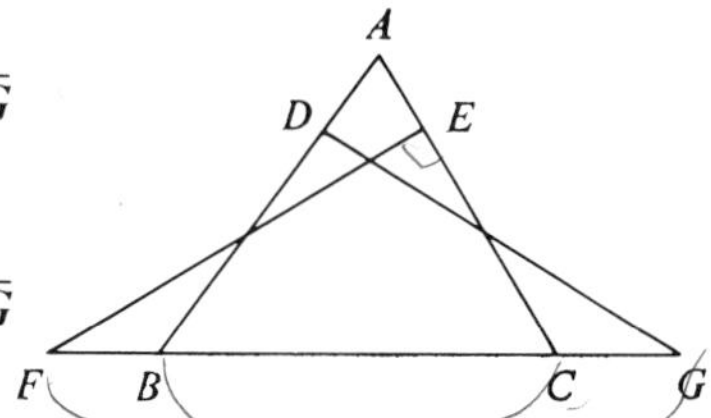

Exercises 7–9

10. **Given** $\overline{DB} \cong \overline{EC}$; $\overline{DF} \perp \overline{BC}$ at F; $\overline{EG} \perp \overline{BC}$ at G; $BG = CF$; $\overline{ADB}$; $\overline{AEC}$; $\overline{BFGC}$
Prove $\triangle ABC$ is isosceles.

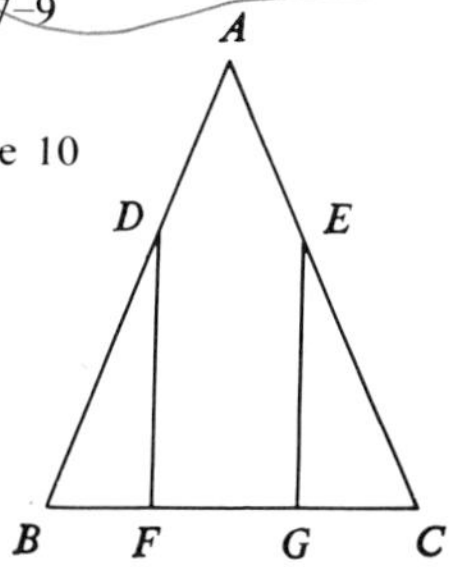

Exercise 10

11. Prove that if two altitudes of a triangle are congruent, the triangle is isosceles.
12. Prove the converse of the statement in Exercise 11.

C

13. **Given** $\overline{BA} \perp \overline{AE}$; $\overline{CD} \perp \overline{ED}$; E is the midpoint of $\overline{BC}$; $\overline{AE} \cong \overline{ED}$;
Prove A, E, and D are collinear.

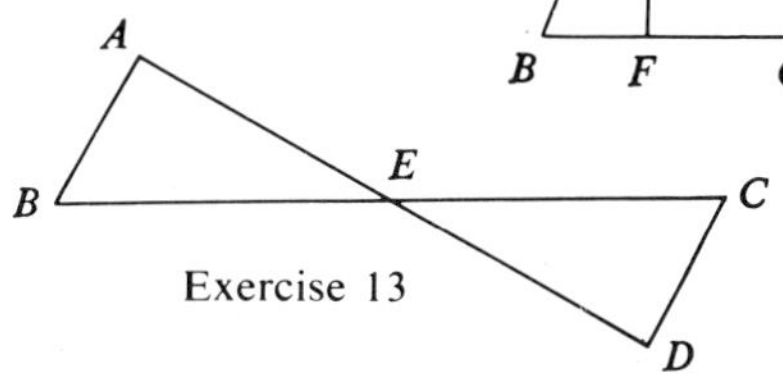

Exercise 13

14. If a quadrilateral has one pair of opposite sides congruent and one pair of opposite angles that are right angles, then both pairs of opposite sides are parallel.

15. Without using Theorem 3-4.1, prove that the altitude to the base of an isosceles triangle is also a median and an angle bisector.
16. Prove Theorem 6-6.2 without using Theorem 6-6.1. Use the pair of diagrams at right. (*Hint*: Show that $\triangle CAP$ and $\triangle CBP$ are isosceles.)

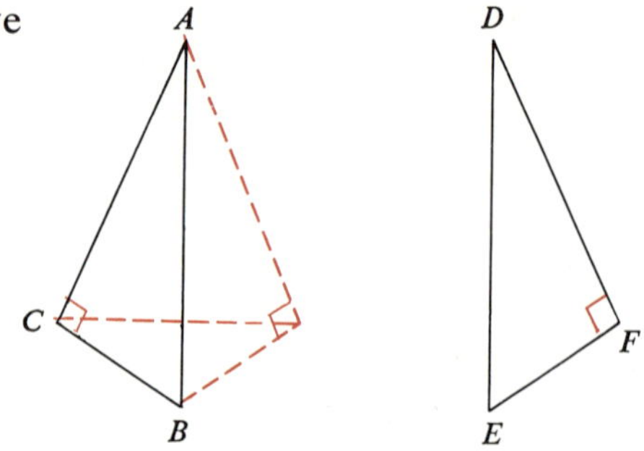

In order to complete the discussion of parallelism, it is necessary to consider parallel planes and lines in parallel planes. Since it would not be applicable until Chapter 13, we have decided not to present it now. If you are interested, you can turn to Section 13-2.

SOMETHING TO THINK ABOUT

Can We Lie with Logic?

Prove that any scalene triangle is isosceles.

GIVEN Scalene $\triangle ABC$

PROVE $\overline{CA} \cong \overline{CB}$

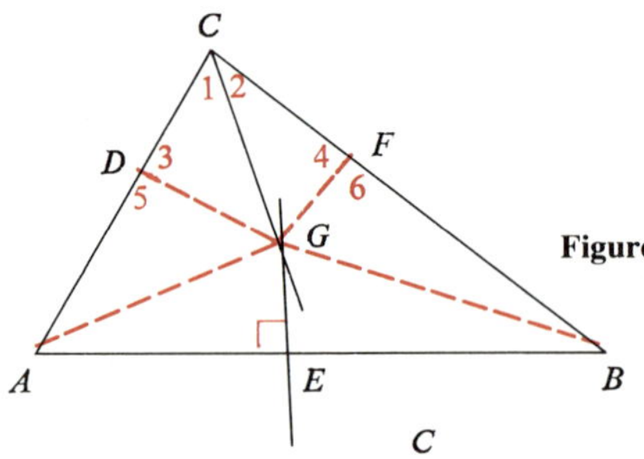

Figure 1

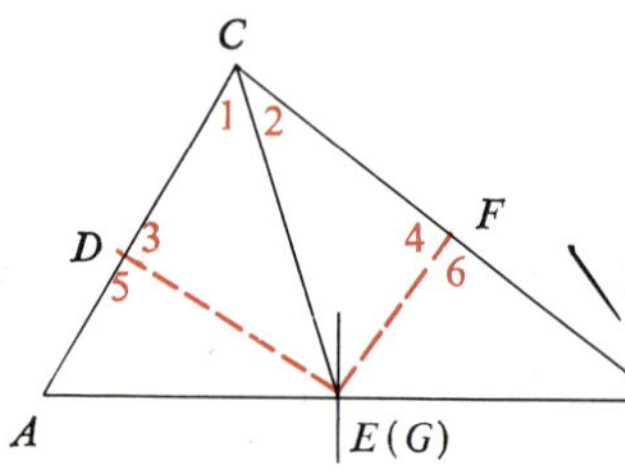

Figure 2

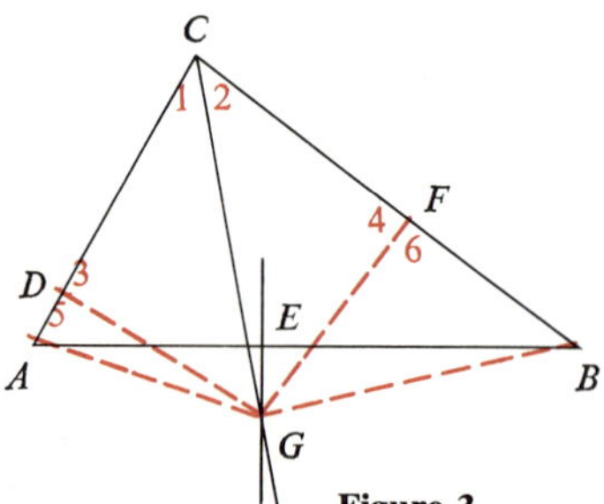

Figure 3

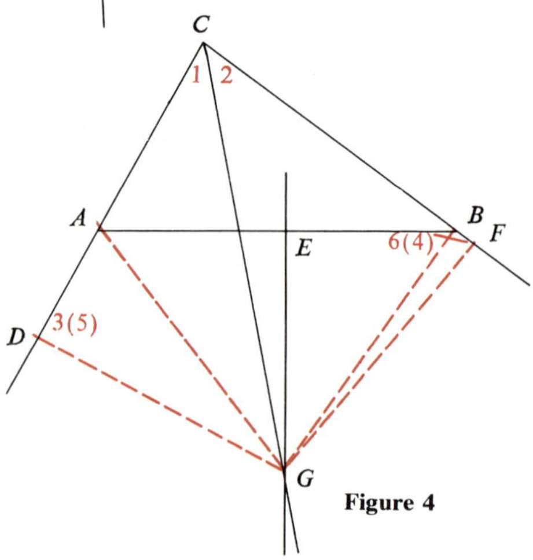

Figure 4

Proof There are four possible situations that will be proved together: where the angle bisector and perpendicular bisector intersect inside the triangle (Figure 1), on the triangle (Figure 2), and outside the triangle (Figures 3 and 4).

STATEMENTS	REASONS
1. Scalene $\triangle ABC$	1. Given
2. Let $\overline{CG}$ be the bisector of $\angle ACB$.	2. Definition of an angle bisector. (1-29)
3. Let $\overleftrightarrow{GE}$ be the perpendicular bisector of $\overline{AB}$ at E; $\overline{GE} \perp \overline{AB}$	3. Definition of a perpendicular bisector. (4-4)
4. Draw $\overline{GD} \perp \overline{AC}$; $\overline{GF} \perp \overline{BC}$; and	4. Through a point external to a line, there is one and only one line perpendicular to the given line. (4-4.6)
5. $\overline{AG}$ and $\overline{GB}$ exist.	5. Definition of a line segment. (1-13)
6. $\angle DCG \cong \angle GCF$	6. Definition of an angle bisector. (1-29)
7. $\angle GDC$ and $\angle GFC$ are right angles.	7. Two lines are perpendicular if and only if they form right angles. (4-4.1)
8. $\angle GDC \cong \angle GFC$	8. All right angles are congruent. (3-1.1).
9. $\overline{CG} \cong \overline{CG}$	9. Every segment is congruent to itself. (3-1.6)
10. $\triangle CDG \cong \triangle CFG$	10. Theorem 6-6.1.
11. $\overline{DG} \cong \overline{FG}$	11. Definition of congruent triangles. (3-3)
12. $\overline{CD} \cong \overline{CF}$	12. Definition of congruent triangles. (3-3)
13. $\angle ADG$ and $\angle GFB$ are right angles.	13. Two lines are perpendicular if and only if they form right angles. (4-4.1)
14. $\angle ADG \cong \angle GFB$	14. All right angles are congruent. (3-1.1)
15. $\overline{AG} \cong \overline{BG}$	15. If two points are both equidistant from the endpoints of a segment, the two points determine the perpendicular bisector of the segment. (4-4.3a)
16. $\triangle DAG \cong \triangle FBG$	16. Theorem 6-6.2.
17. $\overline{DA} \cong \overline{FB}$	17. Definition of congruent triangles. (3-3)
18. $\overline{CA} \cong \overline{CB}$	18. Addition property in Figures 1–3. (See steps 12 and 17.) Subtraction property in Figure 4. (See steps 12 and 17.)

We have just "proved" that all scalene triangles are isosceles. Is this a valid conclusion? When considering the various points at which the bisector of $\angle ACB$ and the perpendicular bisector of $\overline{AB}$ might intersect, we omitted two possibilities: the angle bisector and the perpendicular bisector might be parallel or coincide. What results would we get under these conditions? Can you prove your response?

Judging from the definitions of *isosceles* and *scalene*, do you think that there is some fault with our proof? If you have not found the error yet, search further before reading on.

The fundamental fault in the proof is the failure to consider the Point Betweenness Postulate. It can be proved that the point of intersection of the angle bisector and the perpendicular is outside the triangle. Furthermore, when the perpendiculars to sides $\overleftrightarrow{CA}$ and $\overleftrightarrow{CB}$ are drawn, one and only one may intersect the side *between* the vertices. That is, if F is between C and B, then D is not between C and A.

How does this invalidate the proof? (*Hint*: See step 18.)

HOW IS YOUR MATHEMATICAL VOCABULARY?

The key words and phrases introduced in this chapter are listed below. How many do you know and understand?

alternate interior angles formed by a transversal (6-1)

consecutive angles of a polygon (6-5)

consecutive vertices of a polygon (6-5)

convex polygon (6-5)

corresponding angles formed by a transversal (6-1)

diagonal of a polygon (6-5)

exterior angles of a polygon (6-5)

interior angles of a polygon (6-5)

interior angles on the same side of a transversal (6-1)

parallel lines (6-1)

Parallel Postulate (6-1)

polygon (6-5)

regular polygon (6-5)

transversal (6-1)

vertices of a polygon (6-5)

REVIEW EXERCISES

6-1 Parallel lines

Indicate whether each of the following is true or false. Correct those which are false.

1. Two lines which do not intersect are said to be parallel.
2. If $\overline{AB}$ does not intersect $\overline{CD}$ (where both are in the same plane), then $\overline{AB} \parallel \overline{CD}$.

3. If two lines in the same plane are perpendicular to the same line, then they are parallel.

Complete each of the following:

4. If $\overleftrightarrow{AB} \perp \overleftrightarrow{CD}$ and $\overleftrightarrow{EF} \perp \overleftrightarrow{AB}$, then $\overleftrightarrow{CD} \parallel$ ____.
5. If $\overleftrightarrow{PQ} \parallel \overleftrightarrow{RS}$ and $\overleftrightarrow{MN} \parallel \overleftrightarrow{PQ}$, then $\overleftrightarrow{MN} \parallel$ ____.
6. According to Euclid's Parallel Postulate, there is ________ line parallel to a given line and containing a given point not contained in the given line.
7. GIVEN Quadrilateral $ABCD$; points M and N are the midpoints of $\overline{AB}$ and $\overline{DC}$, respectively; $\angle D \cong \angle C$; $\overline{AD} \cong \overline{BC}$

   PROVE $\overline{AB} \parallel \overline{DC}$

6-2 Proving lines parallel

8. State five ways of proving lines parallel.
9. GIVEN $\overline{BE}$ bisects $\angle ABC$; $\overline{CE}$ bisects $\angle DCB$; $\angle EBC$ is complementary to $\angle ECB$.

   PROVE $\overline{AB} \parallel \overline{CD}$

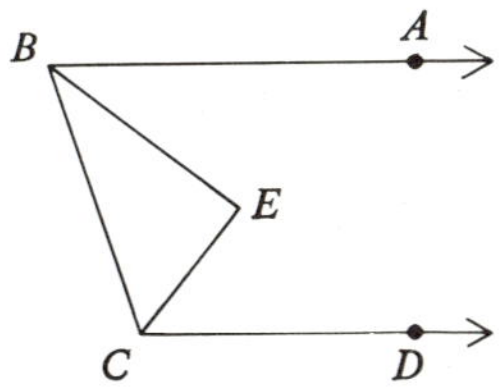

10. GIVEN $BD = CF$; $\angle B \cong \angle F$; $\overline{AB} \cong \overline{FE}$

    PROVE $\overline{AC} \parallel \overline{DE}$

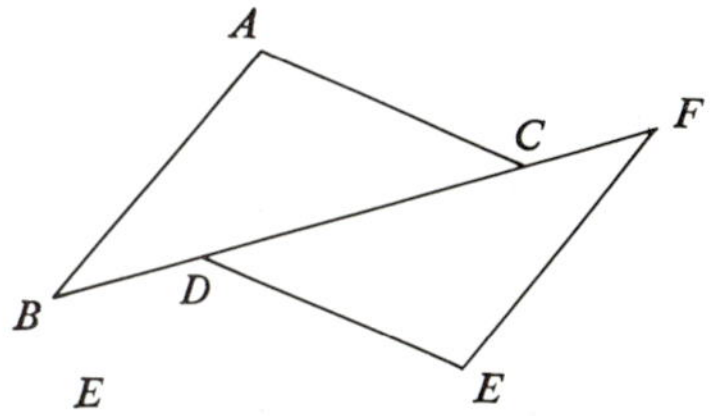

11. GIVEN $\overline{AE}$ and $\overline{BD}$ bisect each other.

    PROVE $\overline{AB} \parallel \overline{DE}$

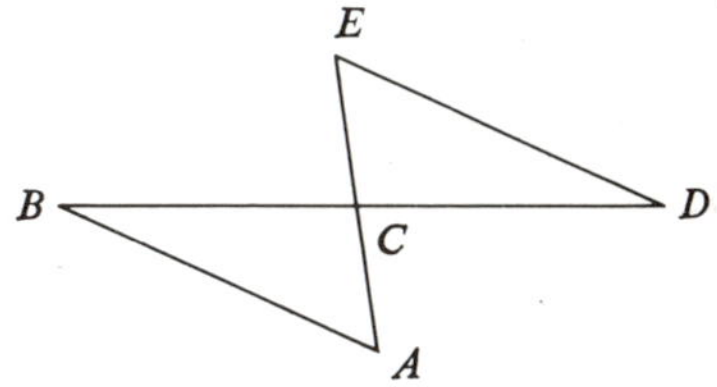

6-3 Properties of angles determined by transversals and parallel lines

12. GIVEN $\overline{AB} \parallel \overline{DC}$; $\overleftrightarrow{BD} \parallel \overleftrightarrow{CE}$; $m\angle DCE = 25$; $m\angle ABC = m\angle ADC = 116$

    FIND $m\angle A$

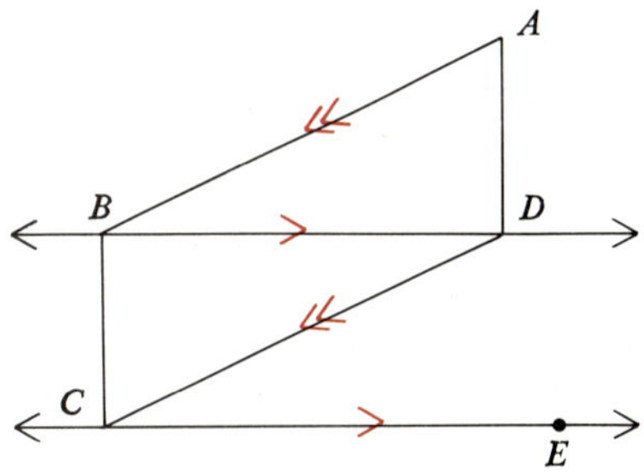

13. GIVEN $\overline{AD} \cong \overline{FC}$; $\overline{ADCF}$;
$\overline{AB} \parallel \overline{FE}$; $\overline{AB} \cong \overline{FE}$
PROVE $\overline{DE} \parallel \overline{BC}$

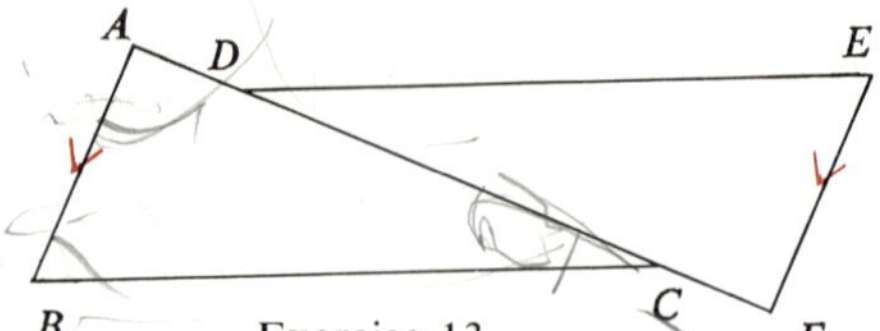

Exercise 13

14. Prove that the opposite angles of a quadrilateral are congruent if both pairs of opposite sides are parallel.

6-4 Angle sum of a triangle

Complete each of the following statements:

15. The measure of each angle of an equilateral triangle is ____.
16. The acute angles of a right triangle are ________.
17. If the measure of one angle of a triangle equals the sum of the remaining two angles, then it is a(n) ________ triangle.
18. When two lines are parallel, the bisectors of a pair of interior angles on the same side of the transversal are ________.
19. GIVEN $\triangle ABC$, $\overrightarrow{CBD}$
$m\angle ACB = 90$;
$m\angle ABD = 140$;
$\overline{EC}$ bisects $\angle ACB$; $\overline{AEB}$
FIND $m\angle AEC$

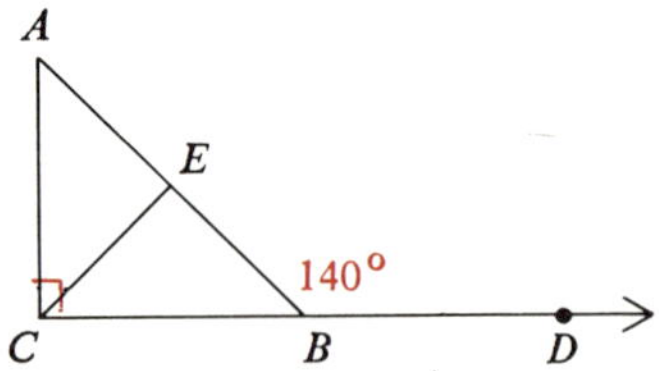

20. The measures of the three angles of a triangle are represented by $4x + 9$, $3x + 18$, and $10x$. Find the measure of each angle, then state precisely what type of triangle it is.

6-5 Polygons

21. How does a convex polygon differ from a concave polygon?
22. What is the measure of an interior angle of a regular n-gon if n equals 26.
23. What is the measure of an exterior angle of a regular n-gon if n equals 720.
24. If the measure of an interior angle of a regular polygon is represented by $27x + 35$ and the measure of an exterior angle of the same polygon is represented by $2x$, how many sides does this polygon have?
25. By how much does the sum of the measures of the interior angles of a 17-sided polygon differ from the sum of the measures of the interior angles of a decagon? Can you answer this question without finding either sum?
26. Prove that if a quadrilateral has congruent opposite angles, then its opposite sides are parallel.

6-6 More triangle congruence theorems

Indicate whether each of the following statements is always true, sometimes true, or never true. Justify your answers.

27. Two triangles are congruent if two sides and a nonincluded angle of one are congruent to the corresponding sides and angle of the other.
28. Two right triangles are congruent if their hypotenuses are congruent and a pair of corresponding acute angles are congruent.
29. GIVEN $\triangle ABC$; $\overline{DC} \cong \overline{EB}$; $\overline{ADB}$; $\overline{AEC}$; $\angle ADC \cong \angle AEB$

    PROVE $\triangle ABC$ is isosceles.

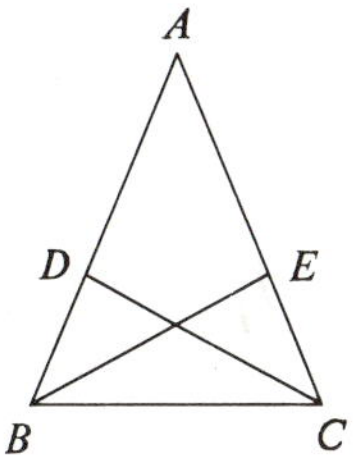

30. GIVEN $m\angle A = m\angle B = 90$; $\overline{AP} \cong \overline{BP}$

    PROVE $\overline{AQ} \cong \overline{BQ}$

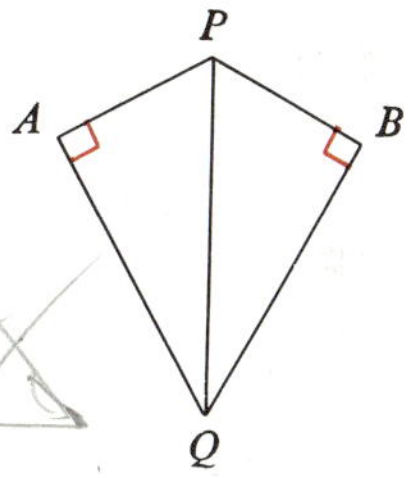

CHAPTER TEST

1. If the measure of the vertex angle of an isosceles triangle is 30, then the measure of a base angle is ____.
2. The difference between the sum of the measures of the exterior angles and the sum of the measures of the interior angles of a quadrilateral is ____.
3. If twice the measure of an angle of a triangle equals the measure of a nonadjacent exterior angle of the triangle, then the triangle is ________.
4. A line parallel to one of two perpendicular lines is ________ to the other line.

Find the measure of $\angle x$ and $\angle y$ in each of the following:

5.

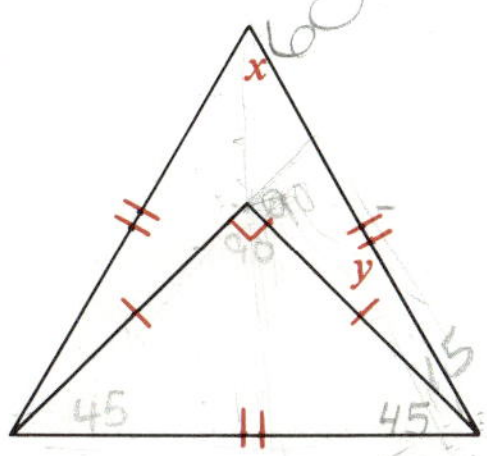

6.

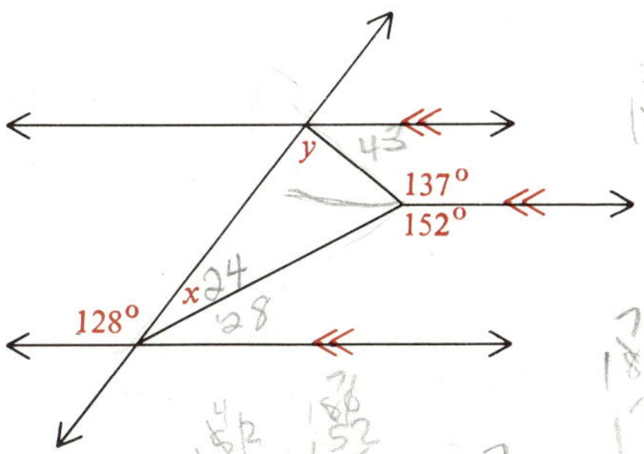

In the figure at right, $\overline{BD} \cong \overline{CE}$; $\overline{AEB}$; and $\overline{ADC}$. Indicate which of the single facts given below allow us to complete a proof that $\overline{CD} \cong \overline{BE}$ and which do not. Justify your answers.

7. $\angle 1 \cong \angle 2$
8. $\angle 3 \cong \angle 4$
9. $\angle 5 \cong \angle 3$
10. $\angle 4 \cong \angle 6$

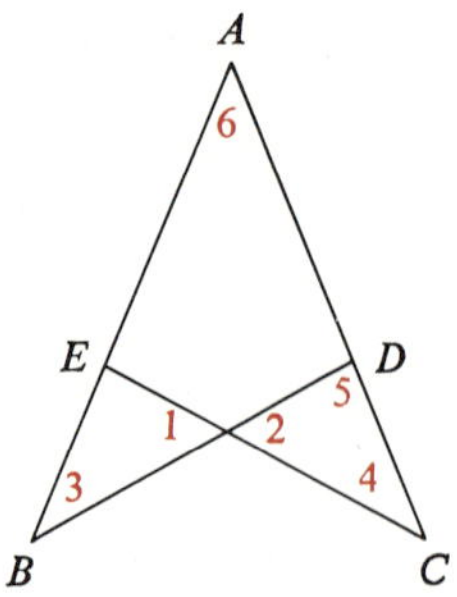

Exercises 7–10

11. In the diagram at right, show that $\overleftrightarrow{AP} \parallel \overline{BC}$ if
$m\angle B = 11x - 7$;
$m\angle C = 3x + 4$;
$m\angle BAC = 10x + 15$; and
$m\angle PAC = 6x - 17$.

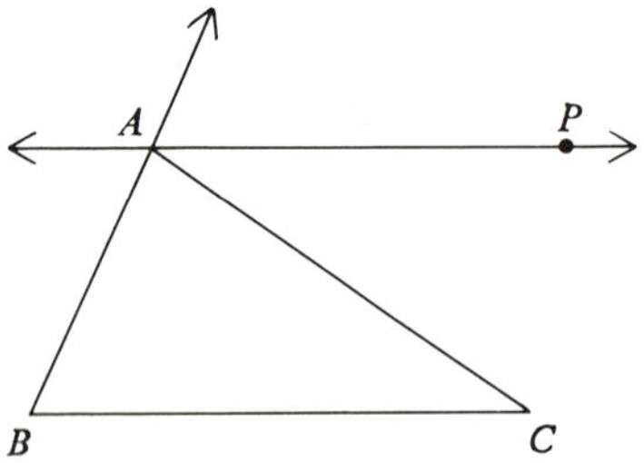

12. **Given** Right $\triangle QPR$ and SPT, with right angles at P; $\overline{PS}$ and $\overline{ST}$ intersect $\overline{QR}$ at points U and M, respectively; $\overline{ST}$ meets $\overline{PR}$ at N; $\overline{QP} \cong \overline{SP}$; $\overline{QR} \cong \overline{ST}$; and $\overline{QR} \parallel \overline{PT}$

Prove $\triangle MNR$ is isosceles.

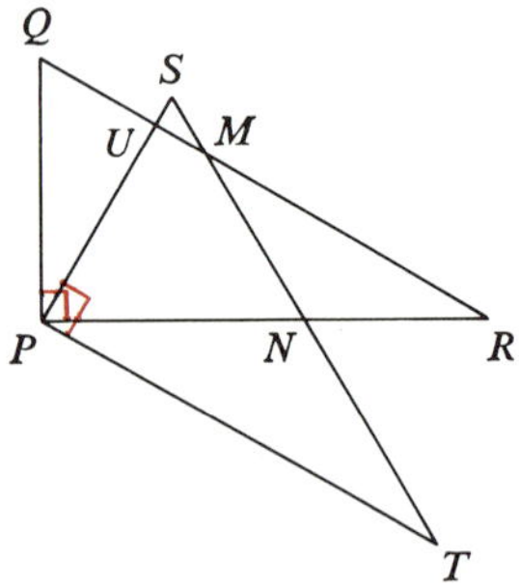

13. Prove that if from any point P on the bisector of $\angle ABC$, $\overline{PR} \perp \overrightarrow{BA}$, and $\overline{PS} \perp \overrightarrow{BC}$, where R and S are on $\overrightarrow{BA}$ and $\overrightarrow{BC}$, respectively, then $\overline{PR} \cong \overline{PS}$.

SUGGESTED RESEARCH

1. Prepare a report about how the ideas learned in this chapter can be used in navigation, sports, carpentry, electricity, surveying, and other fields.
2. Present a report on Morley's Theorem to your class. One possible source for your report is *Challenging Problems in Geometry*, vol. 2, by A. S. Posamentier and C. T. Salkind (New York: Macmillan Co., 1970).
3. A tessellation is a mosaic composed of small squares. Prepare a report on tessellations. Some sources include *Mathematical Recreations*, by

Maurice Kraitchik (New York: Dover Publications, 1953) and *Introduction to Geometry*, by H. S. M. Coxeter (New York: John Wiley and Sons, 1961).

MATHEMATICAL EXCURSION

The Parallel Postulate

The parallel postulate that we stated at the beginning of this chapter is not in the form Euclid used. The form we used is known as Playfair's Axiom. Euclid's statement of the parallel postulate does not even mention parallel lines. In modern terminology, Euclid's postulate states: If two lines are cut by a transversal and the sum of the measures of the interior angles on the same side of the transversal is less than 180, then the two lines meet on that side of the transversal.

In the figure at the right, if $n \perp l$ at A and $m\angle ABC < 90$, then, by the postulate above, m intersects l in the direction of $\overrightarrow{BC}$ and $\overrightarrow{AD}$.

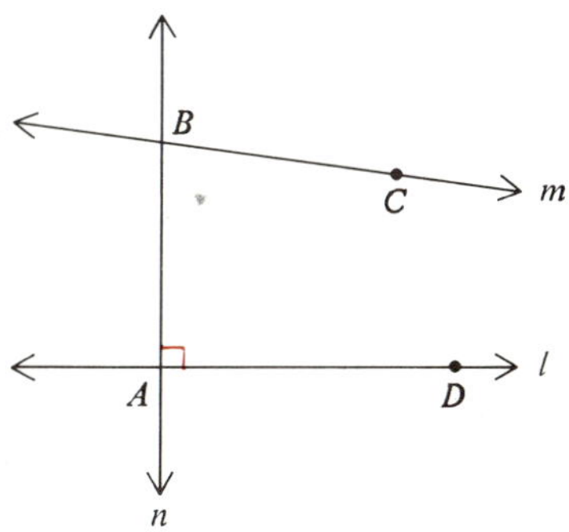

Euclid's postulate and Playfair's Axiom are equivalent. That is, you can assume either and prove the other. The following, though not a complete proof, makes their equivalence seem reasonable. Consider $n \perp l$ at A and m intersecting n at B. If $m \perp n$, we negate the hypothesis of Euclid's Parallel Postulate and the conclusion is negated because m and l are parallel by Theorem 6-1.1. Note carefully that Theorem 6-1.1 does not depend on any assumption of parallels. If m is not perpendicular to n, it must make an acute angle with n on one side or other of n. Thus, there is only one angle that m can make with n that will not require m to intersect l. Essentially, this is Playfair's Axiom.

Theorem 6-1.1 is the opposite of Euclid's Parallel Postulate. That is, it is equivalent to the converse of the postulate. Because of this, mathematicians thought that the postulate should be provable. Most early attempts to prove the postulate failed because an equivalent assumption was inadvertently used. It is easy to prove the postulate if an equivalent assumption is made. For example, using Playfair's Axiom, we can easily prove Euclid's Parallel Postulate. The question that has always concerned mathematicians was whether Euclid's Parallel Postulate could be proved without making an equivalent assumption.

The most significant efforts used the form of indirect proof. Refer to the figure above. Essentially, Euclid said that the only value of $\angle ABC$ for which m does not cut l is a right angle. If it can be shown that (1) the assumption that m does not cut l if $m\angle ABC < 90$ leads to a contradiction and (2) the assumption that m cuts l for any value of $m\angle ABC$ leads to a contradiction, then we can conclude that m does not cut l if and only if $m\angle ABC = 90$. This conclusion is equivalent to Euclid's Parallel Postulate.

The situation can be more easily described using Playfair's terminology. If we consider a line l and a point P not on l, then there are three possible situations.

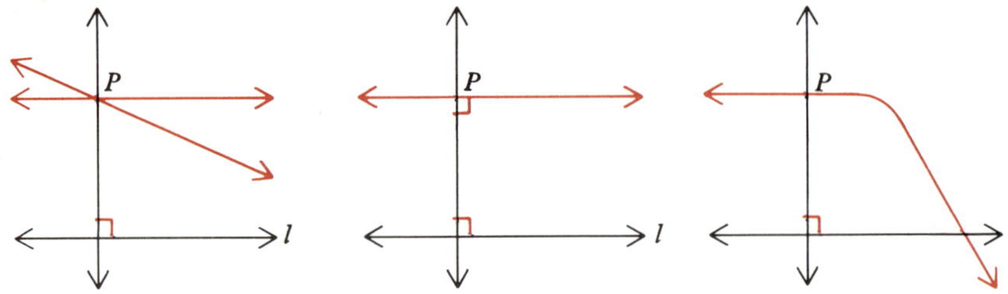

(*a*) (*b*) (*c*)

As shown in (a), there is either more than one line in the plane of P and l through P that does not cut l, or as in (b), there is exactly one line through P that does not cut l, or, as in (c), there are no lines through P that do not cut l. If we assume that there is more than one line through P parallel to l and reach a contradiction, and assume that there are no parallels and reach another contradiction, then the only remaining alternative is the existence of one parallel to l through P. This remaining alternative is Playfair's Axiom.

Using the no-parallel assumption, it is easy to establish a contradiction. Without using any parallel assumption, Theorem 6-1.1 shows that there is a parallel. Historically, all attempts to arrive at a contradiction under the assumption of more than one parallel met with failure. It is now known that no contradiction can be found if we assume that there are no contradictions in Euclidean geometry. This poses a basic question. For this figure is it actually possible, in physical space, for $\overleftrightarrow{AD}$ to be parallel to $\overleftrightarrow{BC}$ and for $m\angle BAD$ to be less than 90? We cannot answer this question. However, if we assume this, we can show that the sum of the measures of the angles of a triangle is less than 180. Why can't we determine which parallel assumption is correct by measuring the angles of a triangle?

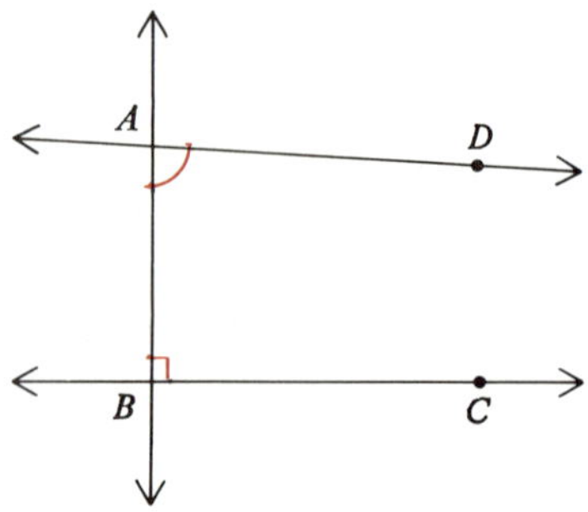

Refer to the proof of Theorem 6-1.1. In this proof we used the idea that the exterior angle of a triangle is greater than either remote interior angle. The proof of this statement depends on the notion that lines are infinite in length. Much of the geometry we are studying would remain valid even if lines were not infinite. However, if we do not use the fact that lines are infinite, the above exterior angle statement would not be valid. If we then accept the no-parallel assumption, we would again fail to reach a contradiction.

If we do not assume that lines are infinite but do assume that there are no parallel lines, we can form a consistent geometry by assuming that two lines always have one point in common. We get another consistent geometry if we assume that two lines always have two points in common. The latter is similar to the geometry of great circles on the surface of a sphere.

EXERCISES

1. Prove that if the sum of the measures of the angles of a right triangle is always 180, then the sum of the measures of the angles of any triangle is 180.
2. Under the hypothesis of no parallels, the sum of the measures of the angles of a triangle must be more than 180. What is the smallest value it must be less than?
3. Under the hypothesis of more than one parallel, the sum of the measures of the angles of a triangle must be less than 180. Prove that under this hypothesis there is no such thing as a square.
4. What can you conclude about the sum of the measures of the exterior angles of a polygon under the hypothesis of more than one parallel?
5. Answer Exercise 4 under the hypothesis of no parallels.

Construct perpendicular lines and measure the same distance on each of the four rays from their intersection. At the ends of these segments, construct perpendiculars.

6. Describe the resulting figure under the hypothesis of one parallel.
7. Describe the resulting figure under the hypothesis of no parallels.
8. Describe the resulting figure under the hypothesis of more than one parallel.

CUMULATIVE REVIEW

Chapters 4, 5, 6

These exercises are designed so that you can check your understanding of some of the principal concepts studied so far. The numbered items in the right column are the correct answers. To complete the review, first cover the answers, read and answer the questions, and check each response by uncovering the corresponding answer.

1. The implication ____ is equivalent to $p \to q$.	1. $\sim q \to \sim p$
2. True or false: If a given implication is true, then the converse must also be true.	2. false
3. Methods of proving statements true include direct proof, existence, enumeration, indirect proof, and ____.	3. contrapositive
4. A projection of a point to a line is the ____ of the perpendicular.	4. foot
5. Given a line, there is a unique ____ to the line at a point of the line.	5. perpendicular
6. No triangle can have more than ____ right angle(s).	6. one
7. There is a(n) ____ perpendicular to a plane from an external point.	7. unique
8. In your classroom, describe the projection of the ceiling corners on the floor.	8. floor corners
9. The set of points common to both faces of a dihedral angle lies on the ____.	9. line of intersection
10. If the supplement of $\angle M$ is greater than the supplement of $\angle N$, what is the relationship between $m\angle M$ and $m\angle N$?	10. $m\angle M < m\angle N$
11. If in $\triangle ABC$, $\angle A$ is an obtuse angle, then $\angle B$ and $\angle C$ are ____.	11. acute
12. True or false: If in $\triangle PQR$, $PQ = RQ$, then $\angle P$ and $\angle Q$ are acute.	12. true
13. The longest side of a right $\triangle ACB$ with $m\angle C = 90$, is side ____.	13. $\overline{AB}$

14. Arrange the angles of $\triangle PQR$ in order of their measure, if $PQ = 7$, $QR = 8$, and $RP = 10$.	14. largest to smallest: $\angle Q$, $\angle P$, $\angle R$
15. True or false: The elements of $\{105, 94, 9\}$ represent the lengths of sides of a triangle.	15. false
16. In $\triangle DEF$, $DE = 1$, $DF = 18$, and EF has what range?	16. $17 < EF < 19$
17. If in $\triangle ABC$ and $\triangle DEF$, $AB = DE$, $BC = EF$, and $m\angle B > m\angle E$, then AC ____ DF.	17. $>$
18. If in $\triangle ABC$ and $\triangle DEF$, $AB = DE$, $BC = EF$, and $AC > DF$, then $m\angle B$ ____ $m\angle E$.	18. $>$
19. In a plane, if a line is perpendicular to one of two parallel lines, then it is ____ to the other line.	19. perpendicular
20. If two lines are cut by a transversal so that the alternate interior angles are ____, then the lines are parallel.	20. congruent
21. If a pair of interior angles on the same side of the transversal are supplementary, then the lines are ____.	21. parallel
22. True or false: Two lines are parallel if the lines are each perpendicular to the same line or to parallel lines.	22. true
23. If in $\triangle ABC$, $\angle C$ is a right angle and $m\angle B = 20$, $m\angle A =$ ____.	23. 70
24. When the interior and adjacent exterior angles of a regular polygon are congruent, there are ____ sides.	24. four
25. If $\triangle ABC$ and $\triangle DEF$ have: $\angle A \cong \angle D$, $\angle B \cong \angle E$, and $BC = EF$, then ____.	25. $\triangle ABC \cong \triangle DEF$
26. If two altitudes of a triangle are congruent, then the triangle is ____.	26. isosceles
27. If in right triangles ABC and DEF, $m\angle C = m\angle F = 90$, $AB = DE$ and $AC = DF$, then ____.	27. $\triangle ABC \cong \triangle DEF$

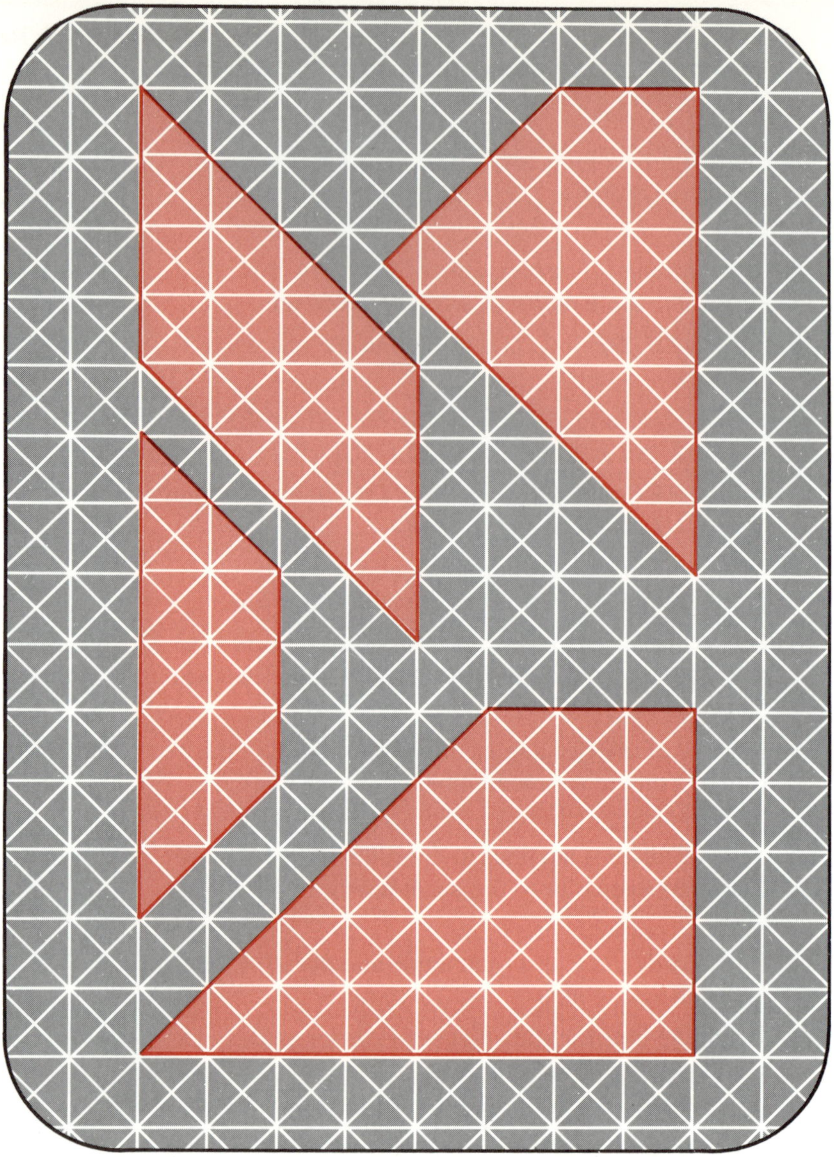

7 Quadrilaterals

The polygon that received the most attention in the first six chapters was the triangle. We shall now concentrate on polygons of four sides. In this chapter, we shall study quadrilaterals, including the general parallelogram, special types of parallelograms, the trapezoid, and other related topics.

7-1 PROPERTIES OF A PARALLELOGRAM

The word *parallel* is contained in the word *parallelogram*. What do you think this indicates about the parallelogram?

D

Definition 7-1 A **parallelogram** is a quadrilateral in which both pairs of opposite sides are parallel.

We shall use the symbol ▱ to denote *parallelogram*. Read ▱$ABCD$, "parallelogram $ABCD$." The consecutive vertices of ▱$ABCD$ are A, B, C, and D. Therefore, we may refer to ▱$ABCD$ as ▱$BCDA$, but not as ▱$ACBD$.

T

Theorem 7-1.1 A diagonal of a parallelogram divides the parallelogram into two congruent triangles.

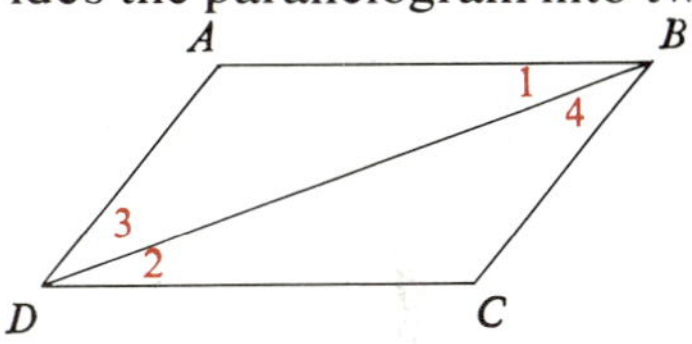

GIVEN ▱$ABCD$ with diagonal $\overline{BD}$
PROVE $\triangle ABD \cong \triangle CDB$
PROOF

STATEMENTS	REASONS
1. ▱$ABCD$ with diagonal $\overline{BD}$	1. Given.
2. $\overline{AB} \parallel \overline{DC}$ and $\overline{AD} \parallel \overline{BC}$	2. Definition of a parallelogram. (7-1)
3. $\angle 1 \cong \angle 2$ and $\angle 3 \cong \angle 4$	3. If two parallel lines are cut by a transversal, then the alternate interior angles are congruent. (6-3.1)
4. $\overline{DB} \cong \overline{DB}$	4. Every segment is congruent to itself. (3-1.6)
5. $\triangle ABD \cong \triangle CDB$	5. ASA Postulate. (3-2)

Using a similar proof, you can show that $\overline{AC}$ also divides ▱$ABCD$ into two congruent triangles. Name the triangles.

Using Theorem 7-1.1, we can prove the following theorems. Use the figure from Theorem 7-1.1 for Theorems 7-1.2 and 7-1.3.

T

Theorem 7-1.2 The opposite sides of a parallelogram are congruent.

GIVEN ▱$ABCD$ with diagonal $\overline{BD}$
PROVE $\overline{AB} \cong \overline{CD}$; $\overline{AD} \cong \overline{CB}$
PROOF

STATEMENTS	REASONS
1. ▱$ABCD$ with diagonal $\overline{BD}$	1. Given.
2. $\triangle ABD \cong \triangle CDB$	2. Theorem 7-1.1.
3. $\overline{AB} \cong \overline{CD}$ and $\overline{AD} \cong \overline{CB}$	3. Definition of congruent triangles (3-3).

T **Theorem 7-1.3** The opposite angles of a parallelogram are congruent.

GIVEN ▱$ABCD$ with diagonal $\overline{BD}$

PROVE $\angle A \cong \angle C$; $\angle ADC \cong \angle ABC$

PROOF

STATEMENTS	REASONS
1. ▱$ABCD$ with diagonal $\overline{BD}$	1. Given.
2. $\triangle ABD \cong \triangle CDB$	2. Theorem 7-1.1.
3. $\angle A \cong \angle C$	3. Definition of congruent triangles. (3-3)
4. $\overline{AB} \parallel \overline{CD}$ and $\overline{AD} \parallel \overline{BC}$	4. Definition of a parallelogram. (7-1)
5. $m\angle 1 = m\angle 2$ and $m\angle 3 = m\angle 4$	5. If two parallel lines are cut by a transversal, then the alternate interior angles are congruent. (6-3.1)
6. $m\angle 1 + m\angle 4 = m\angle 2 + m\angle 3$ (or $\angle ADC \cong \angle ABC$)	6. Addition property.

D **Definition 7-2** A pair of **consecutive angles** of a parallelogram is formed by two angles that have their vertices in the endpoints of the same side of the parallelogram.

T **Theorem 7-1.4** Any two consecutive angles of a parallelogram are supplementary.

PROOF OUTLINE Since the opposite sides of a parallelogram are parallel, each pair of consecutive angles is also a pair of interior angles on the same side of the transversal. Therefore, by Corollary 6-3.1b, each pair of consecutive angles is supplementary.

T **Theorem 7-1.5** The diagonals of a parallelogram bisect each other.

GIVEN ▱$ABCD$ with diagonals $\overline{BD}$ and $\overline{AC}$ intersecting at E

PROVE $\overline{DE} \cong \overline{BE}$; $\overline{AE} \cong \overline{CE}$

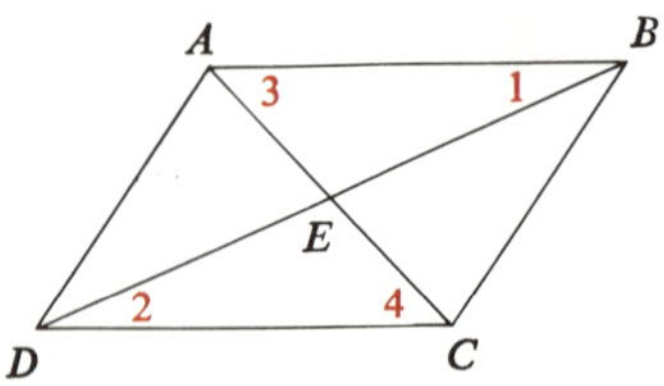

Proof

STATEMENTS	REASONS
1. $\square ABCD$ with diagonals $\overline{BD}$ and $\overline{AC}$ intersecting at E	1. Given.
2. $\overline{AB} \parallel \overline{DC}$	2. Definition of a parallelogram. (7-1)
3. $\angle 1 \cong \angle 2$; $\angle 3 \cong \angle 4$	3. If two parallel lines are cut by a transversal, then the alternate interior angles are congruent. (6-3.1)
4. $\overline{AB} \cong \overline{CD}$	4. Theorem 7-1.2.
5. $\triangle AEB \cong \triangle CED$	5. ASA Postulate. (3-2)
6. $\overline{DE} \cong \overline{BE}$; $\overline{AE} \cong \overline{CE}$	6. Definition of congruent triangles. (3-3)

Class Exercises

In the figure at right, $\overleftrightarrow{AB} \parallel \overleftrightarrow{CD}$.
Complete each of the following:

1. If $\overline{AD} \perp \overleftrightarrow{DC}$ and $\overline{BC} \perp \overleftrightarrow{DC}$, then quadrilateral $ABCD$ is a ________.
2. $AD =$ ____.
3. What is the distance from A to $\overleftrightarrow{DC}$?
4. What is the distance from B to $\overleftrightarrow{DC}$?
5. How do these distances relate?

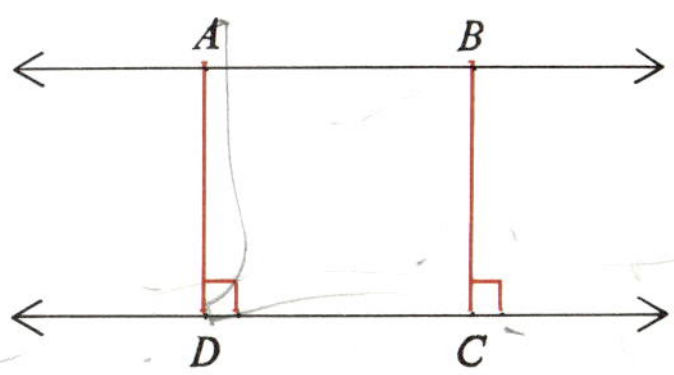

Exercises 1–5

Exercises 1–5 above lead to the following theorem. The proof of this theorem is left as an exercise.

T **Theorem 7-1.6** If two lines are parallel, then all the points of each line are equidistant from the other line.

D **Definition 7-3** The **distance between two parallel lines** is the length of the perpendicular segment from any point of one line to the other line.

Definition 7-3 makes it possible for us to restate Theorem 7-1.6: "Parallel lines are everywhere equidistant." Can you explain why?

D **Definition 7-4** An **altitude of a parallelogram** is the perpendicular segment from any point of a line containing one side of the parallelogram to the line containing the opposite side of the parallelogram.

EXERCISES

A Exercises 1–4 refer to $\square MNKL$.

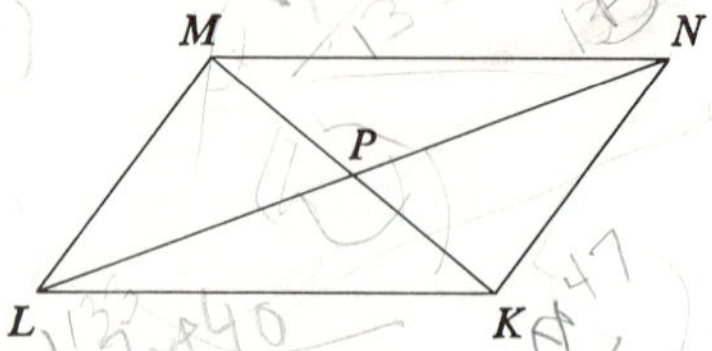

Exercises 1–5

1. Name two pairs of parallel segments and four pairs of congruent segments.
2. How many pairs of congruent angles are there in $\square MNKL$? Name them.
3. Name two angles that are supplementary to $\angle NML$.
4. Name two angles that are supplementary to $\angle MLK$.

5. In $\square WXYZ$, $m\angle W = 47$. Find the measures of the other angles.
6. In $\square ABCD$, $m\angle A = x$ and $m\angle D = 3x + 40$. Find the measures of all the angles of the parallelogram.
7. In $\square EFGH$, $EH = 3x - 5$, $EF = 6x - 11$, and $FG = 7x - 17$. Find the lengths of $\overline{EH}$, $\overline{GH}$, and $\overline{FG}$.
8. The diagonals of $\square WXYZ$ meet at P. $WP = 3x - 5$, $YP = 12x - 41$, and $XP = 3x + 11$. Find the lengths of the diagonals.

B

9. The diagonals of $\square HTAM$ intersect at P. $HP = 5x + y$, $AP = 4y - x$, $MP = 3y + 7$, and $TP = x + 6y$. Find the lengths of the diagonals.
10. Use Theorem 7-1.4 to prove Theorem 7-1.3 without using the diagonals of the parallelogram.
11. Use Theorems 7-1.3 and 6-5.1 to prove Theorem 7-1.4 without using the diagonals of the parallelogram.
12. GIVEN Points M, N, L, and K are the respective midpoints of sides $\overline{AB}$, $\overline{BC}$, $\overline{CD}$, and $\overline{AD}$ of $\square ABCD$.
    PROVE $\overline{MN} \cong \overline{LK}$
13. GIVEN $\square DEFG$; $\overrightarrow{GH}$, the bisector of $\angle DGF$, meets $\overline{DE}$ at H; $\overrightarrow{EJ}$, the bisector of $\angle DEF$, meets $\overline{GF}$ at J.
    PROVE $HE = GJ$
14. GIVEN $\square ABCD$; $\overrightarrow{DH} \perp \overline{AC}$ at H; $\overrightarrow{BG} \perp \overline{AC}$ at G
    PROVE $\overline{HD} \cong \overline{GB}$
15. GIVEN $\square MATH$; $\overline{GO}$ bisects $\overline{MT}$ at E; $\overline{MGA}$; $\overline{HOT}$
    PROVE $\overline{MT}$ bisects $\overline{GO}$ at E.

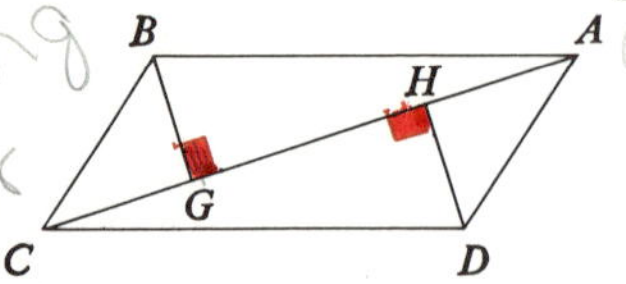

Exercise 14

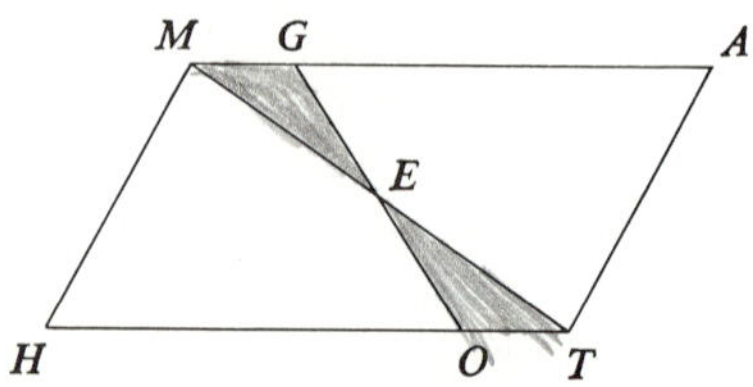

16. Prove the converse of Exercise 15.

18. GIVEN $\square ABCD$; $\overline{ABP}$; $\overline{CDQ}$; $BP = DQ$
PROVE $\overline{AC}$ and $\overline{PQ}$ bisect each other.

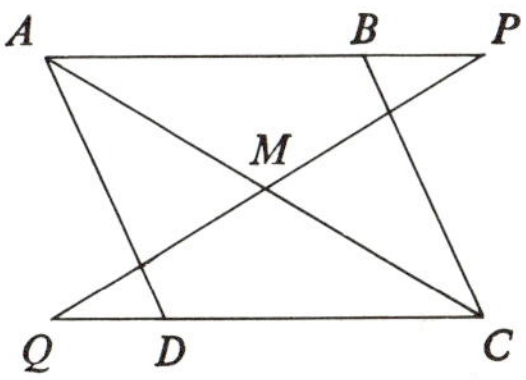

17. GIVEN $\square ABCD$; M and N are the midpoints of $\overline{AB}$ and $\overline{CD}$, respectively; $\overline{MC}$ meets $\overline{DB}$ at E; $\overline{NA}$ meets $\overline{DB}$ at F
PROVE $\overline{DF} \cong \overline{BE}$

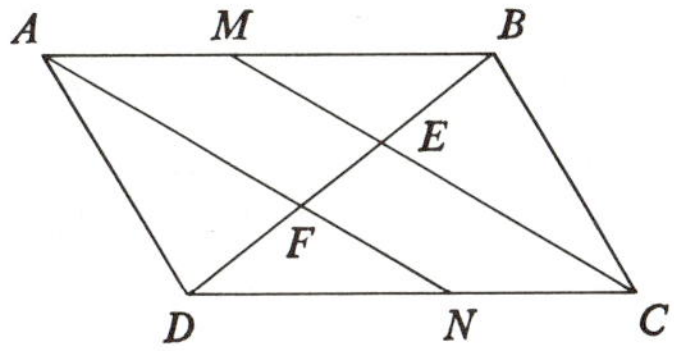

C

19. GIVEN $\square ABCD$; $\triangle AED$ and $\triangle AFB$ are equilateral.
PROVE $\triangle EFC$ is equilateral.

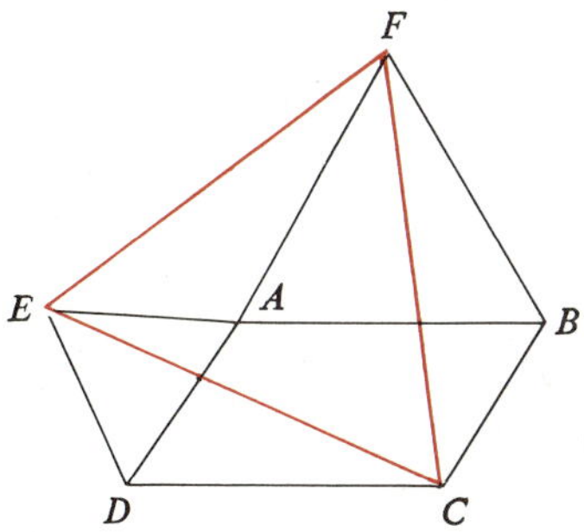

20. GIVEN Isosceles $\triangle ABC$; $\overline{AB} \cong \overline{AC}$; altitude $\overline{CD}$; $\overline{BPC}$; $\overline{PM} \perp \overline{AB}$ at M; $\overline{PN} \perp \overline{AC}$ at N
PROVE $PM + PN = CD$ (*Hint*: Draw $\overline{PE} \perp \overline{DC}$ at E.)

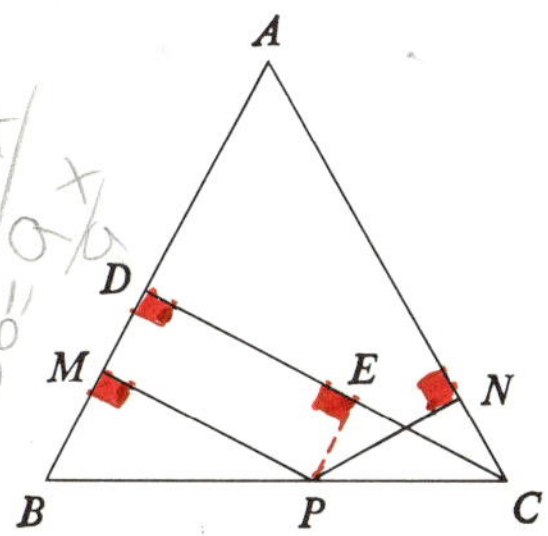

21. Prove that the distance between two parallel lines is the length of the shortest line segment that has an endpoint in each of the parallel lines.
22. State and prove the converse of Exercise 21.

7-2 WHEN A QUADRILATERAL IS A PARALLELOGRAM

With the proper information about the sides, angles, and/or diagonals of a quadrilateral, you can determine whether the quadrilateral is a parallelogram. For example, if the pairs of opposite sides of the given quadrilateral are parallel, then by definition the quadrilateral is a parallelogram.

T **Theorem 7-2.1** A quadrilateral is a parallelogram if both pairs of opposite sides are congruent.

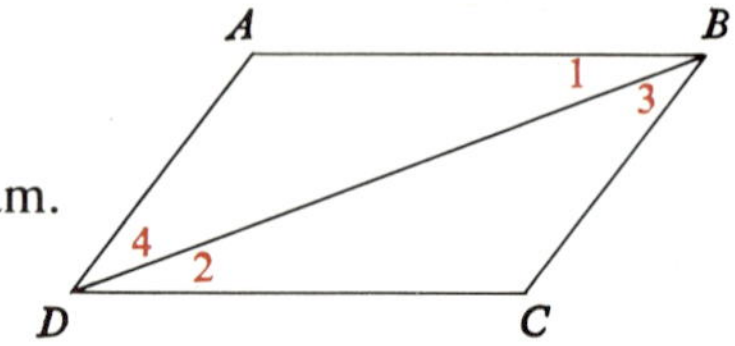

GIVEN Quadrilateral $ABCD$; $\overline{AB} \cong \overline{CD}$; $\overline{AD} \cong \overline{CB}$

PROVE Quadrilateral $ABCD$ is a parallelogram.

PROOF

STATEMENTS	REASONS
1. Quadrilateral $ABCD$; $\overline{AB} \cong \overline{CD}$; $\overline{AD} \cong \overline{CB}$	1. Given.
2. Draw diagonal $\overline{BD}$.	2. Definition of a line segment. (1-13)
3. $\overline{BD} \cong \overline{BD}$	3. Every segment is congruent to itself. (3-1.6)
4. $\triangle ABD \cong \triangle CDB$	4. SSS Postulate. (3-3)
5. $\angle 1 \cong \angle 2$	5. Definition of congruent triangles. (3-3)
6. $\overline{AB} \parallel \overline{CD}$	6. If two lines are cut by a transversal so that the alternate interior angles are congruent, then the lines are parallel. (6-2.1)
7. $\angle 3 \cong \angle 4$	7. Same reason as step 5.
8. $\overline{AD} \parallel \overline{BC}$	8. Same reason as step 6.
9. Quadrilateral $ABCD$ is a parallelogram.	9. Definition of a parallelogram. (7-1)

We now have two ways of proving that a given quadrilateral is a parallelogram. What are they? The next few theorems offer additional ways. Use the figure from Theorem 7-2.1 for proof of Theorem 7-2.2.

T **Theorem 7-2.2** A quadrilateral is a parallelogram if two of its sides are both congruent and parallel.

GIVEN Quadrilateral $ABCD$; $\overline{AB} \cong \overline{CD}$; $\overline{AB} \parallel \overline{CD}$

PROVE Quadrilateral $ABCD$ is a parallelogram.

PROOF

STATEMENTS	REASONS
1. Quadrilateral $ABCD$; $\overline{AB} \cong \overline{CD}$; $\overline{AB} \parallel \overline{CD}$	1. Given.
2. Draw diagonal $\overline{BD}$.	2. Definition of a line segment. (1-13)

(*continued*)

STATEMENTS	REASONS
3. $\angle 1 \cong \angle 2$	3. If two parallel lines are cut by a transversal, then the alternate interior angles are congruent. (6-3.1)
4. $\overline{BD} \cong \overline{BD}$	4. Every segment is congruent to itself. (3-1.6)
5. $\triangle BAD \cong \triangle DCB$	5. SAS Postulate. (3-1)
6. $\angle 3 \cong \angle 4$	6. Definition of congruent triangles. (3-3)
7. $\overline{AD} \parallel \overline{BC}$	7. If two lines are cut by a transversal so that the alternate interior angles are congruent, the lines are parallel. (6-2.1)
8. Quadrilateral $ABCD$ is a parallelogram.	8. Definition of a parallelogram. (7-1)

T

Theorem 7-2.3 A quadrilateral is a parallelogram if the opposite angles are congruent.

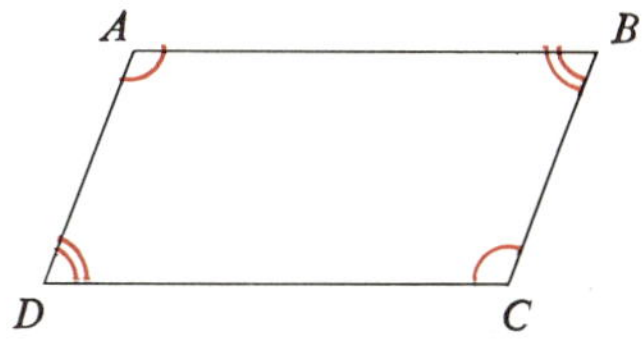

GIVEN Quadrilateral $ABCD$; $\angle A \cong \angle C$; $\angle D \cong \angle B$

PROVE Quadrilateral $ABCD$ is a parallelogram.

PROOF

STATEMENTS	REASONS
1. Quadrilateral $ABCD$; $\angle A \cong \angle C$; $\angle D \cong \angle B$	1. Given.
2. $m\angle A + m\angle B + m\angle C + m\angle D = 360$	2. The sum of the measures of the interior angles of a convex polygon of n sides equals $(n-2)180$. (6-5.1)
3. $2(m\angle A) + 2(m\angle B) = 360$	3. Substitution Postulate. (2-1)
4. $m\angle A + m\angle B = 180$	4. Multiplication property.
5. $\overline{AD} \parallel \overline{BC}$	5. If two lines are cut by a transversal so that the interior angles on the same side of the transversal are supplementary, then the lines are parallel. (6-2.1b)
6. $m\angle C + m\angle B = 180$	6. Substitution Postulate. (2-1)
7. $\overline{AB} \parallel \overline{CD}$	7. Same reason as step 5.
8. Quadrilateral $ABCD$ is a parallelogram.	8. Definition of a parallelogram. (7-1)

When proving the following theorem, use the figure in Theorem 7-2.3.

T **Theorem 7-2.4** A quadrilateral is a parallelogram if the angles of either opposite pair are congruent and the sides of either opposite pair are parallel.

GIVEN Quadrilateral $ABCD$; $\angle B \cong \angle D$; $\overline{AB} \parallel \overline{CD}$

PROVE Quadrilateral $ABCD$ is a parallelogram

PROOF

STATEMENTS	REASONS
1. Quadrilateral $ABCD$; $\angle B \cong \angle D$; $\overline{AB} \parallel \overline{CD}$	1. Given.
2. $\angle A$ is supplementary to $\angle D$.	2. If two parallel lines are cut by a transversal, then the interior angles on the same side of the transversal are supplementary. (6-3.1b)
3. $\angle A$ is supplementary to $\angle B$.	3. Substitution Postulate. (2-1)
4. $\overline{AD} \parallel \overline{BC}$	4. If two lines are cut by a transversal so that the interior angles on the same side of the transversal are supplementary, then the lines are parallel. (6-2.1b)
5. Quadrilateral $ABCD$ is a parallelogram.	5. Definition of a parallelogram. (7-1)

The proof of the next theorem is far less obvious than the proofs of the previous theorems.

T **Theorem 7-2.5** A quadrilateral is a parallelogram if the angles of either opposite pair are congruent and the sides of either opposite pair are congruent.

GIVEN Quadrilateral $ABCD$; $\angle B \cong \angle D$; $\overline{AB} \cong \overline{CD}$

PROVE Quadrilateral $ABCD$ is a parallelogram.

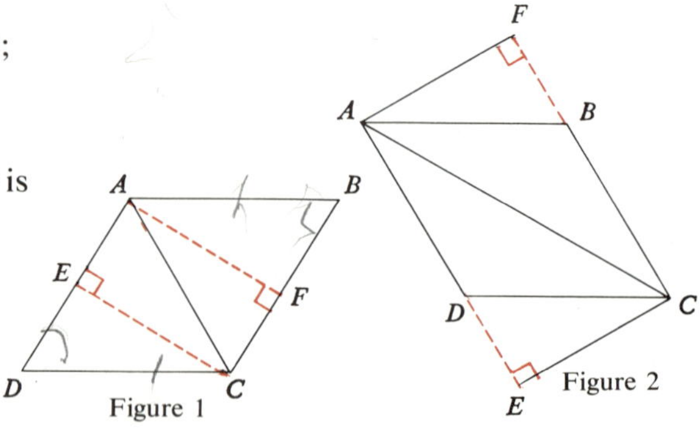

Proof The proof is valid whether the congruent angles are acute (Fig. 1) or obtuse (Fig. 2). We shall prove both cases at the same time.

STATEMENTS	REASONS
1. Quadrilateral $ABCD$; $\angle B \cong \angle D$; $\overline{AB} \cong \overline{CD}$	1. Given.
2. Draw diagonal $\overline{AC}$.	2. Definition of a line segment. (1-13)
3. Draw perpendiculars $\overline{CE}$ and $\overline{AF}$ to $\overleftrightarrow{AD}$ and $\overleftrightarrow{BC}$, respectively.	3. Through a point external to a line, there is one and only one line perpendicular to the given line. (4-4.6)
4. $\angle DEC$ and $\angle AFB$ are right angles.	4. Definition of perpendicular lines. (1-25)
5. $\angle DEC \cong \angle AFB$	5. All right angles are congruent. (3-1.1)
6. $\triangle DEC \cong \triangle BFA$	6. If two angles and a nonincluded side of one triangle are congruent to the two corresponding angles and the nonincluded side of another triangle, then the two triangles are congruent. (6-6.1)
7. $m\angle DCE = m\angle BAF$; $\overline{EC} \cong \overline{FA}$	7. Definition of congruent triangles. (3-3)
8. $\overline{AC} \cong \overline{AC}$	8. Every segment is congruent to itself. (3-1.6)
9. $\triangle AEC$ and $\triangle CFA$ are right triangles.	9. Definition of a right triangle. (1-24)
10. $\triangle AEC \cong \triangle CFA$	10. If the hypotenuse and one leg of a right triangle are congruent to the corresponding hypotenuse and leg of another right triangle, then the two triangles are congruent. (6-6.2)
11. $m\angle ECA = m\angle FAC$	11. Definition of congruent triangles. (3-3)
12. $m\angle BAC = m\angle DCA$	12. In Fig. 1, addition property. In Fig. 2, subtraction property.
13. $\overline{AB} \parallel \overline{CD}$	13. If two lines are cut by a transversal so that the alternate interior angles are congruent, then the lines are parallel. (6-2.1)
14. Quadrilateral $ABCD$ is a parallelogram.	14. Theorem 7-2.2.

The next method of proving that a quadrilateral is a parallelogram involves only its diagonals. What do you recall about the diagonals of a parallelogram?

T **Theorem 7-2.6** If the diagonals of a quadrilateral bisect each other, then the quadrilateral is a parallelogram.

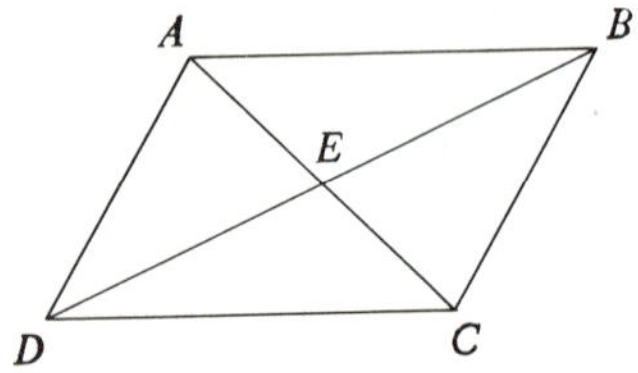

GIVEN Quadrilateral $ABCD$; diagonals $\overline{AC}$ and $\overline{BD}$ bisect each other at E.

PROVE Quadrilateral $ABCD$ is a parallelogram.

PROOF

STATEMENTS	REASONS
1. Quadrilateral $ABCD$; diagonals $\overline{AC}$ and $\overline{ED}$ bisect each other at E.	1. Given.
2. $\overline{AE} \cong \overline{CE}$; $\overline{DE} \cong \overline{BE}$	2. Definition of a midpoint. (1-15)
3. $\angle AEB \cong \angle DEC$	3. Vertical angles are congruent. (3-1.5)
4. $\triangle AEB \cong \triangle CED$	4. SAS Postulate. (3-1)
5. $\overline{AB} \cong \overline{CD}$	5. Definition of congruent triangles. (3-3)
6. $\angle BAC \cong \angle DCA$	6. Definition of congruent triangles. (3-3)
7. $\overline{AB} \parallel \overline{CD}$	7. If two lines are cut by a transversal so that the alternate interior angles are congruent, then the lines are parallel. (6-2.1)
8. Quadrilateral $ABCD$ is a parallelogram	8. Theorem 7-2.2.

Summary. Methods of Proving That a Quadrilateral Is a Parallelogram

To prove that a quadrilateral is a parallelogram, prove that:

1. the opposite sides are parallel.
2. both pairs of opposite sides are congruent.
3. two of its sides are both congruent and parallel.
4. the opposite angles are congruent.
5. the angles of one pair of opposite angles are congruent and the sides of one pair of opposite sides are parallel.
6. the angles of one pair of opposite angles and the sides of one pair of opposite sides are congruent.
7. the diagonals bisect each other.

You should notice one obvious omission from the summary above. If one pair of opposite sides of a quadrilateral is congruent and the other pair of opposite sides is parallel, then we do not necessarily have a parallelogram. This type of quadrilateral will be studied in Section 7-7.

EXERCISES

A The congruent sides and angles of each of the quadrilaterals below are marked. Arrows indicate parallel sides. Indicate which figures are parallelograms and which are not. For the parallelograms, justify your answers by citing a definition, postulate, theorem, or corollary.

1. 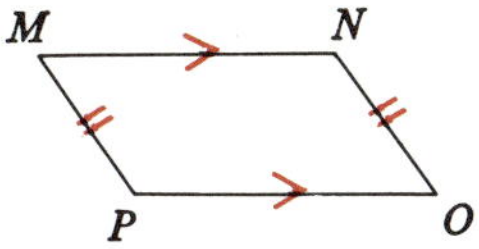

2. 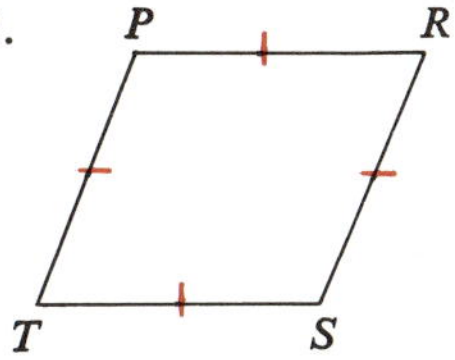

3. 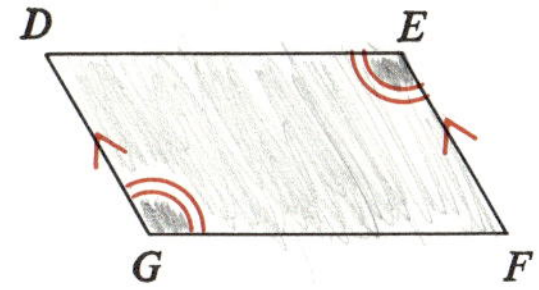

4. Prove that the quadrilateral formed by joining the midpoints of the consecutive sides of a parallelogram is also a parallelogram.

5. GIVEN $\square ABCD$; $FB = DE$; $\overline{AFB}$; $\overline{DEC}$
   PROVE Quadrilateral $AFCE$ is a parallelogram.

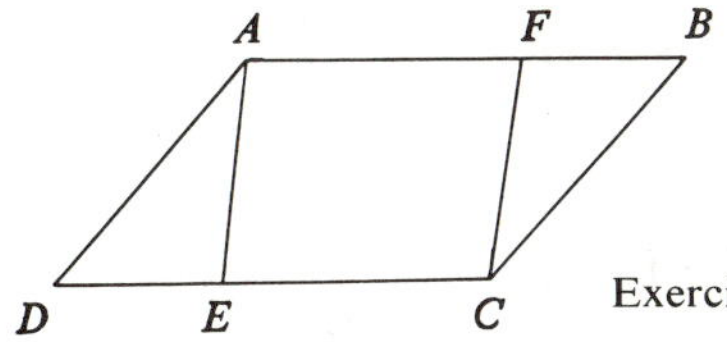

Exercise 5

6. GIVEN $\square ABCD$; $\angle DAE \cong \angle BCF$; $\overline{AFB}$; $\overline{DEC}$
   PROVE Quadrilateral $AFCE$ is a parallelogram.

7. GIVEN $\triangle ABC$ with median $\overline{AM}$; $\triangle ACP$ with median $\overline{CM}$
   PROVE Quadrilateral $ABPC$ is a parallelogram.

8. GIVEN Quadrilateral $PQRS$; $\overrightarrow{PQK}$; $\overrightarrow{RSN}$; $\angle 3 \cong \angle 4$; $\angle 1 \cong \angle 2$
   PROVE Quadrilateral $PQRS$ is a parallelogram.

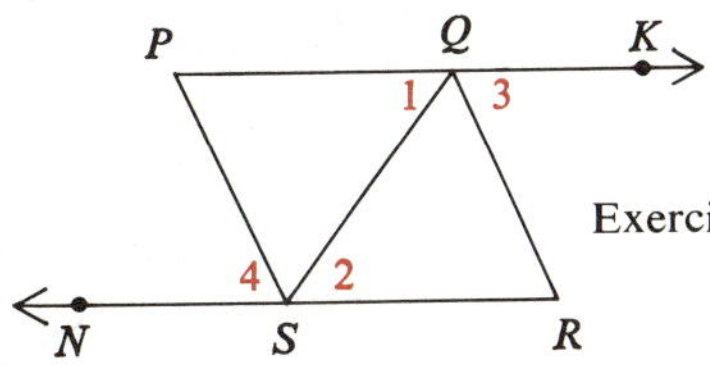

Exercise 8

B

9. GIVEN Quadrilateral $ASPN$; $\overline{NFA}$; $\overline{SRP}$; $\overrightarrow{PND}$; $\overrightarrow{ASE}$; $\angle AND \cong \angle PSE$; $\angle ARS \cong \angle PFN$; $\overline{AR} \cong \overline{PF}$
   PROVE Quadrilateral $ASPN$ is a parallelogram.

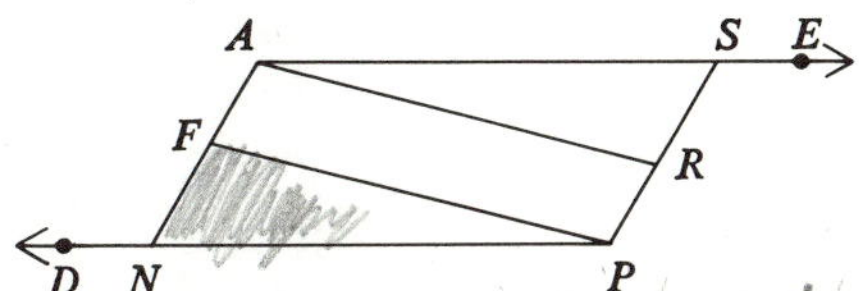

10. **Given** ▱ $PQRS$; $\overline{ST} \perp \overline{PR}$ at T; $\overline{QH} \perp \overline{PR}$ at H
Prove Quadrilateral $TSHQ$ is a parallelogram; $\angle TQH \cong \angle TSH$

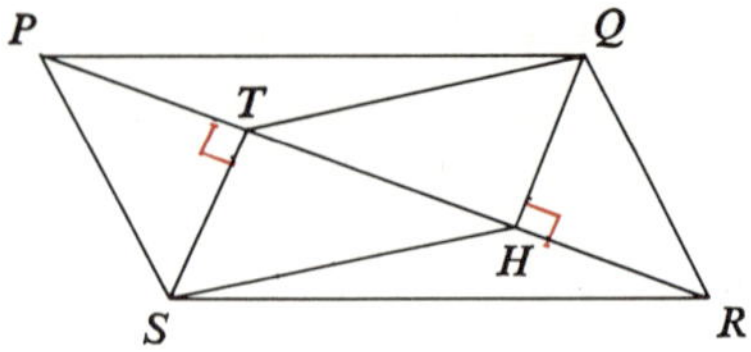

Exercise 10

11. **Given** ▱ $MNRS$ with diagonal $\overline{MPQR}$; $\angle MPN \cong \angle RQS$
Prove Quadrilateral $PNQS$ is a parallelogram; $\overline{PS} \parallel \overline{QN}$

12. **Given** ▱ $MNRS$ with diagonal $\overline{MPQR}$; $MQ = RP$
Prove Quadrilateral $PNQS$ is a parallelogram; $\angle PSQ \cong \angle PNQ$

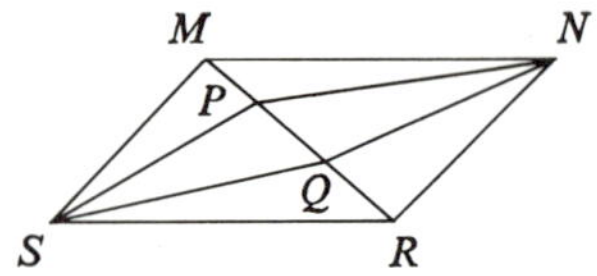

Exercises 11–13

13. **Given** ▱ $PNQS$ with $\overline{MPQR}$; $MQ = RP$
Prove Quadrilateral $MNRS$ is a parallelogram; $\overline{MS} \parallel \overline{NR}$

14. Prove that the segment joining the midpoints of a pair of opposite sides of a parallelogram is parallel to the remaining two sides and has the same measure as these sides.

C

15. Prove that if the bisectors of a pair of consecutive angles of a quadrilateral are not perpendicular, then the quadrilateral is not a parallelogram.

16. **Given** ▱ $PQRS$ with diagonals $\overline{RP}$ and $\overline{QS}$ meeting at A; M and N are midpoints of $\overline{SA}$ and $\overline{QA}$, respectively; $\overline{KAL}$; $\overline{PKQ}$; $\overline{SLR}$
Prove Quadrilateral $MKNL$ is a parallelogram.

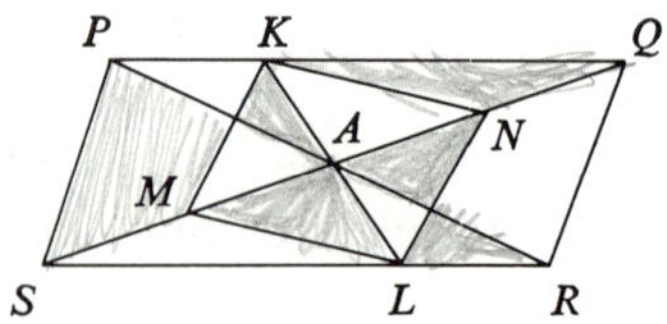

Exercise 16

17. **Given** Quadrilateral $ABCD$; $\overline{AD} \perp \overline{DC}$; $\overline{BC} \perp \overline{DC}$; $\overline{AB} \cong \overline{DC}$
Prove Quadrilateral $ABCD$ is a parallelogram.

18. **Given** $\triangle ABC$ with medians $\overline{AM}$ and $\overline{BN}$; $\overline{BGNP}$ and $\overline{AGMQ}$; $GN = NP$ and $GM = MQ$
Prove Quadrilateral $GCQB$ is a parallelogram; quadrilateral $GCPA$ is a parallelogram; quadrilateral $GPCQ$ is a parallelogram; $GM = \frac{1}{3}AM$; $GN = \frac{1}{3}BN$

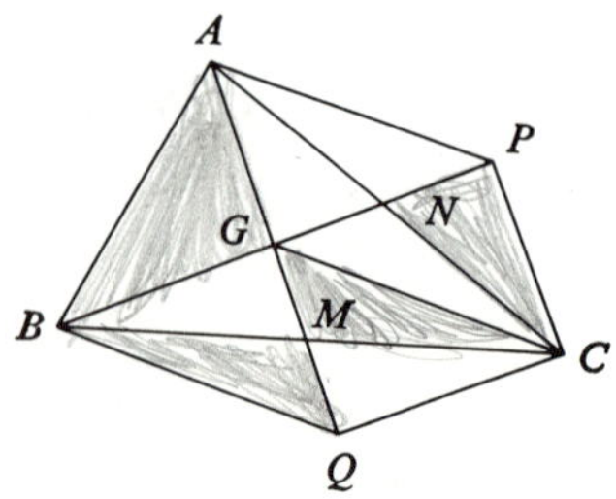

7-3

THE RECTANGLE

The first parallelogram with special properties that we shall define is the rectangle. In order to arrive at a definition of a rectangle as a type of parallelogram, we must determine what is true of a rectangle that is not true of all parallelograms.

D

Definition 7-5 A **rectangle** is a parallelogram with one right angle.

The first theorem follows immediately from the definition. The formal proof is left as an exercise.

T

Theorem 7-3.1 A rectangle has four right angles.

Class Exercises

1. Draw a rectangle and label the vertices A, B, C, and D.
2. Besides being parallel, what relationship exists between $\overline{AD}$ and $\overline{BC}$?
3. What types of angles are $\angle ADC$ and $\angle BCD$?
4. What is their relationship to each other?
5. Why is $\triangle ADC \cong \triangle BCD$?
6. What is the relationship between $\overline{AC}$ and $\overline{BD}$?
7. What is true about the diagonals of a rectangle?

By answering the questions above, we have proved the following theorem:

T

Theorem 7-3.2 The diagonals of a rectangle are congruent.

How can we prove that a quadrilateral is a rectangle? One method would be to use Theorem 7-3.1. However, there are other theorems that can be used, presented below. Their proofs, which are relatively simple, are left as exercises.

T

Theorem 7-3.3 If a quadrilateral has four right angles, then it is a rectangle.

T

Theorem 7-3.4 If a parallelogram has congruent diagonals, then it is a rectangle.

Summary. Methods of Proving That a Quadrilateral Is a Rectangle

To prove that a quadrilateral is a rectangle, prove that:

1. it is a quadrilateral with four right angles.
2. it is a parallelogram with one right angle.
3. it is a parallelogram with congruent diagonals.

Example 1

GIVEN $\overline{BE}$ is a median of right $\triangle ABC$ with $m\angle ABC = 90$; D is in $\overrightarrow{BE}$ so that E is the midpoint of $\overline{BD}$.

PROVE Quadrilateral $ABCD$ is a rectangle.

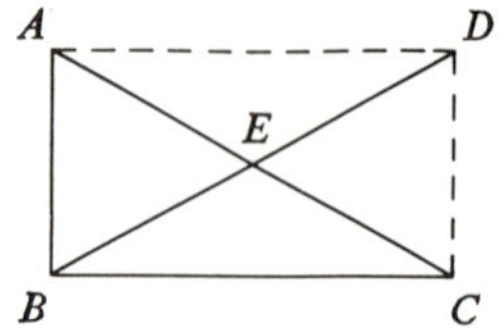

Solution

PROOF

STATEMENTS	REASONS
1. $\overline{BE}$ is a median of right $\triangle ABC$ with $m\angle ABC = 90$; D is in $\overrightarrow{BE}$ so that E is the midpoint of $\overline{BD}$.	1. Given.
2. E is the midpoint of $\overline{AC}$.	2. Definition of a median of a triangle. (3-9)
3. Quadrilateral $ABCD$ is a parallelogram.	3. If the diagonals of a quadrilateral bisect each other, then the quadrilateral is a parallelogram. (7-2.6)
4. ▱ $ABCD$ is a rectangle.	4. Theorem 7-3.1.

Example 2 If diagonals $\overline{AC}$ and $\overline{BD}$ of rectangle $ABCD$ of Example 1 intersect in E, and $AC = 18$, what is the length of $\overline{BE}$?

Solution Since $BE = ED$, or $BE = \frac{1}{2}BD$ (Theorem 7-1.5), and $BD = AC$ (Theorem 7-3.2), we get $BE = \frac{1}{2}AC$. So if $AC = 18$, then $BE = 9$.

By combining the results of Examples 1 and 2 above, we are able to formulate the following theorem. Refer to the figure for Example 1.

T **Theorem 7-3.5** The median to the hypotenuse of a right triangle is half as long as the hypotenuse.

GIVEN Right $\triangle ABC$ with median $\overline{BE}$

PROVE $BE = \frac{1}{2}AC$

PROOF OUTLINE

In Example 1, we showed that by selecting a point D in $\overrightarrow{BE}$ so that E is the midpoint of $\overline{BD}$, we create rectangle $ABCD$. In Example 2, we showed that since the diagonals of a rectangle are congruent, a segment which is half as long as one diagonal is also half as long as the other diagonal. Using both of these examples, write a detailed proof of Theorem 7-3.5.

EXERCISES

A Which of the following quadrilaterals are rectangles and which are not? Justify your answers. Information is marked on each figure.

1. 

2. 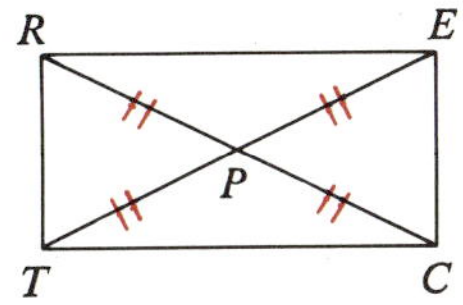

3. 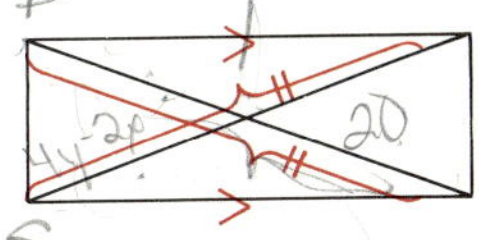

4. The lengths of diagonals $\overline{AC}$ and $\overline{BD}$ of rectangle $ABCD$ are represented by $7x + 5$ and $3x + 17$, respectively. Find the lengths of $\overline{AC}$ and $\overline{BD}$.
5. Diagonals $\overline{RP}$ and $\overline{SQ}$ of rectangle $PQRS$ meet at M. If $PM = x + 3y$, $SM = 4y - 2x$, and $RM = 20$, find x and y.
6. In ▱ $MNQR$, diagonals $\overline{MQ}$ and $\overline{RN}$ meet at P. If $MP = 15x - 11$ and $PQ = 7x + 21$, find MQ. Find RN if $RN = 20x + 18$. What type of parallelogram is $MNQR$? Why?
7. What is the length of the median to the hypotenuse of a right triangle if the hypotenuse has a length of 15?
8. What is the length of the hypotenuse of a right triangle if the median to the hypotenuse has a length of 15?
9. **Given** Rectangle $ASPN$ with F the midpoint of $\overline{AN}$
   Prove $\triangle SFP$ is isosceles.
10. **Given** Rectangle $TCER$; $\overline{CMNE}$; $CM = NE$
    Prove $TN = RM$; $\triangle TPR$ and $\triangle MPN$ are isosceles.

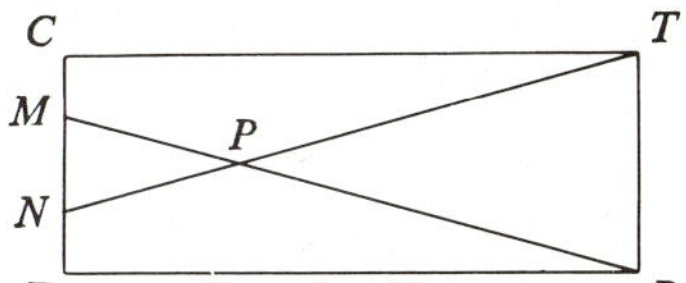

Exercise 10

11. Prove Theorem 7-3.1.
12. Prove Theorem 7-3.3.
13. Prove Theorem 7-3.4.

B

14. **Given** Regular hexagon $ABCDEF$
    Prove Quadrilateral $ABDE$ is a rectangle.
15. **Given** M is the midpoint of $\overline{AB}$; $\angle APB$, $\angle AQB$, $\angle ARB$, and $\angle ASB$ are right angles.
    Prove $PM = QM = RM = SM$

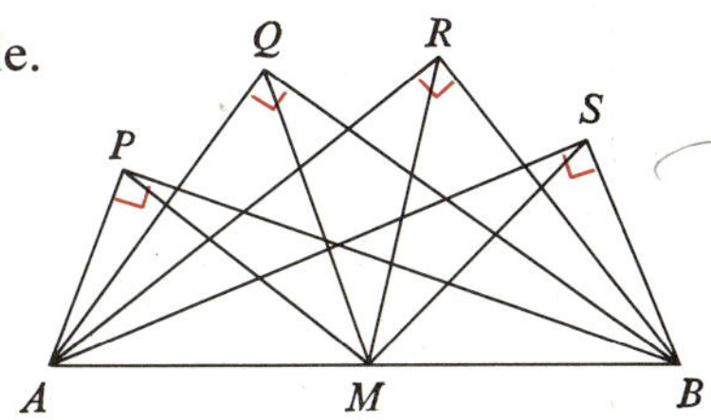

Exercise 15

16. **Given** $\triangle ABC$ with median $\overline{BE}$; $AE = BE$
    Prove $m\angle ABC = 90$
17. **Given** ▱ $ABCD$; $\overline{AB} \cong \overline{AD}$; M, N, P, and R are the midpoints of $\overline{AB}$, $\overline{BC}$, $\overline{CD}$, and $\overline{AD}$, respectively.
    Prove Quadrilateral $MNPR$ is a rectangle.

C

18. **Given** Rectangle $ABCD$; $\overline{AE} \cong \overline{AC}$; $\overline{AC}$ meets $\overline{BD}$ at P; D is the midpoint of $\overline{AE}$.

Prove $\overline{APC} \perp \overline{PE}$ (*Hint*: Use the result of Exercise 16.)

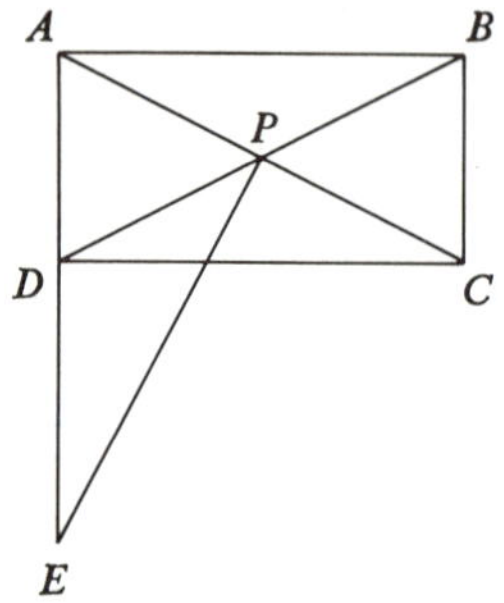

19. **Given** P is any point in equilateral $\triangle ABC$; $\overline{PE} \perp \overline{AC}$ at E; $\overline{PF} \perp \overline{BC}$ at F; $\overline{PD} \perp \overline{AB}$ at D; $\overline{AH}$ is the altitude of $\triangle ABC$.

Prove $PD + PE + PF = AH$ (*Hint*: Consider $\overline{MPN} \parallel \overline{BC}$.)

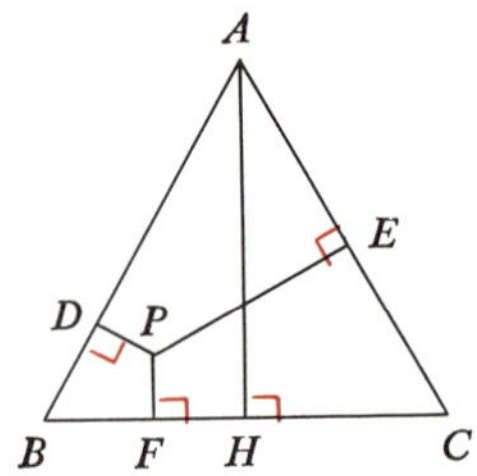

SOMETHING TO THINK ABOUT

An Impossible Situation?

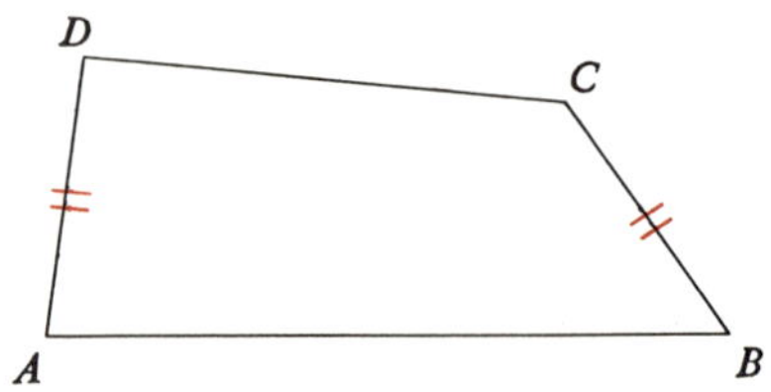

In quadrilateral $ABCD$, one pair of opposite sides is congruent and the other pair of opposite sides is not parallel. But the statement proved below indicates that this situation is impossible. Study the proof of the statement and see if you can determine whether the drawing is misleading or the statement is incorrect.

Statement If one pair of opposite sides of a quadrilateral is congruent, then the other pair of opposite sides is parallel.

Given Quadrilateral $ABCD$; $\overline{AD} \cong \overline{BC}$

Prove $\overline{AB} \parallel \overline{DC}$

Proof There are three cases to be proved. The first two steps apply to all three cases.

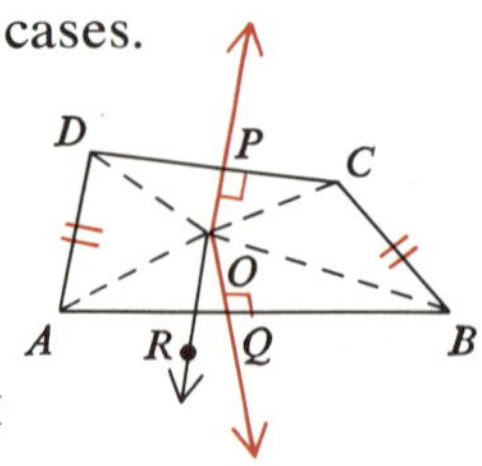

Case I

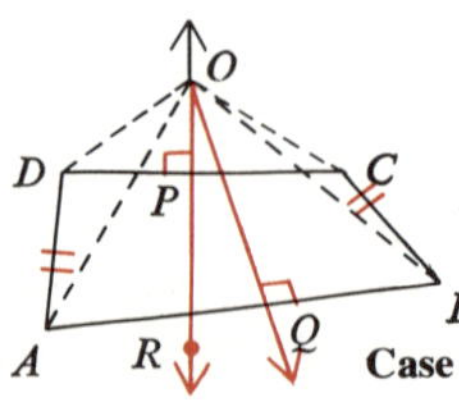

Case II

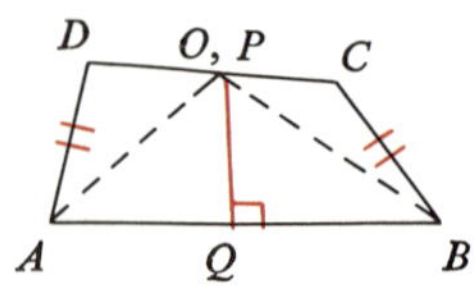

Case III

STATEMENTS	REASONS
1. Quadrilateral $ABCD$; $\overline{AD} \cong \overline{BC}$	1. Given.
2. Let $\overline{QO}$ and $\overline{PO}$, the respective perpendicular bisectors of $\overline{AB}$ and $\overline{DC}$, intersect at O; $\overline{POR}$	2. Definition of a perpendicular bisector. (4-4)
Case I Point O is in the interior of quadrilateral $ABCD$	
3. $\triangle DPO \cong \triangle CPO$	3. SAS Postulate. (3-1) ($\overline{DP} \cong \overline{CP}$; $\angle DPO \cong \angle CPO$; $\overline{PO} \cong \overline{PO}$)
4. $DO = CO$	4. Definition of congruent triangles. (3-3)
5. $\triangle AOQ \cong \triangle BOQ$	5. SAS Postulate. (3-1) ($\overline{AQ} \cong \overline{BQ}$; $\angle AQO \cong \angle BQO$; $\overline{QO} \cong \overline{QO}$)
6. $\overline{AO} \cong \overline{BO}$	6. Definition of congruent triangles. (3-3)
7. $\triangle ADO \cong \triangle BCO$	7. SSS Postulate. (3-3)
8. $m\angle DOA = m\angle BOC$; $m\angle DOP = m\angle COP$	8. Definition of congruent triangles. (3-3)
9. $m\angle AOP = m\angle BOP$	9. Addition property.
10. $\angle AOR$ is supplementary to $\angle AOP$; $\angle BOR$ is supplementary to $\angle BOP$	10. Definition of a linear pair. (1-26)
11. $\angle AOR \cong \angle BOR$	11. Supplements of congruent angles, or of the same angle, are congruent. (3-14)
12. $\overline{OR}$ bisects $\angle AOB$.	12. Definition of an angle bisector. (1-29)
13. But $\angle AOQ \cong \angle BOQ$	13. Definition of congruent triangles, step 5. (3-3)
14. $\overline{OQ}$ bisects $\angle AOB$.	14. Definition of an angle bisector. (1-29)
15. $\overleftrightarrow{POR}$ contains $\overline{OQ}$.	15. Every angle has exactly one bisector. (3-4.1)
16. $\overline{AB} \parallel \overline{CD}$	16. If two distinct lines in the same plane are both perpendicular to the same line, then they are parallel. (6-1.1)

In Case I, the point O is inside quadrilateral $ABCD$. In Case II, O is outside quadrilateral $ABCD$, and in Case III, O is contained in a side of quadrilateral $ABCD$, say $\overline{DC}$. Write informal proofs for Case II and Case III. Then try to state the fallacy. Consider all the postulates carefully.

7-4 THE RHOMBUS

The rectangle is an equiangular parallelogram. The rhombus is an equilateral parallelogram.

D **Definition 7-6** A **rhombus** is a parallelogram with two adjacent sides congruent.

The next theorem follows from the definition:

T **Theorem 7-4.1** The four sides of a rhombus are congruent.

PROOF OUTLINE From Theorem 7-1.2, we know that both pairs of opposite sides of a parallelogram are congruent. Therefore, by the transitive property, we can easily prove that all sides of the rhombus are congruent if one pair of adjacent sides is congruent.

Class Exercises

1. Draw a rhombus. Label the vertices A, B, C, and D.
2. What is the relationship between $\overline{AB}$ and $\overline{BC}$?
3. Why is $\overline{AD} \cong \overline{DC}$?
4. Would this be true if quadrilateral $ABCD$ were not a rhombus?
5. Why is $\triangle ADB \cong \triangle CDB$?
6. As a result of the triangle congruence, $\angle ABD \cong \angle$____.
7. As a further result of the triangle congruence, $\angle ADB \cong \angle$____.
8. What is true about diagonal $\overline{BD}$ in relation to $\angle ABC$ and $\angle ADC$?
9. Based on your answer to Exercise 8, make a general statement about the relationship between the diagonals and the angles of a rhombus.
10. Considering only Exercises 2 and 3, what else can you conclude about $\overline{AC}$ and $\overline{BD}$? State a theorem that justifies your conclusion.
11. Based on your answer to Exercise 10, make a general statement about the relationship between the diagonals of a rhombus.

The statements called for in Exercises 9 and 11 are the next two theorems. The complete proof of each theorem will be left as an exercise.

T **Theorem 7-4.2** The diagonals of a rhombus bisect its angles.

T **Theorem 7-4.3** The diagonals of a rhombus are perpendicular to each other.

Example If quadrilateral $PQRS$ is a rhombus and $m\angle PQS = 52$, find $m\angle PRQ$.

Solution We know from Theorem 7-4.2 that $m\angle SQR = 52$. Since by Theorem 7-4.3 the diagonals of a rhombus are perpendicular to each other, $\triangle QMR$ is a right triangle. By Corollary 6-4.2b, $\angle MRQ$ is complementary to $\angle SQR$. Therefore $m\angle PRQ = 38$.

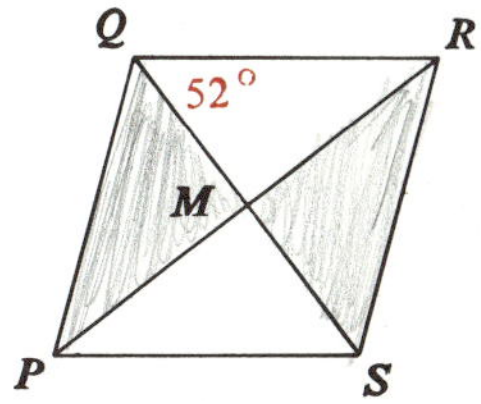

Theorem 7-4.1 provides one method of proving that a quadrilateral is a rhombus. The following theorems provide other methods. Since their proofs are short and simple, the formal proof of each is left as an exercise.

(T) **Theorem 7-4.4** If a quadrilateral has four congruent sides, then it is a rhombus.

(T) **Theorem 7-4.5** If a parallelogram has perpendicular diagonals, then it is a rhombus.

(T) **Theorem 7-4.6** If a diagonal of a parallelogram bisects an angle of the parallelogram, then the parallelogram is a rhombus.

Summary. Methods of Proving That a Quadrilateral Is a Rhombus

To prove that a quadrilateral is a rhombus, prove that:

1. it has four congruent sides.
2. it is a parallelogram with consecutive sides congruent.
3. it is a parallelogram in which a diagonal bisects an angle of the parallelogram.
4. it is a parallelogram with perpendicular diagonals.

EXERCISES

(A) Which of the following quadrilaterals are rhombuses and which are not? Information is marked on the figures. Justify your answers.

1. 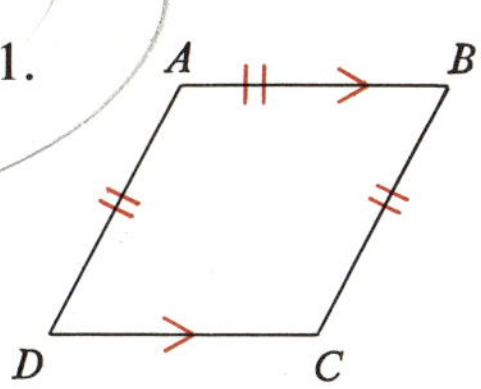

2. 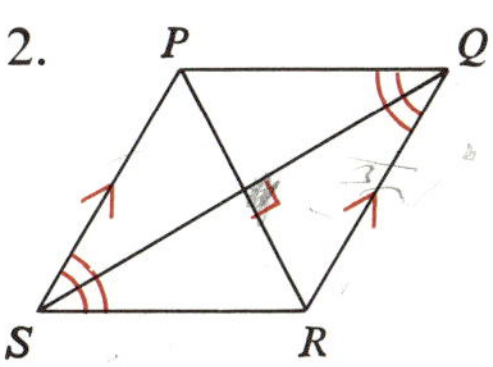

3. 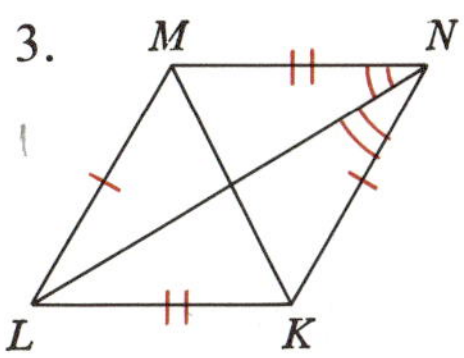

4. In rhombus $ABCD$, $m\angle ABD = 3x - 5$ and $m\angle BAC = 11x - 3$. Find the measures of all of the angles of the rhombus.

5. In $\square ABCD$, $AB = 17x - 3$, $BC = 13x + 5$, and $DC = 4x + 23$. Find the lengths of the sides of $\square ABCD$. What type of parallelogram is $\square ABCD$?

B

6. Prove Theorem 7-4.2.
7. Prove Theorem 7-4.3.
8. Prove Theorem 7-4.4.
9. Prove Theorem 7-4.5.
10. Prove Theorem 7-4.6.
11. GIVEN $\overline{AD}$ is an angle bisector of isosceles $\triangle ABC$; $\overline{AB} \cong \overline{AC}$; E lies on $\overrightarrow{AD}$ so that D is the midpoint of $\overline{AE}$.
    PROVE Quadrilateral $ABEC$ is a rhombus; $\overline{AB} \cong \overline{EB}$
12. GIVEN Rectangle $PQRS$; M, N, K, and L are the midpoints of $\overline{PQ}$, $\overline{QR}$, $\overline{RS}$, and $\overline{SP}$, respectively.
    PROVE Quadrilateral $MNKL$ is a rhombus.

C

13. GIVEN $\overleftrightarrow{AB} \parallel \overleftrightarrow{CD}$ with $\overleftrightarrow{EF}$ meeting the parallel lines in points M and N, respectively; $\overrightarrow{MR}$ bisects $\angle BMF$ and meets $\overleftrightarrow{CD}$ in R; $\overrightarrow{NP}$ bisects $\angle END$ and meets $\overleftrightarrow{AB}$ in P.
    PROVE Quadrilateral $MPRN$ is a rhombus.
14. GIVEN Rhombus $WXYZ$; A, B, and C are the midpoints of $\overline{WX}$, $\overline{XY}$, and $\overline{YX}$, respectively
    PROVE $\triangle ABC$ is a right triangle.
15. GIVEN $\overline{AN}$ is an angle bisector of $\triangle ABC$; $\overrightarrow{CNBH}$; $\overline{HA} \perp \overline{AN}$; $\overrightarrow{CAR}$; $\overrightarrow{HR} \parallel \overleftrightarrow{AB}$; $\overrightarrow{ABP}$; $\overrightarrow{HP} \parallel \overleftrightarrow{CA}$
    PROVE Quadrilateral $APHR$ is a rhombus.

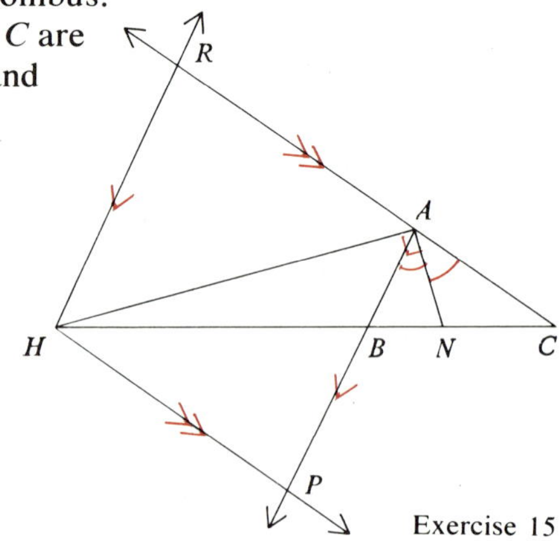

Exercise 15

7-5 THE SQUARE

Rectangles are equiangular parallelograms. Rhombuses are equilateral parallelograms. In this section we shall study parallelograms that are both equiangular and equilateral.

D

Definition 7-7 A **square** is a parallelogram with one right angle and two adjacent sides congruent.

A square is both an equilateral rectangle and an equiangular rhombus. Consequently, it has all the properties of a rectangle and a rhombus.

Example Find m$\angle ABD$ in square $ABCD$.

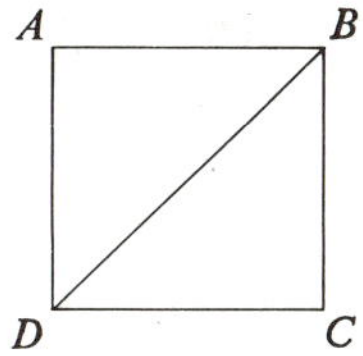

Solution Since square $ABCD$ is a rectangle, m$\angle ABC = 90$. However, square $ABCD$ is also a rhombus. Therefore, by Theorem 7-4.2, $\overline{BD}$ bisects $\angle ABC$. As a result, m$\angle ABD = 45$. Can you solve this problem another way?

Summary. Methods of Proving That a Quadrilateral Is A Square

To prove that a quadrilateral is a square, prove that:

1. it is a rectangle with consecutive sides congruent.
2. it is a rectangle with a diagonal bisecting one of its angles.
3. it is a rectangle with perpendicular diagonals.
4. it is a rhombus with one right angle.
5. it is a rhombus with congruent diagonals.

EXERCISES

A What quadrilaterals have each of the following properties?

1. The diagonals are congruent.
2. Both pairs of opposite angles are congruent.
3. The diagonals are perpendicular and bisect the angles of the quadrilateral.
4. All angles are congruent.
5. The diagonals bisect each other.
6. The diagonals are congruent and perpendicular.
7. GIVEN Square $HIJK$ and equilateral $\triangle ILJ$, with L in the exterior of square $HIJK$

   FIND m$\angle y$

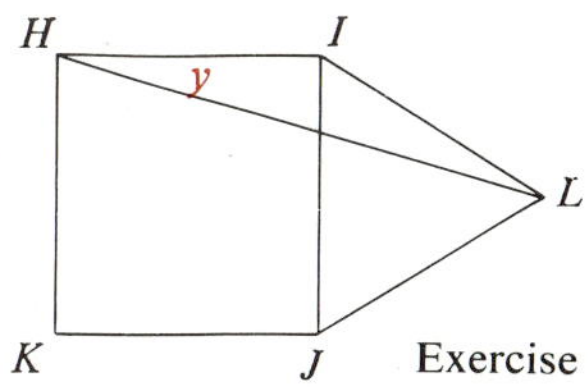

Exercise 7

8. Prove that the diagonals of a square are congruent and perpendicular.
9. Prove that the diagonals of a square bisect the angles of the square.
10. Prove that the diagonals of a square divide the square into four congruent isosceles right triangles.
11. Prove that the quadrilateral formed by joining the midpoints of the consecutive sides of a square is also a square.

B

12. GIVEN Square $ABCD$; $\overline{AQC}$; $\overline{APB}$; $\overline{BC} \cong \overline{QC}$; $\overline{PQ} \perp \overline{AC}$

    PROVE $AQ = PQ = PB$

    (*Hint*: Draw $\overline{PC}$.)

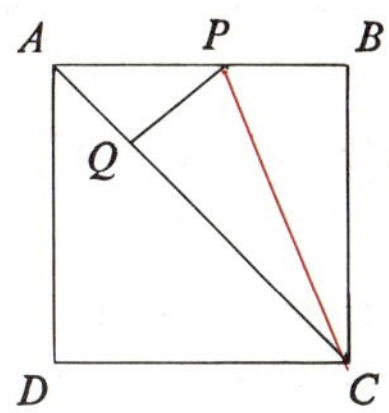

13. GIVEN Square $PQRS$, with N in $\overline{RS}$ so that $\overline{TS} \cong \overline{SN}$
PROVE $m\angle STN = 3(m\angle NTR)$

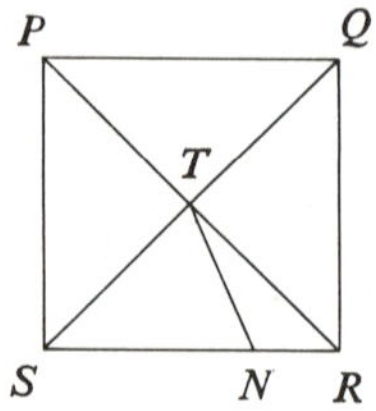

14. GIVEN Square $ABCD$; P is any point of $\overline{AB}$ so that $\overline{APB}$; $\overline{CQ} \perp \overline{PD}$ at R; $\overline{AQD}$
PROVE $\overline{PD} \cong \overline{QC}$

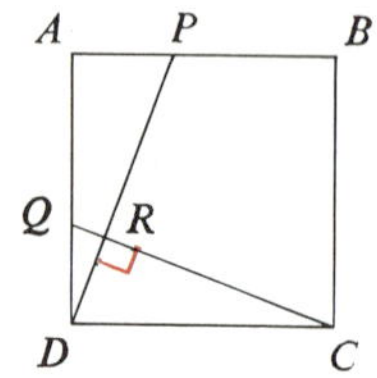

15. GIVEN Square $ABCD$; $\overline{APS}$; $\overline{DSR}$; $\overline{CRQ}$; $\overline{BQP}$; $m\angle 1 = m\angle 2 = m\angle 3 = m\angle 4$
PROVE Quadrilateral $PQRS$ is a square.

16. GIVEN Square $ABCD$; $\overline{APS} \perp \overline{DSR}$; $\overline{DSR} \perp \overline{CRQ}$; $\overline{CRQ} \perp \overline{BQP}$; $\overline{BQP} \perp \overline{APS}$
PROVE Quadrilateral $PQRS$ is a square.

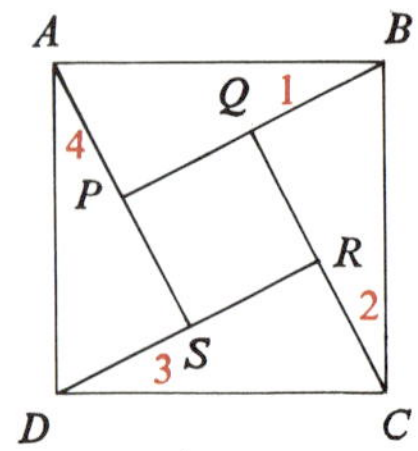

Exercises 15–17

17. GIVEN Square $PQRS$; $\overline{APS}$; $\overline{DSR}$; $\overline{CRQ}$; $\overline{BQP}$; $AP = BQ = CR = DS$
PROVE Quadrilateral $ABCD$ is a square.

C

18. GIVEN Points E and F are exterior to square $ABCD$; $\overline{CF} \cong \overline{CE}$; $\angle ACF \cong \angle ACE$
PROVE $\overline{EF} \parallel \overline{BD}$

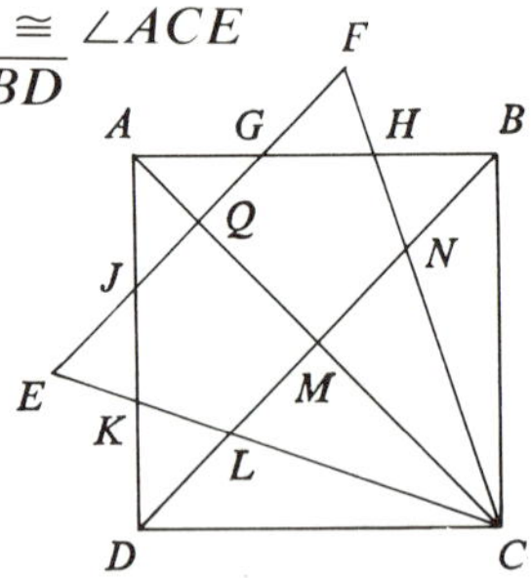

19. GIVEN Square $PQRS$; A is any point of $\overline{QMS}$; $\overline{QBC} \perp \overline{PCA}$; $\overline{PBMR}$
PROVE $\overline{AS} \cong \overline{BP}$

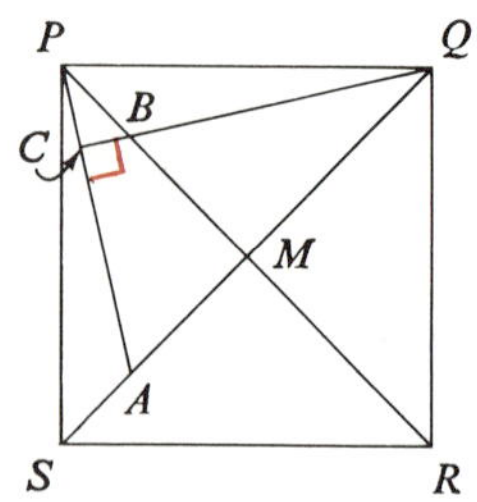

7-6 THE MIDLINE OF A TRIANGLE

The proofs of the following theorems are based upon our study of parallelograms.

D

Definition 7-8 If a transversal intersects two lines m and n in points A and B, then lines m and n **intercept** $\overline{AB}$ on the transversal.

T

Theorem 7-6.1 If three or more parallel lines intercept congruent segments on one transversal, then they intercept congruent segments on any other transversal.

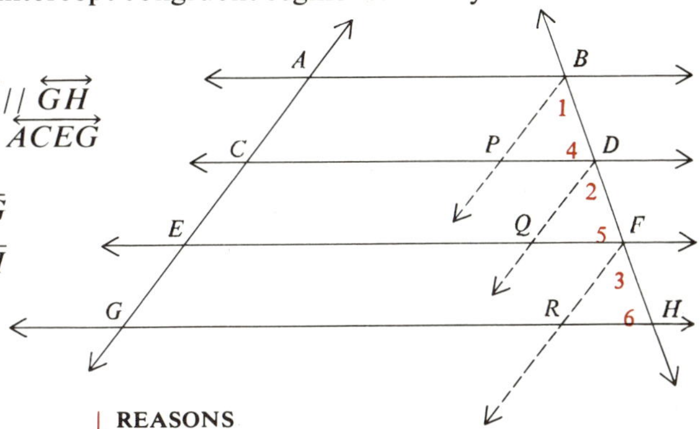

GIVEN $\overleftrightarrow{AB} \parallel \overleftrightarrow{CD} \parallel \overleftrightarrow{EF} \parallel \overleftrightarrow{GH}$ with transversals $\overleftrightarrow{ACEG}$ and $\overleftrightarrow{BDFH}$; $\overline{AC} \cong \overline{CE} \cong \overline{EG}$

PROVE $\overline{BD} \cong \overline{DF} \cong \overline{FH}$

PROOF

STATEMENTS	REASONS
1. $\overleftrightarrow{AB} \parallel \overleftrightarrow{CD} \parallel \overleftrightarrow{EF} \parallel \overleftrightarrow{GH}$ with transversals $\overleftrightarrow{ACEG}$ and $\overleftrightarrow{BDFH}$; $\overline{AC} \cong \overline{CE} \cong \overline{EG}$	1. Given.
2. Consider the lines through B, D, and F which are parallel to $\overleftrightarrow{ACEG}$ and meet $\overleftrightarrow{CD}$, $\overleftrightarrow{EF}$, and $\overleftrightarrow{GH}$ at points P, Q, and R, respectively.	2. Parallel lines exist in any given plane (6-1.1a) and the Parallel Postulate. (6-1)
3. Quadrilaterals $ABPC$, $CDQE$, and $EFRG$ are parallelograms.	3. Definition of a parallelogram. (7-1)
4. $\overline{BP} \cong \overline{AC}$; $\overline{DQ} \cong \overline{CE}$; $\overline{FR} \cong \overline{EG}$	4. The opposite sides of a parallelogram are congruent. (7-1.2)
5. $\overline{BP} \cong \overline{DQ} \cong \overline{FR}$	5. Transitive property.
6. $\angle 1 \cong \angle 2 \cong \angle 3$; $\angle 4 \cong \angle 5 \cong \angle 6$	6. If two parallel lines are cut by a transversal, then the corresponding angles are congruent. (6-3.1a)
7. $\triangle BPD \cong \triangle DQF \cong \triangle FRH$	7. If two angles and a nonincluded side of one triangle are congruent to the two corresponding angles and the nonincluded side of another triangle, then the two triangles are congruent. (6-6.1)
8. $\overline{BD} \cong \overline{DF} \cong \overline{FH}$	8. Definition of congruent triangles. (3-3)

Although the proof of Theorem 7-6.1 considered only four parallel lines, the same proof can be applied to any number of parallel lines greater than two.

D

Definition 7-9 A **midline** of a triangle is the line segment joining the midpoints of two sides of the triangle.

How many midlines does a triangle have?

We shall now establish some properties of the midline of a triangle. Theorems 7-6.2 and 7-6.3 will be proved together.

T

Theorem 7-6.2 The midline of a triangle is parallel to the third side of the triangle.

T

Theorem 7-6.3 The midline of a triangle is half as long as the third side of the triangle.

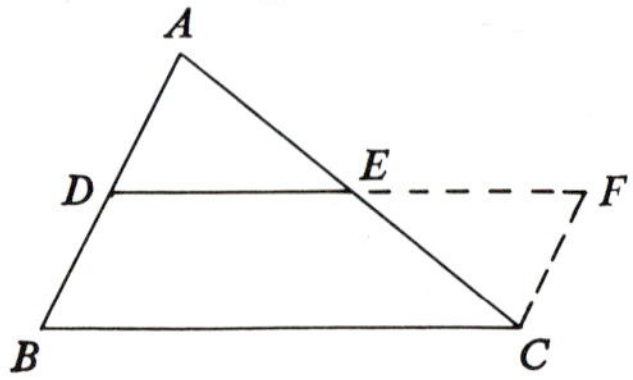

GIVEN $\overline{DE}$ is a midline of $\triangle ABC$, joining D and E, the respective midpoints of $\overline{AB}$ and $\overline{AC}$.

PROVE $\overline{DE} \parallel \overline{BC}$ (Theorem 7-6.2); $DE = \frac{1}{2}BC$ (Theorem 7-6.3)

PROOF

STATEMENTS	REASONS
1. $\overline{DE}$ is a midline of $\triangle ABC$, joining D and E, the respective midpoints of $\overline{AB}$ and $\overline{AC}$.	1. Given.
2. Select point F of $\overrightarrow{DE}$ so that $\overline{DE} \cong \overline{FE}$.	2. Point Uniqueness Postulate. (2-2)
3. $\overline{AE} \cong \overline{CE}$	3. Definition of a midpoint of a line segment. (1-15)
4. $\angle AED \cong \angle CEF$	4. Vertical angles are congruent. (3-1.5)
5. $\triangle AED \cong \triangle CEF$	5. SAS Postulate. (3-1)
6. $\angle A \cong \angle ECF$	6. Definition of congruent triangles. (3-3)
7. $\overleftrightarrow{AB} \parallel \overleftrightarrow{CF}$	7. If two lines are cut by a transversal so that the alternate interior angles are congruent, then the lines are parallel. (6-2.1)
8. $\overline{CF} \cong \overline{AD}$	8. Definition of congruent triangles. (3-3)
9. $\overline{AD} \cong \overline{DB}$	9. Definition of midpoint of a line segment. (1-15)

(continued)

STATEMENTS	REASONS
10. $\overline{CF} \cong \overline{DB}$	10. Transitive property.
11. Quadrilateral $DFCB$ is a parallelogram.	11. A quadrilateral is a parallelogram if two of its sides are both congruent and parallel. (7-2.2)
12. $\overline{DE} \parallel \overline{BC}$	12. Definition of a parallelogram. (7-1)
13. $\overline{DF} \cong \overline{BC}$	13. The opposite sides of a parallelogram are congruent. (7-1.2)
14. $DE = \frac{1}{2}DF$	14. See step 2.
15. $DE = \frac{1}{2}BC$	15. Substitution Postulate. (2-1)

Example

Given P, Q, R, and S are the respective midpoints of sides $\overline{AB}$, $\overline{BC}$, $\overline{CD}$, and $\overline{AD}$ of quadrilateral $ABCD$.

Prove Quadrilateral $PQRS$ is a parallelogram.

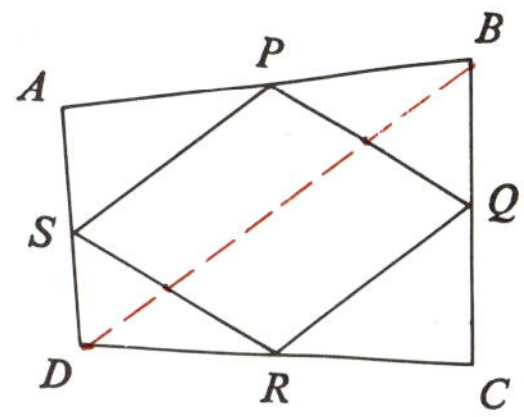

Solution

Proof

STATEMENTS	REASONS
1. P, Q, R, and S are the respective midpoints of sides $\overline{AB}$, $\overline{BC}$, $\overline{CD}$, and $\overline{AD}$ of quadrilateral $ABCD$.	1. Given.
2. $\overline{SP}$ is a midline of $\triangle ABD$.	2. Definition of a midline. (7-9)
3. $\overline{SP} \parallel \overline{DB}$	3. Theorem 7-6.2.
4. $SP = \frac{1}{2}DB$	4. Theorem 7-6.3.
5. $\overline{QR}$ is a midline of $\triangle CDB$.	5. Definition of a midline. (7-9)
6. $\overline{QR} \parallel \overline{DB}$	6. Theorem 7-6.2.
7. $QR = \frac{1}{2}DB$	7. Theorem 7-6.3.
8. $\overline{SP} \parallel \overline{QR}$	8. If each of two lines is parallel to a third line, then they are parallel to each other. (6-1.1c)
9. $SP = QR$	9. Transitive property.
10. Quadrilateral $PQRS$ is a parallelogram.	10. A quadrilateral is a parallelogram if two of its sides are both congruent and parallel. (7-2.2)

Class Exercises

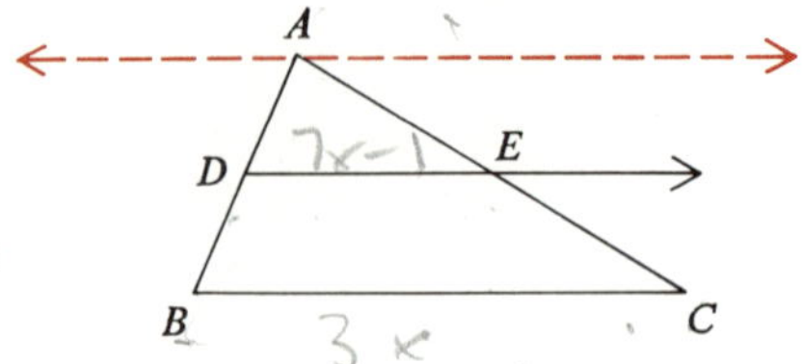

In the figure at right, D is the midpoint of $\overline{AB}$, $\overline{AEC}$, and $\overrightarrow{DE} \parallel \overline{BC}$.

1. How many lines can we draw through point A parallel to $\overline{BC}$? Why?
2. $\overline{AD} \cong$ ____. Why?
3. $\overline{AE} \cong$ ____. State a theorem to justify your answer.
4. Point E is the _______ of $\overline{AC}$.
5. $\overline{DE}$ is a _______ of $\triangle ABC$.
6. How does DE compare with BC? Why?

By answering the questions above, you have proved the following theorem:

T **Theorem 7-6.4** If a line containing the midpoint of one side of a triangle is parallel to a second side of the triangle, then it also contains the midpoint of the third side of the triangle.

EXERCISES

A Find the lengths of midlines of triangles given the following lengths of the side parallel to each midline:

1. 14 2. 18 3. 17 4. $15\frac{1}{2}$
5. If the perimeter of a triangle is 15, what is the perimeter of the triangle formed by the midlines of the first triangle?
6. $\overline{DE}$ is a midline of $\triangle ABC$, where D is in $\overline{AB}$ and E is in $\overline{AC}$. If $DE = 7x - 1$ and $BC = 3x + 20$, find the numerical lengths of $\overline{DE}$ and $\overline{BC}$.
7. In $\triangle ABC$, E is the midpoint of $\overline{AC}$ and $\overline{DE} \parallel \overline{BC}$, where D is in $\overline{AB}$. If $AD = 18x - 31$ and $AB = 7x + 25$, find the length of $\overline{DB}$.

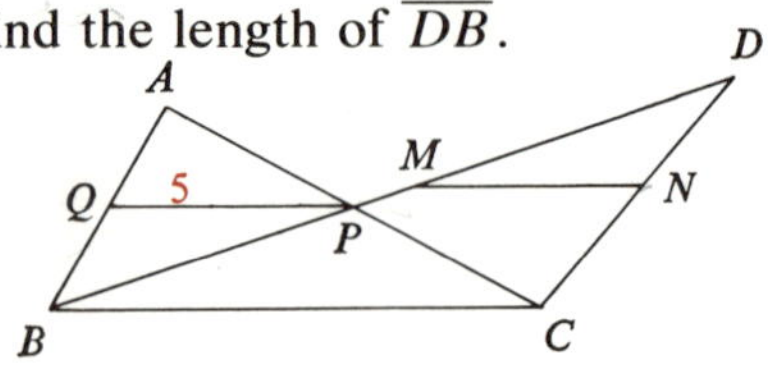

8. Q, P, M, and N are the midpoints of $\overline{AB}$, $\overline{AC}$, $\overline{BD}$, and $\overline{DC}$, respectively. If $QP = 5$, find MN.

In this figure, $\overleftrightarrow{AB} \parallel \overleftrightarrow{CD} \parallel \overleftrightarrow{EF}$; $AC = \frac{3}{5}AE$.

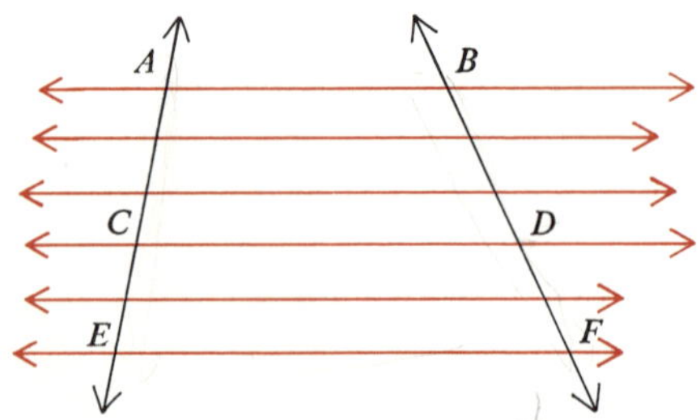

9. What part of BF is BD?
10. What part of BD is DF?
11. If $BF = 15$, find DF.
12. Use Theorem 7-6.2 or 7-6.3 to prove that the triangle formed by the midlines of an equilateral triangle is also equilateral.

13. Use Theorem 7-6.2 or 7-6.3 to prove that the line segments joining the midpoints of consecutive sides of a rectangle form a rhombus.
14. Use Theorem 7-6.2 or 7-6.3 to prove that the line segments joining the midpoints of consecutive sides of a rhombus form a rectangle.
15. Prove that the perimeter of the triangle formed by the three midlines of a triangle is half the perimeter of the original triangle.
16. Prove that the three midlines of a triangle divide the triangle into four congruent triangles.
17. GIVEN M is the midpoint of side $\overline{AB}$ of $\square ABCD$; $\overline{BPQ} \parallel \overline{MD}$; $\overline{DPC}$; $\overline{ADQ}$;
PROVE $\overline{AD} \cong \overline{DQ}$

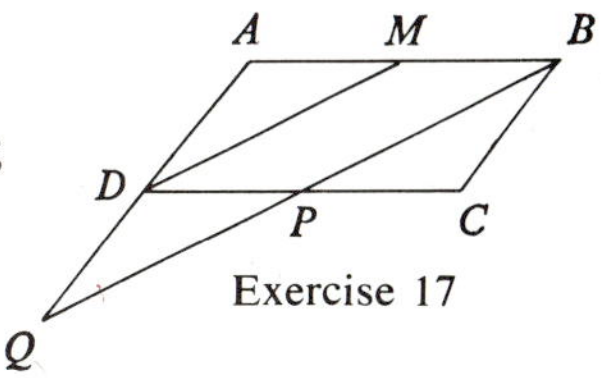

Exercise 17

B

18. Prove that the midline of a triangle bisects the median, angle bisector, and altitude to the third side.
19. GIVEN M, N, K, and L are the respective midpoints of sides $\overline{RP}$, $\overline{PQ}$, $\overline{SQ}$, and $\overline{RS}$ of quadrilateral $PQSR$.
PROVE The perimeter of quadrilateral $MNKL = PS + RQ$

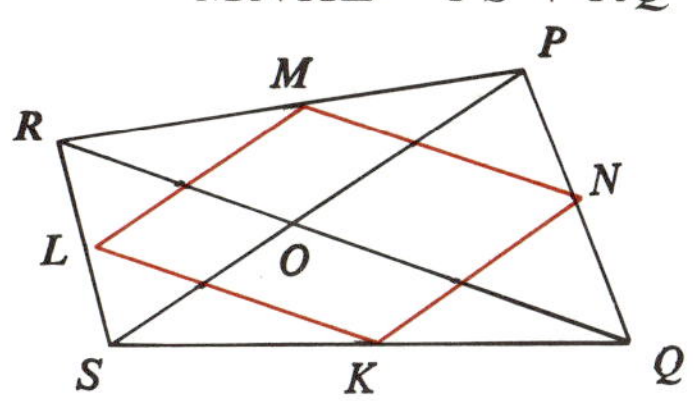

20. GIVEN Medians $\overline{BE}$ and $\overline{CD}$ of $\triangle ABC$ meet at G; M and N are the midpoints of $\overline{BG}$ and $\overline{CG}$, respectively.
PROVE Quadrilateral $DENM$ is a parallelogram.

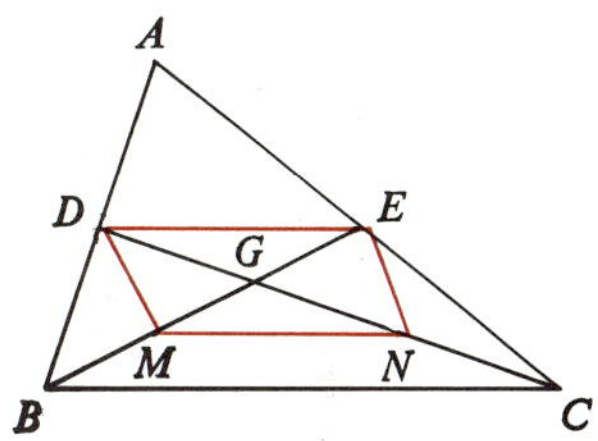

21. GIVEN $\triangle ABC$; $\overline{APMRB}$; $AP = PM = MR = RB$; D and E are the midpoints of $\overline{AC}$ and $\overline{BC}$, respectively.
PROVE $\overline{PD} \cong \overline{RE}$ (*Hint*: Draw $\overline{MC}$.)

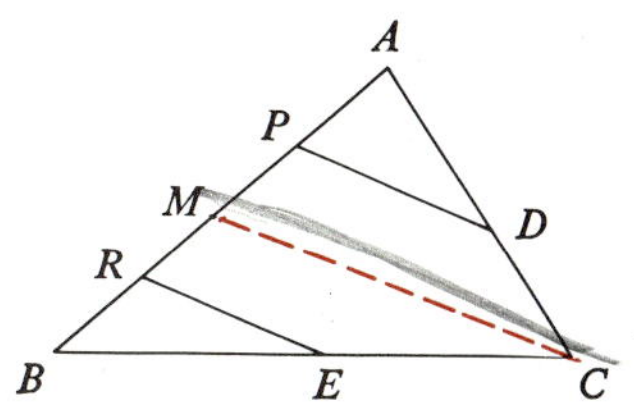

C

22. GIVEN $\triangle ABC$; $\overline{ADB}$; $\overline{AEC}$; $\overleftrightarrow{DE} \parallel \overline{BC}$; $DE = \frac{1}{2} BC$
PROVE $\overline{AD} \cong \overline{DB}$; $\overline{AE} \cong \overline{EC}$
23. GIVEN $\triangle ABC$; M is the midpoint of $\overline{AB}$; N is the midpoint of $\overline{MC}$. $\overleftrightarrow{ML} \parallel \overline{ANK}$; $\overline{BLKC}$
PROVE $BL = LK = KC$

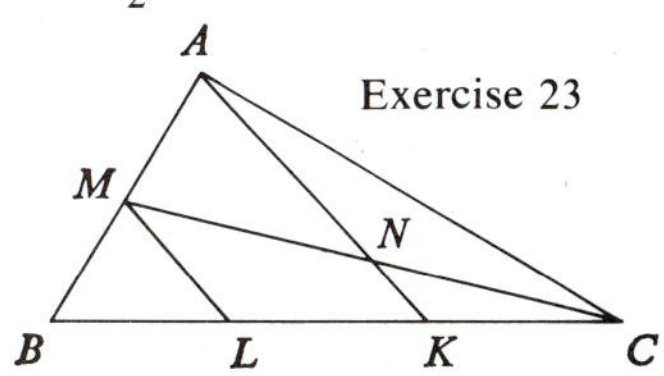

Exercise 23

24. **Given** In quadrilateral $ABCD$, $\overline{AC}$ is the perpendicular bisector of $\overline{DB}$; P, Q, R, and S are the midpoints of $\overline{AD}$, $\overline{AB}$, $\overline{BC}$, and $\overline{DC}$, respectively.
Prove Quadrilateral $PQRS$ is a rectangle.

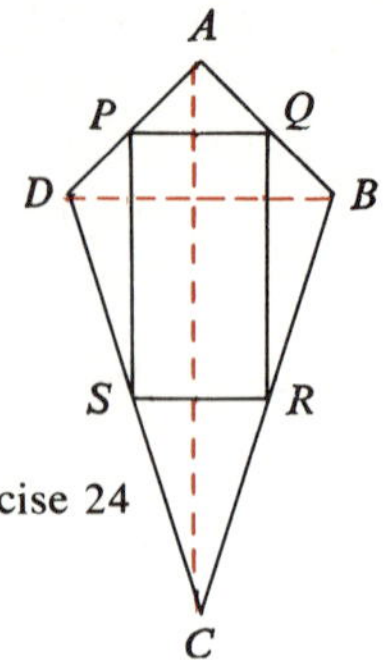

Exercise 24

25. **Given** $\triangle ABC$ with $\overline{ADFB}$ and $\overline{AEGC}$; $AE = \frac{1}{3}AC$; $\overline{DE} \parallel \overline{BC}$
Prove $AD = \frac{1}{3}AB$
(*Hint*: Draw $\overline{FG} \parallel \overline{BC}$, where G is the midpoint of $\overline{EC}$.)

26. **Given** $\square ABCD$; P and Q are the midpoints of $\overline{AD}$ and $\overline{BC}$, respectively; $\overline{PB}$ and $\overline{DQ}$ meet $\overline{AC}$ at points M and N, respectively.
Prove $AM = MN = NC$

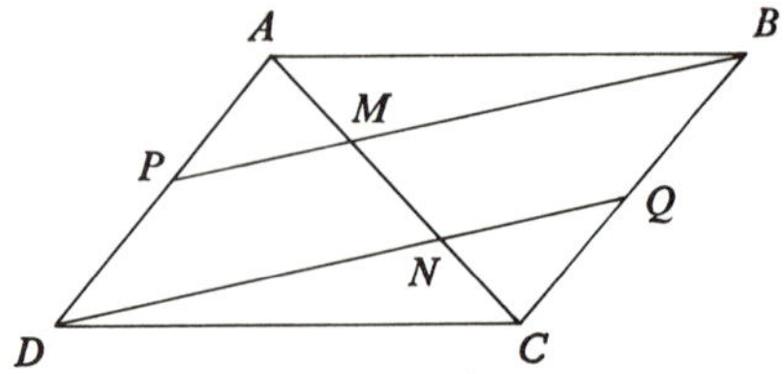

Exercise 26

7-7 THE TRAPEZOID

So far in our study of quadrilaterals we have been concerned only with quadrilaterals with both pairs of opposite sides parallel. We shall now consider quadrilaterals with only one pair of opposite sides parallel.

D **Definition 7-10** A quadrilateral is a **trapezoid** if the sides of exactly one opposite pair are parallel.

The parallel sides of a trapezoid are called the **bases** of the trapezoid. The angles that the nonparallel sides of a trapezoid form with each base are called the **base angles** of a trapezoid. To be considered a pair of base angles, each of the two angles must have one of its rays contained in the same base as the other angle. In the trapezoid for Theorem 7-7.2, in which $\overline{AB} \parallel \overline{DC}$, $\overline{AB}$ and $\overline{DC}$ are the bases. Angles C and D form one pair of base angles, and angles A and B form the other pair.

D **Definition 7.11** The **median of a trapezoid** is the line segment joining the midpoints of the nonparallel sides.

In the trapezoid $ABCD$ for Theorem 7-7.2, the median is $\overline{MN}$.

Theorem 7-7.1 The median of a trapezoid is parallel to the bases.

Theorem 7-7.2 The length of the median of a trapezoid is half the sum of the lengths of the bases.

We shall consider a proof for each theorem. A second proof will be suggested as an exercise.

GIVEN Trapezoid $ABCD$ with median $\overline{MN}$

PROVE $\overline{MN} \parallel \overline{DC}$; $MN = \frac{1}{2}(DC + AB)$

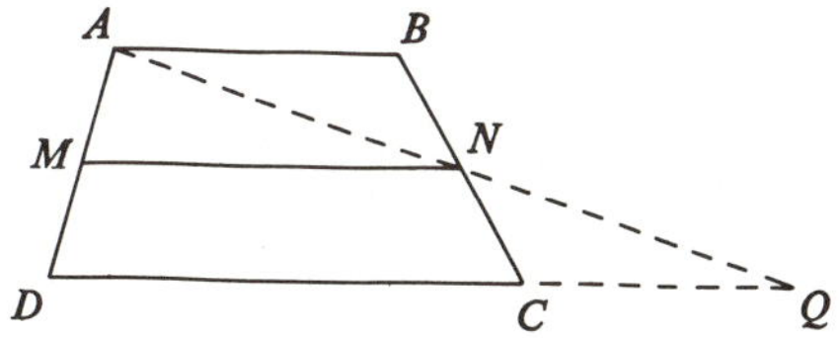

PROOF OUTLINE
Let $\overleftrightarrow{AN}$ meet $\overleftrightarrow{DC}$ at Q. Since N is the midpoint of $\overline{BC}$ and $\overleftrightarrow{AB} \parallel \overleftrightarrow{DC}$, we can easily prove that $\triangle ABN \cong \triangle QCN$. Therefore $\overline{AB} \cong \overline{QC}$. Since $\overline{MN}$ is the midline of $\triangle ADQ$, $\overline{MN} \parallel \overline{DC}$. Why? Also, $MN = \frac{1}{2}DQ$ or $MN = \frac{1}{2}(DC + CQ) = \frac{1}{2}(DC + AB)$.

Example 1 Find the length of the median of a trapezoid whose bases have lengths 9 and 17.

Solution By Theorem 7-7.2 the length of the median equals $\frac{1}{2}(9 + 17) = 13$.

Example 2 If the lengths of the shorter base ($\overline{AB}$), longer base ($\overline{DC}$), and median ($\overline{MN}$) of a trapezoid $ABCD$ are $6x - 11$, $5x + 4$, and $2x + 7$, respectively, find the length of the shorter base.

Solution From Theorem 7-7.2, we are able to establish the following relationship:

$$2x + 7 = \tfrac{1}{2}[(6x - 11) + (5x + 4)]$$
$$4x + 14 = 11x - 7$$
$$7x = 21$$
$$x = 3$$

$$AB = 6x - 11 = 6(3) - 11 = 7$$
$$MN = 2x + 7 = 2(3) + 7 = 13$$
$$DC = 5x + 4 = 5(3) + 4 = 19$$

Before the next example, consider the following definition.

Definition 7-12 An **altitude of a trapezoid** is the perpendicular segment from any point in the line containing one base of the trapezoid to the line containing the other base.

Name the altitude in trapezoid $PQRS$ for Example 3.

Example 3 Prove that the line containing the median of a trapezoid bisects any altitude of the trapezoid.

Solution

GIVEN Trapezoid $PQRS$ with altitude $\overline{PT}$ meeting median $\overline{MN}$ at H

PROVE $\overline{MN}$ bisects $\overline{PT}$.

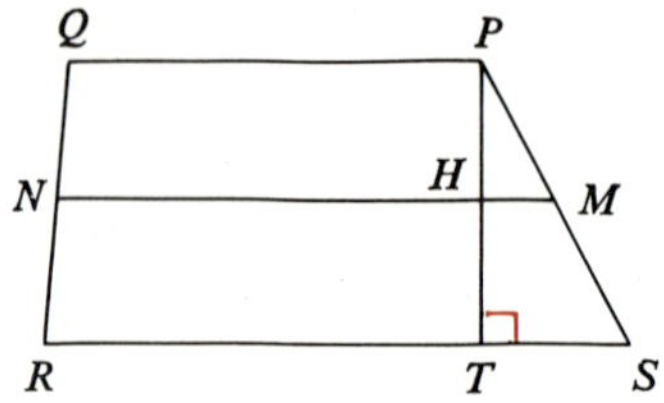

PROOF

STATEMENTS	REASONS
1. Trapezoid $PQRS$ with altitude $\overline{PT}$ meeting median $\overline{MN}$ at H	1. Given.
2. $\overline{MN} \parallel \overline{SR}$	2. Theorem 7-7.1.
3. M is the midpoint of $\overline{PS}$.	3. Definition of the median of a trapezoid. (7-11)
4. $\overline{PM} \cong \overline{MS}$	4. Definition of the midpoint of a line segment. (1-15)
5. $\overline{PH} \cong \overline{HT}$ (or H bisects $\overline{PT}$.)	5. If three or more parallel lines intercept congruent segments on one transversal, then they intercept congruent segments on any other transversal. (7-6.1)

Trapezoids have a variety of shapes. The trapezoid that we shall study in greatest detail is the one whose nonparallel sides are congruent.

D **Definition 7-13** An **isosceles trapezoid** is a trapezoid whose nonparallel sides are congruent.

Properties of an Isosceles Trapezoid

Draw a few different isosceles trapezoids and their diagonals. What do you think is true about the angles in each pair of base angles of an isosceles trapezoid? What appears to be true about the diagonals? Compare your findings with the following theorems:

T **Theorem 7-7.3** The angles in each pair of base angles of an isosceles trapezoid are congruent.

T **Theorem 7-7.4** The diagonals of an isosceles trapezoid are congruent.

We shall prove both of these theorems in one proof. Indicate the statements that conclude the proof of each theorem.

GIVEN Isosceles trapezoid $ABCD$ with $\overline{AD} \cong \overline{BC}$

PROVE $\left.\begin{array}{l}\angle ADC \cong \angle BCD \\ \angle DAB \cong \angle CBA\end{array}\right\}$ (Theorem 7-7.3)

$\overline{AC} \cong \overline{BD}$ (Theorem 7-7.4)

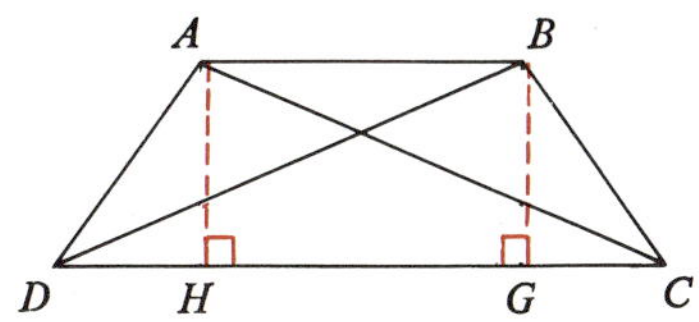

PROOF

STATEMENTS	REASONS
1. Isosceles trapezoid $ABCD$, with $\overline{AD} \cong \overline{BC}$	1. Given.
2. Draw altitudes $\overline{AH}$ and $\overline{BG}$; $\angle AHD$ and $\angle BGC$ are right angles.	2. Definition of an altitude of a trapezoid. (7-12)
3. $AH = BG$	3. Parallel lines are everywhere equidistant. (7-1.6)
4. $\triangle AHD \cong \triangle BGC$	4. If the hypotenuse and one leg of a right triangle are congruent to the corresponding hypotenuse and leg of another right triangle, then the two triangles are congruent. (6-6.2)
5. $\angle ADC \cong \angle BCD$	5. Definition of congruent triangles. (3-3)
6. $m\angle DAH = m\angle CBG$	6. Definition of congruent triangles. (3-3)
7. $m\angle HAB = 90$; $m\angle GBA = 90$	7. Definition of an altitude of a trapezoid (7-12) and, in a plane, if a line is perpendicular to one of two parallel lines, then it is also perpendicular to the other line. (6-1.1b)
8. $m\angle HAB = m\angle GBA$	8. Transitive property.
9. $m\angle DAB = m\angle CBA$	9. Angle Sum Postulate. (2-10)
10. $\overline{AB} \cong \overline{AB}$	10. Every segment is congruent to itself. (3-1.6)
11. $\triangle DAB \cong \triangle CBA$	11. SAS Postulate. (3-1)
12. $\overline{AC} \cong \overline{BD}$	12. Definition of congruent triangles. (3-3)

The proof of the following theorem will be left as an exercise:

T **Theorem 7-7.5** The opposite angles of an isosceles trapezoid are supplementary.

It will be helpful for us to be able to determine whether a given trapezoid is isosceles. The following theorems provide us with some methods:

T **Theorem 7-7.6** A trapezoid is isosceles if the angles in one pair of base angles are congruent.

T **Theorem 7-7.7** A trapezoid is isosceles if the angles in one pair of opposite angles are supplementary.

T **Theorem 7-7.8** A trapezoid is isosceles if its diagonals are congruent.

The proofs of these theorems are similar to the proof of Theorems 7-7.3 and 7-7.4. The proofs will be left as exercises.

Summary. Methods of Proving That a Trapezoid Is Isosceles

To prove that a trapezoid is isosceles, prove that:

1. its nonparallel sides are congruent.
2. the base angles of one pair are congruent.
3. the opposite angles of one pair are supplementary.
4. its diagonals are congruent.

EXERCISES

A

Find the length of the median of a trapezoid whose bases have the following lengths.

1. 3 and 7 2. 14 and 23 3. 51.3 and 16.5 4. $18\frac{2}{3}$ and $16\frac{2}{3}$

Find the length of the longer base of a trapezoid if the median and shorter base have respective lengths as follows:

5. 14 and 11 6. 34 and 23 7. $6\frac{1}{2}$ and 5 8. $26\frac{3}{4}$ and $15\frac{1}{2}$

Find the measures of the other three base angles of an isosceles trapezoid if one of the base angles has the following measure:

9. 25 10. 58 11. 100 12. 170

13. $\overline{AB}$ is the median of trapezoid $PQRS$. $PQ = 16x - 15$, $AB = 5x + 23$, and $SR = 13x + 4$. Find the measures of $\overline{PQ}$, $\overline{AB}$, and $\overline{SR}$.

14. In trapezoid $ABCD$, $\overline{AB} \parallel \overline{DC}$ and $\overline{AC} \cong \overline{BD}$; $m\angle D = 23x - 3$ and $m\angle C = 5x + 33$. Find the measures of all the angles of trapezoid $ABCD$.

15. The length of the altitude of an isosceles trapezoid is 7, the measure of a base angle is 45, and the length of the shorter base is 15. Find the length of the longer base of the trapezoid.

B

16. Prove that the median of a trapezoid bisects each of the diagonals. (*Hint*: Use Theorem 7-6.1 or 7-6.4.)

17. GIVEN $\triangle ABC$; $\overline{ADB} \cong \overline{AEC}$; $\overline{DE} \parallel \overline{BC}$
    PROVE Quadrilateral $DECB$ is an isosceles trapezoid.

18. GIVEN Isosceles trapezoid $ABCD$; $\overline{AD} \cong \overline{BC}$; $DE = CF$; $\overline{EDCF}$
    PROVE Trapezoid $ABFE$ is isosceles

19. Prove Theorem 7-7.5.

20. Prove Theorem 7-7.6.

21. Prove Theorem 7-7.7.

22. Prove Theorem 7-7.8.

C

23. Prove that the line containing the midpoints of the bases of an isosceles trapezoid is perpendicular to the median of the trapezoid. Can you prove this theorem another way?

24. GIVEN Trapezoid $ABCD$; P and Q are the midpoints of diagonals $\overline{BD}$ and $\overline{AC}$, respectively; $\overline{BQ}$ meets $\overline{DC}$ at E.

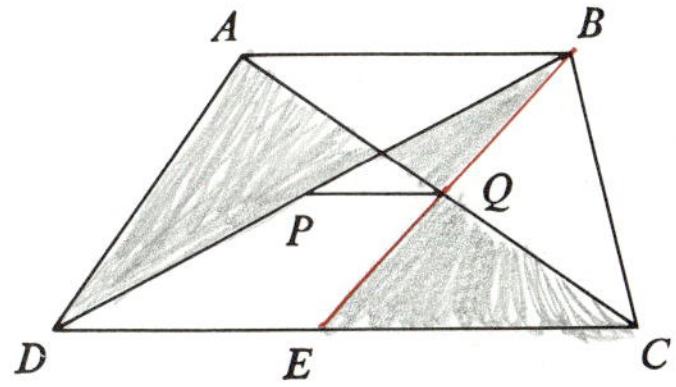

    PROVE $\overline{BQ} \cong \overline{EQ}$; $\overleftrightarrow{PQ} \parallel \overleftrightarrow{AB}$; $PQ = \frac{1}{2}(DC - AB)$. If $\overline{BE} \parallel \overline{AD}$, then E is the midpoint of $\overline{DC}$.

25. Use the figure at right to prove Theorem 7-7.2. $\overline{MN}$ is the median of trapezoid $ABCD$, and $\overline{BE} \parallel \overline{AD} \parallel \overline{NQ}$.

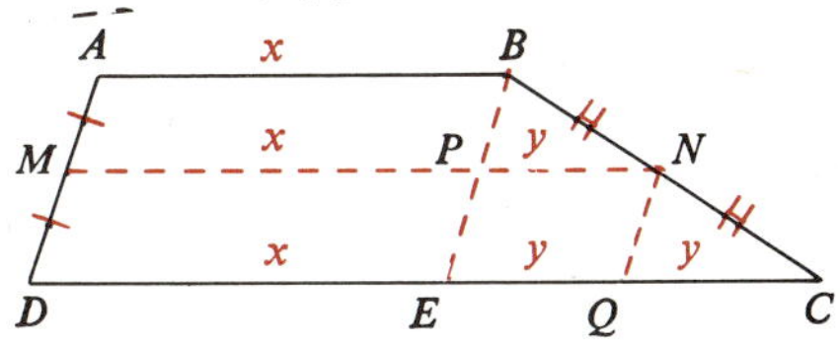

A LOOK AT THE PAST

Euclid's Classification of Quadrilaterals

Before Euclid wrote *The Elements*, about 300 B.C., mathematicians used the same terms to refer to different quadrilaterals. For example, the Greek word used by Pythagoras for a square was used by Aristotle to mean "quadrilateral."

The following definitions from *The Elements* put an end to this confusion: "Of quadrilateral figures, a square is that which is both equilateral and right-angled;

an oblong that which is right-angled but not equilateral; a rhombus that which is equilateral but not right-angled; and a rhomboid that which has its opposite sides and angles equal to one another but is neither equilateral nor right-angled. And let quadrilaterals other than these be called trapezia."

From the way the last sentence is worded, we can assume that Euclid introduced the word *trapezia* (*trapezium* in the singular). All the other words were already in use. Their definitions, however, were not standardized.

Above, Euclid seems to classify all nonparallelograms as trapezia. Further study of *The Elements* indicates, however, that he used *trapezium* to refer to a quadrilateral with one and only one pair of opposite sides parallel. Quadrilaterals with no parallel sides he called trapezoids.

HOW IS YOUR MATHEMATICAL VOCABULARY?

The key words and phrases introduced in this chapter are listed below. How many do you know and understand?

altitude of a parallelogram (7-1)

altitude of a trapezoid (7-7)

base angles of a trapezoid (7-7)

bases of a trapezoid (7-7)

consecutive angles of a parallelogram (7-1)

distance between two parallel lines (7-1)

intercept (7-6)

isosceles trapezoid (7-7)

median of a trapezoid (7-7)

midline of a triangle (7-6)

parallelogram (7-1)

rectangle (7-3)

rhombus (7-4)

square (7-5)

trapezoid (7-7)

REVIEW EXERCISES

7-1 Properties of a parallelogram

Fill in the missing information in each of the following. Explain why the result is true.

1. $\triangle BAD \cong$ ____
2. $AD =$ ____
3. $\angle ADC \cong$ ____
4. $\overline{AE} \cong$ ____
5. $\triangle AED \cong$ ____
6. $\overline{AB} \parallel$ ____

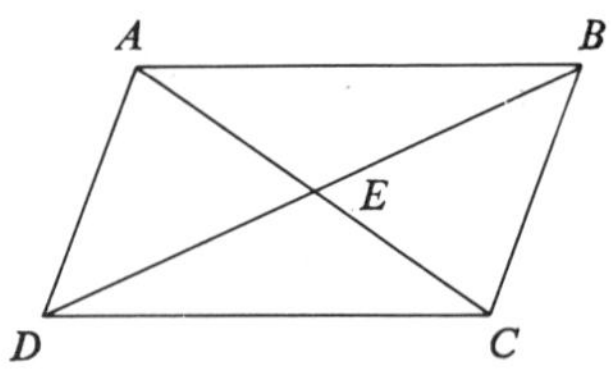

Exercises 1–6

7. Find the measures of the angles of a parallelogram if one angle has a measure of 47.
8. The diagonals of $\square ABCD$ meet at E. Find the length of $\overline{BD}$ if $BE = 7x - 20$ and $DE = 12x - 35$.

9. **Given** $\square PQRS$; $\overline{PQT}$; $\overline{PQ} \cong \overline{QT}$
Prove $\overline{ST}$ and $\overline{QR}$ bisect each other.

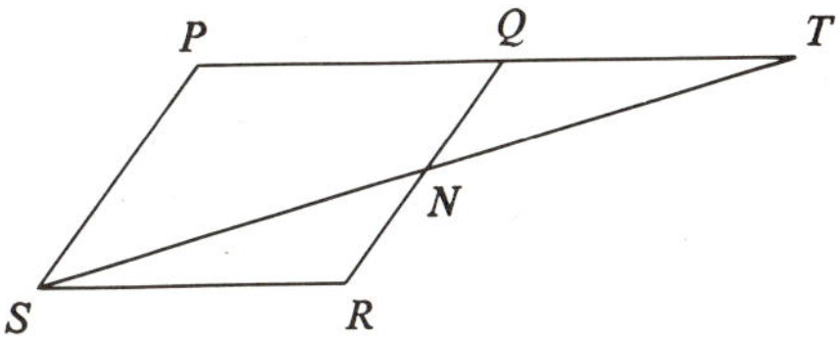

7-2 When a quadrilateral is a parallelogram

Indicate whether each of the following is true or false.
A quadrilateral is a parallelogram if:

10. both pairs of opposite sides are congruent.
11. the diagonals bisect each other.
12. two pairs of consecutive angles are supplementary.

13. **Given** $\overline{AM}$ is a median of $\triangle ABC$; $\overrightarrow{BE} \perp \overleftrightarrow{AM}$ at E; $\overrightarrow{CD} \perp \overleftrightarrow{AM}$ at D.
Prove Quadrilateral $BECD$ is a parallelogram.

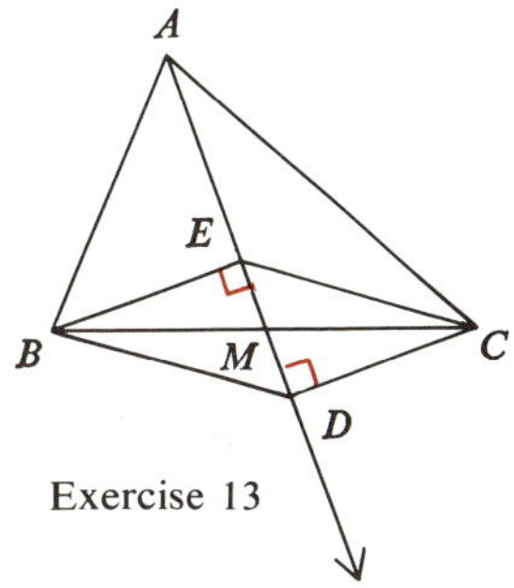

Exercise 13

7-3 The rectangle

14. If the length of the median to the hypotenuse of a right triangle is 7, find the length of the hypotenuse.
15. **Given** Quadrilateral $ABCD$ with $\overline{AD} \perp \overline{DC}$; $\overline{BC} \perp \overline{DC}$; $\overline{AC} \cong \overline{BD}$
Prove Quadrilateral $ABCD$ is a rectangle.
16. **Given** Rectangle $ABCD$; $\overline{AEC}$; $\overline{BED}$; $\overline{EF} \perp \overline{DC}$ at F
Prove $\overline{DF} \cong \overline{CF}$
17. **Given** $\overline{CD}$ and $\overline{BE}$ are altitudes of $\triangle ABC$; M is the midpoint of $\overline{BC}$.
Prove $\triangle DME$ is isosceles.

7-4 The rhombus

18. List three ways of proving that a parallelogram is a rhombus.

For Exercises 19–23, use a rhombus $ABCD$ with diagonals meeting at E.

19. What is the relationship between $\overline{AC}$ and $\overline{BD}$?
20. If $AB = 4x - 7$ and $AD = 2x + 17$, find BC.
21. If $m\angle BDC = 23$, find $m\angle ACD$.
22. If $m\angle BAC = 32$, find $m\angle DAC$.
23. If $m\angle ADB = 7x - 11$ and $m\angle DAC = 2x - 7$, find $m\angle DBC$.
24. **Given** $\overline{AD} \perp \overline{DB}$; $\overline{BC} \perp \overline{DB}$; $\overline{AB} \parallel \overline{DC}$; E is the midpoint of $\overline{AB}$; F is the midpoint of $\overline{DC}$.
Prove Quadrilateral $DEBF$ is a rhombus.

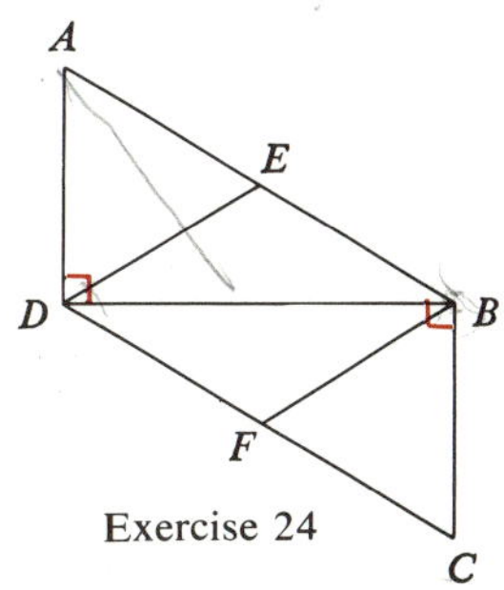

Exercise 24

7-5 *The square*

25. List the properties of a square.
26. List at least three ways of proving that a quadrilateral is a square.
27. Is it correct to say that a square is a rectangular rhombus? Explain.
28. In square $ABCD$, where diagonals $\overline{AC}$ and $\overline{BD}$ meet at E, find $m\angle ABD$.
29. GIVEN P, Q, R, and S are in sides $\overline{AB}$, $\overline{BC}$, $\overline{CD}$, and $\overline{DA}$, respectively, of square $ABCD$, so that $AS = BP = CQ = DR$.
    PROVE Quadrilateral $PQRS$ is a square.

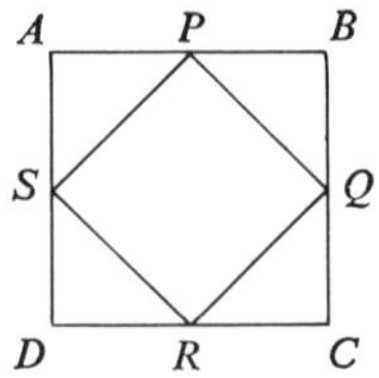

7-6 *The midline of a triangle*

30. Find the length of a line joining the midpoints of two sides of a triangle if the length of the third side is 24.
31. If the perimeter of the triangle formed by the midlines of a given triangle is 15, find the perimeter of the given triangle.
32. If $\overline{MN}$ is the midline of $\triangle ABC$ so that $\overline{MN} \parallel \overline{BC}$, find MN if $BC = 3x + 21$ and $MN = 7x - \frac{1}{2}$.
33. Prove that the midline of a triangle bisects the median drawn to the side that is parallel to the midline.
34. GIVEN $\square ABCD$; M and N are the midpoints of $\overline{AB}$ and $\overline{DC}$, respectively; $\overline{MP} \parallel \overline{AC}$; $\overline{NQ} \parallel \overline{AC}$; $\overline{DQEPB}$
    PROVE Quadrilateral $MPNQ$ is a parallelogram.
35. GIVEN $\triangle PTR$ with $\overline{PQR}$; M is the midpoint of $\overline{PT}$; $QT = QP$; $\overline{MQ} \parallel \overline{TR}$
    PROVE $\angle QTR \cong \angle R$

7-7 *The trapezoid*

36. List the properties of a trapezoid.
37. List the properties of an isosceles trapezoid.
38. State at least four ways of proving that a trapezoid is isosceles.
39. If the length of the median of a trapezoid is 18 and the length of the longer base is 32, find the length of the shorter base.
40. GIVEN $\overline{MN}$ is the median of trapezoid $ABCD$; $\overline{AB} \parallel \overline{DC}$; $\overline{MN}$ meets $\overline{BD}$ at P
    PROVE $DP = BP$
41. GIVEN Isosceles trapezoid $ABCD$; $\overline{AD} \cong \overline{BC}$; $\overline{PB} \cong \overline{QC}$; $\overline{APB}$; $\overline{DQC}$
    PROVE $APQD$ is an isosceles trapezoid.

CHAPTER TEST

1. The diagonals of rectangle $ABCD$ meet at E. If $AE = 17$, find BD.
2. The diagonals of rhombus $ABCD$ meet at M. If $m\angle ABD = 50$, find $m\angle ADB$.
3. If the length of the median of a trapezoid is 10 and the length of the shorter base is 6, find the length of the longer base.
4. Points M and N trisect diagonal $\overline{AC}$ of ▱ $ABCD$; $\overline{PMQ}$; $\overline{RNS}$; $\overline{BQSC}$; $\overline{APRD}$. If $AD = 21$, find SB.

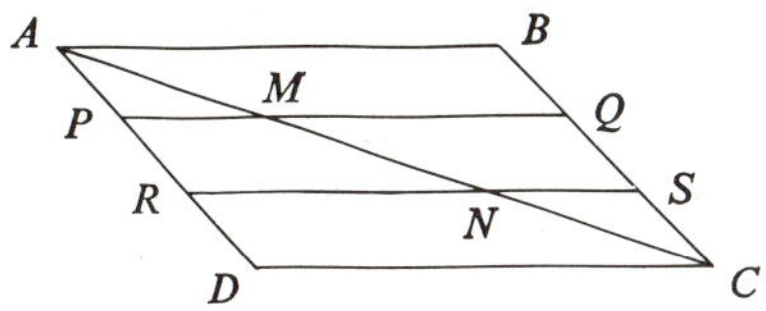

5. $\overline{CD}$ is the median of $\triangle ACB$. If $m\angle ACB = 90$, $CD = 8x - 3$, and $BD = 15x - 17$, find AB.

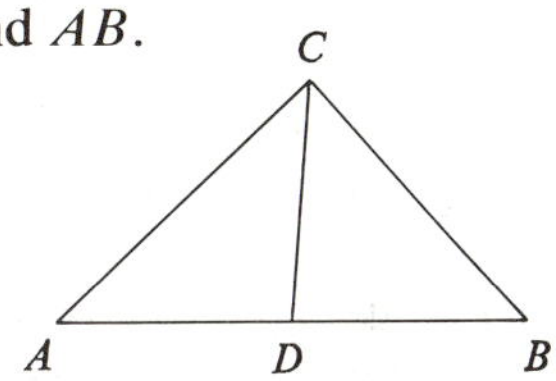

6. The lengths of the bases of an isosceles trapezoid are 17 and 9. If the measure of a base angle is 45, find the length of the altitude of the trapezoid.
7. GIVEN $\overline{AD}$ and $\overline{CE}$ are altitudes of $\triangle ABC$ intersecting at H; F, M, and N are midpoints of $\overline{AH}$, $\overline{AC}$, and $\overline{CB}$, respectively.

   PROVE $\overline{MF} \perp \overline{MN}$

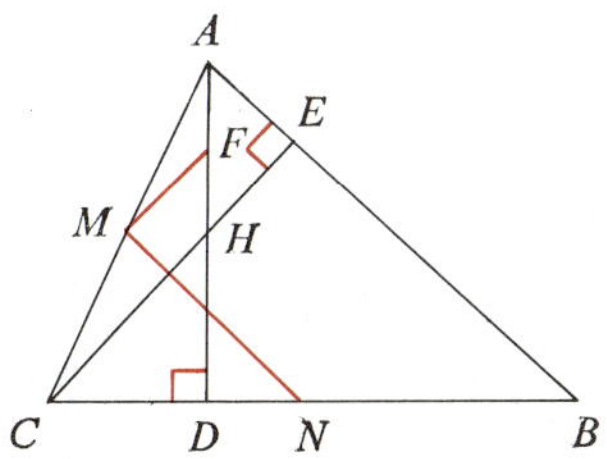

8. GIVEN P, Q, R, and S are the midpoints of $\overline{AB}$, $\overline{AC}$, $\overline{DB}$, and $\overline{DC}$, respectively; P, Q, R, and S are not collinear.

   PROVE Quadrilateral $PQRS$ is a parallelogram.

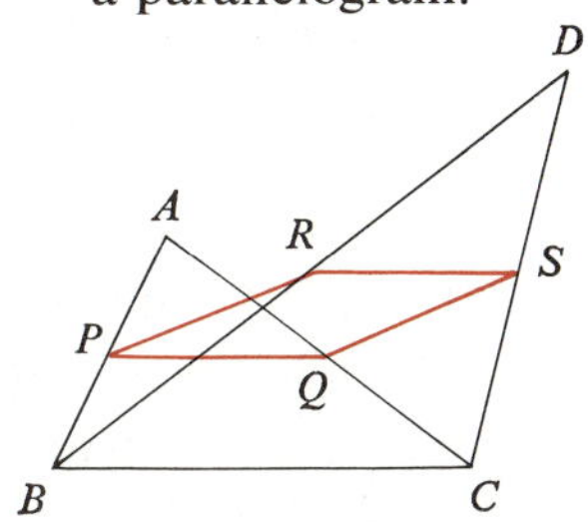

SUGGESTED RESEARCH

1. Induction plays a prominent role in all branches of mathematics. What role does it play in geometry? You may understand better by reading *Induction*

in Geometry, by L. I. Golovina and I. M. Yaglom (Boston: D. C. Heath & Co., 1963).

2. Many instruments and mechanical devices use properties of parallelograms. Investigate some of these instruments and demonstrate their uses to your class, showing how they use parallelogram properties.

MATHEMATICAL EXCURSION

The Complete Quadrilateral

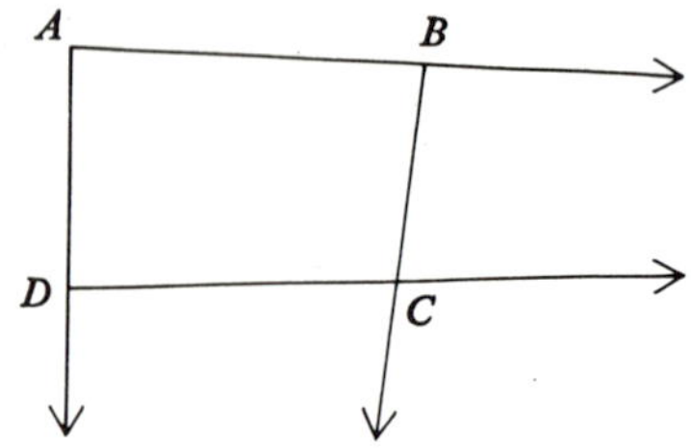

How many points do the sides of a quadrilateral determine? Suppose that no two sides of the quadrilateral are parallel. It should be evident that the lines of each pair of opposite sides, when they intersect, determine a point. Together with the vertices, this makes a total of six points.

D **Definition** Four lines, no three concurrent, and the six points determined by them, form a *complete quadrilateral*.

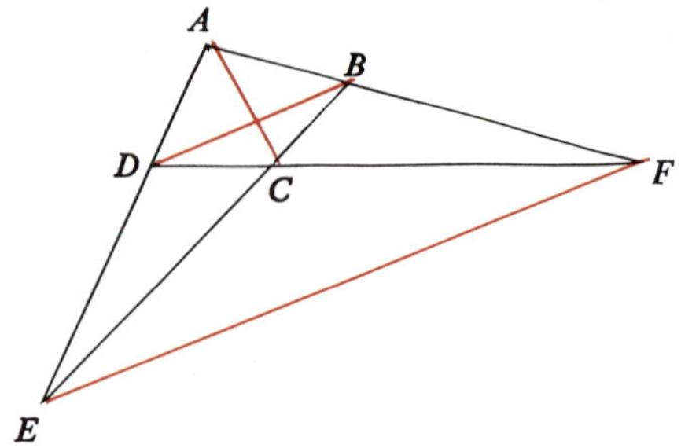

Thus, the complete quadrilateral has six vertices. If we define a line joining two vertices not on the same side as a diagonal, how many diagonals does a complete quadrilateral have? In the figure at right, the quadrilateral has sides $\overline{ABF}$, $\overline{BCE}$, $\overline{DCF}$, $\overline{ADE}$, and vertices A, B, C, D, E, F. A and C are not on a side, but A is on a side with the other four points. Therefore, the only diagonal from A is $\overline{AC}$. Similarly, we have $\overline{DB}$ and $\overline{EF}$ as diagonals.

Suppose one or both pairs of opposite sides of the quadrilateral are parallel. Would we then have a complete quadrilateral? In the geometry we are studying, the answer is no. In projective geometry, the answer would be yes. In projective geometry, parallel lines meet in an *ideal point*. The set of all ideal points is the *ideal line*. The plane of projective geometry is called the *extended plane*. This may all sound contradictory, but it really is not. Playfair's Axiom or Postulate, given in the Mathematical Excursion for Chapter 6, still applies. The definition of parallel lines, however, is slightly modified.

D

Definition *Parallel lines* are two lines that have a common ideal point.

Before you decide that ideal points cannot really be points, remember that points are undefined. "Point" is used to indicate what is common to two lines. When we introduce ideal points, we are merely adopting a common language in describing what is common to any two lines.

In the extended plane, we can still say that two points determine a line. The only way we can show an ideal point is by drawing a line that it is on. We can indicate the ideal point, α, like this:

The line determined by the ordinary point A and the ideal point α is the line through A and parallel to the line that defines α.

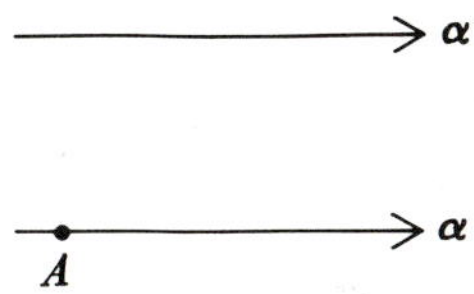

Two ideal points determine the ideal line.

In the extended plane, we can also say that two lines always determine a point: an ordinary point if the lines intersect and an ideal point if they are parallel.

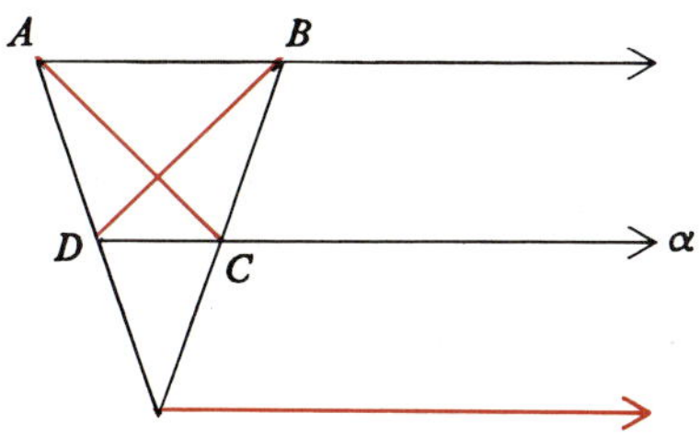

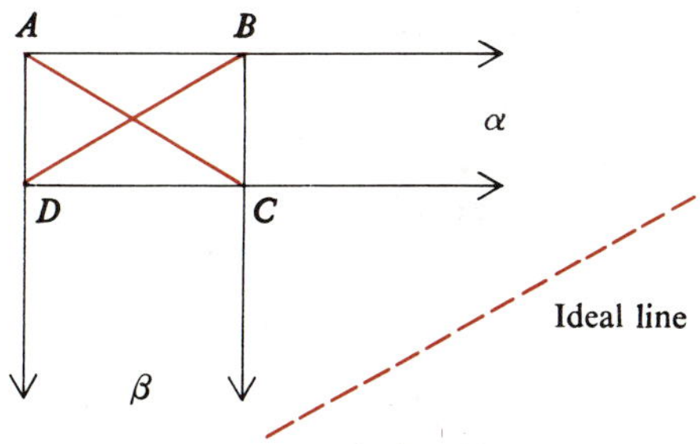

In the extended plane, every quadrilateral is a complete quadrilateral. In the figure at the left above, there is one pair of parallel sides and we have one ideal point for a vertex. One diagonal is parallel to the parallel sides.

In the figure at the right above, the quadrilateral is a parallelogram and there are two ideal vertices and one of the diagonals is the ideal line.

Later in this course you will find that the bisector of an angle of a triangle divides the opposite side of the triangle into two segments. The ratio of the measures of the segments is equal to the ratio of the measures of the other two sides. That is, if $\angle BAD \cong \angle CAD$, then $\frac{BD}{DC} = \frac{AB}{AC}$.

A segment can be considered divided externally by any point collinear with but outside the segment. The point C divides the segment $\overline{AB}$ internally into the segments $\overline{AC}$ and $\overline{CB}$. The point D divides the segment $\overline{AB}$ externally into segments $\overline{AD}$ and $\overline{DB}$. We can generalize the idea of a point separating a segment: If C is collinear with A and B, then C divides $\overline{AB}$ into two segments, $\overline{AC}$ and $\overline{CB}$.

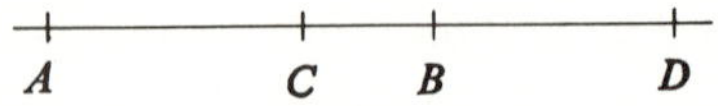

The bisector $\overline{AE}$ of an exterior $\angle BAR$ of $\triangle ABC$ also divides the side $\overline{BC}$ into two segments, $\overline{BE}$ and $\overline{EC}$, such that $\frac{BE}{EC} = \frac{AB}{AC}$. Thus, the bisectors of the interior and exterior angles at a vertex of a triangle divide the opposite side internally and externally into the same ratio. A segment that is divided in this way is *divided harmonically*.

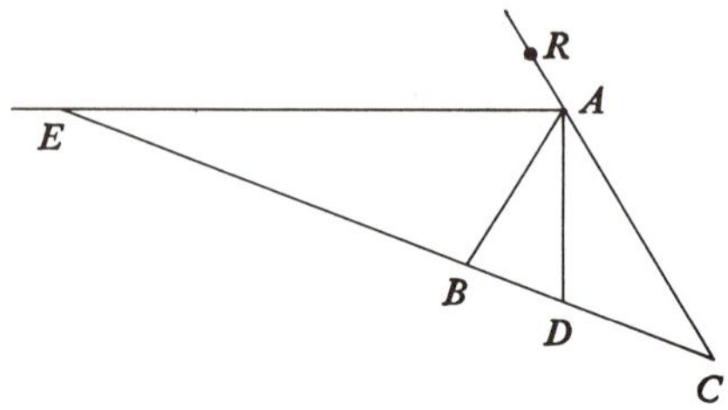

D **Definition** Four collinear points located so that two of the points divide the segment determined by the other two points internally and externally into the same ratio are called a *harmonic range*.

The points of division in a harmonic range are called the *harmonic conjugates* with respect to the other two points.

In the figure below, C and D are harmonic conjugates with respect to A and B. We can show that if C and D are harmonic conjugates with respect to A and B, then A and B are also harmonic conjugates with respect to C and D.

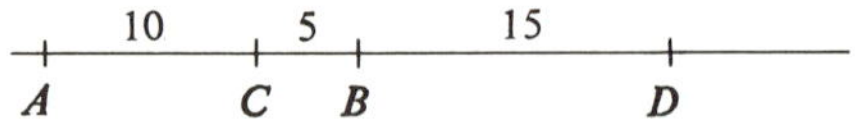

In the figure, $\frac{AC}{CB} = \frac{10}{5}$, $\frac{AD}{DB} = \frac{30}{15}$, $\frac{DB}{BC} = \frac{15}{5}$, and $\frac{DA}{AC} = \frac{30}{10}$. In the general case, we must prove that for collinear points A, B, C, and D, if $\frac{AC}{CB} = \frac{AD}{DB}$, then $\frac{DB}{BC} = \frac{DA}{AC}$.

PROOF

STATEMENTS	REASONS
1. $\frac{AC}{CB} = \frac{AD}{DB}$	1. Given
2. $\frac{AC}{CB} \cdot \frac{CB}{AD} = \frac{AD}{DB} \cdot \frac{CB}{AD}$	2. Multiplication property of equality
3. $\frac{AC}{AD} = \frac{CB}{DB}$	3. Simplify statement 2
4. $\frac{AC}{AD} \cdot \frac{AD}{AC} \cdot \frac{DB}{CB} = \frac{CB}{DB} \cdot \frac{AD}{AC} \cdot \frac{DB}{CB}$	4. Multiplication property of equality
5. $\frac{DB}{CB} = \frac{AD}{AC}$	5. Simplify statement 4
6. $\frac{DB}{BC} = \frac{DA}{AC}$	6. Distance Postulate (1-1)

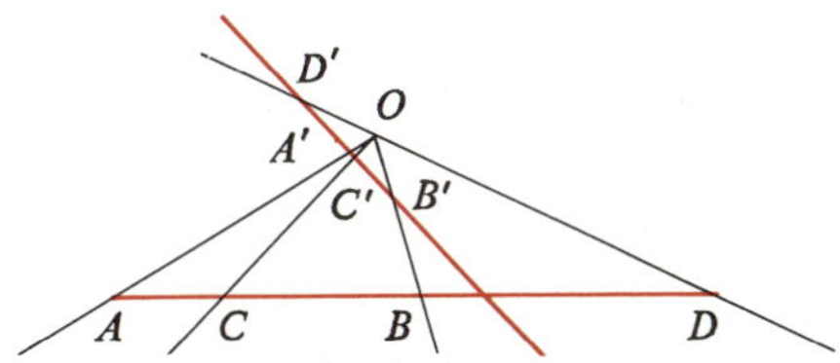

If four concurrent lines cut a fifth line in a harmonic range, then we call the four concurrent lines a *harmonic pencil*. If A, B, C, D is a harmonic range, then O-A, B, C, D is a harmonic pencil. The converse of this is also true. In other words, any other line cutting the pencil will be cut in a harmonic range. In the figure, O-A', B', C', D' is a harmonic pencil, and A', B', C', D' is a harmonic range.

Harmonic division is a central topic in projective geometry. We can now illustrate the importance of the complete quadrilateral in the context of harmonic division. Since we do not yet have the tools to establish the desired results, we shall verify them experimentally.

On a complete quadrilateral, any diagonal is divided harmonically by the other two diagonals.

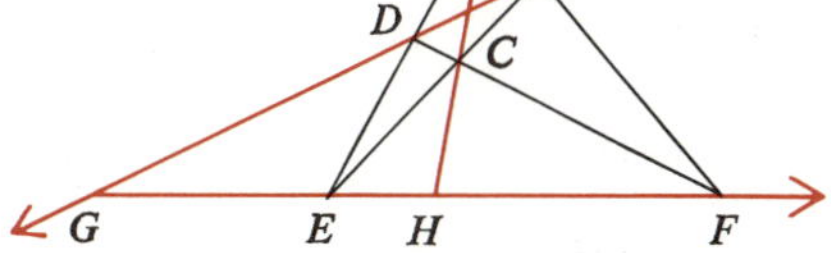

Given complete quadrilateral $ABCDEF$. Diagonal $\overline{EF}$ is divided by $\overleftrightarrow{AC}$ at H and by $\overleftrightarrow{DB}$ at G. H and G divide $\overline{EF}$ harmonically. Also, $\overline{BD}$ intersects $\overline{AC}$ at J. Hence, J and H separate A and C harmonically; J and G separate $\overline{DB}$ harmonically.

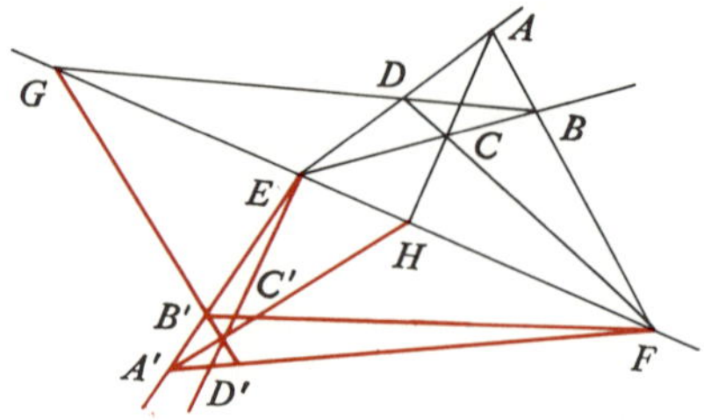

We should be able to verify this by constructing another complete quadrilateral with H at the intersection of $\overleftrightarrow{AC}$ and $\overleftrightarrow{EF}$. Then the other diagonal should cut $\overleftrightarrow{EF}$ at G. In this figure, we construct $ABCDEF$. Then below $\overline{GF}$ we draw two lines from E. Then draw a line from H cutting one line from E at C' and the other at A'. Then draw $\overleftrightarrow{FC'}$ cutting $\overleftrightarrow{EA'}$ at B' and $\overleftrightarrow{FA'}$ cutting $\overleftrightarrow{EC'}$ at D'. Now, if it is true that two diagonals of a complete quadrilateral separate the other diagonal harmonically, then $\overleftrightarrow{D'B'}$ must pass through G. This is true because H divides $\overleftrightarrow{EF}$ internally in a fixed ratio, and one and only one point will divide it externally into the same ratio.

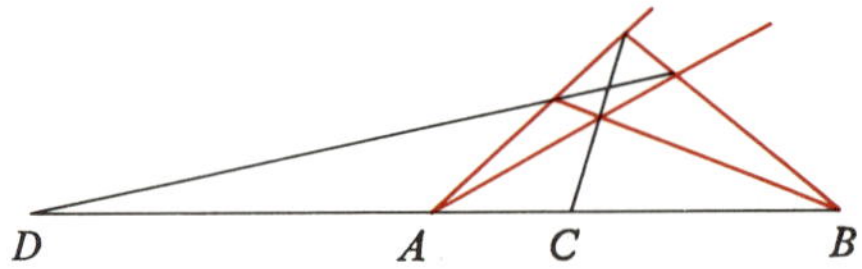

The complete quadrilateral provides a means of separating a segment harmonically without using anything but a straightedge. To divide $\overline{AB}$ harmonically, draw any two rays through A and cut them with two rays through B. These four rays will form a complete quadrilateral with $\overline{AB}$ as one diagonal The other two diagonals will cut $\overline{AB}$ harmonically in C and D.

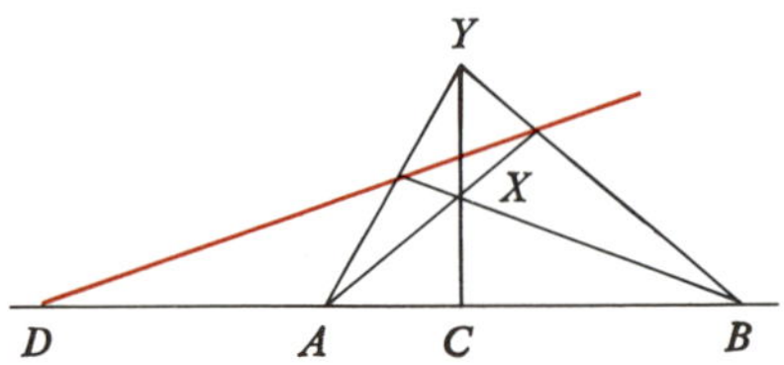

The above discussion yields a harmonic range without concern for the value of the ratio $\frac{AC}{CB}$. Suppose we know one point of division and want to find the other. Here, we draw two rays from A, then one ray from C cutting them in X and Y. Then we draw rays $\overrightarrow{BX}$ and $\overrightarrow{BY}$. This completes the quadrilateral and the third diagonal will cut $\overleftrightarrow{AB}$ in D.

EXERCISES

1. Construct triangle ABC. Construct the bisectors of the interior and exterior angles at A, cutting $\overleftrightarrow{BC}$ in D and E. Measure as closely as you can the segments $\overline{BD}$, $\overline{DC}$, $\overline{BE}$, and $\overline{EC}$. Compare $\frac{BD}{DC}$ with $\frac{BE}{EC}$.
2. Construct the complete quadrilateral $ABCDEF$. Draw the diagonals $\overline{AC}$, $\overline{BD}$, and $\overline{EF}$. Extend $\overline{AC}$ and $\overline{BD}$ cutting $\overline{EF}$ in G and H. Measure as accurately as you can the segments $\overline{EG}$, $\overline{GF}$, $\overline{EH}$, and $\overline{HF}$. Compare $\frac{EG}{GF}$ with $\frac{EH}{HF}$.

3. Divide $\overline{AB}$ harmonically by constructing a complete quadrilateral with $\overline{AB}$ as a diagonal and finding the intersection X and Y of $\overleftrightarrow{AB}$ with the other diagonals. Now divide $\overline{XY}$ harmonically, using A as one of the points of division. B should be the other point of division.
4. In the figure, if A, B, C, D is a harmonic range, so is E, F, G, H. That is, if $\frac{AC}{CB} = \frac{AD}{DB}$, then $\frac{EG}{GF} = \frac{EH}{HF}$. Verify this. Is it necessarily true that $\frac{AC}{CB} = \frac{EG}{GF}$? Determine whether they are equal.

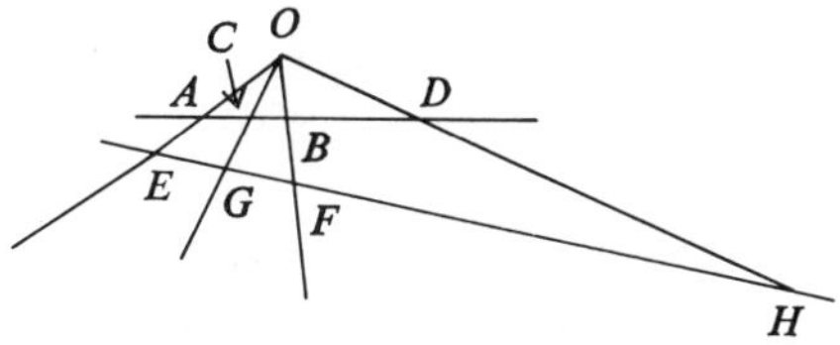

5. Given four concurrent rays, two of them perpendicular. If one of the perpendicular rays bisects the angle formed by the other two rays, prove that the four rays form a harmonic pencil.
6. Four points, no three collinear, and the six lines determined by them, form a complete quadrangle. In a complete quadrangle nonconsecutive sides meet in diagonal points. Draw a picture of a complete quadrangle and its diagonal points. Show how a complete quadrangle differs from a complete quadrilateral.
7. If C is the midpoint of $\overline{AB}$, where is the harmonic conjugate of C with respect to A and B?

8 Geometric Proportions and Similarity

In this chapter we shall consider geometric proportions and the *similarity relation*—the relation that exists between triangles that have the same shape, but not necessarily the same size. Then, using this knowledge, we shall derive many useful theorems dealing with measurement concepts, including the Pythagorean Theorem.

8-1 RATIO AND PROPORTION

In mathematics, as in our daily routines, comparisons of numbers occur frequently. In comparing a weight of 100 pounds to a weight of 150 pounds, you could say that the first weight is two-thirds of the second weight. In other words, the ratio of the first weight to the second weight is $\frac{100}{150}$ or $\frac{2}{3}$.

Ratios also occur in geometry. For segments $\overline{AB}$ and $\overline{CD}$, $AB = 14$ and $CD = 35$.

In comparing the lengths of these segments, we say that the ratio of the length of $\overline{AB}$ to the length of $\overline{CD}$ is $\frac{14}{35}$ or $\frac{2}{5}$. We write this $\frac{AB}{CD} = \frac{2}{5}$.

D **Definition 8-1** For any two positive real numbers a and b, $b \neq 0$, the **ratio** of a to b is the quotient $\frac{a}{b}$. We may also denote this $a : b$.

Any real number can be thought of as a ratio. For example, r can be considered a ratio since it can be expressed as $\frac{r}{1}$.

We may have two ratios which have the same quotient. For example, $\frac{15}{40} = \frac{3}{8}$. By equating these two ratios, we form a proportion.

D **Definition 8-2** An equation expressing the equality of two ratios is called a **proportion**.

We can extend this definition to include more than four numbers. Thus, we can say that positive numbers $a, b, c, d, e, \ldots$ are proportional to positive numbers $p, q, r, s, t, \ldots$ if $\frac{a}{p} = \frac{b}{q} = \frac{c}{r} = \frac{d}{s} = \frac{e}{t} = \ldots$.

In the proportion $\frac{a}{b} = \frac{c}{d}$, d is called the **fourth proportional** because it is considered the fourth term of the proportion. The first term is a; the second, b; the third, c. The first and fourth terms, a and d, are called the **extremes** of the proportion. The second and third terms, b and c, are called the **means** of the proportion.

In the proportion $\frac{a}{b} = \frac{b}{c}$, b is said to be the **mean proportional** between a and c because it appears in both mean positions. The mean proportional between two numbers is sometimes referred to as the **geometric mean** between the two numbers.

Because the algebra of proportions is essentially the same as the algebra of all real numbers, the following theorems can easily be proved using ordinary algebraic methods.

The first theorem is sometimes stated verbally, "A proportion exists if and only if the product of the means equals the product of the extremes."

T **Theorem 8-1.1** For any positive real numbers a, b, c, and d, $\frac{a}{b} = \frac{c}{d}$ if and only if $ad = bc$.

PROOF OUTLINE

Part I Prove that if $\frac{a}{b} = \frac{c}{d}$ and a, b, c, and d are positive real numbers, then $ad = bc$. By multiplying both members of $\frac{a}{b} = \frac{c}{d}$ by bd, we get $\left(\frac{a}{b}\right)bd = \left(\frac{c}{d}\right)bd$ or $ad = bc$.

Part II Prove that if $ad = bc$ and a, b, c, and d are positive real numbers, then $\frac{a}{b} = \frac{c}{d}$. By multiplying both members of $ad = bc$ by $\frac{1}{bd}$, we get $\frac{ad}{bd} = \frac{bc}{bd}$ or $\frac{a}{b} = \frac{c}{d}$.

Example 1 Is the ratio $\frac{51}{57}$ equal to the ratio $\frac{34}{38}$?

Solution By using Theorem 8-1.1, we may test the validity of a proportion. Since $51 \times 38 = 1938$ and since $57 \times 34 = 1938$, we know that $\frac{51}{57} = \frac{34}{38}$. We could also have reduced both fractions to $\frac{17}{19}$ to show this equality.

Example 2 If CD is the mean proportional between AC and BC, find CD when $AC = 3$ and $BC = 12$.

Solution We begin by writing the proportion

$$\frac{AC}{CD} = \frac{CD}{BC}$$

Now substitute the numerical measures.

$$\frac{3}{CD} = \frac{CD}{12}$$

$$CD^2 = 3 \cdot 12 = 36$$

$$CD = 6$$

C **Corollary 8-1.1a The Inversion Corollary** For any positive real numbers a, b, c, and d, $\frac{a}{b} = \frac{c}{d}$ if and only if $\frac{b}{a} = \frac{d}{c}$.

C **Corollary 8-1.1b The Alternation Corollary** For any positive real numbers a, b, c, and d, $\frac{a}{b}=\frac{c}{d}$ if and only if $\frac{a}{c}=\frac{b}{d}$.

The proofs of these corollaries are left as exercises.

We shall now consider two more theorems about proportions.

T **Theorem 8-1.2** For any positive real numbers a, b, c, and d, $\frac{a}{b}=\frac{c}{d}$ if and only if $\frac{a+b}{b}=\frac{c+d}{d}$.

PROOF OUTLINE

Part I Prove that if $\frac{a}{b}=\frac{c}{d}$, then $\frac{a+b}{b}=\frac{c+d}{d}$.

Add 1 in the form $\frac{b}{b}$ to the left ratio of the given proportion and add 1 in the form $\frac{d}{d}$ to the right ratio to obtain the desired result, $\frac{a+b}{b}=\frac{c+d}{d}$.

Part II Prove that if $\frac{a+b}{b}=\frac{c+d}{d}$, then $\frac{a}{b}=\frac{c}{d}$.

Subtract the proper forms of 1 from each part of the proportion. The complete proof is left as an exercise.

T **Theorem 8-1.3** For any positive real numbers a, b, c, and d, $\frac{a}{b}=\frac{c}{d}$ if and only if $\frac{a-b}{b}=\frac{c-d}{d}$.

The proof of this theorem follows the same pattern as the proof of the previous theorem and is left as an exercise.

Class Exercises

Choose three equal ratios and do Exercises 1–4.

1. Add the numerators of the three ratios.
2. Add the denominators of the three ratios.
3. Find the ratio of the sum of the numerators to the sum of the denominators.
4. How does the ratio in Exercise 3 compare with each of the ratios chosen?

5. Now choose any four equal ratios and repeat Exercises 1–4.
6. What do you expect would be the ratio of the sum of the numerators to the sum of the denominators of seven equal ratios?

The Class Exercises above suggest the following theorem:

T **Theorem 8-1.4** For any positive real numbers a, b, c, d, e, and f, if $\frac{a}{b}=\frac{c}{d}=\frac{e}{f}$, then $\frac{a+c+e}{b+d+f}=\frac{a}{b}$.

PROOF OUTLINE Suppose $\frac{a}{b}=k$, where k is any constant real number. Then $a=kb$. Since $\frac{a}{b}=\frac{c}{d}$, $\frac{c}{d}=k$ and $c=kd$. Similarly, $e=kf$. Hence

$$\frac{a+c+e}{b+d+f}=\frac{kb+kd+kf}{b+d+f}=\frac{k(b+d+f)}{b+d+f}=k=\frac{a}{b}.$$

Can you extend this theorem to more than three equal ratios?

To complete our set of theorems on proportions, we shall consider the following:

T **Theorem 8-1.5** For any positive real numbers a, b, c, d, and e, if $\frac{a}{b}=\frac{c}{d}$ and $\frac{a}{b}=\frac{c}{e}$, then $d=e$.

The proof of this theorem is left for you to complete as an exercise.

Summary. Proportion Theorems and Corollaries

For positive real numbers:

1. $\frac{a}{b}=\frac{c}{d}$ if and only if $ad=bc$.
2. $\frac{a}{b}=\frac{c}{d}$ if and only if $\frac{b}{a}=\frac{d}{c}$.
3. $\frac{a}{b}=\frac{c}{d}$ if and only if $\frac{a}{c}=\frac{b}{d}$.
4. $\frac{a}{b}=\frac{c}{d}$ if and only if $\frac{a+b}{b}=\frac{c+d}{d}$.
5. $\frac{a}{b}=\frac{c}{d}$ if and only if $\frac{a-b}{b}=\frac{c-d}{d}$.
6. If $\frac{a}{b}=\frac{c}{d}=\frac{e}{f}$, then $\frac{a+c+e}{b+d+f}=\frac{a}{b}$.
7. If $\frac{a}{b}=\frac{c}{d}$ and $\frac{a}{b}=\frac{c}{e}$, then $d=e$.

EXERCISES

A

Find each of the following ratios assuming that P is a point between points A and B of $\overline{AB}$, where $AB = 24$ and $AP = 6$.

1. $\frac{AP}{AB}$ 2. $\frac{AP}{PB}$ 3. $\frac{PB}{AB}$ 4. $\frac{AB}{AP}$

Complete each of the following:

5. If $\frac{7}{3} = \frac{21}{9}$, then $7 \times 9 = 3 \times$ ____. 6. If $\frac{a}{b} = \frac{7}{11}$, then $11a =$ ____.

Find x in each of the following proportions:

7. $\frac{12}{15} = \frac{x}{7}$ 8. $\frac{5}{x} = \frac{17}{3}$ 9. $\frac{p}{q} = \frac{x}{r}$ 10. $\frac{AB}{x} = \frac{x}{CD}$

Complete each of the following:

11. If $5 \cdot x = 3 \cdot y$, then $\frac{x}{} = \frac{3}{}$ and $\frac{5}{y} =$ ____.

12. If $PQ \cdot PS = PR \cdot RT$, then $\frac{RT}{} = \frac{PS}{}$ and ____ $= \frac{PR}{PQ}$.

Find the fourth proportional for each of the following:

13. 3, 5, 13 14. 2, 4, 6 15. 5, 7, 11 16. 2, 16, 3

Find the mean proportional between each of the following pairs of numbers:

17. 2, 8 18. 6, 54 19. 5, 3 20. $\sqrt{2}, \sqrt{8}$

21. Prove Corollary 8-1.1a. 22. Prove Corollary 8-1.1b.
23. Prove Theorem 8-1.2. 24. Prove Theorem 8-1.3.

B

Find the values of p and q in Exercises 25–28.

25. If $\frac{3}{7} = \frac{21}{p}$, then $\frac{10}{7} = \frac{q}{p}$. 26. If $\frac{p}{33} = \frac{5}{15}$, then $\frac{p}{q} = \frac{5}{10}$.

27. If $\frac{8}{3} = \frac{40}{p}$, then $\frac{5}{3} = \frac{q}{p}$. 28. If $\frac{x}{y} = \frac{z}{w}$, then $\frac{x}{x+y} = \frac{p}{z+q}$.

Complete each of the following, stating the theorem or corollary that justifies the completed statement:

29. If $\frac{AD}{DB} = \frac{AE}{EC}$, then $\frac{DB}{AD} = \frac{}{AE}$.

30. If $\frac{AD}{DE} = \frac{AB}{BC}$, then $\frac{BC}{} = \frac{AB}{AD}$.

31. If $\frac{AE}{DE} = \frac{AC}{BC}$, then $AE \cdot$ ____ $= DE \cdot AC$.

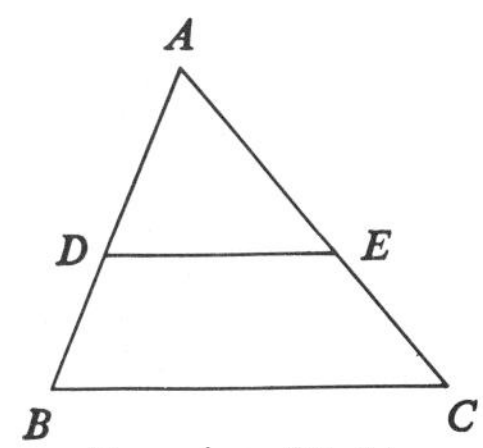

Exercises 29–31

C

32. If $AD \cdot BC = DE \cdot AB$, then $\frac{DE}{AD} = \frac{BC}{}$.

33. Point P divides $\overline{AB}$ into two segments so that $\frac{AP}{PB} = \frac{3}{5}$. Find AP if $AB = 40$.

34. Point Q divides $\overline{CD}$ into two segments so that $\frac{QD}{CD} = \frac{4}{7}$. Find CD if $CQ = 28$.

8-2 PARALLEL LINES AND PROPORTIONS

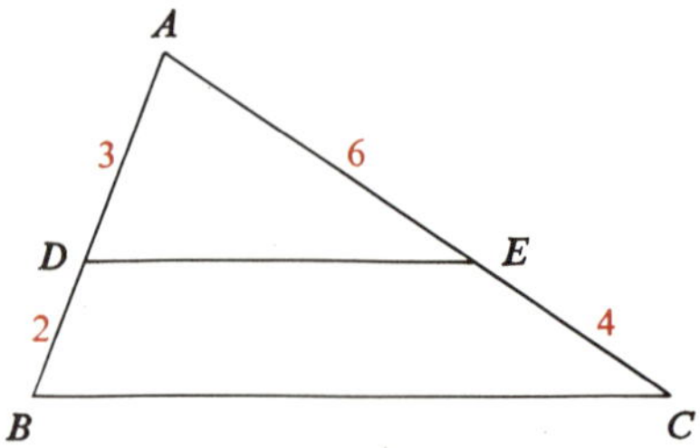

Pairs of segments whose measures or lengths are proportional are called **proportional segments**. For example, in the figure at right, the pair $\overline{AD}$ and $\overline{DB}$, and the pair $\overline{AE}$ and $\overline{EC}$ are proportional segments. That is, AD and DB are proportional to AE and EC, respectively, since $\frac{AD}{DB} = \frac{AE}{EC}$, or $\frac{3}{2} = \frac{6}{4}$. Can you form other proportions with these four segments? If so, name some.

We are now ready to apply our knowledge of proportions to geometry. Let us begin with the following postulate:

P **Postulate 8-1 Proportional Line Segments Postulate** Three or more parallel lines intercept proportional segments on two or more transversals.

Although we shall accept the above statement as a postulate for now, it is possible to prove it. Since the proof involves some rather difficult concepts, we shall not demonstrate it here.

The following theorem is an application of Postulate 8-1.

T **Theorem 8-2.1** If a line segment parallel to one side of a triangle intersects the other two sides, then it divides these two sides into proportional segments.

GIVEN $\triangle ABC$ with $\overline{DE} \parallel \overline{BC}$; $\overline{ADB}$ and $\overline{AEC}$

PROVE $\frac{AD}{DB} = \frac{AE}{EC}$

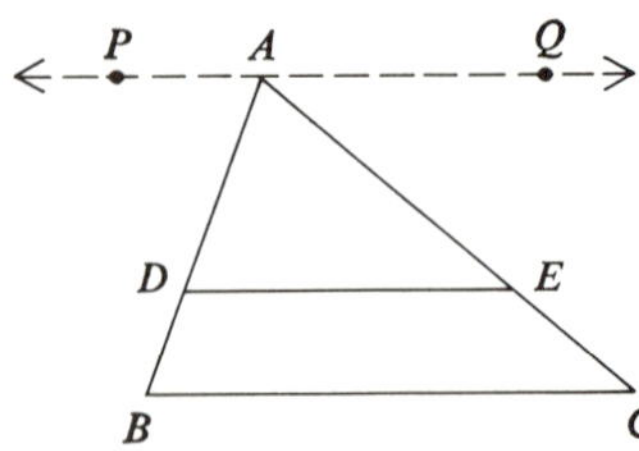

PROOF

STATEMENTS	REASONS
1. $\triangle ABC$ with $\overline{DE} \parallel \overline{BC}$, $\overline{ADB}$ and $\overline{AEC}$	1. Given.
2. Consider $\overleftrightarrow{PAQ} \parallel \overline{BC}$.	2. The Parallel Postulate. (6-1)
3. $\overleftrightarrow{PAQ} \parallel \overline{DE}$	3. If each of two lines is parallel to a third line, then they are parallel to each other. (C6-1.1c)
4. $\frac{AD}{DB} = \frac{AE}{EC}$	4. Proportional Line Segments Postulate. (8-1)

Using Theorem 8-1.2, we can easily establish the following corollary:

C **Corollary 8-2.1a** If a line parallel to one side of a triangle intersects the other two sides, then it divides these two sides proportionally.

If we apply Corollary 8-2.1a to $\triangle ABC$ in Theorem 8-2.1, where $\overline{DE} \parallel \overline{BC}$, $\overline{ADB}$, and $\overline{AEC}$, we can form the proportions $\frac{AB}{AD} = \frac{AC}{AE}$; $\frac{AB}{DB} = \frac{AC}{EC}$; or any of their variations.

State some other proportions of line segments in $\triangle ABC$. How does Theorem 8-2.1 differ from Corollary 8-2.1a? The distinction is minor, but you should be aware of it. Why may we use Corollary 8-2.1a in place of Theorem 8-2.1? Why may we *not* use Theorem 8-2.1 in place of its corollary?

Example 1 In $\triangle PQR$, S and T are points of $\overline{PQ}$ and $\overline{RP}$, respectively, so that $\overline{ST} \parallel \overline{QR}$, $PS = 6$, $SQ = 2$, and $PT = 4$. Find PR.

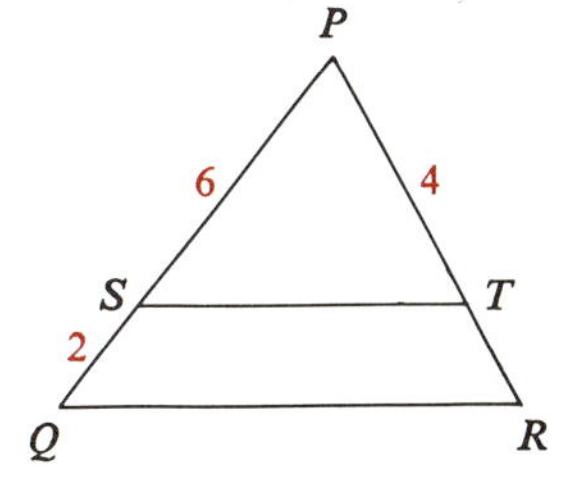

Solution There are two methods of finding PR. Corollary 8-2.1a enables us to form the proportion

$$\frac{PS}{PQ} = \frac{PT}{PR}$$

Substituting the appropriate measures, we get

$$\frac{6}{8} = \frac{4}{PR}$$

By Theorem 8-1.1, $6 \cdot PR = 4 \cdot 8$

$$PR = \frac{16}{3}$$

If we apply Theorem 8-2.1 instead, we get $\frac{PS}{SQ} = \frac{PT}{TR}$ and find that $TR = \frac{4}{3}$.

Since $PR = PT + TR$, $PR = 4 + \frac{4}{3} = \frac{16}{3}$.

Theorem 8-2.2 is the converse of Theorem 8-2.1.

T

Theorem 8-2.2 If a line divides two sides of a triangle into proportional segments, then the line is parallel to the remaining side of the triangle.

GIVEN $\triangle ABC$ with sides $\overline{ADB}$ and $\overline{AEC}$; $\frac{AD}{DB} = \frac{AE}{EC}$

PROVE $\overline{DE} \parallel \overline{BC}$

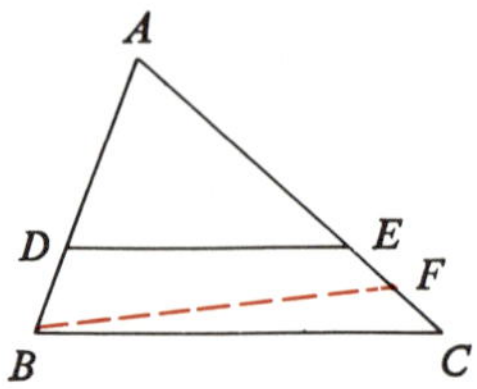

PROOF

STATEMENTS	REASONS
1. $\triangle ABC$ with sides $\overline{ADB}$ and $\overline{AEC}$; $\frac{AD}{DB} = \frac{AE}{EC}$	1. Given.
2. Assume $\overline{DE}$ is not parallel to $\overline{BC}$.	2. Assumption.
3. Let $\overline{BF} \parallel \overline{DE}$, with F in $\overline{EC}$.	3. The Parallel Postulate. (6-1)
4. In $\triangle ABF$, $\frac{AD}{DB} = \frac{AE}{EF}$.	4. Theorem 8-2.1.
5. But $\frac{AD}{DB} = \frac{AE}{EC}$	5. Given.
6. $EF = EC$	6. For any positive real numbers a, b, c, d, and e, if $\frac{a}{b} = \frac{c}{d}$ and $\frac{a}{b} = \frac{c}{e}$, then $d = e$. (8-1.5)
7. But $EF < EC$	7. If P is a point of $\overline{AB}$ between A and B, then $AB > AP$ and $AB > BP$. (C5-1.1a)
8. It is false that $\overline{DE}$ is not parallel to $\overline{BC}$.	8. Step 6 contradicts step 5.
9. $\overline{DE} \parallel \overline{BC}$	9. The contradiction of a false statement is true.

In the above proof, we let F be between points E and C. The proof could just as easily have been completed if point F were a point of $\overrightarrow{EC}$. Could F possibly be on $\overrightarrow{EA}$ when $\overline{BF} \parallel \overline{DE}$? Why?

The following corollary is similar to Corollary 8-2.1a:

C

Corollary 8-2.2a If a line divides two sides of a triangle proportionally, then the line is parallel to the remaining side of the triangle.

As was true for Theorem 8-1.1 and Corollary 8-1.1a, the distinction between Theorem 8-2.2 and Corollary 8-2.2a is minor. We may use Corollary 8-2.2a in place of Theorem 8-2.2, but not the reverse. Why?

Example 2 The lengths of the segments in the figure at right are marked. If $m\angle R = 30$, find $m\angle PNM$.

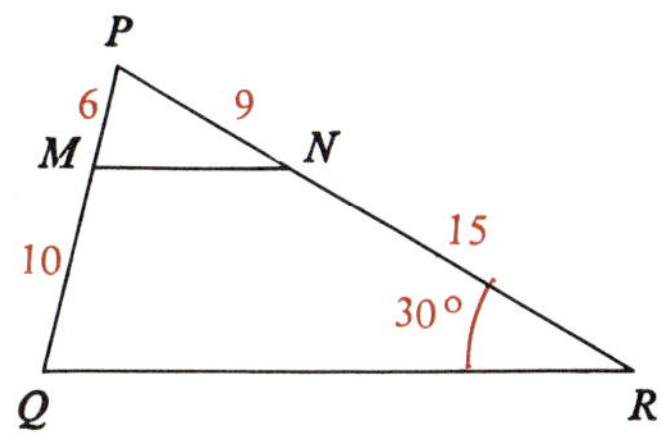

Solution Since $\frac{6}{10} = \frac{9}{15}$,
$\frac{PM}{MQ} = \frac{PN}{NR}$. Therefore $\overline{MN} \parallel \overline{QR}$, by Theorem 8-2.2. By Corollary 6-3.1a, $m\angle R = m\angle PNM$. Thus $m\angle PNM = 30$.

Example 3 D and E are respective points of sides $\overline{AB}$ and $\overline{BC}$ of $\triangle ABC$, so that $\frac{AD}{DB} = \frac{2}{3}$ and $\frac{BE}{EC} = \frac{1}{4}$. If $\overline{AE}$ and $\overline{DC}$ meet at P, find $\frac{PC}{DP}$.

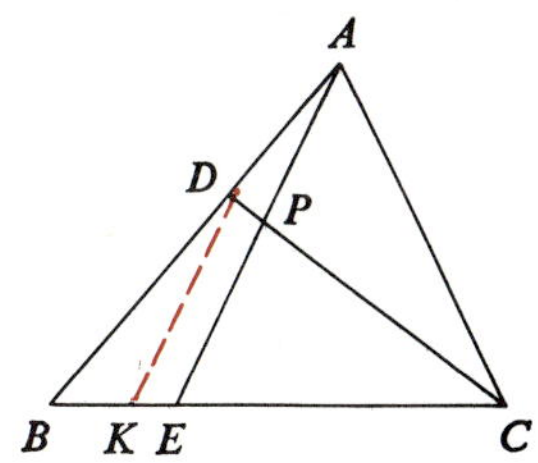

Solution Consider $\overrightarrow{DK} \parallel \overline{AE}$, with K in $\overline{BC}$.
In $\triangle ABE$, since $\overrightarrow{DK} \parallel \overline{AE}$, $\frac{KE}{BK} = \frac{AD}{DB} = \frac{2}{3}$, by Theorem 8-2.1.

By Theorem 8-1.2, $\frac{KE}{BK + KE} = \frac{2}{3+2}$ or $\frac{KE}{BE} = \frac{2}{5}$.

By the multiplication property $\frac{KE}{BE} \cdot \frac{BE}{EC} = \frac{2}{5} \cdot \frac{1}{4}$

$$\frac{KE}{EC} = \frac{1}{10}$$

But since $\overline{PE} \parallel \overrightarrow{DK}$ in $\triangle DCK$, $\frac{KE}{EC} = \frac{DP}{PC}$, by Theorem 8-2.1. Therefore $\frac{DP}{PC} = \frac{1}{10}$.

Example 4 In this trapezoid, $\overline{AB} \parallel \overline{MN} \parallel \overline{DC}$ and $\overline{AMD}$ and $\overline{BNC}$. Find x, referring to the lengths of the line segments indicated.

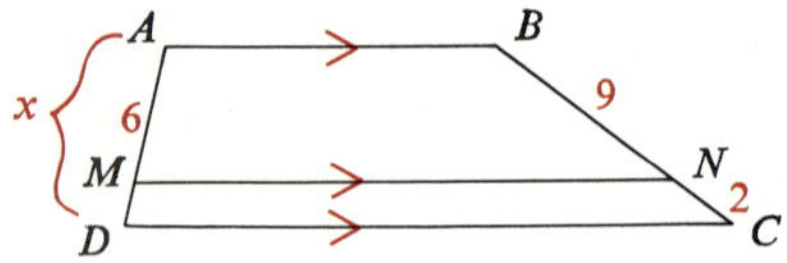

Solution Using Postulate 8-1, we can establish the proportion $\frac{6}{x} = \frac{9}{11}$. Solving the proportion, $x = 7\frac{1}{3}$.

EXERCISES

A Exercises 1–8 refer to $\triangle MNK$, where $\overline{MPN}$ and $\overline{KQN}$.

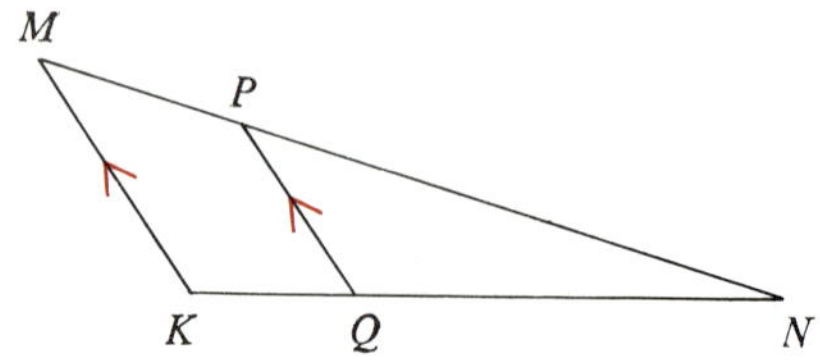

Exercises 1–8

1. If $\overline{PQ} \parallel \overline{MK}$, $PN = 15$, $PM = 12$, and $QN = 10$, find KN.
2. If $\overline{PQ} \parallel \overline{MK}$, $MN = 36$, $KQ = 5$, and $QN = 4$, find NP.
3. If $\overline{PQ} \parallel \overline{MK}$, $KN = 45$, $QN = 5$, and $MP = 8$, find NP.
4. If $\overline{PQ} \parallel \overline{MK}$, $MN = 10$, $KN = 12$, and $NP = 6$, find KQ.
5. If $NP = 8$, $MP = 6$, $NQ = 28$, and $QK = 21$, is $\overline{PQ} \parallel \overline{MK}$? Why?
6. If $MN = 5$, $NP = 2$, $KN = 7$, and $NQ = 4$, is $\overline{PQ} \parallel \overline{MK}$? Why?
7. If $MN = 10$, $MP = 2$, $KN = 20$, and $KQ = 5$, is $\overline{PQ} \parallel \overline{MK}$? Why?
8. If $KN = 33$, $KQ = 15$, $NP = 4\frac{1}{2}$, and $MN = 8\frac{1}{4}$, is $\overline{PQ} \parallel \overline{MK}$?Why?

Complete each of the following, given $\overline{MN} \parallel \overline{HJ}$ and $\overline{HMK}$ and $\overline{KNJ}$. State which theorem or corollary justifies each proportion.

9. $\frac{HK}{MK} =$ ____
10. $\frac{KN}{NJ} =$ ____
11. $\frac{MH}{HK} =$ ____
12. $\frac{KJ}{NJ} =$ ____

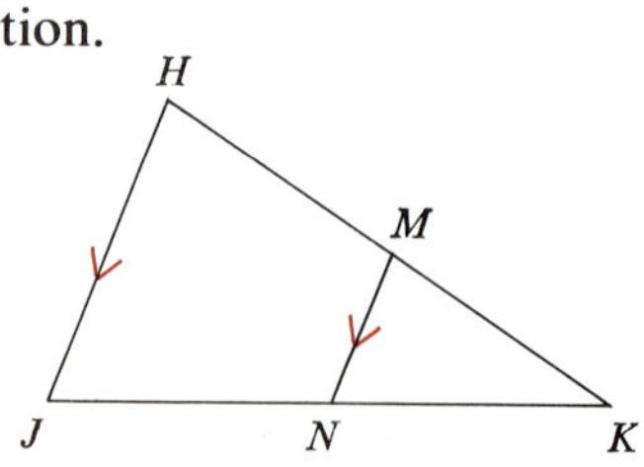

Exercises 9–12

In the figures for Exercises 13–14, $\overline{MN} \parallel \overline{BC}$, $\overline{AMB}$, $\overline{ANC}$, and the lengths of certain line segments are marked. Find x in each figure.

13.

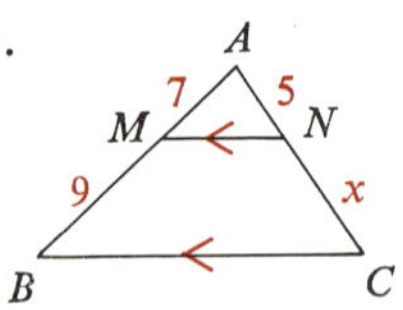

14.

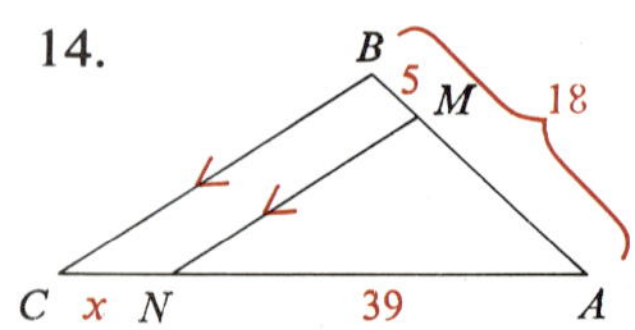

In each of the following trapezoids, $\overline{AB} \parallel \overline{MN} \parallel \overline{DC}$ and $\overline{AMD}$ and $\overline{BNC}$, and lengths of certain line segments are indicated. Find x in each trapezoid.

15.

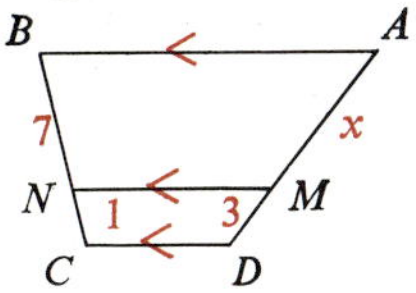

16.

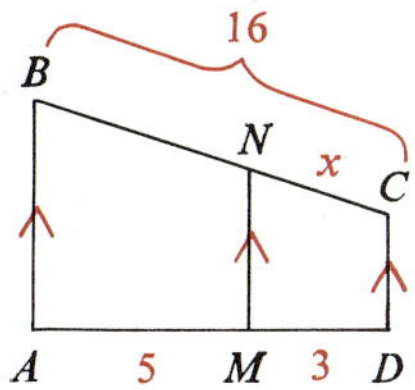

17.

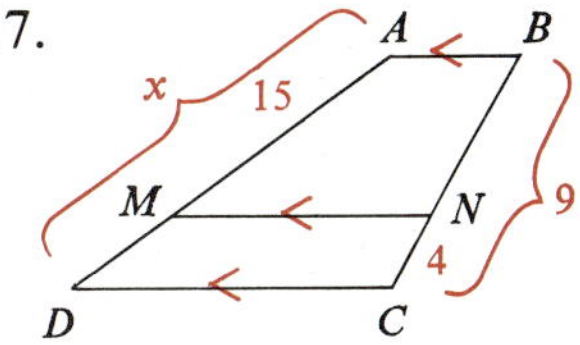

18. In $\triangle ABC$, $\overline{DE} \parallel \overline{FG} \parallel \overline{BC}$, $\overline{ADFB}$, and $\overline{AEGC}$. Lengths of certain line segments are indicated. Find x and y.

19. Prove Corollary 8-2.2a.

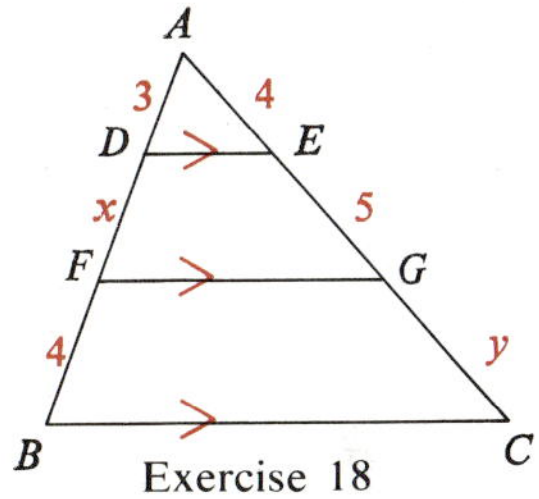

Exercise 18

B

20. GIVEN Trapezoid $ABCD$ with $\overline{AB} \parallel \overline{DC}$; $\overline{AMD}$ and $\overline{BNC}$; $\overline{MN}$ meets $\overline{BD}$ at P; $\dfrac{MD}{AD} = \dfrac{PD}{BD}$

PROVE $\overline{MN} \parallel \overline{DC}$; $\dfrac{PD}{DB} = \dfrac{NC}{CB}$

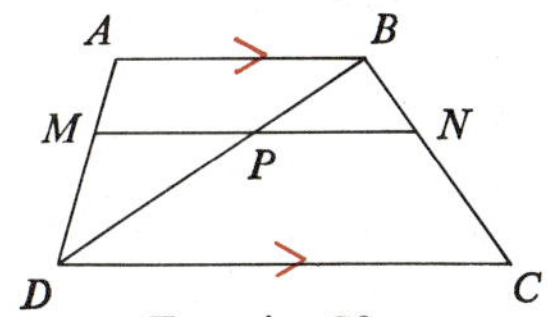

Exercise 20

If $\overline{MN} \parallel \overline{QR}$; $\overline{MK} \parallel \overline{PR}$; $\overline{PMQ}$; $\overline{PNR}$; $\overline{QKR}$; and $QK = \frac{1}{3}KR$; find each of the following ratios.

21. $\dfrac{PM}{MQ}$ 22. $\dfrac{PN}{PR}$ 23. $\dfrac{MQ}{PQ}$ 24. $\dfrac{NR}{PN}$

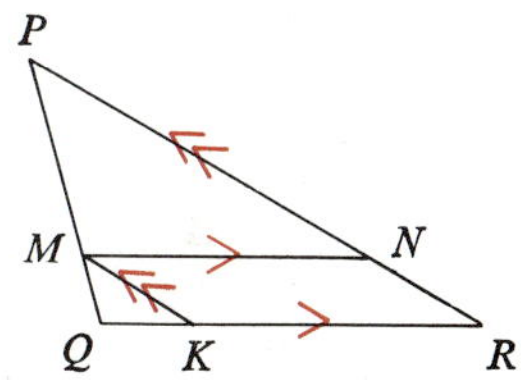

Exercises 21–24

C

25. GIVEN Trapezoid $ABCD$ with $\overline{AB} \parallel \overline{CD}$; diagonals $\overline{AC}$ and $\overline{BD}$ meet at E.

PROVE $\dfrac{AE}{EC} = \dfrac{BE}{ED}$

(*Hint*: Consider $\overline{EP} \parallel \overline{AB}$.)

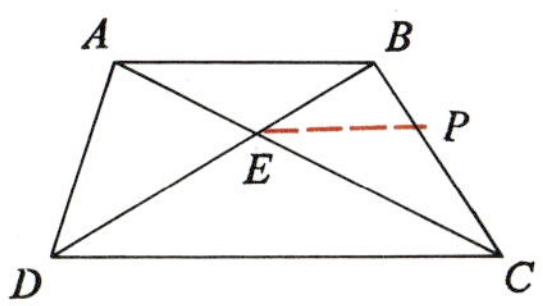

26. In $\triangle ABC$, $\overline{DE} \parallel \overline{BC}$; $\overline{FE} \parallel \overline{DC}$; $AF = 8$; and $DF = 12$; $\overline{AFDB}$; $\overline{AEC}$. Find DB.

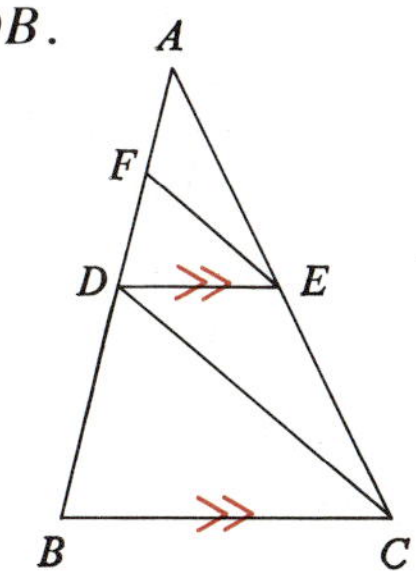

27. GIVEN Points D and F in sides $\overline{AB}$ and $\overline{AC}$, respectively, of $\triangle ABC$, so that $\overline{AD} \cong \overline{AF}$; $\overrightarrow{DF}$ meets $\overrightarrow{BC}$ at point E.

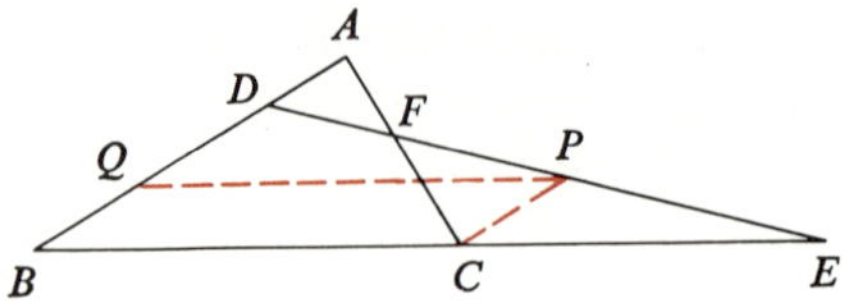

PROVE $BE \cdot CF = BD \cdot CE$ (*Hint*: Consider $\overrightarrow{CP} \parallel \overline{AB}$ where $\overline{DFPE}$, and $\overrightarrow{PQ} \parallel \overline{BE}$ where $\overline{ADQB}$.)

28. GIVEN P is any point in side $\overline{BC}$ of $\triangle ABC$; N is in $\overline{AB}$ so that $\overline{NP} \parallel \overline{AC}$; Q is in side $\overline{BR}$ of $\triangle BRC$ so that $\overline{PQ} \parallel \overline{RC}$.

PROVE $\overline{NQ} \parallel \overline{AR}$

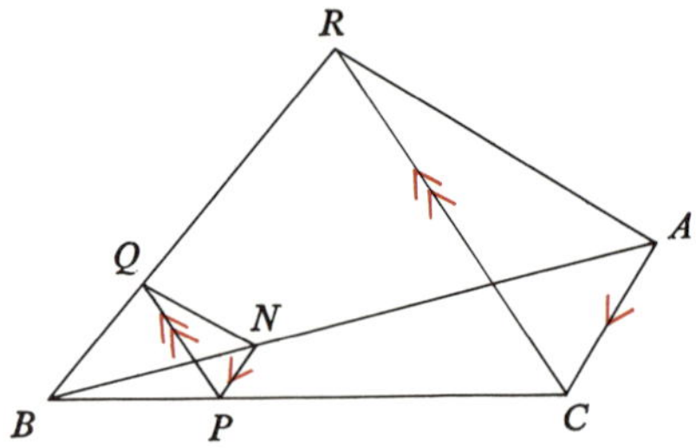

8-3 THE ANGLE BISECTOR THEOREM

Using Theorem 8-2.1, we are able to derive a theorem about the angle bisectors of a triangle. As you already know, an angle bisector of a triangle has one endpoint in the side opposite the bisected angle and the other endpoint at the vertex of the angle. Theorem 8-3.1 is concerned with how the endpoint in the side of the triangle divides that side of the triangle.

T **Theorem 8-3.1** An angle bisector of any triangle divides the side of the triangle opposite the angle into segments proportional to the adjacent sides.

GIVEN $\overline{AD}$ is an angle bisector of $\triangle ABC$.

PROVE $\dfrac{CD}{DB} = \dfrac{CA}{AB}$

PROOF

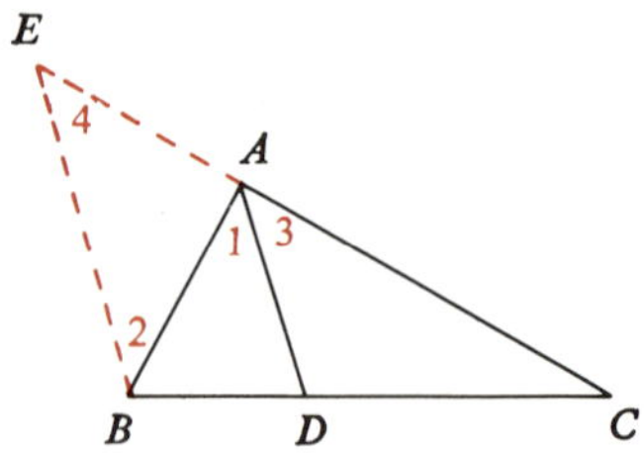

STATEMENTS	REASONS
1. $\overline{AD}$ is an angle bisector of $\triangle ABC$.	1. Given.
2. Consider the line through B parallel to $\overline{AD}$ and meeting $\overrightarrow{CA}$ at E.	2. The Parallel Postulate. (6-1)

(*continued*)

STATEMENTS	REASONS
3. $\angle 1 \cong \angle 2$	3. If two parallel lines are cut by a transversal, then the alternate interior angles are congruent. (6-3.1)
4. $\angle 3 \cong \angle 4$	4. If two parallel lines are cut by a transversal, then the corresponding angles are congruent. (C6-3.1a)
5. $\angle 1 \cong \angle 3$	5. Definition of an angle bisector of a triangle. (3-8)
6. $\angle 2 \cong \angle 4$	6. Substitution Postulate. (2-1)
7. $\overline{AE} \cong \overline{AB}$	7. In a triangle, if two angles are congruent, then the sides opposite those angles are congruent. (3-4.3)
8. In $\triangle ECB$, $\dfrac{CD}{DB} = \dfrac{CA}{AE}$	8. If a line parallel to one side of a triangle intersects the other two sides then it divides these two sides into proportional segments. (8-2.1)
9. $\dfrac{CD}{DB} = \dfrac{CA}{AB}$	9. Substitution Postulate. (2-1)

The following examples show how Theorem 8-3.1 can be applied.

Example 1 The sides of a triangle have lengths 15, 20, and 28. Find the lengths of the segments into which the bisector of the angle with the greatest measure divides the opposite side.

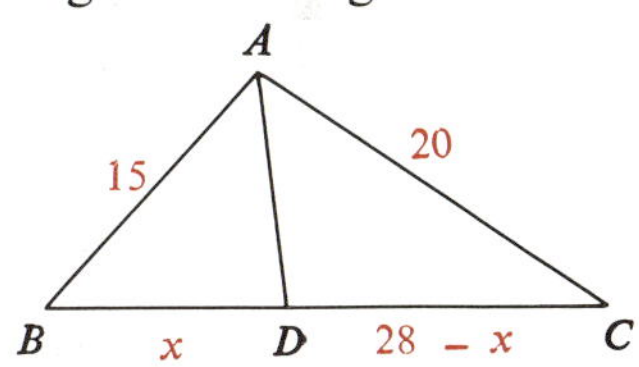

Solution By Theorem 5-3.1, the angle with the greatest measure is $\angle BAC$. The angle bisector $\overline{AD}$ separates $\overline{BC}$ into two segments, $\overline{BD}$ and $\overline{DC}$. We must find BD and DC. From Theorem 8-3.1 we get the proportion $\dfrac{BD}{DC} = \dfrac{AB}{AC}$. If we let $BD = x$, then $DC = 28 - x$. Substituting the appropriate values in the proportion above, we get

$$\frac{x}{28 - x} = \frac{15}{20} \quad \text{or} \quad \frac{x}{28 - x} = \frac{3}{4}$$

Applying Corollary 8-1.1a and Theorem 8-1.2

$$\frac{x}{28} = \frac{3}{7}$$

$$7x = 3 \cdot 28 = 84$$

$$x = 12$$

Therefore, $BD = 12$ and $DC = 28 - 12 = 16$.

Class Exercises

In the figure at right, $\overline{AD}$ bisects exterior $\angle FAB$ of $\triangle ABC$. Also, $\overrightarrow{CBD}$. Consider $\overline{BE} \parallel \overline{AD}$, with E in $\overline{AC}$.

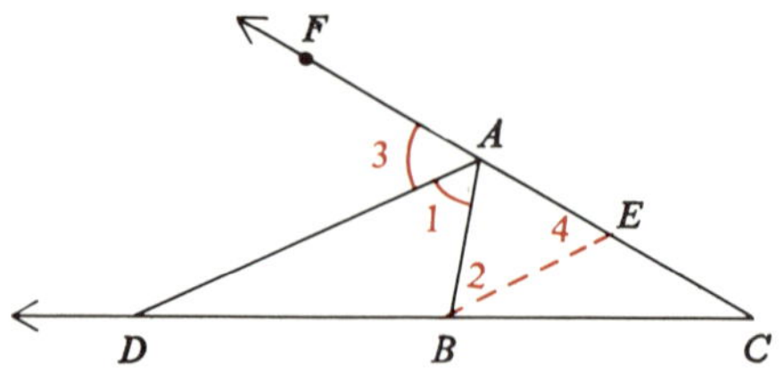

1. Prove that $\dfrac{CD}{DB} = \dfrac{CA}{AB}$ by following each step in the proof of Theorem 8-3.1.
2. Can you draw an analogy between your proof and that of Theorem 8-3.1?
3. Summarize what you have proven with a statement.

In the figure above, $\overline{AD}$ is an exterior angle bisector of $\triangle ABC$.

D **Definition 8-3** An **exterior angle bisector of a triangle** is a segment that bisects an exterior angle of a triangle, and has its endpoints at the vertex of the bisected angle and in the line containing the side of the triangle opposite this angle.

In both Theorem 8-3.1 and in the Class Exercises above, point D is an endpoint of each of the segments which are proportional to two sides of $\triangle ABC$.

In Theorem 8-3.1 we say that point D divides $\overline{BC}$ in the ratio $\dfrac{CD}{DB}$, whereas in the figure for the Class Exercises, we say that point D determines segments in the ratio $\dfrac{CD}{DB}$ along the line containing $\overline{BC}$.

We shall now state formally a corollary to Theorem 8-3.1. You have already proved the corollary in the Class Exercises above.

C **Corollary 8-3.1a** An exterior angle bisector of a triangle determines with each of the other vertices segments along the line containing the opposite side of the triangle which are proportional to the two remaining sides.

In the figure for the Class Exercises, $\overline{AD}$ is an exterior angle bisector of $\triangle ABC$. By Corollary 8-3.1a, $\dfrac{CD}{DB} = \dfrac{CA}{AB}$. Suppose however, that $\overline{AB} \cong \overline{AC}$. What would be true about $\overrightarrow{AD}$ with respect to $\overleftrightarrow{BC}$? Where would point D be? Draw an isosceles triangle and the exterior angle bisector to help answer these questions.

In your drawing it should appear that $\overrightarrow{AD} \parallel \overline{BC}$. Now try to prove it. You will need to use Theorems 6-4.1 and 6-2.1. The proof is short and straightforward.

Example 2 In the figure at right, $\overline{MP}$ is an exterior angle bisector of $\triangle MKN$. The lengths of the sides of $\triangle MKN$ are indicated in the figure. Find PK.

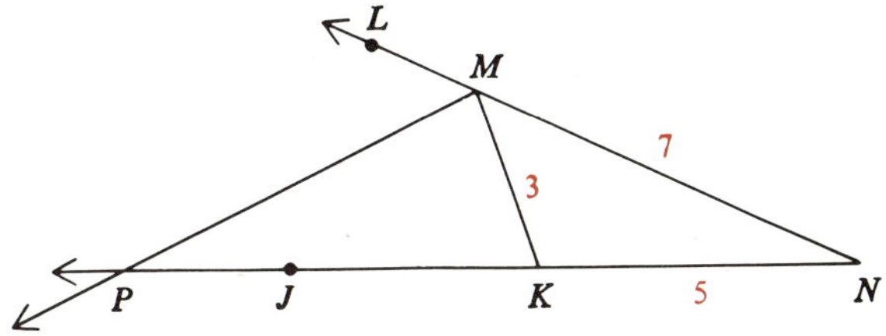

Solution From Corollary 8-3.1a we get

$$\frac{MK}{MN} = \frac{PK}{PN}$$

Let $PK = x$. Then $PN = x + 5$. Substitute the appropriate values in the proportion.

$$\frac{3}{7} = \frac{x}{x+5}$$

$$7x = 3x + 15 \quad \text{(Theorem 8-1.1)}$$

$$4x = 15$$

$$x = 3\tfrac{3}{4} = PK$$

EXERCISES

A Exercises 1–12 refer to the figure at right. $\overline{AT}$ is an angle bisector of $\triangle ABC$.

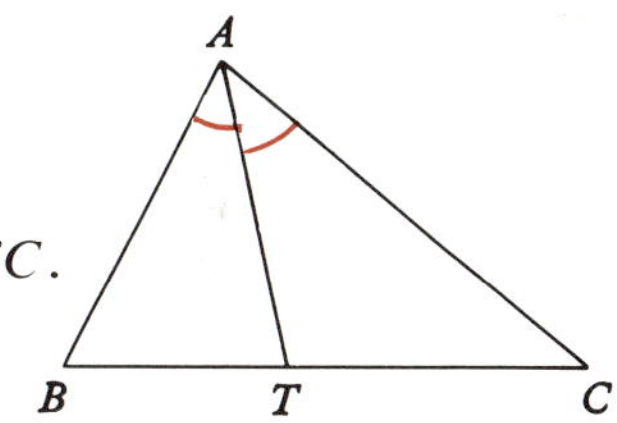

Exercises 1–12

1. If $AB = 16$, $AC = 24$, and $BT = 4$, find TC.
2. If $AB = 20$, $AC = 15$, and $TC = 6$, find BT.
3. If $AB = 12$, $AC = 24$, and $BC = 18$, find BT and TC.
4. If $BT = 3$, $AB = 6$, and $TC = 4$, find AC.
5. If $AB = 5$ and $BT = 3$, find $\frac{TC}{AC}$.
6. If $AC = 7$ and $TC = 4$, find $\frac{BC}{AB + AC}$.
7. If $AB = 11$ and $AC = 11$, find $\frac{BT}{TC}$.
8. If $AB = 5$, $AC = 6$, and $BC = 7$, find BT and TC.
9. If $AB = 13$, $AC = 14$, and $TC = 5$, find BC.
10. If $AB = 3x$, $TC = 2x$, $AC = 6$, and $BT = 4$, find AB and TC.
11. If $AB = 2a$, $TC = 4d$, $AC = 5d$, and $BT = 3b$, find $\frac{a}{b}$.
12. If the lengths of the sides of a triangle are 9, 12, and 18, find the lengths of the segments into which the three angle bisectors divide the sides.

Exercises 13–18 refer to the figure at right, where $\overline{PT}$ is an exterior angle bisector of $\triangle PQR$.

13. If $PQ = 6$, $PR = 10$, and $TQ = 9$, find TR.
14. If $PQ = 8$, $PR = 11$, and $TR = 15$, find TQ.
15. If $PQ = 16$, $PR = 24$, and $TQ = 18$, find RQ.
16. If $PQ = 24$, $PR = 32$, and $QR = 40$, find TR.

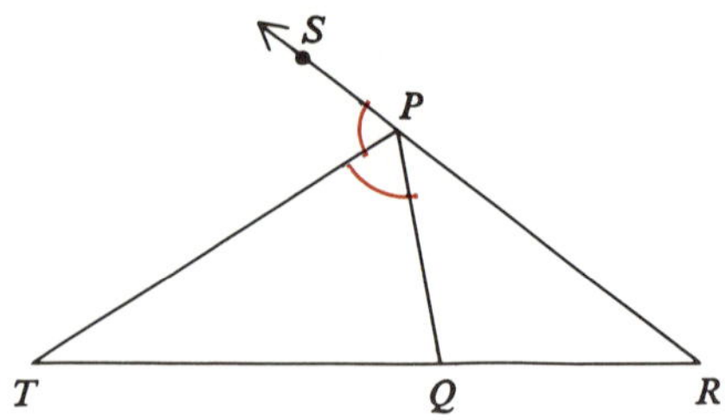

Exercises 13–18

17. If $TQ = 10$, $QR = 7$, and $PQ = 4$, find PR.
18. If $TQ = PQ$, is $\triangle PQR$ isosceles?

B

19. GIVEN $\overline{AF}$ is a median of $\triangle ABC$; $\overline{FD}$ bisects $\angle AFB$; $\overline{FE}$ bisects $\angle AFC$; $\overline{ADB}$ and $\overline{AEC}$
    PROVE $\overline{DE} \parallel \overline{BC}$

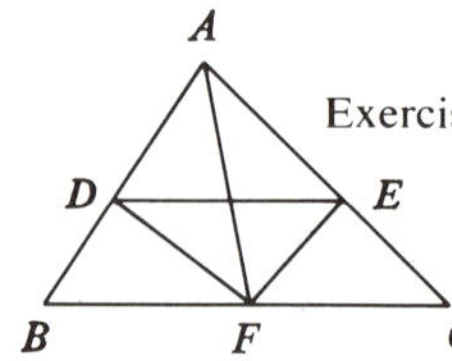

Exercise 19

20. GIVEN $\overline{AT}$ is an angle bisector of $\triangle ABC$; points P and Q are in $\overline{AB}$ and $\overline{AC}$, respectively, so that $\overline{BP} \cong \overline{BT}$ and $\overline{CQ} \cong \overline{CT}$.
    PROVE $\overline{PQ} \parallel \overline{BC}$
21. Use Theorem 8-3.1 to prove that if an angle bisector of a triangle bisects a side, then the triangle is isosceles.
22. Suppose that a line segment joins the vertex of a triangle with a point that divides the opposite side of the triangle into segments proportional to the remaining two sides of the triangle. Prove that this line segment bisects an angle of the triangle.
23. A triangle has sides of lengths 12, 24, and 32. The interior and exterior angle bisectors at the vertex of the angle of greatest measure intersect the opposite side in points P and Q. Find the distances of P and Q from the vertex of the angle of smallest measure of the triangle.

C

24. GIVEN $\overline{AD}$, $\overline{BE}$, and $\overline{CF}$ are the angle bisectors of $\triangle ABC$.
    PROVE $AF \cdot BD \cdot EC = BF \cdot DC \cdot AE$
25. GIVEN $\square ABCD$ is not a rhombus; $\overrightarrow{AP}$ bisects $\angle DAB$; $\overrightarrow{DQ}$ bisects $\angle ADC$; P and Q are on $\overline{BD}$ and $\overline{AC}$, respectively.
    PROVE $\overline{PQ} \parallel \overline{AD}$
    (*Hint*: Use Theorems 8-3.1 and 8-2.2.)

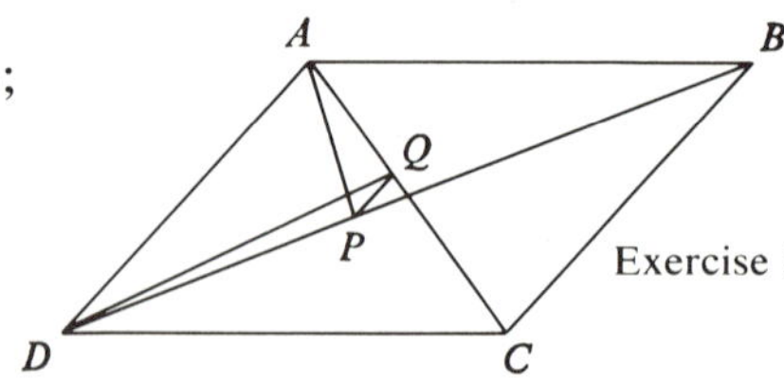

Exercise 25

26. GIVEN In $\triangle ABC$, $AC = 2BC$; D is the point of $\overrightarrow{AB}$ (distinct from A) where $DB = AB$; E is the point in $\overrightarrow{AB}$ where $\overrightarrow{CE} \perp \overline{DC}$

PROVE $\frac{BE}{AD} = \frac{1}{3}$

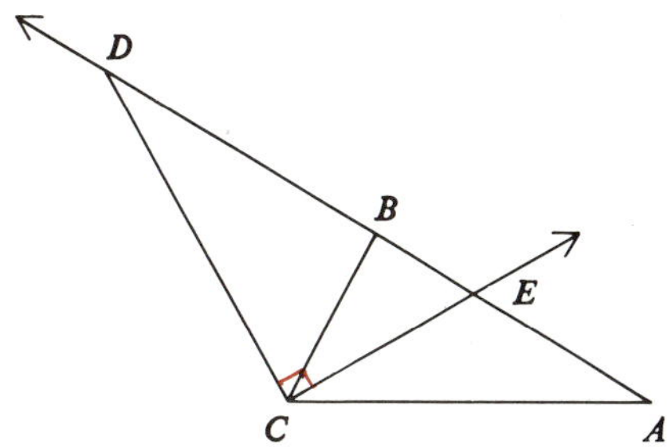

8-4 SIMILAR POLYGONS

Suppose that we make an enlargement of a picture of a polygon. If you compare the shapes of the two polygons, what would you think is the relationship between their corresponding angles? If you were to compare the sizes of the two polygons, do you think that the corresponding sides would be congruent? What other relationship is possible between the corresponding sides?

When two polygons have the same shape, we say that they are similar to each other. What do we call the relationship between two polygons if they are also the same size?

D **Definition 8-4** **Similar polygons** are polygons whose corresponding angles are congruent and whose corresponding sides are proportional.

D **Definition 8-5** A **ratio of similitude** of two similar polygons is the ratio of the measures of any pair of corresponding sides.

In our discussion of similarity, most of our attention will be focused on one type of polygon, the triangle. We can apply the definition of similar polygons to form a definition of similar triangles.

D **Definition 8-6** **Similar triangles** are triangles whose corresponding angles are congruent and whose corresponding sides are proportional.

We shall use the symbol $\sim$ to denote "is similar to." For example, $\triangle ABC \sim \triangle DEF$ is read "$\triangle ABC$ is similar to $\triangle DEF$." As with the congruence relation, vertices A, B, and C correspond to vertices D, E, and F, respectively.

To crystallize your thoughts about similar triangles, consider the following questions: Are congruent triangles similar? What is the ratio of similitude of a pair of congruent triangles?

Example 1 If $\triangle ABC \sim \triangle PRQ$, with side lengths as marked, find x and y.

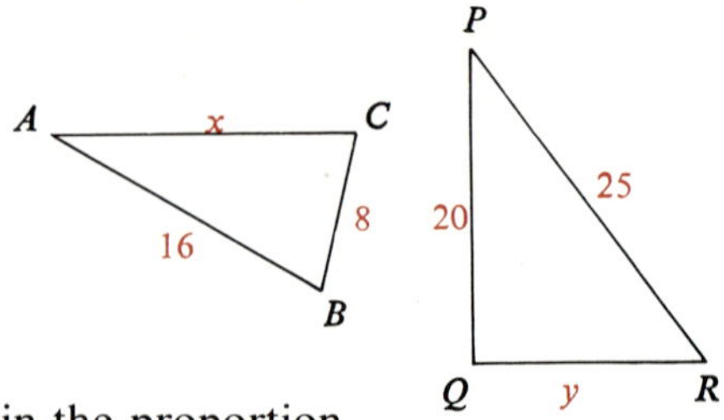

Solution Since the triangles are similar, the corresponding sides are proportional. Therefore, $\frac{AB}{PR} = \frac{AC}{PQ}$. Substitute the proper values in the proportion.

$$\frac{16}{25} = \frac{x}{20}$$

$$x = 12\tfrac{4}{5}$$

To find y, substitute the proper values in the proportion $\frac{AB}{PR} = \frac{BC}{RQ}$.

$$\frac{16}{25} = \frac{8}{y}$$

$$y = 12\tfrac{1}{2}$$

Example 2

GIVEN D and E are the respective midpoints of sides $\overline{AB}$ and $\overline{AC}$ of $\triangle ABC$.

PROVE $\triangle ADE \sim \triangle ABC$

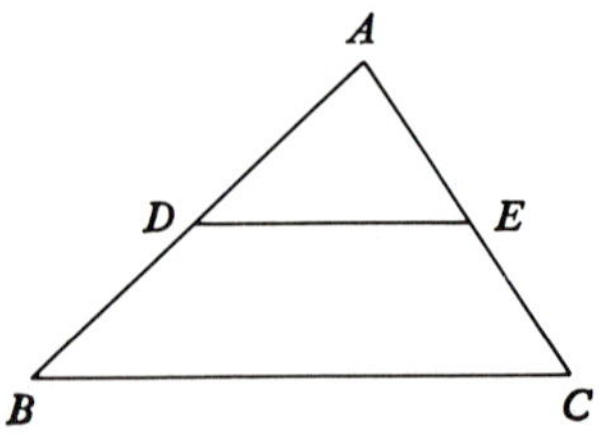

Solution

PROOF

STATEMENTS	REASONS
1. $\triangle ABC$; D is the midpoint of $\overline{AB}$; E is the midpoint of $\overline{AC}$.	1. Given.
2. $AD = \frac{1}{2}AB$ and $AE = \frac{1}{2}AC$	2. Definition of a midpoint. (1-15)
3. $\frac{AD}{AB} = \frac{1}{2}$ and $\frac{AE}{AC} = \frac{1}{2}$	3. Division property.
4. $\overline{DE}$ is a midline of $\triangle ABC$.	4. Definition of a midline. (7-9)
5. $DE = \frac{1}{2}BC$	5. The midline of a triangle is half as long as the third side of the triangle. (7-6.3)

(continued)

STATEMENTS	REASONS
6. $\frac{DE}{BC} = \frac{1}{2}$	6. Division property.
7. $\frac{AD}{AB} = \frac{AE}{AC} = \frac{DE}{BC}$	7. Transitive property.
8. $\angle A \cong \angle A$	8. Reflexive property.
9. $\overline{DE} \parallel \overline{BC}$	9. The midline of a triangle is parallel to the third side of the triangle. (7-6.2)
10. $\angle ADE \cong \angle B$; $\angle AED \cong \angle C$	10. If two parallel lines are cut by a transversal, then the alternate interior angles are congruent. (6-3.1)
11. $\triangle ADE \sim \triangle ABC$	11. Definition of similar triangles. (8-6)

The simple proofs of the following theorem and corollary are left as exercises. Theorem 8-4.1 states the transitive property of similarity.

T

Theorem 8-4.1 If two triangles are similar to the same triangle, or to similar triangles, then the triangles are similar to each other.

C

Corollary 8-4.1a If a given triangle is similar to a triangle that is congruent to a third triangle, then the given triangle is also similar to the third triangle.

Corollary 8-4.1a means that if $\triangle ABC \sim \triangle DEF$, and $\triangle DEF \cong \triangle MNP$, then $\triangle ABC \sim \triangle MNP$.

EXERCISES

A

The exercises below refer to similar $\triangle PQR$ and HJK.

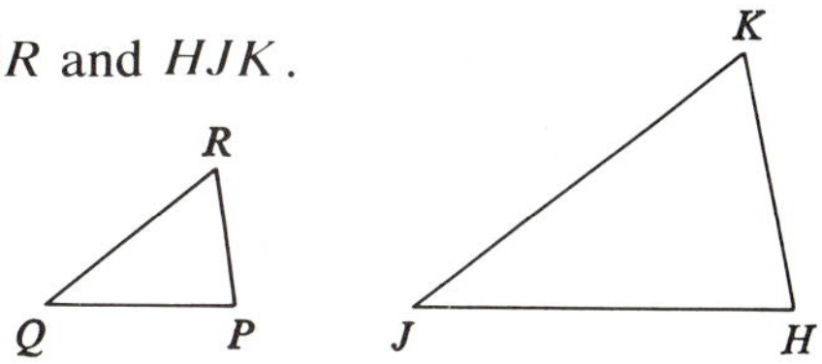

Exercises 1–6

1. $\angle Q \cong \angle$ ____
2. $\frac{PQ}{HJ} = \frac{\quad}{KJ}$ 3. $\frac{KH}{JH} =$ ____
4. If $RQ = 6$, $RP = 4$, and $KJ = 15$, find HK.
5. If $RQ = 6$, $JK = 15$, and $HJ = 12$, find PQ.
6. What is the ratio of similitude of $\triangle PQR$ and $\triangle HJK$?

Indicate whether each of the statements below is true or false. Justify each true statement by stating a theorem, corollary, postulate, or definition.

7. If $\triangle ASP \cong \triangle NRP$, then $\triangle ASP \sim \triangle NRP$.
8. If $\triangle HWP \sim \triangle EJP$, then $\triangle HWP \cong \triangle EJP$.
9. If $\triangle GEL \sim \triangle WRP$, and $\triangle WRP \sim \triangle VSB$, then $\triangle GEL \sim \triangle VSB$.

If $\triangle HMB \sim \triangle ASP$ and $HM = 8$, $MB = 9$, $HB = 10$, and $SP = 3$, find each of the following:

10. AS 11. AP
12. The ratio of similitude of $\triangle HMB$ to $\triangle ASP$.
13. The ratio of the perimeter of $\triangle HMB$ to that of $\triangle ASP$.
14. How does the ratio of similitude of two similar polygons compare to the ratio of their perimeters? Present an informal proof outline.
15. If the ratio of similitude of two similar quadrilaterals is $\frac{5}{7}$, what is the ratio of their perimeters?

Find the following if the lengths of the sides of a triangle are 13, 14, and 15:

16. the length of the shortest side of a similar triangle whose longest side has length 21.
17. the perimeter of the similar triangle described in Exercise 16.
18. Prove Theorem 8-4.1. 19. Prove Corollary 8-4.1a.

B

20. Prove that the triangle formed by the three midlines of a triangle is similar to the original triangle.
21. What is the ratio of similitude of the two triangles in Exercise 20?
22. Prove that any two equilateral triangles are similar.
23. GIVEN Points D, E, and F are in sides $\overline{AB}$, $\overline{AC}$, and $\overline{BC}$, respectively, of $\triangle ABC$; $\triangle ADE \sim \triangle EFC$

    PROVE $\overline{EF} \parallel \overline{AB}$; $\angle B \cong \angle DEF$

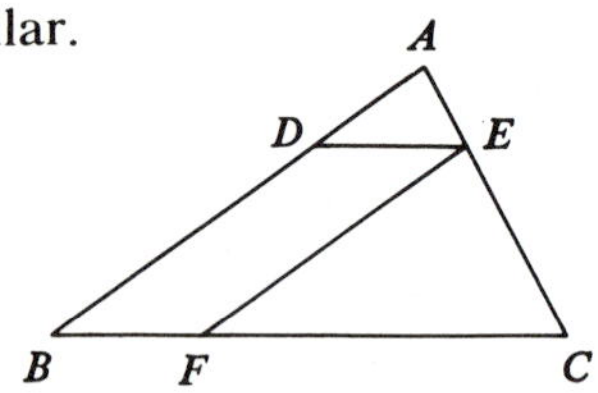

24. GIVEN Diagonals of trapezoid $TRAP$ meet at point O; $\overline{TR} \parallel \overline{PA}$; $\triangle TAP \sim \triangle RPA$

    PROVE $\overline{TP} \cong \overline{RA}$

C

25. Prove that any two regular polygons with the same number of sides are similar.
26. GIVEN $\triangle ABC$ with $\overline{ADB}$ and $\overline{AEC}$; $\frac{AD}{AB} = \frac{DE}{BC} = \frac{AE}{AC}$

    PROVE $\triangle ADE \sim \triangle ABC$

8-5 PROVING TRIANGLES SIMILAR

We shall now consider similar triangles in the same way we considered congruent triangles. We shall first develop theorems that will provide us with short cuts for proving triangles similar. Once we prove the triangles similar, we automatically know certain facts about them, because of the properties of similar triangles. What are these properties?

The first method of proving a pair of triangles similar is to apply the definition of similar triangles. Theorem 8-5.1 provides us with a second method.

Theorem 8-5.1 AAA Similarity Theorem If the corresponding angles of two triangles are congruent, then the two triangles are similar.

To prove this theorem, we must show that two triangles with corresponding angles congruent have proportional corresponding sides. Since congruent triangles are always similar, we shall use noncongruent triangles in the proof.

GIVEN $\triangle ABC$ and $\triangle DEF$ where $\angle A \cong \angle D$, $\angle B \cong \angle E$, and $\angle C \cong \angle F$

PROVE $\triangle ABC \sim \triangle DEF$

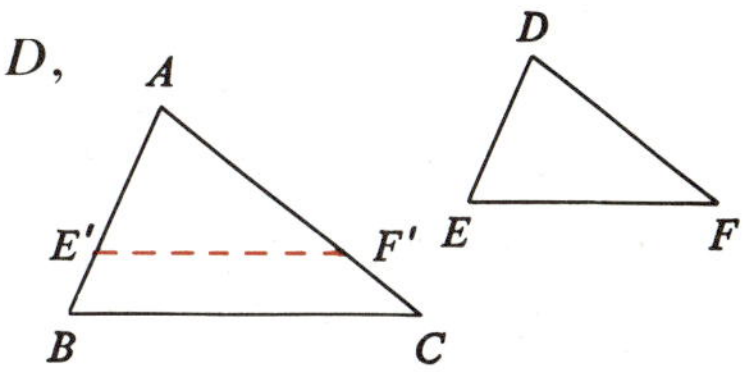

PROOF

STATEMENTS	REASONS
1. $\triangle ABC$; $\triangle DEF$; $\angle A \cong \angle D$; $\angle B \cong \angle E$; $\angle C \cong \angle F$	1. Given.
2. Points E' and F' are in $\overrightarrow{AB}$ and $\overrightarrow{AC}$, respectively, so that $\overline{AE'} \cong \overline{DE}$ and $\overline{AF'} \cong \overline{DF}$.	2. Point Uniqueness Postulate. (2-2)
3. $\triangle E'AF' \cong \triangle EDF$	3. SAS Postulate. (3-1)
4. $\angle AE'F' \cong \angle E$	4. Definition of congruent triangles. (3-3)
5. $\angle AE'F' \cong \angle B$	5. Substitution Postulate. (2-1)
6. $\overline{E'F'} \parallel \overline{BC}$	6. If two lines are cut by a transversal so that the corresponding angles are congruent, then the lines are parallel. (C6-2.1a)
7. $\dfrac{AE'}{AB} = \dfrac{AF'}{AC}$	7. If a line parallel to one side of a triangle intersects the other two sides, then it divides these two sides proportionally. (C8-2.1a)
8. $AE' = DE$ and $AF' = DF$	8. Definition of congruent triangles. (3-3)
9. $\dfrac{DE}{AB} = \dfrac{DF}{AC}$	9. Substitution Postulate. (2-1)
10. $\dfrac{DF}{AC} = \dfrac{EF}{BC}$	10. Steps 1–9, after selecting D' and E' on $\overline{AC}$ and $\overline{BC}$, respectively, so that $\overline{CD'} \cong \overline{FD}$ and $\overline{CE'} \cong \overline{FE}$.
11. $\dfrac{DE}{AB} = \dfrac{DF}{AC} = \dfrac{EF}{BC}$	11. Transitive property, statements 9 and 10.
12. $\triangle ABC \sim \triangle DEF$	12. Definition of similar triangles. (8-6)

To prove that two triangles are similar, can we prove that fewer than three corresponding angles are congruent? Why? After reviewing Corollary 6-4.1a, you will see why we can state the following corollary:

C **Corollary 8-5.1a AA Similarity Corollary** If two pairs of corresponding angles of two triangles are congruent, then the triangles are similar.

The next two corollaries follow from Corollary 8-5.1a.

C **Corollary 8-5.1b** If two right triangles have a congruent pair of corresponding acute angles, then the triangles are similar.

C **Corollary 8-5.1c** If a line parallel to one side of a triangle intersects the other two sides, then it cuts off a triangle similar to the original triangle.

The proofs of these corollaries are left as exercises.

Example 1

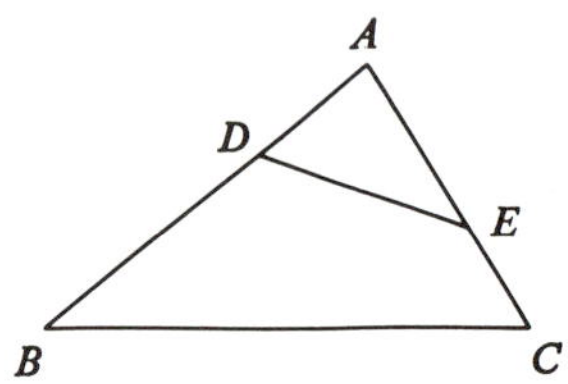

GIVEN $\triangle ABC$ with $\overline{ADB}$ and $\overline{AEC}$ such that $\angle EDB$ is supplementary to $\angle C$

PROVE $\dfrac{AE}{AB} = \dfrac{AD}{AC}$

Solution

PROOF

STATEMENTS	REASONS
1. $\triangle ABC$; $\overline{ADB}$ and $\overline{AEC}$; $\angle EDB$ is supplementary to $\angle C$.	1. Given.
2. $\angle EDB$ and $\angle EDA$ form a linear pair.	2. Definition of a linear pair. (1-26)
3. $\angle EDB$ is supplementary to $\angle EDA$.	3. Supplementary Angles Postulate. (1-6)
4. $\angle C \cong \angle EDA$	4. Supplements of congruent angles, or of the same angle, are congruent. (3-1.4)
5. $\angle A \cong \angle A$	5. Every angle is congruent to itself. (3-1.7)
6. $\triangle ADE \sim \triangle ACB$	6. Corollary 8-5.1a.
7. $\dfrac{AE}{AB} = \dfrac{AD}{AC}$	7. Definition of similar triangles. (8-6)

Class Exercises

Explain why each pair of triangles below is or is not similar. The information you will need is indicated on the figures.

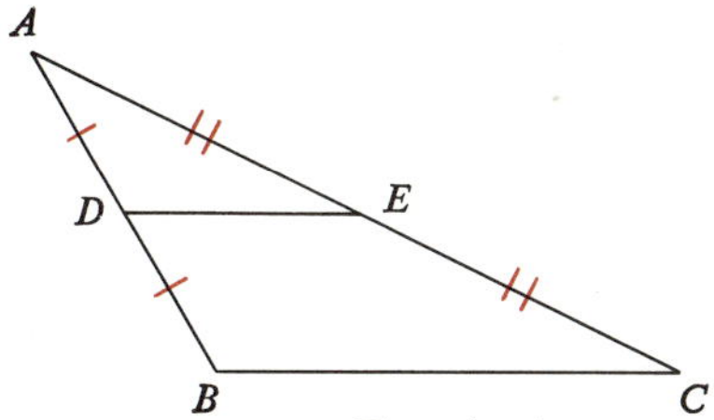

Exercise 1

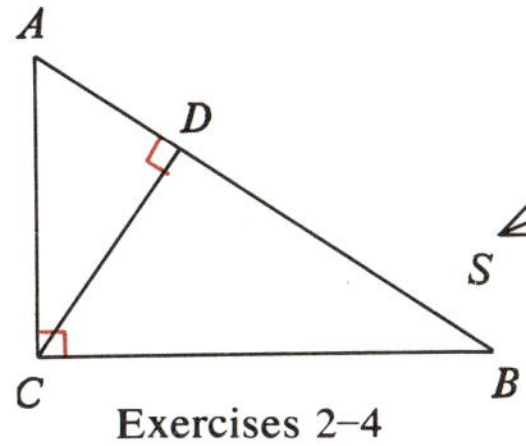

Exercises 2–4

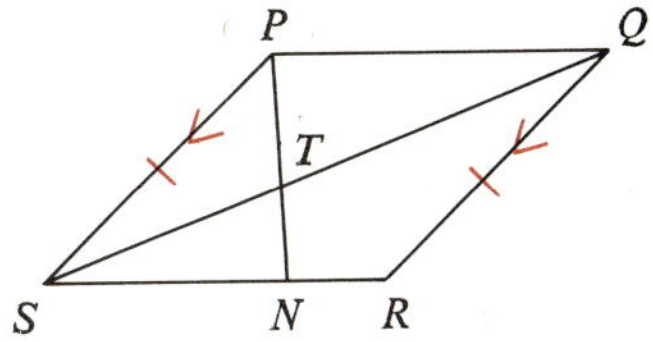

Exercise 5

1. $\triangle ADE$ and $\triangle ABC$
2. $\triangle ADC$ and $\triangle ACB$
3. $\triangle BDC$ and $\triangle ACB$
4. $\triangle DCA$ and $\triangle DBC$
5. $\triangle STN$ and $\triangle QTP$

EXERCISES

A In $\triangle ABC$, $m\angle B = 60$ and $m\angle C = 50$. In $\triangle DEF$, $m\angle D = 70$ and $m\angle F = 60$.

1. Is $\triangle ABC \sim \triangle DEF$? Why?
2. Is $\triangle BCA \sim \triangle FED$? Why?
3. Are two isosceles triangles similar if their vertex angles are congruent? Why or why not?

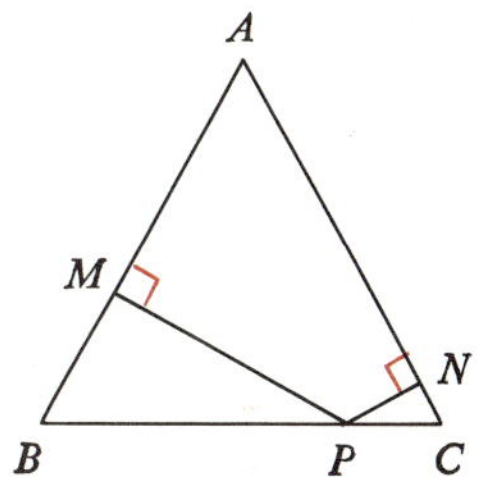

Exercise 4

4. GIVEN Isosceles $\triangle ABC$ with $\overline{AB} \cong \overline{AC}$, $\overline{PM} \perp \overline{AMB}$, $\overline{PN} \perp \overline{ANC}$, and $\overline{BPC}$
   PROVE $\triangle MPB \sim \triangle NPC$
   FIND The proportion resulting from this similarity.

5. GIVEN $\square ABCD$; $\overline{DCE}$; $\overline{AE}$ intersects $\overline{BC}$ in F.
   PROVE $\triangle FAB \sim \triangle FEC$; $\frac{FE}{AF} = \frac{FC}{FB}$; $\frac{FE}{AE} = \frac{FC}{BC}$; $FE \cdot BC = AE \cdot FC$

6. GIVEN $\triangle QPR$ and $\triangle STU$, where $\angle P$ and $\angle T$ are right angles; $\overline{PR} \parallel \overline{ST}$; $\overline{QSRU}$
   PROVE $\triangle PQR \sim \triangle TUS$; $\frac{PQ}{TU} = \frac{RQ}{SU}$; $PQ \cdot SU = RQ \cdot TU$

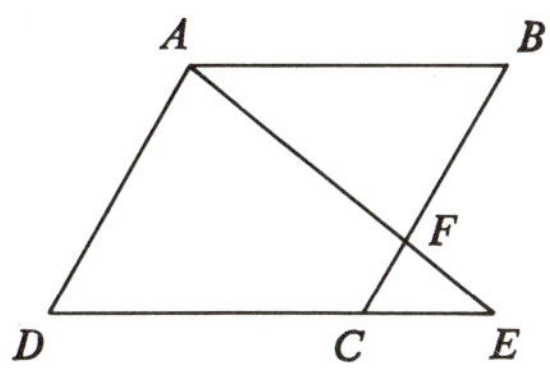

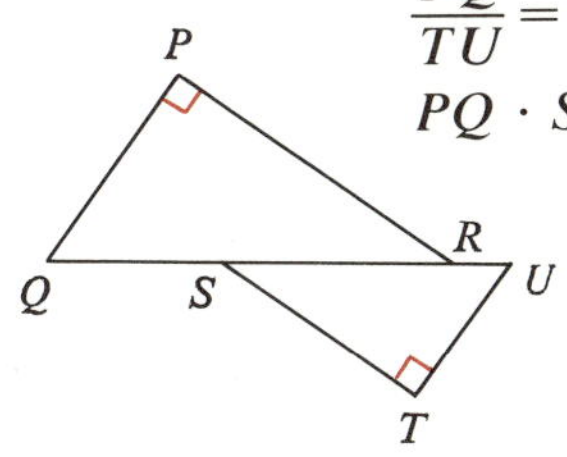

7. Prove Corollary 8-5.1a. 8. Prove Corollary 8-5.1b.
9. Prove Corollary 8-5.1c.
10. Prove that two isosceles triangles are similar if a base angle of one triangle is congruent to a base angle of the other triangle.
11. Use Theorem 8-5.1 to prove that all equilateral triangles are similar.
12. Prove that the diagonals of a trapezoid divide each other proportionally.

B

13. Prove that the ratio of the corresponding altitudes of two similar triangles equals the ratio of similitude.
14. Prove that the ratio of the corresponding angle bisectors of two similar triangles equals the ratio of similitude.
15. GIVEN $\triangle ABC$ with right $\angle C$; $\overline{ED} \perp \overline{ADB}$; $\overline{CEB}$

PROVE $\frac{EB}{AB} = \frac{BD}{BC}$

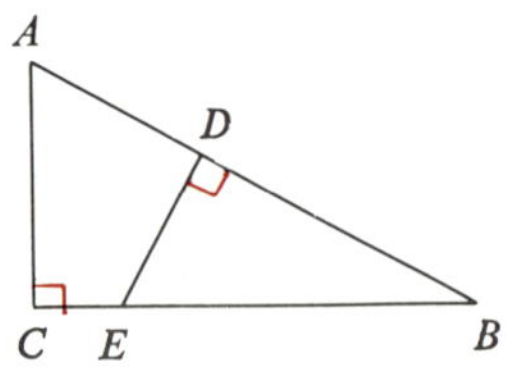

16. GIVEN $\square ABCD$, with $\overline{APB}$, $\overline{BRC}$, $\overline{CSD}$, and $\overline{DQA}$, so that $\overline{PQ} \parallel \overline{RS}$

PROVE $\frac{AQ}{RC} = \frac{QP}{SR}$.

(*Hint*: Consider $\overline{PS}$.)

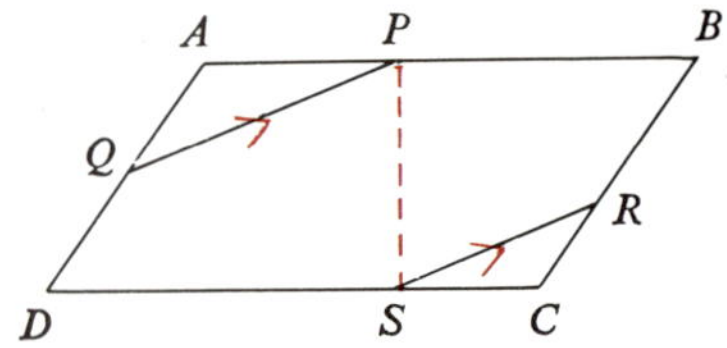

17. GIVEN Point P is in side $\overline{AB}$ of rectangle $ABCD$; $m\angle DPC = 90$

PROVE $\frac{AP}{BC} = \frac{AD}{PB}$; $AP \cdot PB =$ ____ $\cdot$ ____.

18. GIVEN Point P is in side $\overline{AB}$ of rectangle $ABCD$; $m\angle DPC = 90$

PROVE $\frac{PD}{AP} = \frac{DC}{PD}$; $PD^2 = AD \cdot DC$.

19. GIVEN $\triangle ABC$ with right $\angle ACB$ and altitude $\overline{CD}$

PROVE $AC \cdot CB = CD \cdot AB$

20. GIVEN $\triangle PQR$ with altitudes $\overline{QT}$ and $\overline{RS}$

PROVE $QT \cdot RP = RS \cdot PQ$

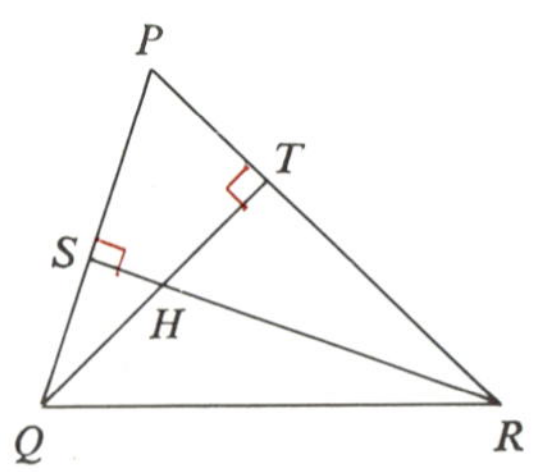

21. GIVEN $\square ABCD$ with $\overline{AFD}$; $\overrightarrow{BF}$ intersects $\overrightarrow{CD}$ in E.

PROVE $AF \cdot BE = BF \cdot BC$

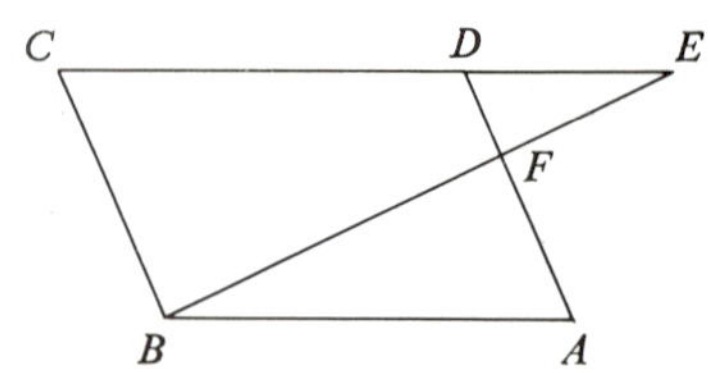

C

22. GIVEN $\overline{AFC} \perp \overline{BCE}$;
$\overline{DF} \perp \overline{AFC}$;
$\overline{DE} \perp \overline{BCE}$;
$\overline{BD} \perp \overline{AD}$

PROVE $AF \cdot BD = BE \cdot AD$;
$\frac{AC}{AB} = \frac{DE}{BD} \cdot \frac{BD}{AB} + \frac{BE}{BD} \cdot \frac{AD}{AB}$

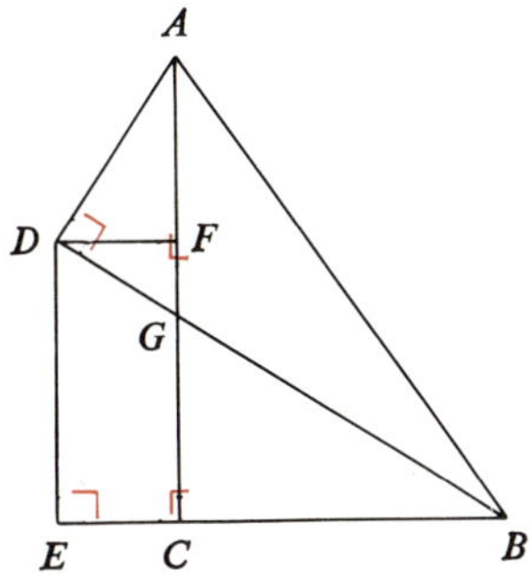

23. GIVEN $\square ABCD$;
$\overleftrightarrow{AE}$ intersects $\overline{BD}$, $\overline{BC}$, and $\overrightarrow{DC}$ in points G, F, and E, respectively.

PROVE AG is the mean proportional between GE and GF.

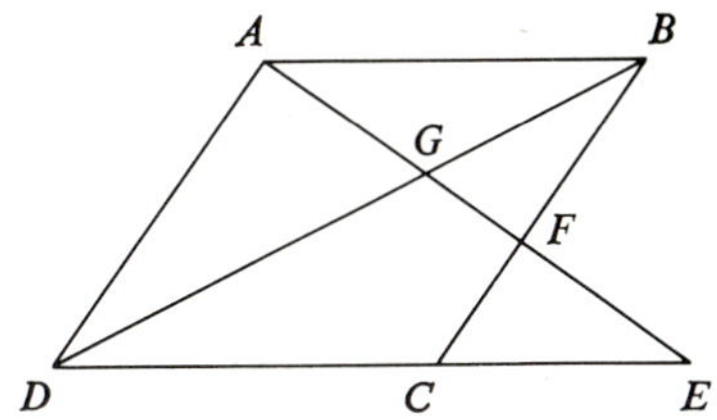

24. GIVEN $\triangle ABC$;
$\overline{AX} \parallel \overline{CZ} \parallel \overline{YB}$;
$\overline{AZB}$, $\overline{ACY}$, and $\overline{BCX}$

PROVE $\frac{1}{AX} + \frac{1}{BY} = \frac{1}{CZ}$

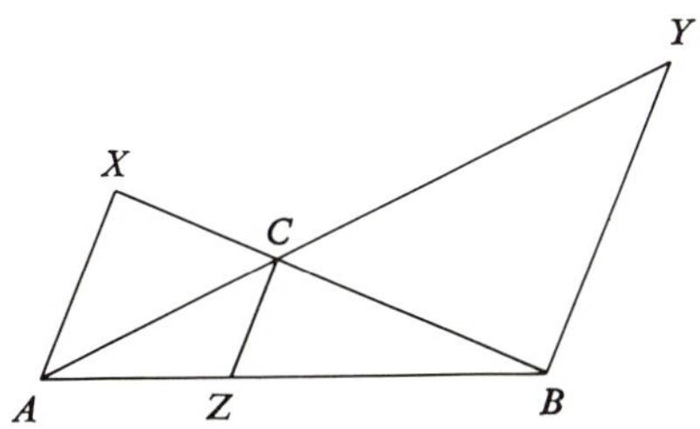

25. GIVEN $\overline{AD}$ is an angle bisector of $\triangle ABC$;
$\overline{BDC}$; E is in $\overrightarrow{DC}$ so that $\overline{BD} \cong \overline{DE}$;
$\overline{EF} \parallel \overline{AC}$; F is in $\overrightarrow{DA}$.

PROVE $\overline{FE} \cong \overline{AB}$
(*Hint*: Use Theorem 8-3.1 and C8-5.1c.)

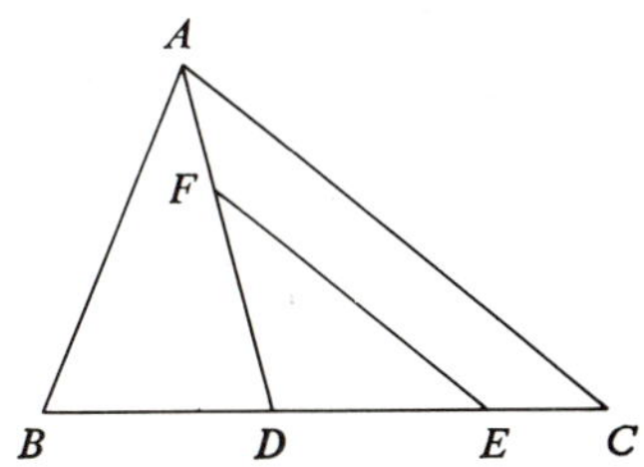

8-6 MORE TRIANGLE SIMILARITY THEOREMS

What methods of proving triangles similar did we derive in the last section? We shall now develop two more methods. Should these new methods involve more than one pair of congruent angles? What would you expect these new methods to be? You may find a clue if you recall the congruence theorems for general triangles that involve fewer than two pairs of congruent angles.

T

Theorem 8-6.1 SAS Similarity Theorem If two sides of one triangle are proportional to two sides of another triangle, and the angles included by those sides are congruent, then the triangles are similar.

GIVEN $\triangle ABC$ and $\triangle DEF$; $\dfrac{DE}{AB} = \dfrac{DF}{AC}$; $\angle A \cong \angle D$

PROVE $\triangle ABC \sim \triangle DEF$

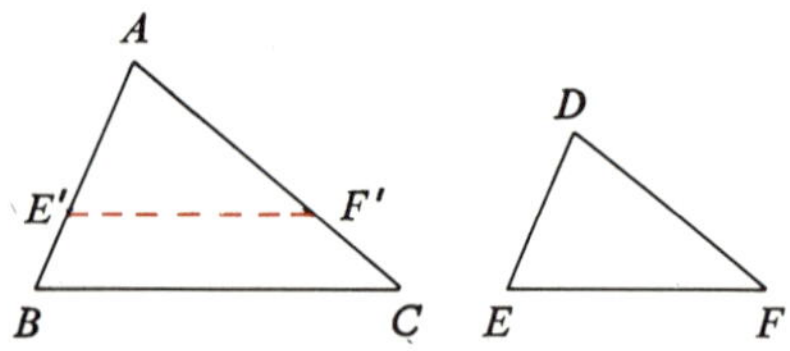

PROOF

STATEMENTS	REASONS
1. $\triangle ABC$ and $\triangle DEF$; $\dfrac{DE}{AB} = \dfrac{DF}{AC}$	1. Given.
2. Select points E' and F' in $\overline{AB}$ and $\overline{AC}$, respectively, so that $\overline{AE'} \cong \overline{DE}$ and $\overline{AF'} \cong \overline{DF}$.	2. Point Uniqueness Postulate. (2-2)
3. $\dfrac{AE'}{AB} = \dfrac{AF'}{AC}$	3. Substitution Postulate. (2-1)
4. $\overline{E'F'} \parallel \overline{BC}$	4. If a line divides two sides of a triangle proportionally, the line is parallel to the remaining side of the triangle. (C8-2.2a)
5. $\angle 1 \cong \angle B$ and $\angle 2 \cong \angle C$	5. If two parallel lines are cut by a transversal, then the corresponding angles are congruent. (C6-3.1a)
6. $\angle A \cong \angle D$	6. Given.
7. $\triangle AE'F' \cong \triangle DEF$	7. SAS Postulate. (3-1)
8. $\angle 1 \cong \angle E$ and $\angle 2 \cong \angle F$	8. Definition of congruent triangles. (3-3)
9. $\angle B \cong \angle E$ and $\angle C \cong \angle F$	9. Transitive property.
10. $\triangle ABC \sim \triangle DEF$	10. If the corresponding angles of two triangles are congruent then the two triangles are similar. (8-5.1)

Example 1 In the figure at right, $\overline{AB}$ intersects $\overline{CD}$ in E. $AE = 3$, $EB = 4\frac{1}{2}$, $CE = 4$, and $ED = 6$. Find BD if $AC = 3\frac{1}{2}$.

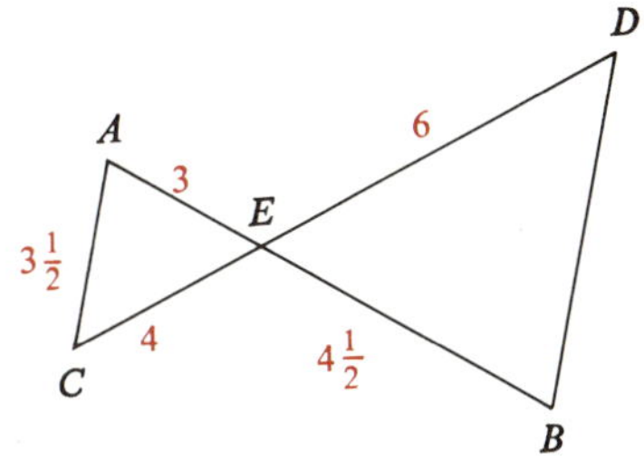

Solution In order to find BD, we must be able to write a proportion. This is possible if $\triangle AEC \sim \triangle BED$.
$\angle AEC \cong \angle BED$. Why?
$\frac{AE}{EB} = \frac{CE}{ED}$ because appropriate substitutions yield the true proportion $\frac{3}{4\frac{1}{2}} = \frac{4}{6}$.

Verify this. By Theorem 8-6.1, $\triangle AEC \sim \triangle BED$. Therefore, $\frac{AC}{BD} = \frac{CE}{DE}$.
Substitute the appropriate lengths.

$$\frac{3\frac{1}{2}}{BD} = \frac{4}{6} \text{ or } \frac{3\frac{1}{2}}{BD} = \frac{2}{3}$$

$$2 \cdot BD = 3 \cdot 3\tfrac{1}{2}$$

$$BD = 5\tfrac{1}{4}$$

Example 2

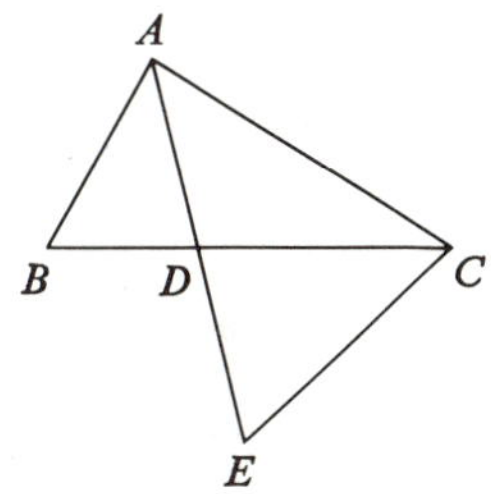

GIVEN $\overline{AD}$ is an angle bisector of $\triangle ABC$; point E is in $\overrightarrow{AD}$ so that $AB \cdot AC = AD \cdot AE$.

PROVE $\angle B \cong \angle AEC$

Solution

PROOF

STATEMENTS	REASONS
1. $\overline{AD}$ bisects $\angle BAC$.	1. Given.
2. $\angle BAD \cong \angle DAC$	2. Definition of an angle bisector. (1-29)
3. $AB \cdot AC = AD \cdot AE$	3. Given.
4. $\frac{AB}{AE} = \frac{AD}{AC}$	4. $\frac{a}{b} = \frac{c}{d}$ if and only if $ad = bc$. (8-1.1)
5. $\triangle ABD \sim \triangle AEC$	5. Theorem 8-6.1.
6. $\angle B \cong \angle AEC$	6. Definition of similar triangles. (8-6)

We shall now consider another theorem that provides a method of proving triangles similar.

T

Theorem 8-6.2 SSS Similarity Theorem If the corresponding sides of two triangles are proportional, then the two triangles are similar.

GIVEN $\triangle ABC$ and $\triangle DEF$; $\frac{DE}{AB} = \frac{DF}{AC} = \frac{EF}{BC}$

PROVE $\triangle ABC \sim \triangle DEF$

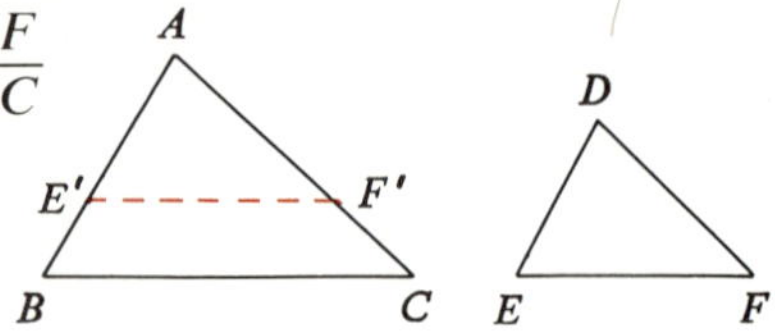

PROOF

STATEMENTS	REASONS
1. $\triangle ABC$ and $\triangle DEF$; $\frac{DE}{AB} = \frac{DF}{AC} = \frac{EF}{BC}$	1. Given.
2. Select points E' and F' so that $\overline{AE'B}$, $\overline{AF'C}$, $\overline{AE'} \cong \overline{DE}$, and $\overline{AF'} \cong \overline{DF}$.	2. Point Uniqueness Postulate. (2-2)
3. $\frac{AE'}{AB} = \frac{AF'}{AC}$	3. Substitution Postulate. (2-1)
4. $\overline{E'F'} \parallel \overline{BC}$	4. If a line divides two sides of a triangle proportionally, then the line is parallel to the remaining side of the triangle. (C8-2.2a)
5. $\triangle AE'F' \sim \triangle ABC$	5. If a line parallel to one side of a triangle intersects the other two sides, then it cuts off a triangle similar to the original triangle. (C8-5.1c)
6. $\frac{E'F'}{BC} = \frac{AE'}{AB}$	6. Definition of similar triangles. (8-6)
7. $\frac{E'F'}{BC} = \frac{DE}{AB}$	7. Substitution Postulate. (2-1)
8. $\frac{EF}{BC} = \frac{DE}{AB}$	8. Statement 1.
9. $E'F' = EF$	9. For any positive real numbers a, b, c, d, and e, if $\frac{a}{b} = \frac{c}{d}$ and $\frac{a}{b} = \frac{c}{e}$, then $d = e$. (8-1.5)
10. $\triangle AE'F' \cong \triangle DEF$	10. SSS Postulate. (3-3)
11. $\triangle ABC \sim \triangle DEF$	11. If a given triangle is similar to a triangle that is congruent to a third triangle, then the given triangle is also similar to the third triangle. (C8-4.1a)

Example 3 The lengths of the sides of one triangle are 6, 8, and 12, and the lengths of the sides of a second triangle are $1\frac{1}{2}$, 2, and 3. Are the two triangles similar?

Solution Since $\frac{6}{1\frac{1}{2}} = \frac{8}{2} = \frac{12}{3}$, the triangles are similar, according to Theorem 8-6.2.

Example 4

GIVEN Quadrilateral $ABCD$; $\frac{AB}{BD} = \frac{BD}{DC} = \frac{AD}{BC}$

PROVE Quadrilateral $ABCD$ is a trapezoid.

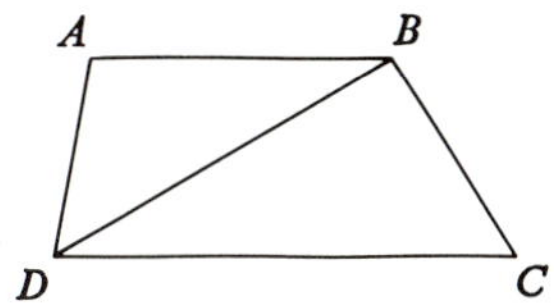

Solution

PROOF

STATEMENTS	REASONS
1. Quadrilateral $ABCD$; $\frac{AB}{BD} = \frac{BD}{DC} = \frac{AD}{BC}$	1. Given.
2. $\triangle ABD \sim \triangle BDC$	2. Theorem 8-6.2.
3. $\angle ABD \cong \angle BDC$	3. Definition of similar triangles. (8-6)
4. $\overline{AB} \parallel \overline{DC}$	4. If two lines are cut by a transversal so that the alternate interior angles are congruent, then the lines are parallel. (6-2.1)
5. Quadrilateral $ABCD$ is a trapezoid.	5. Definition of a trapezoid. (7-10)

Class Exercises

Complete the exercises below, referring to the measures of angles and line segments indicated in the figures.

1. $\frac{AE}{BE} =$ ——. Why?
2. $\angle AEB \cong \angle$ ____
3. $\triangle ABC \sim \triangle$ ____. Why?
4. $\angle B \cong$ ____
5. $\frac{CD}{AC} =$ ____
6. $\frac{AB}{} = \frac{}{CD} = \frac{AC}{}$

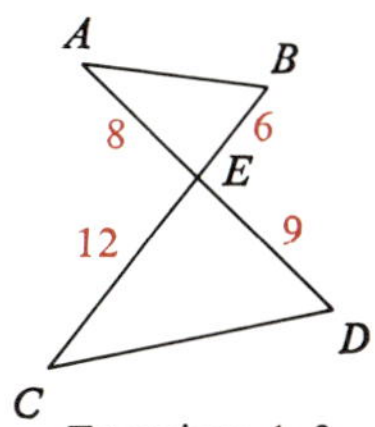

Exercises 1–3

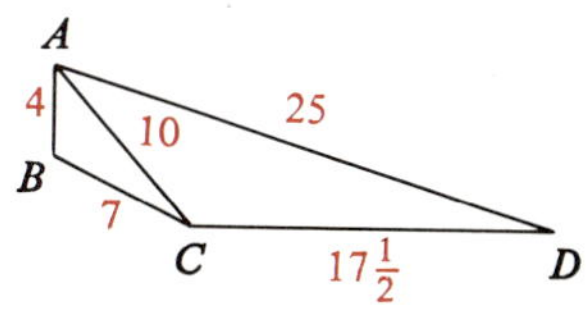

Exercises 4–8

7. AC is the mean proportional between ____ and ____.

8. $\triangle ABC \sim \triangle$____

9. $\angle BAC \cong \angle$____

10. $\dfrac{AE}{\quad} = \dfrac{AD}{\quad}$

11. $\triangle AED \sim \triangle$____

12. $\angle AED = \angle$____

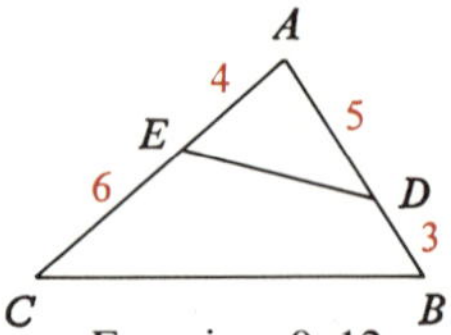

Exercises 9–12

Summary. Methods of Proving Triangles Similar

1. If at least two angles of one triangle are congruent to the corresponding angles of the other triangle (AA).
2. If two sides of one triangle are proportional to two sides of the other triangle, and the angles included by those sides are congruent (SAS).
3. If the corresponding sides of the two triangles are proportional (SSS).

EXERCISES

A

1. In the figure at right, $\overline{MTR}$, $\overline{NPR}$, $MT = 1$, $TR = 5$, $NP = 7$, and $PR = 3$. Name a pair of similar triangles and tell why they are similar.

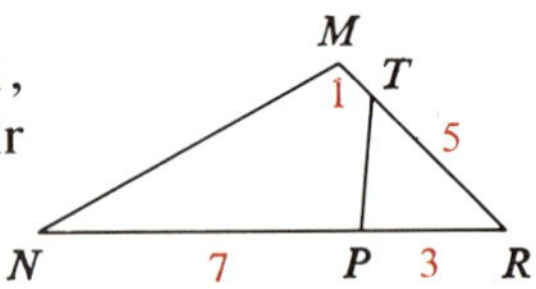

Indicate whether or not each of the following pairs of triangles is similar. State the theorem which proves each similarity.

2.

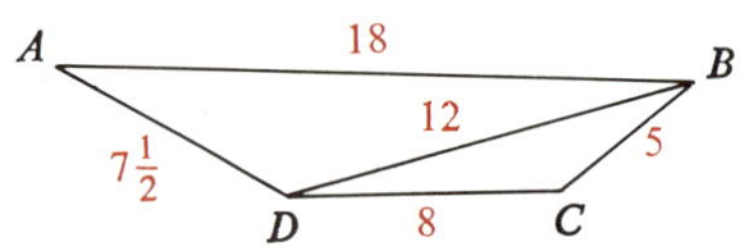

3.

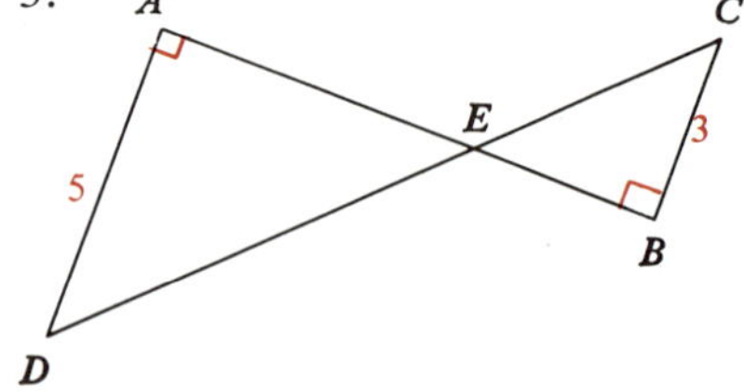

4.

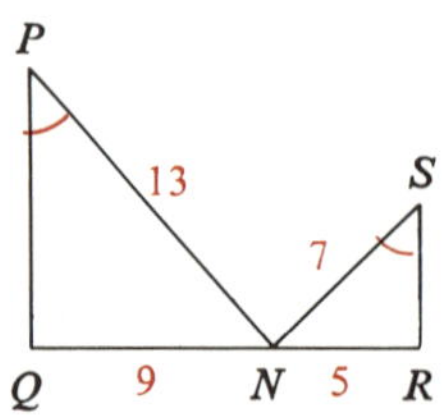

5.

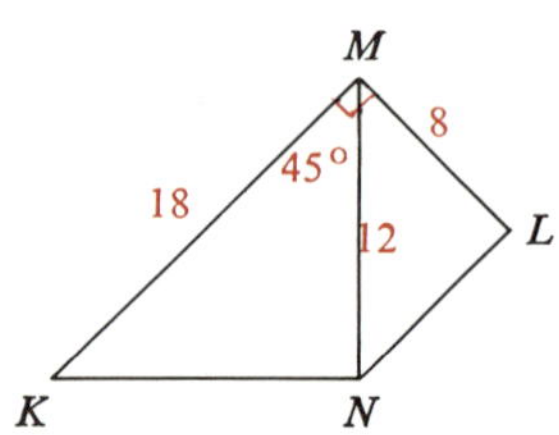

6.

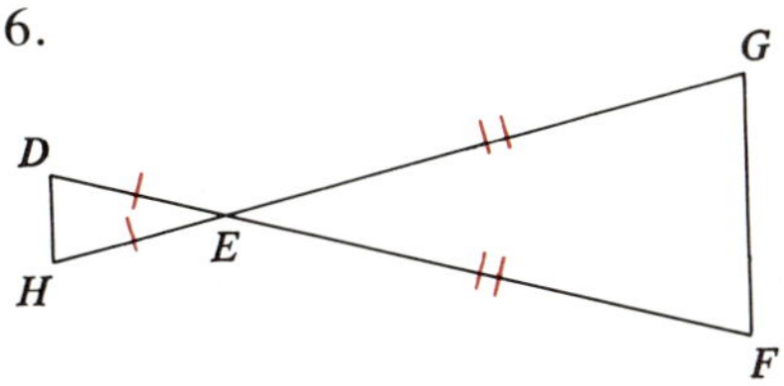

7.

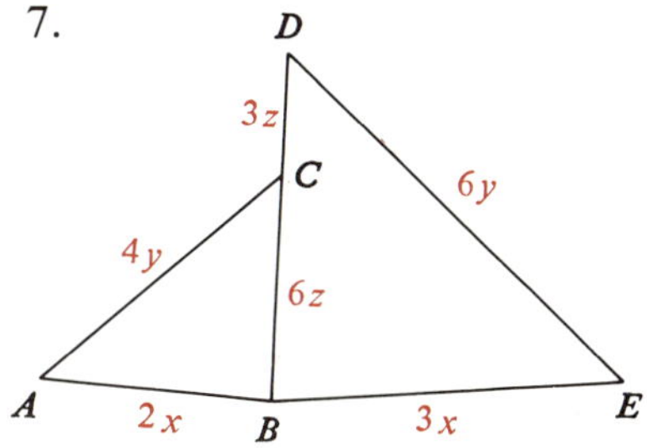

8. GIVEN $\triangle ABC$, where $\overline{AEC}$ and $\overline{ADB}$; $AE = \frac{1}{3}AB$ and $AD = \frac{1}{3}AC$

PROVE $m\angle BDE + m\angle C = 180$

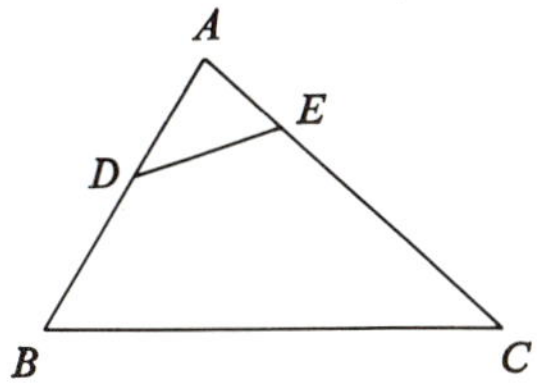

9. GIVEN Quadrilaterals $ABCD$ and $PQRS$; $\triangle ABC \sim \triangle PQR$; $\triangle ADC \sim \triangle PSR$

PROVE $\triangle BDC \sim \triangle QSR$

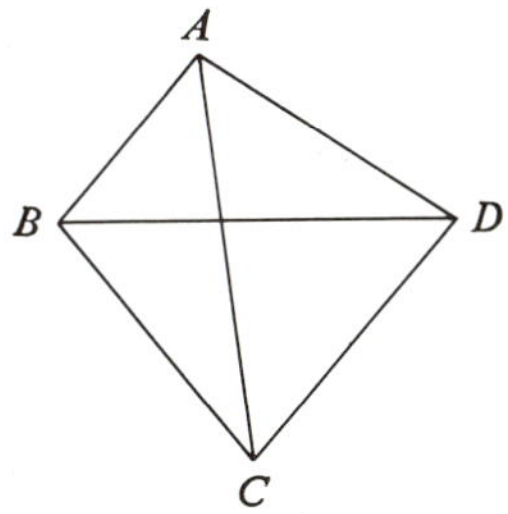

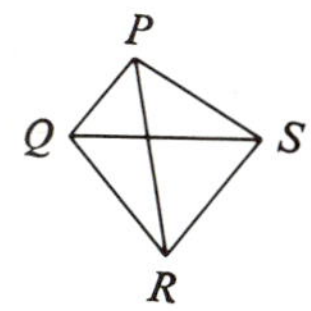

B

10. Prove that if the legs of one right triangle are proportional to the legs of another right triangle, the two triangles are similar.
11. Prove that the ratio of any pair of corresponding medians of two similar triangles equals the ratio of similitude of those two triangles.
12. GIVEN $\overline{AM}$ is a median of $\triangle ABC$; points D and E are in $\overleftrightarrow{AB}$ and $\overleftrightarrow{AC}$, respectively, so that $\overline{DE} \parallel \overline{BC}$; $\overleftrightarrow{AM}$ intersects $\overline{DE}$ at N.

    PROVE $\overline{AN}$ is a median of $\triangle ADE$.
13. Prove that if two sides of one triangle and the median on one of these sides are proportional to the corresponding parts of another triangle, the triangles are similar.
14. Prove that two right triangles are similar if the hypotenuse and leg of one triangle are proportional to the hypotenuse and leg of the other triangle. (*Hint*: Consider the median to the hypotenuse of each triangle.)
15. GIVEN $\overline{BD}$ and $\overline{AE}$ are altitudes of $\triangle ABC$.

    PROVE $\triangle ABC \sim \triangle EDC$
16. GIVEN Point D is in side $\overline{BC}$ of $\triangle ABC$ so that $\frac{DC}{BD} = \frac{AC}{AB}$; point E is in $\overrightarrow{AD}$ so that $\frac{AE}{AB} = \frac{AC}{AD}$.

    PROVE $\triangle BEC$ is isosceles

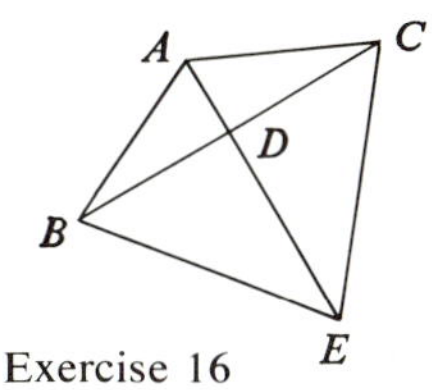

Exercise 16

17. GIVEN Square $ABCD$, where P is in $\overline{BC}$ so that $PC = \frac{1}{4}BC$; Q is the midpoint of $\overline{DC}$.
PROVE $AQ = 2PQ$; $\overline{PQ}$ bisects $\angle APC$.

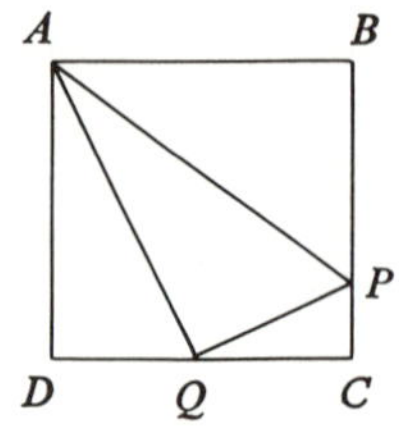

Exercise 17

18. GIVEN $\triangle ABC$ is isosceles, with $\overline{AB} \cong \overline{AC}$; point P is in the interior of $\triangle ABC$ so that PQ is the mean proportional between PR and PS; $\overline{PQ} \perp \overline{BC}$; $\overline{PR} \perp \overline{AC}$; and $\overline{PS} \perp \overline{AB}$.
PROVE $\triangle SPQ \sim \triangle QPR$
(*Hint*: Use Theorem 6-5.1.)

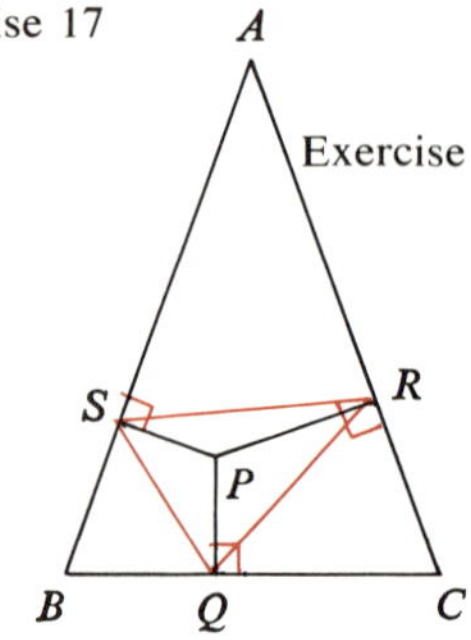

Exercise 18

19. GIVEN $\triangle ABC$ with $\overline{AB} \cong \overline{AC}$; $\angle A$ a right angle; quadrilateral $BCDE$ is a rectangle where $BC = 2DC$; M is the midpoint of $\overline{AC}$.
PROVE $m\angle BMD = 90$

20. GIVEN $\triangle APB$; points K and L are in $\overline{AB}$ so that AL is the mean proportional between AK and AB; $\overline{AP} \cong \overline{AL}$
PROVE $\overline{PL}$ bisects $\angle KPB$.

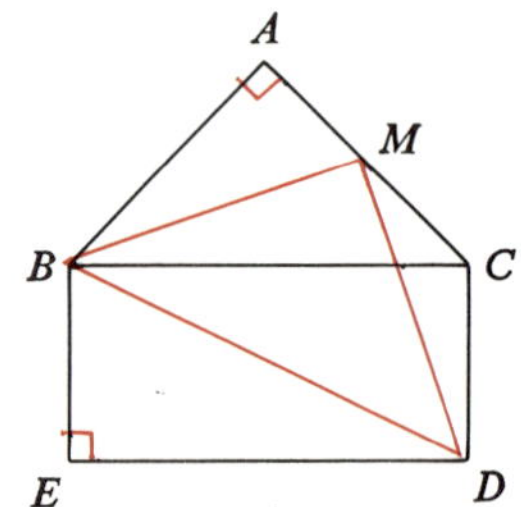

Exercise 19

8-7 SIMILARITY AND PROPORTION IN RIGHT TRIANGLES

Similarity in right triangles involves many interesting relationships. The following theorem is basic to these relationships:

T

Theorem 8-7.1 The altitude to the hypotenuse of a right triangle separates the triangle into two triangles that are similar to each other and to the original triangle.

GIVEN $\overline{AD}$ is an altitude of right $\triangle ABC$ with right $\angle ACB$.

PROVE $\triangle ADC \sim \triangle ACB$;
$\triangle CDB \sim \triangle ACB$;
$\triangle ADC \sim \triangle CDB$

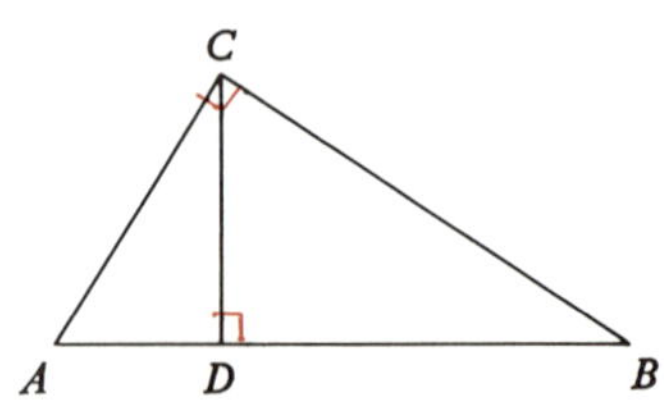

PROOF

STATEMENTS	REASONS
1. $\overline{AD}$ is an altitude of right $\triangle ABC$ with right $\angle ACB$.	1. Given
2. $\overline{CD} \perp \overline{AB}$	2. Definition of an altitude of a triangle. (3-10)
3. $\angle ADC$ is a right angle.	3. Definition of perpendicular lines. (1-25)
4. $\angle ABC \cong \angle ADC$	4. All right angles are congruent. (3-1.1)
5. $\angle A \cong \angle A$	5. Every angle is congruent to itself. (3-1.7)
6. $\triangle ADC \sim \triangle ACB$	6. If two pairs of corresponding angles of two triangles are congruent, then the triangles are similar. (C8-5.1a)
7. $\angle CDB$ is a right angle.	7. Same reason as for step 3.
8. $\angle ACB \cong \angle CDB$	8. Same reason as for step 4.
9. $\angle B \cong \angle B$	9. Same reason as for step 5.
10. $\triangle CDB \sim \triangle ACB$	10. Same reason as for step 6.
11. $\triangle ADC \sim \triangle CDB$	11. If two triangles are similar to the same triangle, or to similar triangles, then the triangles are similar to each other. (8-4.1)

Class Exercises

Complete each of the following with the measure of the appropriate line segment. Each exercise refers to a statement in the proof of Theorem 8-7.1.

1. Since $\triangle ADC \sim \triangle ACB$, $\dfrac{AD}{AC} = \dfrac{}{AB}$. Why?
2. ____ is the mean proportional between AD and AB.
3. Since $\triangle CDB \sim \triangle ACB$, $\dfrac{BD}{} = \dfrac{}{AB}$. Why?
4. BC is the mean proportional between ____ and ____.
5. Since $\triangle ADC \sim \triangle CDB$, $\dfrac{AD}{} = \dfrac{CD}{}$. Why?
6. CD is the mean proportional between ____ and ____.

By restating Exercises 2, 4, and 6, we have the following corollaries:

C **Corollary 8-7.1a** The altitude to the hypotenuse of a right triangle divides the hypotenuse so that either leg is the mean proportional between the hypotenuse and the segment of the hypotenuse adjacent to that leg.

Corollary 8-7.1b The altitude to the hypotenuse of a right triangle is the mean proportional between the segments of the hypotenuse.

By completing the Class Exercises above, you have actually proved Corollaries 8-7.1a and 8-7.1b. The formal proofs of these corollaries are left as exercises.

Example 1 $\overline{AD}$ is the altitude on hypotenuse $\overline{BC}$ of right $\triangle ABC$. If $BD = 2$ and $DC = 8$, find AD.

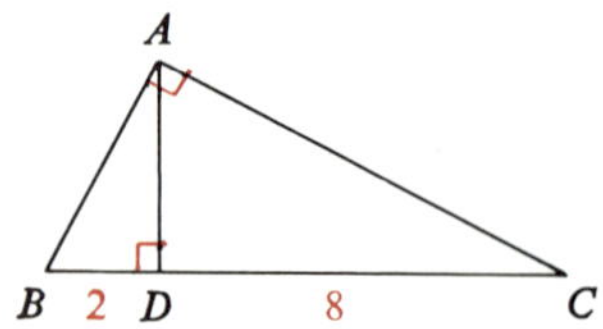

Solution Corollary 8-8.1b indicates that AD is the mean proportional between BD and DC. Therefore $\frac{BD}{AD} = \frac{AD}{DC}$, or $AD^2 = BD \cdot DC$ (Theorem 8-1.1). By substituting the appropriate lengths, we get $AD^2 = 2 \cdot 8 = 16$. Therefore $AD = \sqrt{16} = 4$.

Example 2 Using the information and the figure for Example 1, find AB.

Solution From Corollary 8-7.1a, we get the proportion $\frac{BD}{AB} = \frac{AB}{BC}$. By substituting the appropriate lengths, we get $\frac{2}{AB} = \frac{AB}{10}$, or $AB^2 = 2 \times 10 = 20$ (Theorem 8-1.1). Therefore $AB = \sqrt{20}$.

To simplify this answer, recall a theorem from algebra that states that if $a \geq 0$ and $b \geq 0$, then $\sqrt{a} \cdot \sqrt{b} = \sqrt{ab}$. To simplify $\sqrt{20}$, we must find the largest perfect square that is a factor of 20, namely 4. Therefore $\sqrt{20} = \sqrt{4(5)}$, which by the above theorem equals $\sqrt{4} \cdot \sqrt{5}$, or $2\sqrt{5}$. Thus $\sqrt{20} = 2\sqrt{5}$, a more workable form for the answer.

Example 3 In the figure for Example 1, find AC if $BD = 2$ and $DC = 8$.

Solution Corollary 8-7.1a gives us

$$\frac{DC}{AC} = \frac{AC}{BC}$$

$$AC^2 = DC \cdot BC. \text{ Why?}$$

Substituting, we get

$$AC^2 = 8 \cdot 10 = 80$$

then

$$AC = \sqrt{80}.$$

The largest perfect square factor of 80 is 16. Therefore, $\sqrt{80} = \sqrt{16(5)} = \sqrt{16} \cdot \sqrt{5} = 4\sqrt{5}$.

EXERCISES

A

1. Name the similar triangles in the figure at right. Perpendicular and parallel lines are indicated.

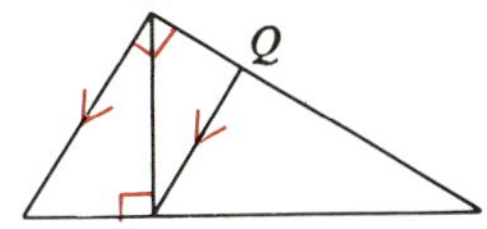

Using Examples 2 and 3 as models, simplify each of the following radicals:

2. $\sqrt{12}$ 3. $\sqrt{32}$ 4. $\sqrt{96}$ 5. $\sqrt{8y^2}$ 6. $\sqrt{108x^3y^2}$

Find the value of x in each of the following:

7. $\frac{3}{x} = \frac{x}{6}$ 8. $\frac{x}{\sqrt{8}} = \frac{\sqrt{18}}{x}$ 9. $\frac{3}{x} = \frac{x}{27}$ 10. $\frac{12}{x} = \frac{x}{16}$ 11. $\frac{4n}{x} = \frac{x}{27n}$

Find the mean proportional between each of the following pairs of numbers:

12. 6, 8 13. 3, 7 14. 12, 27 15. $\sqrt{3}, \sqrt{7}$

16. Prove Corollary 8-7.1a. 17. Prove Corollary 8-7.1b.

B

18. Find the lengths of the three altitudes of a right triangle, the measures of whose sides are 5, 12, and 13. How long is the hypotenuse? Why?

19. GIVEN $\overline{CD}$ is an altitude of right $\triangle ABC$ with $m\angle ACB = 90$.
PROVE $AC^2 - DC^2 = (AD \cdot AB) - (AD \cdot DB)$; $AC^2 - DC^2 = AD^2$

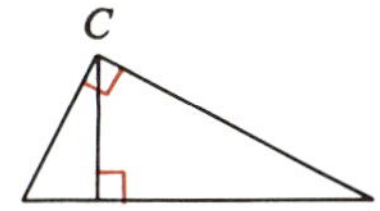

Exercises 19–25

Copy and complete the following chart. The segment measures refer to the figure at right above. Simplify all radicals.

	AC	BC	AB	AD	DB	CD
20.	2					$\sqrt{3}$
21.		3				$2\sqrt{2}$
22.	$\sqrt{3}$		3			
23.			5			2
24.		$\sqrt{2}$	2			
25.				$x - 2$		$x + 3$

C

26. An altitude and an angle bisector are drawn to the hypotenuse of a right triangle whose sides are 8, 15, and 17 inches long. Find the length of the segment joining the points where the altitude and angle bisector intersect the hypotenuse. Express your answer to the nearest hundredth.

27. GIVEN $\overline{CD}$ is an altitude and $\overline{CE}$ is an angle bisector of $\triangle ABC$; $\angle ACB$ is a right angle.

PROVE $\frac{AD}{DB} = \frac{AE^2}{EB^2}$

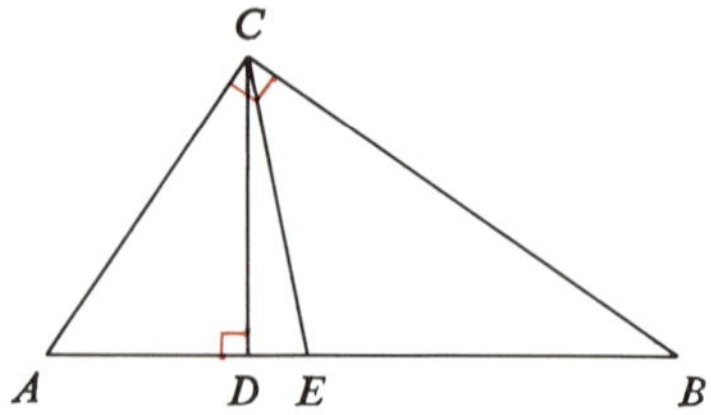

8-8 THE PYTHAGOREAN THEOREM

We are now ready to prove one of the most famous and useful theorems of geometry. This theorem is said to have been first proved by the Greek mathematician Pythagoras in about 525 B.C., although it was known to the Babylonians over 1,000 years earlier. Unfortunately, we have no record of Pythagoras's proof.

Since the days of Pythagoras, many people have offered different proofs of the Pythagorean Theorem. The proof of Pres. James A. Garfield, based on areas, is presented in Chapter 12. Today, there are more than 360 different proofs of this theorem. One such proof is presented below. Some of the exercises in later chapters will provide you with opportunities to try other proofs.

By completing the following Class Exercises, you will prove the Pythagorean Theorem in a manner similar to that believed to have been used by Pythagoras. This proof is based on Corollary 8-7.1a. Try to state the Pythagorean Theorem after you have completed the exercises.

Class Exercises

In the figure at right, $\overline{CD}$ is an altitude of right $\triangle ABC$ with right angle at C. The lengths of the segments are marked. Referring to the figure, complete each of the following:

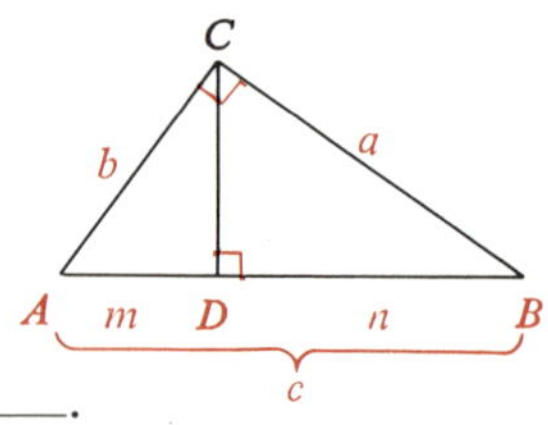

1. AC is the mean proportional between ____ and ____.
2. Therefore $\frac{c}{\ } = \frac{b}{\ }$, or $b^2 =$ ____. Why?
3. ____ is the mean proportional between AB and BD.
4. Therefore $\frac{\ }{a} = \frac{\ }{n}$; or $a^2 =$ ____. Why?
5. Adding the results of Exercises 2 and 4, we get $a^2 + b^2 =$ ____ $+$ ____ $=$ ____$(m + n)$.
6. But $m + n =$ ____.
7. Therefore $a^2 +$ ____ $=$ ____.

Now try to state the Pythagorean Theorem.

T

Theorem 8-8.1 The Pythagorean Theorem The sum of the squares of the lengths of the legs of a right triangle equals the square of the length of the hypotenuse.

Example 1 Can a circular table top with a diameter of 16 feet fit through a rectangular door whose dimensions are 8 feet by 15 feet?

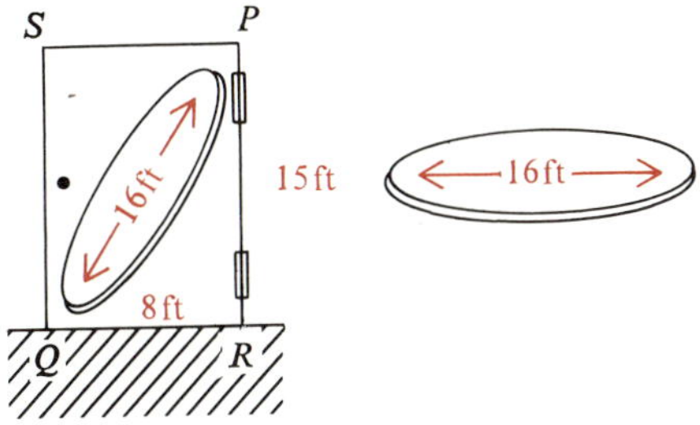

Solution Since the height of the door is less than the diameter of the table top, we cannot fit the table through the door while it is standing vertically. To determine whether the table top will fit if we tilt it, we need to know the length of the longest line segment joining two points on the door. Therefore, we must find the length of a diagonal of the door ($\overline{PQ}$). Since $\angle PRQ$ is a right angle, we can use the Pythagorean Theorem, whereby $PQ^2 = RQ^2 + RP^2$. Substituting the appropriate lengths, we get

$$PQ^2 = 8^2 + 15^2 = 64 + 225 = 289$$
$$PQ = \sqrt{289} = 17.$$

Therefore, the table top will fit through the door.

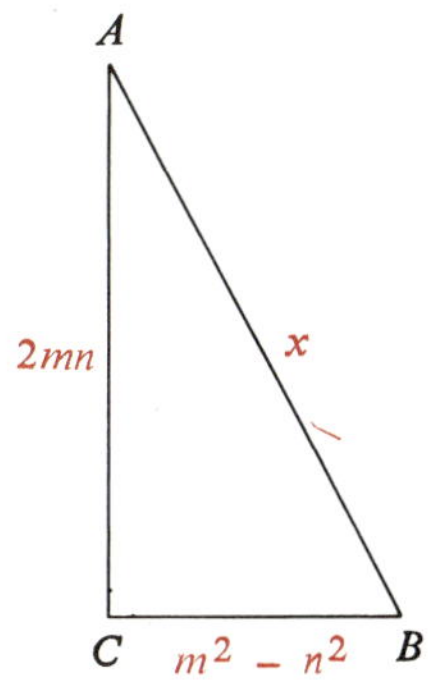

Example 2 Find x in terms of m and n in right $\triangle ABC$, where $m > n$.

Solution By the Pythagorean Theorem,

$$AB^2 = CB^2 + CA^2$$

or

$$x^2 = (m^2 - n^2)^2 + (2mn)^2$$
$$x^2 = m^4 - 2m^2n^2 + n^4 + 4m^2n^2$$
$$x^2 = m^4 + 2m^2n^2 + n^4$$
$$x^2 = (m^2 + n^2)^2$$
$$x = \sqrt{(m^2 + n^2)^2} = m^2 + n^2$$

Theorem 8-8.2 provides a method of proving that a given triangle is a right triangle. What is the relationship of Theorem 8-8.2 to the Pythagorean Theorem?

T

Theorem 8-8.2 If the sum of the squares of the lengths of two sides of a triangle equals the square of the length of the third side, then the angle opposite this third side is a right angle.

GIVEN $\triangle ABC$ with sides of length a, b, and c; $a^2 + b^2 = c^2$

PROVE $\angle C$ is a right angle.

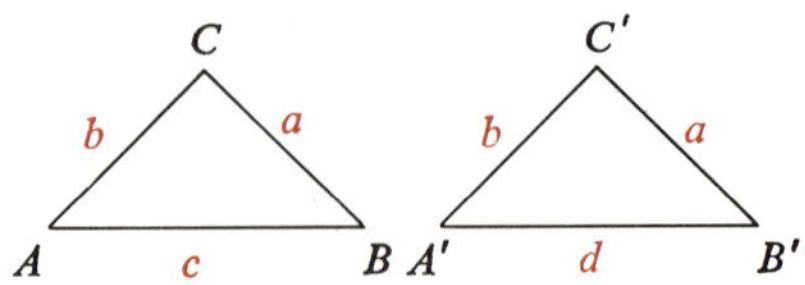

PROOF

STATEMENTS	REASONS
1. $\triangle ABC$ with sides a, b, and c; $a^2 + b^2 = c^2$	1. Given.
2. Consider $\triangle A'B'C'$ where $A'C' = AC = b$; $B'C' = BC = a$; and $\angle C'$ is a right angle.	2. Definition of a triangle (1-31)
3. In $\triangle A'B'C'$, $a^2 + b^2 = d^2$	3. Theorem 8-8.1
4. $c^2 = d^2$, or $c = d$	4. Transitive property in steps 1 and 3
5. $\triangle ABC \cong \triangle A'B'C'$	5. SSS Postulate (3-3)
6. $\angle C \cong \angle C'$	6. Definition of congruent triangles (3-3)
7. $\angle C$ is a right angle.	7. Substitution Postulate (2-1)

Example 3 The lengths of some of the segments in the figure at right are indicated. Find DC.

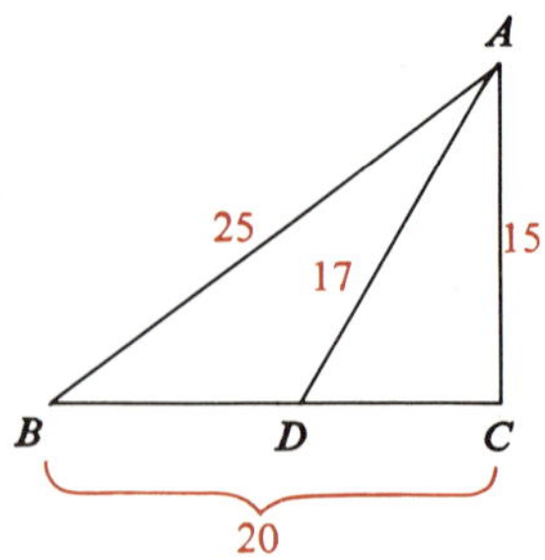

Solution Use the Pythagorean Theorem.

$$DC^2 + AC^2 = AD^2$$

$$DC^2 + 15^2 = 17^2$$

$$DC^2 = 17^2 - 15^2 = 289 - 225 = 64$$

$$DC = 8$$

Example 4 Show that if d is the length of a diagonal of a rectangular solid in which every pair of intersecting edges is perpendicular, then if we represent the three possible lengths of the edges by a, b, and c, $a^2 + b^2 + c^2 = d^2$.

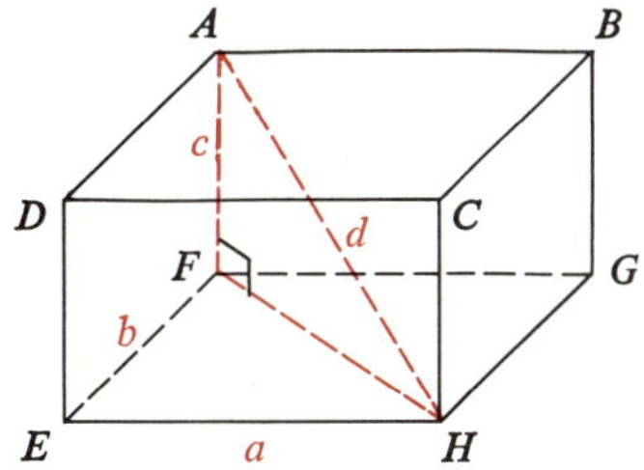

Solution In the rectangular solid, let $EH = a$, $EF = b$, $AF = c$, and $AH = d$. Apply the Pythagorean Theorem to right $\triangle FEH$ to get $EH^2 + EF^2 = FH^2$, or $a^2 + b^2 = FH^2$. Now apply the Pythagorean Theorem to right $\triangle AFH$ to get $FH^2 + AF^2 = AH^2$. Appropriate substitution gives us $a^2 + b^2 + c^2 = d^2$.

EXERCISES

A Find x in each of the following figures:

1.
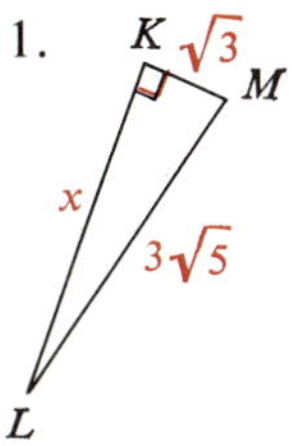

2.
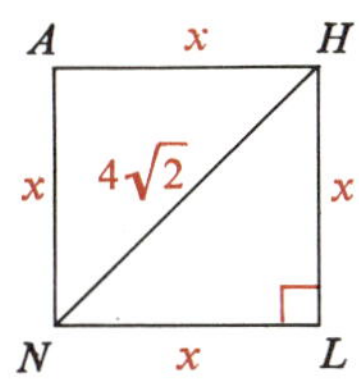

3.
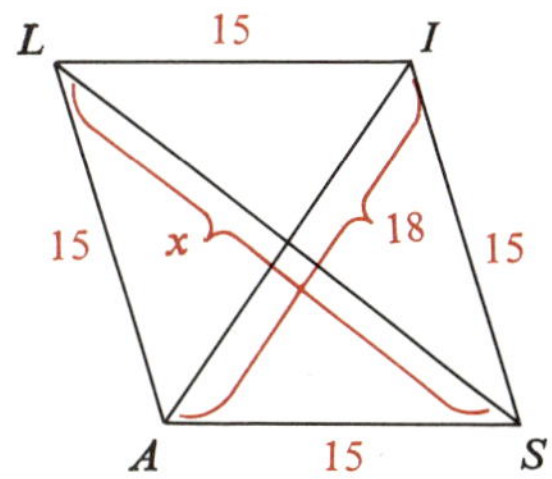

4.
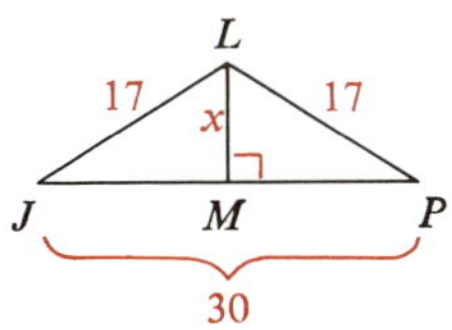

5.
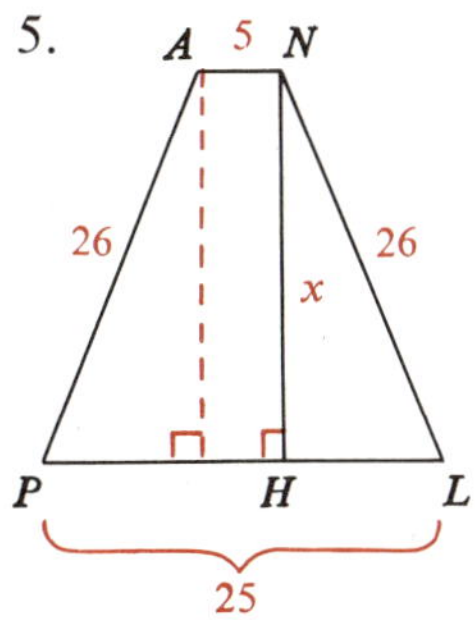

6.
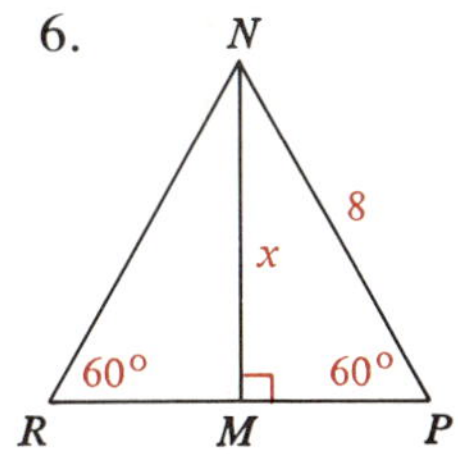

Which of the following sets may represent the lengths of the sides of a right triangle?

7. $\{\sqrt{5}, \sqrt{12}, \sqrt{13}\}$
8. $\{\sqrt{7}, 2\sqrt{5}, \sqrt{69}\}$
9. $\{8, 10, 12\}$
10. $\{1, 1, \sqrt{2}\}$
11. $\{4, \sqrt{65}, 9\}$
12. $\{6\sqrt{2}, 8\sqrt{2}, 10\sqrt{2}\}$

13. The length of altitude $\overline{MH}$ of $\triangle MNR$ is 8. If $MN = 17$ and $MR = 10$, find NR.

14. GIVEN $\overline{PS}$ is an altitude of $\triangle PQR$.
PROVE $PQ^2 - RP^2 = QS^2 - SR^2$

15. GIVEN $\overline{AM}$ is a median of right $\triangle ABC$ with right $\angle B$; $\overline{MN} \perp \overline{AC}$ at N
PROVE $AN^2 + BM^2 = NC^2 + AM^2$

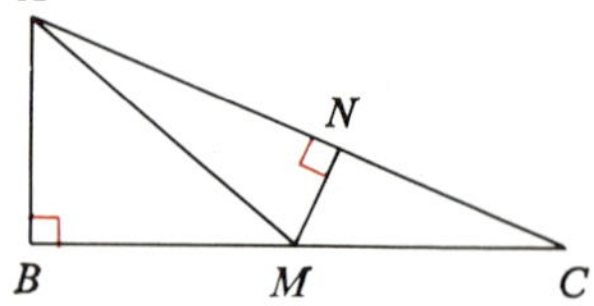

Exercise 15

B

16. Find the length of the median to the longest side of a triangle whose sides have lengths 7, 7, and 8.

17. In the figure at right, $\overline{PC} \perp \overline{BD}$ at C and $\overline{PB} \perp \overline{AB}$. $AP = 17, AB = 8, BC = 9$, and $CD = 3\frac{1}{2}$. Find PD.

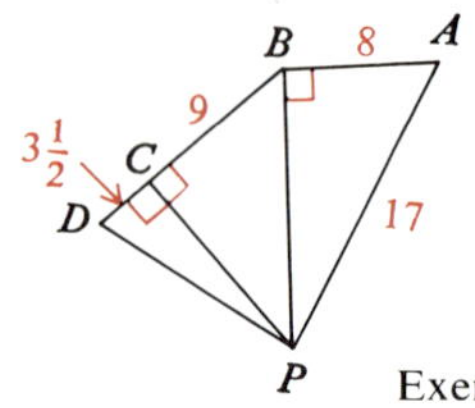

Exercise 17

18. Find the diagonal of a rectangular solid whose edges have lengths 3, 4, and 5.

19. If two edges and a diagonal of a rectangular solid have lengths 4, 12, and 13, respectively, find the length of the shortest edge.

20. Prove that the length of the diagonal of a cube with edge length e is $e\sqrt{3}$.

21. A 25-foot ladder leaning against the side of a house touches the house at a point 24 feet above the ground. If the top of the ladder slips down 4 feet, how far would the bottom of the ladder slide out along the ground?

22. Prove that the sum of the squares of the lengths of the diagonals of a rhombus equals four times the square of the length of a side.

23. GIVEN Right $\triangle ABC$ with right $\angle C$; $\overline{CMNB}$; $CM = MN = NB$
PROVE $3AB^2 + 5AM^2 = 8AN^2$

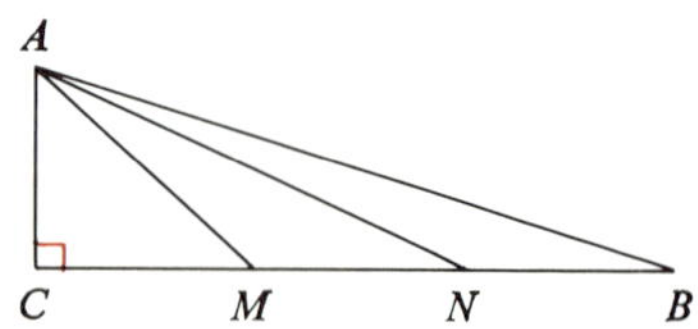

24. GIVEN $\triangle ABC$, where $\overline{PS} \perp \overline{AB}$ at S, $\overline{PR} \perp \overline{BC}$ at R, and $\overline{PQ} \perp \overline{AC}$ at Q
PROVE $AS^2 + BR^2 + CQ^2 = AQ^2 + BS^2 + CR^2$
(*Hint:* Consider $\overline{AP}$, $\overline{BP}$, and $\overline{CP}$.)

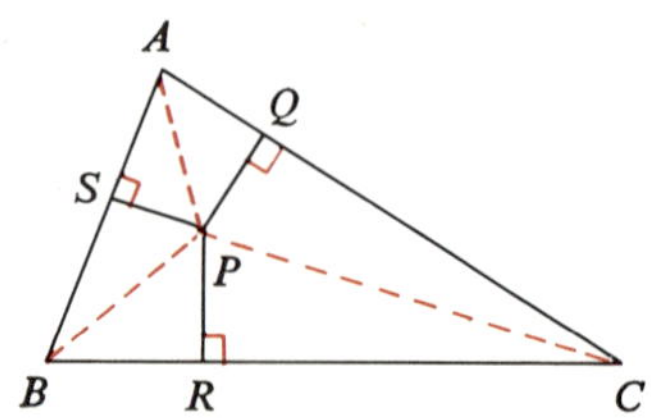

25. GIVEN In quadrilateral $KLMN$, $\overline{KM} \perp \overline{NL}$ at P.
PROVE $ML^2 + NK^2 = KL^2 + MN^2$

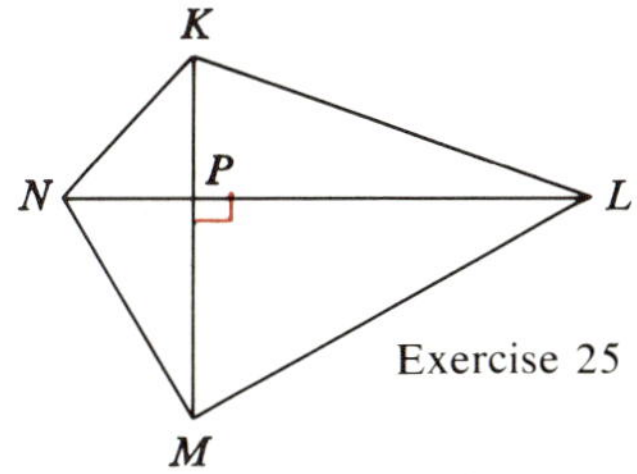

Exercise 25

C

26. GIVEN $\overline{AD}$ is an altitude of $\triangle ABC$; $AD^2 = BD \cdot DC$
PROVE $m\angle BAC = 90$

27. Show that a rectangular solid whose edges have lengths n, $n + 1$, and $n^2 + n$, where n is a positive rational number, has a diagonal whose length is a rational number.

28. The lengths of the sides of an isosceles triangle are 9, 9, and $6\sqrt{2}$. Find the length of the segment representing the shortest distance from the point of intersection of the altitudes to the base of the isosceles triangle.

29. If the medians to the legs of a right triangle have lengths 10 and 15, find the length of the hypotenuse.

30. Find the length of the segment determined by the endpoints of the altitude and angle bisector to the longest side of a triangle if the sides of the triangle have lengths 10, 24, and 26.

31. The sides $\overline{AB}$, $\overline{BC}$, and $\overline{AC}$ of $\triangle ABC$ have lengths 13, 14, and 15, respectively. Find the length of altitude $\overline{AD}$.

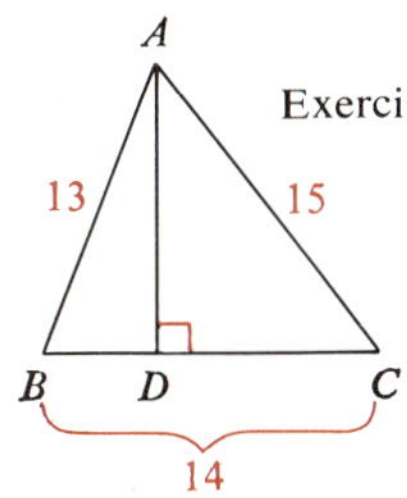

Exercise 31

32. GIVEN $\overline{BE}$ is a median of $\triangle ABC$.
PROVE $AB^2 + BC^2 = 2(AE^2 + BE^2)$

SOMETHING TO THINK ABOUT

An Extension of the Converse of the Pythagorean Theorem

Theorem 8-8.2 provides a method of proving that a given triangle is a right triangle when the lengths of its three sides are known. There is also a method of determining whether a nonright triangle is acute or obtuse.

To make this determination knowing only the lengths of the sides of a triangle, we must consider the angle with the greatest measure. The angle with the greatest measure in a triangle is the angle opposite the longest side, by Theorem 5-3.1. If this angle is acute, then the triangle is acute, by definition. If it is obtuse, then the triangle is obtuse. We must develop a method of finding whether this angle is acute or obtuse.

First, we shall establish that in an obtuse triangle, the square of the length of the longest side is greater than the sum of the squares of the lengths of the two shorter sides. We shall also establish that in an acute triangle, the square of the length of the longest side is less than the sum of the squares of the lengths of the two shorter sides.

Consider $\triangle ABC$, in which c is the measure of the longest side, and $\angle C$ has the greatest measure. We must consider two cases: $\angle C$ obtuse (on the left below), and $\angle C$ acute (on the right below). Since the two cases are similar, we shall prove them at the same time.

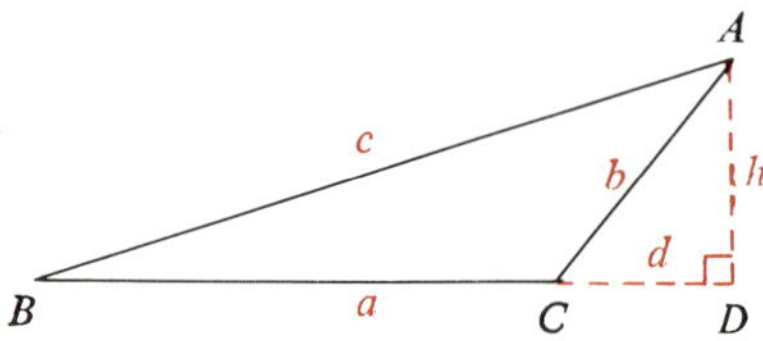

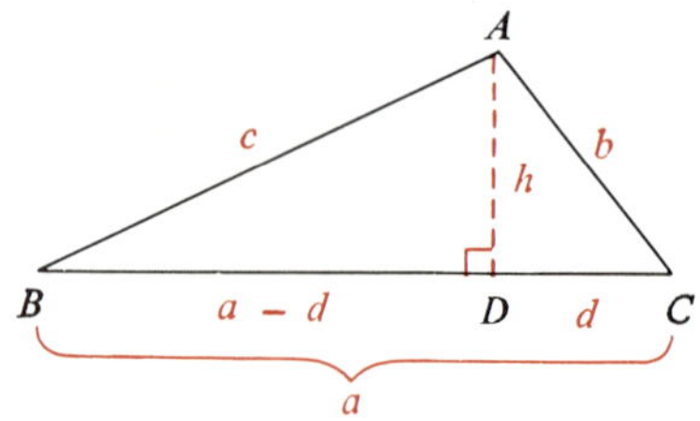

GIVEN $\triangle ABC$; $\angle C$ is obtuse.	GIVEN $\triangle ABC$; $\angle C$ is acute.
PROVE $c^2 > a^2 + b^2$	PROVE $c^2 < a^2 + b^2$

PROOF

1. $\overline{AD}$ is an altitude of $\triangle ABC$. Let $DC = d$.

2. $BD = a + d$	2. $BD = a - d$

Apply the Pythagorean Theorem to $\triangle ABD$. Let $AD = h$.

3. $c^2 = h^2 + (a + d)^2$	3. $c^2 = h^2 + (a - d)^2$
4. $c^2 = h^2 + a^2 + d^2 + 2ad$	4. $c^2 = h^2 + a^2 + d^2 - 2ad$

5. Apply the Pythagorean Theorem to $\triangle ACD$; $b^2 = h^2 + d^2$ in both cases.

Substitute the result in step 5 into the equations in step 4:

6. $c^2 = a^2 + b^2 + 2ad$	6. $c^2 = a^2 + b^2 - 2ad$
7. $c^2 > a^2 + b^2$ (Theorem 5-1.1)	7. $c^2 < a^2 + b^2$ (Why?)

We have now established that in an obtuse triangle, the square of the length of the longest side is greater than the sum of the squares of the lengths of the two shorter sides.

We have also established that in an acute triangle, the square of the length of the longest side is less than the sum of the squares of the lengths of the two shorter sides.

In what type of triangle is the square of the length of the longest side equal to the sum of the squares of the lengths of the two shorter sides? Why?

Using the above discussion, convince yourself that:
In $\triangle ABC$, where c is the length of the longest side,

if $c^2 < a^2 + b^2$, then $\angle C$ is acute;
if $c^2 = a^2 + b^2$, then $\angle C$ is right;
if $c^2 > a^2 + b^2$, then $\angle C$ is obtuse.

How can we modify these rules for isosceles triangles where $AB = AC$?

8-9 SPECIAL RIGHT TRIANGLES

Usually we must know the lengths of two sides of a right triangle in order to use the Pythagorean Theorem to find the third side. In this section, we shall discuss two special types of right triangles, for which we need to know only the length of one side in order to find the lengths of the remaining sides of the triangle. These triangles are the isosceles right triangle and the right triangle whose acute angles have measures of 30 and 60. We shall call the second type of triangle a 30-60-90 triangle.

T **Theorem 8-9.1** In an isosceles right triangle, the hypotenuse is $\sqrt{2}$ times as long as a leg.

For example, in isosceles right $\triangle ABC$, with right $\angle C$ and $AC = BC = a$, Theorem 8-9.1 tells us that $c = a\sqrt{2}$.

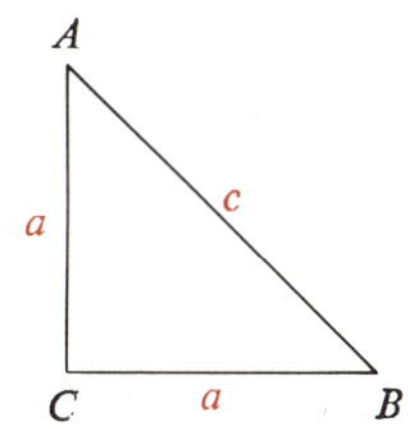

Proof Outline By applying the Pythagorean Theorem to the isosceles right triangle at right, we get $a^2 + a^2 = c^2$, or $2a^2 = c^2$. Therefore $c = \sqrt{2a^2} = a\sqrt{2}$.

C **Corollary 8-9.1a** In an isosceles right triangle, either leg is $\frac{\sqrt{2}}{2}$ times as long as the hypotenuse.

This corollary tells us that in the above right $\triangle ABC$,

$$a = c \cdot \frac{1}{2}\sqrt{2} = \frac{c\sqrt{2}}{2}$$

Proof Outline By dividing both sides of the resulting equation in the proof of Theorem 8-9.1 by $\sqrt{2}$, we get $\frac{c}{\sqrt{2}} = \frac{a\sqrt{2}}{\sqrt{2}} = a$.

Rationalizing the denominator, we get

$$a = \frac{c}{\sqrt{2}} \cdot \frac{\sqrt{2}}{\sqrt{2}} = \frac{c\sqrt{2}}{2}$$

Example 1 Find the hypotenuse of an isosceles right triangle whose legs are 8 inches long.

Solution By Theorem 8-9.1, we find that the length of the hypotenuse is $8\sqrt{2}$ inches.

Example 2 Find the length of a leg of an isosceles right triangle whose hypotenuse is 8 inches long.

Solution By Corollary 8-9.1a, we find that the length of either leg is

$8 \times \frac{\sqrt{2}}{2} = 4\sqrt{2}.$

By completing the following Class Exercises you can infer relationships between the sides of the other special right triangle, the 30-60-90 triangle.

Class Exercises

$\overline{CM}$ is a median of right $\triangle ABC$; $m\angle A = 60$; and $m\angle B = 30$.

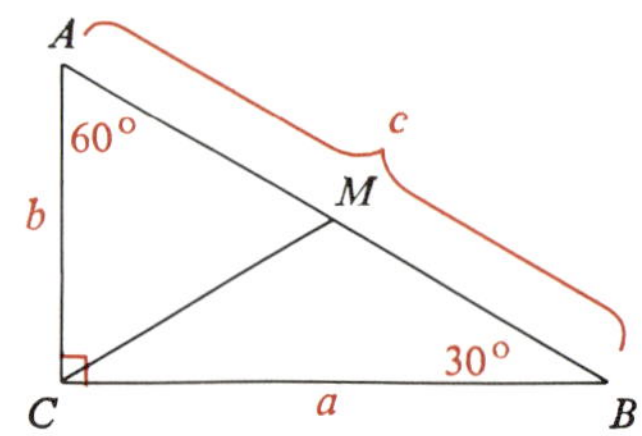

1. Why is $AM = MC$?
2. $m\angle ACM =$ ____
3. $m\angle AMC =$ ____
4. What type of triangle is $\triangle AMC$? Why?
5. Explain why $AC = \frac{1}{2}AB$.
6. $a^2 + b^2 =$ ____
7. $b = \frac{1}{2} \times$ ____
8. $a^2 + (\frac{1}{2} \cdot \text{\_\_\_\_})^2 =$ ____
9. $a^2 +$ ____ $=$ ____
10. $a^2 = (\text{\_\_\_\_})c^2$
11. $a = \sqrt{(\text{\_\_\_\_})c^2} = \frac{\sqrt{3}}{2}(\text{\_\_\_\_})$
12. State the results of Exercises 5 and 11 using general desciptions rather than algebraic designations. A restatement of Exercise 5 gives us the following theorem:

T **Theorem 8-9.2** In a 30-60-90 triangle, the side opposite the angle of measure 30 is half as long as the hypotenuse.

We may also say that in a 30-60-90 triangle, the hypotenuse is twice as long as the shorter leg.

The Class Exercises above also indicate a relationship between the hypotenuse and the longer leg of a 30-60-90 triangle. Which exercise states this relationship? Which theorem justifies the relationship? It is restated by the following theorem:

T **Theorem 8-9.3** In a 30-60-90 triangle, the side opposite the angle of measure 60 is $\frac{\sqrt{3}}{2}$ times as long as the hypotenuse.

C **Corollary 8-9.3a** In a 30-60-90 triangle, the hypotenuse is $\frac{2\sqrt{3}}{3}$ times as long as the side opposite the angle of measure 60.

C **Corollary 8-9.3b** In a 30-60-90 triangle, the longer leg is $\sqrt{3}$ times as long as the shorter leg.

Example 3 Find the legs of 30-60-90 $\triangle ABC$, if the hypotenuse $AB = 10$ and $m\angle A = 60$.

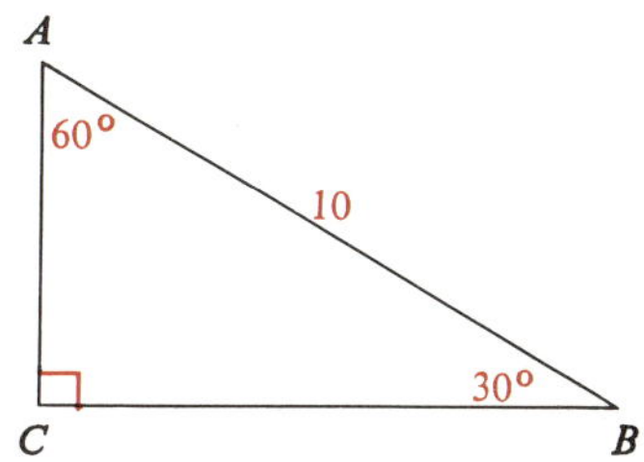

Solution By Theorem 8-9.2, $AC = \frac{1}{2} \cdot 10 = 5$.

By Theorem 8-9.3, $BC = \frac{\sqrt{3}}{2} \cdot 10 = 5\sqrt{3}$.

EXERCISES

A

Find the length of the longer leg of a 30-60-90 triangle if the hypotenuse has the following lengths:

1. 12 2. 30 3. 17 4. $23\frac{1}{2}$ 5. $6\sqrt{3}$ 6. $5\sqrt{2}$

Find the length of the hypotenuse of a 30-60-90 triangle if the longer leg has the following lengths:

7. 15 8. 36 9. 25 10. $13\frac{1}{2}$ 11. $12\sqrt{3}$ 12. $10\sqrt{2}$

Find the length of the altitude of an equilateral triangle if a side has the following lengths:

13. 16 14. 20 15. 27 16. $11\frac{1}{2}$ 17. $8\sqrt{3}$ 18. $5\sqrt{5}$

Find the length of the diagonal of a square if the following are the lengths of a side:

19. 5 20. 6 21. 11 22. $35\frac{1}{2}$ 23. $9\sqrt{2}$ 24. $7\sqrt{8}$

Find the lengths of the longer leg and hypotenuse of a 30-60-90 triangle if the shortest side has the following lengths:

25. 3 26. 1 27. 5 28. $7\frac{1}{2}$ 29. $11\sqrt{3}$ 30. $3\sqrt{5}$

Find the length of a side of the square whose diagonal has the following lengths:

31. 1 32. 6 33. 5 34. $12\frac{1}{2}$ 35. $8\sqrt{2}$ 36. $5\sqrt{3}$

37. If the altitude of an equilateral triangle is 5 inches long, find the perimeter of the triangle.
38. Find the length of the base of an isosceles triangle if the measure of the vertex angle is 120 and if the sides that include the angle are both 16 inches long.

Find the lengths of the altitude from the vertex angle and the base of an isosceles triangle whose equal sides are 10 inches long and whose base angles have these measures:

39. 30 40. 45 41. 60

B

42. In trapezoid $ABCD$, $\overline{AB} \parallel \overline{CD}$, $AB = 8$, $BC = 8$, $m\angle A = 135$, and $m\angle B = 120$. Find the length of altitude $\overline{EF}$ and base $\overline{DC}$.
43. In $\triangle ABC$, $m\angle B = 60$ and $m\angle A = 75$. If the length of altitude $\overline{AD} = 6$, find the perimeter of the triangle.
44. In $\triangle PQR$, $m\angle Q$ and $m\angle P = 105$. Find the perimeter of $\triangle PQR$ if $PQ = 10$. (*Hint*: Consider altitude $\overline{PN}$.)
45. Find the perimeter of a rhombus whose longer diagonal has length 24 and one of whose angles is 120.
46. In trapezoid $ABCD$, $\overline{AB} \parallel \overline{CD}$, $BC = 20$, $DC = 24$, $m\angle B = 120$, and $\overline{AD} \perp \overline{DC}$. Find the perimeter of the trapezoid.
47. In trapezoid $ABCD$, $\overline{AB} \parallel \overline{CD}$, $m\angle A = m\angle B = 135$, $AD = 8\sqrt{2}$, and $AB = 12$. Find the perimeter of the trapezoid.
48. $\overline{AD}$ is the altitude to hypotenuse $\overline{BC}$ of right $\triangle ABC$, where $m\angle B = 60$. If $AD = 5\sqrt{3}$, find the perimeter of $\triangle ABC$.

C

49. GIVEN Right $\triangle ABC$ with $m\angle C = 90$ and $AB = AC\sqrt{2}$
    PROVE $\triangle ABC$ is isosceles
50. GIVEN $\triangle ABC$ where $BC = AC\sqrt{3}$ and $2AC = AB$
    PROVE $m\angle C = 90$
51. $\overline{CE}$, $\overline{CG}$, and $\overline{EG}$ are the diagonals of faces of cube $ABCDEFGH$. $\overline{CN} \perp \overline{EG}$. Find CN if $AB = 8$.

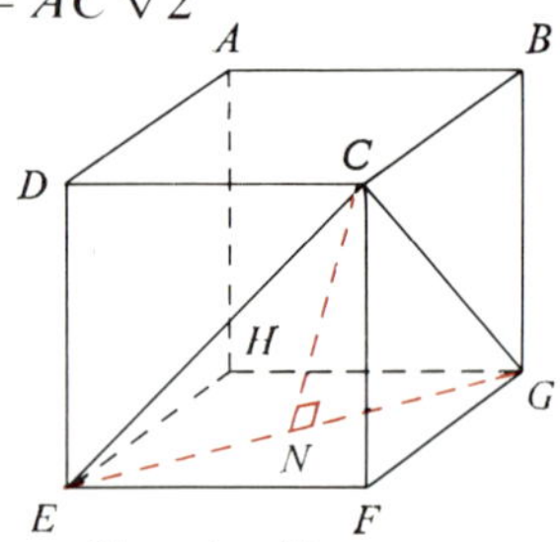

Exercise 51

52. GIVEN $\overline{PM} \perp$ plane of square $ABCD$; $m\angle PBM = 45$; $\overline{PE} \perp \overline{BC}$; $\overline{DMB}$; $\overline{AMC}$
    PROVE $PE = \frac{1}{2}AB\sqrt{3}$

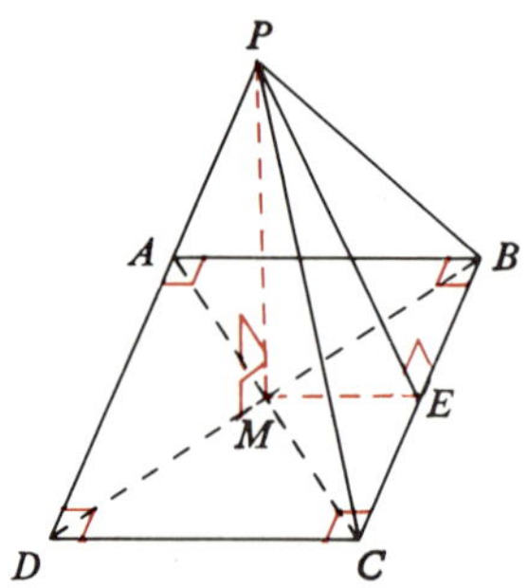

53. GIVEN $\overline{AB} \perp$ plane of $\triangle BCD$; $\overline{CD} \perp \overline{BD}$; $m\angle BAD = 45$; $m\angle BCD = 30$; $CD = 6$
    FIND AC
    PROVE $\overline{AD} \perp \overline{CD}$

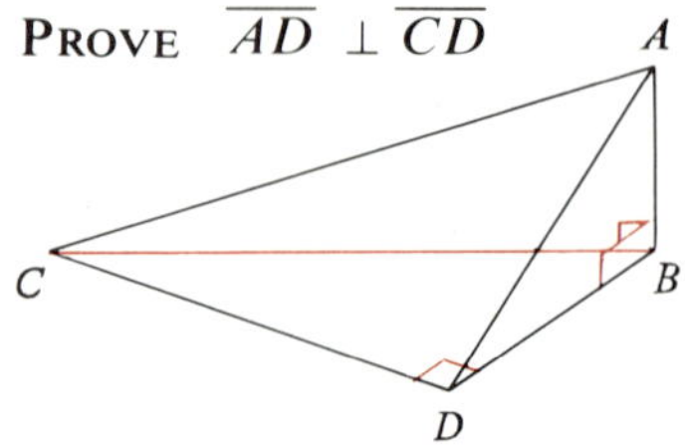

A LOOK AT THE PAST

The Pleasing Rectangle

Which of the two rectangles below seems more pleasing to you?

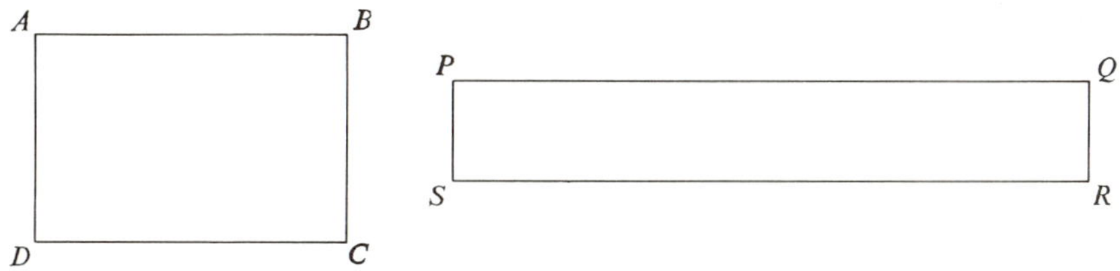

Psychologists have conducted experiments showing that rectangles like rectangle *ABCD*, or any rectangles whose sides are in the same proportion, are the most pleasing rectangles. A glance at rectangle *ABCD* catches the entire rectangle at once, while looking at rectangle *PQRS* requires a scanning motion of the eyes by most people.

The beauty of rectangle *ABCD* is not a new discovery; ancient civilizations were very much aware of it. The pictures below illustrate the use of this Golden Rectangle, as it is frequently called, in art and in architecture.

Parthenon

Cathedral of Chartres

8-10 TRIGONOMETRIC RATIOS

Consider below $\triangle ABC$ at right. How do the triangles below compare to $\triangle ABC$? Why?

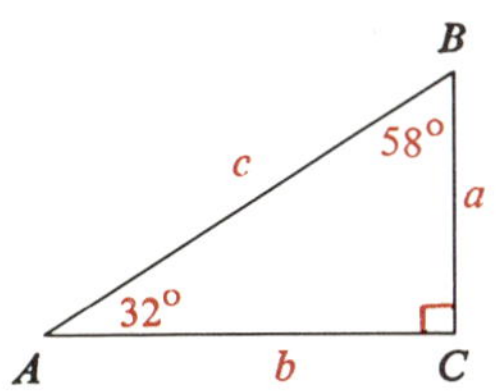

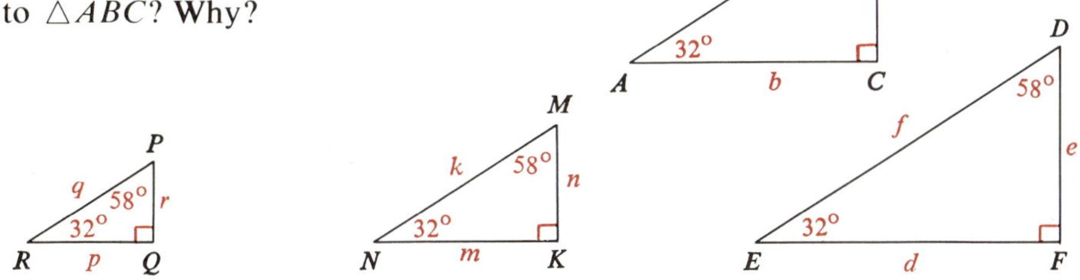

Since each of these triangles is similar to $\triangle ABC$, we can form the following equal ratios between the measures of their sides. Why are the ratios equal?

$$\frac{a}{c} = \frac{r}{q} = \frac{n}{k} = \frac{e}{f} \qquad \frac{b}{c} = \frac{p}{q} = \frac{m}{k} = \frac{d}{f} \qquad \frac{a}{b} = \frac{r}{p} = \frac{n}{m} = \frac{e}{d}$$

The equality of such ratios in similar triangles of different sizes led mathematicians of ancient times to establish a branch of mathematics called **trigonometry**. In Greek, *trigonon* means "triangle" and *metria* means "measurement." Trigonometry deals with the measurement of triangles. Specifically, it deals with the relations between the sides and angles of triangles.

The equality of the ratios above depends only on the size of the angles. We can name the ratios between the sides of a right triangle in relation to its acute angles. In the definitions below, $\angle A$ is either acute angle in a right triangle.

D **Definition 8-7** The **sine of $\angle A$** $= \dfrac{\text{length of side opposite } \angle A}{\text{length of hypotenuse}} = \dfrac{a}{c}$

The abbreviation for *sine* is *sin*. Thus, we may write "$\sin \angle A = \frac{a}{c}$."

Furthermore, if $m\angle A = x$, in degrees, we may write "$\sin x° = \frac{a}{c}$." (We must show by symbol that x represents degrees because there is another unit of measure for angles, called radian measure.)

D **Definition 8-8** The **cosine of $\angle A$** $= \dfrac{\text{length of side adjacent to } \angle A}{\text{length of hypotenuse}} = \dfrac{b}{c}$

Using the abbreviation *cos*, we write "$\cos \angle A = \frac{b}{c}$." As before, if $m\angle A = x$ degrees, we may write "$\cos x° = \frac{b}{c}$."

D **Definition 8-11** The **tangent of $\angle A$** $= \dfrac{\text{length of side opposite } \angle A}{\text{length of side adjacent } \angle A} = \dfrac{a}{b}$

Using the abbreviation *tan*, we write "$\tan \angle A = \frac{a}{b}$." If $m\angle A = x$ degrees, we may write "$\tan x° = \frac{a}{b}$."

These three ratios are called **trigonometric ratios.**

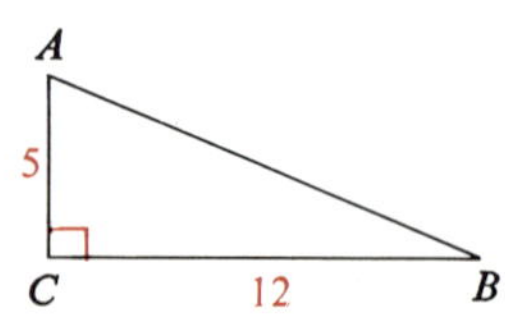

Example 1 Find the three trigonometric ratios of $\angle B$ of $\triangle ABC$.

Solution By the Pythagorean Theorem, $AB = 13$.

$$\sin \angle B = \frac{AC}{AB} = \frac{5}{13} \qquad \cos \angle B = \frac{BC}{AB} = \frac{12}{13} \qquad \tan \angle B = \frac{AC}{BC} = \frac{5}{12}$$

Class Exercises

1. Find x and y.

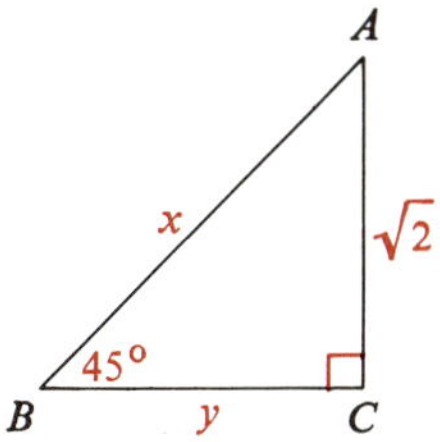

2. Find x and y.

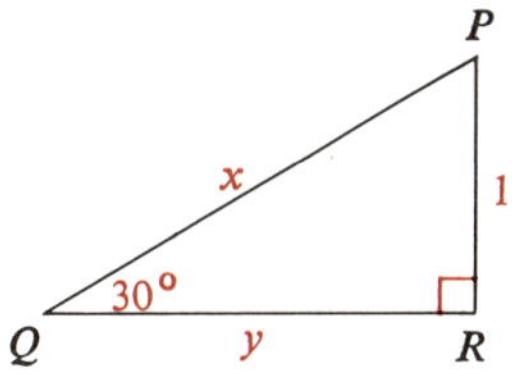

3. Complete the following chart by using the trigonometric ratios in the triangles in Exercises 1 and 2.

	$b = 30$	$b = 45$	$b = 60$
$\sin b°$			
$\cos b°$			
$\tan b°$			

4. Evaluate $\dfrac{\sin 45° \cdot \cos 45°}{\sin 30°}$

5. Evaluate $(\sin 60°)^2 + (\cos 60°)^2$.

In the Class Exercises above, we found that $\sin 45° = \dfrac{\sqrt{2}}{2} \approx \dfrac{1.4142}{2} = .7071$.

We also found that $\cos 60° = \frac{1}{2} = .5000$. Can you find the value of sin 31°, or cos 43°, or tan 79°? This would be very difficult, so we shall use the following table, which gives the values of the trigonometric ratios correct to the nearest ten-thousandth. Why does the table give no values for angles measuring more than 90°?

The first table of trigonometric ratios is believed to have been compiled by Hipparchus of Nicea (circa 150 B.C.), who is now referred to as the Father of Trigonometry.

How do the entries in the chart in the Class Exercises compare with the corresponding entries in the table of trigonometric ratios? The following examples will clarify the use of this table.

Values of Trigonometric Ratios

Angle	Sin	Cos	Tan	Angle	Sin	Cos	Tan
1°	.0175	.9998	.0175	46°	.7193	.6947	1.0355
2°	.0349	.9994	.0349	47°	.7314	.6820	1.0724
3°	.0523	.9986	.0524	48°	.7431	.6691	1.1106
4°	.0698	.9976	.0699	49°	.7547	.6561	1.1504
5°	.0878	.9962	.0875	50°	.7660	.6428	1.1918
6°	.1045	.9945	.1051	51°	.7771	.6293	1.2349
7°	.1219	.9925	.1228	52°	.7880	.6157	1.2799
8°	.1392	.9903	.1405	53°	.7986	.6018	1.3270
9°	.1564	.9877	.1584	54°	.8090	.5878	1.3764
10°	.1736	.9848	.1763	55°	.8192	.5736	1.4281
11°	.1908	.9816	.1944	56°	.8290	.5592	1.4826
12°	.2079	.9781	.2126	57°	.8387	.5446	1 5399
13°	.2250	.9744	.2309	58°	.8480	.5299	1.6003
14°	.2419	.9703	.2493	59°	.8572	.5150	1.6643
15°	.2588	.9659	.2679	60°	.8660	.5000	1.7321
16°	.2756	.9613	.2867	61°	.8746	.4848	1.8040
17°	.2924	.9563	.3057	62°	.8829	.4695	1.8807
18°	.3090	.9511	.3249	63°	.8910	.4540	1.9626
19°	.3256	.9455	.3443	64°	.8988	.4384	2.0503
20°	.3420	.9397	.3640	65°	.9063	.4226	2.1445
21°	.3584	.9336	.3839	66°	.9135	.4067	2.2460
22°	.3746	.9272	.4040	67°	.9205	.3907	2.3559
23°	.3907	.9205	.4245	68°	.9272	.3746	2.4751
24°	.4067	.9135	.4452	69°	.9336	.3584	2.6051
25°	.4226	.9063	.4663	70°	.9397	.3420	2.7475
26°	.4384	.8988	.4877	71°	.9455	.3256	2.9042
27°	.4540	.8910	.5095	72°	.9511	.3090	3.0777
28°	.4695	.8829	.5317	73°	.9563	.2924	3.2709
29°	.4848	.8746	.5543	74°	.9613	.2756	3.4874
30°	.5000	.8660	.5774	75°	.9659	.2588	3.7321
31°	.5150	.8572	.6009	76°	.9703	.2419	4.0108
32°	.5299	.8480	.6249	77°	.9744	.2250	4.3315
33°	.5446	.8387	.6494	78°	.9781	.2079	4.7046
34°	.5592	.8290	.6745	79°	.9816	.1908	5.1446
35°	.5736	.8192	.7002	80°	.9848	.1736	5.6713
36°	.5878	.8090	.7265	81°	.9877	.1564	6.3138
37°	.6018	.7986	.7536	82°	.9903	.1392	7.1154
38°	.6157	.7880	.7813	83°	.9925	.1219	8.1443
39°	.6293	.7771	.8098	84°	.9945	.1045	9.5144
40°	.6428	.7660	.8391	85°	.9962	.0872	11.4301
41°	.6561	.7547	.8693	86°	.9976	.0698	14.3007
42°	.6691	.7431	.9004	87°	.9986	.0523	19.0811
43°	.6820	.7314	.9325	88°	.9994	.0349	28.6363
44°	.6947	.7193	.9657	89°	.9998	.0175	57.2900
45°	.7071	.7071	1.0000	90°	1.0000	.0000	

Example 2 Find, to the nearest foot, the height of the Empire State Building if the angle of elevation ($\angle TPB$) from a point 478 feet from the foot of the building is 72°.

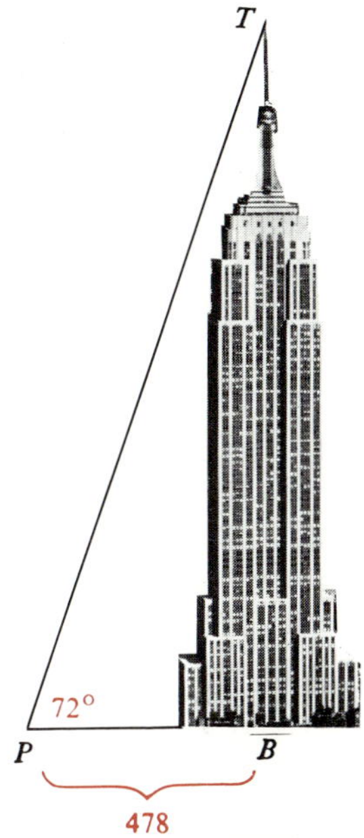

Solution Let $h =$ the height of the building, in feet.

$$\tan 72° = \frac{h}{478}$$
$$h = 478 \cdot \tan 72°$$
$$h = 478 \cdot 3.0777$$
$$h = 1471.1406 \approx 1471 \text{ feet}$$

Example 3 Find, correct to the nearest degree, the angle that a 50-foot ladder leaning against a wall forms with the ground when it is 20 feet away from the base of the wall.

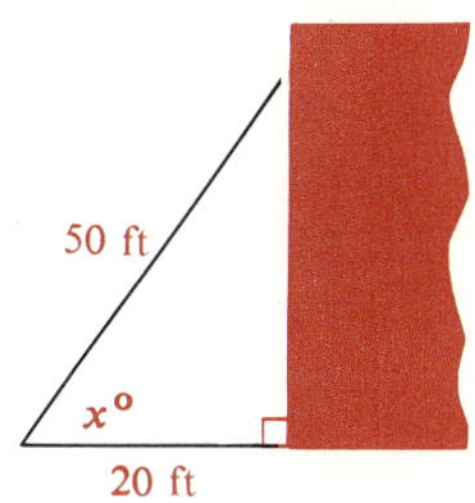

Solution Let $x =$ the measure of the required angle.

$$\cos x = \frac{20}{50} = .4000$$

Since cos 66° (.4067) is the closest approximation to .4000 on our trigonometric ratio table, $x = 66°$, to the nearest degree.

EXERCISES

A Find the approximate value of each of the following. Use the table of trigonometric ratios shown in this section.

1. sin 57° 2. cos 79° 3. tan 10° 4. cos 18° 5. tan 45°

Find θ (the Greek letter *theta*, pronounced "thay-ta") to the nearest degree for each of the following:

6. $\sin \theta = \frac{3}{5}$ 7. $\tan \theta = \frac{8}{3}$ 8. $\sin \theta = .5678$ 9. $\cos \theta = .3500$

Using the figure at right, find the following:

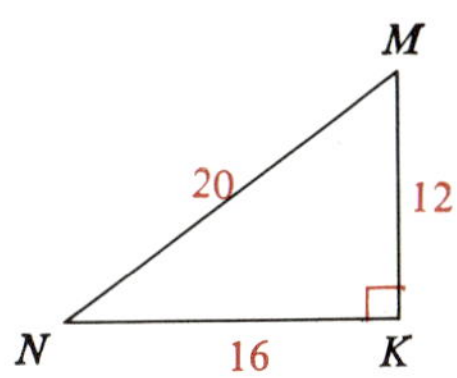

10. $\cos \angle M$ 11. $\tan \angle N$ 12. $\sin \angle N$
13. $\tan \angle M$ 14. $\sin \angle M$ 15. $\cos \angle N$
16. $\tan \angle N \cdot \tan \angle M$ 17. $\tan \angle M - \dfrac{\sin \angle M}{\cos \angle M}$
18. $(\sin \angle N)^2 + (\cos \angle N)^2$ 19. $1 - (\sin \angle M)^2$

In each of the following figures, find x correct to the nearest tenth:

20. 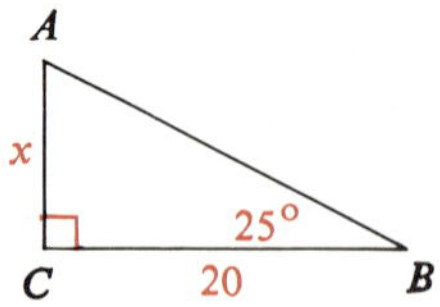

21. 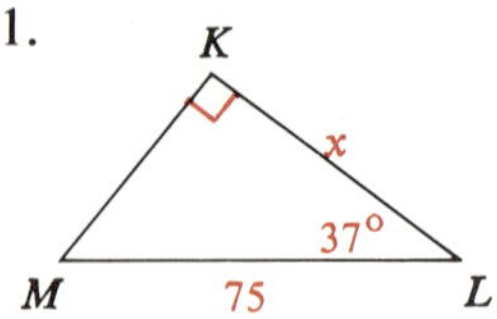

22. 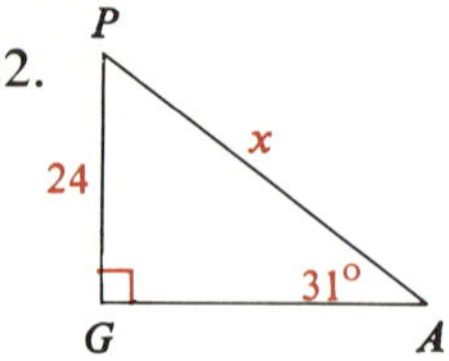

In each of the following figures, find θ to the nearest degree:

23. 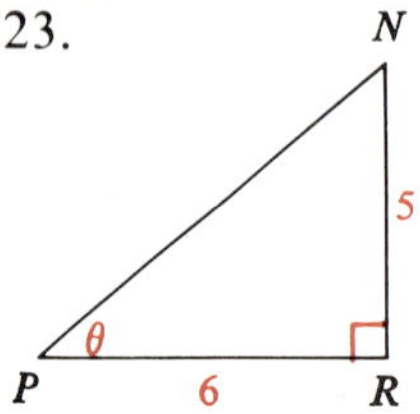

24. 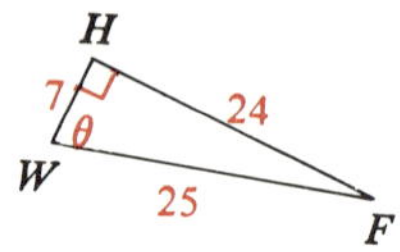

B

25. Find the measures of the acute angles, correct to the nearest degree, of a triangle whose side lengths are 5, 12, and 13.
26. Find the measures of the angles, correct to the nearest degree, of an isosceles triangle whose side lengths are 5, 5, and 6.

In the figure at right, the measures of line segments and angles are marked.

27. Find x and y.
28. Mention a method, independent of trigonometry, for checking your answers to Exercise 27. Why is this method inaccurate?

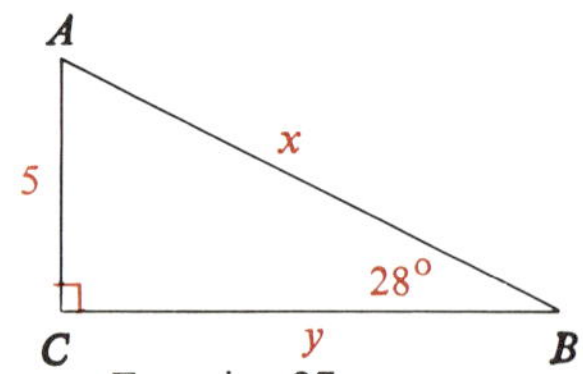

Exercise 27

29. In trapezoid $ABCD$, $\overline{AB} \parallel \overline{DC}$, and the measures of angles and line segments are marked. Find the length of altitude $\overline{BE}$ and $m\angle C$.

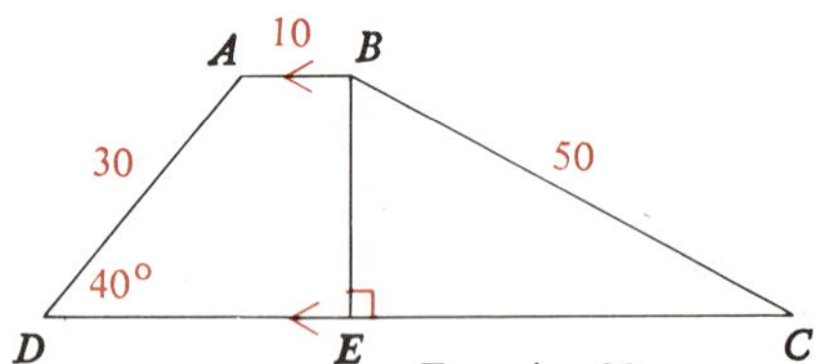

Exercise 29

30. In measuring across a lake between points A and B, a surveyor measures 10 feet along the ground to point P, in a direction perpendicular to $\overline{AB}$. He then measures $\angle APB$ and finds it measures 85°. What is the width of the lake between points A and B, correct to the nearest tenth of a foot?

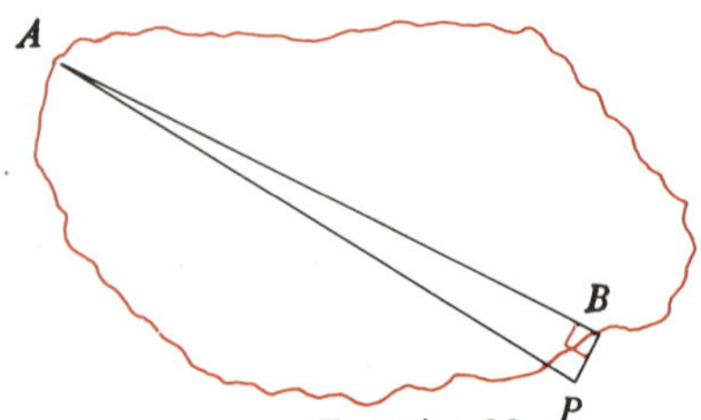

Exercise 30

31. In guiding a ship past dangerously shallow water, a lighthouse attendant must determine the distance of the ship from the lighthouse. The lighthouse attendant sights the ship 15° down from the horizontal. This 15° angle is called the angle of depression. If the lighthouse is 200 feet above sea level, how far is the ship from the lighthouse? Express your answer to the nearest foot.

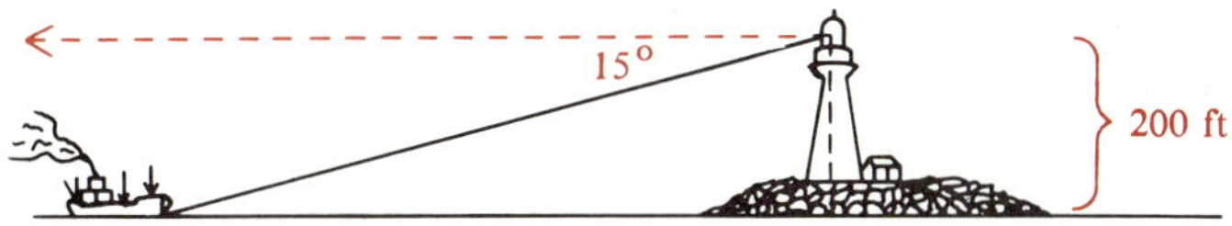

C

32. An airplane at an altitude of 24,000 feet approaches an airport that is at sea level. The pilot receives orders from the airport's traffic controllers to descend at an angle of 4° when coming in for a landing. How far from the airport should the airplane begin its descent? Express your answer to the nearest tenth of a mile. (1 mile ... 5,280 feet.)

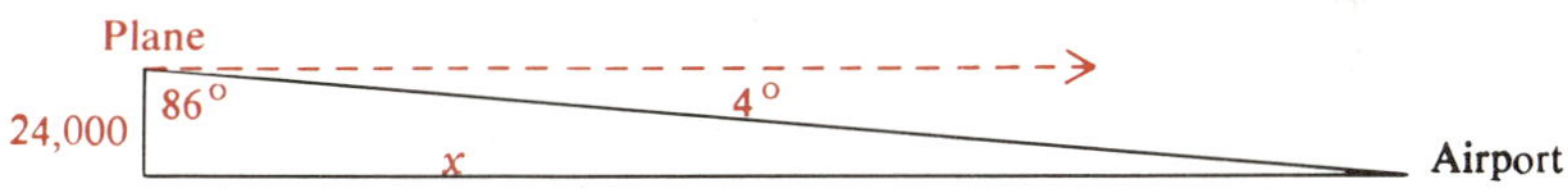

33. Two vertical cliffs are on opposite sides of a 100-foot-wide river. From an observation point P on the top edge of the shorter cliff, the angle of elevation to the top of the higher cliff (point A) is 32°, while the angle of depression to the bottom of the cliff (point B) is 41°. Find, correct to the nearest foot, the height of each cliff.

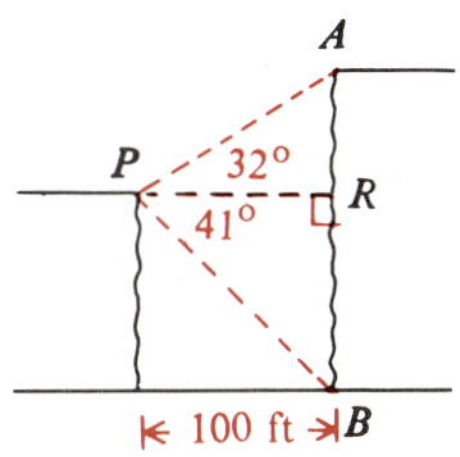

HOW IS YOUR MATHEMATICAL VOCABULARY?

The key words and phrases introduced in this chapter are listed below. How many do you know and understand?

cosine (8-10)
exterior angle bisector (8-3)
extremes (8-1)
fourth proportional (8-1)
geometric mean (8-1)
mean proportional (8-1)
means (8-1)
proportion (8-1)
Proportional Line Segments Postulate (8-2)
proportional segments (8-2)
ratio (8-1)
ratio of similitude (8-4)
similar polygons (8-4)
similar triangles (8-4)
sine (8-10)
tangent (8-10)
trigonometry (8-10)

REVIEW EXERCISES

8-1 *Ratio and proportion*

1. If point P is between R and S of $\overline{RS}$ so that $RS = 36$ and $PS = 12$, find $\frac{PS}{RS}$, $\frac{RP}{RS}$, and $\frac{RP}{PS}$.
2. Find the value of x in the proportion $\frac{x-3}{5} = \frac{x+2}{6}$.
3. Find the mean proportional between 3 and 12, 5 and 12, and $\sqrt{3}$ and $\sqrt{27}$.
4. If $\frac{p+q}{q} = \frac{a}{b}$, then $\frac{p}{q} =$ ——.
5. If $\frac{r}{s} = \frac{m}{n}$, then $\frac{n}{s} =$ ——.
6. If $\frac{a}{b} = \frac{3}{2}$, then $\frac{b}{a} =$ ——.
7. If $\frac{m-n}{n} = \frac{7}{15}$, then $\frac{n}{m} =$ ——.

8-2 *Parallel lines and proportions*

In Exercises 8–10, parallel lines and the lengths of line segments are marked. Find x in each figure.

8. 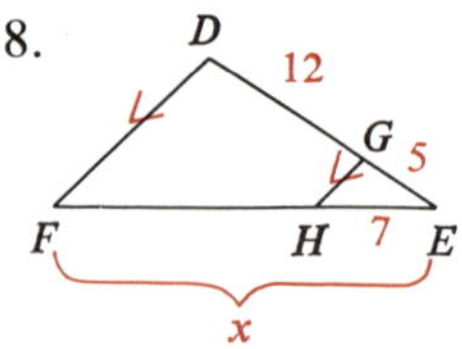

9. 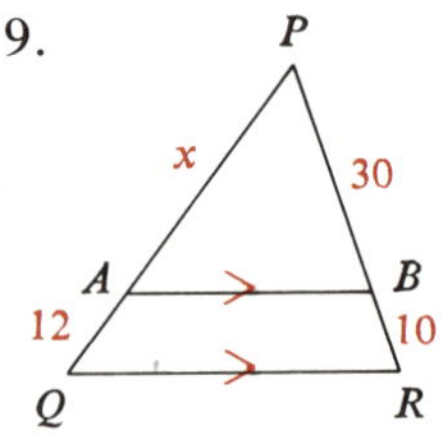

10. 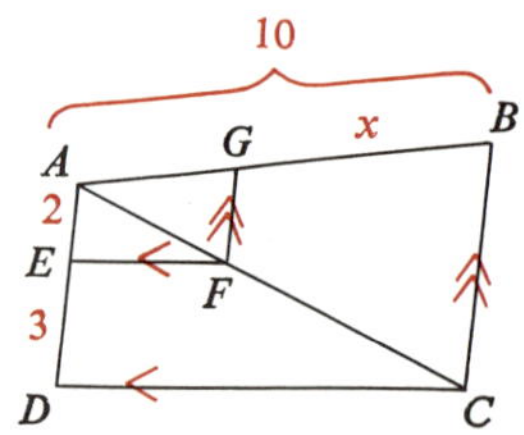

11. $\overline{AN}$ meets median $\overline{BM}$ of $\triangle ABC$ at P, $\overline{BNC}$, $BN = 21$, and $NC = 14$. Find $\frac{AP}{PN}$. (*Hint:* Draw $\overline{NR} \parallel \overline{BM}$.)

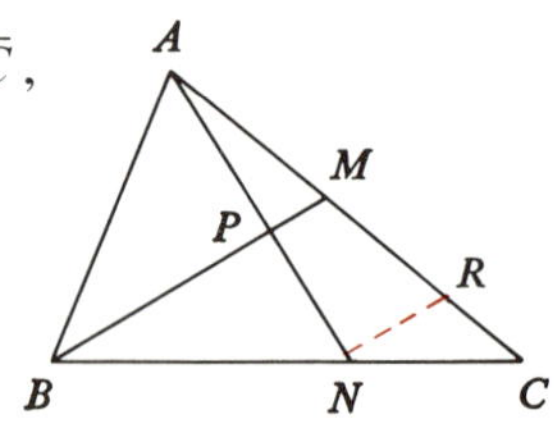

Exercise 11

8-3 *The angle bisector theorem*

12. Find the lengths of the segments into which the bisector of the angle of greatest measure of a triangle divides the opposite side if the sides are 14, 21, and 25 inches long.
13. The perimeter of a triangle is 50 inches. The bisector of one angle divides the opposite side into segments that are 8 and 12 inches long. Find the lengths of the remaining two sides of the triangle.

14. If $\overline{AD}$ is an angle bisector of $\triangle ABC$, and $AC = 6$, $BC = 7$, and $AB = 8$, find CD.

15. If $\overline{AP}$ is an exterior angle bisector of $\triangle ABC$, $AB = a$ and $AC = b$, express $\frac{PC}{PB}$ in terms of a and b.

8-4 Similar polygons

16. If the lengths of the sides of one triangle are 10, 12, and 14, find the length of the shortest side and the perimeter of a similar triangle whose longest side is 35.
17. If the lengths of the sides of a triangle are 9, 12, and 18, find the lengths of the sides of a similar triangle whose perimeter is 52.
18. Prove that the ratio of the perimeters of two similar polygons equals their ratio of similitude.
19. GIVEN D, E, and F are the respective midpoints of $\overline{BC}$, $\overline{AC}$, and $\overline{AB}$ of $\triangle ABC$.
PROVE $\triangle ABC \sim \triangle DEF$

8-5 Proving triangles similar

20. GIVEN $\square ABCD$; $\overline{AFE}$ intersects $\overrightarrow{BC}$ and $\overline{DC}$ in points E and F, respectively.
PROVE $\triangle ADF \sim \triangle ECF$

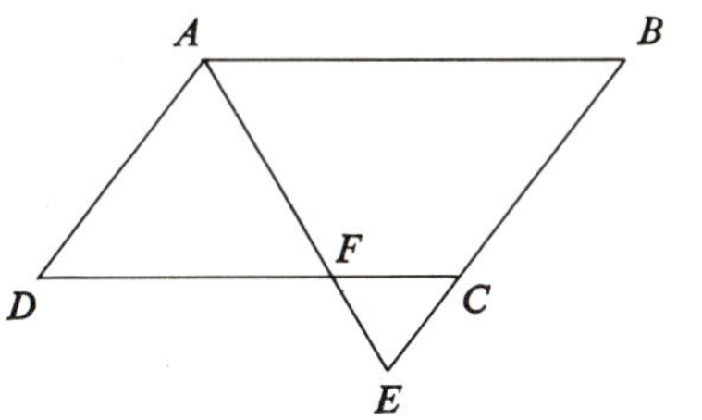

21. GIVEN Right $\triangle ABC$ with right $\angle C$; $\overline{DE} \perp \overline{AC}$ at D; $\overline{AEB}$
PROVE $\triangle ADE \sim \triangle ACB$

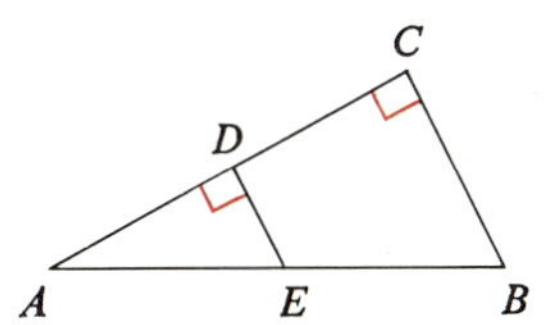

22. Prove that two isosceles triangles having congruent vertex angles are similar.

23. GIVEN $\overline{CH}$ is an altitude of right $\triangle ABC$ with right $\angle ACB$; $\overline{HD}$ bisects $\angle CHB$; $\overline{CE}$ bisects $\angle ACB$; $\overline{AHEB}$; $\overline{CDB}$
PROVE $\frac{HD}{CE} = \frac{CB}{AB}$

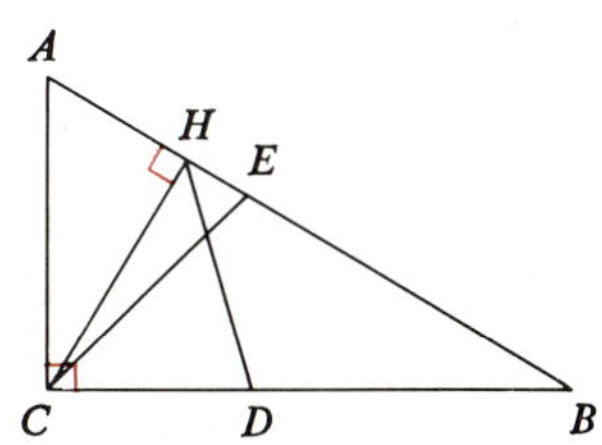

8-6 More triangle similarity theorems

24. State three methods of proving triangles similar.

State the theorem that proves each of the following pairs of triangles similar:

25. 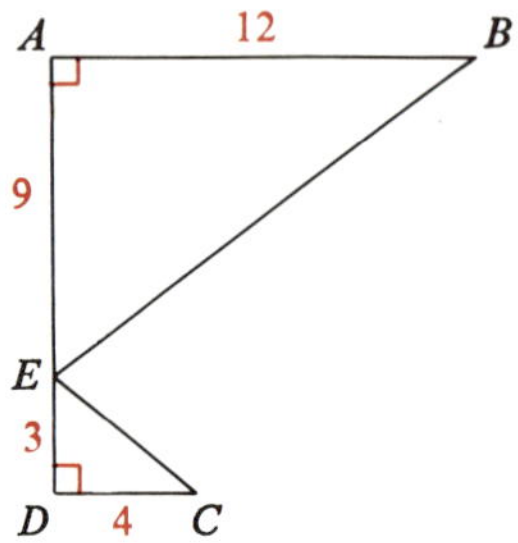

26. 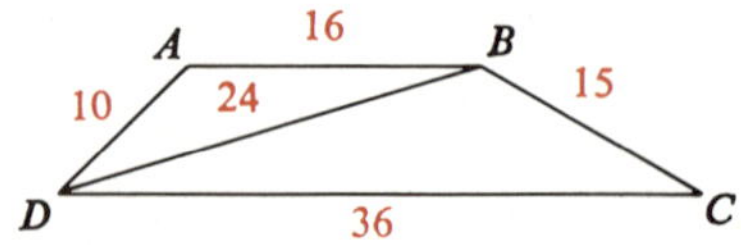

27. 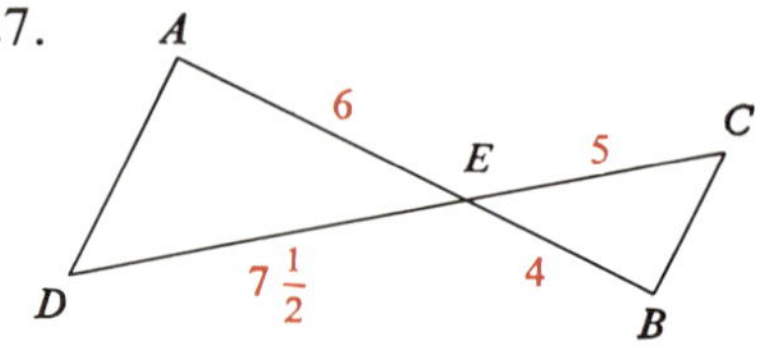

28. GIVEN $\overline{AD}$ intersects $\overline{BC}$ in E; $AE \cdot ED = BE \cdot EC$

PROVE $\angle B \cong \angle D$

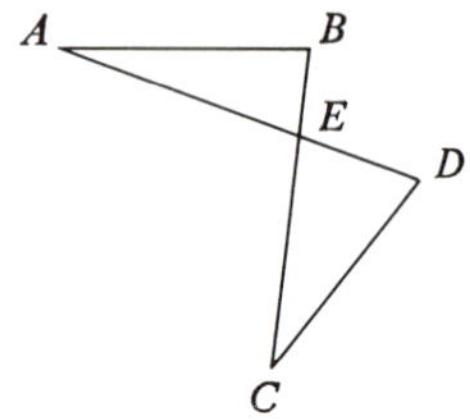

8-7 Similarity and proportion in right triangles

29. In the figure at right, E and F are respective points of sides $\overline{AC}$ and $\overline{CB}$ of right $\triangle ABC$ so that quadrilateral $CEDF$ is a rectangle and $\overline{CD}$ is an altitude of $\triangle ABC$. List all pairs of similar triangles.

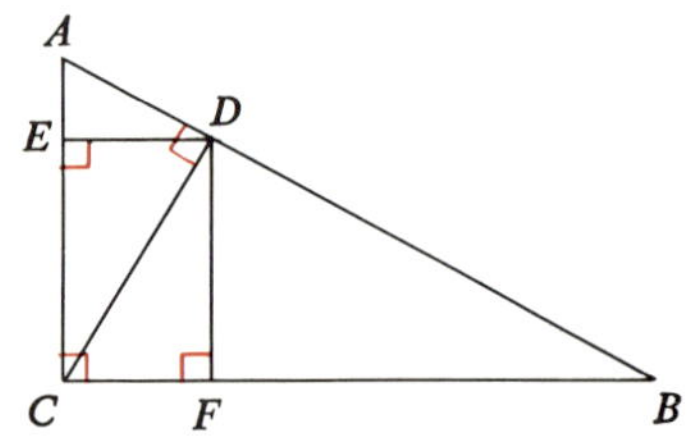

Find x in each of the following figures:

30. 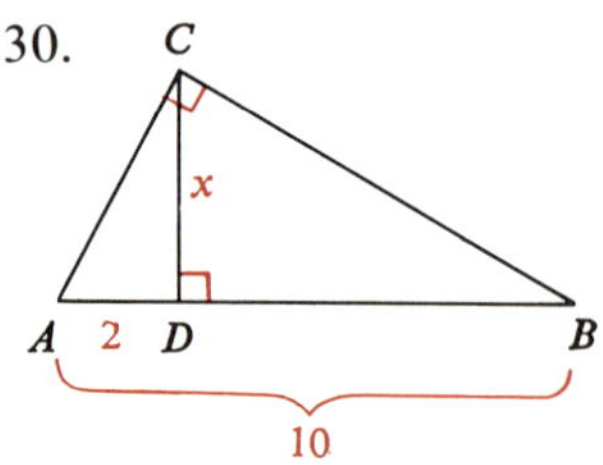

31. 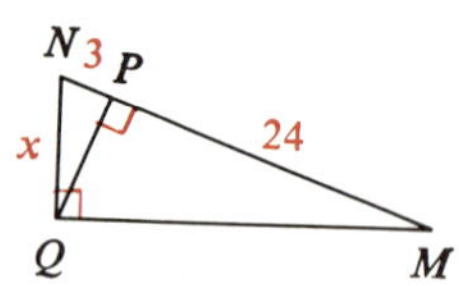

32. 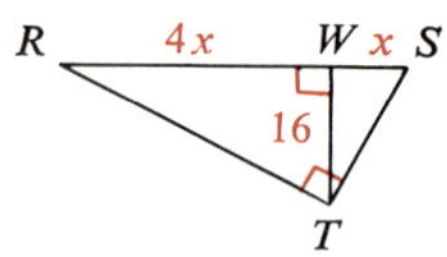

33. GIVEN $\overline{AD} \perp \overline{AB}$; $\overline{BC} \perp \overline{AB}$; $\overline{AEC} \perp \overline{DEB}$

PROVE $\dfrac{CA}{BC} = \dfrac{BC}{CE}$

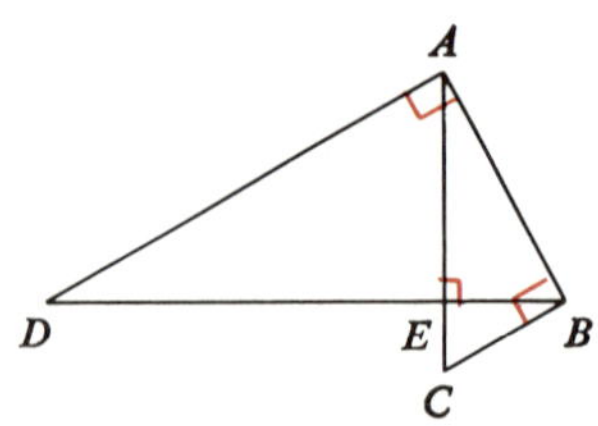

34. GIVEN $\overline{CD}$ is an altitude to hypotenuse $\overline{AB}$ of right $\triangle ABC$.
PROVE $CD \cdot AB = AC \cdot BC$

8-8 The Pythagorean theorem

Find x in each of the following right triangles:

35.

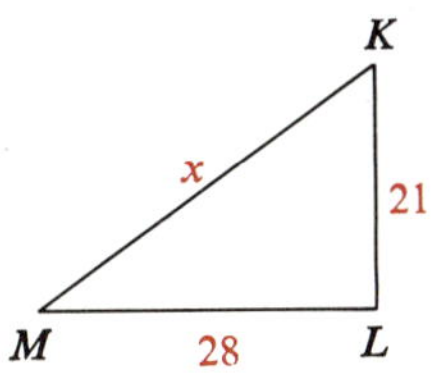

36.

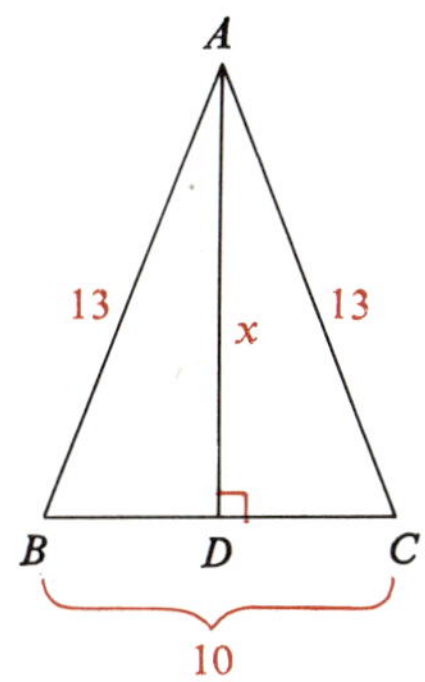

37.

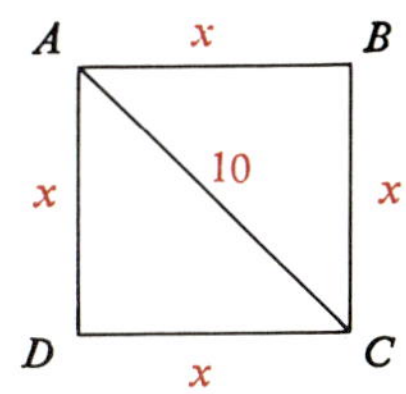

8-9 Special right triangles

38. Find the length of the longer leg of a 30-60-90 triangle if $7\sqrt{2}$ represents the length of its hypotenuse.

39. Find the length of the hypotenuse of a 30-60-90 triangle if the length of the longer leg is 1.

40. Find the lengths of the sides of a 30-60-90 triangle if the length of the shortest side is 3.

41. Find the length of the side of a square if the length of a diagonal is $3\sqrt{2}$.

42. Find the length of a diagonal of the square whose side length is $\sqrt{3}$.

43. Find the perimeter of the equilateral triangle whose altitude measure is $3\sqrt{3}$.

8-10 Trigonometric ratios

Without the use of a table of the values of trigonometric ratios, find each of the following:

44. $\sin 30°$ 45. $\cos 45°$ 46. $\tan 60°$
47. $\cos 60°$ 48. $\sin 60°$ 49. $\tan 45°$

Find x to the nearest tenth for each of the following right triangles:

50. 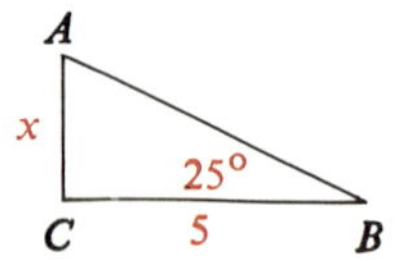

51. 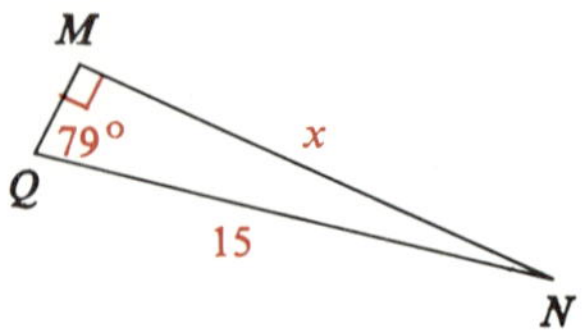

52. Find the measures of the angles of a triangle whose sides are 25, 25, and 14 inches long.

CHAPTER TEST

Find x in each of the following figures:

1. 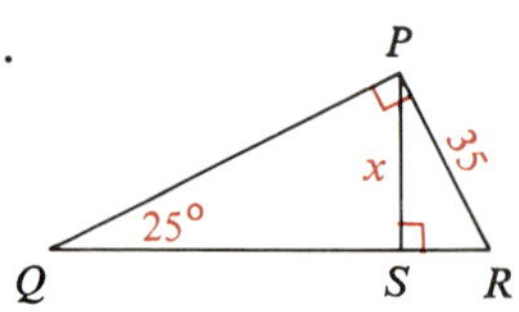

2. 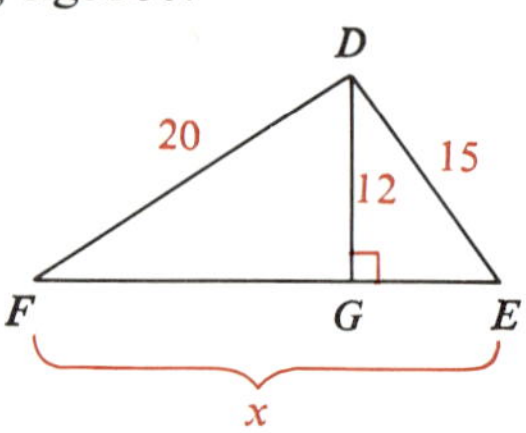

3. 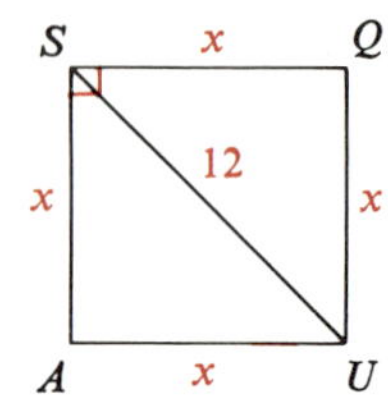

4. 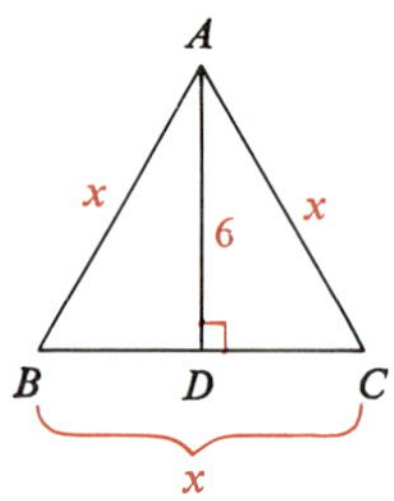

5. 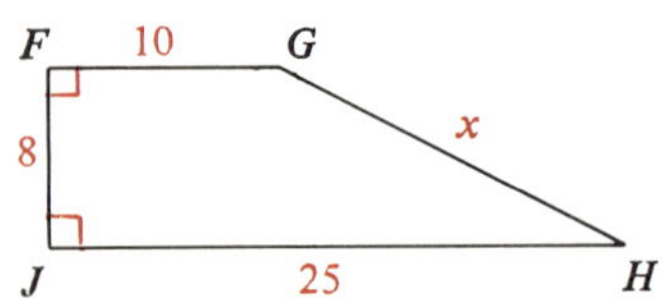

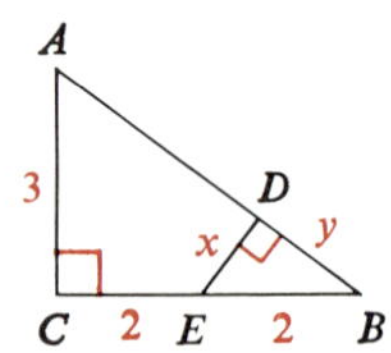

Exercise 6

6. Find x and y in right $\triangle ABC$, where $\overline{ED} \perp \overline{AB}$ at D.

7. GIVEN $\overline{BD}$ is an angle bisector of $\triangle ABC$; $AG = AD$; $HC = CD$

   FIND AG, HC, and HG

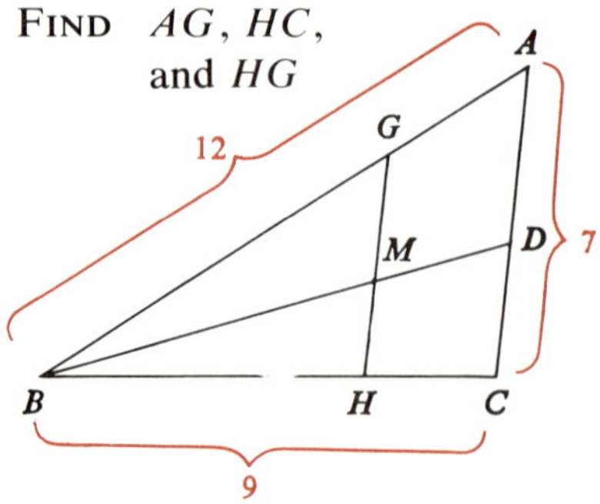

8. GIVEN M is a point in side $\overline{AP}$ of $\triangle APB$; $\overline{MN} \parallel \overline{AB}$, where N is in $\overline{BP}$; $\overline{NL} \parallel \overline{BC}$, where L is in $\overline{CP}$.

   PROVE $\triangle MNL \sim \triangle ABC$

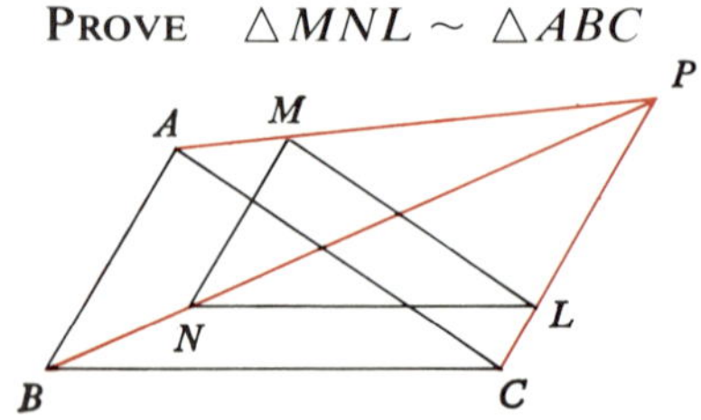

SUGGESTED RESEARCH

1. A popular source of proofs of the Pythagorean Theorem is *The Pythagorean Proposition*, by Elisha S. Loomis (Washington, D.C.: The National Council of Teachers of Mathematics, 1968). Present several of these proofs to your geometry class.

2. Stewart's Theorem is one of the neglected theorems of geometry. Refer to *Challenging Problems in Geometry*, Vol. 2, by Alfred S. Posamentier and Charles T. Salkind (New York: Macmillan Co., 1970), and prepare a report on this very useful theorem.

3. Prepare a report on the relationships among the members of a Pythagorean Triple. One source of information is Waclaw Sierpinski's *Pythagorean Triangles*, The Scripta Mathematica Studies (New York: Yeshiva University, 1962).

$\pi \approx$ 3.14159265358979323846264338327950288419716939937510
58209749445923078164062862089986280348253421170679
82148086513282306647093844609550582231725359408128
48111745028410270193852110555964462294895493038196
44288109756659334461284756482337867831652712019091
45648566923460348610454326648213393607260249141273
72458700660631558817488152092096282925409171536436
78925903600113305305488204665213841469519415116094
33057270365759591953092186117381932611793105118548
07446237996274956735188575272489122793818301194912
98336733624406566430860213949463952247371907021798
60943702770539217176293176752384674818467669405132
00056812714526356082778577134275778960917363717872
14684409012249534301465495853710507922796892589235
42019956112129021960864034418159813629774771309960
51870721134999999837297804995105973173281609631859
50244594553469083026425223082533446850352619311881
71010003137838752886587533208381420617177669147303
59825349042875546873115956286388235378759375195778
18577805321712268066130019278766111959092164201989
38095257201065485863278865936153381827968230301952
03530185296899577362259941389124972177528347913151
55748572424541506959508295331168617278558890750983
81754637464939319255060400927701671139009848824012
85836160356370766010471018194295559619894676783744
94482553797747268471040475346462080466842590694912
93313677028989152104752162056966024058038150193511
25338243003558764024749647326391419927260426992279
67823547816360093417216412199245863150302861829745
55706749838505494588586926995690927210797509302955
32116534498720275596023648066549911988183479775356
63698074265425278625518184175746728909777727938000
81647060016145249192173217214772350141441973568548
16136115735255213347574184946843852332390739414333
45477624168625189835694855620992192221842725502542
56887671790494601653466804988627232791786085784383
82796797668145410095388378636095068006422512520511
73929848960841284886269456042419652850222106611863
06744278622039194945047123713786960956364371917287
46776465757396241389086583264599581339047802759009
94657640789512694683983525957098258...

9 Circles and Spheres

In this chapter, we shall discuss some of the properties of circles and spheres, on which we shall base some new theorems.

9-1 BASIC DEFINITIONS

Before studying the properties of circles and spheres, we must define some basic terms.

D **Definition 9-1** A **circle** is the set of all points in a plane that are the same distance from a given point in the plane.

We use the symbol ⊙ for *circle*, and name a given circle by naming its **center**, the point coplanar with and equidistant from all points of the circle. In ⊙P at right, point P is the center. Distance r at right is called the **radius** of the circle.

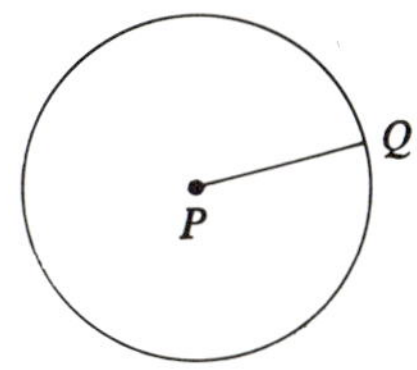

We can also refer to $\overline{PQ}$ as a radius. The term *radius* may refer either to the segment joining the center of a circle with a point on the circle, or to the length of that segment. The context in which the term is used will indicate the intended meaning. The plural of *radius* is *radii*. Based on the definition of a circle and the definition of a radius, what can you say about all radii of a circle?

If we substitute *space* for *plane* in the definition of a circle, we have the definition of a sphere.

D **Definition 9-2** A **sphere** is the set of all points in space that are the same distance from a given point.

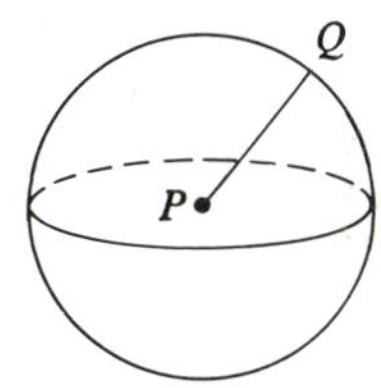

P is the center of the sphere at right and r, or $\overline{PQ}$, the radius. We call a sphere with its center at P, sphere P.

D **Definition 9-3** A **chord of a circle** is a line segment whose endpoints are points of the circle.
A **secant** of the circle is a line that contains a chord.
A **secant ray** is a ray that contains a chord of the circle and whose endpoint is one of the endpoints of the chord.

In the figure at right, $\overline{AB}$ is a chord of ⊙P; $\overleftrightarrow{AB}$, or $\overleftrightarrow{ABC}$, is a secant of the circle; and $\overrightarrow{ABC}$ is a secant ray of the circle.

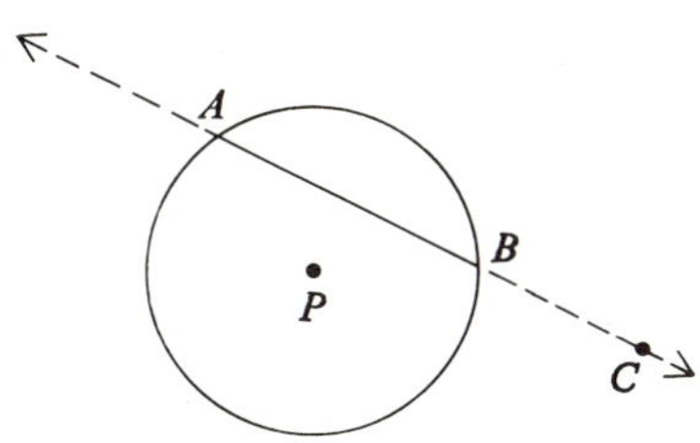

A segment whose endpoints are points of a sphere is a chord of the sphere. A line that contains a chord of a sphere is a secant of the sphere. A ray of the line that contains two points of the sphere is a secant ray of the sphere. In sphere P, $\overline{AB}$ is a chord, $\overleftrightarrow{AB}$ is a secant, and $\overrightarrow{ABC}$ is a secant ray.

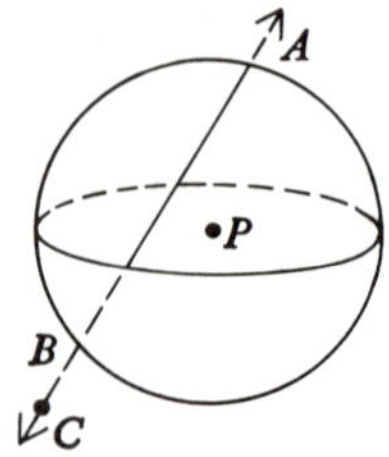

If a chord is a segment joining any two points on a circle or a sphere, can a chord contain the center of the circle or sphere?

D

Definition 9-4 A **diameter** of a circle or sphere is any chord containing the center of the circle or sphere, or the length of such a chord.

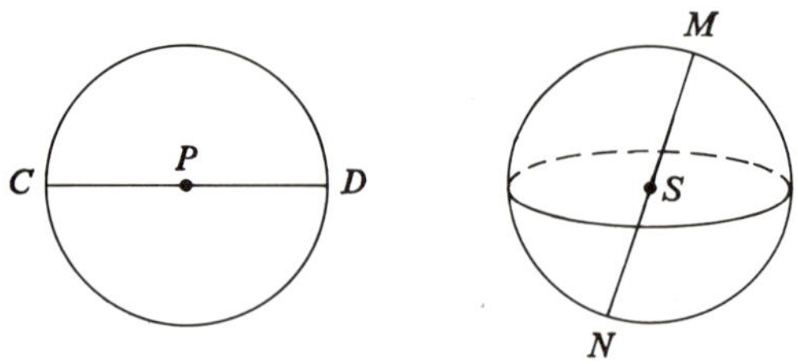

Since chord $\overline{CD}$ in the circle above contains point P, $\overline{CD}$ is a diameter of $\odot P$. The word *diameter* may refer either to $\overline{CD}$ or to CD. The context will indicate the intended meaning. Chord $\overline{MSN}$ is a diameter of sphere S, above.

Since all radii of a circle are congruent, what property does the midpoint of a diameter of a circle have? Is this also true for the midpoint of a diameter of a sphere?

D

Definition 9-5 Two or more circles or spheres are **concentric** if they have a common center.

The **interior of a circle** consists of the points in the plane of the circle whose distance from the center of the circle is less than the radius. The **exterior of a circle** consists of the points in the plane of the circle whose distance from the center is greater than the radius. The points whose distance from the center of the circle is equal to the radius are points of the circle. By analogy, you should be able to state the meanings of "interior of a sphere," "exterior of a sphere," and "points of a sphere."

EXERCISES

A

Indicate whether each of the statements in Exercises 1-17 is true or false. Correct the false statements.

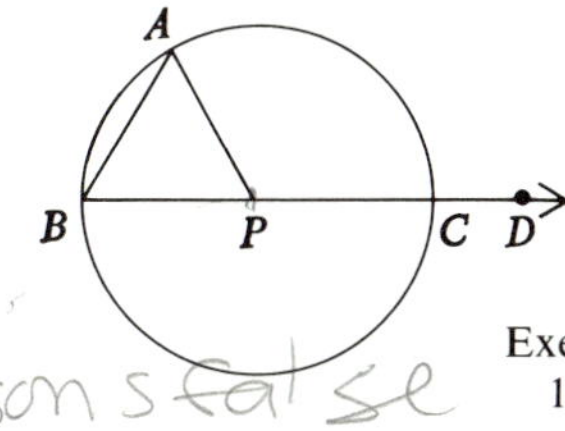

Exercises 1–10

Exercises 1-10 refer to $\odot P$ at right. Points of the circle are indicated.

1. Point P is a point of the circle.
2. $\overleftrightarrow{AB}$ is a chord of $\odot P$.
3. $AP = BP = CP$
4. $\overline{BD}$ is a chord of $\odot P$.
5. Point D is in the exterior of $\odot P$.
6. $\triangle APB$ is isosceles.
7. $\overline{CB}$ is a chord of $\odot P$.
8. Point A is a point of the circle.
9. $\overline{BP}$ is a radius of $\odot P$.
10. $\overleftrightarrow{CB}$ is a diameter of $\odot P$.

11. The diameter of a sphere is a secant of the sphere.
12. A chord of a circle contains fewer points of a circle than a secant of the same circle.
13. A sphere has only one segment that can be called a radius.
14. The midpoint of a chord of a circle is always in the interior of the circle.
15. A secant ray contains one and only one ray whose endpoint is a point of the circle, with the rest of its points in the exterior of the circle.
16. The diameter of a circle is twice the radius of the circle.
17. Any chord of a circle containing the center of the circle is a diameter of the circle.

B

18. $\overline{AB}$ is a diameter of $\odot P$. Find AP if $AB = 5x + 6$ and $BP = x + 12$.
19. $\overline{MQN}$ is a chord of $\odot Q$. If $MQ = 5x - 13$ and $MN = 7x - 5$, find NQ.
20. Prove that all diameters of the same circle are congruent.
21. Prove that all diameters of a sphere are congruent.
22. Prove that the center of a circle is the midpoint of the diameters of the circle.
23. Prove the statement in Exercise 22 for spheres.
24. GIVEN $\overline{AB}$ is a diameter of $\odot P$; $\overrightarrow{AD} \perp \overline{AB}$ and $\overrightarrow{BC} \perp \overline{AB}$; $\overline{AD} \cong \overline{BC}$
    PROVE $\overline{DP} \cong \overline{CP}$

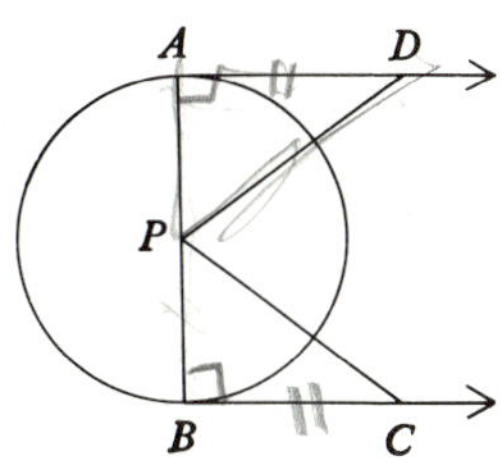

25. GIVEN $\overline{AC}$ and $\overline{BD}$ are diameters of the same sphere and meet at E.
    PROVE Quadrilateral $ABCD$ is a rectangle.

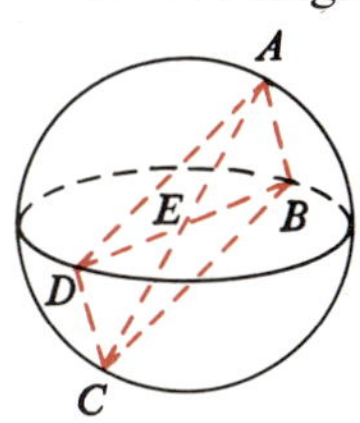

26. Given $\overline{AB}$ is a diameter of $\odot M$; P is any point of $\odot M$; $\overrightarrow{MN}$ bisects $\angle AMP$.
Prove $\overrightarrow{MN} \parallel \overline{BP}$

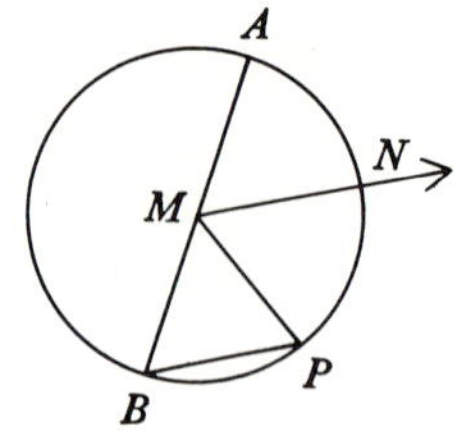

C 27. Given $\overline{AB}$ is a diameter of $\odot Q$; $\overline{CG} \perp \overline{AB}$ at G; $\overline{DH} \perp \overline{AB}$ at H; $\overline{AG} \cong \overline{BH}$; $\overline{AC}$ and $\overline{BD}$ are chords of $\odot Q$.
Prove $\overline{CG} \cong \overline{DH}$
(*Hint*: Consider $\overline{CQ}$ and $\overline{DQ}$.)

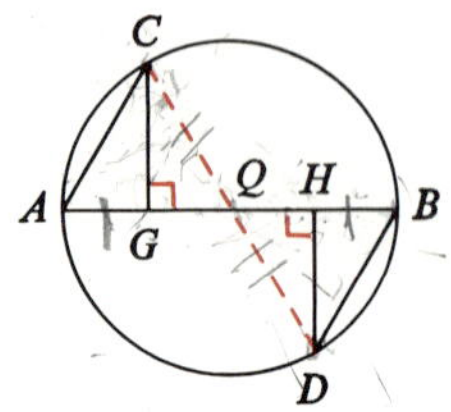

28. Given N is an interior point of $\odot P$; M is an exterior point of $\odot P$.
Prove $m\angle N > m\angle M$

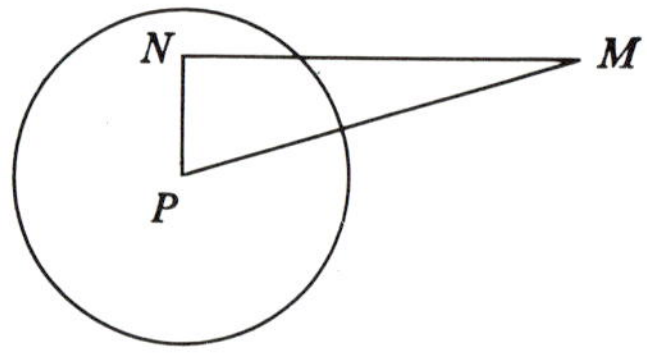

29. Given $\overline{AB}$ is a diameter of $\odot P$; $\overline{BD}$ and $\overline{AC}$ are chords; $\overline{AC} \cong \overline{BD}$
Prove $\overline{AC} \parallel \overline{BD}$

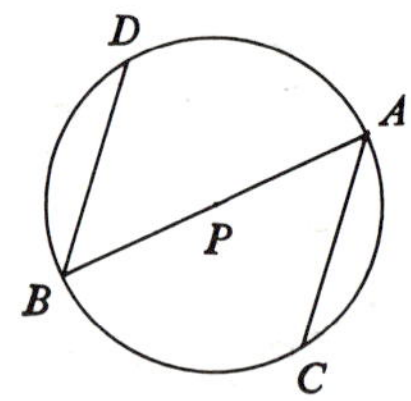

9-2 CHORDS OF A CIRCLE

The following Class Exercises will help you to derive three theorems about the chords of a circle.

Class Exercises

Exercises 1–14 refer to the figure at right. For Exercises 1–4, assume that M is the midpoint of $\overline{AB}$.

1. $\overline{AM} \cong$ ____.
2. $\overline{AP} \cong$ ____. Why?
3. ____ and P are two points that are ________ from the endpoints of $\overline{AB}$.
4. What is the relationship between $\overline{AB}$ and $\overleftrightarrow{MP}$? Why?

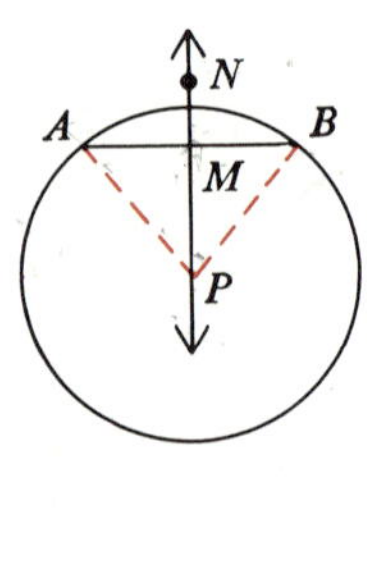

For Exercises 5–10, assume $\overleftrightarrow{MP} \perp \overline{AB}$ at M.

5. $\triangle AMP$ and $\triangle$____ are ________ triangles.
6. ____ $\cong \overline{BP}$
7. By the ________ property, $\overline{MP} \cong$ ____.
8. $\triangle AMP \cong \triangle$____. Why?
9. $\overline{AM} \cong$ ____. Why?
10. Consider Exercises 5–9 and explain how to find the midpoint of a chord of a circle.

Consider M the midpoint of $\overline{AB}$ and $\overleftrightarrow{NM} \perp \overline{AB}$ for Exercises 11–14.

11. What is true about every point of the perpendicular bisector of a line segment?
12. Is point P equidistant from A and B? Why?
13. Explain why point P is contained in the perpendicular bisector of $\overline{AB}$.
14. Summarize Exercises 1–4, 5–10, and 11–13 in three separate statements.

Your response to Exercise 14 above should be similar to the following theorems:

T **Theorem 9-2.1** The line containing the center of a circle and the midpoint of a chord of the circle that is not a diameter is perpendicular to the chord.

T **Theorem 9-2.2** The line containing the center of a circle and perpendicular to a chord of the circle bisects the chord.

T **Theorem 9-2.3** In the plane of a circle, the perpendicular bisector of a chord of the circle contains the center of the circle.

The following corollary follows directly from Theorem 9-2.3. The proof is left as an exercise.

C **Corollary 9-2.3a** No circle contains three collinear points.

In order to prove Corollary 9-2.3a, we must show that if three points are collinear, then the perpendicular bisectors of the two noncoinciding segments determined are parallel. This cannot be true for any three points on the circle since perpendicular bisectors of the segments must contain the center of the circle. (9-2.3) Hence three collinear points cannot all be on a circle.

Explain why this corollary implies that any three noncollinear points are contained in a circle.

Example 1

Given $\overline{AB}$ and $\overline{CD}$ are chords of $\odot P$; $\overline{PM} \perp \overline{AB}$ at M; $\overline{PN} \perp \overline{CD}$ at N; $\overline{PM} \cong \overline{PN}$

Prove $\overline{AB} \cong \overline{CD}$

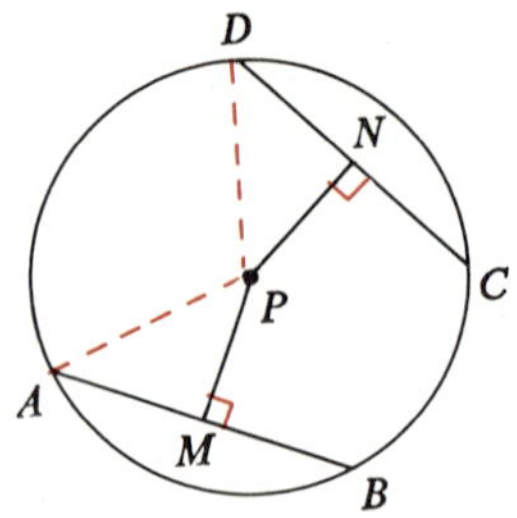

Solution

Proof

STATEMENTS	REASONS
1. $\overline{AB}$ and $\overline{CD}$ are chords of $\odot P$; $\overline{PM} \perp \overline{AB}$ at M; $\overline{PN} \perp \overline{CD}$ at N; $\overline{PM} \cong \overline{PN}$	1. Given.
2. $\overline{AP} \cong \overline{DP}$	2. Definition of a circle. (9-1)
3. $\angle AMP$ and $\angle DNP$ are right angles.	3. Definition of perpendicular lines. (1-25)
4. $\triangle AMP$ and $\triangle DNP$ are right triangles.	4. Definition of a right triangle. (1-32)
5. $\triangle AMP \cong \triangle DNP$	5. If the hypotenuse and one leg of a right triangle are congruent to the corresponding hypotenuse and leg of another right triangle, then the two triangles are congruent. (6-6.2)
6. $AM = DN$	6. Definition of congruent triangles. (3-3)
7. M is the midpoint of $\overline{AB}$; N is the midpoint of $\overline{CD}$.	7. Theorem 9-2.2.
8. $AB = 2AM$; $CD = 2DN$	8. Property of a midpoint.
9. $AB = CD$, or $\overline{AB} \cong \overline{CD}$	9. Multiplication property and Substitution Postulate (2-1).

D **Definition 9-6** **Circles are congruent** if and only if their radii are congruent.

Example 1 leads to the following theorem:

T **Theorem 9-2.4** In the same circle, or in congruent circles, chords are congruent if and only if they are equidistant from the center of the circle.

The proof of one part of this theorem is similar to the proof in Example 1. However, we must also prove the converse. This part of the proof is left as an exercise.

If congruent chords are equidistant from the center of a circle, how do distances from the center of a circle to chords of unequal measures compare?

T

Theorem 9-2.5 In the same circle, or in congruent circles, if two chords are not congruent, then the longer chord is nearer the center of the circle than the shorter chord.

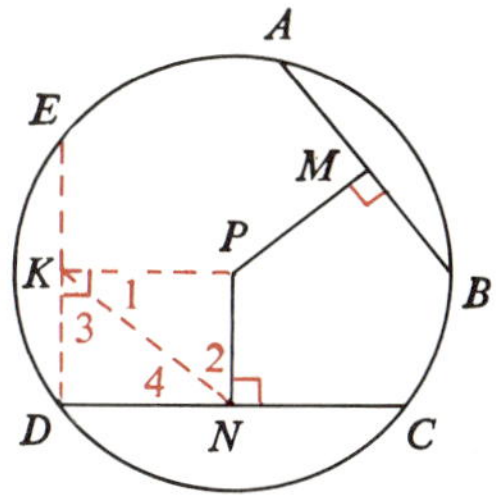

GIVEN $\overline{AB}$ and $\overline{CD}$ are chords of $\odot P$; $\overline{PM} \perp \overline{AB}$ at M; $\overline{PN} \perp \overline{CD}$ at N; $CD > AB$

PROVE $PN < PM$

PROOF OUTLINE We must first introduce chord $\overline{DE}$ so that $\overline{DE} \cong \overline{AB}$. By Theorem 9-2.4, we know that $\overline{PM} \cong \overline{PK}$, where $\overline{PK} \perp \overline{DE}$ at K. We must therefore show that $PN < PK$.

$DC > DE$ and $DN > DK$. Why? Therefore, in $\triangle DNK$, $m\angle 3 > m\angle 4$. Since $m\angle PKD = m\angle PND$, $m\angle 1 < m\angle 2$. Why? Therefore, from $\triangle PNK$, we find $PN < PM$.

C

Corollary 9-2.5a The diameter of a circle is the longest chord of the circle.

T

Theorem 9-2.6 In the same circle or in congruent circles, if two chords are not congruent, then the chord nearer the center of the circle is the longer of the two chords.

This theorem can easily be proved by reversing the reasoning in the proof of Theorem 9-2.5. The proof of this theorem is left as an exercise.

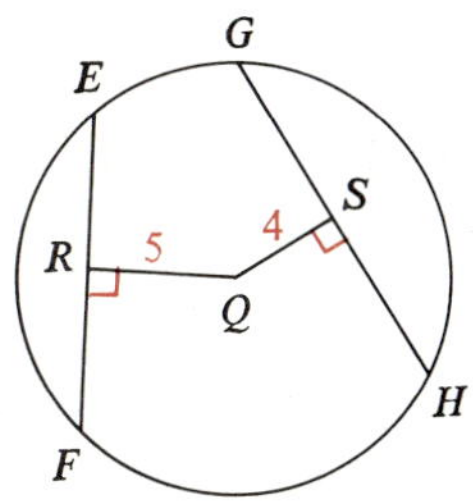

Example 2 In the figure at right, $\overline{QR} \perp \overline{EF}$ at R, and $\overline{QS} \perp \overline{GH}$ at S. If $RQ = 5$ and $QS = 4$, is $EF > GH$?

Solution No. Since $RQ > QS$, then $EF < GH$, by Theorem 9-2.6.

D

Definition 9-7 The **common chord** of two intersecting circles is the segment whose endpoints are the points of intersection of the two circles.

If circles P and Q intersect at points A and B, then $\overline{AB}$ is their common chord. $\overleftrightarrow{PQ}$ is called the **line of centers** of the two circles.

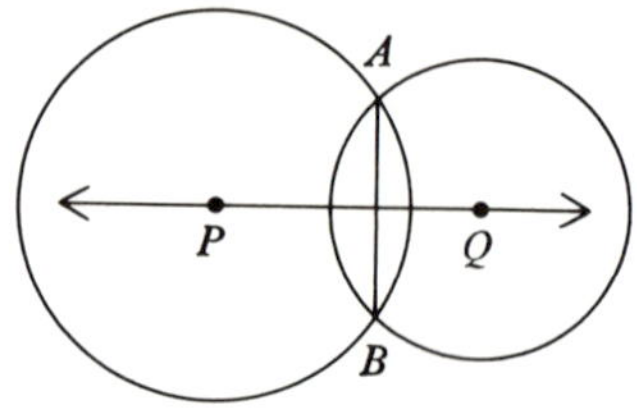

D

Definition 9-8 An **inscribed polygon** is a polygon whose vertices are all points of a circle.

EXERCISES

A

Indicate whether the following statements are true or false. Correct the false statements.

1. Two circles are congruent if their diameters are congruent.
2. If a line contains the center of a circle and the midpoint of one of the chords of the circle, and is perpendicular to another chord, then the two chords are perpendicular.
3. The perpendicular bisector of one chord of a circle is also the perpendicular bisector of only one other chord.
4. The distance from the center of a circle to a chord is the length of the segment whose endpoints are the center of the circle and the midpoint of the chord.
5. The closer a chord is to the center of a circle, the shorter it is.
6. GIVEN M is the midpoint of chord $\overline{AB}$ of $\odot Q$; $\overleftrightarrow{MQ}$ intersects $\odot Q$ at C.
   PROVE $\triangle ABC$ is isosceles.

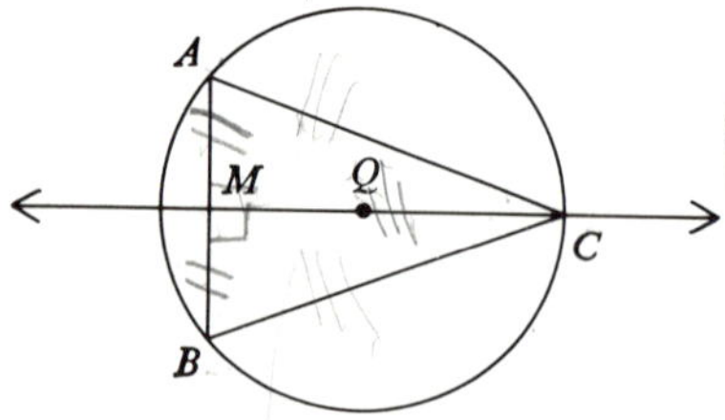

7. GIVEN $\overline{AM}$ is a median of $\triangle ABC$, which is inscribed in $\odot Q$; $\angle B \cong \angle C$
   PROVE $\overleftrightarrow{AM}$ contains point Q.

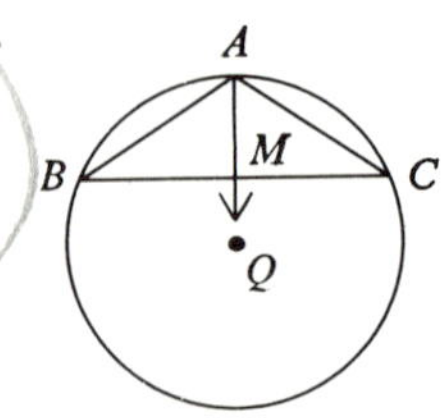

8. Prove that the midpoints of all the chords of a circle congruent to a given chord determine a circle concentric with the original circle.

Prove the following:

9. Theorem 9-2.1 10. Theorem 9-2.2. 11. Theorem 9-2.3.
12. Theorem 9-2.4 (second part) 13. Theorem 9-2.5
14. Corollary 9-2.5a

B

15. Present an indirect proof of Theorem 9-2.6.
16. Find the length of a chord of a circle of radius 13 if its distance from the center of the circle is 5.
17. Find the radius of a circle if a 14-inch chord is 24 inches from the center of the circle.
18. How far from the center of a circle of radius 10 is a chord of length 16?
19. Prove Corollary 9-2.3a.
20. Explain how we may use Theorem 9-2.3 to locate the center of a given circle.
21. $\overline{RS}$ is perpendicular to chords $\overline{ARB}$ and $\overline{DSC}$ of $\odot P$. If $AB = 6$, $RS = 7$, and the radius of $\odot P$ is 5, find DC.
22. GIVEN $\overleftrightarrow{ABP}$, $\overleftrightarrow{CDP}$, and $\overleftrightarrow{FQEP}$ are secants of $\odot Q$; A, B, C, and D are points of the circle; $\angle APF \cong \angle CPF$

    PROVE $\overline{AB} \cong \overline{CD}$; $\overline{BP} \cong \overline{DP}$; $\overleftrightarrow{QP} \perp \overleftrightarrow{BD}$; (*Hint*: Consider $\overline{QM} \perp \overline{AP}$; and $\overline{QN} \perp \overline{CP}$.)

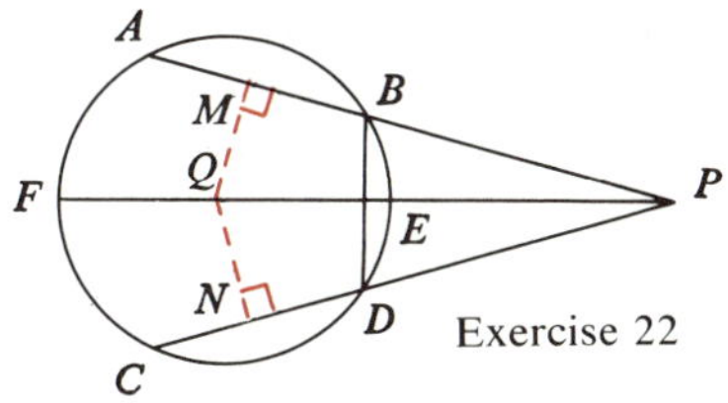

Exercise 22

C

23. GIVEN $\overleftrightarrow{ABP}$, $\overleftrightarrow{CDP}$, and $\overleftrightarrow{FQEP}$ are segments of $\odot Q$; A, B, C, D are points of the circle; $\overline{AP} \cong \overline{CP}$

    PROVE $\angle APF \cong \angle CPF$ and $\overline{AB} \cong \overline{CD}$; (*Hint*: Consider $\overline{AQ}$ and $\overline{CQ}$; $\overline{QM} \perp \overline{AP}$; and $\overline{QN} \perp \overline{CP}$.)
24. The length of the longest chord of a circle passing through interior point P is 10 inches, while the length of the shortest chord through point P is 6. Find the radius of the circle and the distance from point P to the center of the circle.
25. GIVEN Circles P and Q intersect in points C and D; $\overleftrightarrow{ACB} \parallel \overleftrightarrow{PQ}$; A and B are points of circles P and Q, respectively.

    PROVE $PQ = \frac{1}{2}AB$

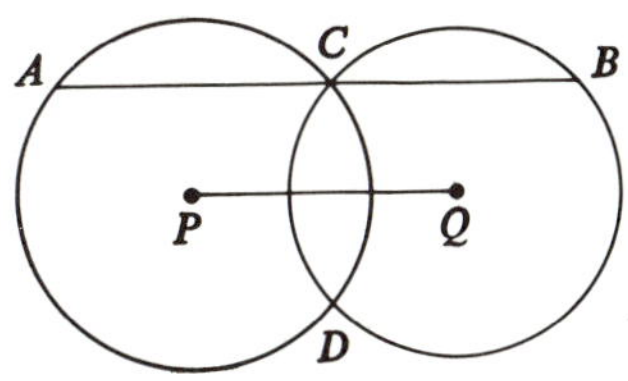

26. GIVEN Circles P and Q intersect in A and C; $\overrightarrow{PQ}$ meets $\overrightarrow{BA}$ and $\overrightarrow{DC}$ at T, a point of $\odot P$.

PROVE $\overline{AB} \cong \overline{CD}$

Exercise 26

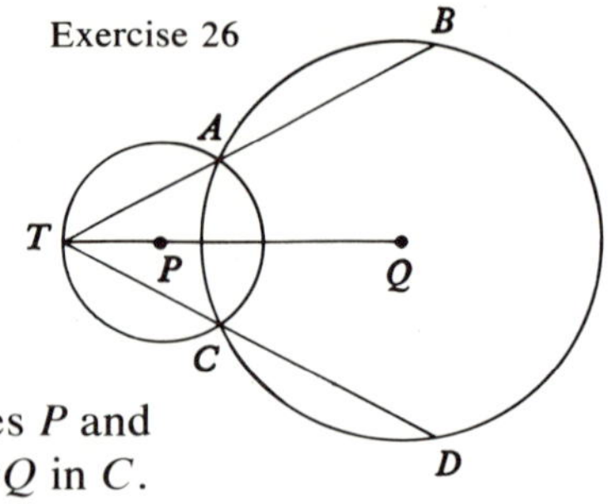

27. $\overline{AB}$ is the common chord of circles P and Q; $\overleftrightarrow{PQ}$ intersects $\odot P$ in D and $\odot Q$ in C. Find AQ if $AP = 13$, $AB = 10$, and $CD = 3$.

28. GIVEN Point Q is in the interior of $\odot P$; $\overrightarrow{QP}$ intersects $\odot P$ in R; S is any point of $\odot P$.

PROVE $RQ > SQ$ (*Hint*: Consider $\overline{PS}$)

9-3 TANGENT LINES AND CIRCLES

We defined a secant as a line that intersects a circle in two points. We shall now consider a line that intersects a circle in exactly one point.

D **Definition 9-9** A **tangent** to a circle is a line in the plane of the circle that intersects the circle in exactly one point.

The point of intersection is called the **point of tangency** or point of contact. The line and the circle are said to be tangent at this point.

In this section, we shall be concerned with circles and tangents in one plane.

How many different segments may be radii of a given circle? How many tangents does a circle have? Perhaps the following theorem will help you answer these questions:

T **Theorem 9-3.1** A line perpendicular to a radius at the endpoint on the circle is tangent to the circle.

To prove this theorem, we must show that not more than one point of the perpendicular line is a point of the circle. To do this we select a random point in line l and show that it is in the exterior of the circle.

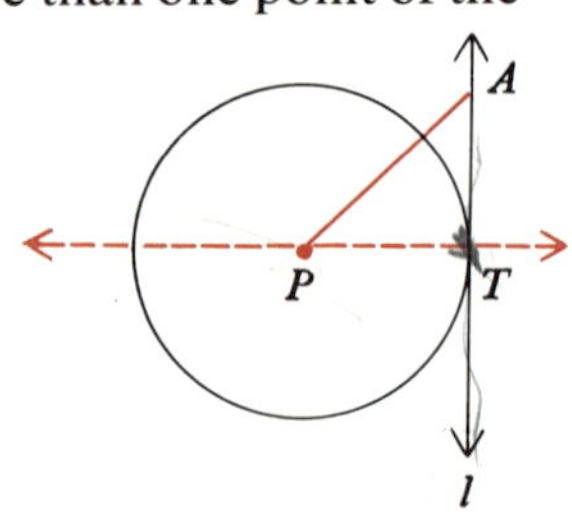

GIVEN $\overline{PT}$ is a radius of $\odot P$; $\overline{PT} \perp l$ at T; A is a point of l distinct from T.

PROVE A is in the exterior of $\odot P$.

PROOF OUTLINE Since the hypotenuse of a right triangle is the longest side, $AP > PT$. Therefore, by the definition of the exterior of a circle, point A is in the exterior of $\odot P$. Thus, the only point of l which is a point of $\odot P$ is T. By definition, line l is tangent to $\odot P$ at T.

In the figure above, consider line l tangent to $\odot P$ at T. Select any point A of l, distinct from point T. Why is $PT < PA$? What is the shortest distance from P to line l? Why? By applying Theorem 5-4.2 we get the following theorem:

T **Theorem 9-3.2** The radius of a circle is perpendicular to a tangent at the point of tangency.

The proof implied above is left as an exercise. Copy the figure above and draw $\overleftrightarrow{PT}$. How do you know that $\overleftrightarrow{PT}$ is perpendicular to line l at T? Can you name a special point in the interior of $\odot P$ that is in $\overleftrightarrow{PT}$? Your answer forms the basis of the following theorem.

T **Theorem 9-3.3** A line perpendicular to a tangent of a circle at the point of tangency contains the center of the circle.

The proof of this theorem is left as an exercise.

Class Exercises

Draw a diagram to illustrate your answer to each of the following questions:

1. Can a line be tangent to more than one circle at the same time?
2. Can a line be tangent to more than one circle at the same point?
3. If two circles are tangent to a line at the same point, assuming they do not coincide, how many points do these circles have in common?
4. What relationship exists between the line containing the centers of the circles in Exercise 3 and the tangent? Why?
5. Can two concentric circles of unequal radii be tangent to the same line?
6. Can two intersecting circles be tangent to the same line?

Your answers to these Class Exercises should suggest the following two definitions:

D **Definition 9-10** A **common tangent** of two circles is a line that is tangent to two coplanar circles.

A common tangent is a common *internal* tangent if the centers of the circles are on opposite sides of the tangent. A common tangent is a common *external* tangent if the centers of the circles are on the same side of the tangent.

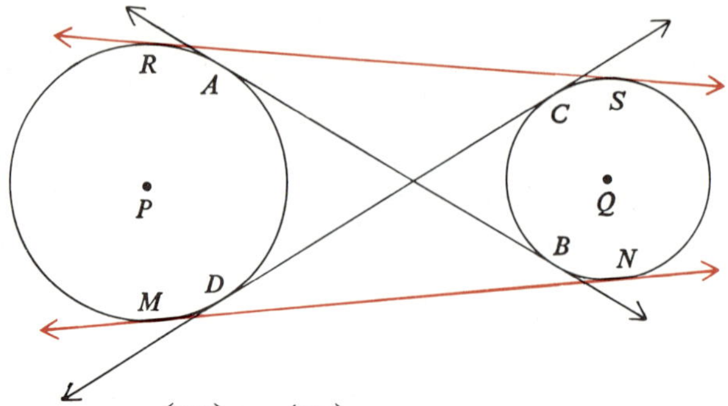

In the figure above $\overleftrightarrow{AB}$ and $\overleftrightarrow{CD}$ are common internal tangents of circles P and Q, while $\overleftrightarrow{MN}$ and $\overleftrightarrow{RS}$ are common external tangents of the circles.

D

Definition 9-11 **Tangent circles** are two coplanar circles that intersect in exactly one point.

They are *externally tangent* if their centers are on opposite sides of the common tangent. They are *internally tangent* if their centers are on the same side of their common tangent.

Circles A and B are externally tangent at point T. The line l containing point T and no other points of $\odot A$ or $\odot B$ is the common tangent.

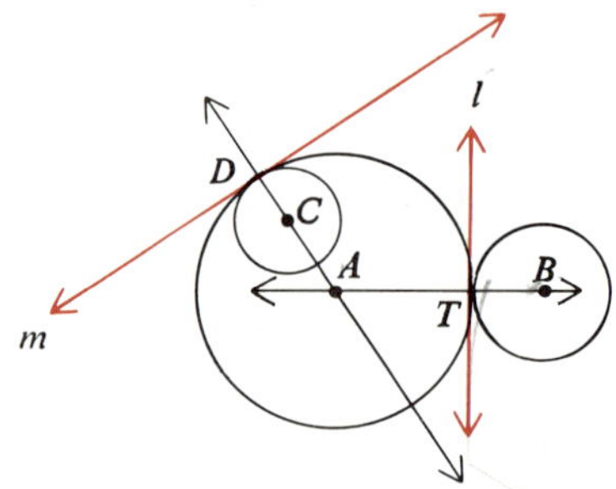

Circles A and C are internally tangent at D. What is their line of centers? If line m contains D and no other points of $\odot A$ and $\odot C$, what do we call line m?

Example 1 How many common tangents does each of the following pairs of circles have: (a) concentric circles, (b) internally tangent circles, (c) nontangent intersecting circles, (d) externally tangent circles, and (e) externally nonintersecting circles?

Solution

(a) Concentric circles have no common tangents.
(b) Internally tangent circles have one common external tangent.
(c) Nontangent intersecting circles have two common external tangents.
(d) Externally tangent circles have three common tangents: two external tangents and one internal tangent.
(e) Externally nonintersecting circles have four common tangents: two external and two internal common tangents.

EXERCISES

A

Indicate whether Exercises 1–3 are true or false. Correct the false statements.

1. A tangent is always perpendicular to some radius of a circle.
2. A tangent is a segment intersecting a circle in one point.
3. Only one tangent can be drawn to a circle from an external point.
4. Two circles are externally tangent. The radius of one circle is 11 and the radius of the other circle is 7. Find the length of the line segment whose endpoints are the centers of the circles.
5. The diameters of two internally tangent circles are 8 and 12. Find the length of the segment joining the centers of the circles.

The radii of two coplanar circles are 5 and 2. Find the number of points of intersection of the two circles if the length of the segment joining their centers is:

6. 2 7. 3 8. 4 9. 7 10. 8

11. Find the number of common tangents in each of Exercises 6–10.
12. If two or more circles are tangent to line l at T, prove that their line of centers is perpendicular to l.
13. Prove that the tangents to a circle at the endpoints of a diameter are parallel.

14. GIVEN $\overleftrightarrow{ATB}$ is tangent to $\odot P$ at T; M is the midpoint of chord $\overline{KL}$ of $\odot P$; $\overline{CMPT}$ is a diameter of $\odot P$.

 PROVE $\overline{KL} \parallel \overleftrightarrow{AB}$

15. GIVEN $\odot P \cong \odot Q$; l is tangent to circles P and Q at point T; A is any point of l.

 PROVE $\overline{AP} \cong \overline{AQ}$

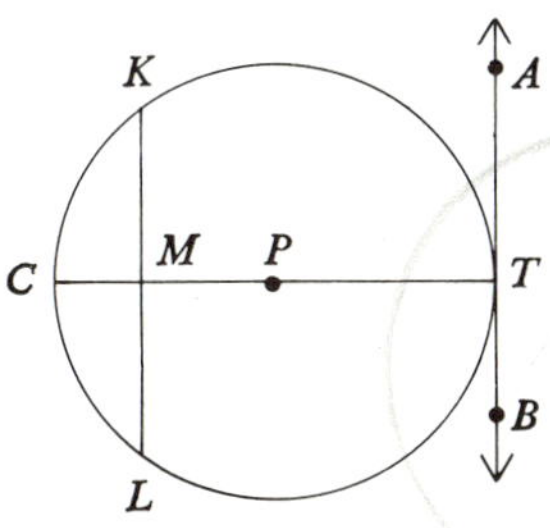

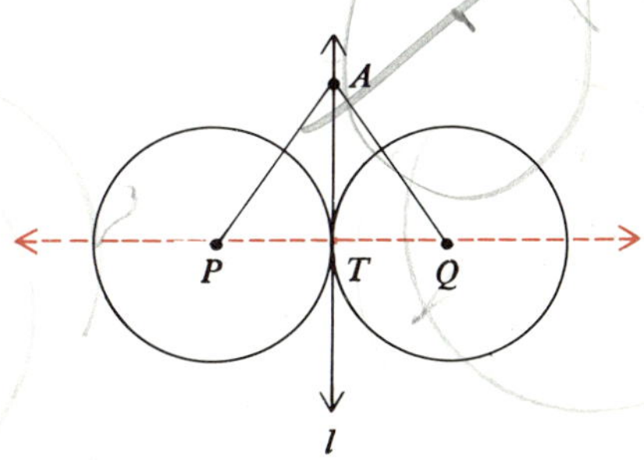

B

16. GIVEN $\overleftrightarrow{TA}$ and $\overleftrightarrow{TB}$ are tangent to $\odot P$ at points A and B, respectively; M is the midpoint of $\overline{AB}$.

 PROVE $\triangle APB$ is isosceles; $\triangle ATB$ is isosceles.

17. **Given** $\overline{CD}$ is tangent to $\odot P$ at M; $\overline{AB}$ is a diameter of $\odot P$; $\overline{AC} \perp \overline{CD}$ and $\overline{BD} \perp \overline{CD}$.
Prove M is the midpoint of $\overline{CD}$.

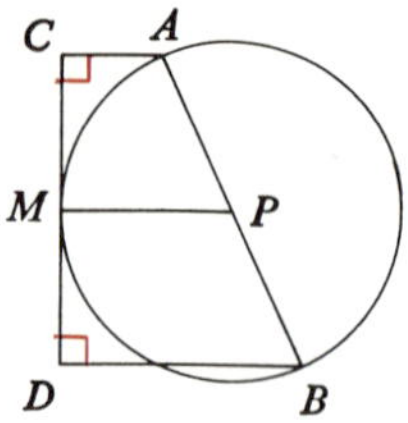

Exercise 17

18. **Given** $\overleftrightarrow{AB}$ is tangent to $\odot P$ at A and to $\odot Q$ at B; $\overleftrightarrow{AB}$ meets $\overline{PRQ}$ at R.
Prove $AR \cdot QR = BR \cdot PR$
19. Present a formal proof of Theorem 9-3.1.
20. Prove Theorem 9-3.2, using a direct proof.
21. Prove Theorem 9-3.2, using an indirect proof.
22. Prove Theorem 9-3.3, using an indirect proof.

C

23. Circles A, B, and C are tangent to one another. Find the radii of the three circles if $AB = 7$, $AC = 5$, and $BC = 9$.
24. **Given** $\overleftrightarrow{AB}$ and $\overleftrightarrow{CD}$ are the common external tangents of circles P and Q; A, B, C, and D are the points of tangency.
Prove $\overline{AB} \cong \overline{CD}$ (*Hint*: Consider the dashed lines in the figure.)

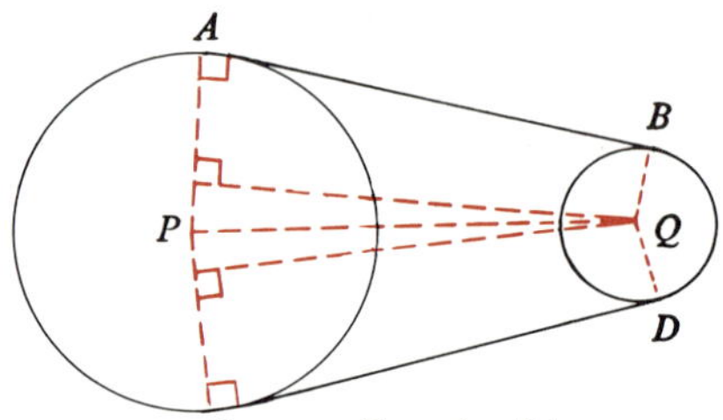

Exercise 24

25. The diameters of two externally tangent circles are 27 and 12. Find the length of the segment of the common external tangent whose endpoints are the points of tangency.
26. The distance between the centers of two circles whose radii are 10 and 2 is 17. Find the length of the segment of the common external tangent whose endpoints are the points of contact.
27. **Given** $\overline{MN}$ is a diameter of $\odot P$; $\overline{MR}$ is a chord of $\odot P$; $\overline{TR}$ is tangent to $\odot P$ at R; $\overleftrightarrow{PT}$ is a secant of $\odot P$ so that $\overleftrightarrow{PT} \parallel \overline{MR}$.
Prove $\overleftrightarrow{NT}$ is tangent to $\odot P$. (*Hint*: Consider $\overline{RP}$.)

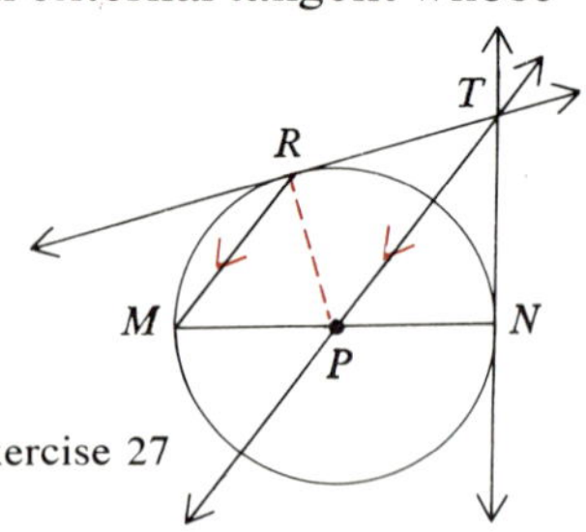

Exercise 27

9-4 PLANES AND SPHERES

There are three ways in which a plane and a sphere may intersect.

Case I A plane may contain the center of the sphere.

Case II A plane may contain an interior point of the sphere but not the center of the sphere.

Case III A plane may contain exactly one point of the sphere.

Each of these cases has different properties.
Consider plane P containing the center of sphere Q.

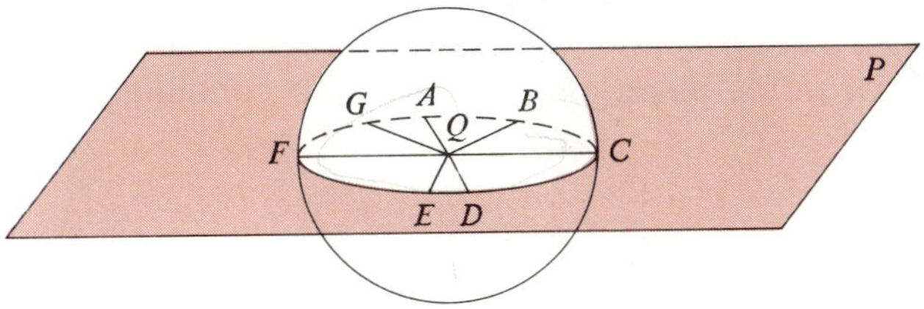

The radii of sphere Q are all congruent to each other. Therefore, the points A, B, C, D, E, F, and G, which are in the intersection of plane P and sphere Q, are equidistant from Q. Are there any other points in plane P that are the same distance as point A from Q? Since all the points in this intersection are equidistant from a point in their plane, we can state the following theorem:

T **Theorem 9-4.1** The intersection of a sphere with a plane containing the center of the sphere is a circle whose center and radius are the same as those of the sphere.

This circle is called a **great circle of the sphere**. The great circle in spherical geometry (see the Mathematical Excursion at the end of this chapter) is comparable to the line in plane geometry.

Suppose that a sphere is intersected by a plane that does not contain its center, but does contain other interior points of the sphere. What do you think the intersection is?

T **Theorem 9-4.2** The intersection of a sphere with a plane that contains points in the interior of the sphere is a circle of the sphere.

This circle is called the **circle of intersection.** When the center of the circle of intersection is not the center of the sphere, we call the circle a **small circle of the sphere**, in contrast to the great circle mentioned before.

T **Theorem 9-4.3** If a line contains the center of a sphere and the center of the circle of intersection of the sphere with a plane that does not contain the center of the sphere, then the line is perpendicular to the intersecting plane.

Theorem 9-4.4 If a line contains the center of a sphere and is perpendicular to a plane that intersects the sphere and contains interior points of the sphere other than the center, then the line also contains the center of the circle of intersection.

Theorem 9-4.5 If a line contains the center of the circle of intersection of a sphere with a plane that does not contain its center and if the line is perpendicular to the plane, then the line also contains the center of the sphere.

The proofs of these theorems are left for you to complete as exercises.

How do these theorems compare with Theorems 9-2.1, 9-2.2, and 9-2.3?

Example 1 What is the radius of the circle of intersection of a plane with a sphere of radius 25 if the plane is 24 units from the center of the sphere?

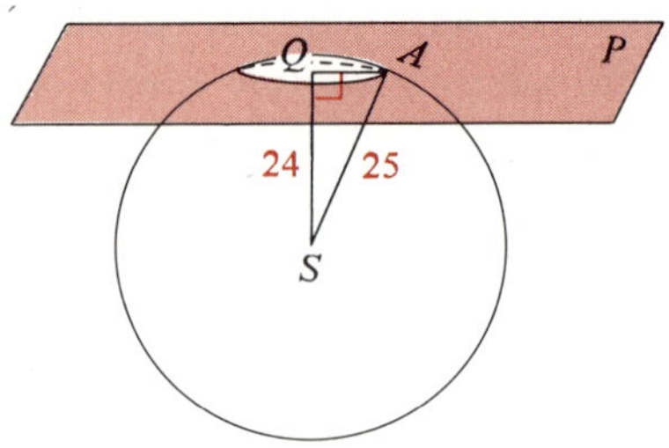

Solution In the figure at right, plane P intersects sphere S to form $\odot Q$. Point A is a point of $\odot Q$. $\overline{SQ} \perp$ plane P by Theorem 9-4.3; hence $\overline{SQ} \perp \overline{QA}$. Why? By using the Pythagorean Theorem, we find that since $SQ = 24$ and $SA = 25$, radius $QA = 7$.

Example 2

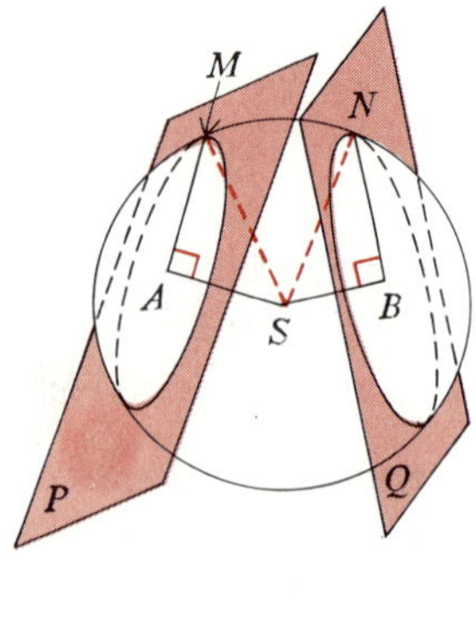

Given Planes P and Q intersect sphere S in $\odot A$ and $\odot B$, respectively; M is a point of $\odot A$; N is a point of $\odot B$; $\overline{AS} \cong \overline{BS}$

Prove $\odot A \cong \odot B$

Solution

Proof

STATEMENTS	REASONS
1. Planes P and Q intersect sphere S in $\odot A$ and $\odot B$, respectively; M is a point of $\odot A$; N is a point of $\odot B$; $\overline{AS} \cong \overline{BS}$	1. Given.
2. $\overline{MS} \cong \overline{NS}$	2. Definition of a sphere. (9-2)
3. $\overline{AS} \perp$ plane P; $\overline{BS} \perp$ plane Q	3. Theorem 9-4.3.
4. $\overline{AS} \perp \overline{AM}$; $\overline{BS} \perp \overline{BN}$	4. A plane perpendicular to a line contains every line that is perpendicular to the line and that also intersects the line at the same point as the plane. (4-5.2)
5. $\angle A$ and $\angle B$ are right angles.	5. Definition of perpendicular lines. (1-25)
6. $\triangle MAS$ and $\triangle NBS$ are right triangles.	6. Definition of a right triangle. (1-32)
7. $\triangle MAS \cong \triangle NBS$	7. If the hypotenuse and one leg of a right triangle are congruent to the corresponding hypotenuse and leg of another right triangle, then the two triangles are congruent. (6-6.2)
8. $\overline{MA} \cong \overline{NB}$	8. Definition of congruent triangles. (3-3)
9. $\odot A \cong \odot B$	9. Definition of congruent circles. (9-6)

D

Definition 9-12 A **tangent plane** of a sphere is a plane that contains exactly one point of the sphere.

The point of intersection is called the **point of tangency** or the point of contact.

By adapting Theorems 9-3.1, 9-3.2, and 9-3.3 and applying them to spheres instead of circles, we create the following new theorems:

T

Theorem 9-4.6 A plane perpendicular to a radius of a sphere at its intersection with the sphere is tangent to the sphere.

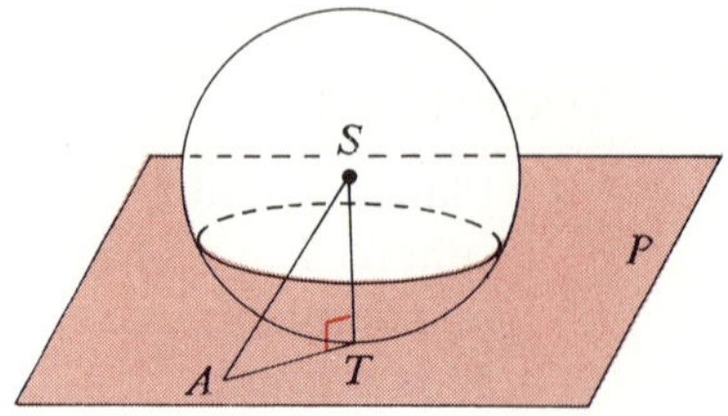

GIVEN Radius $\overline{ST}$ of sphere S is perpendicular to plane P at T.

PROVE Plane P is tangent to sphere S at T.

PROOF

STATEMENTS	REASONS
1. Radius $\overline{ST}$ of sphere S is perpendicular to plane P at T.	1. Given.
2. Select any point A in plane P.	2. Point Uniqueness Postulate. (2-2)
3. $\overline{ST} \perp \overline{AT}$	3. A plane perpendicular to a line contains every line that is perpendicular to the line and that also intersects the line at the same point as the plane. (4-5.2)
4. $\angle STA$ is a right angle.	4. Definition of perpendicular lines. (1-25)
5. $\triangle STA$ is a right triangle.	5. Definition of a right triangle. (1-32)
6. $m\angle T > m\angle A$	6. If a triangle has one right angle, then the other two angles must be acute (C5-2.1a); and definition of an acute angle. (1-24)
7. $SA > ST$	7. If two angles of a triangle are not congruent, then the sides opposite those angles are not congruent, the longer side being opposite the angle with greater measure. (5-3.2)
8. Point A is in the exterior of sphere S.	8. Definition of the exterior of a sphere
9. Point T is the only point of intersection of plane P and sphere S.	9. The only remaining possibility, since all other points of plane P, represented by A, are in the exterior of sphere S.
10. Plane P is tangent to sphere S at point T.	10. Definition of a tangent plane. (9-12)

T **Theorem 9-4.7** The radius of a sphere is perpendicular to a tangent plane at the point of tangency.

T **Theorem 9-4.8** A line perpendicular to a tangent plane of a sphere at the point of tangency contains the center of the sphere.

The proofs of these theorems, similar to the proofs of Theorems 9-3.2 and 9-3.3, are left as exercises.

EXERCISES

Complete each of the following statements:

A

1. If the circle of intersection of a plane with a sphere has the same radius as the sphere, then the circle also has the same ________ as the sphere.
2. The intersection of a plane with a sphere is either a(n) ________ or a(n) ________.
3. A plane which contains exactly one point of a sphere is called a(n) ________.
4. If the intersection of a plane with a sphere contains more than one point, then the intersection is a ________.
5. Two distinct tangent planes of a sphere that intersect the same diameter of the sphere are ________.
6. A line perpendicular to a plane containing the circle of intersection of the plane with a sphere contains the center of the sphere if it also contains the ________.
7. Prove that the radius of a great circle is longer than the radius of a small circle of the same sphere.
8. Prove that all great circles of the same sphere are congruent.
9. Prove that any two great circles of a sphere intersect in the endpoints of a diameter of the sphere.
10. Prove that the line perpendicular to a chord of a sphere and containing the center of a sphere bisects the chord of the sphere that it intersects.
11. Prove Theorem 9-4.2.
12. Prove Theorem 9-4.3.
13. Prove Theorem 9-4.4.
14. Prove Theorem 9-4.5.
15. Prove Theorem 9-4.7.
16. Prove Theorem 9-4.8.

B

17. If the diameter of a sphere is 13 inches, find the diameter of a circle of intersection in a plane 6 inches from the center of the sphere.
18. Find the radius of a sphere if a plane 7 inches from its center intersects the sphere in a circle whose diameter is 14 inches.
19. How far from the center of a sphere of radius 15 is a plane whose circle of intersection with the sphere has a diameter of 24?

C

20. **Given** $\overleftrightarrow{TA}$ and $\overleftrightarrow{TB}$ are tangent to sphere S at points A and B, respectively.
Prove $TA \cong TB$

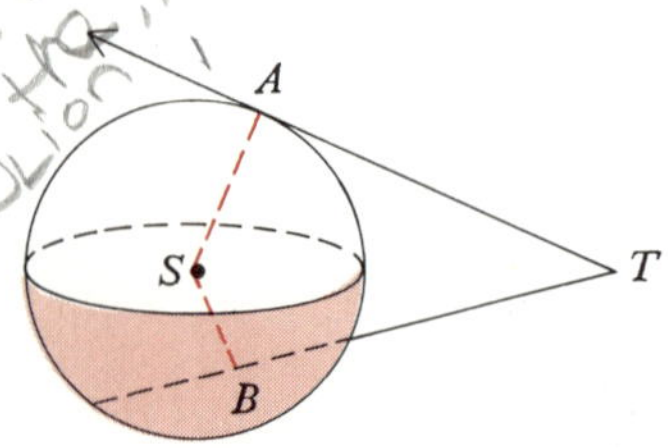

21. **Given** Plane P intersects sphere S in $\odot Q$; A and B are points in $\odot Q$; $\overline{SQ} \cong \overline{AQ}$; $\overline{AQ} \perp \overline{BQ}$
Prove $\triangle ABS$ is equilateral.

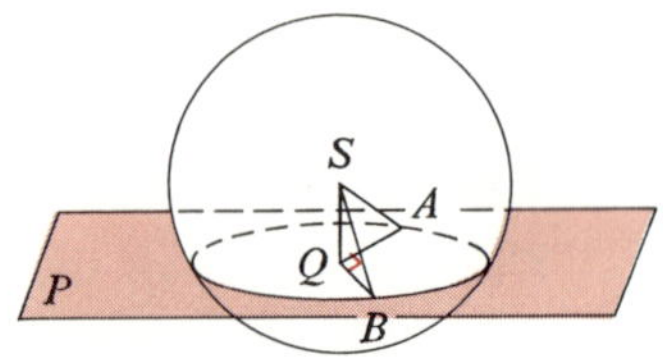

22. Explain how we can locate the center of a sphere using Theorem 9-4.5.
23. Prove that if two planes intersect a sphere in two noncongruent circles, then the plane nearer the center of the sphere contains the circle with the longer diameter.

9-5 ARCS OF CIRCLES

In this section, we shall devote our attention exclusively to circles and the properties of their parts.

D **Definition 9-13** A **central angle** of a circle is an angle whose vertex is at the center of the circle.

In the figure at right, $\angle APB$ is a central angle of $\odot P$. The part of the circle in color is called *minor arc AB*. The remainder of the circle is called *major arc AB*. "Arc AB" is denoted by $\overset{\frown}{AB}$. The **endpoints of an arc** are the two points of intersection of a central angle with the circle.

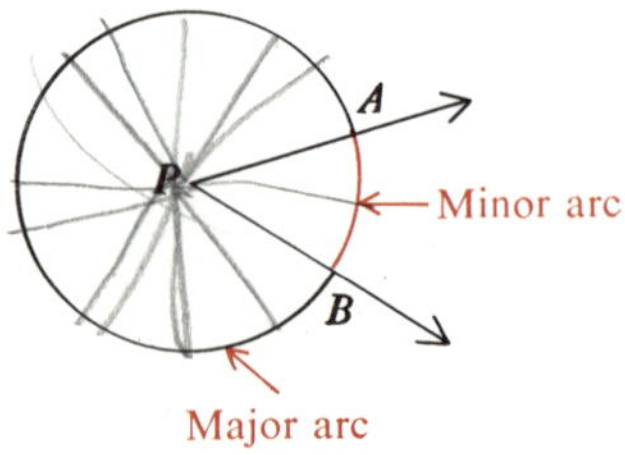

D **Definition 9-14** A **minor arc** of a circle is the part of the circle intersected by a central angle and included in the angle's interior.

D **Definition 9-15** A **major arc** of a circle is the part of the circle intersected by a central angle and included in the angle's exterior.

Suppose that the endpoints of an arc of a circle are also the endpoints of a diameter of the circle. The two arcs so determined are called *semicircles*. $\overline{AB}$ is a diameter of $\odot P$ at right. $\overset{\frown}{AWB}$ and $\overset{\frown}{AVB}$ are both semicircles.

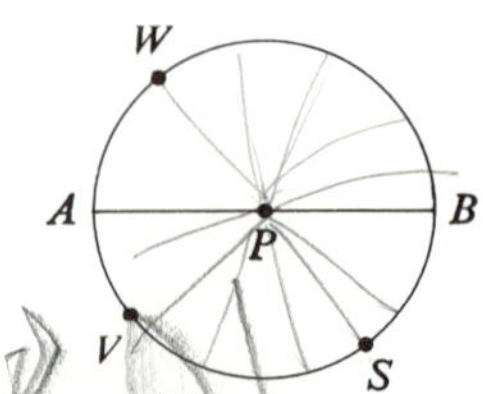

D **Definition 9-16** A **semicircle** is an arc of a circle whose endpoints are the endpoints of a diameter of the circle.

$\overset{\frown}{AB}$ denotes the arc with endpoints A and B. Sometimes, to avoid ambiguity, we shall label another point on the circle and refer to the arc using three letters, as we did with $\overset{\frown}{AWB}$ and $\overset{\frown}{AVB}$ above.
We shall use the notation m $\overset{\frown}{ASB}$ for the degree measure of $\overset{\frown}{ASB}$.

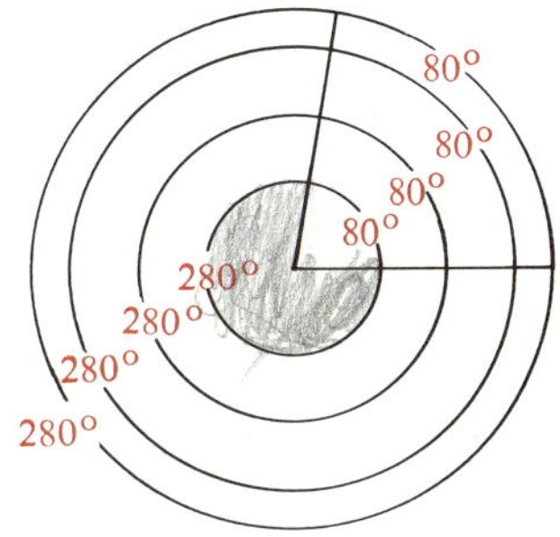

You know that the scale of a protractor is in the shape of a semicircle. You should also recall that, regardless of its size, every protractor measures 180°. This suggests that all semicircles, regardless of size, have a degree measure of 180. The diagram at right suggests that the degree measure of any arc does not depend on the size of its circle, but rather on the measure of its central angle.

D **Definition 9-17** The **degree measure of a minor arc** is the measure of its corresponding central angle.
The **degree measure of a semicircle** is 180.
The **degree measure of a major arc** is equal to 360 minus the degree measure of its corresponding minor arc.

From now on, we shall refer to the degree measure of an arc simply as the **measure of an arc**.

Although the following statement is easy to understand, its proof is difficult. Hence, we shall accept it as a postulate.

P **Postulate 9-1 The Arc Addition Postulate** If P is a point of $\overset{\frown}{AB}$, distinct from A and B, then $m\overset{\frown}{APB} = m\overset{\frown}{AP} + m\overset{\frown}{PB}$.

An alternate form of this postulate is "$m\overset{\frown}{AP} = m\overset{\frown}{APB} - m\overset{\frown}{PB}$."

D **Definition 9-18** **Congruent arcs** are arcs of the same or congruent circles whose degree measures are equal.

From the above definition and the definition of the degree measure of an arc, we derive the next theorem.

T **Theorem 9-5.1** In the same circle or in congruent circles, two arcs are congruent if and only if their corresponding central angles are congruent.

The proof of this theorem is left as an exercise.

D

Definition 9-19 The **chord of an arc** of a circle is the chord of the circle whose endpoints are the same as those of the arc.

Theorems 9-5.2 and 9-5.3 relate a chord to its arc.

T

Theorem 9-5.2 In the same circle, or in congruent circles, if two chords are congruent, then their arcs are congruent.

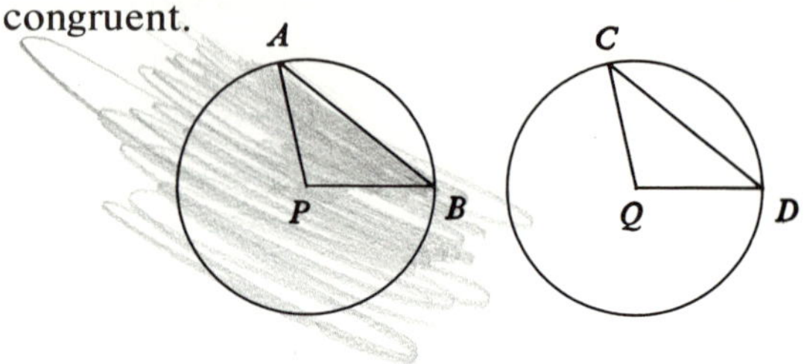

GIVEN $\odot P \cong \odot Q$
$\overline{AB} \cong \overline{CD}$

PROVE $\widehat{AB} \cong \widehat{CD}$

PROOF

STATEMENTS	REASONS
1. $\odot P \cong \odot Q$; $\overline{AB} \cong \overline{CD}$	1. Given.
2. $\overline{AP} \cong \overline{CQ}$; $\overline{BP} \cong \overline{DQ}$	2. Definition of congruent circles (9-6)
3. $\triangle ABP \cong \triangle CDQ$	3. SSS Postulate (3-3)
4. $\angle P \cong \angle Q$	4. Definition of congruent triangles (3-3)
5. $\widehat{AB} \cong \widehat{CD}$	5. Theorem 9-5.1

Example 1

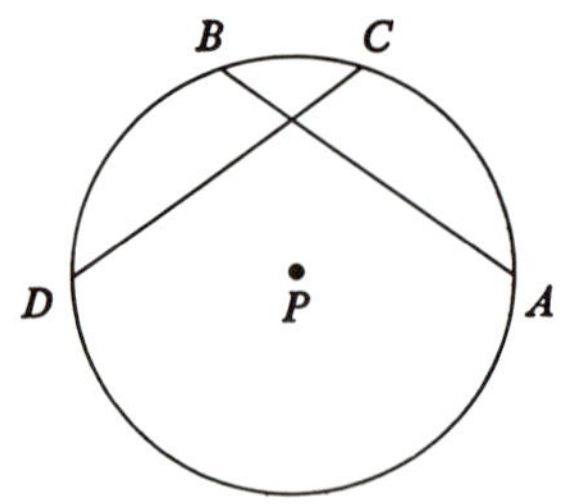

GIVEN $\overline{AB}$ and $\overline{CD}$ are chords of $\odot P$;
$\overline{AB} \cong \overline{CD}$

PROVE $\widehat{AC} \cong \widehat{BD}$

Solution

PROOF

STATEMENTS	REASONS
1. $\overline{AB}$ and $\overline{CD}$ are chords of $\odot P$; $\overline{AB} \cong \overline{CD}$	1. Given.
2. $\widehat{ACB} \cong \widehat{DBC}$	2. Theorem 9-5.2
3. $m\widehat{ACB} = m\widehat{DBC}$	3. Definition of congruent arcs (9-18)
4. $m\widehat{BC} = m\widehat{BC}$	4. Reflexive property
5. $m\widehat{AC} = m\widehat{BD}$	5. Arc Addition Postulate (9-1)
6. $\widehat{AC} \cong \widehat{BD}$	6. Definition of congruent arcs (9-18)

The converse of Theorem 9-5.2 is as follows:

T **Theorem 9-5.3** In the same circle, or in congruent circles, if two arcs are congruent, then their chords are congruent.

The proof of this theorem entails reasoning that reverses the proof of Theorem 9-5.2, and is left as an exercise.

Example 2

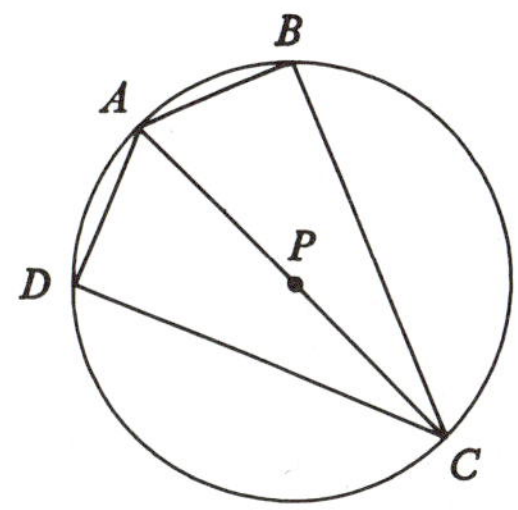

GIVEN Points A, B, C, and D are in $\odot P$; $\overset{\frown}{AB} \cong \overset{\frown}{AD}$; $\overset{\frown}{BC} \cong \overset{\frown}{DC}$

PROVE $\angle B \cong \angle D$

Solution

PROOF

STATEMENTS	REASONS
1. Points A, B, C, and D are in $\odot P$; $\overset{\frown}{AB} \cong \overset{\frown}{AD}$; $\overset{\frown}{BC} \cong \overset{\frown}{DC}$	1. Given.
2. $\overline{AB} \cong \overline{AD}$; $\overline{BC} \cong \overline{DC}$	2. Theorem 9-5.3.
3. $\overline{AC} \cong \overline{AC}$	3. Every segment is congruent to itself. (3-1.6)
4. $\triangle ABC \cong \triangle ADC$	4. SSS Postulate. (3-3)
5. $\angle B \cong \angle D$	5. Definition of congruent triangles. (3-3)

Class Exercises

Complete each of the following exercises for $\odot P$ at right. Justify each of your answers.

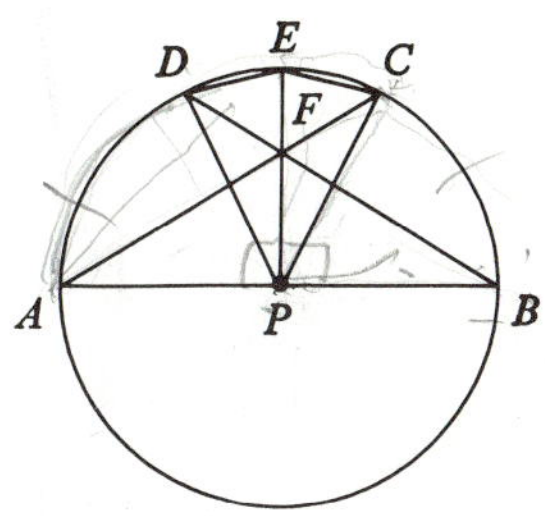

1. If $\overline{AB}$ is a diameter of $\odot P$, then $\overset{\frown}{ADB}$ is a ________.
2. m$\angle CPB$ = ________. (arc)
3. By the Arc Addition Postulate, m$\overset{\frown}{AD}$ = m$\overset{\frown}{ADC}$ − m ________.
4. If $\overset{\frown}{AD} \cong \overset{\frown}{BC}$ and $\overline{EP} \perp \overline{APB}$, then $\angle DPE \cong$ ____.
5. If $\angle DPE \cong \angle CPE$, then $\overset{\frown}{DE} \cong$ ____.
6. If $\overline{AC} \cong \overline{BD}$, then $\overset{\frown}{ADC} \cong$ ____.
7. If $\overline{EP} \perp \overline{APB}$, then $\overset{\frown}{AE} \cong$ ____.
8. If $\overset{\frown}{AD} \cong \overset{\frown}{BC}$, then $\overset{\frown}{ADC} \cong$ ____.
9. If $\overset{\frown}{AEC} \cong \overset{\frown}{BED}$, then $\overline{AC} \cong$ ____.
10. If $\overset{\frown}{DE} \cong \overset{\frown}{CE}$, then $\triangle DPE \cong$ ____.

EXERCISES

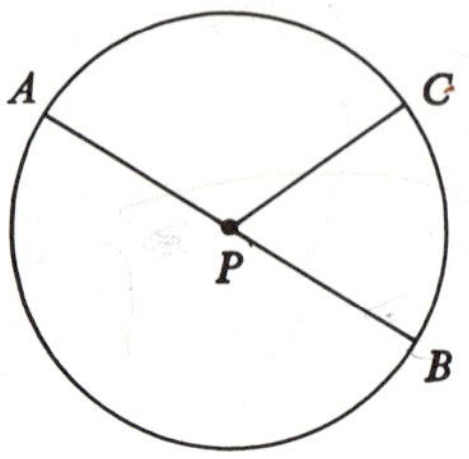

Exercises 1–5

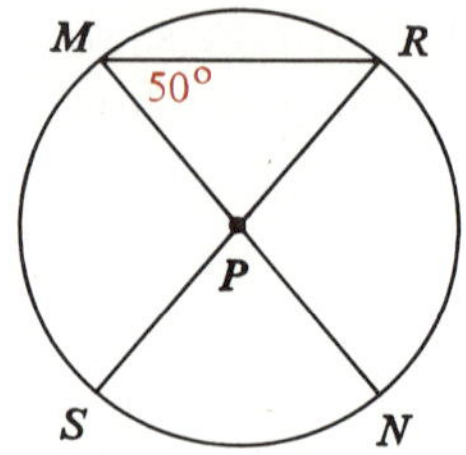

Exercises 6–11

A

In the figure at left above, $\overline{APB}$ is a diameter of $\odot P$; $\overline{PC}$ is a radius.

1. Name a semicircle.
2. Name two minor arcs.
3. Name two major arcs.
4. Name the corresponding arc of $\angle APC$.
5. Name the corresponding central angle of $\widehat{BC}$.

In the figure at right above, $\overline{MN}$ and $\overline{RS}$ are diameters of $\odot P$ and $m\angle RMN = 50$.

6. Find $m\angle R$.
7. Find $m\widehat{MR}$.
8. Find $m\widehat{NS}$.
9. Find $m\widehat{MS}$.
10. Find $m\angle MPS$.
11. Find $m\widehat{RNS}$.

12. GIVEN Equilateral $\triangle ABC$ inscribed in $\odot P$
    PROVE $m\widehat{AB} = m\widehat{BC} = m\widehat{AC}$

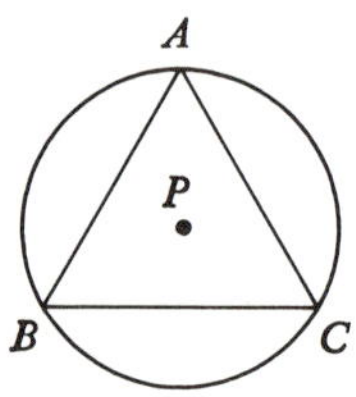

13. GIVEN Regular pentagon $DEFGH$ inscribed in $\odot Q$
    PROVE $\widehat{DEF} \cong \widehat{HGF}$

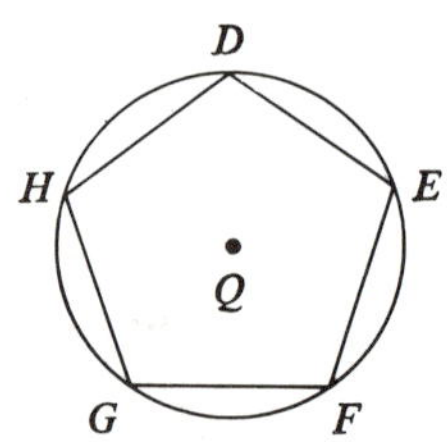

14. GIVEN A, M, and B are in $\odot P$; $\overline{MR} \perp \overline{ARP}$ and $\overline{MS} \perp \overline{BSP}$; $\overline{MR} \cong \overline{MS}$
    PROVE $\widehat{AM} \cong \widehat{MB}$
15. GIVEN A, P, and B are in $\odot Q$; $\overline{PQ} \perp \overline{AB}$ at M.
    PROVE $\widehat{AP} \cong \widehat{BP}$
16. GIVEN A and B are in $\odot Q$; M is the midpoint of $\overline{AB}$; $\overline{QM}$ intersects $\odot Q$ at P.
    PROVE $\widehat{AP} \cong \widehat{BP}$
17. GIVEN A, P, and B are points of $\odot Q$; M is the midpoint of $\overline{AB}$; $\overline{MP} \perp \overline{AB}$
    PROVE $\widehat{AP} \cong \widehat{BP}$

18. Prove Theorem 9-5.1.
19. Prove Theorem 9-5.3.
20. GIVEN $\overline{AD}$ is a diameter of $\odot P$; $\overline{AB}$ and $\overline{AC}$ are chords; $\overline{AB} \cong \overline{AC}$
PROVE $\overline{BD} \cong \overline{CD}$

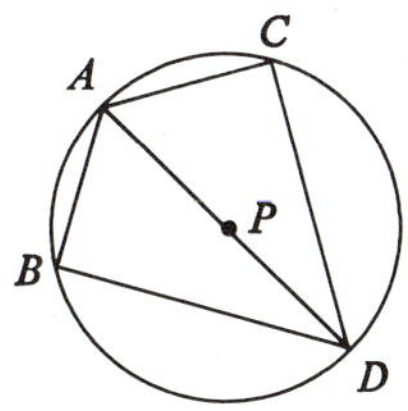

Exercise 20

B

21. GIVEN Points A, D, B, and C are points of $\odot Q$; $m\angle CEB = m\angle DEB$; $\overline{CE} \cong \overline{DE}$
PROVE $\overparen{CB} \cong \overparen{BD}$; $\overline{AB}$ is a diameter of $\odot Q$.

22. GIVEN Points A, N, B, and M are points of $\odot P$; $\angle ABM \cong \angle ABN$; $\overline{AB}$ is a diameter of $\odot P$.
PROVE $\overparen{MA} \cong \overparen{NA}$

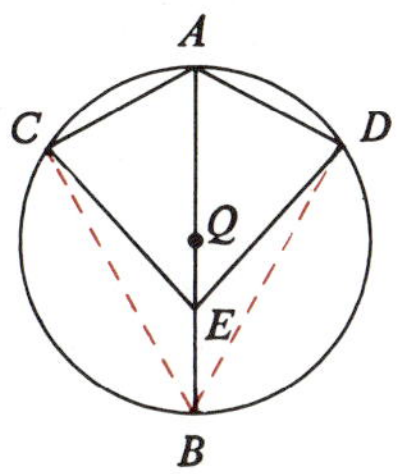

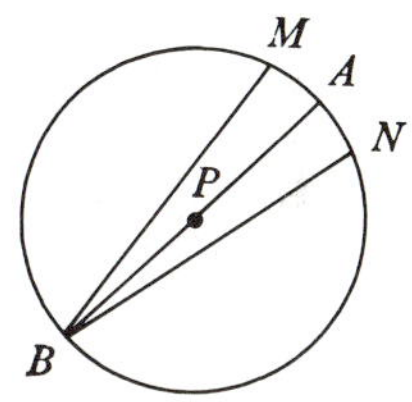

23. GIVEN Points A, B, C, and D are points of $\odot P$; $\overline{AB} \cong \overline{CD}$; R is the midpoint of $\overline{AC}$; S is the midpoint of $\overline{BD}$.
PROVE $\overline{PR} \cong \overline{PS}$

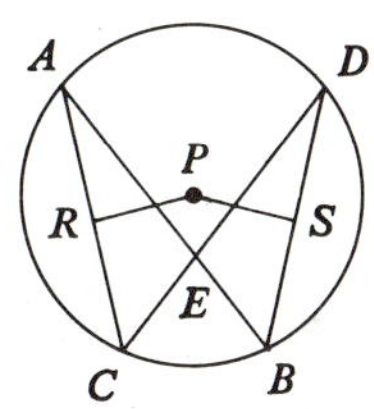

Exercise 23

24. GIVEN Quadrilateral $ABCD$ is inscribed in $\odot P$; $\overparen{AB} \cong \overparen{DC}$; $\overparen{AD} \cong \overparen{BC}$
PROVE Quadrilateral $ABCD$ is a rectangle.

C

25. Prove that the line containing the midpoints of the major and minor arcs of a chord of a circle is the perpendicular bisector of the chord.

26. GIVEN Points A, B, C, and D are points of $\odot Q$; $\overline{AB} \perp \overline{AC}$; $\overline{CD} \perp \overline{AC}$
PROVE $\overparen{AB} \cong \overparen{CD}$

27. GIVEN $\overline{AB}$ is a chord of $\odot Q$; $\overleftrightarrow{CD}$ is tangent to $\odot Q$ at C; $\overline{AB} \parallel \overleftrightarrow{CD}$
PROVE $\overparen{AC} \cong \overparen{BC}$

9-6 INSCRIBED ANGLES OF A CIRCLE

Using our knowledge of the relationship between the measure in degrees of a central angle and its arc, we can study many relationships between the measures of arcs and other types of angles. First, however, we must consider some definitions. In the figure at right, we say that $\angle F$ intercepts $\overparen{AC}$ and $\overparen{BED}$. Also, $\angle BAE$ intercepts $\overparen{BE}$.

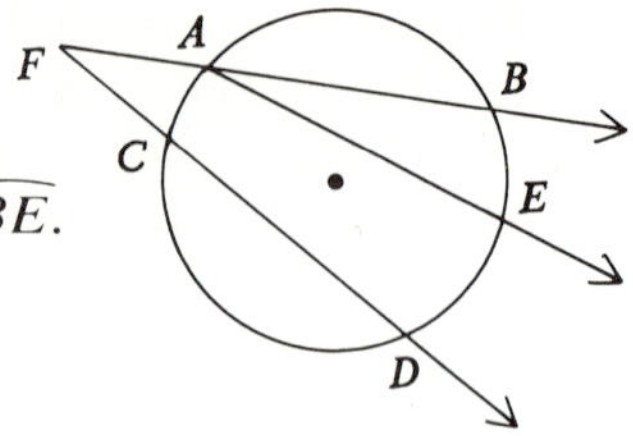

D **Definition 9-20** An angle **intercepts** an arc if:

1. each ray of the angle contains exactly one endpoint of the arc;
2. all other points of the arc lie in the interior of the angle.

This arc is called the **intercepted arc** of the angle.

The red arcs are the *intercepted arcs* of the angles below.

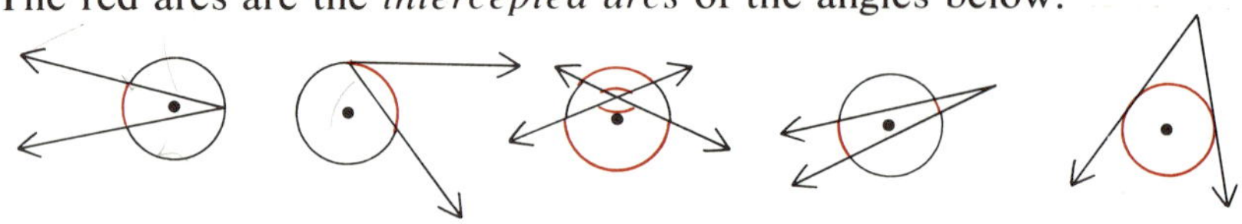

State why each of the red arcs is *not* an intercepted arc of the angle.

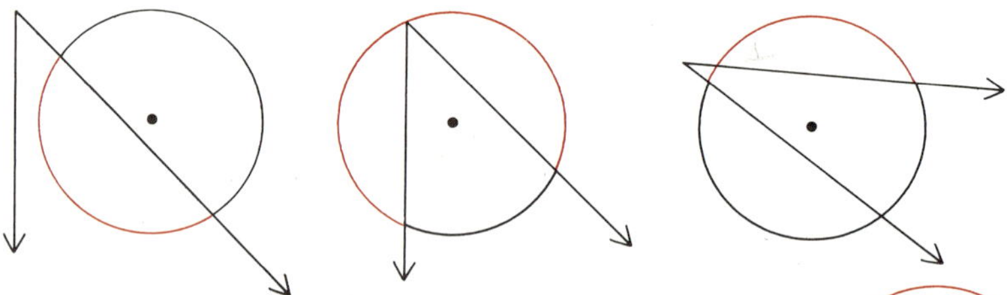

We stated earlier that a polygon is inscribed in a circle if all its vertices are in the circle. We shall now describe an angle that is inscribed in an arc.

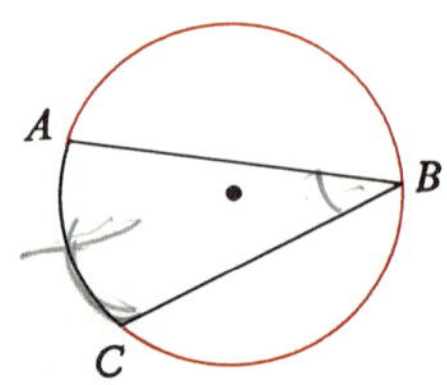

D **Definition 9-21** An **inscribed angle** of a circle is an angle whose vertex is a point of the circle and whose rays contain two other points of the circle.

The figure above shows that the rays of the inscribed angle contain the endpoints of ABC, in which the angle is inscribed, and that vertex B is another point of the arc.

T **Theorem 9-6.1** The measure of an inscribed angle is one-half the measure of its intercepted arc.

GIVEN $\angle ABC$ is inscribed in $\odot P$.

PROVE $m\angle ABC = \frac{1}{2} m\overparen{AC}$

Proof We must consider three cases.

Case I The center of the circle is *contained in a ray* of the angle.

Case II The center of the circle is *in the interior* of the angle.

Case III The center of the circle is *in the exterior* of the angle.

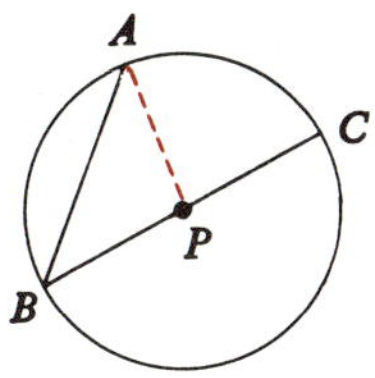

Case I

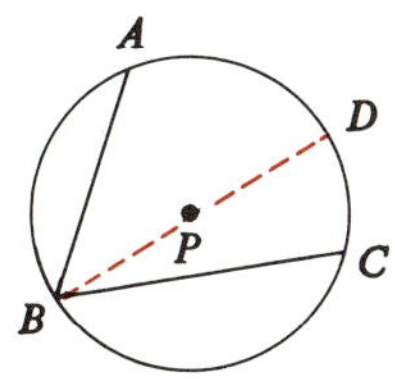

Case II

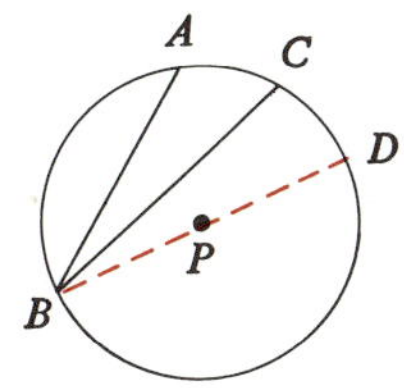

Case III

Proof

STATEMENTS	REASONS
Case I	
1. $\angle ABC$ is inscribed in $\odot P$ and contains center P.	1. Given.
2. Consider $\overline{AP}$.	2. Definition of a line segment. (1-13)
3. $\overline{AP} \cong \overline{BP}$	3. Definition of a circle. (9-1)
4. $\angle BAP \cong \angle ABC$	4. The base angles of an isosceles triangle are congruent. (3-4.2)
5. $m\angle APC = m\angle BAP + m\angle ABC$	5. The measure of the exterior angle of a triangle is equal to the sum of the measures of the two remote interior angles. (6-4.1)
6. $m\angle APC = 2m\angle ABC$	6. Substitution Postulate. (2-1)
7. $m\angle ABC = \frac{1}{2}m\angle APC$	7. Multiplication property.
8. $m\angle APC = m\,\widehat{AC}$	8. Definition of the measure of a minor arc. (9-1.7)
9. $m\angle ABC = \frac{1}{2}m\,\widehat{AC}$	9. Substitution Postulate. (2-1)
Case II	
1. Draw diameter $\overline{BPD}$.	1. Definition of a diameter. (9-4)
2. $m\angle ABD = \frac{1}{2}m\,\widehat{AD}$, $m\angle DBC = \frac{1}{2}m\,\widehat{DC}$	2. Proved in Case I.
3. $m\angle ABC = \frac{1}{2}m\,\widehat{AC}$	3. Angle Sum Postulate (2-10) and Arc Addition Postulate. (9-1)

(*continued*)

STATEMENTS	REASONS
Case III	
1. Draw diameter $\overline{BPD}$.	1. Definition of a diameter. (9-4)
2. $m\angle ABD = \frac{1}{2}m\,\widehat{ACD}$; $m\angle DBC = \frac{1}{2}m\,\widehat{DC}$	2. Proved in Case I.
3. $m\angle ABC = \frac{1}{2}m\,\widehat{AC}$	3. Angle Difference Postulate (2-11) and Arc Addition Postulate. (9-1)

Example 1 Points K, L, and M are in $\odot P$. If $m\widehat{KL} = 100$ and $m\widehat{LM} = 130$, find $m\angle KLM$.

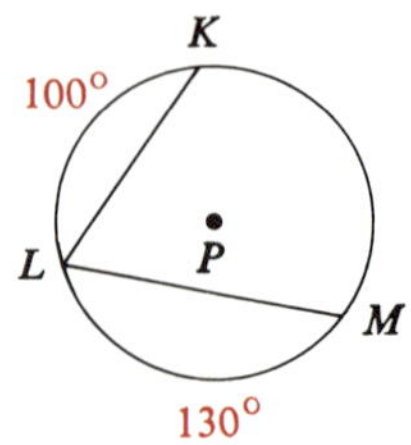

Solution Since $m\widehat{KL} + m\widehat{LM} + m\widehat{KM} = 360$, $m\widehat{KM} = 360 - (100 + 130) = 360 - 230 = 130$. $\angle KLM$ is an inscribed angle; therefore, by Theorem 9-6.1, its measure is $\frac{1}{2}m\widehat{KM}$, or 65.

Example 2

GIVEN A, B, C, and D are points of $\odot Q$.

PROVE $\angle ABD \cong \angle ACD$

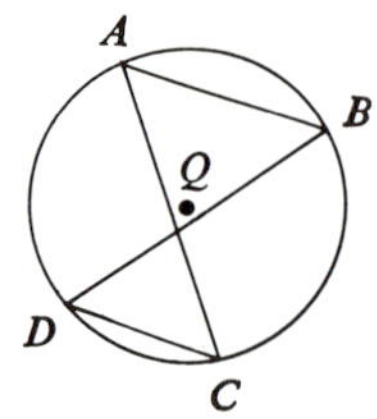

Solution

PROOF

STATEMENTS	REASONS
1. A, B, C, and D are points of $\odot Q$.	1. Given.
2. $m\angle ABD = \frac{1}{2}m\widehat{AD}$; $m\angle ACD = \frac{1}{2}m\widehat{AD}$	2. Theorem 9-6.1.
3. $m\angle ABD = m\angle ACD$	3. Transitive property.
4. $\angle ABD \cong \angle ACD$	4. Definition of congruent angles. (1-30)

Suppose we inscribe more angles in $\widehat{ABD}$ of Example 2. Angles AED, AFD, and AGD are all inscribed in $\widehat{ABD}$. Since they intercept the same arc, $\widehat{AD}$, a proof like the one in Example 2 would indicate that
$m\angle AED = m\angle AFD = m\angle AGD$.
We may therefore state the following corollary to Theorem 9-6.1:

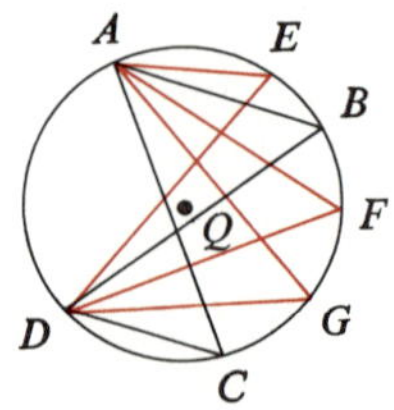

C **Corollary 9-6.1a** Angles inscribed in the same arc are congruent.

Following are two other corollaries of Theorem 9-6.1:

C **Corollary 9-6.1b** If two inscribed angles intercept congruent arcs, then the angles are congruent.

C **Corollary 9-6.1c** An angle inscribed in a semicircle is a right angle.

The proofs of these corollaries are left as exercises.

Class Exercises

In the figure at right, quadrilateral $ABCD$ is inscribed in $\odot Q$.

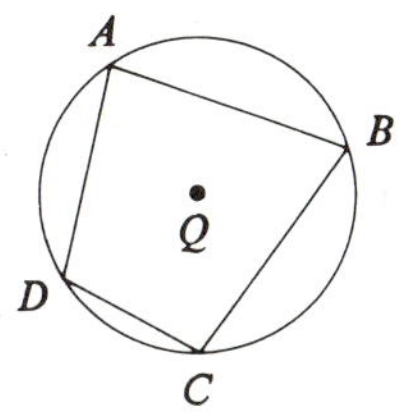

1. $m\angle B = \frac{1}{2}$ m ____
2. $m\angle D = \frac{1}{2}$ m____
3. $m\angle B + m\angle D = \frac{1}{2}$m (____ + ____)
4. $m\angle B + m\angle D = \frac{1}{2}$ (____) (Express answer in degree measure.)
5. $m\angle B + m\angle D =$ ____ (Express answer in degree measure.)
6. Therefore, we may say that $\angle B$ and $\angle D$ are ________.
7. The sum of the measures of the angles of a quadrilateral equals ____.
8. $m\angle A + m\angle B + m\angle C + m\angle D =$ ____ (Express answer in degree measure.)
9. $m\angle A + m\angle C =$ ____ (Express answer in degree measure.)
10. Therefore, we may say that $\angle A$ and $\angle C$ are ________.

Having completed the above Class Exercises, you have actually proved the next theorem.

T **Theorem 9-6.2** The opposite angles of an inscribed quadrilateral are supplementary.

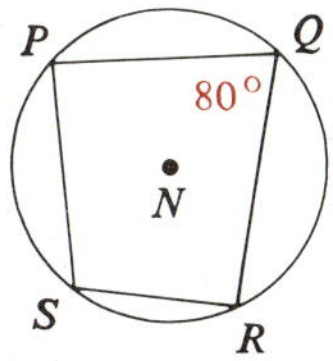

Example 3 If quadrilateral $PQRS$ is inscribed in $\odot N$ and $m\angle Q = 80$, find $m\angle S$.

Solution By Theorem 9-6.2, $m\angle Q + m\angle S = 180$. Therefore $m\angle S = 180 - 80 = 100$.

Another useful theorem whose proof depends on Theorem 9-6.1 deals with parallel chords of a circle.

T

Theorem 9-6.3 In any circle, parallel chords intercept congruent arcs.

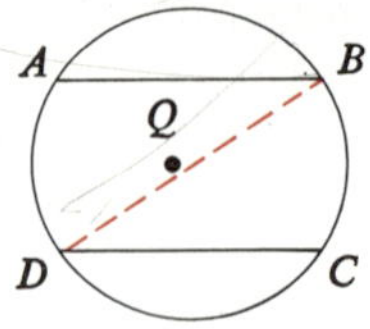

GIVEN $\overline{AB}$ and $\overline{CD}$ are chords of $\odot Q$; $\overline{AB} \parallel \overline{CD}$

PROVE $\overset{\frown}{AD} \cong \overset{\frown}{BC}$

PROOF

STATEMENTS	REASONS
1. $\overline{AB}$ and $\overline{CD}$ are chords of $\odot Q$; $\overline{AB} \parallel \overline{CD}$.	1. Given.
2. Consider $\overline{BD}$.	2. Definition of a line segment. (1-13)
3. $m\angle ABD = \frac{1}{2} m\overset{\frown}{AD}$ $m\angle BDC = \frac{1}{2} m\overset{\frown}{BC}$	3. Theorem 9-6.1.
4. $\angle ABD \cong \angle BDC$	4. If two parallel lines are cut by a transversal, then the alternate interior angles are congruent. (6-3.1)
5. $m\overset{\frown}{AD} = m\overset{\frown}{BC}$	5. Transitive property.
6. $\overset{\frown}{AD} \cong \overset{\frown}{BC}$	6. Definition of congruent arcs. (9-18)

T

Theorem 9-6.4 In any circle, a tangent and a chord parallel to it intercept congruent arcs.

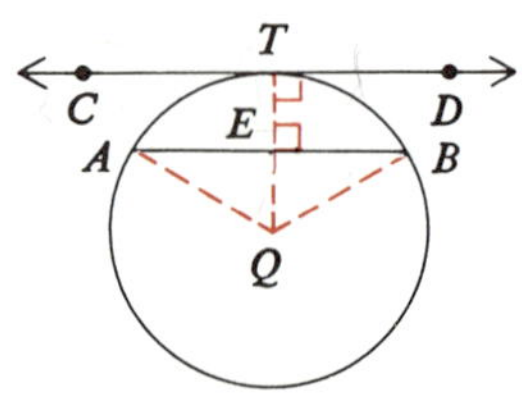

GIVEN $\overleftrightarrow{CTD}$ is tangent to $\odot Q$ at T; chord $\overline{AB} \parallel \overline{CD}$

PROVE $\overset{\frown}{AT} \cong \overset{\frown}{BT}$

PROOF OUTLINE To prove that $\overset{\frown}{AT} \cong \overset{\frown}{BT}$, we must first prove that $\angle AQT \cong \angle BQT$. $\overline{QT} \perp \overleftrightarrow{CTD}$ at T and $\overline{QT} \perp \overline{AB}$ at E. Why is E the midpoint of $\overline{AB}$? $\triangle AEQ \cong \triangle BEQ$, and $\angle AQT \cong \angle BQT$. Therefore $\overset{\frown}{AT} \cong \overset{\frown}{BT}$.

EXERCISES

A

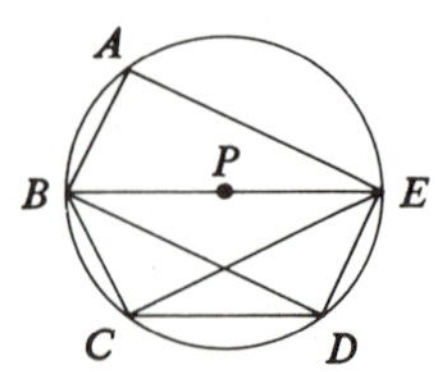

Exercises 1–12

Points A, B, C, D, and E are points of $\odot P$.

1. Name the arc intercepted by $\angle BDE$.
2. Name the arc in which $\angle ABE$ is inscribed.
3. Name the intercepted arc of $\angle CDE$.
4. Which angle is inscribed in $\overset{\frown}{ADC}$?
5. $\overset{\frown}{BCE}$ is the intercepted arc of which angle?

6. Name two angles inscribed in $\widehat{BCE}$.
7. What is the relationship between the two angles referred to in Exercise 6?
8. Why is $\angle BEC \cong \angle BDC$?
9. $\angle EBD \cong$ ____.
10. If $\overline{BE}$ is a diameter of $\odot P$, then $m\angle BDE =$ ____.
11. What is the sum of $m\angle ABC + m\angle AEC$?
12. If $\overline{CD} \parallel \overline{BE}$, then $\widehat{BC} \cong$ ____.

Find the measure of an angle inscribed in an arc of each of these measures.

13. 200 14. 60 15. 340 16. $3x + 5$ 17. $360 - 2x$

Find the measure of an inscribed angle that intercepts an arc of each of these measures.

18. 80 19. 75 20. 145 21. $5x - 7$ 22. $180 - 6x$

Find the values of x and y in each of the following figures. Assume that the center of each circle, Q, is contained in the lines that appear to contain it.

23.
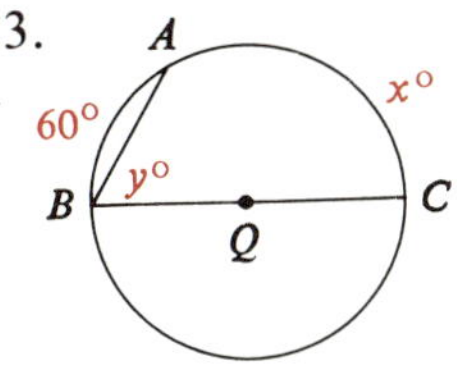

24.
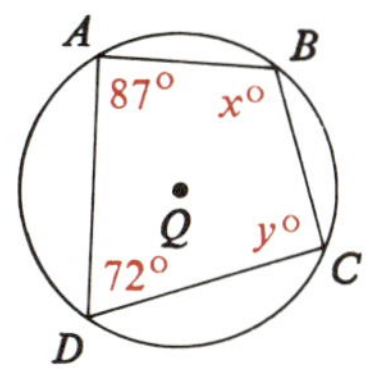

25.
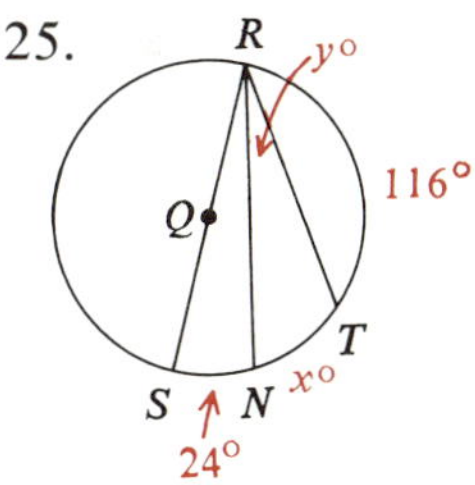

26.
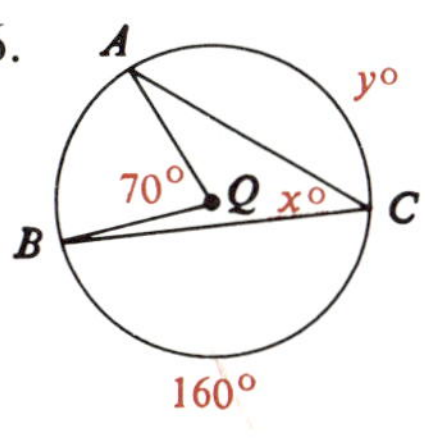

27.
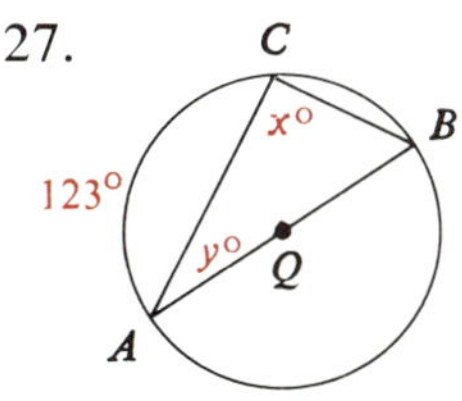

28.
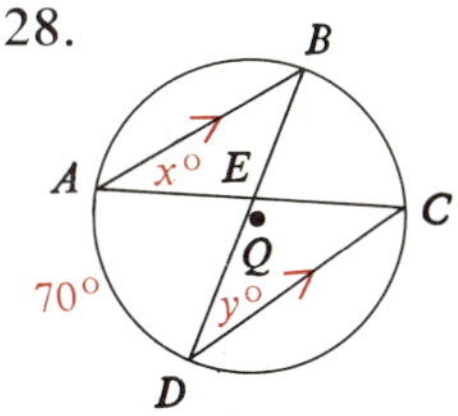

29. Prove Corollary 9-6.1a.
30. Prove Corollary 9-6.1b.
31. Prove Corollary 9-6.1c.
32. Prove Theorem 9-6.4.
33. Prove that a parallelogram inscribed in a circle is a rectangle.

B

34. Using the figure at right for reference, prove that the median to the hypotenuse of a right triangle is half as long as the hypotenuse (Theorem 7-3.5).

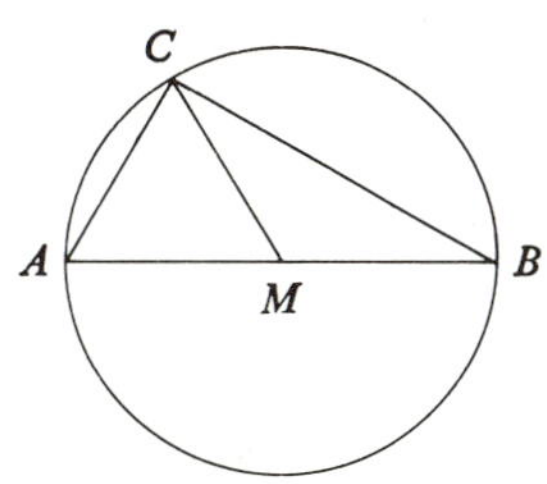

35. Circles P and Q are internally tangent at point T; chord $\overline{TA}$ of $\odot Q$ meets $\odot P$ at B; chord $\overline{TC}$ of $\odot Q$ meets $\odot P$ at D. If $m\widehat{AC} = 106$, find $m\widehat{BD}$.

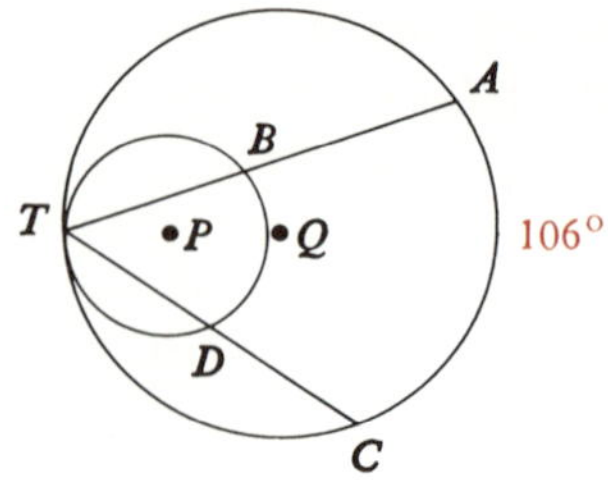

Exercise 35

36. **Given** $\overline{AB}$ is a diameter of $\odot Q$; $\overline{BC}$ intersects $\odot Q$ in D; $\overline{BD} \cong \overline{CD}$

Prove $\overline{AB} \cong \overline{AC}$

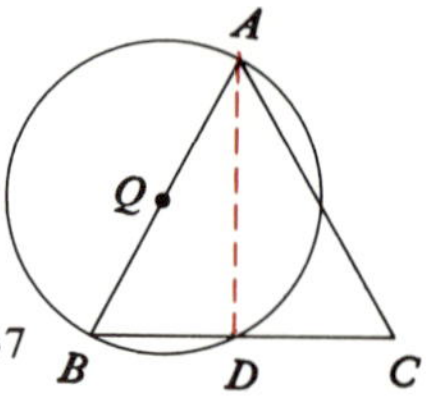

Exercises 36–37

37. **Given** $\overline{AB}$ is a diameter of $\odot Q$; $\overline{BC}$ intersects $\odot Q$ in D; $\overline{AB} \cong \overline{AC}$

Prove $\overline{BD} \cong \overline{CD}$

Find the values of x and y in each of the following figures:

38.

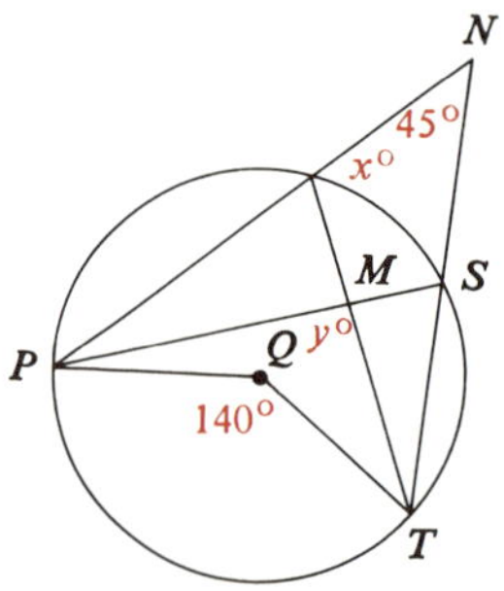

39.

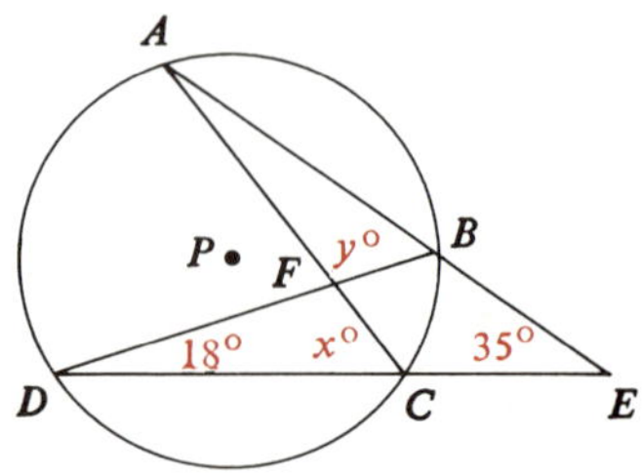

C

40. **Given** Circles P and Q are internally tangent at T; $\odot Q$ contains point P; $\overline{TA}$ is a chord of $\odot P$; $\odot Q$ intersects $\overline{TA}$ in M.

Prove M is the midpoint of $\overline{TA}$.

41. **Given** Quadrilateral $ABCD$ is inscribed in $\odot Q$; $\overleftrightarrow{AB}$ meets $\overleftrightarrow{DC}$ at P.

Prove $\dfrac{BP}{DP} = \dfrac{CP}{AP}$

42. **Given** Quadrilateral $ABCD$ is inscribed in $\odot Q$; $\overleftrightarrow{AB}$ meets $\overleftrightarrow{DC}$ at P; $\angle ABD \cong \angle PBC$

Prove $AD \cdot BP = CP \cdot BD$

43. **Given** $\triangle ABC$ is inscribed in $\odot P$; $\overline{AD}$ is an altitude of $\triangle ABC$; $\overline{APE}$ is a diameter of $\odot P$.

Prove $AB \cdot AC = AD \cdot AE$

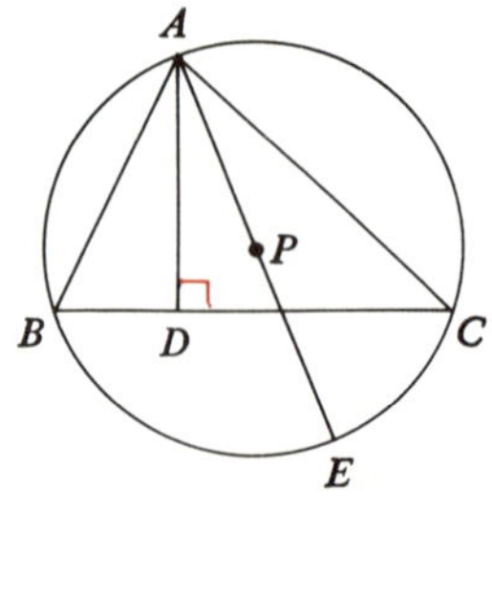

9-7 CONCYCLIC POINTS (Optional)

Points contain in the same circle are called **concyclic points.** Do we already know that any three noncollinear points are concyclic? Can you prove this, using Theorem 9-2.3? When are four points concyclic? The answer to this question is the main concern of this section.

Not all sets of four noncollinear points are concyclic. For instance, the vertices of many parallelograms are not concyclic.

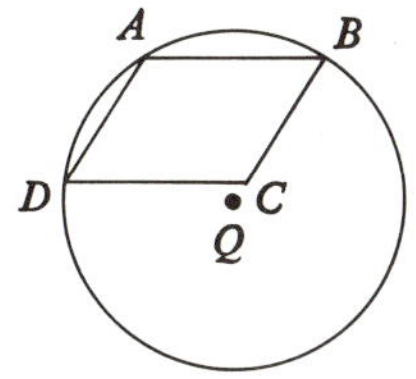

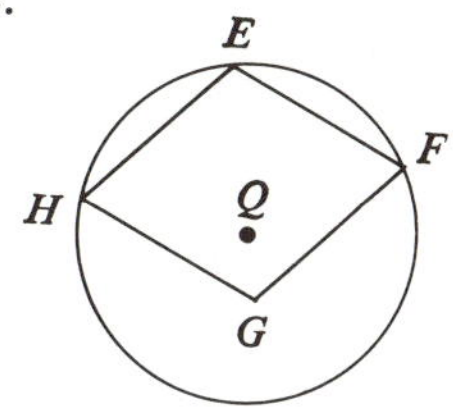

Definition 9-22 A **cyclic quadrilateral** is a quadrilateral whose vertices are concyclic.

The following theorem is the converse of Corollary 9-6.1a:

Theorem 9-7.1 If one side of a quadrilateral subtends congruent angles at the two nonadjacent vertices, then the quadrilateral is cyclic.

GIVEN Side $\overline{AB}$ of quadrilateral $ABCD$ subtends congruent angles ACB and ADB.

PROVE Quadrilateral $ABCD$ is cyclic.

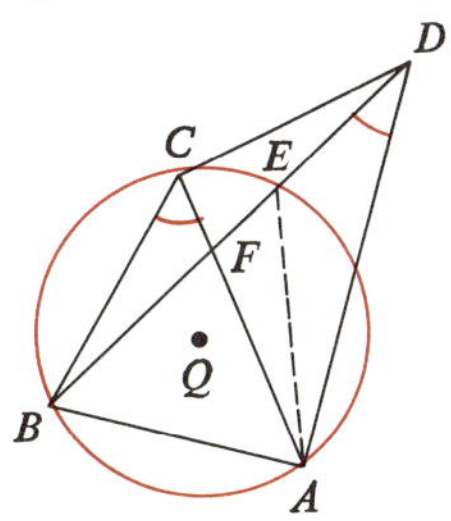

Case I

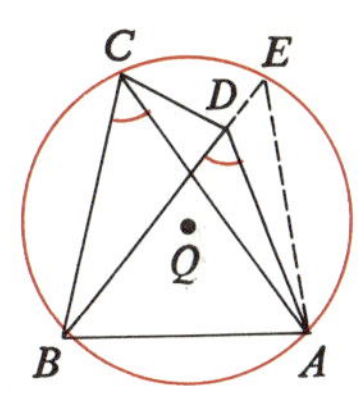

Case II

PROOF PLAN Assume that $\odot Q$ contains points A, B, and C. There are three possibilities for point D: it is an exterior point of $\odot Q$; it is an interior point of $\odot Q$; or it is a point of $\odot Q$. We shall eliminate the first two possibilities.

PROOF Assume point D is an exterior point of $\odot Q$. We are given that $\angle ACB \cong \angle ADB$. By Corollary 9-6.1a, $\angle AEB \cong \angle ACB$. Hence, $\angle AEB \cong \angle ADB$ by the Transitive property. But, by Theorem 5-2.1, $m\angle AEB > m\angle ADB$. This contradiction tells us point D is not in the exterior of $\odot Q$. If we assume point D is an interior point of $\odot Q$, we have $\angle AEB \cong \angle ADB$ and $m\angle ADB > m\angle AEB$. Why? Again, our assumption is invalid. Therefore, D is a point of $\odot Q$. Quadrilateral $ABCD$ is cyclic by Definition 9-22.

Example 1 In the figure at right, the measures of angles are indicated. Why are points B, C, D, and E concyclic? What is $m\angle DEB$?

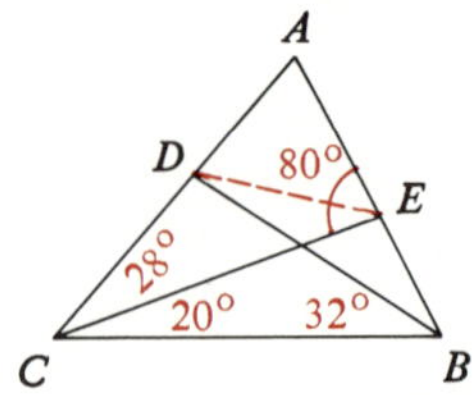

Solution $\angle AEC$ is an exterior angle of $\triangle CEB$.
Therefore, $m\angle AEC = m\angle ECB + m\angle EBC = m\angle ECB + m\angle DBC + m\angle EBD$.
Therefore, $m\angle ABD = 28$ and $\angle ABD \cong \angle ACE$. As a result, by Theorem 9-7.1, we conclude that quadrilateral $BCDE$ is cyclic, or that points B, C, D, and E are concyclic. Then by Theorem 9-6.2, we find $m\angle DEB = 180 - 48 = 132$.

The next theorem presents another method of proving that a quadrilateral is cyclic. How is this theorem related to Theorem 9-6.2?

T

Theorem 9-7.2 If a pair of opposite angles of a quadrilateral are supplementary, then the quadrilateral is cyclic.

GIVEN Quadrilateral $ABCD$ with $\angle B$ supplementary to $\angle D$

PROVE Quadrilateral $ABCD$ is cyclic.

PROOF

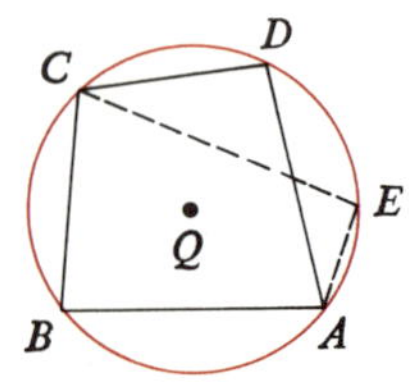

STATEMENTS	REASONS
1. In quadrilateral $ABCD$, $\angle B$ supplementary to $\angle D$.	1. Given.
2. Assume that $\odot Q$ contains points A, B, and C, and select a point E of $\odot Q$, on the same side of $\overline{AC}$ as point D.	2. Definition of a circle. (9-1)
3. $\angle E$ is supplementary to $\angle B$.	3. The opposite angles of an inscribed quadrilateral are supplementary. (9-6.2)
4. $\angle E \cong \angle D$	4. Supplements of congruent angles, or of the same angle, are congruent. (3-1.4)
5. Points C, D, E, and A are concyclic.	5. Theorem 9-7.1.
6. Quadrilateral $ABCD$ is a cyclic quadrilateral.	6. Definition of a cyclic quadrilateral. (9-22)

Why is D concylic with points A, B, and C?

Example 2

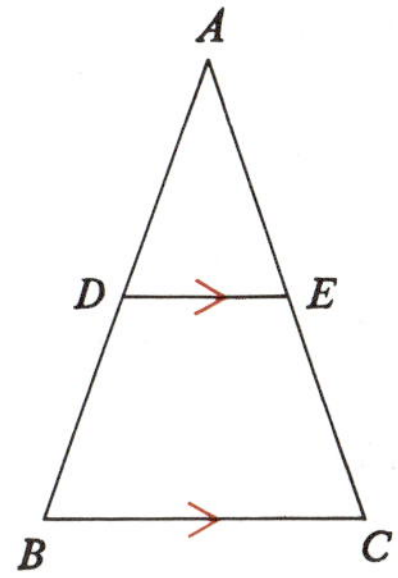

Given $\triangle ABC$ is isosceles, with $\overline{AB} \cong \overline{AC}$; $\overline{ADB}$ and $\overline{AEC}$; $\overline{DE} \parallel \overline{BC}$

Prove Quadrilateral $EDBC$ is cyclic.

Solution

Proof

STATEMENTS	REASONS
1. $\triangle ABC$ is isosceles, with $\overline{AB} \cong \overline{AC}$; $\overline{ADB}$ and $\overline{AEC}$; $\overline{DE} \parallel \overline{BC}$	1. Given.
2. $\angle B$ is supplementary to $\angle EDB$.	2. If two parallel lines are cut by a transversal, then the interior angles on the same side of the transversal are supplementary. (C6-3.1b)
3. $\angle B \cong \angle C$	3. The base angles of an isosceles triangle are congruent. (3-4.2)
4. $\angle C$ is supplementary to $\angle EDB$.	4. Substitution Postulate. (2-1)
5. Quadrilateral $EDBC$ is cyclic.	5. Theorem 9-7.2.

EXERCISES

A Explain why quadrilateral $ABCD$ is cyclic in each of the following exercises. Angle measures are indicated.

1.

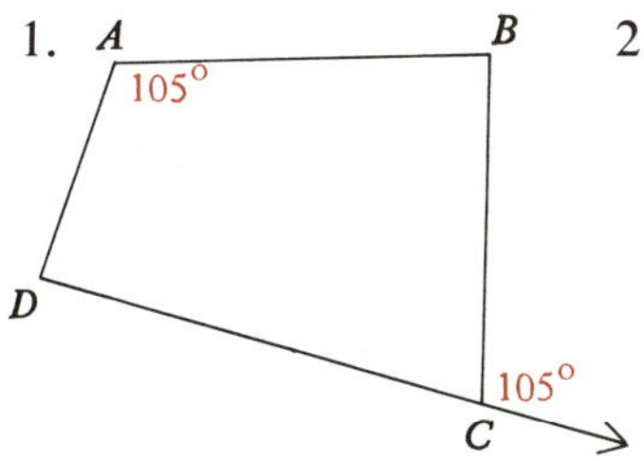

2.

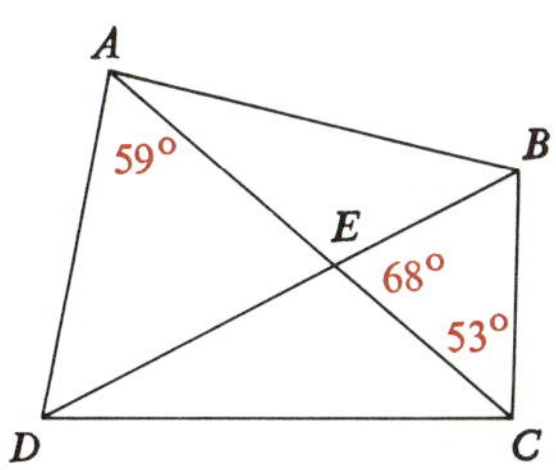

3.

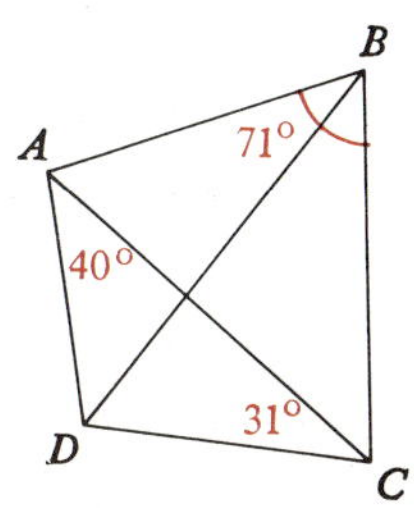

4. Prove that all rectangles are cyclic quadrilaterals.
5. Prove that any isosceles trapezoid can be inscribed in a circle.
6. State what we proved in Example 2 of this section in theorem form.
7. **Given** $\overline{AC}$ intersects $\overline{BD}$ in E; $\overline{AE} \cong \overline{EB}$; $\overline{DE} \cong \overline{EC}$

   Prove Points A, B, C and D are concyclic.

B

8. GIVEN $\overline{AB} \cong \overline{AD}$; $\overline{BD} \cong \overline{CD}$; $m\angle ABD = \frac{1}{2} m\angle C$
   PROVE Quadrilateral $ABCD$ is cyclic.

9. GIVEN Isosceles $\triangle DEC$ with $\overline{DE} \cong \overline{DC}$; $\overrightarrow{ADC}$; $\overline{AB}$ intersects $\overline{DE}$ in F; $m\angle A = \frac{1}{2} m\angle ABE$
   PROVE Points A, D, B, and E are concyclic.

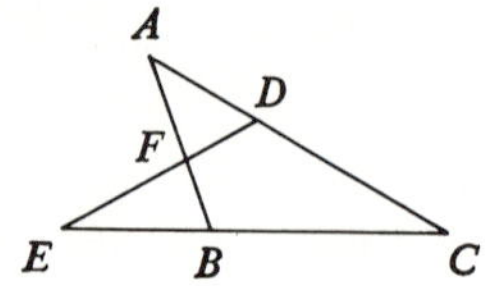

10. GIVEN $\overline{AB}$ is a diameter of $\odot Q$; $\overline{CQ} \perp \overline{AB}$ at Q; $\overline{AC}$ intersects $\odot Q$ in D; $\overline{BC}$ intersects in F.
    PROVE Quadrilateral $CDQB$ is cyclic; $\angle QDB \cong \angle QCA$

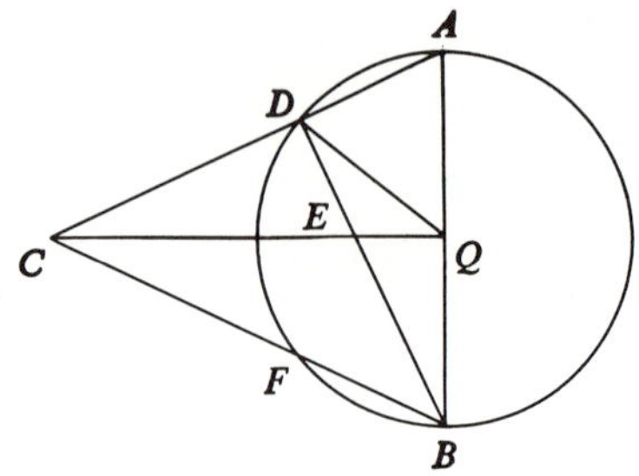

11. GIVEN Circles P and Q are internally tangent at T; $\overrightarrow{RABCD}$ and $\overrightarrow{RMNLK}$; A, D, K, and M are points of $\odot Q$ and B, C, L, and N are points of $\odot P$; quadrilateral $ABNM$ is cyclic.
    PROVE Quadrilateral $CDKL$ is cyclic.

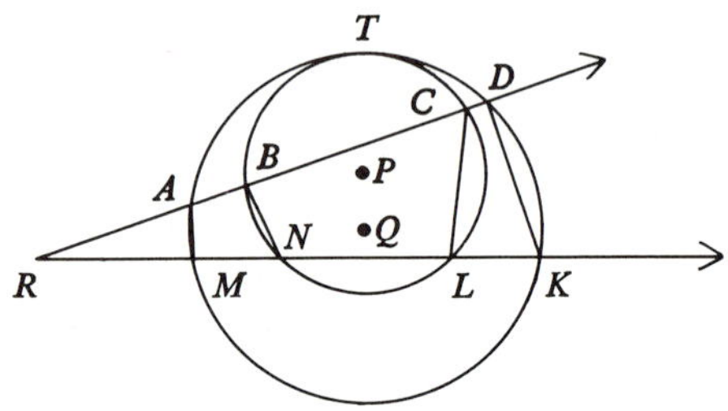

C

12. GIVEN Circles P and Q intersect in points M and N; $\overline{AMB}$, with A and B points of $\odot P$ and $\odot Q$, respectively; $\overline{AC}$ intersects $\odot P$ in D; $\overline{BC}$ intersects $\odot Q$ in E.
    PROVE Quadrilateral $DNEC$ is cyclic. (*Hint*: Consider $\overline{MN}$.)

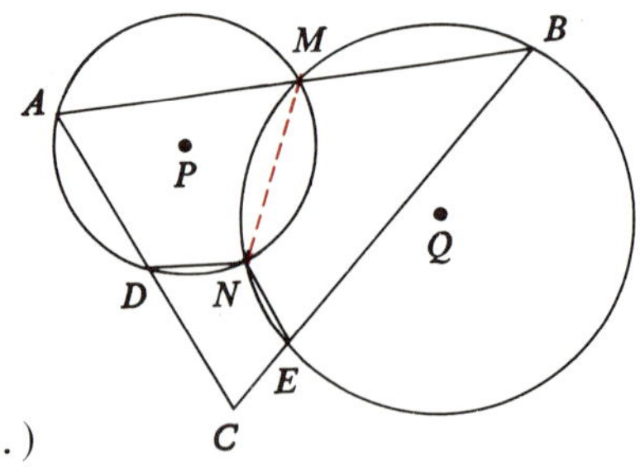

13. GIVEN Right $\triangle ABC$ with $m\angle ACB = 90$; $\overline{CF}$ is an altitude of $\triangle ABC$; $\overline{ADC}$ and $\overline{BEC}$; $\overline{CF}$ bisects $\overline{DE}$ at M.
    PROVE Quadrilateral $ADEB$ is cyclic.

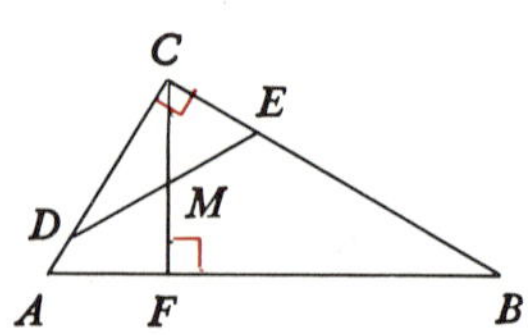

14. GIVEN Quadrilateral $ABCD$, where $\overline{APS}$ bisects $\angle A$; $\overline{BRS}$ bisects $\angle B$; $\overline{CRQ}$ bisects $\angle C$; $\overline{DPQ}$ bisects $\angle D$.
PROVE Quadrilateral $PQRS$ is cyclic.

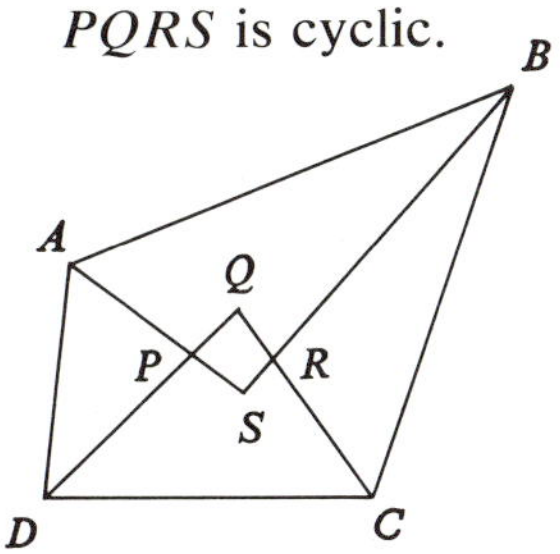

15. GIVEN Altitudes $\overline{AD}$ and $\overline{CE}$ of $\triangle ABC$ intersect in H; $\overline{CHEF}$; $\overline{FE} \cong \overline{HE}$
PROVE Quadrilateral $AFBC$ is cyclic.

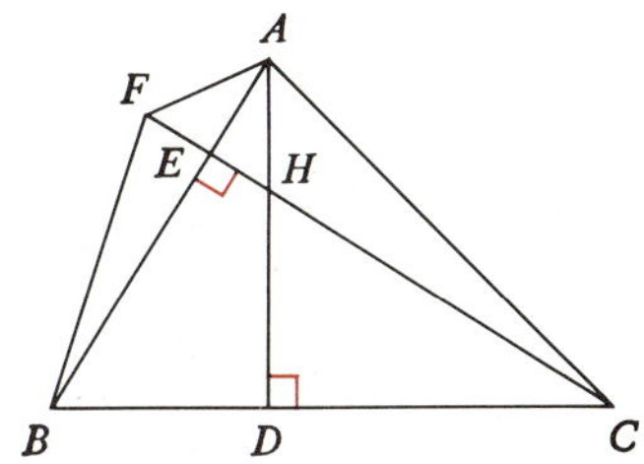

SOMETHING TO THINK ABOUT

Ptolemy's Theorem

One of the more helpful theorems about cyclic quadrilaterals was first published in the second century, by the Greek mathematician Ptolemy (Claudius Ptolemaeus). The theorem appears in his famous work, *The Almagest*, and states that the product of the lengths of the diagonals of a cyclic quadrilateral equals the sum of the products of the lengths of the opposite sides.

GIVEN Quadrilateral $ABCD$ inscribed in $\odot Q$.

PROVE $AC \cdot BD = AB \cdot CD + AD \cdot BC$

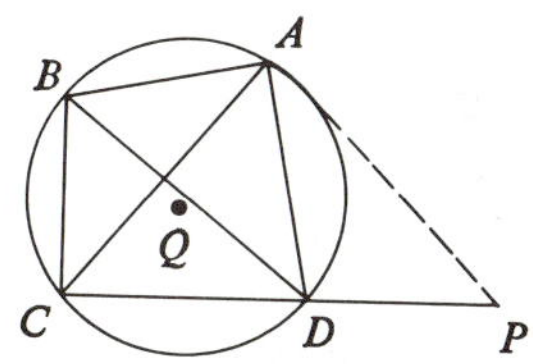

PROOF Quadrilateral $ABCD$ is inscribed in $\odot Q$. Select point P of $\overline{CD}$ so that $m\angle BAC = m\angle DAP$. Then, by Theorem 9-6.2, $\angle ABC$ is supplementary to $\angle ADC$. Apply Postulate 1-6 and Theorem 3-1.4 to show that $m\angle ABC = m\angle ADP$. Then, by Corollary 8-5.1a, $\triangle BAC \sim \triangle DAP$ and $\frac{DP}{BC} = \frac{AD}{AB}$ or $DP = \frac{AD \cdot BC}{AB}$. Since $m\angle BAD = m\angle CAP$ (Postulate 2-10) and $\frac{AB}{AD} = \frac{AC}{AP}$ (Definition 8-6), $\triangle ABC \sim \triangle ACP$ by Theorem 8-6.1. Thus, $\frac{CP}{BD} = \frac{AC}{AB}$ or $CP = \frac{AC \cdot BD}{AB}$. Substituting $CD + DP$ for CP and simplifying, we get $AC \cdot BD = AB \cdot CD + AD \cdot BC$.

Since all rectangles are cyclic quadrilaterals, we can apply Ptolemy's Theorem to any rectangle. In rectangle $ABCD$, the following relationship exists: $AC \cdot BD = AB \cdot CD + AD \cdot BC$.

Since the diagonals of a rectangle are of equal length, as are the opposite sides, the members of each product above are equal. Thus, for quadrilateral $ABCD$, we can express the equation above as $AC^2 = CD^2 + AD^2$. What famous theorem is this?

9-8 OTHER ANGLES RELATED TO CIRCLES

The inscribed angle and the central angle are the only angles related to circles that we have studied so far. In this section, we shall examine the properties of other angles that are related to circles. We shall consider angles whose vertices are points of a circle, in the interior, and in the exterior of a circle.

By completing the Class Exercises below, you will discover many interesting relationships between angles and circles.

Class Exercises

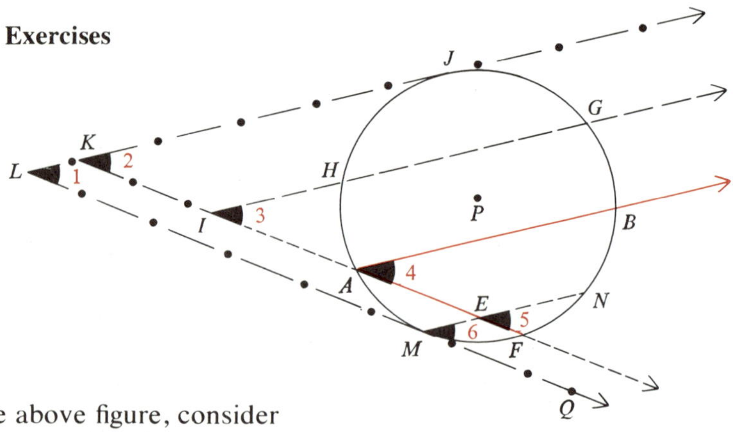

In the above figure, consider all lines parallel that appear to be so. All answers are arcs.

1. Why are angles 1–6 congruent?
2. $m\angle BAF = \frac{1}{2}m$ ____
3. $m\widehat{BN} = m$____ $= m$____
4. $m\angle NMQ = m\angle BAF = \frac{1}{2}(m\widehat{BN} + m$____$) = \frac{1}{2}(m\widehat{MF} + m$____$) = \frac{1}{2}m$____
5. $m\angle NEF = m\angle BAF = \frac{1}{2}(m\widehat{BN} + m$____$) = \frac{1}{2}(m\widehat{AM} + m$____$)$
6. $m\angle GIF = m\angle BAF = \frac{1}{2}m$____ $= \frac{1}{2}(m$____ $+ m\widehat{BG} - m\widehat{BG})$
   $= \frac{1}{2}(m$____ $+ m\widehat{BG} - m$____$) = \frac{1}{2}(m\widehat{GBF} - m$____$)$
7. $m\angle JKF = m\angle BAF = \frac{1}{2}m$____ $= \frac{1}{2}(m$____ $+ m\widehat{JB} - m\widehat{JB})$
   $= \frac{1}{2}(m$____ $+ m\widehat{JB} - m$____$) = \frac{1}{2}(m\widehat{JBF} - m$____$)$

8. $m\angle JLM = m\angle NMQ = \frac{1}{2}m\_\_\_\_ = \frac{1}{2}(m\_\_\_\_ + m\widehat{JN} - m\widehat{JN})$
$= \frac{1}{2}(m\_\_\_\_ + m\widehat{JN} - m\_\_\_\_) = \frac{1}{2}(m\widehat{JBM} - m\_\_\_\_)$

9. Express the statements that appear in color above in general terms.

Your response to Exercise 9 above should suggest the theorems presented in the rest of this section.

T

Theorem 9-8.1 The measure of the angle formed by a tangent and a chord of a circle is one-half the measure of its intercepted arc.

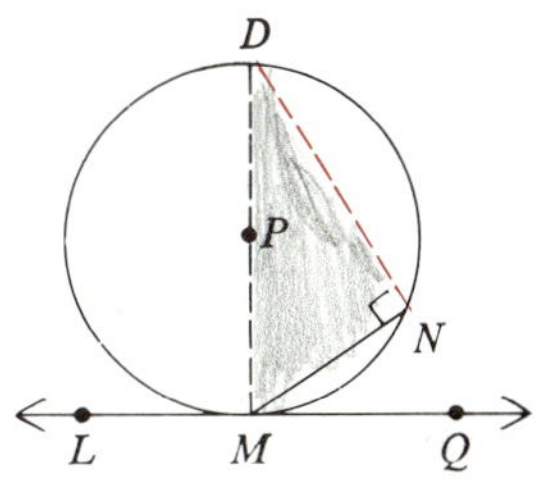

GIVEN $\overline{NM}$ is a chord of $\odot P$; $\overleftrightarrow{LMQ}$ is tangent to $\odot P$ at M.

PROVE $m\angle NMQ = \frac{1}{2}m\widehat{MN}$

PROOF OUTLINE Since $\overline{DPM}$ is a diameter of $\odot P$, $\angle N$ is a right angle. Why is $\overline{DM} \perp \overleftrightarrow{LMQ}$? Since both $\angle D$ and $\angle NMQ$ are complementary to $\angle DMN$, $\angle D \cong \angle NMQ$. However, $m\angle D = \frac{1}{2}m\widehat{MN}$. Therefore, $m\angle NMQ = \frac{1}{2}m\widehat{MN}$.

Example 1 In the figure above, find $m\angle NMQ$ if $m\widehat{MDN} = 310$.

Solution $m\widehat{MN} = 360 - 310 = 50$. By Theorem 9-8.1, $m\angle NMQ = \frac{1}{2}m\widehat{MN} = \frac{1}{2}(50) = 25$.

T

Theorem 9-8.2 The measure of an angle formed by two chords intersecting in a point in the interior of a circle is one-half the sum of the measures of the arcs intercepted by the angle and its vertical angle.

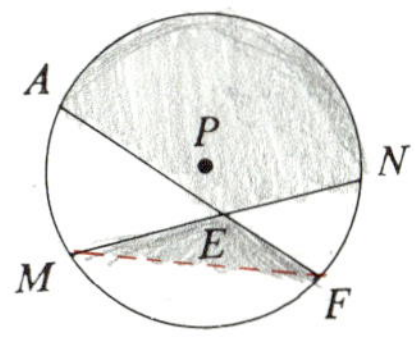

GIVEN Chords $\overline{AF}$ and $\overline{MN}$ of $\odot P$ intersect in E.

PROVE $m\angle NEF = \frac{1}{2}(m\widehat{AM} + m\widehat{NF})$

PROOF OUTLINE For $\triangle EMF$, $m\angle NEF = m\angle M + m\angle F$. However, $m\angle F = \frac{1}{2}m\widehat{AM}$, and $m\angle M = \frac{1}{2}m\widehat{NF}$. By substitution, $m\angle NEF = \frac{1}{2}(m\widehat{AM} + m\widehat{NF})$.

Example 2 Find $m\widehat{NF}$ in the figure above, if $m\widehat{AM} = 50$ and $m\angle NEF = 35$.

Solution By Theorem 9-8.2, $m\angle NEF = \frac{1}{2}(m\widehat{AM} + m\widehat{NF})$. By substitution, $35 = \frac{1}{2}(50 + m\widehat{NF})$. Multiplying by 2, $70 = 50 + m\widehat{NF}$. Therefore $m\widehat{NF} = 20$.

T **Theorem 9-8.3** The measure of an angle formed by two secants of a circle intersecting in a point in the exterior of the circle is equal to one-half the difference of the measures of the intercepted arcs.

GIVEN $\overleftrightarrow{GHI}$ and $\overleftrightarrow{FAI}$ are secants of $\odot P$; G, H, A, and F are points of the circle; I is an exterior point of $\odot P$.

PROVE $m\angle I = \frac{1}{2}(m\widehat{GF} - m\widehat{AH})$

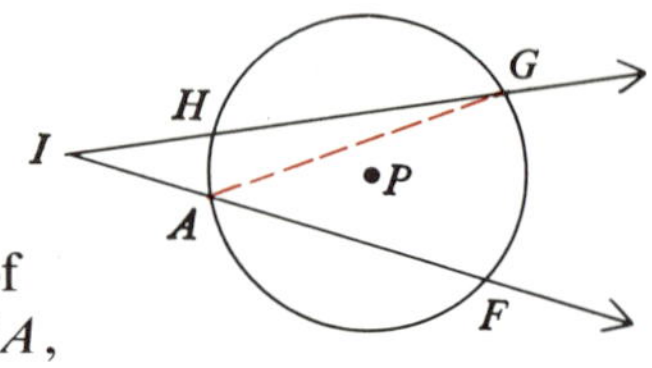

PROOF OUTLINE $\angle GAF$ is an exterior angle of $\triangle AIG$. Therefore, $m\angle GAF = m\angle I + m\angle IGA$, or $m\angle I = m\angle GAF - m\angle IGA$. By Theorem 9-6.1, $m\angle GAF = \frac{1}{2}m\widehat{GF}$ and $m\angle IGA = \frac{1}{2}m\widehat{AH}$. Substituting, $m\angle I = \frac{1}{2}(m\widehat{GF} - m\widehat{AH})$.

Example 3 In the above figure, if $\overline{HG} \cong \overline{AF}$, $m\widehat{GF} = 130$, and $m\widehat{AF} = 80$, find $m\angle I$.

Solution Since $\overline{HG} \cong \overline{AF}$, $\widehat{HG} \cong \widehat{AF}$. Therefore $m\widehat{HG} = 80$. Thus
$m\widehat{AH} = 360 - (80 + 80 + 130) = 70$. By Theorem 9-8.3,
$m\angle I = \frac{1}{2}(130 - 70) = 30$.

T **Theorem 9-8.4** The measure of an angle formed by a secant and a tangent to a circle intersecting in a point in the exterior of the circle is equal to one-half the difference of the measures of the intercepted arcs.

GIVEN $\overleftrightarrow{KJ}$ is tangent to $\odot P$ at J; $\overleftrightarrow{KAF}$ is a secant of the circle; A and F are points of $\odot P$.

PROVE $m\angle K = \frac{1}{2}(m\widehat{JF} - m\widehat{JA})$

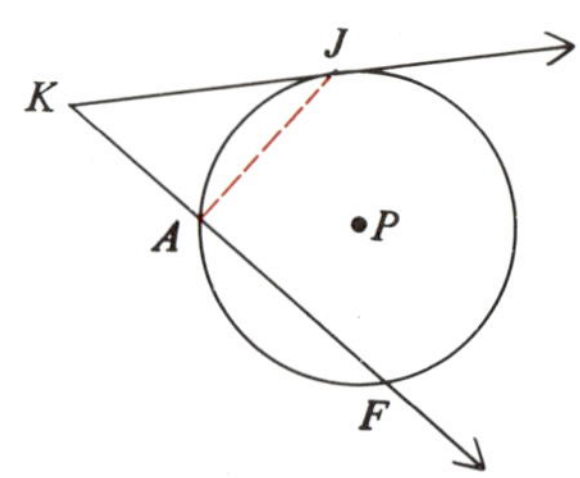

PROOF OUTLINE For $\triangle KJA$,
$m\angle JAF = m\angle K + m\angle KJA$, or $m\angle K = m\angle JAF - m\angle KJA$. Since $m\angle JAF = \frac{1}{2}m\widehat{JF}$ by Theorem 9-6.1, and $m\angle KJA = \frac{1}{2}m\widehat{JA}$ by Theorem 9-8.1, we find that $m\angle K = \frac{1}{2}(m\widehat{JF} - m\widehat{JA})$.

Can you prove Theorem 9-8.4 by considering $\overline{JF}$ instead of $\overline{JA}$?

Example 4 In the figure above, if $m\widehat{JA} = 85$ and $m\angle K = 40$, find $m\widehat{JF}$.

Solution Theorem 9-8.4 tells us that $m\angle K = \frac{1}{2}(m\widehat{JF} - m\widehat{JA})$; therefore, $40 = \frac{1}{2}(m\widehat{JF} - 85)$. Then $80 = m\widehat{JF} - 85$, and $m\widehat{JF} = 165$.

T

Theorem 9-8.5 The measure of an angle formed by two tangents to a circle is equal to one-half the difference of the measures of the intercepted arcs.

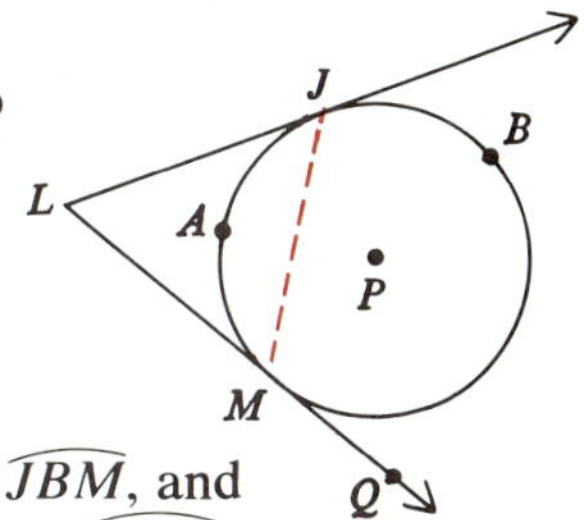

GIVEN $\overleftrightarrow{LJ}$ is tangent to $\odot P$ at J; $\overleftrightarrow{LMQ}$ is tangent to $\odot P$ at M.

PROVE $m\angle L = \frac{1}{2}(m\widehat{JBM} - m\widehat{JAM})$

PROOF OUTLINE
For $\triangle JLM$, $m\angle QMJ = m\angle L + m\angle LJM$, or $m\angle L = m\angle QMJ - m\angle LJM$. Since $m\angle QMJ = \frac{1}{2}m\widehat{JBM}$, and $m\angle LJM = \frac{1}{2}m\widehat{JAM}$, we find that $m\angle L = \frac{1}{2}(m\widehat{JBM} - m\widehat{JAM})$.

Example 5 In the figure above, find $m\angle L$, if $m\widehat{JAM} = 120$.

Solution Since $m\widehat{JAM} = 120$, $m\widehat{JBM} = 360 - 120 = 240$. By Theorem 9-8.5, $m\angle L = \frac{1}{2}(240 - 120) = 60$.

The solution of Example 5 suggests the following corollary to Theorem 9-8.5:

C

Corollary 9-8.5a The sum of the measure of an angle formed by two tangents to a circle and the measure of the closer intercepted arcs is 180.

You will notice that in Example 5, $m\angle L + m\widehat{JAM} = 180$. The proof of the above corollary is left as an exercise.

Summary. Related Measures of Angles and Intercepted Arcs

1. When the vertex of an angle is a point of a circle, the measure of the angle is one-half the measure of the intercepted arc.
2. When the vertex of an angle is in the interior of a circle, the measure of the angle is one-half the sum of the measures of the intercepted arcs.
3. When the vertex of an angle is in the exterior of a circle, the measure of the angle is one-half the difference of the measures of the intercepted arcs.

EXERCISES

A Find the value of x in each of the following figures:

1.

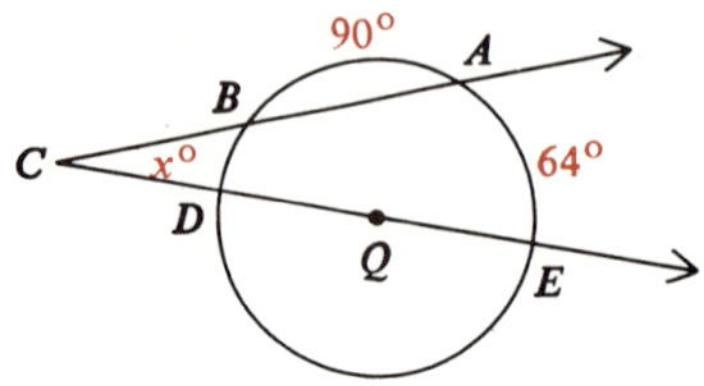

2.

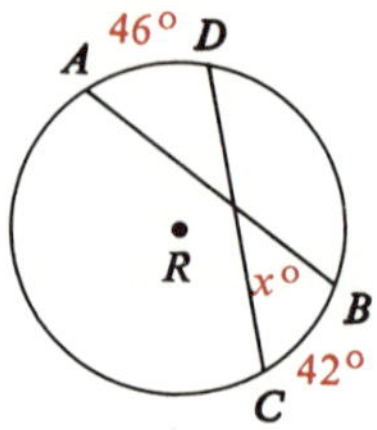

3.

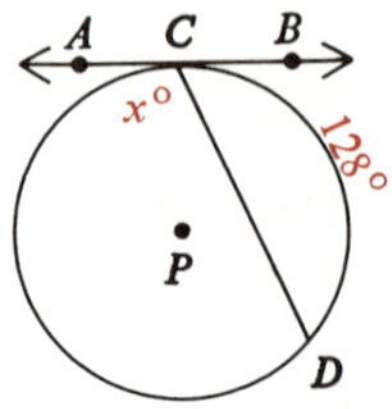

4.

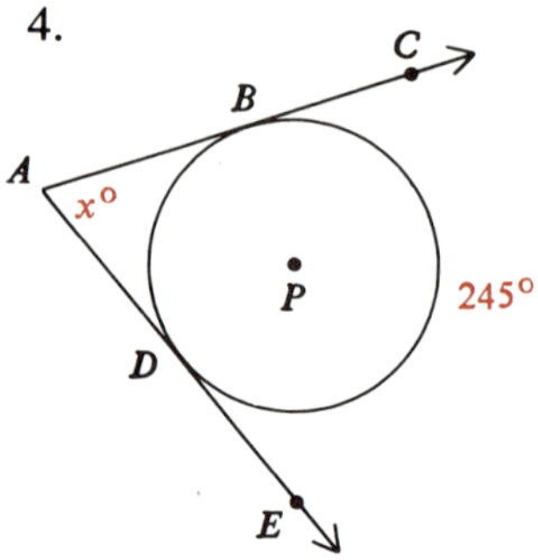

5.

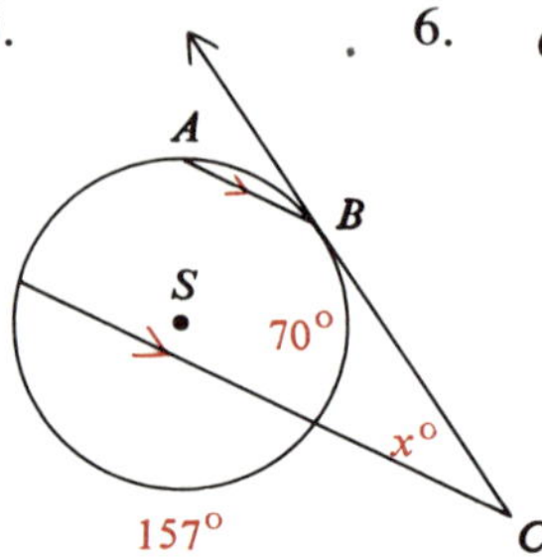

6.

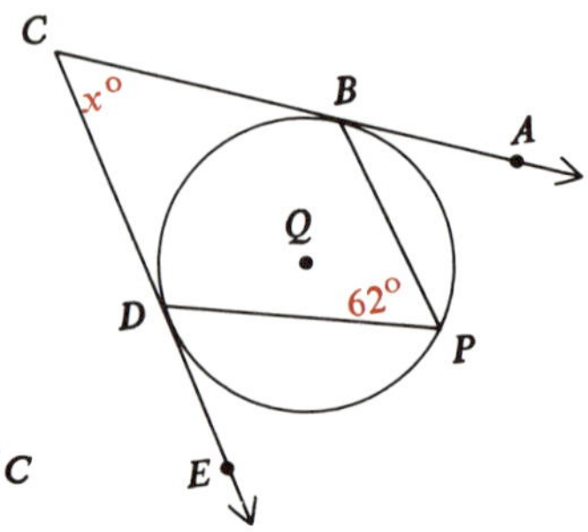

7. Prove that two parallel tangents of the same circle intercept congruent arcs on the circle.
8. Prove Theorem 9-8.1.
9. Prove Theorem 9-8.2.
10. Prove Theorem 9-8.3.
11. Prove Theorem 9-8.4.
12. Prove Theorem 9-8.5.
13. Circles P and Q are internally tangent at B. $\overline{AB}$ is a chord of $\odot Q$ and intersects $\odot P$ in C; $\overleftrightarrow{TB}$ is the common external tangent of the circles at B. If $m\widehat{AB} = 105$, find $m\widehat{BC}$.

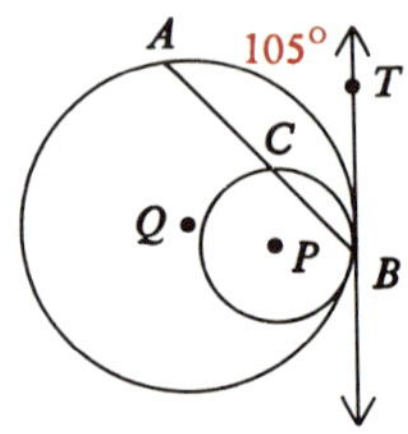

B Find the values of x, y, or z in each of the following figures:

14.

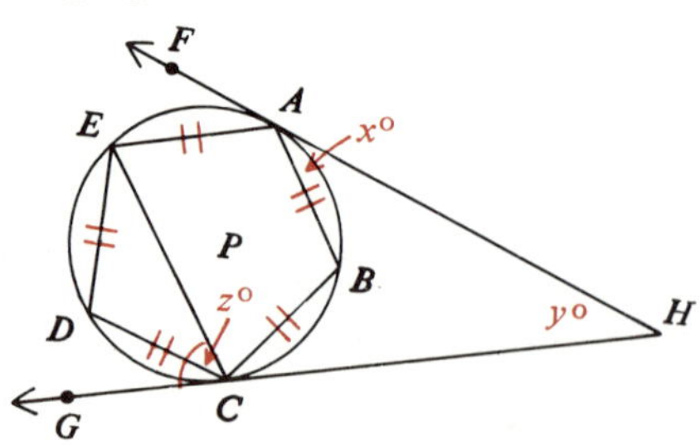

15.

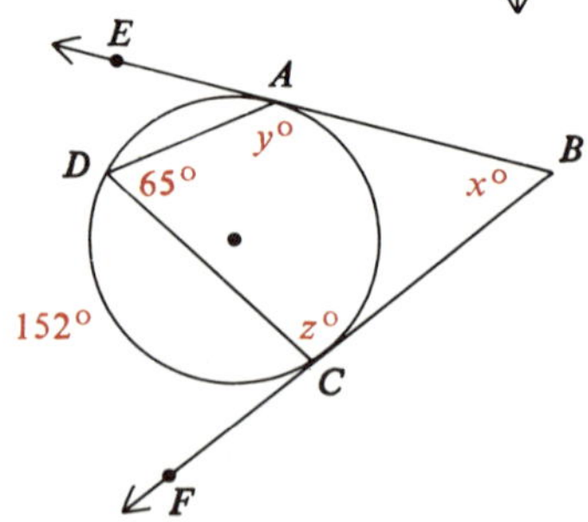

16.

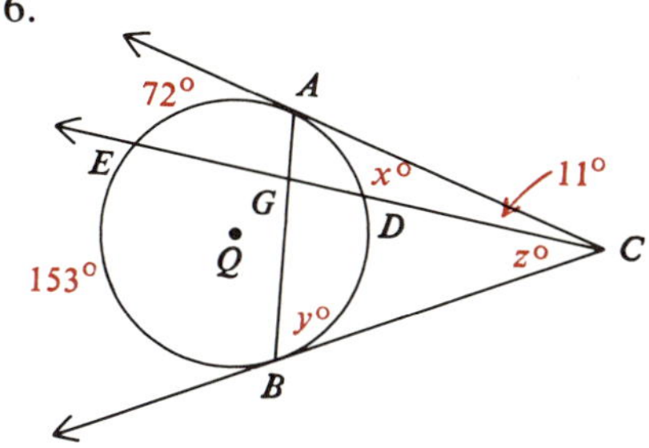

17.

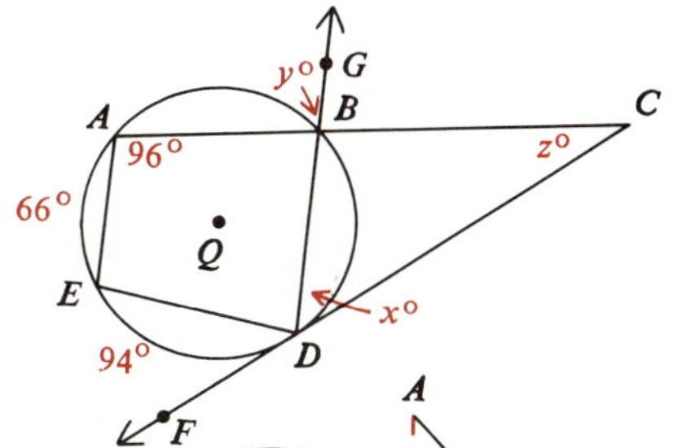

18. **Given** $\odot P$ is externally tangent to $\odot Q$ at T; $\overline{DTC}$; D is a point of $\odot P$, C is a point of $\odot Q$; $\overleftrightarrow{ACE}$ is tangent to $\odot Q$ at C; $\overleftrightarrow{BDF}$ is tangent to $\odot P$ at D.

Prove $\overleftrightarrow{ACE} \parallel \overleftrightarrow{BDF}$ (*Hint*: Consider the common internal tangent.)

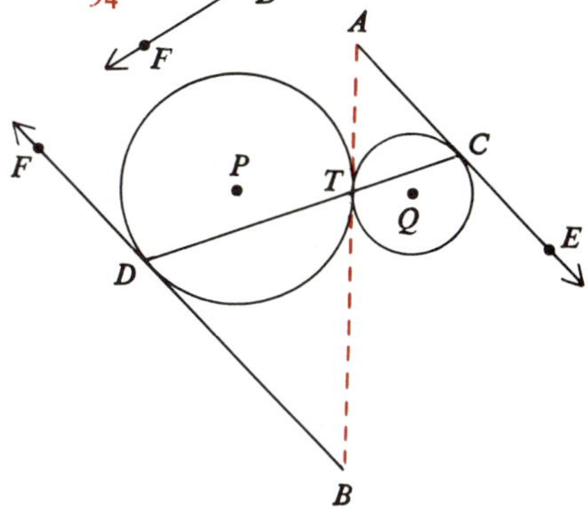

Using simultaneous equations, find the values of x and y in the figures below.

19.

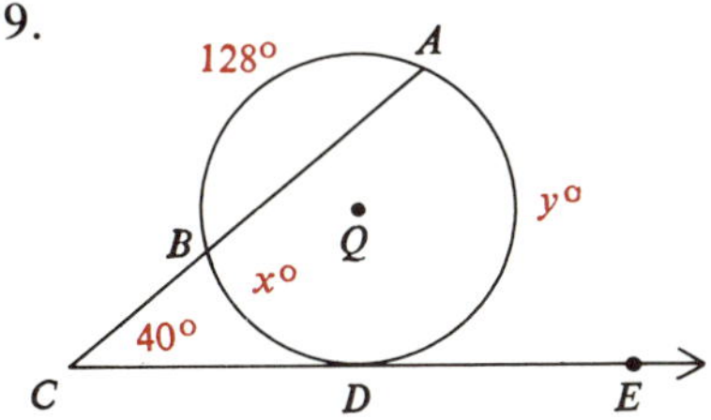

20.

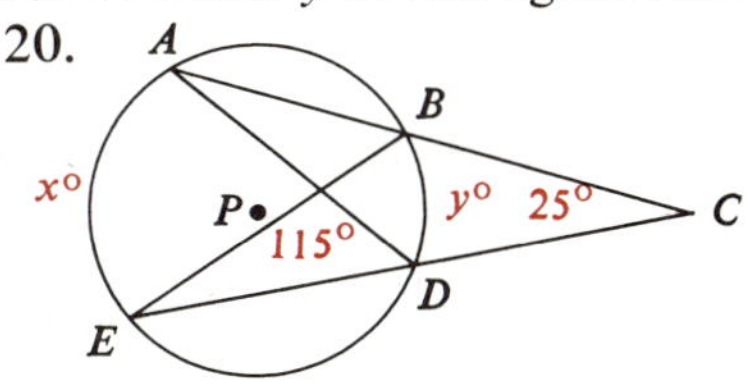

21. Prove Corollary 9-8.5a.

C

22. Prove Theorem 9-8.1 using the information in the figure at right.

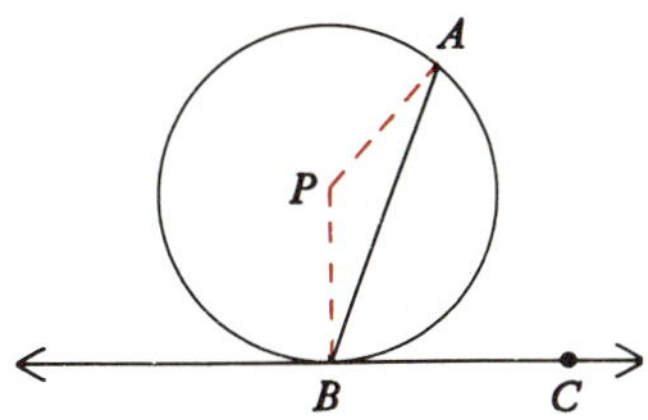

Exercise 22

23. **Given** Circles P and Q intersect in points A and C; chord $\overline{AB}$ of $\odot P$ is tangent to $\odot Q$; chord $\overline{AD}$ of $\odot Q$ is tangent to $\odot P$; E is a point of $\odot Q$; F is a point of $\odot P$.

Prove $\triangle ABC \sim \triangle DAC$; AC is the mean proportional between BC and DC.

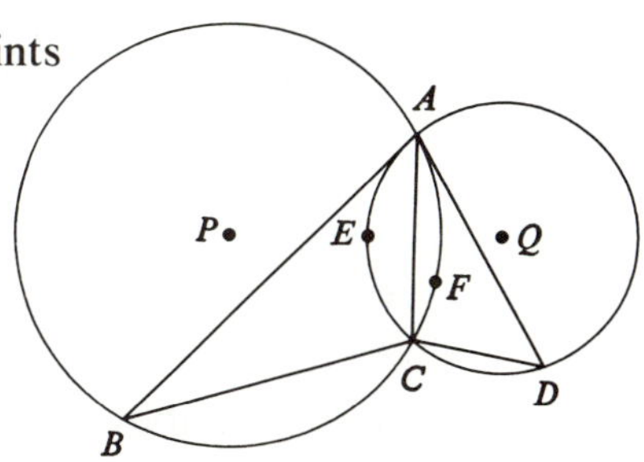

24. Prove Theorem 9-6.4, using only the information indicated in the figure at right.

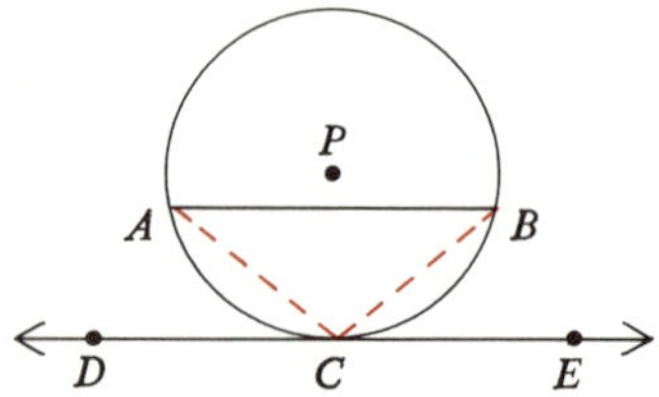

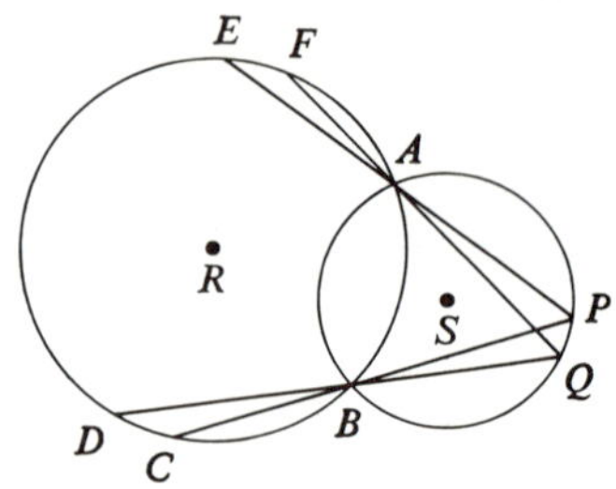

25. GIVEN $\odot R$ intersects $\odot S$ in points A and B; E, F, D, and C are points of $\odot R$; P and Q are points of $\odot S$; $\overline{EAP}$, $\overline{FAQ}$, $\overline{DBQ}$, and $\overline{CBP}$.
PROVE $\overline{EC} \cong \overline{FD}$

9-9 TANGENT, SECANT, AND CHORD SEGMENTS

In this section, we shall develop theorems that deal with segments contained in the tangents, secants, and chords of a circle. We shall first consider segments called tangent segments.

D **Definition 9-23** A **tangent segment** is a segment of a tangent to a circle, one of whose endpoints is the point of tangency.

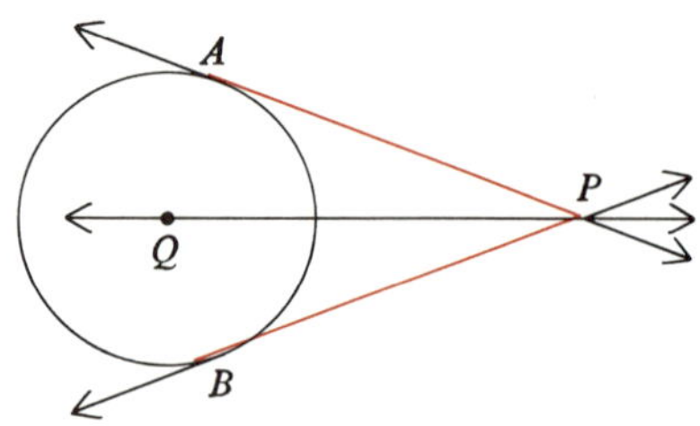

In the figure at right, $\overleftrightarrow{PA}$ and $\overleftrightarrow{PB}$ are tangent to $\odot Q$ at A and B, respectively. $\overline{PA}$ and $\overline{PB}$ are tangent segments. How do the lengths of $\overline{AP}$ and $\overline{PB}$ compare? In Example 1 of Section 9-3, we proved that $\overline{AP} \cong \overline{PB}$. We shall restate this as the first theorem of this section.

T **Theorem 9-9.1** Two tangent segments that have the same endpoint in the exterior of the circle to which they are tangent are congruent.

C **Corollary 9-9.1a** Two tangents to a circle that intersect in an exterior point of the circle form congruent angles with the line containing both the exterior point and the center of the circle.

Thus, in the figure above, $\angle APQ \cong \angle BPQ$. This corollary was proved in Example 1 of Section 9-3.

D **Definition 9-24** A **circle is inscribed** in a polygon if it is tangent to each side of the polygon. The polygon is a **circumscribed polygon** and the circle is an **inscribed circle**.

Example 1 $\odot Q$ is inscribed in $\triangle ABC$, intersecting the triangle in points D, E, and F. Find the perimeter of $\triangle ABC$, if $AB = 5$, $BF = 2$, and $EC = 4$.

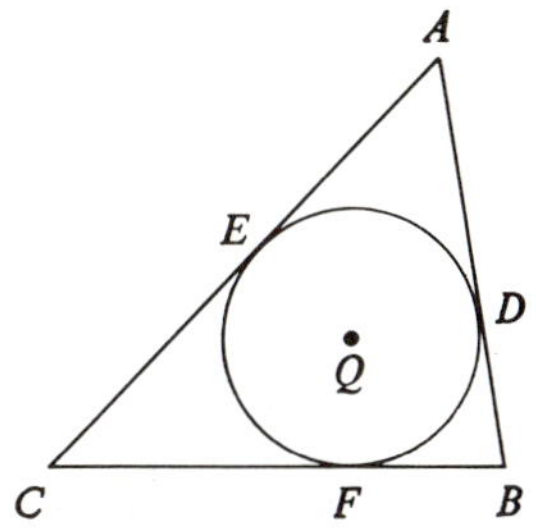

Solution By Theorem 9-9.1, $BD = BF = 2$. Then $AD = AB - BD = 5 - 2 = 3$. Applying Theorem 9-9.1 again, $AE = AD = 3$. Similarly, $FC = EC = 4$. By addition, we find that the perimeter of $\triangle ABC$ equals 18.

D **Definition 9-25** A **secant segment** is a segment that intersects a circle in two distinct points, exactly one of which is an endpoint of the segment.

In the figure below $\overline{PCD}$ is a secant segment of $\odot Q$. $\overline{PC}$ is called the *external segment* of the secant segment.

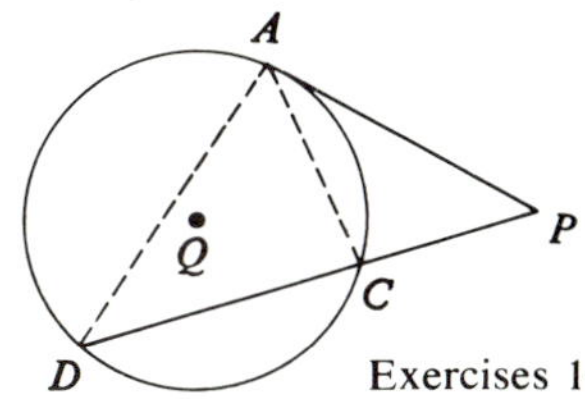

Exercises 1–7

Class Exercises

Exercises 1–7 refer to $\odot Q$ with tangent segment $\overline{AP}$ and secant segment $\overline{PCD}$, where C and D are points of $\odot Q$.

1. $m\angle PAC = \frac{1}{2}m$ __(arc)__.
2. $m\angle ADC = \frac{1}{2}m$ __(arc)__.
3. Why is $\angle P \cong \angle P$?
4. $\triangle APC \sim \triangle$ ____.
5. $\frac{\quad}{AP} = \frac{AP}{\quad}$.
6. We may then say that AP is the mean proportional between ____ and ____.
7. From Exercise 5, we know that $(AP)^2 =$ ____ $\cdot$ ____.

For Exercises 8–10, consider additional secant segment $\overline{PEF}$ of $\odot Q$, where E and F are points of $\odot Q$.

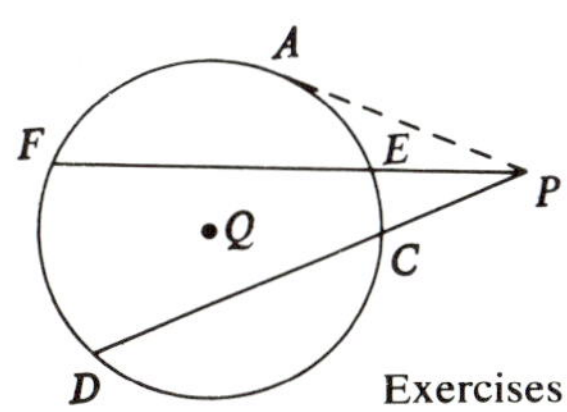

Exercises 8–10

8. Considering Exercise 6, we can determine that $\frac{PE}{AP} = \frac{\quad}{PF}$.
9. From Exercise 8, we know that ____ $= PE \cdot PF$.
10. Applying the ______ property to Exercises 7 and 9, we find that ____ $\cdot$ ____ $= PE \cdot PF$.

The theorems on the next page summarize the Class Exercises.

T

Theorem 9-9.2 If a secant segment and a tangent segment to the same circle share an endpoint in the exterior of the circle, then the length of the tangent segment is the mean proportional between the length of the secant segment and the length of its external segment.

In $\odot Q$ above, AP is the mean proportional between PD and PC.

An alternate form of this theorem follows:

C

Corollary 9-9.2a If a secant segment and a tangent segment to the same circle share an endpoint in the exterior of the circle, then the square of the length of the tangent segment equals the product of the lengths of the secant segment and its external segment.

This corollary tells us that in $\odot Q$ above, $(AP)^2 = PC \cdot PD$.

T

Theorem 9-9.3 If two secant segments of the same circle share an endpoint in the exterior of the circle, then the product of the lengths of one secant segment and its external segment equals the product of the lengths of the other secant segment and its external segment.

In the figure above, where $\overline{PCD}$ and $\overline{PEF}$ are secant segments, Theorem 9-9.3 tells us that $PD \cdot PC = PF \cdot PE$.

The proofs of the above theorems are outlined in the Class Exercises. Complete proofs of these theorems are left as exercises.

Example 2 $\overline{APBE}$ and $\overline{DCE}$ are secant segments of $\odot P$. Find the radius of $\odot P$ if $BE = 3$, $EC = 4$, and $CD = 2$.

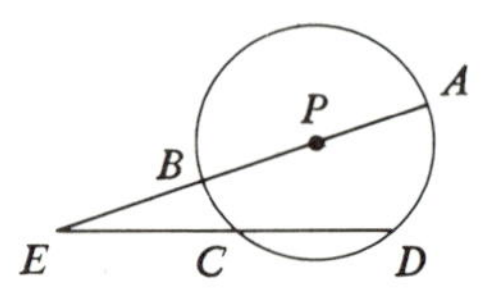

Solution By Theorem 9-9.3, $AE \cdot EB = DE \cdot EC$. Therefore, $AE \cdot 3 = 6 \cdot 4$; $AE = 8$. Since $AB = AE - BE = 8 - 3 = 5$, the radius equals $\frac{1}{2}AB = 2\frac{1}{2}$.

Theorem 9-9.4 concerns the segments determined by the intersection of two chords of a circle.

T

Theorem 9-9.4 If two chords intersect in the interior of a circle, thus determining two segments in each chord, the product of the lengths of the segments of one chord equals the product of the lengths of the segments of the other chord.

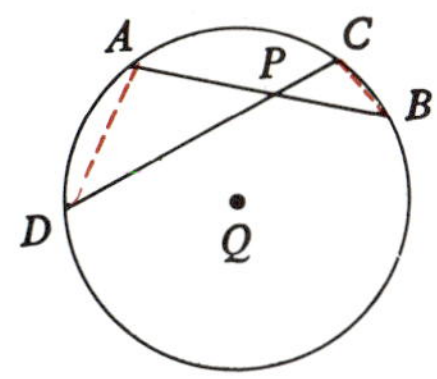

Given Chords $\overline{AB}$ and $\overline{CD}$ of $\odot Q$ intersect in point P.

Prove $AP \cdot PB = CP \cdot PD$

Proof

STATEMENTS	REASONS
1. Chords $\overline{AB}$ and $\overline{CD}$ of $\odot Q$ intersect in point P.	1. Given.
2. $\angle ADC \cong \angle ABC$	2. Angles inscribed in the same arc are congruent. (C9-6.1a)
3. $\angle DAB \cong \angle DCB$	3. Same reason as for step 2.
4. $\triangle ADP \sim \triangle CBP$	4. If two pairs of corresponding angles of two triangles are congruent, then the triangles are similar. (C8-5.1a)
5. $\frac{AP}{PD} = \frac{CP}{PB}$	5. Definition of similar triangles. (8-6)
6. $AP \cdot PB = CP \cdot PD$	6. For any positive real numbers a, b, c, and d, $\frac{a}{b} = \frac{c}{d}$ if and only if $ad = bc$. (8-1.1)

Example 3 Find the length of the chord whose distance from the center of a circle of diameter 10 is 3.

Solution In the figure at right, the distance of chord $\overline{AB}$ from the center of $\odot Q$ is 3; diameter $\overline{CD}$ meets chord $\overline{AB}$ at M; $CD = 10$; therefore, $QD = 5$. Since $MD = 8$, $CM = 2$. By Theorem 9-9.4 $AM \cdot MB = CM \cdot MD$. Why is $AM = MB$? Let $AM = x$. Then $x^2 = 2 \cdot 8 = 16$, or $x = 4$. Thus, the length of chord $\overline{AB}$ is 8.

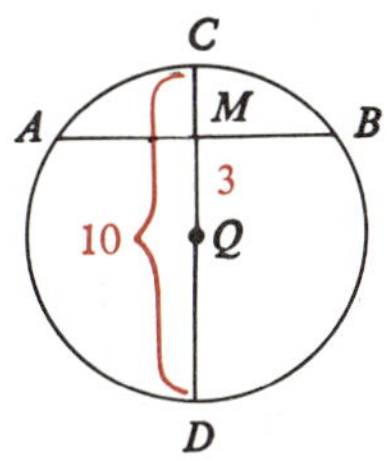

EXERCISES

A Find the value of x in each of the following figures:

1.

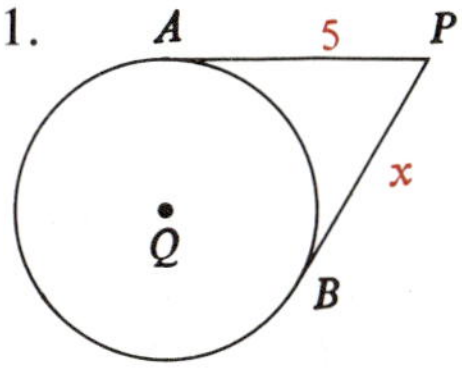

2.

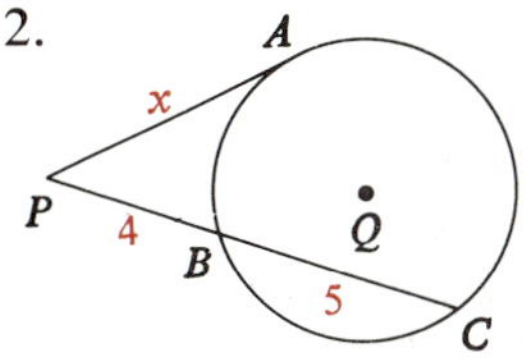

3.

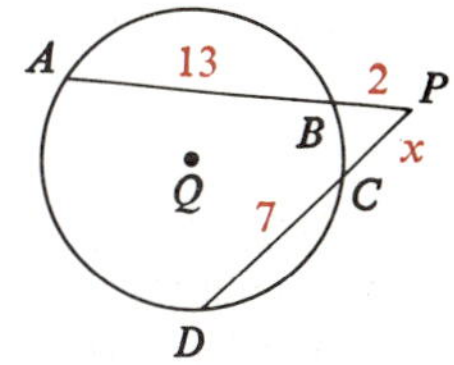

Find the values of x and y in each of the following figures:

4.

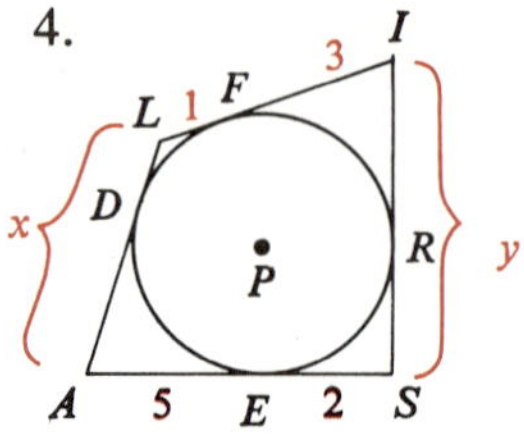

5.

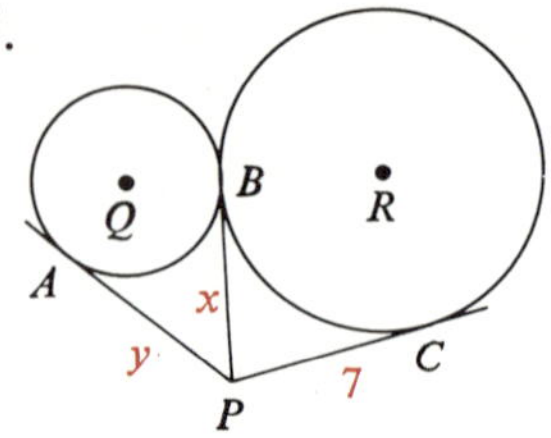

6.

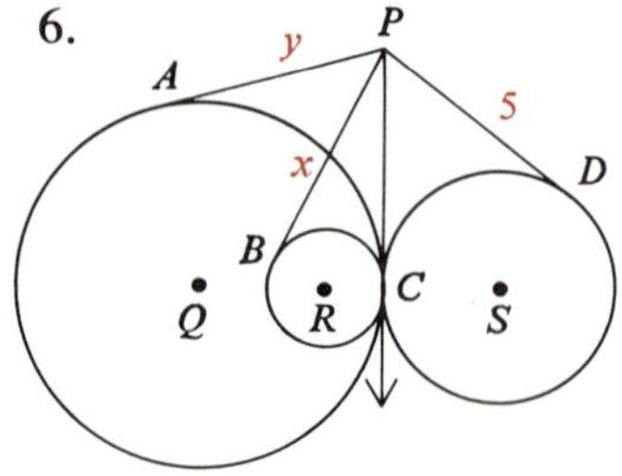

Exercises 7–9 refer to $\odot Q$, with chords $\overline{AB}$ and $\overline{CD}$ intersecting in point E.

7. If $AE = 3$, $AB = 12$, and $CE = 4$, find DC.
8. If $DE = 6$, $BE = 8$, and $AB = 11$, find CE.
9. If $AB = 18$, $DC = 14$, and $CE = 5$, find BE and AE.

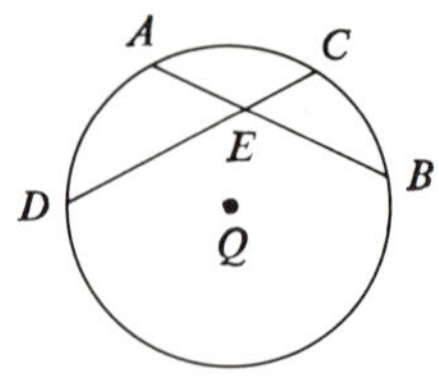

Exercises 7-9

Exercises 10–12 refer to $\odot Q$, with secant segments $\overline{PCB}$ and $\overline{PED}$, and tangent segment $\overline{PA}$.

10. If $PC = 4$, and $BC = 5$, find AP.
11. If $PE = 6$, $ED = 2$, and $PC = 3$, find BC.
12. If $AP = 12$, and $ED = 7$, find PD.

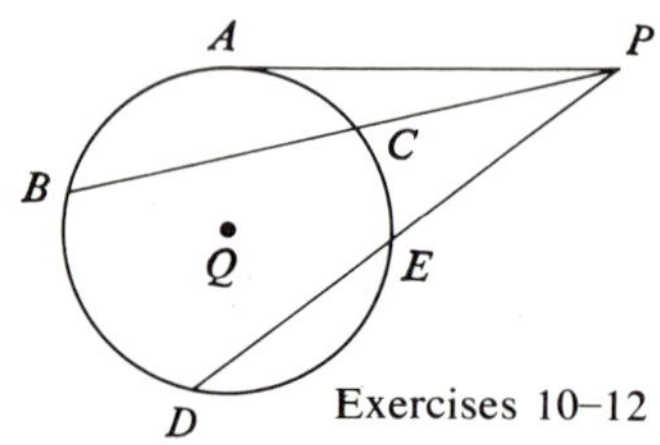

Exercises 10–12

13. GIVEN $\overleftrightarrow{AMB}$ and $\overleftrightarrow{CMD}$ are common internal tangents of nonintersecting circles P and Q; A and D are points of $\odot P$; B and C are points of $\odot Q$.

PROVE $\overline{AB} \cong \overline{CD}$

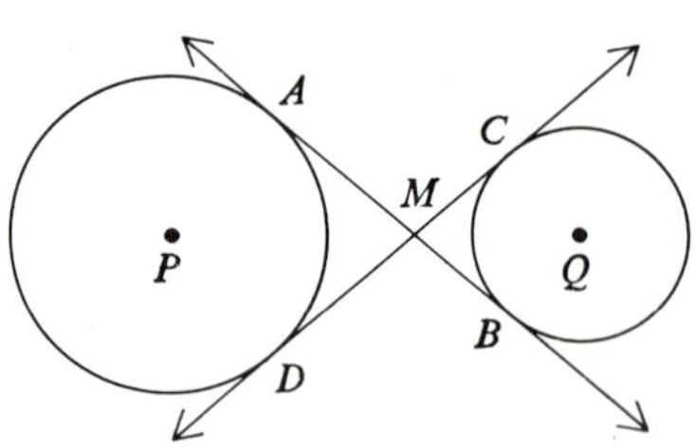

14. GIVEN $\odot Q$ intersects $\odot R$ in points C and D; $\overleftrightarrow{PA}$ is tangent to $\odot Q$ at A; $\overleftrightarrow{PB}$ is tangent to $\odot R$ at B; $\overleftrightarrow{DCP}$

PROVE $PA = PB$

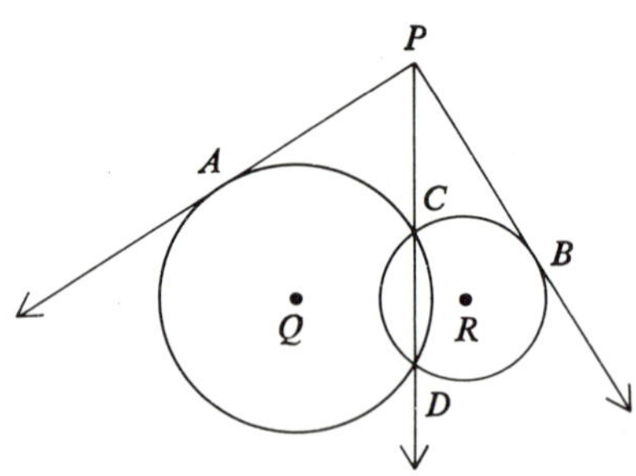

15. Prove Theorem 9-9.2.
16. Prove Corollary 9-9.2a.
17. Prove Theorem 9-9.3.

B

18. A point P is 17 inches from the center of a circle whose diameter is 16. Find the lengths of the tangent segments from point P to the circle.

19. **Given** $\odot Q$ is inscribed in quadrilateral $ABCD$, meeting the sides at points E, F, G, and H.
Prove $AB + CD = AD + CB$

20. In the figure at right, P is 8 units from the center of $\odot Q$. If the radius of $\odot Q$ is 10, find the product of the two segments determined by P in each chord containing P.

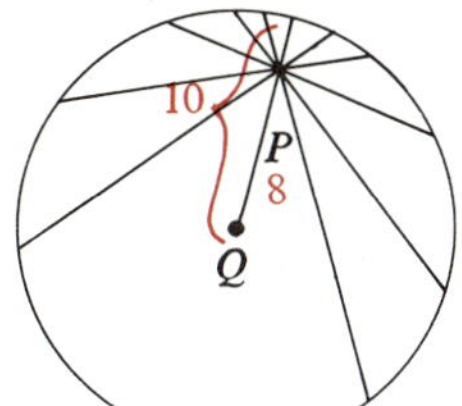

Find the values of x and y in each of the following figures:

21.

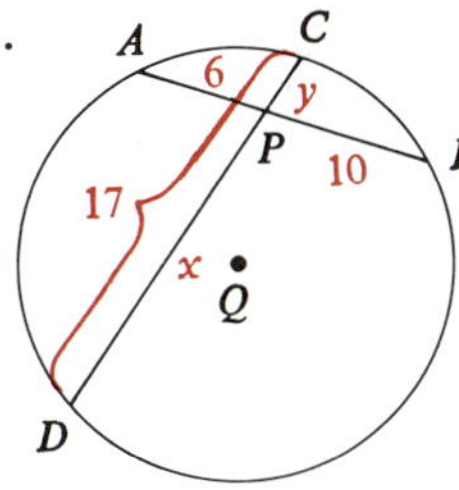

22.

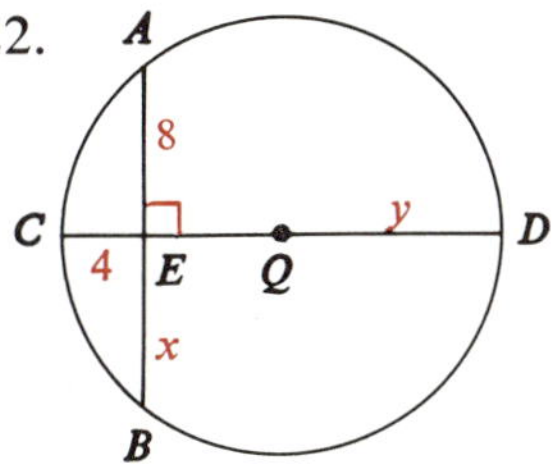

23.

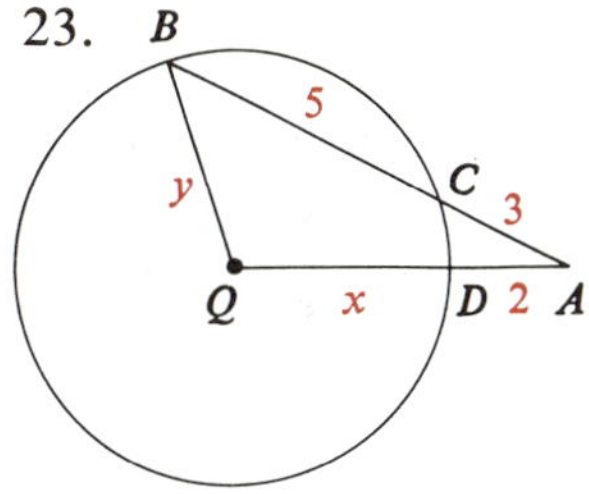

24. **Given** $\odot Q$ is externally tangent to $\odot R$ at T; $\overleftrightarrow{PAB}$ and $\overleftrightarrow{PCD}$ are the common external tangents of the circles; $\overline{MTN}$ is the common internal tangent; A and C are points of $\odot Q$; B and D are points of $\odot R$; $\overline{AMB}$; $\overline{CND}$
Prove $\overline{AB} \cong \overline{CD}$; $\overline{MTN} \cong \overline{AB}$

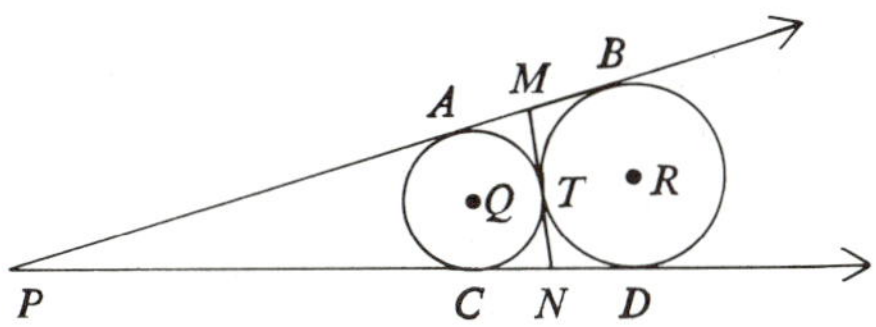

C

25. Prove Theorem 9-9.3, using the information indicated in the figure at right.

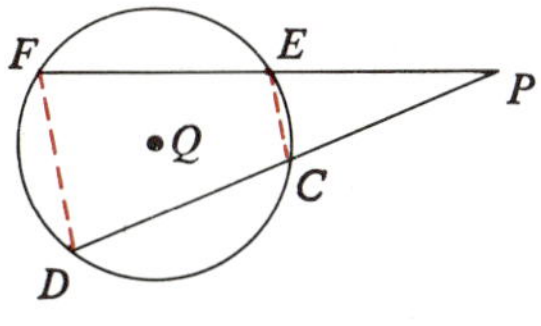

26. Find the radius of a circle inscribed in a triangle whose sides have lengths 3, 4, and 5.

27. **Given** $\overline{DC}$ is an angle bisector of $\triangle ABC$.
Prove $(DC)^2 = AC \cdot BC - AD \cdot DB$
(*Hint*: Consider the dashed lines in the figure.)

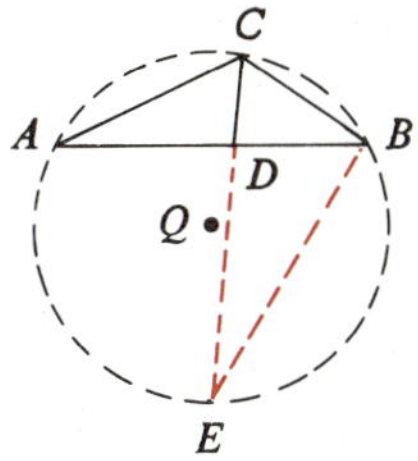

28. GIVEN $\overline{AB}$ and $\overline{CD}$ intersect in P; $AP \cdot BP = CP \cdot DP$
PROVE Points A, B, C, and D are concyclic.

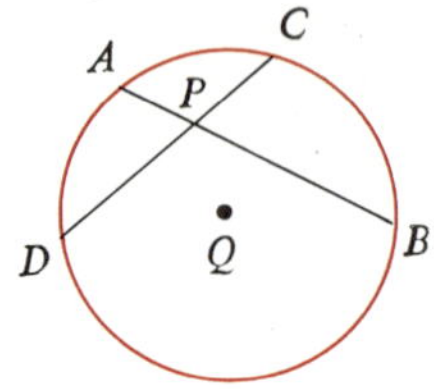

29. GIVEN $\overline{ABP}$ and $\overline{DCP}$; $AP \cdot BP = CP \cdot DP$
PROVE Points A, B, C, and D are concyclic.

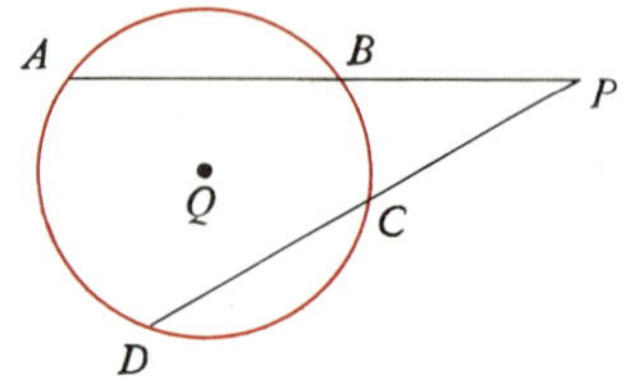

9-10 CIRCUMFERENCE AND LENGTHS OF ARCS

"The circumference of a circle is the perimeter of the circle." Although there is nothing wrong with this definition of *circumference*, one that you probably learned in an earlier mathematics course, we shall develop a more rigorous definition in this section.

Consider a circle, $\odot Q$, with an equilateral triangle inscribed in it. Suppose we take the same circle and inscribe in it, in turn, a square, a regular pentagon, and a regular hexagon.

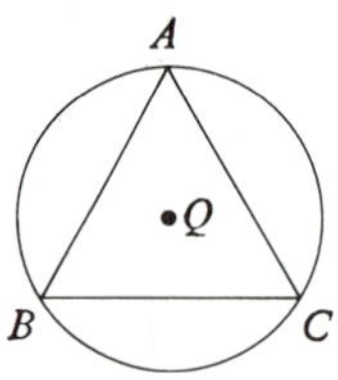

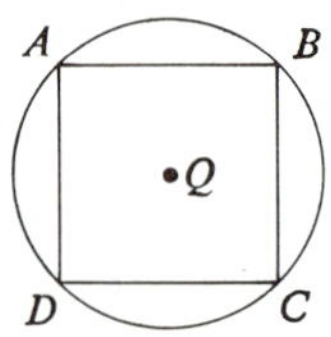

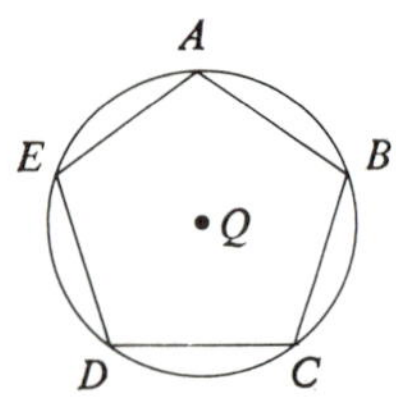

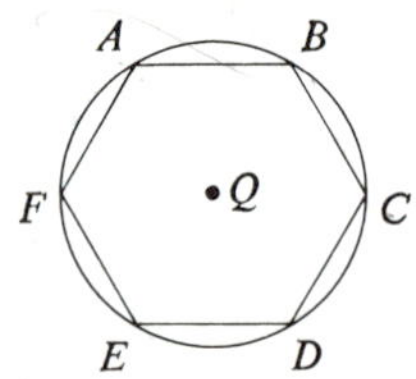

How do you suppose the perimeters of these inscribed regular polygons compare to the circumference of the circle? Which of these regular polygons has a perimeter closest to the circumference of the circle? As we increase the number of sides of the polygons we inscribe in $\odot Q$, how does the perimeter of the polygon compare to the circumference of the circle? What is the largest possible perimeter of an inscribed polygon?

The figure at right, a composite of the four figures above, illustrates what happens when we keep increasing the number of sides of the regular polygons we inscribe in circle Q. As the number of sides increases, notice that the perimeters of the polygons become closer and closer to the circumference of the circle. When this happens, we say that the perimeters of these

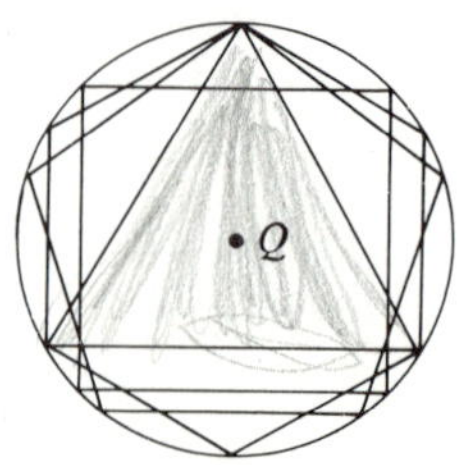

polygons approach the circumference of the circumscribed circle as a limit. In other words, if each new regular polygon inscribed has a greater number of sides than the one before, the perimeter of each new polygon will be closer to the circumference. Yet no matter how many times we repeat this process the perimeter of the polygon will never actually reach the circumference. There will always be some difference between the perimeter of the regular polygon and the circumference of its circumscribed circle. Can you explain why?

In future mathematics courses, you will have an opportunity to study the concept of limit in greater depth. However, for our purposes, this introduction to limit will suffice. You may wish to think of the circumference of a circle as the perimeter of a regular polygon with a boundless number of sides.

Definition 9-26 The **circumference** of a circle is the limit of the perimeters of the inscribed regular polygons as the number of sides of the polygons increases without bound.

Associated with the term *circumference*, you probably remember the symbol π. We shall define π as the ratio of the circumference of a circle to its diameter. That is, $\pi = \frac{c}{d}$.

Theorem 9-10.1 The ratio, π, of the circumference of a circle to its diameter is the same for all circles.

PROOF OUTLINE To show that the ratio π is the same for all circles, we shall choose any two circles and show that in each circle this ratio has the same value.

Consider $\odot P$ with radius r and circumference c, and $\odot P'$ with radius r' and circumference c'. In each circle, a regular n-gon is inscribed. The figures at right show only a part of each n-gon.

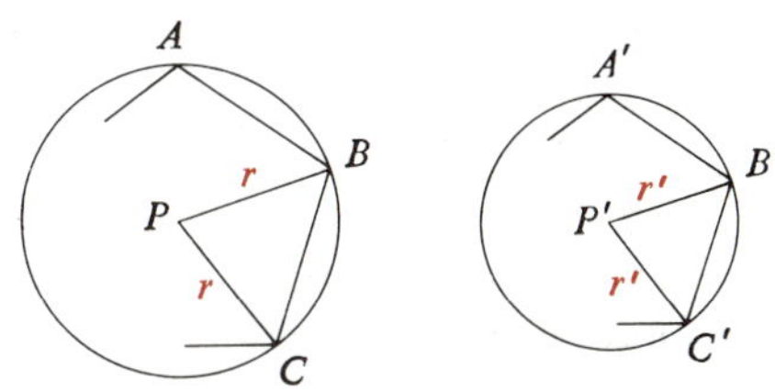

The measure of each central angle whose rays intersect the vertices of these n-gons is $\frac{360}{n}$. Therefore $\angle BPC \cong \angle B'P'C'$. Since $\frac{r}{r'} = \frac{r}{r'}$, $\triangle BPC \sim \triangle B'P'C'$ by Theorem 8-6.1.

Let $BC = s$ and $B'C' = s'$. Therefore $\frac{s}{r} = \frac{s'}{r'}$, and $\frac{ns}{r} = \frac{ns'}{r'}$. However, $ns = p$, the perimeter of regular n-gon $ABC \cdots$, and $ns' = p'$, the perimeter of regular n-gon $A'B'C' \cdots$. Therefore $\frac{p}{r} = \frac{p'}{r'}$.

By the definition of circumference, as the number of sides of each n-gon increases without bound, p approaches c as a limit, while p' approaches c' as a limit. In symbols we write this as

$$p \to c \qquad \text{and} \qquad p' \to c'$$

It then follows that

$$\frac{p}{r} \to \frac{c}{r} \qquad \text{and} \qquad \frac{p'}{r'} \to \frac{c'}{r'}$$

But since $\frac{p}{r} = \frac{p'}{r'}$, we get $\frac{c}{r} = \frac{c'}{r'}$. Hence, $\frac{c}{2r} = \frac{c'}{2r'}$ or $\frac{c}{d} = \frac{c'}{d'}$, where d and d' are the diameters of the two circles. Thus, the ratio π is the same for both circles.

We now accept that the ratio $\pi = \frac{c}{d}$ is the same for all circles. Thus $\pi = \frac{c}{2r}$ is the same for all circles. From this second ratio we obtain the formula $c = 2\pi r$.

Corollary 9-10.1a The circumferences of any two circles are proportional to their radii.

The proof of this corollary is left as an exercise.

Class Exercises

The following exercises refer to the figure at right. The radius of $\odot Q$ is 6 and the measures of the central angles are indicated.

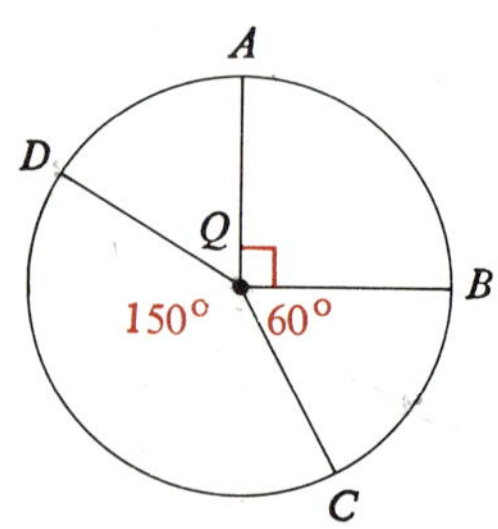

1. Find the circumference of $\odot Q$.
2. What part is $m\widehat{AB}$ of the measure of $\odot Q$?
3. What fraction of the circumference of the circle is $\widehat{AB}$?
4. How long is $\widehat{AB}$?
5. What part of 360 is $m\widehat{BC}$?
6. Find the length of $\widehat{BC}$.
7. What fractional part of $\odot Q$ is $\widehat{DC}$?
8. Find the length of $\widehat{DC}$.

From the Class Exercises above, we find that the length of an arc of degree measure n is $\frac{n}{360}$ of the circumference of the circle. This forms the basis for Theorem 9-10.2:

T

Theorem 9-10.2 If an arc of a circle of radius r has measure n, then the length of the arc is $\frac{n}{360} \cdot 2\pi r$.

Example A and B are points of $\odot Q$ such that $\triangle AQB$ is equilateral. If $AB = 12$, find the length of $\overset{\frown}{AB}$.

Solution Since $\triangle AQB$ is equilateral, the radius of $\odot Q$ is 12. Also $m\angle AQB = 60$. Therefore, $m\overset{\frown}{AB} = 60$. Applying Theorem 9-10.2,

$$\begin{aligned} L &= \tfrac{60}{360} \cdot 2\pi(12) \\ &= \tfrac{1}{6} \cdot 24\pi \\ &= 4\pi \end{aligned}$$

EXERCISES

A

Leave all answers in terms of π unless otherwise indicated.

Which of the following is the closest approximation of π? Which is the least accurate approximation?

1. 3.14 2. $\frac{22}{7}$ 3. 3 4. $\frac{754}{240}$ 5. $\frac{355}{113}$

Find the circumference of the circle whose radius is:

6. 4 7. 7 8. $2r$ 9. x 10. $3x - 5$

Find the radius of the circle whose circumference is:

11. 6π 12. 13π 13. 18 14. c 15. $2x + 1$

Find the diameter of a circle whose circumference is:

16. 5π 17. 12π 18. 20 19. c 20. $4x + 7$

21. State a formula that can be used to find the length of a semicircle of radius r.

Using the approximation $\pi = 3.14$, find the length of the semicircles contained in circles with the following radii:

22. 1 23. 4 24. 9 25. $2x$ 26. $3x - 2$

Using the approximation $\pi = \frac{22}{7}$, find the length of an arc of measure 90 of circles with these diameters.

27. 14 28. 7 29. 35 30. $56x$ 31. $14x - 1$

32. If the wheels of a bicycle have 26-inch diameters, how far will the bicycle have traveled when the front wheel has made 70 revolutions? (Use $\pi = \frac{22}{7}$.)

Find the length of $\overset{\frown}{AB}$ in each of the following figures:

33.

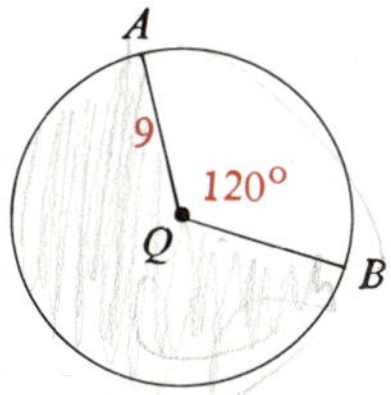

34.

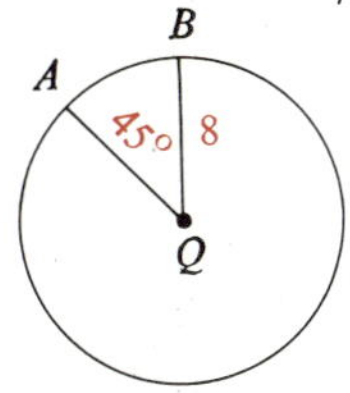

35. Prove Corollary 9-10.1a

36. Prove Theorem 9-10.2.

B

Find the value of x in each of the figures:

37.

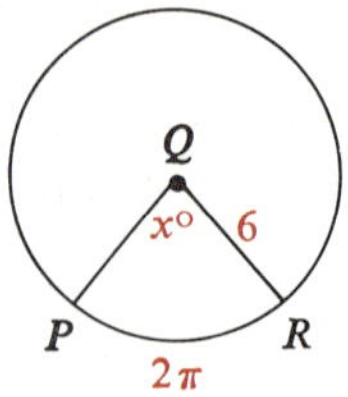

38.

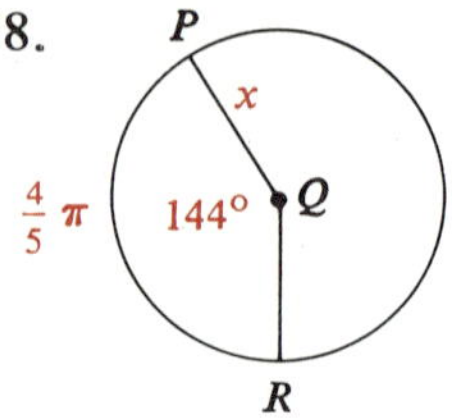

39.

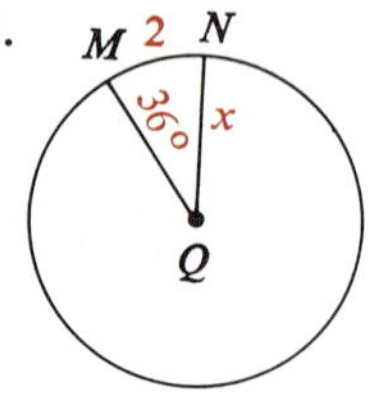

40. Find the circumference of the circle inscribed in a square with 20-inch sides.
41. If the perimeter of a regular hexagon is 18, find the circumference of its circumscribed circle.
42. Prove that the circumference and diameter of a circle cannot both be rational.
43. By how much does the circumference of a circle increase when its radius, r, is doubled? tripled?
44. By how much does the radius of a circle increase when the circumference, c, is doubled? tripled?

Find the ratios of the circumferences of pairs of circles whose radii are in the following ratios:

45. $\frac{2}{3}$ 46. $\frac{5}{6}$ 47. 8:1 48. 2:7

Find the radius of the smaller of pairs of circles with circumferences in the ratio of $\frac{5}{9}$ when the radii of the larger circles are as follows:

49. 9 50. 27 51. 24 52. 81

C

53. Find the circumference of the circle circumscribed about an equilateral triangle with perimeter of 24.
54. Find the circumference of the circle inscribed in an equilateral triangle with a perimeter of 24.
55. By how much does the radius of a circle increase when the circumference increases from 20 inches to 25 inches?

56. In the figure at right, $\odot P$ contains point Q and $\odot Q$ contains point P. Find the perimeter of the shaded region if $PQ = 12$.

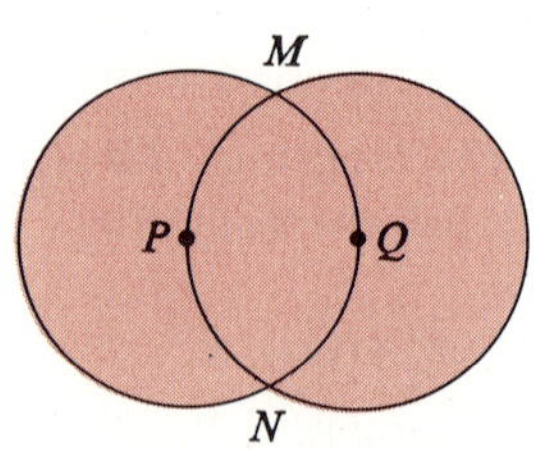

57. In the figure below, $\triangle ABC$ is equilateral. $\widehat{DE}$ is contained in $\odot A$, $\widehat{DF}$ is contained in $\odot B$, and $\widehat{EF}$ is contained in $\odot C$. If the perimeter of $\triangle ABC$ is 24, find the perimeter of the shaded region.

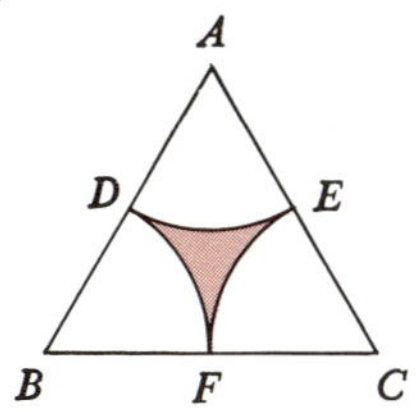

58. $\odot P$ and $\odot Q$ are externally tangent at T. $PT = 6$, and $QT = 2$. Find the length of a belt that would fit exactly once around the two circles. Assume that there is no slack.

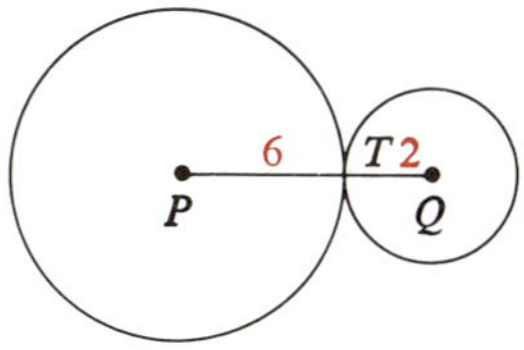

A LOOK AT THE PAST

The Value of π

One of the earliest references to the ratio of the circumference and diameter of a circle can be found in the Old Testament. The passage in Kings 7:23 gives a description of the pool in the courtyard of Solomon's Temple. The pool is a circle whose diameter is 10 cubits and whose circumference is 30 cubits. According to these measurements, $\pi = \frac{30}{10} = 3$.

The Hebrews were not the only ancient civilization to adopt this approximation of π; it was also used by the Babylonians, Chinese, and Hindus.

A more accurate approximation was recorded around 1800 B.C., in the Egyptian Ahmes Papyrus, in which π is said to be $\frac{(16)^2}{9}$, or about 3.1604 . . .

The Greeks refined the calculation even more. Euclid determined only that $3 < \pi < 4$, but in the second century Ptolemy used $\pi = 3.141552$. Archimedes evaluated π by measuring the perimeters of many-sided inscribed and circumscribed regular polygons. He found that $3.1408 < \pi < 3.1428$.

Since π is an irrational number, it cannot be expressed exactly in decimal fraction form, but mathematicians have continually attempted to correct the existing approximations. In 1579, the French algebraist François Viete worked with a polygon of $6(2^{16})$, or 393,216, sides and found that $3.1415926535 < \pi < 3.1415926537$, which gives π correct to the nine decimal places.

The most celebrated calculation of π was completed by William Shanks in 1873. It took Shanks 20 years to calculate π to 707 decimal places.

In contrast, in 1961 it took Daniel Shanks and John W. Wrench Jr. only 8 hours and 43 minutes to calculate the value of π correct to 100,265 decimal places, using an electronic calculating device. It was not until this calculation was made that an error was found in the 528th decimal place of the 1873 calculation.

The Greek letter π was not always used to represent the ratio of a circle's circumference and diameter. It was first used to represent this ratio in a book by the English writer William Jones, in 1706; but it was not until 1737 that the current symbol was made popular by the famous Swiss mathematician Leonhard Euler. Refer to the introduction to this chapter for the value of π expressed to about 2,800 decimal places.

HOW IS YOUR MATHEMATICAL VOCABULARY?

The key words and phrases introduced in this chapter are listed below. How many do you know and understand?

- Arc Addition Postulate (9-5)
- center (9-1)
- central angle (9-5)
- chord of an arc (9-5)
- chord of a circle (9-1)
- circle (9-1)
- circle of intersection (9-4)
- circumference (9-10)
- circumscribed circle (9-9)
- circumscribed polygon (9-9)
- common chord (9-2)
- common tangent (9-3)
- concentric (9-1)
- concyclic points (9-7)
- congruent arcs (9-5)
- congruent circles (9-2)
- cyclic quadrilateral (9-7)
- diameter (9-1)
- endpoints of an arc (9-5)
- exterior of a circle (9-1)
- great circle of a sphere (9-4)
- inscribed angle (9-6)
- inscribed circle (9-9)
- inscribed polygon (9-2)
- intercepted arc (9-6)
- interior of a circle (9-1)
- line of centers (9-2)
- major arc (9-5)
- measure of an arc (9-5)
- minor arc (9-5)
- π (9-10)
- point of tangency (9-3)
- radius (9-1)
- secant (9-1)
- secant ray (9-1)
- secant segment (9-9)
- semicircle (9-5)
- small circle of a sphere (9-4)
- sphere (9-1)
- tangent (9-3)
- tangent circles (9-3)
- tangent plane (9-4)
- tangent segment (9-9)

REVIEW EXERCISES

9-1 Basic definitions

Indicate whether each of the following statements is true or false. Correct those that are false.

1. A circle is the set of all points at a given distance from a given point.
2. Two circles that have the same center and are in the same plane are called concentric circles.
3. If P is a point of $\odot Q$, then $\overline{QP}$ is a radius of $\odot Q$.
4. If P is a point of $\odot Q$, then we may say that P is in the interior of $\odot Q$.
5. The plane containing the center of a sphere contains exactly one diameter of the sphere.

9-2 Chords of a circle

6. Prove that the line that contains the midpoints of two parallel chords of a circle also contains the center of the circle.
7. Two parallel chords of a circle with a 10-inch radius are 12 and 16 inches long. What is the distance between them?
8. Find the length of a chord of a circle of radius 17, if its distance from the center is 8.
9. GIVEN $\overline{BD}$ is a diameter of $\odot P$; $\overline{AB}$ and $\overline{BC}$ are chords; $\angle ABD \cong \angle CBD$
   PROVE $\overline{AB} \cong \overline{BC}$

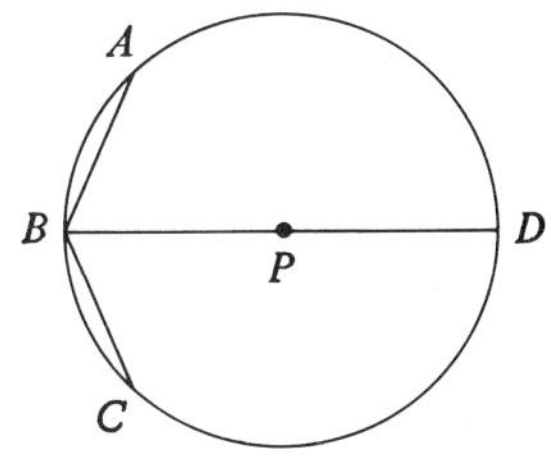

9-3 Tangent lines and circles

10. How many common tangents do two intersecting circles have? How many do two externally tangent circles have? two concentric circles? two internally tangent circles?
11. What do we call a line that is perpendicular to the radius of a circle at its intersection with the circle?
12. $\overleftrightarrow{TP}$ is tangent to $\odot Q$ at T. $\overline{PQ}$ meets $\odot Q$ at R. If the radius of $\odot Q$ is 10 and $RP = 16$, find PT.
13. Two concentric circles with centers Q have radii of 9 and 15. Find the length of a chord of the larger circle which is tangent to the smaller circle.

9-4 Planes and spheres

Indicate whether each of the following statements is true or false. Correct the false statements.

14. The diameter of a great circle of a sphere equals the diameter of the sphere.
15. The intersection of a sphere with a plane is always a circle.

16. A line joining the center of a sphere with the center of the circle of intersection formed by the sphere and a plane, is perpendicular to the plane.
17. When a plane intersects a sphere, the line perpendicular to the plane, at the center of its circle of intersection with the sphere, contains the center of the sphere.
18. GIVEN Planes P and Q intersect sphere S in $\odot A$ and $\odot C$, respectively; B is a point of $\odot A$; D is a point of $\odot C$; $\overline{AB} \cong \overline{CD}$

    PROVE Planes P and Q are equidistant from S.

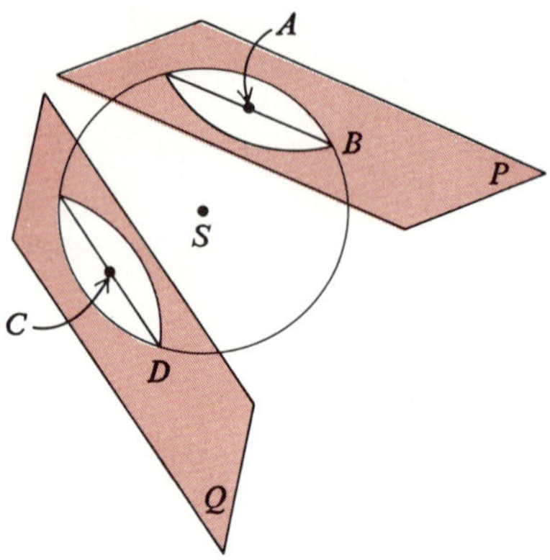

9-5 Arcs of circles

19. GIVEN $\overline{AB}$ is a chord of $\odot P$; radius $\overline{MP}$ of $\odot P$ intersects $\overline{AB}$ in N; $\overline{AN} \cong \overline{NB}$

    PROVE $AM \cong MB$

20. GIVEN Points D, A, C, and B are respective points of $\odot P$; $\widehat{DAC} \cong \widehat{ACB}$

    PROVE $\overline{AD} \cong \overline{BC}$

21. The measure of minor arc $\widehat{AB}$ of $\odot P$ is represented by $7x + 44$. The measure of major arc $\widehat{AB}$ of $\odot P$ is represented by $12x - 7$. Find the measure of each arc.

9-6 Inscribed angles of a circle

22. Prove that the bisector of an angle inscribed in a circle also bisects the arc intercepted by the angle.

Find the value of x in each of the following figures:

23.

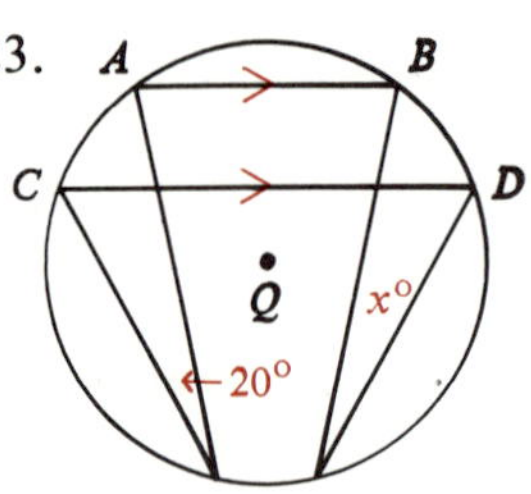

24.

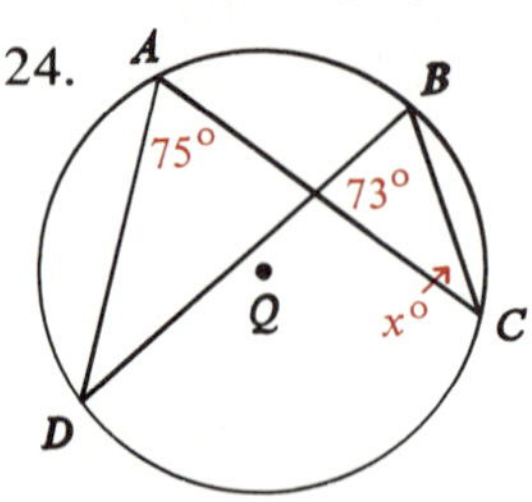

9-7 Concyclic points (Optional)

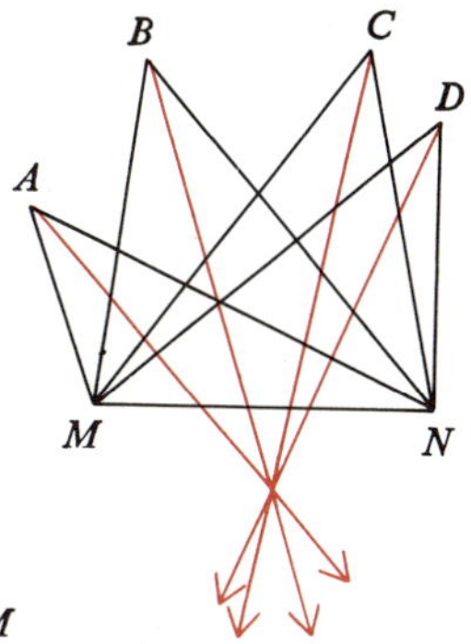

25. Prove that the bisectors of the mutually congruent angles, $\angle A$, $\angle B$, $\angle C$, and $\angle D$ of $\triangle MAN$, $\triangle MBN$, $\triangle MCN$, and $\triangle MDN$, respectively, intersect at a point.
26. What type of parallelogram is a cyclic parallelogram?
27. GIVEN Quadrilateral $AMND$ is cyclic; $\overline{BC} \parallel \overline{AD}$; $\overline{AMB}$; $\overline{DNC}$

    PROVE Quadrilateral $BMNC$ is cyclic.

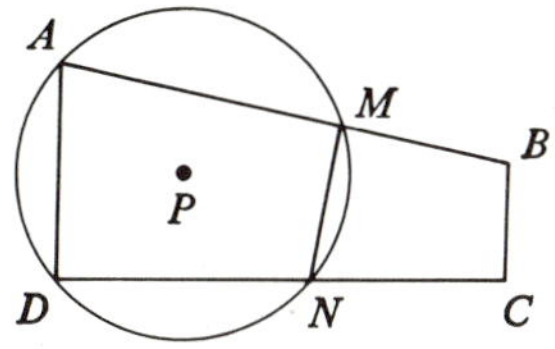

Exercise 27

9-8 Other angles related to circles

Find the value of x in each of the following figures:

28.

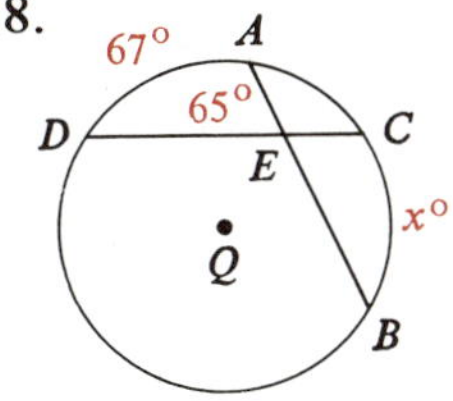

29.

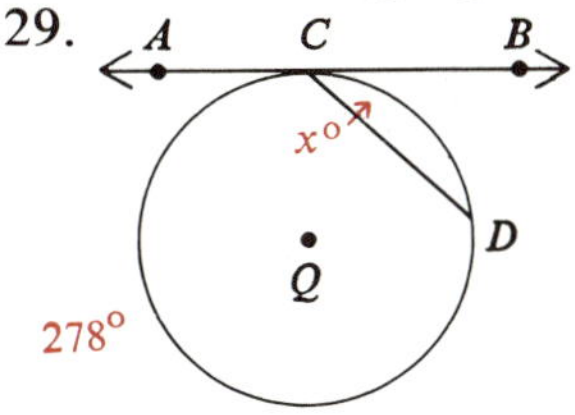

30.

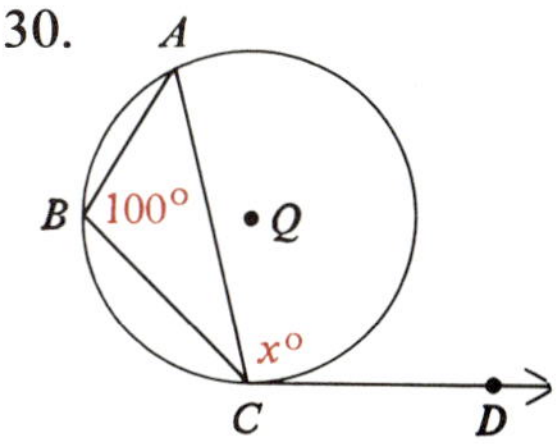

31. GIVEN N and M are the respective midpoints of $\overset{\frown}{AB}$ and $\overset{\frown}{AC}$ of $\odot Q$; $\overline{MN}$ intersects $\overline{AB}$ and $\overline{AC}$ in points D and E, respectively. B and C are on opposite sides of $\overleftrightarrow{AQ}$.

    PROVE $\overline{AD} \cong \overline{AE}$ (*Hint*: Use Theorem 9-8.2.)

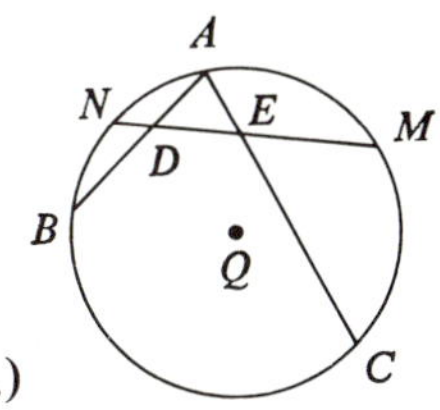

9-9 Tangent, secant, and chord segments

Find the value of x in each of the following figures:

32.

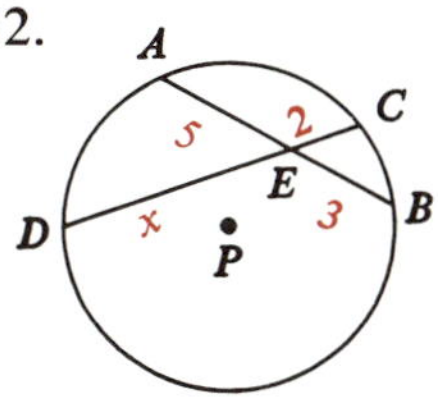

33.

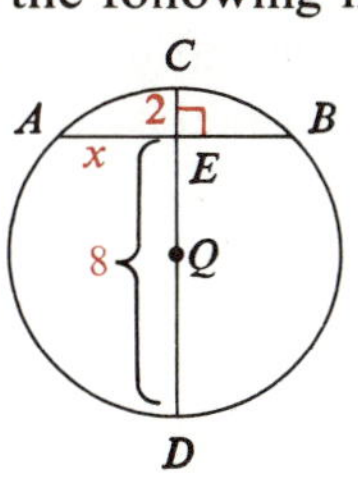

34.

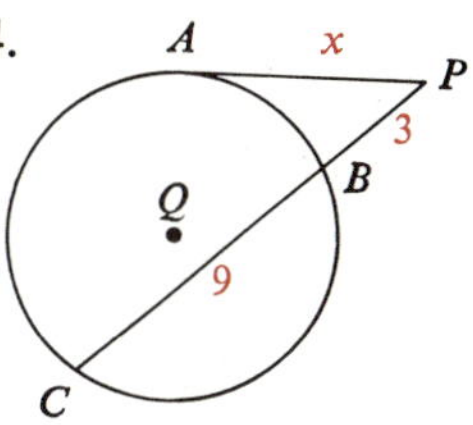

35. Prove that the sum of the lengths of the legs of a right triangle equals the sum of the length of the hypotenuse and the diameter of the inscribed circle.

36. GIVEN $\overline{ADP}$ and $\overline{BCP}$ are the common external tangents of $\odot Q$ and $\odot R$; A and B are points of $\odot Q$; D and C are points of $\odot R$.

PROVE $\overline{AD} \cong \overline{BC}$

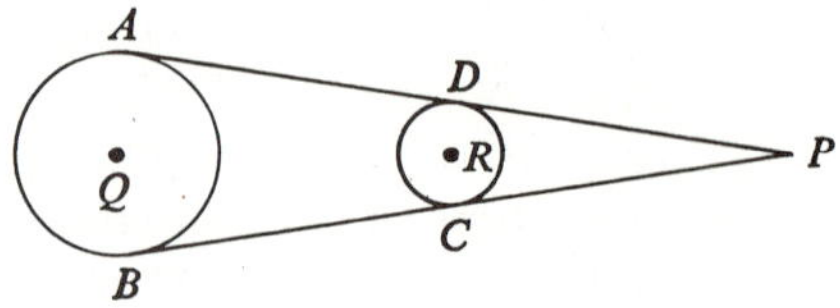

9-10 Circumference and length of arcs

37. What is the ratio represented by π?
38. Find the radius of a circle whose circumference is 18π.
39. Find the circumference of a circle whose diameter is 18.
40. Find the diameter of a semicircle whose arc length is 15.

Find the value of x in each of the following figures:

41.

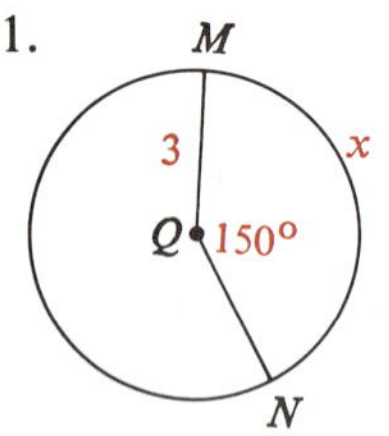

42.

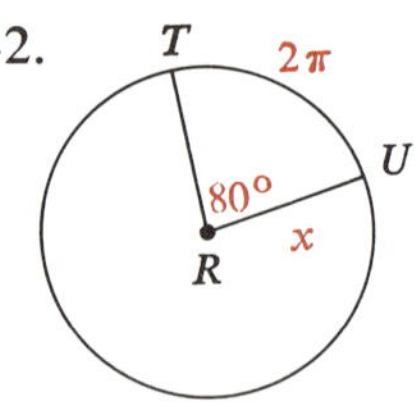

43.

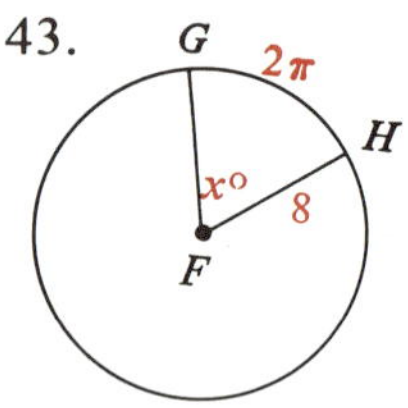

CHAPTER TEST

Complete each of the following sentences:

1. A line that contains exactly one point of a circle is ________ to the circle.
2. Two externally tangent circles have ________ common tangents.
3. If chord $\overline{AB}$ is closer to the center of $\odot Q$ than chord $\overline{CD}$, then AB is ________ than CD.
4. Opposite angles of an inscribed quadrilateral are always ________.
5. The measure of an angle formed by two chords intersecting in the interior of a circle is one-half the ________ of the measures of its intercepted arcs.
6. Two tangent segments to the same circle from the same external point are ________.
7. The ratio of the length of a semicircle to its radius is ________.

Find the value of x in each of the following:

8.

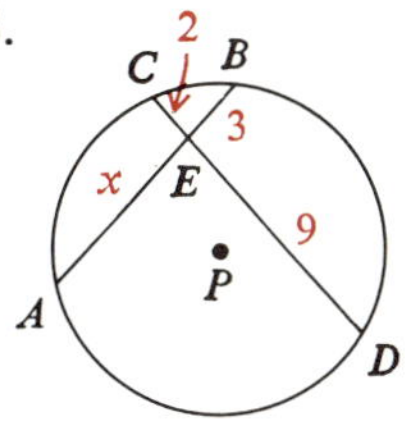

9.

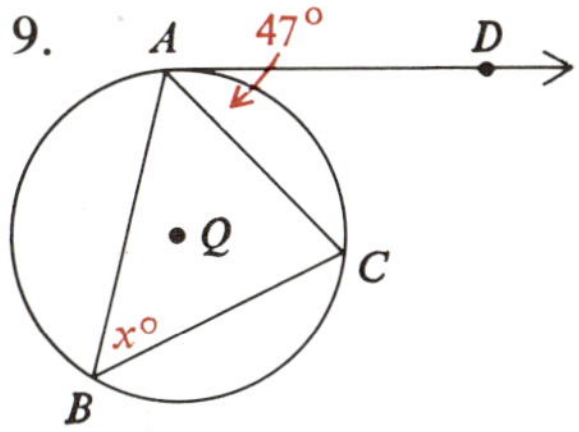

10.

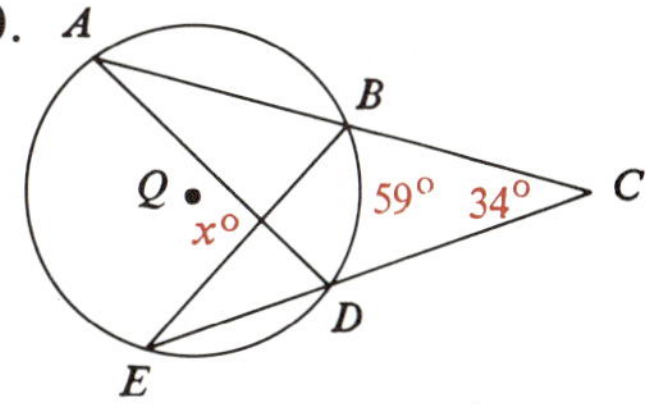

11. Find the circumference of the circle of intersection of a sphere and a plane if the plane is 6 inches from the center of the sphere and the circumference of a great circle of the sphere is 20π.

SUGGESTED RESEARCH

1. Prepare a report on the history of π. You may use "A Look at the Past" on page 413 as a guide; however your report should be more detailed and should discuss more of the attempts to derive π.
2. Present to your class a few methods used to compute π.
3. Find out what Simson's Theorem is and present several of its proofs. Show how Simson's Theorem may be used to prove Ptolemy's Theorem. You may find it helpful to read pages 40-45 of *Geometry Revisited*, by Samuel L. Greitzer and H. S. M. Coxeter (New York: L. W. Singer Co., 1967). After you have become familiar with Simson's Theorem, try a few problems which use this theorem. One possible source is *Challenging Problems in Geometry*, vol. 2, by Alfred S. Posamentier and Charles T. Salkind (New York: Macmillan Co., 1970).
4. Prepare a report on the *radical axis* of two circles. You may wish to tie in a discussion of *coaxial circles*. Two books which would be useful to you are *Advanced Euclidean Geometry*, by Roger H. Johnson (New York: Dover Publications, 1960) and *College Geometry*, by Nathan Altshiller Court (New York: Barnes & Noble, 1952).

MATHEMATICAL EXCURSION

Spherical Geometry

The shortest path connecting two points in a plane is the line segment that they determine. The shortest path connecting two points of a sphere is the shorter arc of the great circle that passes through the two points. Thus, we may consider great circles of a sphere analogous to lines of a plane.

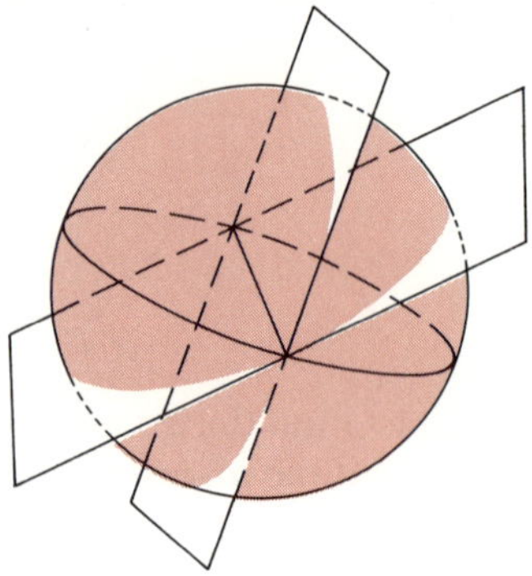

Unfortunately, the analogy is not perfect. Two points in a plane determine a unique line. Two points of a sphere do not always determine a unique great circle. A great circle is the intersection of a sphere with a plane through the center of the sphere. The line of intersection of any two such planes contains one of the diameters of the sphere. The endpoints of the diameter are two points common to the two great circles formed. In fact, any plane containing this diameter will intersect the sphere in a great circle through these points.

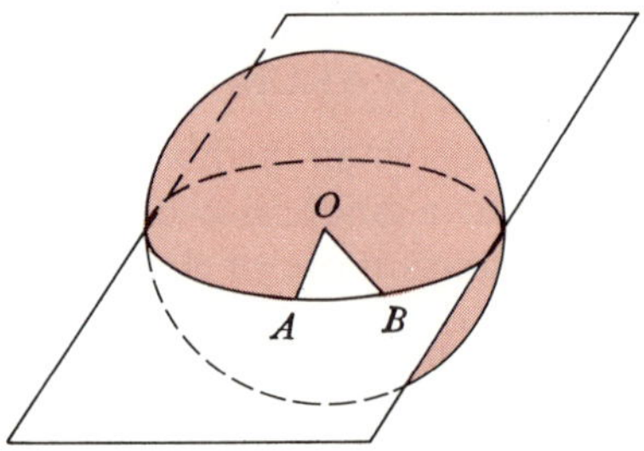

It follows from this that we cannot say that two points that are at opposite ends of a common diameter are points of only one great circle. What can we say about any other two points of the sphere? The great circle through A and B, any two points of the sphere, must lie in a plane through O, the center of the sphere. If A and B are not endpoints of a diameter, the three points A, B, and O are not collinear. Thus, they determine exactly one plane. Hence, one and only one great circle passes through A and B.

We can conclude that if two points of a sphere are not endpoints of the same diameter, exactly one great circle passes through both points. If two points of a sphere are endpoints of the same diameter, then an infinite number of great circles pass through both points.

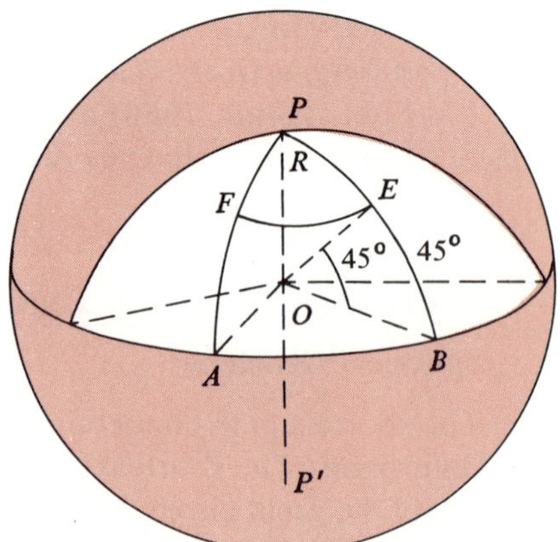

We call the diameter of a sphere that is perpendicular to the plane containing any circle of the sphere the *axis* of the circle. We call the endpoints of the diameter the *poles* of the circle. In the figure at right, we shall refer to P and P' as the poles of the great circle with center O and containing $\overset{\frown}{AB}$. P and P' are also the poles of the small circles with center R and containing $\overset{\frown}{EF}$. The diameter $\overline{PP'}$ is the axis of both circles.

We measure the spherical distance between two points of a sphere that are not the endpoints of a diameter by the minor arc of the great circle through the two points. We use the degree as a unit of measure. In the figure above, where $m\angle AOB = 45$ and $\angle AOB$ is the central angle of $\overset{\frown}{AB}$, the spherical distance from A to B is 45°. Why?

The *polar distance* of a circle is the spherical distance from a point of the circle to its pole. In the figure above, the polar distance of the small circle containing $\overset{\frown}{EF}$ is 45°. The polar distance of the great circle containing $\overset{\frown}{AB}$ is 90°. We call an arc of 90° a *quadrant*. The polar distance of all great circles is a quadrant.

When the arcs of two great circles of a sphere meet at a point, the figure formed is called a *spherical angle*. The great circle arcs are the *sides* of the angle, and their point of intersection is the *vertex*. A spherical angle has the same measure as the angle formed by the two tangents to the great circles at the vertex of the angle. If $\overset{\frown}{PA}$ and $\overset{\frown}{PB}$ are arcs of great circles meeting at P, then the measure of spherical $\angle APB$ is equal to $m\angle A'PB'$, where $\overrightarrow{PA'}$ is tangent to the great circle of $\overset{\frown}{AP}$ and $\overrightarrow{PB'}$ is tangent to the great circle of BP. Tangent rays $\overrightarrow{PA'}$ and $\overrightarrow{PB'}$ are perpendicular to the diameter at P and lie in the dihedral angle formed by the planes of the two great circles. Thus we can say that the spherical angle has the same measure as the dihedral angle, since its measure is equal to that of the angle formed by the tangent rays.

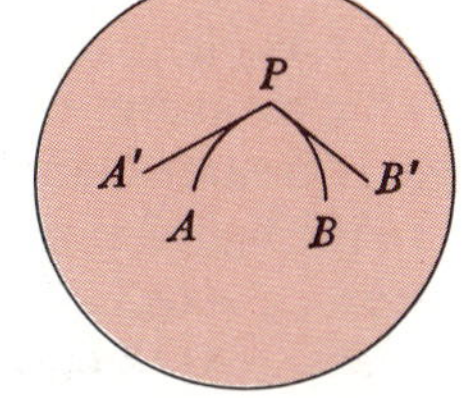

Spherical Polygonal Regions

A region enclosed by three or more minor arcs of great circles is called a *spherical polygonal region*. The boundary of a spherical polygonal region is called a *spherical polygon*. When a spherical polygon is formed by three such arcs, we have a *spherical triangle*. The figure formed by $\overset{\frown}{AB}$, $\overset{\frown}{BC}$, and $\overset{\frown}{CA}$ is spherical $\triangle ABC$.

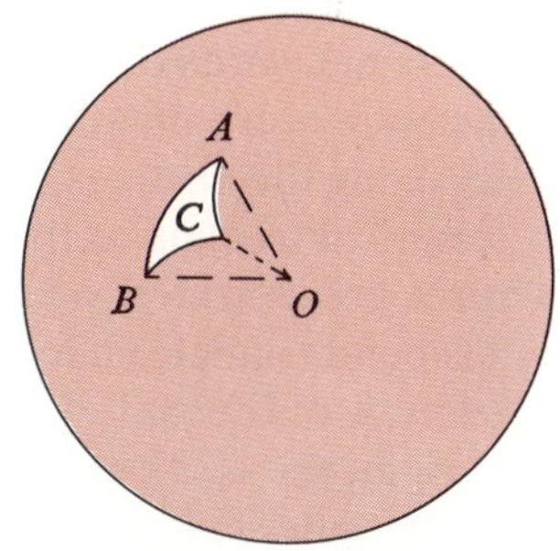

The sides of a spherical triangle are measured in terms of the spherical distance between vertices. The angles are spherical angles. Thus, we can see that both the sides and the angles of spherical triangles can be measured in degrees.

If two points of a sphere that are not the endpoints of the same diameter both determine quadrants with a third point, the third point is a pole of the great cirlce passing through the first two points.

In the figure at right, $m\widehat{PA} = m\widehat{PB} = 90$. Hence, $\overline{PO} \perp \overline{OA}$ and $\overline{PO} \perp \overline{OB}$. Therefore, $\overline{PO}$ is perpendicular to the plane of the great circle of AB. The measure of the spherical $\angle APB$ is the same as $m\angle AOB$, which is the measure of the dihedral angle A-OP-B. But the measure of central $\angle AOB$ is the same as $m\widehat{AB}$. Thus, by the transitive property, the measure of spherical $\angle APB$ is the same as $m\widehat{AB}$.

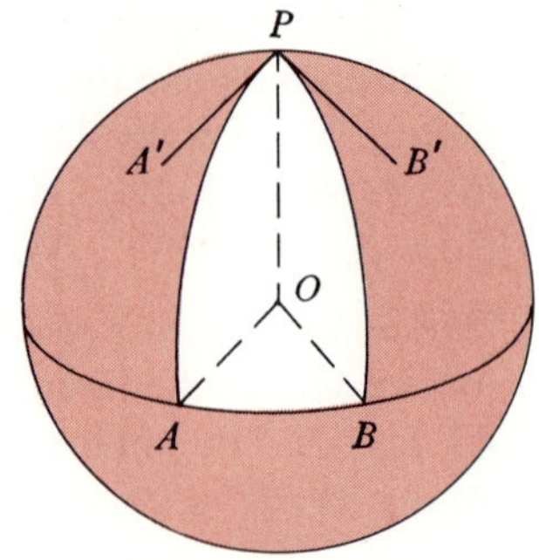

Spherical triangles have many of the same properties as plane triangles. For example, the sum of the measures of two sides of a spherical triangle is greater than the measure of the third side. In the figure at right, $\widehat{AC}$ has the same measure as $\angle AOC$. Similarly, $\widehat{AB}$ has the same measure as $\angle AOB$ and $\widehat{CB}$ has the same measure as $\angle COB$, where O is the center of the sphere of which A, B, and C are points. It can be shown that $m\angle BOC < m\angle BOA + m\angle AOC$. Therefore, the measure of side $\widehat{BC}$ is less than the sum of the measures of sides $\widehat{BA}$ and $\widehat{AC}$. This is consistent with the statement that the shortest path between two points of a sphere is the minor arc of the great circle connecting them.

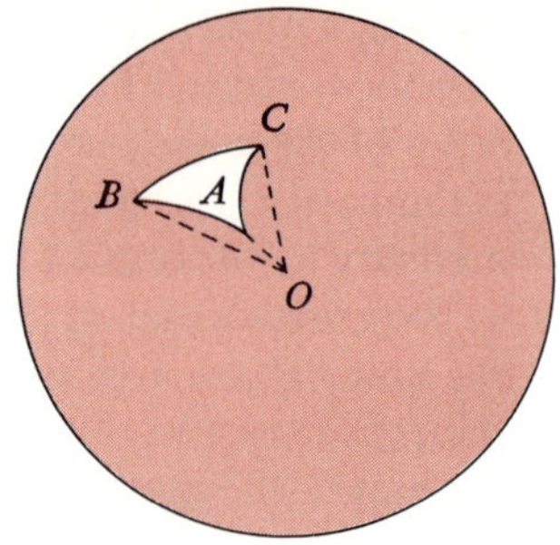

It can also be shown that the sum of the measures of the face angles of a polyhedral angle is less than 360. By considering the polyhedral angle formed by the planes of the sides of a spherical polygon, we can conclude that the sum of the measures of the sides of any spherical polygon is less than 360. If the measure were 360, the "polygon" would have to be a great circle.

By using the vertices of a spherical triangle as the poles of three great circles, we can form a second triangle called the *polar triangle* of the first. In the figure at right, the vertices A, B, and C of the spherical $\triangle ABC$ are the poles of three great circles. These three circles intersect to form eight triangles on the surface of the sphere. In one of them, spherical $\triangle PQR$, two vertices (Q and R) are points of the great circle with pole A; two (P and Q) are points of the great circle with pole B; and two

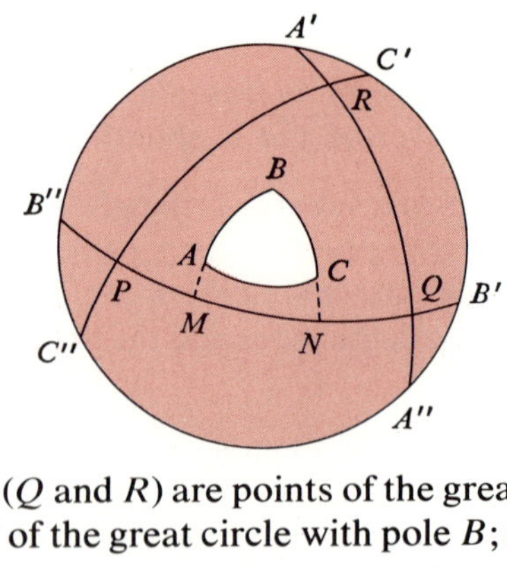

(P and R) are points of the great circle with pole C. Thus, spherical $\triangle PQR$ is the polar triangle of spherical $\triangle ABC$, since the arcs that form the sides of spherical $\triangle PQR$ have as poles the vertices of $\triangle ABC$.

If spherical $\triangle PQR$ is the polar triangle of $\triangle ABC$, then $\triangle ABC$ is also the polar triangle of $\triangle PQR$. To show that this is true, note that since C is the pole of $\overset{\frown}{PR}$ and B is the pole of $\overset{\frown}{PQ}$, $\overset{\frown}{PC}$ and $\overset{\frown}{PB}$ are both quadrants. Hence, P is the pole of $\overset{\frown}{BC}$, by the theorem explained four paragraphs above. Similarly, Q is the pole of $\overset{\frown}{AB}$ and R is the pole of $\overset{\frown}{AC}$. Thus, we can conclude that $\triangle ABC$ is the polar triangle of $\triangle PQR$.

The measure of a side of a spherical triangle plus the measure of the corresponding angle of its polar triangle equals 180. By "corresponding angle" we mean the angle whose vertex is the pole of the given side. To prove this, refer to the figure above. Let the extension of $\overset{\frown}{AB}$ cut $\overset{\frown}{PQ}$ at M and let the extension of $\overset{\frown}{BC}$ cut $\overset{\frown}{PQ}$ at N. Then we have,

$$\mathrm{m}\overset{\frown}{MN} = \mathrm{m}\angle B$$

$$\mathrm{m}\overset{\frown}{PN} = 90 = \mathrm{m}QM$$

Hence $$\mathrm{m}\overset{\frown}{PQ} = 180 - \mathrm{m}\angle B$$

That is, $$\mathrm{m}\angle B + \mathrm{m}\overset{\frown}{PQ} = 180$$

This result can be used to show that the sum of the measures of the angles of a spherical triangle is greater than 180 and less than 540. Justify each statement in the proof below.

We have, using the accompanying figure, with $\triangle ABC$ and $\triangle A'B'C'$ polar triangles,

$$\mathrm{m}\angle A + \mathrm{m}\overset{\frown}{a'} = 180$$
$$\mathrm{m}\angle B + \mathrm{m}\overset{\frown}{b'} = 180$$
$$\mathrm{m}\angle C + \mathrm{m}\overset{\frown}{c'} = 180$$

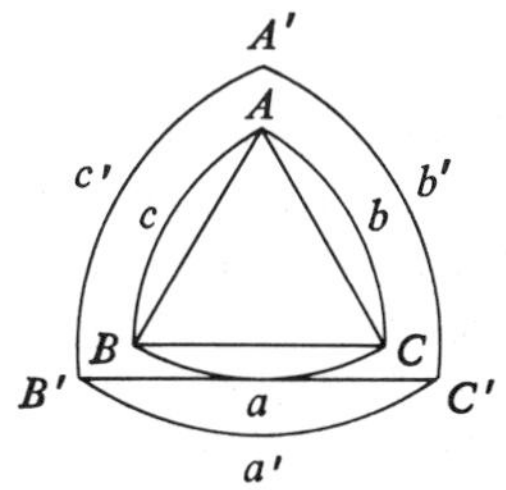

or $$(\mathrm{m}\angle A + \mathrm{m}\angle B + \mathrm{m}\angle C) + (\mathrm{m}\overset{\frown}{a'} + \mathrm{m}\overset{\frown}{b'} + \mathrm{m}\overset{\frown}{c'}) = 540$$

But $$(\mathrm{m}\overset{\frown}{a'} + \mathrm{m}\overset{\frown}{b'} + \mathrm{m}\overset{\frown}{c'}) < 360$$

Hence $$\mathrm{m}\angle A + \mathrm{m}\angle B + \mathrm{m}\angle C > 180$$

We also know $$\mathrm{m}\overset{\frown}{a'} + \mathrm{m}\overset{\frown}{b'} + \mathrm{m}\overset{\frown}{c'} > 0$$

And hence $$\mathrm{m}\angle A + \mathrm{m}\angle B + \mathrm{m}\angle C < 540$$

Congruent Spherical Triangles

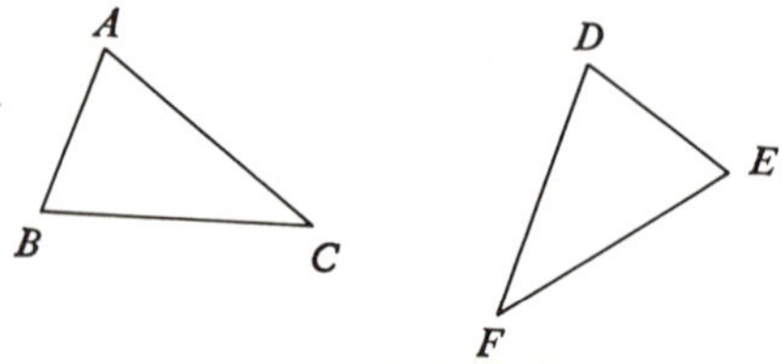

In the accompanying figure, if $\angle A \cong \angle D$, $\angle B \cong \angle E$, $\angle C \cong \angle F$, $\overline{AB} \cong \overline{DE}$, $\overline{AC} \cong \overline{DF}$ and $\overline{BC} \cong \overline{EF}$, then we can say that $\triangle ABC \cong \triangle DEF$. Even if we could move the triangles in the plane without changing their shape or size, we could never make one coincide with the other. But, if we could pick one of them up out of the plane, turn it over, and put it back in the plane, then the triangles could be made to coincide. This situation does not apply to spherical triangles. If the corresponding parts of two spherical triangles are congruent, but the triangles are arranged so that if we move clockwise around one, we must move counterclockwise around the other to reach the corresponding parts, then one triangle cannot be made to coincide with the other whether we slide the triangle around on the surface of the sphere or remove it from the sphere, turn it over, and put it back. When the triangle is turned over it will not fit back on the surface of the sphere. The two triangles are related like the right hand and left hand of a pair of gloves. They are called *symmetric*, rather than congruent. However, two triangles in a sphere with congruent corresponding parts are congruent, if they can be made to coincide by sliding one triangle around on the sphere.

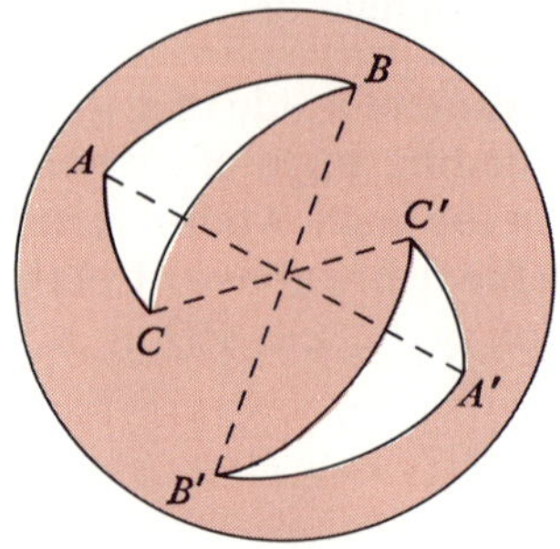

If we consider their congruence and symmetry relations, spherical triangles are much like plane triangles. Two spherical triangles on the same sphere are congruent or symmetric if:

1. two sides and the included angle of one are congruent to the corresponding parts of the other;
2. two angles and the included side of one are congruent to the corresponding parts of the other;
3. three sides of one are congruent to the corresponding parts of the other.

Spherical triangles, however, have one major property that plane triangles do not have: Two spherical triangles on the same sphere are congruent or symmetric if the angles of one triangle are congruent to the corresponding angles of the other.

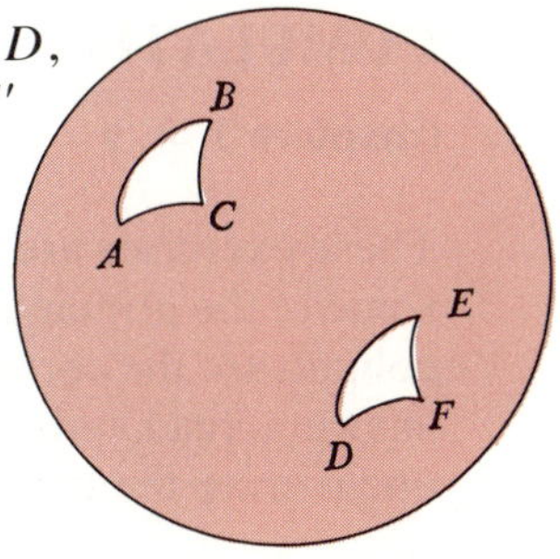

Consider spherical $\triangle ABC$ and DEF, with $\angle A \cong \angle D$, $\angle B \cong \angle E$, and $\angle C \cong \angle F$. Let spherical $\triangle A'B'C'$ be the polar triangle of spherical $\triangle ABC$ and spherical $\triangle D'E'F'$ be the polar triangle of spherical $\triangle DEF$. The two polar triangles are congruent, by the SSS Postulate. For example, $\widehat{B'C'}$ is the supplement of m $\angle A$ and m $\widehat{E'F'}$ is the supplement of m $\angle D$. Since $\angle A \cong \angle D$ it follows that m $\widehat{B'C'}$ = m $\widehat{E'F'}$. Similar reasoning applies to the other sides of the polar triangles. Conversely, where $\angle A' \cong \angle D'$, $\angle B' \cong \angle E'$, $\angle C' \cong \angle F'$, these angles are the supplements of the corresponding sides of the polar triangles ABC and DEF. Then applying the same reasoning as before, $\triangle ABC \cong \triangle DEF$, by the SSS Postulate.

Angle bisectors, medians, and altitudes of spherical triangles have definitions analogous to those of plane triangles. Spherical triangles may also be classified as acute, right, or obtuse. But a spherical triangle may have two, or even three, right angles or obtuse angles.

The definitions for spherical isosceles and equilateral triangles are the same as those for plane triangles. Also, the base angles of an isosceles spherical triangle are congruent; and equilateral spherical triangles are equiangular.

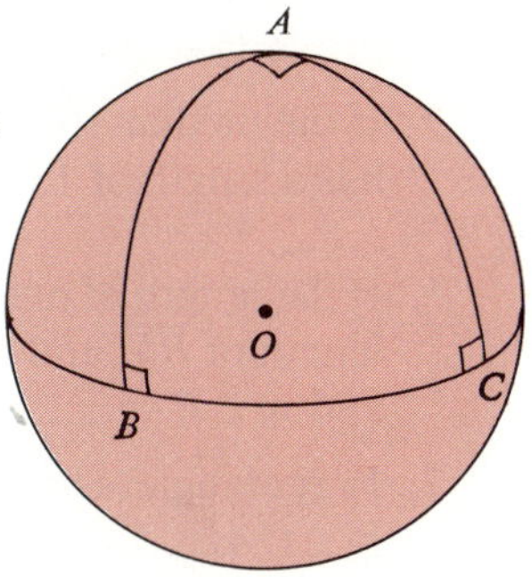

EXERCISES

1. List as many properties as you can that are common to plane and spherical triangles.
2. List as many ways as you can in which plane and spherical triangles are different.
3. Is it possible for a spherical triangle to be self-polar?
4. Prove that if two angles of a spherical triangle are equal in measure, the sides opposite these angles are equal in measure. (*Hint*: Consider the polar triangle.)
5. Prove that a spherical triangle that is equilateral is also equiangular.
6. Prove the converse of Exercise 5.
7. What do we know about a spherical triangle if the altitude of one or more sides is not unique?

CUMULATIVE REVIEW

Chapters 7, 8, 9

These exercises are designed so that you can check your understanding of some of the principal concepts studied so far. The numbered items in the right column are the correct answers. To complete the review, first cover the answers, read and answer the questions, and check each response by uncovering the corresponding answers.

1. In ▱ $PQRS$, $m\angle P = 4(m\angle Q)$. The measures of angles P, Q, R, and S are ____.	1. 144, 36, 144, 36
2. The diagonals of ▱ $ABCD$ meet at E. If $BE = 7x - 20$ and $DE = 12x - 35$, then $BD =$ ____.	2. 3
3. True or false: A quadrilateral is a parallelogram if the diagonals bisect each other.	3. true
4. To prove that a quadrilateral is a parallelogram, prove that both pairs of opposite angles are ____.	4. congruent
5. If the median to the hypotenuse of a right triangle has a length of 15, the length of the hypotenuse is ____.	5. 30
6. If the length of median $\overline{AM}$ of $\triangle ABC$ equals the length of $\overline{BM}$ and $m\angle C = 27$, then $m\angle A =$ ____ and $m\angle B =$ ____.	6. 90, 63
7. In rectangle $ABCD$, $m\angle BAC = 32$. Find $m\angle DAC$.	7. 58
8. In square $ABCD$, $\overline{AC}$ and $\overline{BD}$ meet at E. $m\angle ABD =$ ____.	8. 45
9. If all angles of a quadrilateral are congruent, the quadrilateral is a ____.	9. rectangle
10. The diagonals of a square are ____.	10. congruent
11. If $\overline{MN}$ is the midline of $\triangle ABC$, $\overline{MN} \parallel \overline{BC}$, $BC = 3x + 21$, and $MN = 7x - \frac{1}{2}$, then MN equals ____.	11. $\frac{27}{2}$
12. If the perimeter of $\triangle ABC$ is 15, find the perimeter of the triangle formed by the midlines of $\triangle ABC$.	12. $\frac{15}{2}$
13. If the bases of a trapezoid measure $\frac{56}{3}$ and $\frac{50}{3}$, the median measures ____.	13. $\frac{53}{3}$

14. In trapezoid $ABCD$, $\overline{AB} \parallel \overline{DC}$ and $AC = BD$, $m\angle D = 23x - 3$ and $m\angle C = 5x + 33$, the measures of angles A, B, C, and D are ____.	14. 137, 137, 43, 43
15. If the median and longer base of a trapezoid measure 18 and 32, respectively, the shorter base is ____.	15. 4
16. If $PQ \cdot PS = PR \cdot RT$, then $\frac{RT}{\quad} = \frac{PS}{\quad}$.	16. PQ, PR
17. 2 is to 16 as 3 is to ____.	17. 24
18. Three or more parallel lines intercept ____ segments on two or more transversals.	18. proportional
19. If in $\triangle ABC$ with sides $\overline{ADB}$ and $\overline{AEC}$, $\frac{AD}{DB} = \frac{AE}{EC}$, then ____.	19. $\overline{DE} \parallel \overline{BC}$
20. If a line divides two sides of $\triangle ABC$ proportionally, then the line is ____ to the third side.	20. parallel
21. If $\overline{AD}$ is an angle bisector of $\triangle ABC$, with $AC = 6$, $BC = 7$, and $AB = 8$, then $CD =$ ____.	21. 3
22. True or false: If $\triangle HWP \sim \triangle EJP$, then $\triangle HWP \cong \triangle EJP$.	22. false
23. If the ratio of similitude of two similar quadrilaterals is $\frac{5}{7}$, the ratio of their perimeters is ____.	23. $\frac{5}{7}$
24. If the corresponding angles of two triangles are congruent, then the two triangles are ____.	24. similar
25. If $\triangle JBL \sim \triangle MRN$, then $\frac{JL}{BL} =$ ____.	25. $\frac{MN}{RN}$
26. If two pairs of corresponding angles of two triangles are congruent, the two triangles are ____.	26. similar
27. If two ____ triangles have a pair of corresponding acute angles congruent, the triangles are similar.	27. right
28. If in $\triangle ABC$ and $\triangle DEF$, $\frac{AB}{DE} = \frac{BC}{EF} = \frac{AC}{DF}$, then the pairs of congruent angles are ____.	28. A and D, B and E, C and F
29. If corresponding sides of two triangles are ____, then the triangles are similar.	29. proportional
30. The mean proportional between 12 and 27 is ____.	30. 18
31. True or false: $6\sqrt{2}$, $8\sqrt{2}$, and $10\sqrt{2}$ represent the lengths of the sides of a right triangle.	31. true

32. If the altitude of an equilateral triangle is 5 inches long, the perimeter measures ____.	32. $10\sqrt{3}$
33. If $\sin A = \frac{3}{5}$, then $\cos A =$ ____.	33. $\frac{4}{5}$
34. Without tables, find the value of cos 60°.	34. $\frac{1}{2}$
35. The diameter of a sphere lies on a ____ of the sphere.	35. secant
36. Any chord of a circle containing the center of the circle is a ____ of the circle.	36. diameter
37. Find the radius of a circle if a 14-inch chord is 24 inches from the center of the circle.	37. 25
38. A ____ tangent is a line tangent to two coplanar circles.	38. common
39. Find the radius of a sphere if a plane 7 inches from its center intersects the sphere in a circle of diameter 14 inches.	39. $7\sqrt{2}$
40. A ____ of a circle is the intersection of a circle with a central angle and its exterior.	40. major arc
41. The measure of an angle inscribed in an arc of measure 200 is ____.	41. 100
42. An angle inscribed in a semicircle is a ____ angle.	42. right
43. A circle is ____ in a polygon if it is tangent to each side of the polygon.	43. inscribed
44. The circumference of a circle with radius 4 is ____.	44. 8π

10 Coordinate Geometry

In the study of algebra, and again in Chapter 1, we discussed the correspondence between the points of a line and various sets of numbers. We agreed that to each real number, there corresponds exactly one point of the line and to each point of the line, there corresponds exactly one real number. We called this real number the coordinate of the point, and each point the graph of the number. In this chapter, we shall apply these concepts of correspondence to points in a plane and to plane figures.

10-1 THE EUCLIDEAN PLANE

When we discussed the graphing of various sets of numbers in Chapter 1, we introduced sets and the set operations of union and intersection. We shall now examine another set operation.

Assume that we are given sets X and Y, where $X = \{1, 2\}$ and $Y = \{3, 4\}$. We can form a new set whose members are ordered pairs of numbers from X and Y. We call this set the Cartesian product of X and Y. The notation for the Cartesian product of X and Y is $X \times Y$ and is read "X cross Y." The members of $X \times Y$ are called ordered pairs because the order of the elements is essential to the accuracy of the notation. The pair (1, 3) is not the same as the pair (3, 1). The first element of each pair (x, y) must be a member of X and the second must be a member of Y. Thus, for $X = \{1, 2\}$ and $Y = \{3, 4\}$, $X \times Y = \{(1, 3), (1, 4), (2, 3), (2, 4)\}$.

D

Definition 10-1 The **Cartesian product** of X and Y is the set of all ordered pairs (x, y), where x belongs to X and y belongs to Y.

To draw the graph of $X \times Y$, we use a pair of perpendicular lines called the **coordinate axes**. One line is usually vertical, the other, horizontal. The vertical line, or axis, is labeled the ***y*-axis**; the horizontal line is the ***x*-axis**. The point of intersection of the two axes, the **origin**, is the zero point of each axis. Points to the right of the origin on the x-axis and above the origin on the y-axis represent positive real numbers. Points to the left on the x-axis or below the origin on the y-axis represent negative numbers.

If we choose a convenient scale to be used on each axis, then for each ordered pair of numbers in $X \times Y$, we can locate a point on the graph. For example, to locate any point such as (2, 4), we start at the origin. The first number of the pair (the element x in our definition above) indicates horizontal motion; thus, for (2, 4), we first move 2 units to the right of the origin, along the x-axis. The second number (y) indicates vertical motion; we then move 4 units up, parallel to the y-axis. The numbers of the ordered pair (2, 4) are called the **coordinates** of a point in a plane; 2 is the x-coordinate and 4 is the y-coordinate.

Example 1 Draw the graph of $X \times Y$ if $X = \{1, 2\}$ and $Y = \{3, 4\}$.

Solution We first draw the axes. We can then draw the graph of $X \times Y$ by locating all the members of the Cartesian product. Thus, we must locate the points with coordinates (1, 3), (1, 4), (2, 3), and (2, 4).

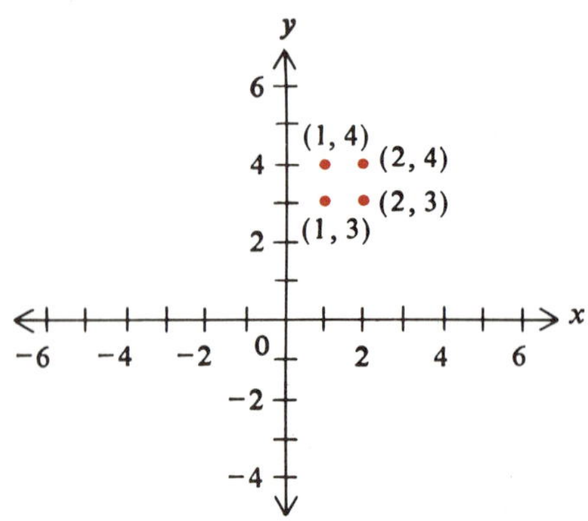

We can also form the Cartesian product of a single set by "crossing" the set with itself. That is, if $A = \{a, b\}$, then we can form $A \times A$.

$$A \times A = \{(a, a), (a, b), (b, a), (b, b)\}$$

Each of the sets above is a finite set. Now consider R, the set of real numbers. The Cartesian product $R \times R$ is an infinite set of ordered pairs. Can we represent an infinite set on our graph?

How many planes are determined by a pair of intersecting lines? If we use real number lines as the axes of our graph, then they both contain an infinite number of points. Using coordinates, we can then describe any point of a plane by an ordered pair of real numbers. Conversely, for every pair of real numbers, we can locate a unique point in the plane. Thus, we can graph any point of $R \times R$.

The graph of $R \times R$ has many different names, but it is most often called the **Cartesian coordinate plane**, after the French mathematician-philosopher René Descartes (1596–1650). Descartes is credited with being the first person to use two-dimensional coordinates to study mathematics. The graph may also be referred to as the **Euclidean plane**, since it can be used to study the properties of Euclidean geometry.

The system of points in a Cartesian plane is called the **rectangular coordinate system**. To name the coordinates of a point, we find the projection of the point onto each of the coordinate axes. Each of the lines showing a projection is parallel to one of the axes and perpendicular to the other. Do you see why we use the name "rectangular coordinate system"?

As we said before, the x-axis and the y-axis intersect at the origin. The axes separate the rest of the points of the plane into four regions or **quadrants**. We number these quadrants I, II, III, and IV, as shown in the figure following.

Example 2 Give the coordinates of points A, B, C, D, E, and F in the figure below right, and determine the quadrants in which they lie.

Solution

Point	Coordinate	Quadrant
A	(3, 2)	I
B	(−3, −3)	III
C	(−2, 0)	on the x-axis
D	(−4, 3)	II
E	(0, −2)	on the y-axis
F	(4, −3)	IV

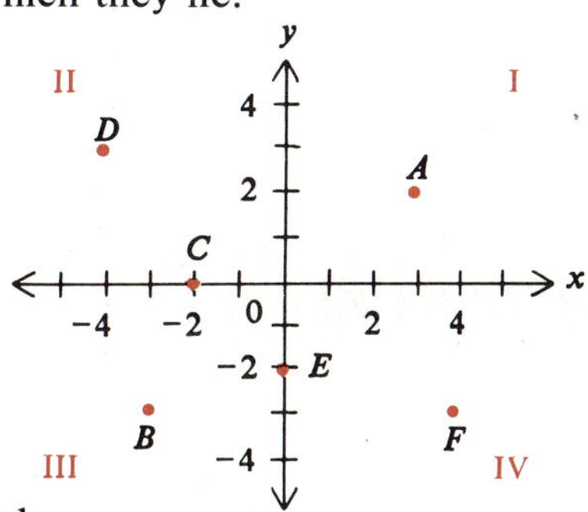

Are points C and E in any quadrant? Can you make a general statement about the positive and negative values of the points in each quadrant?

Each coordinate of a point has a name. The x-coordinate is called the abscissa. The y-coordinate is called the ordinate. Thus, in Example 2, for point F, the abscissa is 4 and the ordinate is -3.

D

Definition 10-2 To each point in a plane, there corresponds a unique ordered pair of real numbers—the abscissa and the ordinate of the point. The **abscissa** is the coordinate of the projection of the point onto the x-axis. The **ordinate** is the coordinate of the projection of the point onto the y-axis.

Class Exercises

Let $A = \{a, b, c, d\}$, $B = \{-1, 0\}$, and $C = \{1, 2\}$. Find each of the following Cartesian products:

1. $A \times B$ 2. $B \times A$ 3. $A \times C$ 4. $B \times C$

Use the figure at right for Exercises 5–11.

5. Name the abscissa for C.
6. Name the ordinate for E.

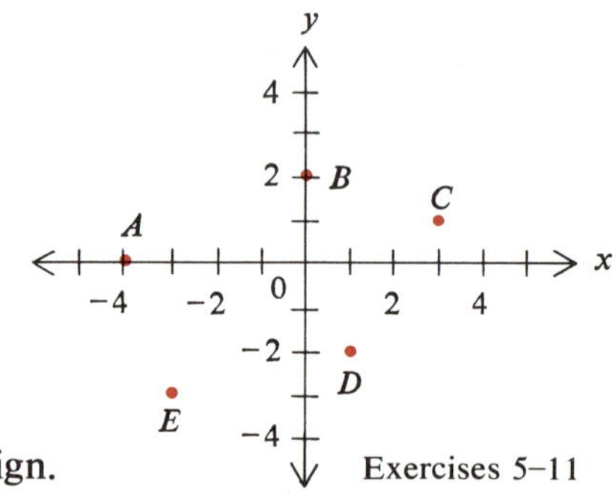

Exercises 5–11

Name the coordinates of the following points:

7. A 8. B 9. C
10. D 11. E
12. Name the quadrants in which the coordinates of a point have the same sign.

Graphing

The rectangular coordinate system can be used to show the relationship between two variables. If the two variables are related by a statement of equality or inequality, we can graph the ordered pairs that can validly be substituted into the statement. This set of ordered pairs is the **solution set** of the statement. The values of x we use in finding the solution set are taken from the **replacement set**.

Example 3 Graph the set $\{(x, y): y = 3\}$ if the replacement set for x is $\{-3, -2, -1, 0, 1, 2, 3\}$.

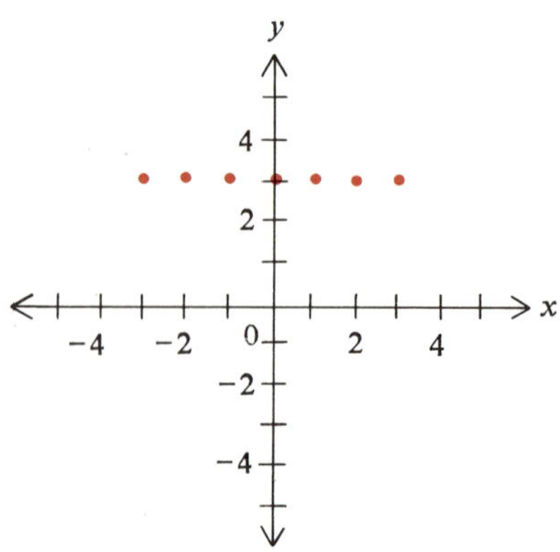

Solution Since $y = 3$ for any value of x, the table of ordered pairs is as shown at right.

x	-3	-2	-1	0	1	2	3
y	3	3	3	3	3	3	3

Example 4 Graph $\{(x, y): y \geq |x|\}$, where x and y are members of $\{-3, -2, -1, 0, 1, 2, 3\}$.

Solution First sketch the graph of $y = |x|$ for this replacement set. The graph of the relation $y > |x|$ consists of the points in which the ordinate is greater than the absolute value of the abscissa: that is, the points $(-2, 3)$, $(-1, 2)$, $(-1, 3)$, $(0, 1)$, $(0, 2)$, $(0, 3)$, $(1, 2)$, $(1, 3)$, and $(2, 3)$.

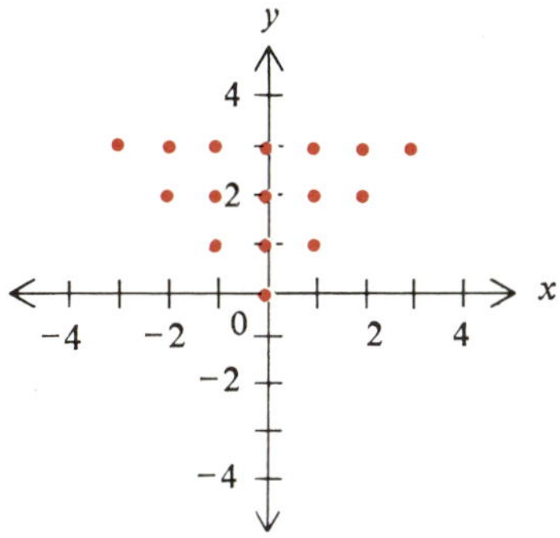

If the possible replacement set for x and y were the set of real numbers, then the graph would contain an infinite set of points. We would have time and space to locate only a few of the points. Thus, from the graph of these points, we would predict the graph of the remaining points. Unless we state a limitation for x or y, we can assume that they are any real numbers.

Example 5 Graph $\{(x, y): y = x + 1\}$.

Solution Make a table for a few values of x chosen in consecutive order: $x = \{-2, -1, 0, 1, 2\}$.

x	-2	-1	0	1	2
y	-1	0	1	2	3

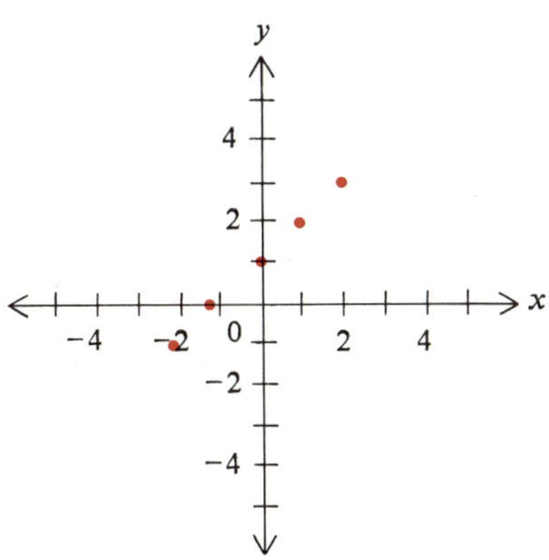

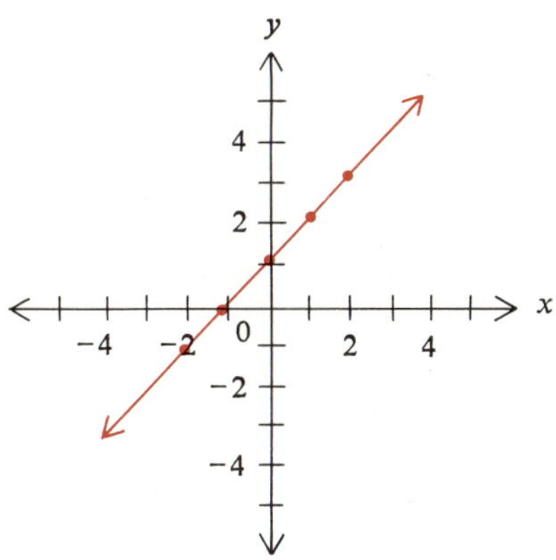

The graph of these pairs is shown at left above. Their graph suggests that the graph of the given set is a line, for if we connect $(-2, -1)$ and $(2, 3)$ with a segment, the other points graphed lie on this segment. The graph for all real values of x continues in both directions. To indicate that the graph is a line, extend the segment and put arrows at the ends, as shown in the figure at right above.

EXERCISES

A

Refer to the graph at right to determine the coordinates of each of the following points:

1. A 2. B 3. C
4. D 5. E

Name the points with the following coordinates:

6. $(0, -2)$ 7. $(-4, 0)$
8. $(5, 2\frac{3}{4})$ 9. $(3, -3)$
10. $(0, 3)$

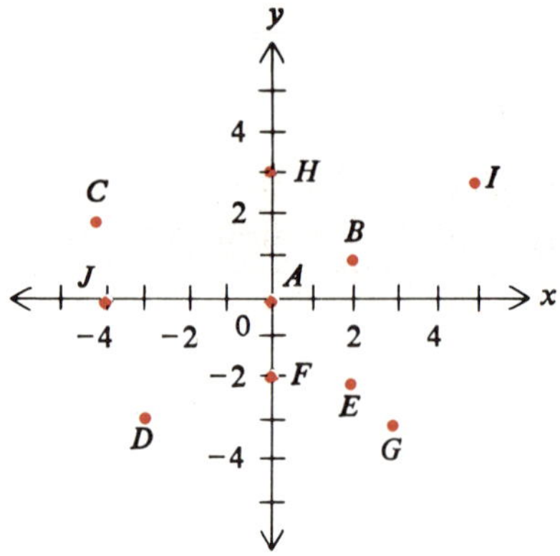

Exercises 1–11

11. Name the points in each quadrant of the graph above.

Name the quadrants in which (x, y) falls, under the following conditions:

12. $x > 0$ and $y > 0$ 13. $x > 0$ 14. $x < 0$ and $y < 0$
15. $y < 0$ 16. $x = 0$ 17. $y = 0$
18. $x > 0$ and $y = 0$ 19. $x = 0$ and $y < 0$

Copy and complete the following tables for the equations given:

20. $y = 2x$

x	-3	-2	-1	0	1	2	3
y							

21. $2x - 3y = 6$

x	-3	-2	-1	0	1	2	3
y							

If x is any member of $\{-2, -1, 0, 1, 2\}$, determine the solution set for each of the following equations:

22. $3y = 2x + 1$ 23. $y = 3x + 9$ 24. $3x - y + 4 = 0$
25. $y = -3x + 4$ 26. $y = 6 - 2x$ 27. $y = |x| + 1$

Draw the graph of $\{(x, y):\ y = 1 - 3x\}$ for each of the following replacement sets for x:

28. $\{-1, 0, 1\}$ 29. $\{-3, -2, -1, 0, 1, 2, 3\}$ 30. all real numbers

B

If $\{-2, -1, 0, 1, 2\}$ is the replacement set for x, graph each of the following:

31. $y = 3x - 1$ 32. $y > 3x - 1$ 33. $y < 3x - 1$
34. $y = 2|x - 1|$ 35. $y > |x| - 1$ 36. $y > |x - 1|$

The replacement set for x is the set of real numbers. Graph each of the following:

37. $y = |x| - 2$ 38. $y = |x - 1|$ 39. $x + y = 2$
40. $\{(x, y): y = 0\}$ 41. $\{(x, y): x = 2\}$ 42. $\{(x, y): x = |y|\}$

C Using the real numbers, graph each of the following:

43. $\{(x, y): y \geq 0\}$ 44. $\{(x, y): x \geq -1 \text{ and } x \leq 1\}$
45. $\{(x, y): |x| = |y|\}$ 46. $\{(x, y): x = 0 \text{ or } y = 0\}$

Using the set of all real numbers, graph enough points to be able to predict and draw the graph of the following equations:

47. $y = \frac{x}{|x|}, x \neq 0$ 48. $y = -\frac{x}{|x|}, x \neq 0$
49. $y = x^2$ 50. $y = -x^2$

10-2 A DISTANCE AND A MIDPOINT FORMULA

In Chapter 1, we stated that to each point of the number line there corresponds exactly one real number. We called this line the real number line. Since establishing this correspondence, we have expanded our discussion from graphing on a number line to graphing in a plane by using two real number lines. To each point in this plane, there corresponds a unique ordered pair of numbers.

We shall now use the Distance Postulate in the coordinate system to find the distance between any two points in a plane. If the two points are in a line parallel to the axes, we can simply count the units between the two points.

In the figure, since both points A and B lie 3 units from the y-axis, $\overleftrightarrow{AB}$ is a vertical line parallel to the y-axis. $\overleftrightarrow{CD}$ is also a vertical line. Similarly, since points C and B are one unit above the x-axis, $\overleftrightarrow{CB}$ is a horizontal line, parallel to the x-axis. $\overleftrightarrow{DQ}$ is another horizontal line. Counting the units along the dashed lines, we see that the distance between A and B is 3; between C and D, 4; between C and B, 6; and between D and Q, 5.

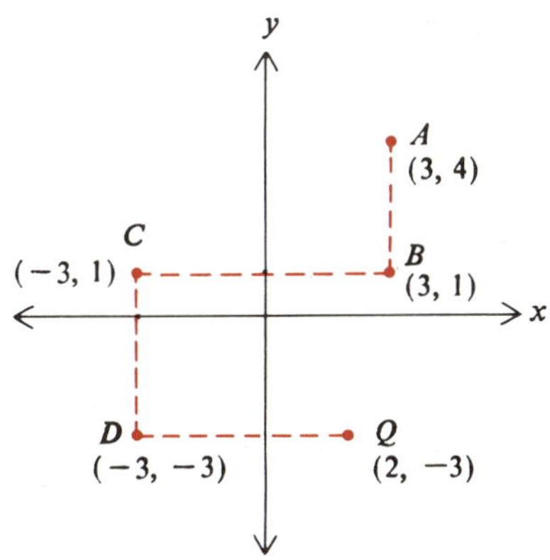

There is another method of determining such distances. In Chapter 1, we said that the distance between two distinct points of a line is the absolute value of the difference of the coordinates of the points. A similar situation arises for two distinct points of a line parallel to an axis. When two distinct points are contained in a horizontal line, their ordinates are the same. Since their abscissas are different, the distance between such points is the absolute value of the difference between their abscissas. For example, in the figure above, the distance CB is $|-3-3| = |-6| = 6$ or $|3-(-3)| = 6$. Unless we are told otherwise, when we refer to two points, we shall assume that they are distinct points.

D **Definition 10-3** The distance between $P(x_1, y_1)$ and $Q(x_2, y_1)$, two points of a line parallel to the x-axis, is PQ, where $PQ = |x_2 - x_1|$ or $|x_1 - x_2|$.

The symbol x_1 is read "x sub one." The 1 is called a **subscript**; it is used to distinguish one value of x from another value of x (for instance, x_2). We can also represent the difference between two values of x symbolically, combining Δ, the capital Greek letter d, with x. This resulting symbol, Δx, is read "delta x." It does not mean "delta times x." It merely indicates a difference between two x values. Thus, $PQ = |\Delta x|$.

Example 1 Find the distance between $R(-3, 4)$ and $S(5, 4)$.

Solution Since the ordinates are the same, $\overline{RS}$ is parallel to the x-axis. Therefore, by Definition 10-3,

$$RS = |5 - (-3)| = |8| = 8 \quad \text{or} \quad RS = |-3 - (5)| = |-8| = 8$$

We can use a similar method to find the distance between two points of a line parallel to the y-axis. Points of a line parallel to the y-axis have equal abscissas but different ordinates.

D **Definition 10-4** The distance between $P(x_1, y_1)$ and $Q(x_1, y_2)$, two points of a line parallel to the y-axis, is PQ, where $PQ = |y_2 - y_1|$ or $|y_1 - y_2|$.

As with Δx, we can denote the difference between y_2 and y_1 by Δy. Thus, $PQ = |\Delta y|$.

Example 2 Find MN for $M(5, -1)$ and $N(5, -7)$.

Solution Since the abscissas of M and N are equal, $MN = |-7 - (-1)| = 6$.

Example 3 In $\triangle ABC$ at right, $\overline{AP}$ is perpendicular to $\overline{BC}$ at P. Find AP.

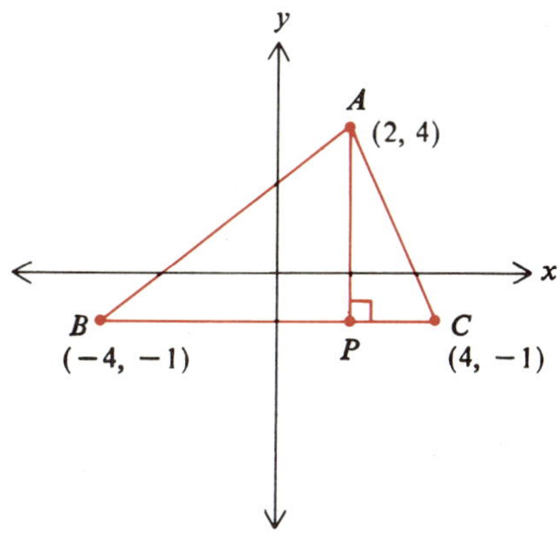

Solution Since the ordinates of B and C are both -1, $\overline{BC}$ is parallel to the x-axis. Thus, the ordinate of P must be -1. Since $\overline{AP}$ is perpendicular to $\overline{BC}$, $\overline{AP}$ is parallel to the y-axis. Thus, the abscissas of A and P are the same, and the coordinates of P are $(2, -1)$. Since $\overline{AP}$ is a vertical line, $AP = |4 - (-1)| = 5$.

Class Exercises

Referring to the figure at right, complete the following:

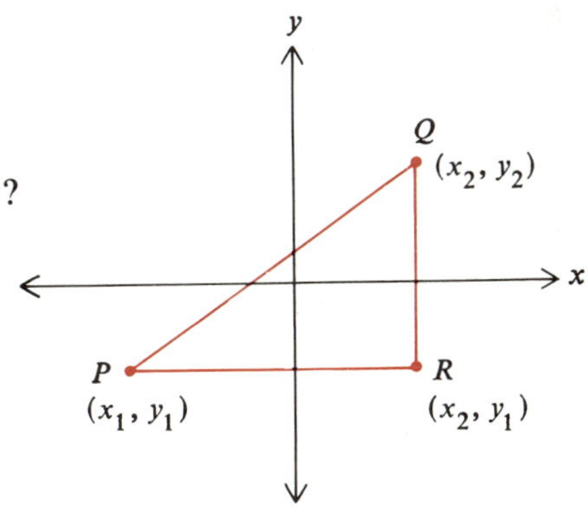

1. $\overline{PR}$ is parallel to ____.
2. $\overline{QR}$ is parallel to ____.
3. $\overline{PR}$ is ________ to $\overline{QR}$. Why?
4. $\triangle PQR$ is a ________ triangle.
5. By Definition 10-3, $PR =$ ____; thus $(PR)^2 =$ ____.
6. By Definition 10-4, $QR =$ ____; thus, $(QR)^2 =$ ____.
7. By the Pythagorean Theorem $(PQ)^2 = (\_\_\_\_)^2 + (\_\_\_\_)^2$.
8. $PQ = \sqrt{\_\_\_\_}$.

Exercises 1–8 above demonstrate a method of finding the distance between two points of a line not parallel to one of the axes. This method is summarized in the following theorem:

T **Theorem 10-2.1 The Distance Formula** The distance PQ between any two points, $P(x_1, y_1)$ and $Q(x_2, y_2)$, is

$$\sqrt{(x_2 - x_1)^2 + (y_2 - y_1)^2} \quad \text{or} \quad \sqrt{(\Delta x)^2 + (\Delta y)^2}$$

GIVEN Points $P(x_1, y_1)$ and $Q(x_2, y_2)$

PROVE $PQ = \sqrt{(x_2 - x_1)^2 + (y_2 - y_1)^2}$ (*Proof continued on next page*)

Proof Outline Since P and Q are any points, we must prove three cases.

Case I If $x_1 = x_2$ and $y_1 = y_2$, P and Q are the same point and $PQ = 0$. Since $\Delta x = 0$ and $\Delta y = 0$, then $PQ = \sqrt{0 + 0} = 0$. The theorem is valid.

Case II If the two points have one of the corresponding coordinates equal, but not both, we can write $x_2 - x_1 = 0$ or $y_2 - y_1 = 0$, but not both. When either of these equations occurs, the points are in a line parallel to one of the axes. Thus,

$$PQ = \sqrt{(x_2 - x_1)^2 + 0} \qquad \text{or} \qquad PQ = \sqrt{0 + (y_2 - y_1)^2}$$

The resulting distance is the same as if we had used either Definition 10-3 or 10-4. Remember, by definition, $\sqrt{x^2} = |x|$. Thus, the formula holds and the theorem is valid.

Case III If P and Q are distinct points of a line not parallel to one of the axes, we can locate a third point R with an abscissa equal to that of Q and an ordinate equal to that of P. We then have $R(x_2, y_1)$. $\overline{PR}$ is parallel to the x-axis and $\overline{RQ}$ is parallel to the y-axis. If we connect the three points, we form right $\triangle PRQ$. By Definitions 10-3 and 10-4, $PR = |x_2 - x_1|$ and $RQ = |y_2 - y_1|$. To find PQ, we apply the Pythagorean Theorem and write $(PQ)^2 = (|x_2 - x_1|)^2 + (|y_2 - y_1|)^2$. But $(|\Delta x|)^2 = (\Delta x)^2$ and $(|\Delta y|)^2 = (\Delta y)^2$. Thus, $(PQ)^2 = (x_2 - x_1)^2 + (y_2 - y_1)^2$.

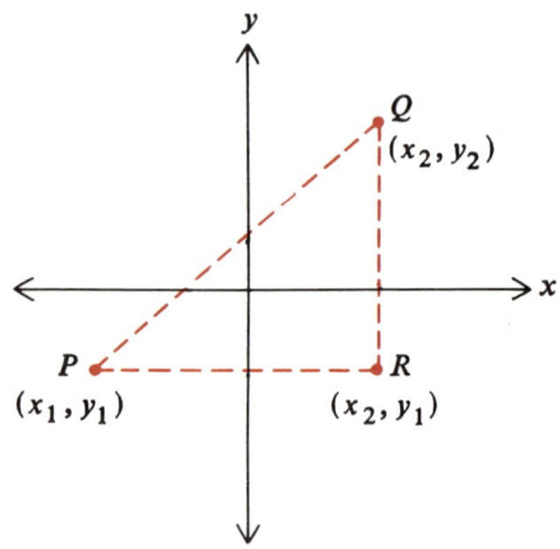

Therefore,

$$PQ = \sqrt{(x_2 - x_1)^2 + (y_2 - y_1)^2}$$

The distance formula has many applications. One of them is the determination of the collinearity of three points. We know that if three points are not collinear, they determine a triangle. By Theorem 5-4.1, we also know that the sum of the lengths of any two sides of a triangle is greater than the length of the third side.

Example 4 Determine if $A(3, 4)$, $B(5, 3)$, and $C(7, 6)$ are collinear.

Solution Use the distance formula to find AB, AC, and BC.

$$AB = \sqrt{(5-3)^2 + (3-4)^2} = \sqrt{4+1} = \sqrt{5}$$
$$AC = \sqrt{(7-3)^2 + (6-4)^2} = \sqrt{16+4} = \sqrt{20} = 2\sqrt{5}$$
$$BC = \sqrt{(7-5)^2 + (6-3)^2} = \sqrt{4+9} = \sqrt{13}$$

Because the sum of the lengths of any two segments is greater than the length of the third segment, the points form a triangle. Thus, they are not collinear.

If the sum of two of the lengths had been equal to a third length, then the three points would have been collinear.

Example 5 Show that the three points $A(-1, 1)$, $B(3, 4)$, and $C(6, 0)$ form an isosceles right triangle.

Solution To show that $\triangle ABC$ is an isosceles right triangle, we must show that two of the sides are congruent and that the Pythagorean Theorem can be applied to the lengths of the sides.

$$AB = \sqrt{(3-(-1))^2 + (4-1)^2} = \sqrt{16+9} = \sqrt{25} = 5$$
$$AC = \sqrt{(6-(-1))^2 + (0-1)^2} = \sqrt{49+1} = \sqrt{50} = 5\sqrt{2}$$
$$BC = \sqrt{(6-3)^2 + (0-4)^2} = \sqrt{9+16} = \sqrt{25} = 5$$

Since $AB = BC = 5$, $\triangle ABC$ is isosceles.

$(AC)^2 = (AB)^2 + (BC)^2$, since $50 = 25 + 25$. Therefore, $\triangle ABC$ is also a right triangle. Thus $\triangle ABC$ is an isosceles right triangle.

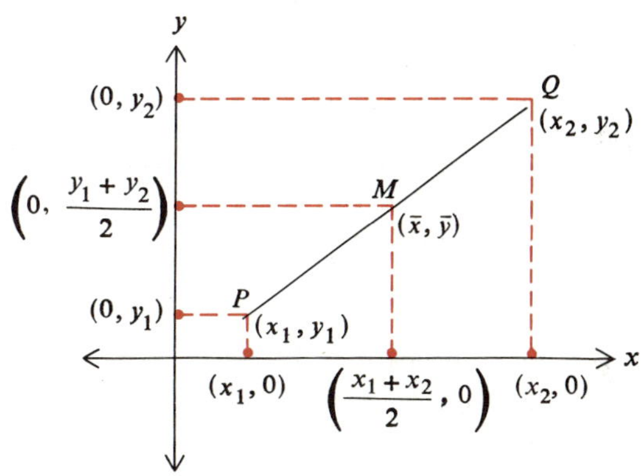

We know that every segment has a length and a unique midpoint. If M is the midpoint of $\overline{PQ}$, M is one-half the distance from one endpoint to the other. The coordinates of M are $(\bar{x}, \bar{y})$. If $P(x_1, y_1)$ and $Q(x_2, y_2)$ are endpoints, we can find $M(\bar{x}, \bar{y})$ by adding half the distance between the coordinates to the coordinate of P. That is, for the figure above,

$\bar{x} = x_1 + \frac{1}{2}(x_2 - x_1)$

or $\bar{x} = \dfrac{x_1 + x_2}{2}$

and $\bar{y} = y_1 + \frac{1}{2}(y_2 - y_1)$ or $\bar{y} = \dfrac{y_1 + y_2}{2}$.

T **Theorem 10-2.2 The Midpoint Formula** The midpoint of the segment determined by points $P(x_1, y_1)$ and $Q(x_2, y_2)$ is the point $M\left(\frac{x_1 + x_2}{2}, \frac{y_1 + y_2}{2}\right)$.

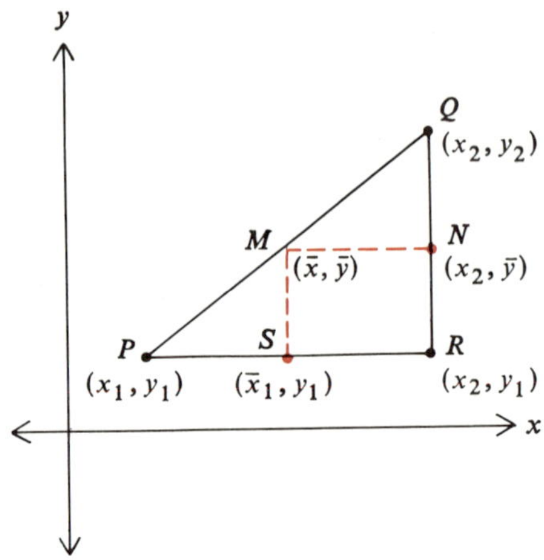

GIVEN $\overline{PQ}$ with endpoints $P(x_1, y_1)$ and $Q(x_2, y_2)$.

PROVE The midpoint of $\overline{PQ}$ is $M\left(\frac{x_1 + x_2}{2}, \frac{y_1 + y_2}{2}\right)$.

PROOF OUTLINE Locate a third point $R(x_2, y_1)$ to form right $\triangle PRQ$. Let $M(\bar{x}, \bar{y})$ be the midpoint of PQ. The line through the midpoint of one side of a triangle and parallel to a second side bisects the third side. Thus, N and S are the midpoints of $\overline{QR}$ and $\overline{PR}$, respectively. Since N is the midpoint of $\overline{QR}$, $y_2 - \bar{y} = \bar{y} - y_1$. Thus, $2\bar{y} = y_1 + y_2$ or $\bar{y} = \frac{y_1 + y_2}{2}$. Likewise, since S is the midpoint of $\overline{PR}$, $x_2 - \bar{x} = \bar{x} - x_1$. Thus, $\bar{x} = \frac{x_1 + x_2}{2}$.

Summary. Distance and Midpoint Formulas

1. The distance from $P(x_1, y_1)$ to $Q(x_2, y_2)$ is $\sqrt{(x_2 - x_1)^2 + (y_2 - y_1)^2}$.
2. The midpoint of $\overline{PQ}$ is $M\left(\frac{x_1 + x_2}{2}, \frac{y_1 + y_2}{2}\right)$.

EXERCISES

A Find the distance between each of the following pairs of points:

1. (1, 2), (2, 4)
2. (2, 2), (5, 7)
3. (7, 6), (4, −7)
4. (5, 1), (14, 0)
5. (3, 0), (0, −11)
6. $(\frac{3}{4}, -\frac{1}{2})$, $(\frac{1}{2}, -2)$
7. (x, y), $(-x, -y)$
8. $(x, 2y)$, $(2x, y)$
9. $(-x, y)$, $(2x, 2y)$

Find the midpoint of the segment formed by each pair of points below.

10. $(\frac{7}{2}, -3)$, $(\frac{4}{3}, 3)$
11. (3, −2), (−5, −5)
12. (0, 2), (1, −6)
13. (3, 5), (7, 2)
14. (x, y), $(x, -y)$
15. $(2x, y)$, $(4x, y)$

16. Are (2, 5), (−1, 8), and (−3, 10) the vertices of a triangle?

Determine whether or not each of the following groups of points is collinear:

17. (2, 5), (−1, 8), (−3, 10)
18. (0, 0), (2, 1), (4, 4)
19. $(2x, y)$, $(0, -3y)$, $(x, -y)$
20. $(x, -y)$, $(x+2, 2-y)$, $(x-2, -y-2)$

B

21. If (−2, 9) is the midpoint of a segment with (−3, 10) as one endpoint, find the coordinates of the other endpoint.
22. Determine whether or not $A(1, -1)$, $B(-3, 5)$, and $C(5, 5)$ are the vertices of an isosceles triangle.
23. Determine if $\triangle ABC$ is a right triangle when A is (−4, 0), B is (0, 4), and C is (4, 0).
24. Determine if quadrilateral $PQRS$ is a parallelogram, given $P(0, 0)$, $Q(8, 0)$, $R(11, 3)$, and $S(3, 3)$.
25. Determine if quadrilateral $WXYZ$ is a rhombus, given $W(0, 0)$, $X(6, -2)$, $Y(11, 0)$, and $Z(5, 2)$.
26. Find the center and radius of a circle if (3, 2) and (5, 2) are the endpoints of a diameter.
27. Find the center and radius of a circle if (3, −4) and (−3, 4) are the endpoints of a diameter.

C

Assume that quadrilateral $ABCD$ at right is a parallelogram and that $\triangle AEF$ is a right triangle.

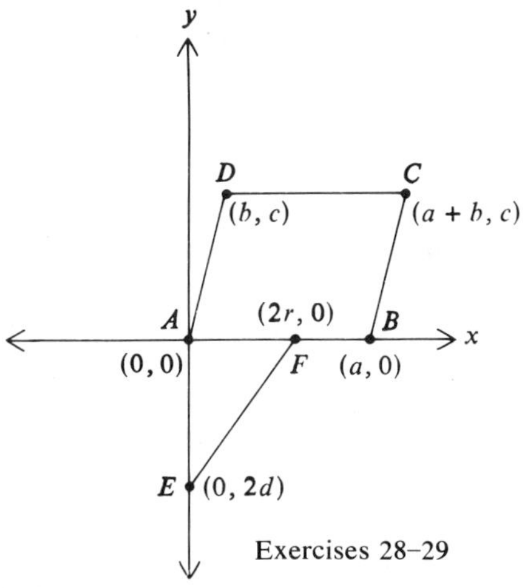

Exercises 28–29

28. Prove that the opposite sides of ▱$ABCD$ are congruent.
29. In $\triangle AEF$ prove that the midpoint of the hypotenuse, $\overline{EF}$, is equidistant from A and E.
30. If (3, 2) is the center of a circle tangent to line m at (7, 0), and the point (5, −4) is in m, show that the radius of the circle is perpendicular to m at (7, 0).
31. Find the perimeter of the triangle with vertices (2, 3), (14, 3), and (7, 8).
32. Find the perimeter of the quadrilateral with vertices (0, 0), (6, 0), (8, 3), and (3, 3).
33. Find the perimeter of the quadrilateral with vertices (0, 0), (8, 0), (11, 3), and (5, 2).
34. Determine if ▱ $PQRS$ is a square, given $P(-2, 3)$, $Q(2, 7)$, $R(8, 5)$, and $S(4, 1)$.

A LOOK AT THE PAST

La Géométrie

When asked if there were a shortcut to determining the properties of geometric figures, Euclid is said to have answered, "There is no royal road to geometry." He was mistaken, however, for coordinate geometry *is* this royal road. By considering geometric figures in terms of two variables (the coordinates of the points), we can treat geometric proofs as algebraic problems, using graphs as aids, as you will see in the rest of this chapter.

Although the Cartesian coordinate system we use bears his name, René Descartes did not actually invent the use of coordinates. Their use in map-making by travelers and surveyors dates back to ancient times. Thus, the rectangular system of coordinates was first used to locate real geographical positions. Coordinates were not used to represent mathematical points in a plane, however, until the seventeenth century, when Descartes and another French mathematician, Pierre de Fermat, working independently of each other, discovered the application of the coordinate system to the study of geometric relations. Although the two men made their discoveries almost simultaneously, Descartes is credited with the invention of coordinate, or analytic, geometry, because he published his ideas before Fermat. His work "La Géométrie" appeared in 1637, as the third appendix to his philosophical treatise, *Discourse on Method*.

10-3 SLOPE

The driver of a car knows by the response of the car whether or not a road is flat. On flat stretches, little additional pressure on the gas pedal is needed to maintain a constant speed. Going up a hill, however, the driver has to apply more pressure; the greater the incline of the hill, the greater the pressure he must apply. We sometimes speak of the steepness of a hill. But what is steepness? What is steep to some people may not be steep to others. We need to agree on some way of measuring it. Over the years, people have agreed to describe steepness as the amount of rise a road makes over a given horizontal distance. We call the measure of steepness the **slope**. The slope is the ratio of the amount of **rise**, the vertical distance attained, to **run**, the horizontal distance covered.

$$\text{Slope} = \frac{\text{rise}}{\text{run}}$$

Looking at the figures at right, note that the greater the rise over the same amount of run, the greater the slope. If a road rises 1 foot for a run of 1,000 feet, then the slope is 1:1,000 or $\frac{1}{1000}$.

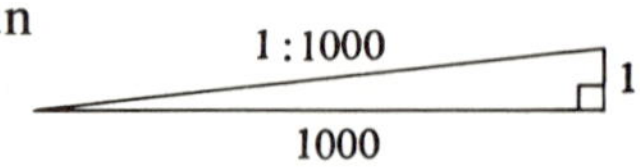

If the road rises 50 feet for the run of 1,000 feet, then the slope is 50 : 1,000 or $\frac{50}{1000}$. If the road inclines downward for 50 feet for the same run, the ratio is $-\frac{50}{1000}$. In this case, the slope has a negative value. We call this a **negative slope**.

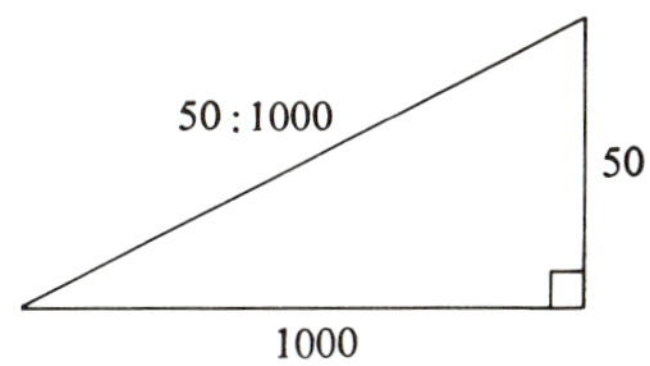

In the coordinate system, lines not parallel to the x-axis are inclined in relation to the x-axis. In the figure at right, points A, B, and C are three points of the graph of $y = 2x + 1$. To measure the rise and run between two points—for example, $A(-2, -3)$ and $B(1, -3)$—we locate a third point, as we did when we were finding distance. Using point $D(1, -3)$ to form the right $\triangle ADB$, we find that the rise from the horizontal $\overline{AD}$ is the change in the y-coordinates from D to B; that is, $DB = 3 - (-3) = 6$. Thus, the rise is 6. The run is the change in the x-coordinates from A to D; that is, $AD = 1 - (-2) = 3$. Thus, the run is 3. Did you notice that these changes in coordinates are also the changes in the corresponding coordinates of A and B?

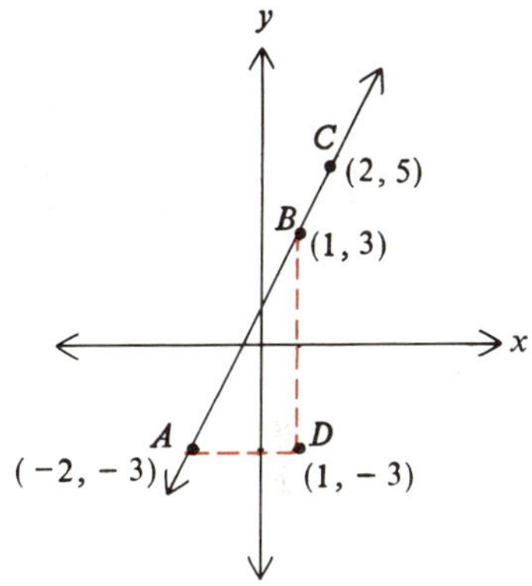

The slope from A to B is

$$\frac{\text{rise}}{\text{run}} = \frac{\text{change in } y\text{-coordinates}}{\text{change in } x\text{-coordinates}} = \frac{3 - (-3)}{1 - (-2)} = \frac{6}{3} = 2$$

If we write the changes in the coordinates using the delta notation, then the slope from B to A is

$$\frac{\Delta y}{\Delta x} = \frac{\text{change in } y\text{-coordinates}}{\text{change in } x\text{-coordinates}} = \frac{-3 - 3}{-2 - 1} = \frac{-6}{-3} = 2$$

What is the slope from B to C? from A to C?

The order in which we consider the difference is very important: the difference between x values must be considered in the same direction as the difference between the y values. If you considered the differences correctly, you should have found that the slope from B to C is 2 and from A to C is also 2.

D **Definition 10-5** Given $P_1(x_1, y_1)$ and $P_2(x_2, y_2)$ such that $x_2 \neq x_1$, the slope of the segment $\overline{P_1P_2}$ is the number m, where $m = \dfrac{\Delta y}{\Delta x} = \dfrac{y_2 - y_1}{x_2 - x_1}$.

Example 1 Find the slope between any two points of the graph of $y = 2x + 1$.

Solution For any x, say x_1, we have $y_1 = 2x_1 + 1$. For a different x, say x_2, we have $y_2 = 2x_2 + 1$. The slope from (x_1, y_1) to (x_2, y_2) is m.

$$\begin{aligned} m = \frac{\Delta y}{\Delta x} &= \frac{y_2 - y_1}{x_2 - x_1} \\ &= \frac{(2x_2 + 1) - (2x_1 + 1)}{x_2 - x_1} \\ &= \frac{2x_2 - 2x_1}{x_2 - x_1} \\ &= \frac{2(x_2 - x_1)}{x_2 - x_1} \\ &= 2 \end{aligned}$$

Thus, $m = 2$ between (x_1, y_1) and (x_2, y_2). Since these can be any two points of the line whose graph is $y = 2x + 1$, we say that the slope of the line is 2.

D **Definition 10-6** The slope of the line determined by (x_1, y_1) and (x_2, y_2), such that $x_1 \neq x_2$, is the same as m, the slope of any segment of the line.

What happens when $x_1 = x_2$? How do we describe a segment whose endpoints have equal abscissas? What happens if we use the formula for slope on two points of a vertical line? The following example should help you answer these questions.

Example 2 Describe the slope of the line determined by the two distinct points (x_1, y_1) and (x_1, y_2).

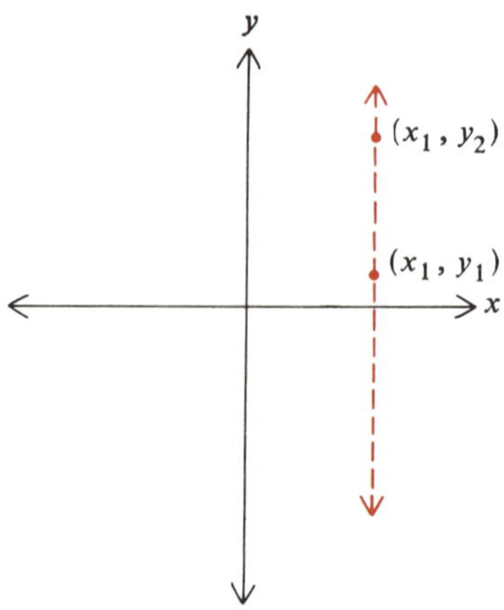

Solution If $y_1 \neq y_2$ but the points have the same abscissas, the line determined is a vertical line parallel to the y-axis. If we use the coordinates to find the value of the slope m, we get

$$m = \frac{\Delta y}{\Delta x} = \frac{y_2 - y_1}{x_1 - x_1} = \frac{y_2 - y_1}{0}$$

Because the denominator is 0, this result is not defined in the real number system. Thus, there is no slope for a vertical line.

Example 3 Find the slope of a line parallel to the x-axis.

Solution If the points determine a line parallel to the x-axis, then they have the same ordinate. Thus,

$$m = \frac{\Delta y}{\Delta x} = \frac{y_1 - y_1}{x_2 - x_1} = \frac{0}{x_2 - x_1}$$

Since the numerator is 0, the slope of a horizontal line must be 0.

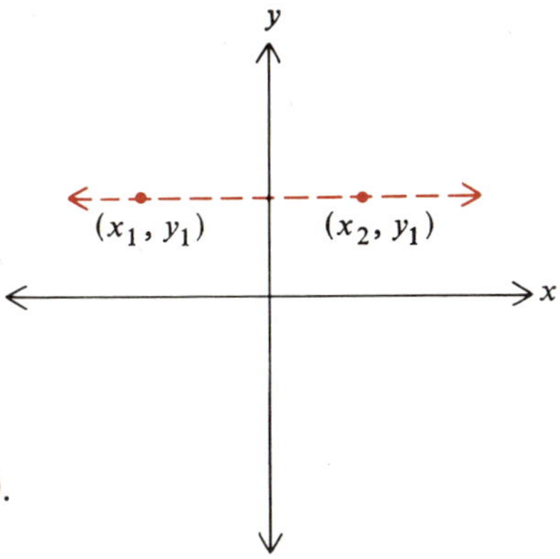

The method of locating two specific points of a line to find its slope can be used even when we are given only the equation of the line, as you can see from Example 1 above.

Example 4 Find the slope of $y = 3x - 1$.

Solution Locate two specific points of the line, by picking two values of x and finding the corresponding values for y. If $x = 0$, then $y = -1$, and $(0, -1)$ is a point of the line. If $x = 1$, then $y = 2$, and $(1, 2)$ is a point of the line. Thus, $m = \frac{\Delta y}{\Delta x} = \frac{2 - (-1)}{1 - 0} = \frac{3}{1} = 3$. The slope is 3.

Given a point of a line and the slope of the line, we should be able to locate and name additional points of the line. Before studying the example of this operation below, examine the following comments on Δx and Δy. If $\Delta x > 0$, then $x_2 - x_1 > 0$, or $x_2 > x_1$. Thus x_2 is to the right of x_1. However, if $\Delta x < 0$, x_2 is to the left of x_1. If $\Delta y > 0$, then $y_2 - y_1 > 0$, or $y_2 > y_1$, and y_2 is above y_1. Similarly, if $\Delta y < 0$, y_2 is below y_1.

Example 5 Locate two additional points of a line with a slope of $-\frac{2}{3}$, if the line contains (1, 2).

Solution If $m = -\frac{2}{3}$, then $\frac{\Delta y}{\Delta x} = -\frac{2}{3}$. This means that if $\Delta y = -2$, then $\Delta x = 3$; and if $\Delta y = 2$, then $\Delta x = -3$. That is, as x increases, y decreases; and as x decreases, y increases.

Case I $\Delta y = -2$ and $\Delta x = -3$. This means that if x moves 3 units to the right, y moves down 2 units. Thus we can locate a new point in relation to (1, 2): $(1 + 3, 2 - 2)$, or (4, 0).

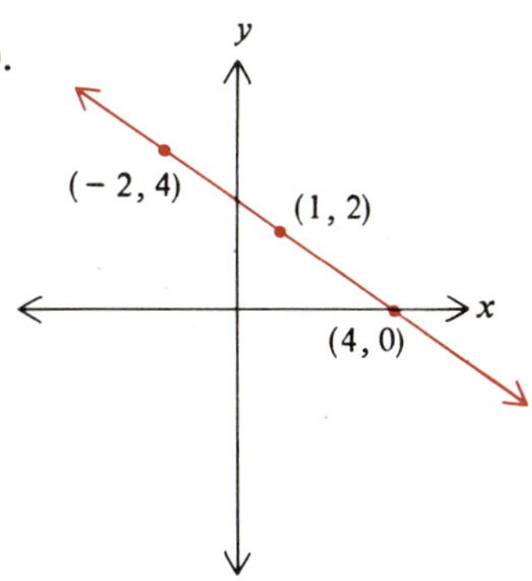

Case II $\Delta y = 2$ and $\Delta x = -3$. This means that if x moves 3 units to the left, y moves up 2 units. This locates a second new point in relation to (1, 2): $(1-3, 2+2)$, or $(-2, 4)$. As x increases, what happens to the graph of this line?

Can you see why lines with negative slopes always fall as x increases?

Class Exercises

Find the slope of the line determined by each of the following pairs of points, where possible:

1. (3, 9), (0, 5)
2. (5, 2), (4, 5)
3. (4, 3), (8, 5)
4. (2, −1), (5, −1)
5. (−2, 1), (2, 5)
6. (4, −1), (4, 12)

Find the slope of the graph of each of the following:

7. $y = 2x + 2$
8. $y = \frac{1}{2}x + 4$
9. $y = -2x - 3$
10. $x - y + 2 = 0$
11. $y - 2x + 5 = 0$
12. $y - \frac{1}{3}x + 2 = 0$

Slope and Perpendicular and Parallel Lines

Every line not parallel to the x-axis intersects the x-axis, forming four angles. The angle we are interested in is the one formed by the half-line of the x-axis to the right of the intersection and the half-line of the intersecting line above the x-axis. We call this the **angle of inclination** of the line. In the figure at left below, $\angle ABC$ is the angle of inclination. In the figure at right, $\angle PQR$ is the angle of inclination.

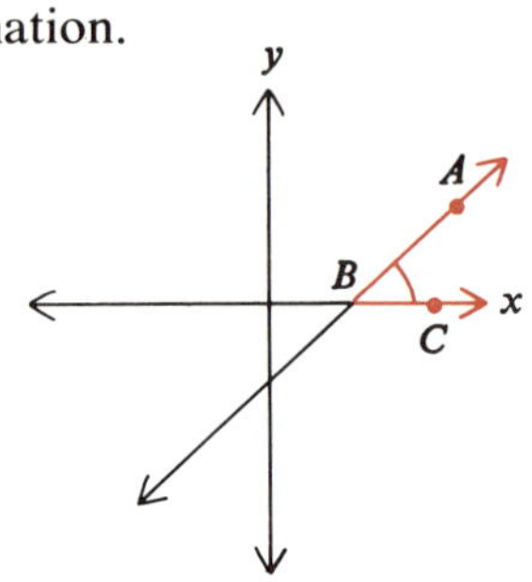

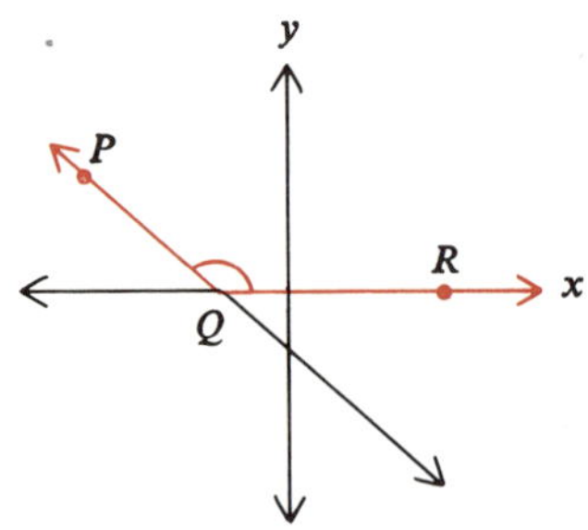

When the angle of inclination is acute, is the slope of the line positive or negative? When the angle of inclination is obtuse, is the slope of the line positive or negative? If a line is parallel to the x-axis, it does not intersect the axis, so we can say that the measure of its angle of inclination is 0.

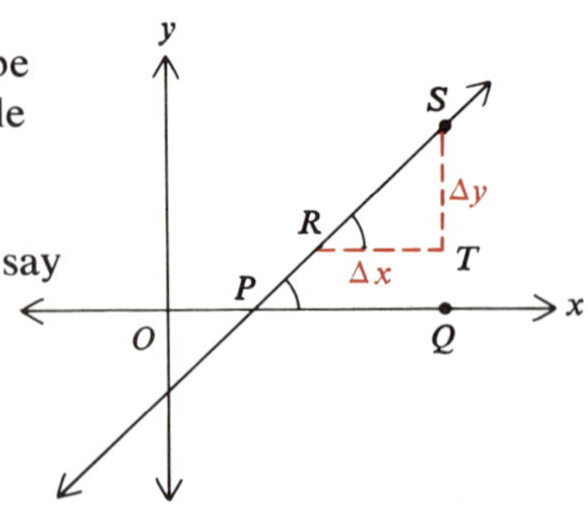

In the figure at right, $\angle RPQ$ is an angle of inclination. Since

$\overline{RT}$ and $\overleftrightarrow{PQ}$ are parallel, and $\overleftrightarrow{PRS}$ is a transversal, can you see that $\angle SRT$ is congruent to $\angle RPQ$?

In Chapter 8, we introduced the trigonometric ratios. Using the tangent ratio, we can write

$$\tan R = \frac{\Delta y}{\Delta x} = m$$

Since $\tan R = \tan P$, we can say that the tangent of the angle of inclination is equal to the slope of the line.

When two parallel lines are cut by a transversal, the corresponding angles are congruent. In the figure at right, l_1 is parallel to l_2 and $\angle 1$ and $\angle 2$, the angles of inclination of the lines, are congruent. Since these angles are congruent, their tangents are equal. Thus, the slopes of the lines are equal. This fact is the basis of our next theorem.

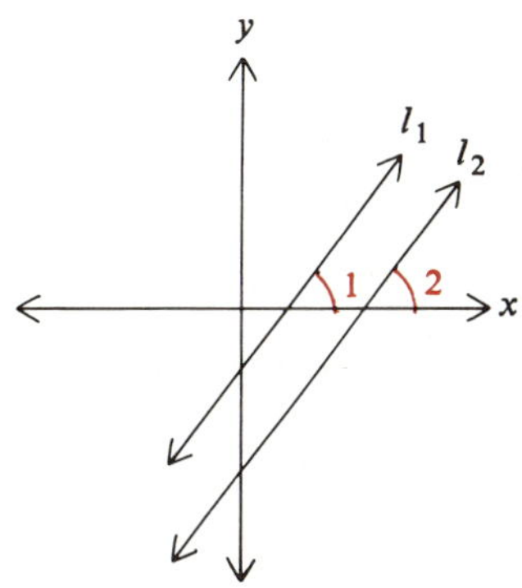

T **Theorem 10-3.1** Two nonvertical lines are parallel if and only if they have equal slopes.

To prove this theorem, we must prove the following statements:
(1) If two nonvertical lines are parallel, their slopes are equal.
(2) If two nonvertical lines have equal slopes, they are parallel.

PROOF OUTLINE OF (1)
We are given two parallel nonvertical lines, $l_1 \parallel l_2$. If these lines are parallel to the x-axis, then each has a slope of zero. Thus, the slopes are equal. If the lines intersect the x-axis, then their slopes are also equal, as we have seen in the discussion of the figure at right above.

The proof of the converse, statement (2), is left as an exercise.

Example 7 Using slopes, show that $A(-1, 3)$, $B(1, -1)$, and $C(3, -5)$ are collinear.

Solution The slope of $\overline{AB} = \dfrac{3+1}{-1-1} = -2$; the slope of $\overline{AC} = \dfrac{3+5}{-1-3} = -2$.
Since the slopes are equal, the segments are either parallel or parts of the same line. Since $\overline{AB}$ and $\overline{AC}$ have point A in common, they cannot be parallel. Thus they are two segments of the same line and the points are collinear.

Example 8 Using slopes, show that quadrilateral $ABCD$ is a parallelogram, given $A(-4, -2)$, $B(1, -3)$, $C(3, 1)$, and $D(-2, 2)$.

Solution A parallelogram is a quadrilateral whose opposite sides are parallel. We shall find the four slopes of $ABCD$.

Slope of $\overline{AB} = \dfrac{-2-(-3)}{-4-1} = -\dfrac{1}{5}$ Slope of $\overline{CD} = \dfrac{1-2}{3-(-2)} = -\dfrac{1}{5}$

Slope of $\overline{BC} = \dfrac{-3-1}{1-3} = 2$ Slope of $\overline{DA} = \dfrac{2-(-2)}{-2-(-4)} = 2$

Since the slopes of $\overline{AB}$ and $\overline{CD}$ are equal, $\overline{AB}$ is parallel to $\overline{CD}$. Since the slopes of $\overline{BC}$ and $\overline{DA}$ are equal, $\overline{BC}$ is parallel to $\overline{DA}$. Thus, $ABCD$ is a parallelogram.

Besides the slopes of parallel lines, we can also compare the slopes of nonvertical perpendicular lines.

T **Theorem 10-3.2** Two nonvertical lines with slopes m_1 and m_2 are perpendicular if and only if $m_1 = -\dfrac{1}{m_2}$; that is, $m_1 \cdot m_2 = -1$.

We must first prove that if two nonvertical lines with slopes m_1 and m_2 are perpendicular, then $m_1 = -\dfrac{1}{m_2}$.

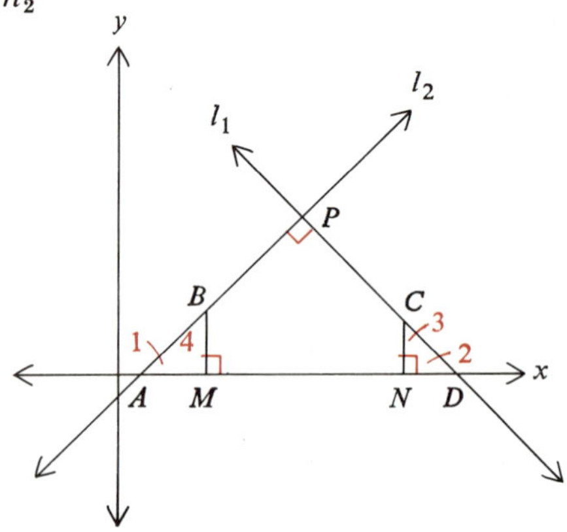

GIVEN Nonvertical perpendicular lines l_1 and l_2, with slopes m_1 and m_2.

PROVE $m_1 = -\dfrac{1}{m_2}$

PROOF OUTLINE Since we are given two nonvertical perpendicular lines, the lines intersect each other and the x-axis, thus forming right $\triangle APD$ with the hypotenuse on the x-axis. Let B be a point of $\overleftrightarrow{AP}$ between A and P. On $\overleftrightarrow{DP}$ locate point C such that $AB = DC$. Let the perpendiculars from B and C to $\overleftrightarrow{AD}$ intersect $\overleftrightarrow{AD}$ in M and N. Two right triangles, $\triangle AMB$ and $\triangle CND$, are formed. Angles 1 and 4 are complementary, as are $\angle 2$ and $\angle 3$, and $\angle 1$ and $\angle 2$. Thus, $\angle 2 \cong \angle 4$ and $\angle 1 \cong \angle 3$. Therefore $\triangle AMB \cong \triangle CND$, by the ASA Postulate.

$$m_1 = \frac{BM}{AM} = -\frac{ND}{CN}$$

This is a negative slope, since the direction of $\overline{BM}$ is opposite that of $\overline{ND}$.

$$m_1 = -\frac{1}{\frac{CN}{ND}}$$

But

$$\frac{CN}{ND} = m_2$$

Thus,

$$m_1 = -\frac{1}{m_2}$$

In effect, this part of the theorem says that if two lines are perpendicular and each has a slope, their slopes are negative reciprocals of each other.

The converse is proved by reversing the steps in the proof outline above. The proof of the converse is left as an exercise.

EXERCISES

A

Find the slope of the line joining each of the following pairs of points:

1. $(3, -2), (5, 7)$
2. $(3, 7), (-2, -1)$
3. $(0, 5), (5, 0)$
4. $(0, 0), (x, y)$
5. $(-3x, 0), (0, -2y)$
6. $(3x, -4y), (5x, 4y)$

Find the slope of the graph of each of the following equations:

7. $y = 2x - 3$
8. $y = -4x + 1$
9. $y = 7x - 5$
10. $y = -2x + 5$
11. $y + 3x - 7 = 0$
12. $y = -5x + 1$

Find two additional points of each of the lines for which one point and the slope are given below.

13. $m = \frac{3}{2}$, $P(1, 1)$
14. $m = -\frac{2}{5}$, $P(0, 1)$
15. $m = -0.5$, $P(-2, -3)$
16. $m = \frac{3}{4}$, $P(5, 2)$

17. Prove the second part of Theorem 10-3.1.
18. Prove the second part of Theorem 10-3.2.

Using slopes, determine whether each of the following sets of three points is collinear:

19. $(-3, 4), (3, 8), (0, 6)$
20. $(-2, 1), (1, -1), (2, 2)$
21. $(3, 1), (4, -1), (5, -3)$
22. $(1, -1), (4, 6), (3, 4)$
23. $(-1, 8), (7, 0), (1, 6)$
24. $(3, -2), (-2, -3), (6, 9)$

B

25. Find the value of k that makes the slope of the line through $(2, k)$ and $(3, 2k)$ equal to $\frac{1}{2}$.
26. Find the slope of a line parallel to the line determined by $(2, 7)$ and $(6, 3)$.
27. Find the slope of a line perpendicular to the line through $(4, 6)$ and $(-2, 1)$.

28. Using slopes, determine if $(1, 1)$, $(0, -4)$, and $(3, 0)$ are the vertices of a right triangle.
29. Using slopes, show that $(-2, 3)$, $(2, 7)$, $(8, 5)$, and $(4, 1)$ are the vertices of a parallelogram.

C

30. Find the slope of the line tangent to a circle at the point $(3, 7)$ if the center of the circle is $(-3, 2)$.
31. $A(3, 1)$, $B(9, 5)$, and $C(5, 7)$ are the vertices of $\triangle ABC$. Show that the line connecting the midpoints of $\overline{AB}$ and $\overline{AC}$ is parallel to $\overline{BC}$.

Using the techniques of coordinate geometry, select convenient coordinates to represent the figures described below and prove each of the following statements:

32. The line joining the midpoints of two sides of a triangle is one-half the length of the third side.
33. If the diagonals of a quadrilateral bisect each other, then the opposite sides of the quadrilateral are congruent.
34. The median to the base of an isosceles triangle is perpendicular to the base.
35. An equilateral quadrilateral has perpendicular diagonals.
36. The median of a trapezoid is parallel to both bases.

10-4 EQUATION OF A LINE

Very early in this course, we accepted the fact that two distinct points determine one and only one line. Earlier in this chapter, we noted that two points also determine the slope of the line. In our discussion of slope in the last section, we showed that knowing the slope and a point of a line enables us to locate other points of the line. If we continued locating points indefinitely, we could determine the line exactly. This process cannot be used, however, to determine a vertical line. Why?

Example 1 Describe the line through the point $(3, 1)$ for which there is no slope.

Solution If the line has no slope, then it is a vertical line. Every point of this line has the same value for its abscissa. That is, all the points on this line have the form $(3, y)$. We can describe the line by $\{(x, y):\ x = 3\}$, or simply by $x = 3$.

In general, the vertical line through the point (a, b) can be described by the equation $x = a$.

Example 2 Describe the horizontal line through the point $(3, 1)$.

Solution Horizontal lines are parallel to the x-axis. This means that all the points of the line have equal ordinates. Thus, all points of the horizontal line through (3, 1) have the form $(x, 1)$. We can describe this line by $\{(x, y): y = 1\}$ or by $y = 1$.

In general, the horizontal line through the point (a, b) can be described by the equation $y = b$.

If we know the slope and a point of any nonvertical line, we can write an equation for this unique line using the formula given in the following theorem. This equation is called the **point-slope form** of the equation of a line, since it describes the line in terms of its slope and a point.

T

Theorem 10-4.1 The Point-Slope Theorem An equation of the line that contains the point (x_1, y_1) and has a slope m is $y - y_1 = m(x - x_1)$, where (x, y) is any other point of the line.

GIVEN Line l contains (x_1, y_1) and has slope m.

PROVE Line l is represented by the equation $y - y_1 = m(x - x_1)$.

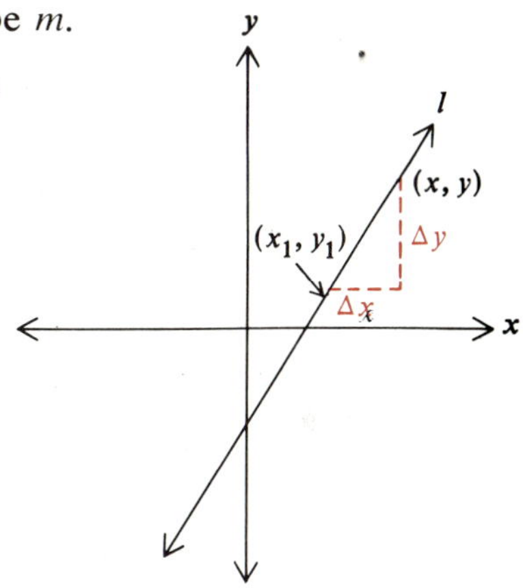

PROOF OUTLINE We have a line l, with a slope m, that contains the point (x_1, y_1). Let (x, y) be any other point of l. The slope between these two points of the line is also the slope of the line. Thus,

$$\frac{y - y_1}{x - x_1} = m$$

Multiplying both sides of the equation by $(x - x_1)$, we have

$$y - y_1 = m(x - x_1)$$

The points of the coordinate axes intersected by a graph of a line are two very special points. If (2, 0) is a point in the graph of $x + y = 2$, we say that 2 is the **x-intercept** of the graph, because it names the point where the graph of $x + y = 2$ crosses the x-axis. To find the x-intercept, let $y = 0$ in the equation and solve for x.

If $(0, -5)$ is a point in the graph of $x - y = 5$, we say that -5 is the **y-intercept** of the graph because it names the point where the graph of $x - y = 5$ crosses the y-axis. Can you tell how to find the y-intercept?

Example 3 Name the x-intercept and the y-intercept of $2x + 3y = 6$.

Solution To find the x-intercept let $y = 0$. Thus, $2x = 6$, or $x = 3$. The x-intercept is 3. To find the y-intercept let $x = 0$. Thus, $3y = 6$, or $y = 2$. The y-intercept is 2.

By substituting the point $(0, b)$ into the Point-Slope Theorem, where b is the y-intercept of the line, we derive the following corollary:

C

Corollary 10-4.1a The Slope-Intercept Theorem An equation of the line with a slope m and y-intercept b is $y = mx + b$, where (x, y) is any other point of the line.

This equation is called the **slope-intercept form** of the equation of a line.

GIVEN Line l with slope m and y-intercept b.

PROVE Line l is represented by $y = mx + b$.

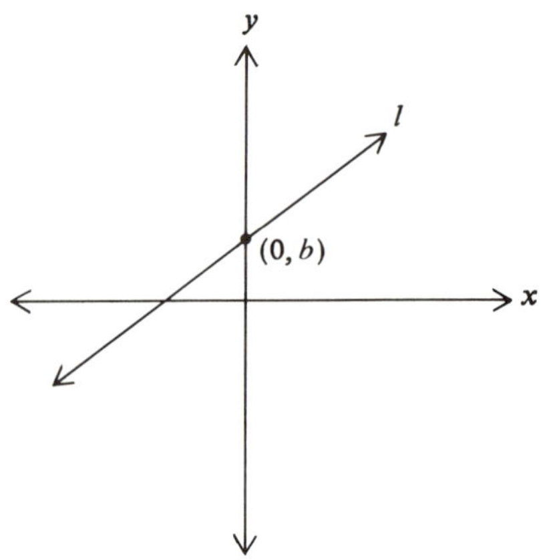

PROOF OUTLINE Since the line has a y-intercept of b, it passes through the point $(0, b)$. Thus, by the last theorem, we can write

$$y - b = m(x - 0)$$

$$y - b = mx$$

$$y = mx + b$$

Example 4 Write the equation for the line that contains the point $(-2, 5)$ and has a slope of $\frac{1}{3}$.

Solution The information given allows us to write an equation in the point-slope form. Thus,

$$y - 5 = \tfrac{1}{3}[x - (-2)]$$

Example 5 Find an equation for the line that has a y-intercept of 3 and is perpendicular to the line whose equation is $y = -\frac{1}{3}x + 5$.

Solution The equation $y = -\frac{1}{3}x + 5$ is in the slope-intercept form. Thus, the slope for this line is $-\frac{1}{3}$. Any line perpendicular to $y = -\frac{1}{3}x + 5$ has a slope of 3, by Theorem 10-3.2. This means that we now know the slope and y-intercept of the desired line. We can write its equation as $y = 3x + 3$.

At the beginning of this section, we said that two distinct points determine a unique line. The next example shows how to use two points of a line to find an equation for the line.

Example 6 Write an equation of the line determined by the points (3, 1) and (2, 5).

Solution The two points determine the slope of the line.

$$m = \frac{\Delta y}{\Delta x} = \frac{5-1}{2-3} = \frac{4}{-1} = -4$$

Using either point, we can write an equation in the point-slope form:

$$y - 1 = -4(x - 3) \qquad \text{or} \qquad y - 5 = -4(x - 2)$$

We shall simplify both forms to show that they are equivalent.

$$y - 1 = -4x + 12 \qquad\qquad y - 5 = -4x + 8$$

$$4x + y = 13 \qquad\qquad 4x + y = 13$$

We can see from Example 6 that using two given points to find the slope of a line enables us to use the point-slope form. We should use this technique to find the equation of a line whenever we are given two points, unless the points are the x-intercept and y-intercept of the line. For these two points, there is a special form of the equation of the line, known as the **two-intercept form**.

T **Theorem 10-4.2 The Two-Intercept Theorem** An equation of the line with x-intercept a and y-intercept b is $\frac{x}{a} + \frac{y}{b} = 1$.

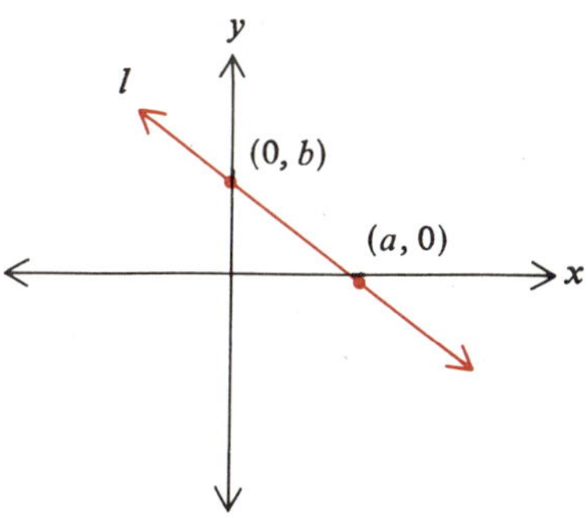

GIVEN Line l has x-intercept a and y-intercept b.

PROVE An equation of l is $\frac{x}{a} + \frac{y}{b} = 1$.

PROOF OUTLINE Since l has an x-intercept of a and y-intercept of b, l passes through the points $(a, 0)$ and $(0, b)$. The slope of this line is $m = \frac{\Delta y}{\Delta x} = \frac{0-b}{a-0} = -\frac{b}{a}$. Since we already know the y-intercept, we can use the slope-intercept form of the equation of a line to write the equation $y = -\frac{b}{a}x + b$. If we multiply all terms of the equation by $\frac{1}{b}$, we get $\frac{y}{b} = -\frac{x}{a} + 1$. Adding $-\frac{x}{a}$ to both sides, $\frac{x}{a} + \frac{y}{b} = 1$.

Summary. Forms for the Equation of a Line

1. Point-Slope form $y - y_1 = m(x - x_1)$
2. Slope-Intercept form $y = mx + b$
3. Two-Intercept form $\frac{x}{a} + \frac{y}{b} = 1$

EXERCISES

A In each of the following exercises, write the equation of the line through the given point and with the given slope:

1. $(1, 9)$, $m = 3$
2. $(0, 5)$, $m = -4$
3. $(-8, 8)$, $m = 5$
4. $(4, 7)$, $m = \frac{2}{7}$
5. $(3, 9)$, $m = 2$
6. $(1, -5)$, $m = 0$

Find the equation of the line passing through each of the following pairs of points:

7. $(2, 3), (-4, 3)$
8. $(3, -6), (0, 0)$
9. $(3, 5), (4, 6)$
10. $(1, -3), (7, 1)$
11. $(4, -1), (0, -9)$
12. $(-4, -13), (5, -1)$

Write the equation for each of the following, where m is the slope and b is the y-intercept.

13. $m = \frac{3}{4}$, $b = 2$
14. $m = 3$, $b = 17$
15. $m = -1$, $b = -9$
16. $m = -\frac{1}{2}$, $b = 10$
17. $m = -\frac{3}{4}$, $b = -8$
18. $m = -\frac{1}{5}$, $b = 6$

Write the equation for each of the lines whose x-intercept (a) and y-intercept (b) are given below.

19. $a = 7$, $b = 21$
20. $a = 2$, $b = 5$
21. $a = 4$, $b = -3$
22. $a = 6$, $b = 6$
23. $a = 3$, $b = -1$
24. $a = -2$, $b = -4$

B Rewrite each of the following equations in the two-intercept form:

25. $x + 2y = 4$
26. $3x + 2y = 6$
27. $y - 2x = 4$

Find the slope and the intercepts for each of the lines represented by the following equations:

28. $y - 2x = 15$
29. $3x + 2y = 6$
30. $x - 7y = 14$
31. $3x - y = 6$
32. $4x - 5y = 20$
33. $2x + y = 5$
34. $x + 2y + 10 = 0$
35. $4x - 2y = 12$
36. $3x + 7y = 9$
37. Write the equation of the line through $(1, 7)$ and perpendicular to the line $2x + y = 6$.
38. Write the equation of the line through $(7, -2)$ and parallel to the line $x = 2y + 4$.
39. Write the equation of the line through $(5, 6)$ and perpendicular to the line $y = 2$.

C

40. Write the equation for the perpendicular bisector of the segment whose endpoints are (1, 1) and (2, −3).
41. Find the equation for the perpendicular bisector of $\overline{AB}$, with A (2, 3) and B (6, −7).
42. If (2, 1) is the center of a circle and C (−1, 5) is a point of the circle, write the equation of the tangent to the circle at C.
43. Explain why the line through the points (2, −5) and (−2, 5) cannot be written in the two-intercept form.
44. The vertices of a triangle are A (3, 2), B(2, 1), and C(5, 7). Write the equation for the median from A to $\overline{BC}$.

SOMETHING TO THINK ABOUT

General Forms of Equations

The equation of a line can be written in many general forms. In the last section, we found that one general form of the equation of a line expresses the relation between the slope and a point of the line and another form expresses the relation between the slope and an intercept.

To find an equation of the line through (1, 1) and parallel to $y - 2x = 1$, we first must find the slope of $y - 2x = 1$, since parallel lines that have slopes have equal slopes. By writing the equation in the slope-intercept form, $y = 2x + 1$, we determine that the slope is 2. Thus, an equation of the line through (1, 1) is $y - 1 = 2(x - 1)$, or $y - 1 = 2x - 2$. This can also be written $y - 2x = -1$. Compare this with the equation of the first line.

We can also write the equations of the two lines above as $2x - y = 1$ and $2x - y = -1$. From this we can predict that if we write the equation of a line in the general form $ax + by = c$, then any line parallel to this line can be written in the form $ax + by = d$. Thus, the line through (2, 3) and parallel to $x + 2y = 1$ has the form $x + 2y = d$. Since (2, 3) is on the graph of $x + 2y = d$, we can write $2 + 2(3) = d$. Thus $d = 8$ and the equation is $x + 2y = 8$.

Perpendicular lines have slopes that are negative reciprocals of each other. If $ax + by = c$ is an equation of a line, we can show that $bx - ay = d$ is an equation of a line perpendicular to $ax + by = c$. Write an equation of the line through (5, 2) perpendicular to $3x + 2y = 7$.

10-5 CIRCLES AND ANALYTIC PROOFS

There are many uses for the formulas and ideas we have developed in this chapter. One of the most basic uses is in describing an equation of a circle. By definition, all the points of a circle are coplanar points equidistant from a fixed point. The fixed point is the center of the circle. Each point lies r units from the center, where r is the radius. If we know the center and radius, we can use the distance formula to write an equation of the circle.

Example 1 Write an equation of the circle with center C at the origin and with radius 7.

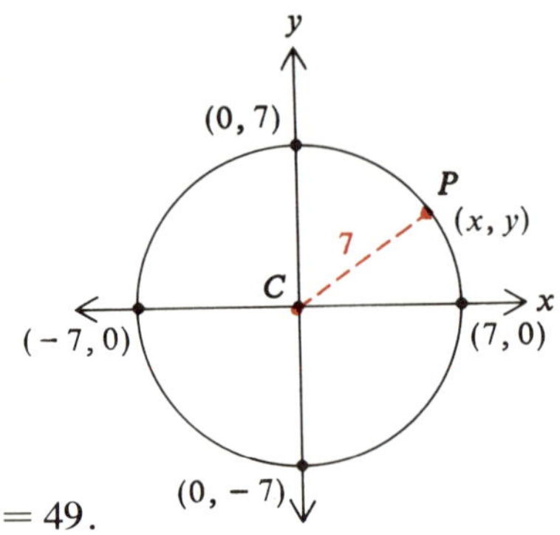

Solution Let $P(x, y)$ be any point of the circle. Thus, $PC = 7$. By the distance formula, we can write

$PC = \sqrt{(x-0)^2 + (y-0)^2}$
$PC = \sqrt{x^2 + y^2}$

Thus, $\sqrt{x^2 + y^2} = 7$, or $x^2 + y^2 = 49$.

Example 2 Write an equation of the circle whose center is $C(0, 0)$ and whose radius is any positive real number r.
Solution Let $P(x, y)$ be any point of the circle. Thus, $PC = r$. But

$$PC = \sqrt{(x-0)^2 + (y-0)^2}$$
$$= \sqrt{x^2 + y^2}$$

Thus, $\sqrt{x^2 + y^2} = r$, or $x^2 + y^2 = r^2$.

Any equation of the form $x^2 + y^2 = r^2$ is an equation of a circle with center at (0, 0) and with radius r. If we use a center other than (0, 0), the equation changes; but the process for developing the equation remains the same.

Example 3 Write an equation for the circle with center $A(-3, -2)$ and radius 3.

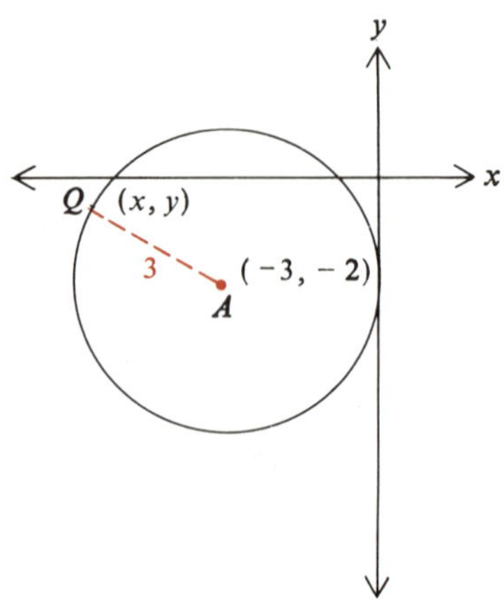

Solution Let $Q(x, y)$ be a point of the circle. Thus, $QA = 3$.

$QA = \sqrt{(x+3)^2 + (y+2)^2}$.

Therefore,

$\sqrt{(x+3)^2 + (y+2)^2} = 3$, or

$(x+3)^2 + (y+2)^2 = 9$.

The location of the center or the size of the radius is unimportant as long as r is positive.

Example 4 If we let any point $D(h, k)$ be the center of a circle with radius r, where $r > 0$, write an equation for the circle.

Solution Let $Q(x, y)$ be any point of the circle. Thus,

$QD = r = \sqrt{(x-h)^2 + (y-k)^2}$,
or $r^2 = (x-h)^2 + (y-k)^2$.

This is called the **standard form of an equation of a circle.**

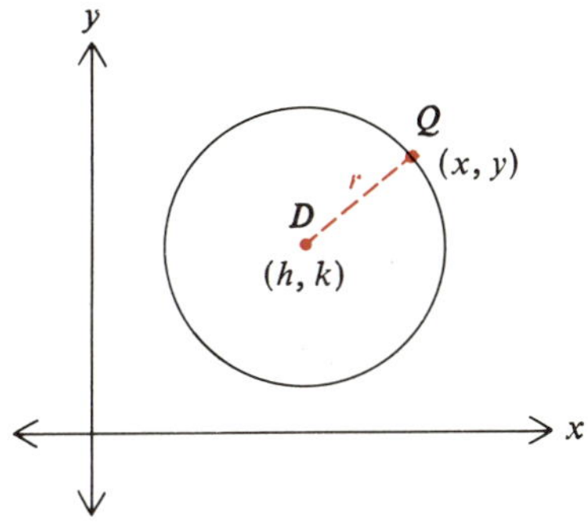

Class Exercises

Write an equation of each of the circles with the radii below, given that the center of each is at the origin.

1. 3 2. $\sqrt{3}$ 3. $\frac{1}{4}$ 4. 1 5. 6

Write an equation of the circle of radius 3 whose center is at:

6. (2, 5) 7. (0, 3) 8. $(\frac{1}{2}, -\frac{1}{2})$
9. (−2, 5) 10. (1, 1)

Analytic Proofs

The most important use of the coordinate system and the formulas we have developed thus far is to aid in proving many geometric theorems. These coordinate proofs, also called **analytic proofs**, combine deductive reasoning with algebraic operations on formulas such as those we have studied for distance, midpoint, and slope.

Much of your success with analytic proofs will depend upon your positioning a given figure on the coordinate system. You should try to place the figure so that the coordinates of as many points as possible are simplified and calculations are minimized. For example, if we want to find a midpoint, it helps to use coordinates such as $(2a, 2b)$ and $(2c, 2d)$, instead of (a, b) and (c, d), since the midpoint coordinates as stated in the Midpoint Formula have a denominator of 2. The midpoint $(a + c, b + d)$ is easier to use than the midpoint $\left(\frac{a+c}{2}, \frac{b+d}{2}\right)$.

Example 5 Prove that the segment formed by joining the midpoints of two sides of a triangle is parallel to the third side and has a length equal to half that of the third side.

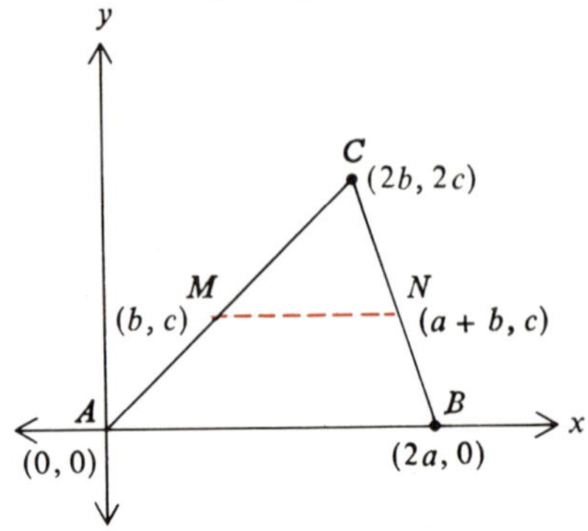

Solution Using the figure at right, we have $\triangle ABC$ with $A(0, 0)$, $B(2a, 0)$, and $C(2b, 2c)$. Note that we chose a scalene triangle with convenient coordinates. Using the midpoint formula for $\overline{AC}$, we get $M(b, c)$; for $\overline{BC}$ we get $N(a+b, c)$. Using M and N, we have to prove that:
(1) $\overline{MN}$ is parallel to $\overline{AB}$; (2) $MN = \frac{1}{2}(AB)$.

Proof of (1) The slope of $\overline{MN} = 0$ equals the slope of $\overline{AB}$. Thus, $\overline{MN}$ is parallel to $\overline{AB}$ (Theorem 10-3.1).

Proof of (2) Applying the Distance Formula,

$$MN = \sqrt{(a+b-b)^2 + (c-c)^2} = \sqrt{a^2 + 0} = |a|$$

$$AB = |2a - 0| = |2a|$$

Thus, $$MN = \tfrac{1}{2}(AB)$$

Example 6 Prove that the midpoint of the hypotenuse of a right triangle is equidistant from all the vertices of the triangle.

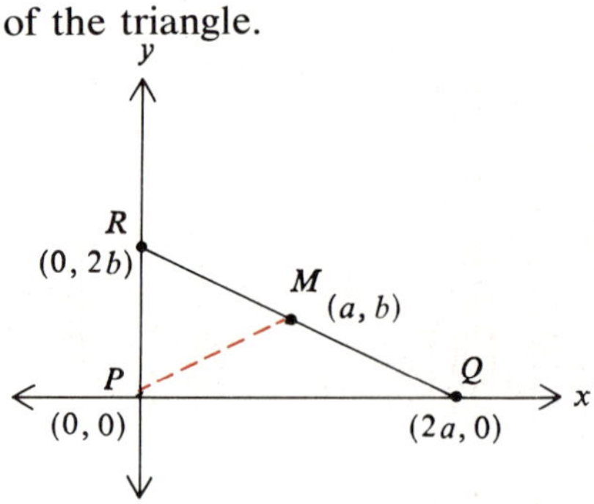

Solution Since the coordinate axes are perpendicular to each other, we form right $\triangle PQR$ by using the origin and a point of each coordinate axis. We choose $(0, 2b)$ and $(2a, 0)$ for convenience. Applying the Midpoint Formula to $\overline{RQ}$ we determine $M(a, b)$. We must prove $PM = RM = QM$.

Proof

$$PM = \sqrt{(a-0)^2 + (b-0)^2} = \sqrt{a^2 + b^2}$$

$$RM = \sqrt{(a-0)^2 + (b-2b)^2} = \sqrt{a^2 + b^2}$$

$$QM = \sqrt{(2a-a)^2 + (0-b)^2} = \sqrt{a^2 + b^2}$$

Thus, $$PM = RM = QM$$

When proving properties of parallelograms, we usually place one side of the parallelogram on the positive x-axis with one vertex at the origin and the other at $(a, 0)$. These are points A and B at right. D is any point (b, c), but since $\overline{DC}$ is parallel to $\overline{AB}$, the ordinate of C must be c also. We can label C as (d, c). However, we also know that the slope of $\overline{AD}$ equals the slope of $\overline{BC}$, since $\overline{AD} \parallel \overline{BC}$. That is,

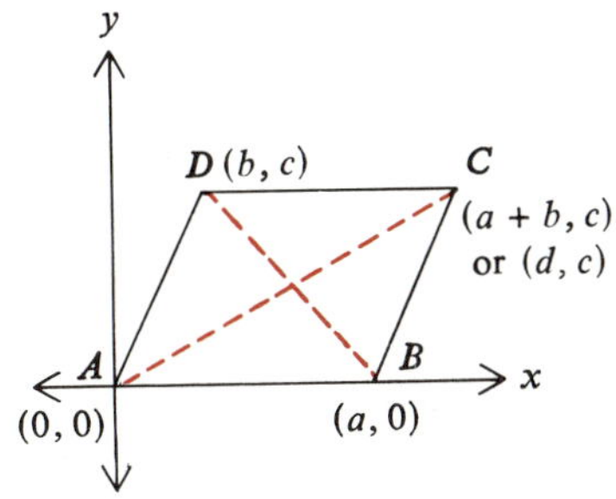

$$\frac{c-0}{b-0} = \frac{c-0}{d-a}$$

This means that $\frac{c}{b} = \frac{c}{d-a}$, or $b = d - a$. This indicates that $d = a + b$.

Example 7 Prove that the diagonals of a parallelogram bisect each other.

Solution We can use the figure above with vertices $A(0, 0)$, $B(a, 0)$, $C(a + b, c)$, and $D(b, c)$. We must prove that $\overline{AC}$ and $\overline{BD}$ bisect each other. Do you see that by showing that the midpoints of the two segments coincide, we will prove the statement? The midpoint of $\overline{AC}$ is $\left(\frac{a+b}{2}, \frac{c}{2}\right)$. The midpoint of $\overline{BD}$ is also $\left(\frac{a+b}{2}, \frac{c}{2}\right)$. Thus, the diagonals bisect each other.

EXERCISES

A Write the equation of each circle whose radius is given below, if its center is at the origin:

1. 2 2. $\frac{5}{2}$ 3. $\sqrt{3}$ 4. 12 5. $2\sqrt{5}$

Determine whether or not the following are points of the circle whose equation is $x^2 + y^2 = 49$.

6. $(-5, 12)$ 7. $(3, 2\sqrt{10})$ 8. $(0, -7)$ 9. $(48, 1)$ 10. $(-1, 4\sqrt{3})$

Write an equation of each circle with the center and radius given below.

11. $(-1, 5)$, $\frac{1}{2}$ 12. $(8, 5)$, 9 13. $(0, 3)$, 4
14. $(-5, 4)$, $\sqrt{11}$ 15. $(0, 0)$, $\frac{2}{3}$ 16. $(6, 5)$, 6

17. Write an equation of the circle with $(8, 5)$ as center, if $(2, 1)$ is a point of the circle.

18. Write an equation of the circle with $(3, -4)$ as center, if $(-3, 0)$ is a point of the circle.
19. If $(-5, 3)$ and $(3, 6)$ are the endpoints of a diameter of a circle, find the center and radius of the circle.

B

20. Write an equation of the circle tangent to the y-axis whose center is at $(5, 13)$.
21. Write an equation of the circle tangent to the x-axis whose center is at $(5, 6)$.

Prove each of the following statements using analytic methods:

22. If the diagonals of a quadrilateral bisect each other, the quadrilateral is a parallelogram.
23. Two coplanar lines perpendicular to the same line are parallel to each other.
24. In a plane, if a line is perpendicular to one of two parallel lines, then it is also perpendicular to the other line.
25. If each of two lines is parallel to a third line, then they are parallel to each other.
26. If both pairs of opposite sides of a quadrilateral are congruent, then the quadrilateral is a parallelogram.
27. If one pair of opposite sides of a quadrilateral is both congruent and parallel, then the quadrilateral is a parallelogram.
28. The diagonals of a rectangle are congruent.
29. The median to the hypotenuse of a right triangle is half as long as the hypotenuse.
30. If a line parallel to one side of a triangle bisects a second side of the triangle, then this line also bisects the third side.
31. The median of a trapezoid is parallel to the bases.
32. The length of the median of a trapezoid is half the sum of the lengths of the bases.
33. The diagonals of an isosceles trapezoid are congruent.

C

Prove each of the following statements using analytic methods:

34. In an isosceles right triangle, the hypotenuse is $\sqrt{2}$ times as long as a leg.
35. An angle inscribed in a semicircle is a right angle.
36. The segments connecting the midpoints of the sides of an isosceles trapezoid, in order, form a rhombus.
37. If the diagonals of a parallelogram are both congruent and perpendicular to each other, then the parallelogram is a square.
38. The segments joining the midpoints of the sides of a triangle divide the triangle into four congruent triangles.
39. The segments joining the midpoints of the sides of a rhombus form a rectangle.

HOW IS YOUR MATHEMATICAL VOCABULARY?

The key words and phrases introduced in this chapter are listed below. How many do you know and understand?

abscissa (10-1)
analytic proof (10-5)
angle of inclination (10-3)
Cartesian coordinate plane (10-1)
Cartesian product (10-1)
coordinate (10-1)
coordinate axes (10-1)
delta (10-2)
Distance Formula (10-2)
Euclidean plane (10-1)
Midpoint Formula (10-2)
negative slope (10-3)
ordered pairs (10-1)
ordinate (10-1)
origin (10-1)
point-slope form (10-4)
quadrants (10-1)
rectangular coordinate system (10-1)
replacement set (10-1)
rise (10-3)
run (10-3)
slope (10-3)
slope-intercept form (10-4)
solution set (10-1)
standard form of a circle (10-5)
subscript (10-2)
two-intercept form (10-4)
x-axis (10-1)
x-intercept (10-4)
y-axis (10-1)
y-intercept (10-4)

REVIEW EXERCISES

10-1 The Euclidean plane

Determine the coordinates of each of the following points on the graph at right:

1. A 2. D 3. F 4. C

Name the points on the graph with the following coordinates:

5. $(\frac{3}{2}, \frac{5}{2})$ 6. $(4, 3)$
7. $(-2.5, -2.5)$ 8. $(-\frac{7}{2}, \frac{7}{2})$

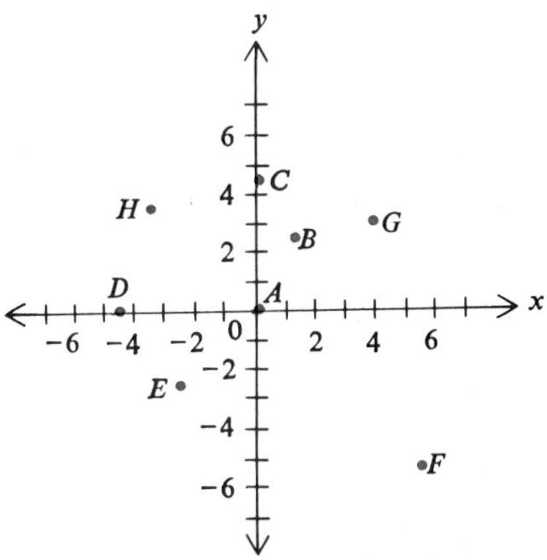

10-2 A distance and a midpoint formula

Find the distance between each of the following pairs of points:

9. $(8x, 2y)$, $(2x, 6y)$ 10. $(\frac{1}{2}, \frac{1}{3})$, $(-\frac{1}{2}, -\frac{1}{3})$ 11. $(-0.5, 2.1)$, $(1.5, -3.1)$.

12. Find the perimeter of the triangle with vertices at $(2, 3)$, $(6, 6)$, and $(-7, 5)$.

13. Determine if the points (5, 2), (8, −1), and (10, −3) are collinear.
14. Show that $A(0, -2)$, $B(6, -2)$, and $C(3, 6)$ are the vertices of an isosceles triangle.
15. In $\triangle ABC$, find the length of the median from $A(0, 4)$ to $\overline{BC}$, with $B(5, 3)$ and $C(3, 2)$.
16. Find the center and radius of the circle with (−5, 13) and (3, 7) as endpoints of a diameter.

10-3 Slope

Find the slope of each of the following lines or pairs of points:

17. (−1, 1), (0, −2)
18. (−3, 5), (5, 0)
19. (−2, −1), (5, 7)
20. $y = 2x$
21. $y - 3 = \frac{1}{2}x$
22. $3y - 1 = x$

23. Using slopes, determine if (−2, 3), (−3, −2), and (9, 6) are collinear.
24. Using slopes, determine if $P(3, 5)$, $Q(0, 6)$, and $R(-3, -5)$ are the vertices of a right triangle.
25. Find the slope of the line perpendicular to the line $6x + 5y - 7 = 0$.

10-4 Equation of a line

In each of the following exercises, write the equation of the line using the information given:

26. $m = \frac{2}{3}$, (2, −8)
27. (−7, 6), (3, 9)
28. $m = -\frac{2}{5}$, $b = -\frac{1}{2}$
29. $a = 7$, $b = -2$
30. $m = 2$, (3, 0)
31. $(\frac{1}{2}, 0)$, $(0, -\frac{1}{2})$

Find the slope and intercepts of the lines with the following equations:

32. $2x - 3y = 6$
33. $5x - 3y = 15$
34. $7x - 12y = 16$
35. $5x + \frac{1}{2}y - 2 = 0$
36. $5y = 7x - 1$
37. $3y + 7 = 2x$

38. Write the equations of the lines containing the diagonals of the quadrilateral $ABCD$, with $A(1, 7)$, $B(5, 3)$, $C(3, 0)$, and $D(-1, 5)$.
39. Write the equation of the perpendicular bisector of $\overline{AB}$ for $A(2, 3)$ and $B(-2, 5)$.

10-5 Circles and analytic proofs

In each exercise below, write an equation for the circle with center at (0, 0) and with the radius given.

40. 3.5
41. $2\sqrt{3}$
42. $\frac{1}{2}$
43. 1

In each of the exercises below, write an equation for the circle with the center and radius given.

44. (7, 4), 2
45. (0, −2), 3
46. (−2, −3), 5
47. (0, 0), $\frac{1}{4}$

48. Prove that the square of the hypotenuse of a right triangle is equal to the sum of the squares of the other two sides.

CHAPTER TEST

1. Draw the graphs of $y = 3x - 5$ and $y = \frac{1}{3}x + 5$. Name their point of intersection.
2. Locate the points that divide the segment whose endpoints are $(-12, 0)$ and $(12, 0)$ into four congruent segments.
3. Using slope, prove that any angle subtended in a semicircle is a right angle.
4. Write the equation of the line through $(1, 1)$ that has the same x-intercept as the line $2x + 3y = 1$.
5. Write the equation of the circle with center at $(3, -2)$, if $(0, 5)$ is a point of the circle.
6. If line l through $(3, 2)$ has an angle of inclination of $45°$, such that $\tan 45° = 1$, write the equation for l.
7. Show in two different ways that $(1, 1)$, $(3, -\frac{1}{3})$, and $(-2, 3)$ are collinear.
8. Find the perimeter of the triangle formed by connecting the midpoints of the sides of $\triangle ABC$, with $A(5, 0)$, $B(-3, 2)$, and $C(-2, -6)$.
9. Prove that the sum of the squares of the sides of a parallelogram equals the sum of the squares of the diagonals of the parallelogram.

SUGGESTED RESEARCH

1. The graphs of the second-degree equations usually studied in a course of analytic geometry date back to Apollonius, a Greek who lived about 250 B.C. Explore the history-of-mathematics books in your library and prepare a report on the work of Apollonius.

2. Research the comparative contributions of Pierre de Fermat and René Descartes to the study of analytic geometry. Report your findings to your class.

3. Describe the conic sections developed by Apollonius. Collect pictures showing examples of these curves in architecture and share them with your class.

11 Loci and Constructions

Throughout the year we have referred to sets of points satisfying given conditions. For example, the set of all points equidistant from the endpoints of a segment lies in the perpendicular bisector of the segment. This set of points forms a locus of points. We have also made reference to forming or constructing a required geometric figure. In this chapter, we shall formally study loci (plural of *locus*) and geometric constructions.

11-1 LOCUS

Consider the tip of the sweep hand of a stopwatch. As the hand travels, the tip passes through many different positions. The different positions it passes through in one full sweep form the circumference of a circle. To each position there corresponds a point. The set of all points corresponding to these positions is the locus determined by the tip of the sweep hand.

D **Definition 11-1** A **locus** is the set of points, and only those points, that satisfy a given condition.

Example 1 The Perpendicular Bisector Theorem states that the set of all points in a plane equidistant from the endpoints of a segment is the perpendicular bisector of the segment. What is the locus described by this theorem and what condition must the points of the locus satisfy?

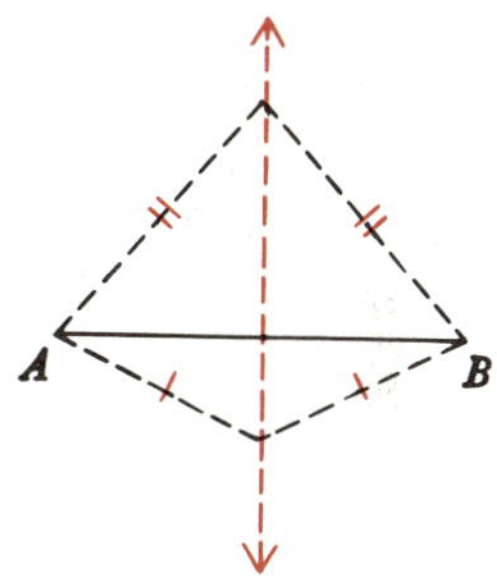

Solution The locus, or set of points, is the perpendicular bisector of the segment.

Example 2 Describe the locus which is the set of points in a plane lying a distance of 3 units from a line in the plane. What is the locus if we remove the condition that the given points and line lie in the same plane?

Solution All points 3 units from the line on one side and all points 3 units from the line on the other side satisfy the given condition. The locus is a pair of lines lying on opposite sides of the given line, each parallel to the given line.

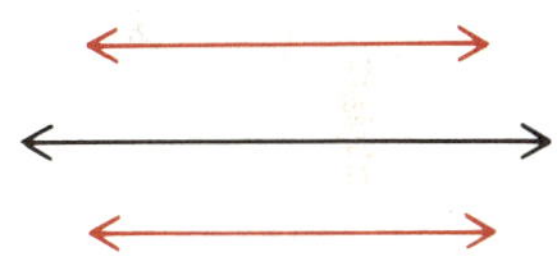

If the points are not restricted to a plane, all the points in space 3 units from the given line are the locus. The locus is then a cylindrical surface around the line.

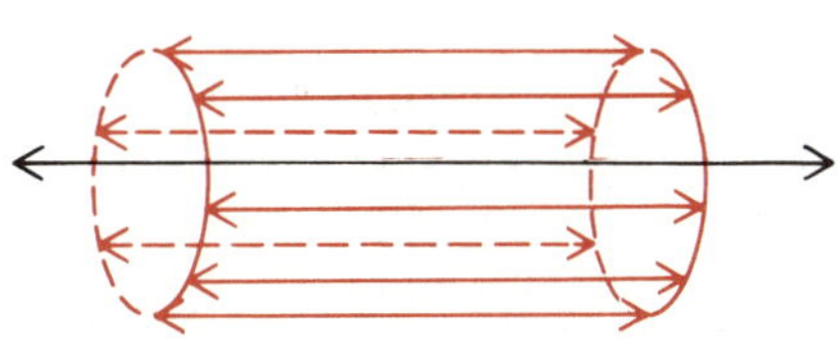

Theorems about loci require the proof of a statement and its converse. The general form of the two parts is:

1. Any point of the locus satisfies certain given conditions;
2. Any point satisfying the given conditions is in the locus.

T

Theorem 11-1.1 The set of points in the interior of an angle and equidistant from the sides is the bisector of the angle, exclusive of the vertex.

PROOF OF (1)

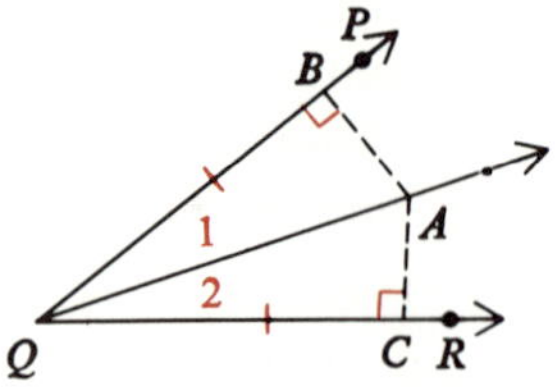

GIVEN A is any point in the interior of $\angle PQR$ equidistant from $\overrightarrow{PQ}$ and $\overrightarrow{QR}$.

PROVE A lies in the bisector of $\angle PQR$.

Using Definition 4-5, we draw $\overline{AB} \perp \overrightarrow{QP}$ and $\overline{AC} \perp \overrightarrow{QR}$. $AC = AB$. Why? By Theorems 2-6.5 and 3-1.1, $\angle ABQ \cong \angle ACQ$. By Theorem 3-1.6 we know that $\overline{AQ} \cong \overline{AQ}$. Hence, $\triangle ABQ \cong \triangle ACQ$ by Theorem 6-6.2. From Definition 3-3, we conclude that $\angle 1 \cong \angle 2$, and hence $\overrightarrow{QA}$ is the bisector of $\angle PQR$.

PROOF OF (2)

GIVEN Point A, distinct from Q, lies in the bisector of $\angle PQR$.

PROVE Point A is in the interior of $\angle PQR$ and is equidistant from $\overrightarrow{QP}$ and $\overrightarrow{QR}$.

The fact that point A is in the interior of $\angle PQR$ follows immediately from the given and Definition 1-29. We draw $\overline{AB} \perp \overrightarrow{QP}$ and $\overline{AC} \perp \overrightarrow{QR}$ using Definition 4-5. Theorems 2-6.5 and 3-1.1 tell us that $\angle ABQ \cong \angle ACQ$. $\overline{QA} \cong \overline{QA}$ by Theorem 3-1.6. Thus, by Theorem 6-6.1, $\triangle ABQ \cong \triangle ACQ$. We know that $\overline{AB} \cong \overline{AC}$ by Definition 3-3, so point A is equidistant from $\overrightarrow{QP}$ and $\overrightarrow{QR}$.

Class Exercises

Draw a model of the descriptions in the exercises and describe the loci obtained. Assume that the points of each locus lie in a plane.

1. The points equidistant from two intersecting lines.
2. The points equidistant from two parallel lines.
3. The points equidistant from the vertices of a rectangle.
4. The points equidistant from the centers of two circles.
5. The points equidistant from the three vertices of a triangle.
6. In the Cartesian coordinate system, the points 7 units from $(-1, -3)$.

Connectives

The points in many loci must satisfy several conditions or alternate conditions. These **compound loci** are described by statements of conditions connected by *and* or *or*. Where *and* connects the statements, the locus can be formed by:

1. drawing the locus of the first condition;
2. drawing the locus of the second condition; and
3. describing the locus that satisfies both conditions, namely the intersection of the two sets of points.

Example 3 Describe the points of a plane that lie 3 units from a given line and that are also equidistant from two distinct points of the line.

Solution Considering the first condition, the locus for points 3 units from a given line is a pair of lines parallel to m, one on each side of m.

To find the points satisfying the second condition, let A and B be any two distinct points on m. From Example 1, we know that the locus for points equidistant from A and B is the perpendicular bisector of $\overline{AB}$.

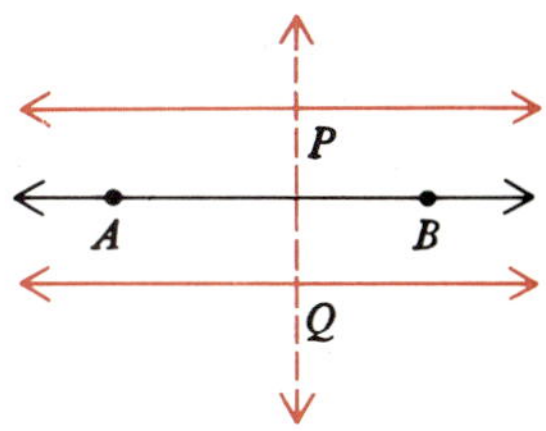

The compound locus is the intersection of the two loci. Thus, the points in the locus described are points P and Q.

Where *or* connects two statements of condition, the compound locus can be found by:

1. drawing the locus for the first condition;
2. drawing the locus for the second condition; and
3. describing the locus that satisfies either condition, namely, the union of the two sets of points.

Example 4 Sketch on a coordinate graph the locus of points whose x-coordinate is greater than 5 or less than -5.

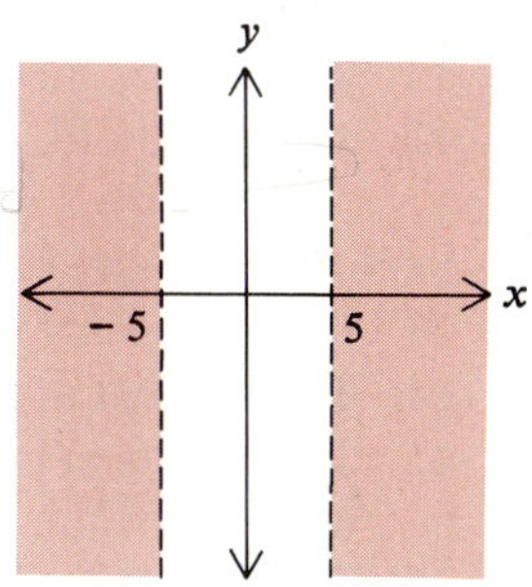

Solution (1) We sketch the line $x = -5$; (2) we then sketch the line $x = 5$. The points we need for the compound locus lie to the left of the line $x = -5$ and to the right of the line $x = 5$. Both sets of points belong to the locus described.

In Example 4, two conditions are specified for the locus, but individual points do not have to satisfy both conditions to be in the compound locus. Each point must, however, be in at least one of the loci.

Using what we learned in the last chapter, we can also describe loci by equations.

Example 5 Write an equation for the locus of points equidistant from $(-3, 2)$ and $(7, 6)$.

Solution From the examples above, we know that this locus is the perpendicular bisector of the segment determined by the points $(-3, 2)$ and $(7, 6)$. The slope of this segment is $\frac{6-2}{7+3}$ or $\frac{2}{5}$. Therefore the slope of the locus is $-\frac{5}{2}$. The locus includes the midpoint of the segment. This midpoint is $\left(\frac{-3+7}{2}, \frac{2+6}{2}\right)$ or $(2, 4)$. Using the point-slope form of the linear equation, we can write the locus as $y - 4 = -\frac{5}{2}(x - 2)$

EXERCISES

A

Draw and describe the locus of each set of points in a plane described below.

1. The points equidistant from two parallel lines.
2. The points equidistant from two perpendicular lines.
3. The points in the interior of a circle with a radius of 7.
4. The points in the exterior of a circle with a radius of 2.
5. The points satisfying both the conditions of Exercise 3 and Exercise 4, when the circles have the same center.
6. The third vertex of a triangle with base $\overline{AB}$ and a height of 3 units.
7. The centers of the circles tangent to $\overline{AB}$ at C.

Sketch the locus of each set of points in a plane described below.

8. The points 3 units from the origin.
9. The points less than 3 units from the origin.
10. The centers of the circles tangent to two intersecting lines.
11. The points more than 6 units from point P and less than 6 units from P.
12. The points of $\overline{AB}$ that are equidistant from A and B.

B

Sketch the locus of each set of points below.

13. The centers of the circles that have a radius 4 and pass through the origin.
14. The centers of the circles that touch a given circle at a given point.
15. The points in the coordinate plane 2 units from the x-axis and less than 3 units from the y-axis.
16. The points in the coordinate plane whose abscissas are greater than 5 and whose ordinates are less than -3.
17. The points in the interior of the circle with point A as center and radius 5 and the points in the exterior of the circle with A as center and radius 4.
18. The points in the coordinate plane 2 units from the x-axis or less than 3 units from the y-axis.
19. The points (x, y) in the coordinate plane whose abscissas are $|x| \geq 3$ or whose ordinates are $|y| \geq 3$.

Write the equation for the locus of each of the following:

20. The points equidistant from (2, 4) and (9, 6).
21. The points of the circle with center $(7, -2)$ where (8, 0) is on the circle.
22. The points of the circle with $(3, -5)$ and (7, 1) as endpoints of a diameter.

C

23. Draw the locus of the points in a plane passed through by a point of a small circle rolling around the interior edge of a circle whose radius is at least three times that of the smaller circle.
24. Describe how you would locate a point in $\overleftrightarrow{PQ}$ of $\triangle PQR$ that is equidistant from P and R.
25. Describe how you would locate a point of side $\overline{PQ}$ of $\triangle PQR$ equidistant from $\overline{QR}$ and $\overline{PR}$.
26. Describe the position of the point $(-5, 12)$ to the locus of points lying more than 13 units from the origin.
27. Describe how you might locate the midpoint of a line segment using geometric procedures.

A LOOK AT THE PAST

Euclidean Constructions

In *The Elements*, Euclid specified three constructions to be used in drawing geometric figures: the drawing of a line segment between any two distinct points; the extension of this segment to form a straight line; and the construction of a circle with a given center and a given radius. While Euclid does not mention the tools used for these constructions, the philosopher Plato tells us that only a compass and a straightedge were used. The compass apparently used by Euclid had one drawback—it could not be locked into one position, but was collapsible. Yet from one of his constructions in the first book of *The Elements*, we can conclude that his collapsible compass could be used to do the same thing our compass will.

Using only the compass and straightedge, the Greek mathematicians succeeded in performing almost every possible construction. They discovered only four that they could not perform:

1. The inscription of any regular polygon of n sides in a given circle. They were able to inscribe polygons for certain values of n—for example, for $n = 3, 4, 5$, and 6—but not for any n;
2. The trisection of any given angle;
3. The construction of a square with an area equal to that of a given circle;
4. The construction of a cube with twice the volume of a given cube.

11-2 COMPASS AND STRAIGHTEDGE

The ruler is a tool for measuring and drawing segments. When we use the ruler merely for drawing, we need only the straight edge, not the scale. Hence, when we refer to a **straightedge**, we mean a ruler without any scale.

Likewise we do not need the scale of a ruler to show that segments are congruent. One other instrument we can use is the **divider**. Dividers have two metal legs. We can adjust the distance between these legs and then hold the distance.

A similar instrument is the **compass**. The difference between the compass and the divider is that the compass has a sharpened pencil point as one of its legs.

To use either the compass or the divider to show that two segments are congruent, place one leg at an endpoint of a given segment. Adjust the legs until the tip of the other leg or pencil rests on the opposite endpoint of the segment. If we lock the compass or divider in this position, then any segment whose endpoints match the tips of the adjusted legs is congruent to the given segment.

We can also use the compass to draw circles and arcs of circles with various radii. The following examples show how we can use the compass in various constructions.

Example 1 Draw a segment on $\overleftrightarrow{AB}$ congruent to $\overline{CD}$.

Solution Adjust the legs of the compass so that the metal tip rests on C and the pencil point rests on D. Rest the metal tip of the adjusted compass on A and make an arc with the pencil tip on the right or left side of A. Label the intersection of the arc with $\overleftrightarrow{AB}$ as P. Then $\overline{AP}$ is congruent to $\overline{CD}$.

Example 2 Draw a circle using the length RS as radius and R as center.

Solution Adjust your compass so that the metal tip rests on R and the pencil tip rests on S. Then rotate the pencil leg around the metal leg at R, holding the metal leg stationary.

Example 3 Construct a line segment whose length is equal to $AB + CD$.

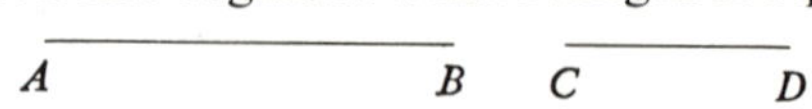

Solution On any line $\overleftrightarrow{PQ}$ select a point and call it A'. To the right of A', locate point B' with your compass so that $\overline{A'B'} \cong \overline{AB}$. Next, with B' as an endpoint, use your compass to locate D' to the right of B' such that $\overline{B'D'} \cong \overline{CD}$.

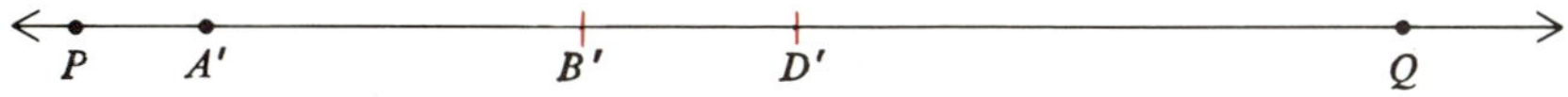

Points A', B', and D' are distinct points of $\overline{PQ}$ such that B' is between A' and D'. Thus, $A'B' + B'D' = A'D'$. But $A'B' = AB$ and $B'D' = CD$. Thus

$$AB + CD = A'D'$$

EXERCISES

A All the exercises below refer to the segments shown in Exercises 1–6. Construct a segment congruent to each of the segments below.

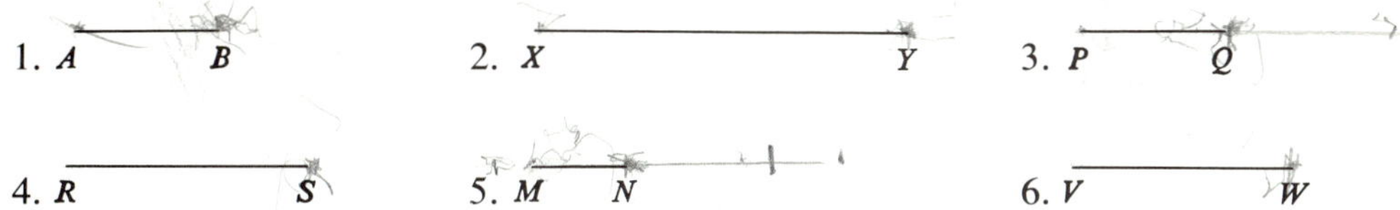

Construct a segment with a length equal to each of the following:

7. $XY + PQ$
8. $XY + MN + PQ$
9. $MN + AB + RS$
10. $VW + AB + PQ$
11. $XY - AB$
12. $XY - AB + MN$

Construct the circles with the following radii:

13. AB
14. XY
15. PQ

B Construct the segments with the following lengths:

16. $2(AB)$
17. $3(MN)$
18. $2(AB) - 2(MN)$
19. $3(AB + MN)$
20. $4(MN) - 2(MN)$

C Construct a circle for each of the following radii:

21. $2(PQ)$
22. $3(MN)$
23. $AB - MN$
24. $PQ + MN$

11-3 BASIC CONSTRUCTIONS

Before considering other constructions, we must introduce a new term. We call the distance between the tips of the legs of a compass the **radius of the compass**. Besides this term, we need the following statement. While it can be proved, the proof would draw too much attention from our main concern. Therefore, we shall postulate the statement and forego the proof.

P

Postulate 11-1 Two-Circle Postulate Given circle A with radius a, circle B with radius b, and the length of their line-segment of centers $\overline{AB}$, such that $AB = c$; if each of the numbers a, b, c is less than the sum of the other two, then the circles intersect in exactly two points which lie on opposite sides of $\overline{AB}$.

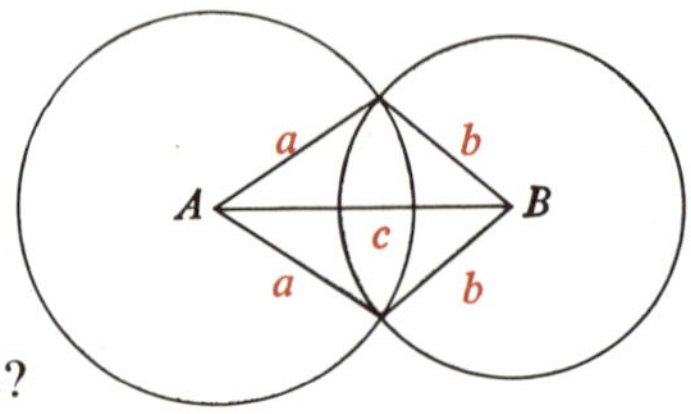

$a + b > c$

$a + c > b$

$b + c > a$

If $a + b = c$, what can we say about the circles?

Example 1 Construct $\triangle ABC$ given sides with the lengths of the following three segments.

Solution On any line, use your compass to construct a segment congruent to c. Label this $\overline{AB}$. Next, using A as center and a compass radius congruent to b, construct an arc above $\overline{AB}$.

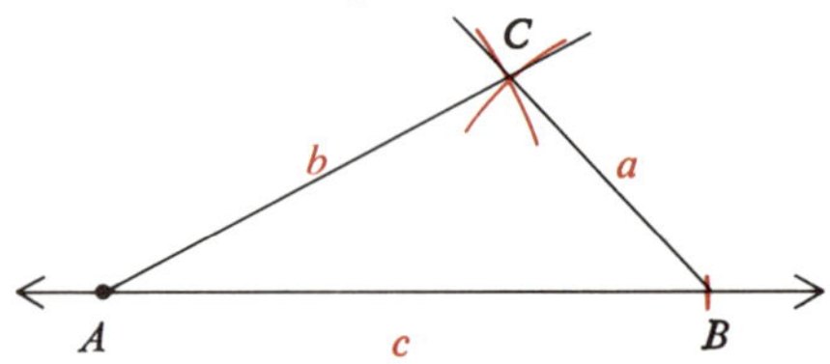

Then, using B as center and a compass radius congruent to a, construct an arc above $\overline{AB}$ towards A. Label the point where these arcs intersect C. If you draw $\overline{AC}$ and $\overline{BC}$ you will have constructed $\triangle ABC$.

Construction 1 Construct the perpendicular bisector of a given segment.

GIVEN $\overline{XY}$

WANTED Perpendicular bisector of $\overline{XY}$

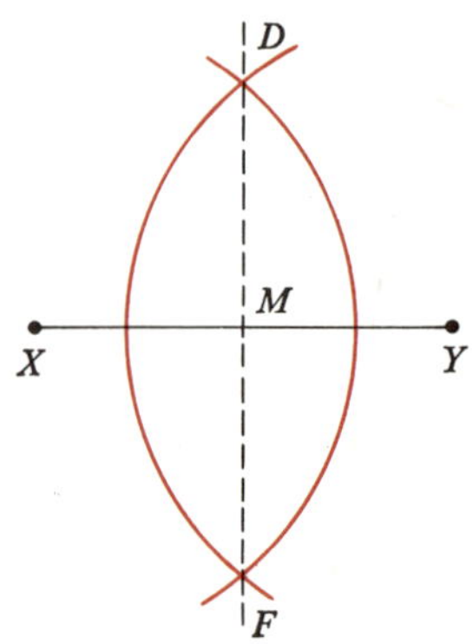

CONSTRUCTION

1. Adjust the compass so that the radius of the compass is more than half $\overline{XY}$.
2. With X as center construct an arc of the circle extending to each side of $\overline{XY}$.
3. Repeat step 2 using Y as center. By the Two-Circle Postulate (11-1), since the sum of the radii is greater than XY, the two circles intersect in two points, one on each side of $\overline{XY}$. Label these points D and F.
4. By the Line Postulate (2-3), D and F determine $\overleftrightarrow{DF}$. $\overleftrightarrow{DF}$ is the perpendicular bisector of $\overline{XY}$ since D and F are both equidistant from the endpoints of $\overline{XY}$. (C4-4.3a)

The next construction adds only one more step to the last construction. Use the illustration for Construction 1.

Step 5. Label the point of intersection between $\overline{XY}$ and $\overleftrightarrow{DF}$, M. M is the midpoint of $\overline{XY}$.

Construction 2 Bisect a given segment.

The following constructions follow from Construction 1. Supply the steps of the construction after studying the figure.

Construction 3 Construct a line perpendicular to a given line through a given point of the line.

GIVEN Point R on line m

WANTED The perpendicular to m at R

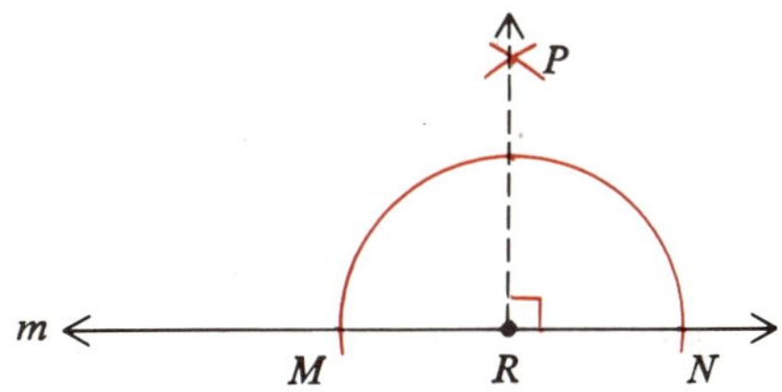

Construction 4 Construct a perpendicular to a line from a point not in the line.

GIVEN Point P not in line m

WANTED The perpendicular to m through P

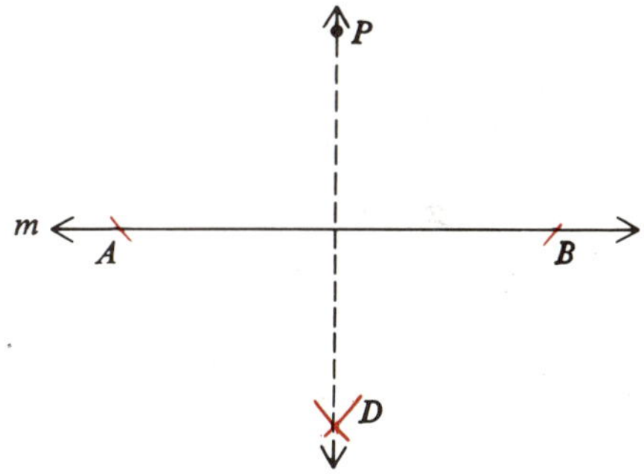

Example 2 Construct the perpendicular to $\overline{AB}$ at B.

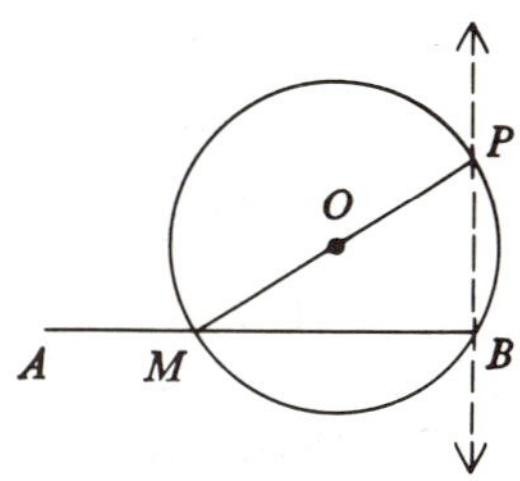

Solution Using a point O above $\overline{AB}$ as center and $\overline{OB}$ as radius, construct $\odot O$. Point O should be located so that $\odot O$ intersects $\overline{AB}$ at a point M between A and B. From M, draw $\overrightarrow{MO}$. Label the second intersection of $\overrightarrow{MO}$ and circle O, point P. Since chord $\overline{MP}$ passes through the center of the circle, $\overline{MP}$ is a diameter. Angle MBP is thus inscribed in a semicircle. Hence, $\angle MBP$ is a right angle. Therefore, $\overleftrightarrow{BP}$ is the only perpendicular to $\overline{AB}$ at B.

EXERCISES

A All constructions in the following exercises are to be performed using only compass and straightedge.

Construct each of the following segments on your paper and then construct the perpendicular bisector of each:

1. 

2. 

3. Construct the perpendicular to $\overline{RS}$ at T.
4. Construct the perpendicular to $\overline{PQ}$ at Q.
5. Using the segment $\overline{RT}$ in Exercise 2 as 1 unit, draw a line and construct a segment 13 units long, marking each unit with your compass.

Using the unit scale constructed in Exercise 5, construct the triangles with sides of the following lengths:

6. 7, 5, 6 7. 5, 5, 7
8. Construct a square with $\overline{PQ}$ from Exercise 1 as a side.

Construct each of the following triangles and then construct a median of each triangle from any vertex of the triangle:

9. 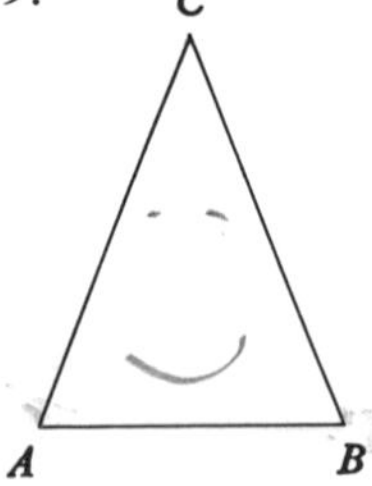

10. 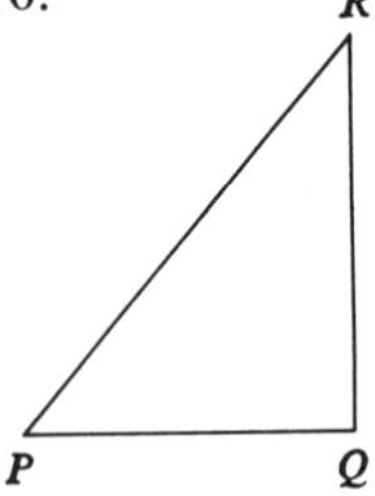

11. 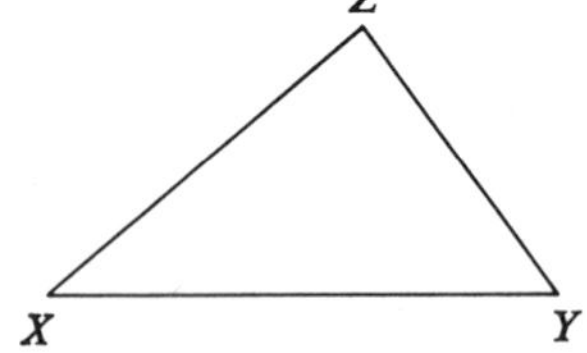

Using your copy of the triangles from Exercises 9–11, construct an altitude from the vertex denoted by the last letter in the name of each triangle below.

12. $\triangle XYZ$ 13. $\triangle PQR$ 14. $\triangle ABC$

B Use the unit scale constructed in Exercise 5 for each of the following:

15. Construct an isosceles triangle with a base 5 units long and congruent sides 7 units long.
16. Construct an isosceles triangle with a base 9 units long and congruent sides 6 units long.
17. Construct an isosceles triangle with a base 5 units long and an altitude 6 units long.
18. Construct an equilateral triangle with sides 3 units long.
19. Construct rhombus $ABCD$, if $AC = 9$ and $AD = 5$.
20. Construct an isosceles right triangle.

C

21. Given any triangle, construct the three medians.
22. Repeat Exercise 21, using a different triangle.
23. What conclusion can you draw from Exercises 21 and 22?
24. Given an acute triangle, construct the three altitudes.
25. Repeat the construction of Exercise 23, using an obtuse triangle. You may have to extend two of the sides.
26. What conclusion can you draw from Exercises 24 and 25?

Using the constructions of this section and your knowledge of the relationships between the sides and angles of certain triangles, construct angles with the following measures:

27. 45 28. 60 29. 30

11-4 ANGLE CONSTRUCTIONS

We have already demonstrated several basic constructions involving lines and segments. To continue our study, we must learn how to construct an angle congruent to a given angle.

Construction 5 Construct an angle congruent to a given angle.

GIVEN $\angle ABC$

WANTED An angle congruent to $\angle ABC$

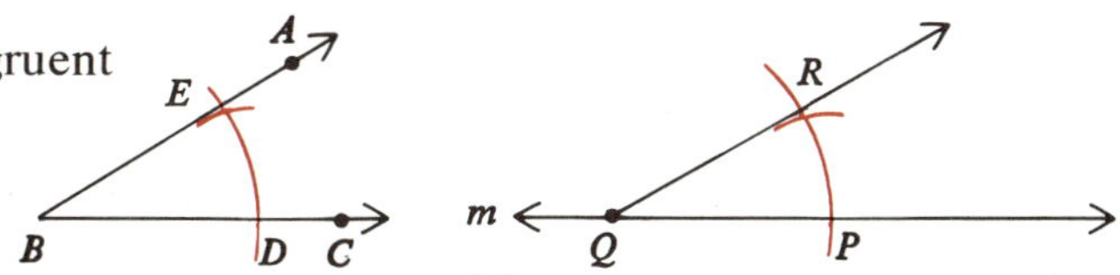

CONSTRUCTION

1. Using B as center and any radius that allows you room to work, construct an arc intersecting $\overrightarrow{BC}$ at D and $\overrightarrow{BA}$ at E.
2. Choose any point Q on any line m.
3. Using Q as center and the same radius as in Step 1, construct an arc above and through line m, intersecting m at P.
4. Using D as center, adjust your compass to make an arc through E.
5. Use this new radius and P as center to construct an arc above and through m, intersecting the first arc at R.
6. Draw $\overrightarrow{QR}$. From the methods of construction, $\overline{BE} \cong \overline{QR}$, $\overline{BD} \cong \overline{QP}$, and $\overline{DE} \cong \overline{PR}$. Thus, by the SSS Postulate (3-3), $\triangle BDE \cong \triangle QPR$. By the definition of congruent triangles (3-3), $\angle RQP \cong \angle ABC$.

We can also demonstrate angle addition by a construction.

Construction 6 Construct an angle whose measure is equal to the sum of the measures of two given angles.

GIVEN $\angle X$ and $\angle Y$

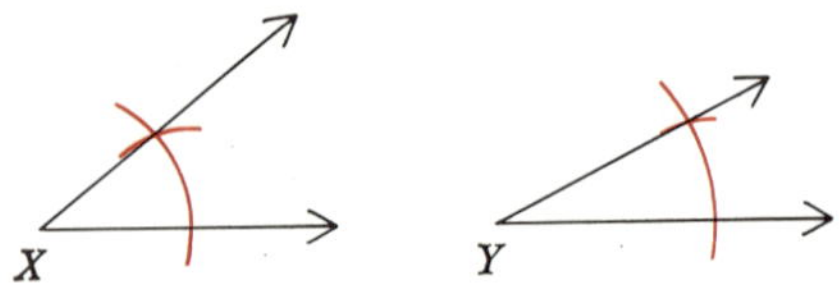

WANTED An angle with a measure of $m\angle X + m\angle Y$

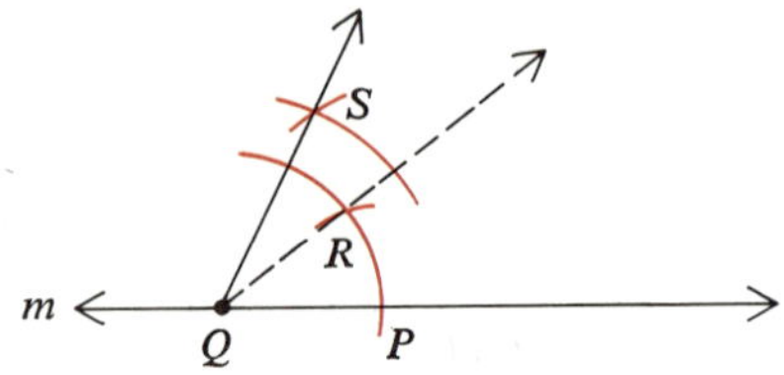

CONSTRUCTION

1. On any line m choose a point Q.
2. At Q construct $\angle PQR \cong \angle X$.
3. Using $\overrightarrow{QR}$, construct $\angle RQS \cong \angle Y$.
4. Since $\overrightarrow{QR}$ is in the interior of $\angle PQS$, and $m\angle PQR + m\angle RQS = m\angle PQS$, $\angle PQS$ is the required angle.

Class Exercises

For Exercises 1–4, we are given line m and point P not in m.

1. Let M and Q be any two points of m with Q to the right of M. Draw $\overleftrightarrow{MP}$.
2. Let R be any point of $\overleftrightarrow{MPR}$ on the opposite side of P from M. On $\overrightarrow{PR}$ construct $\angle RPN$ with N to the right of P, such that $\angle RPN \cong \angle PMQ$.

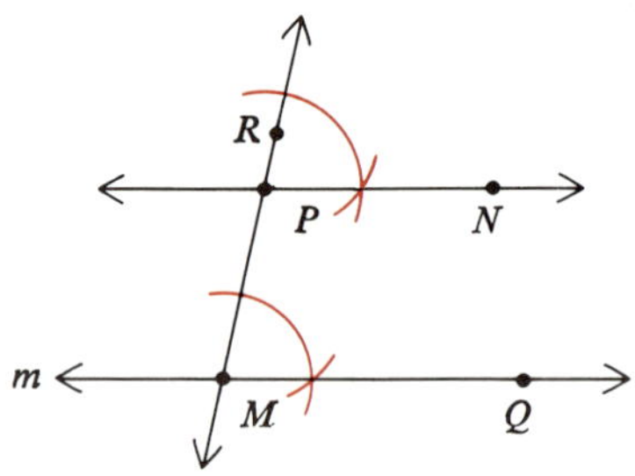

Exercises 1–4

3. $\overleftrightarrow{PN}$ and $\overleftrightarrow{MQ}$ have $\overleftrightarrow{MPR}$ as transversal. What is the angle relationship between $\angle RPN$ and $\angle PMQ$?
4. Since $\angle RPN \cong \angle PMQ$, $\overleftrightarrow{MQ}$ is parallel to $\overleftrightarrow{PN}$. Why?

Construction 7 Construct a line parallel to a given line through a point not on the line.

Exercises 1–4 form the four steps this construction.

Construction 8 Bisect a given angle.

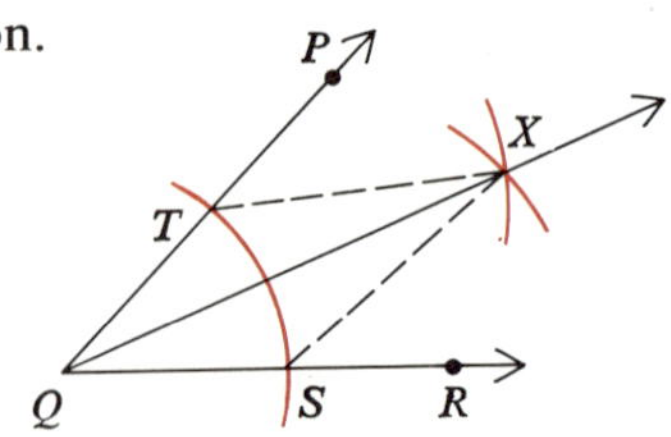

GIVEN $\angle PQR$

WANTED Bisect $\angle PQR$

Construction

1. Using Q as center and any radius, construct an arc of a circle that intersects $\overrightarrow{QR}$ at S and $\overrightarrow{QP}$ at T. Since radii of a circle are equal, $QS = QT$.
2. With S as center and a radius of more than half ST, make an arc in the interior of $\angle PQR$.
3. Repeat step 2, using the same radius and T as center. Label the point of intersection of the two arcs X. Using the same radius, $SX = TX$.
4. Draw $\overrightarrow{QX}$. This ray bisects $\angle PQR$. We know this since, by our construction, $QS = QT$, $TX = SX$, and $QX = QX$. Thus, by the SSS Postulate (3-3), $\triangle QTS \cong \triangle QSX$. Therefore, by the definition of congruent triangles (3-3), $\angle XQT \cong \angle XQS$.

In the last section, we constructed a triangle given its three segments. The next two examples show us how to construct a triangle given a combination of angles and sides, totalling three given parts.

Example 2 Construct the triangle with sides of lengths a and b and the included angle $\angle C$.

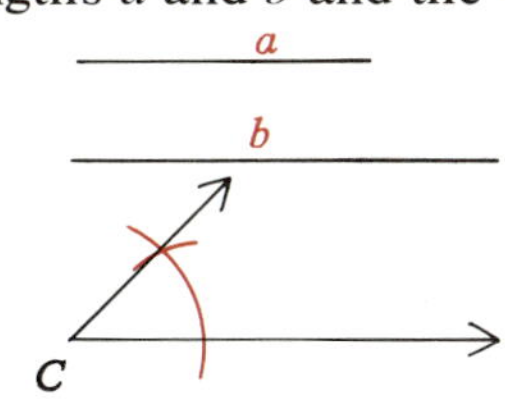

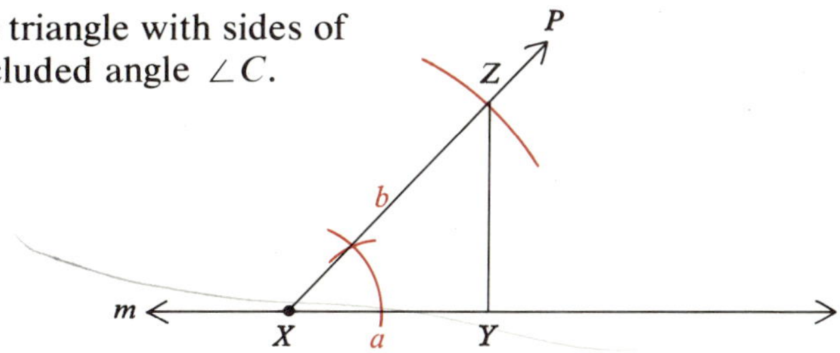

Solution On any horizontal line m choose a point and call it X. To the right of X, construct $\overline{XY}$ such that $XY = a$. Next, at X on m, construct an angle, $\angle YXP$, such that $\angle C \cong \angle YXP$. On $\overrightarrow{XP}$ construct $\overline{XZ}$ such that $XZ = b$. By drawing $\overline{YZ}$, we form the required triangle.

Example 3 Construct the triangle with $\angle A$ and $\angle B$ and the included side of length c.

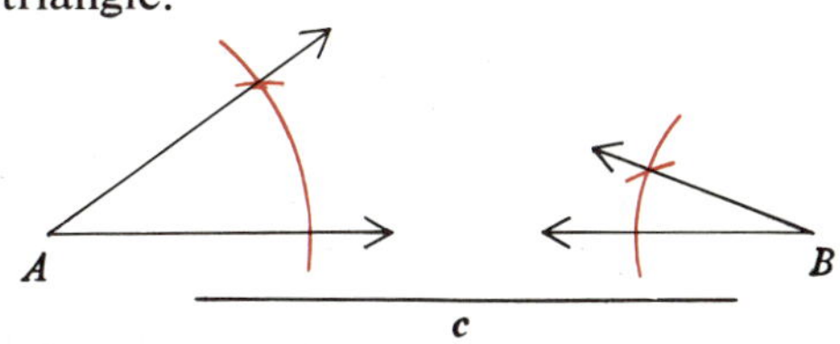

Solution On any line m, construct $\overline{XY}$ such that $XY = c$. At X construct $\angle YXP \cong \angle A$, and at Y construct $\angle XYR \cong \angle B$. Extend $\overrightarrow{XP}$ and $\overrightarrow{YR}$ until they intersect, and call the intersection Z. $\triangle XYZ$ is the required triangle.

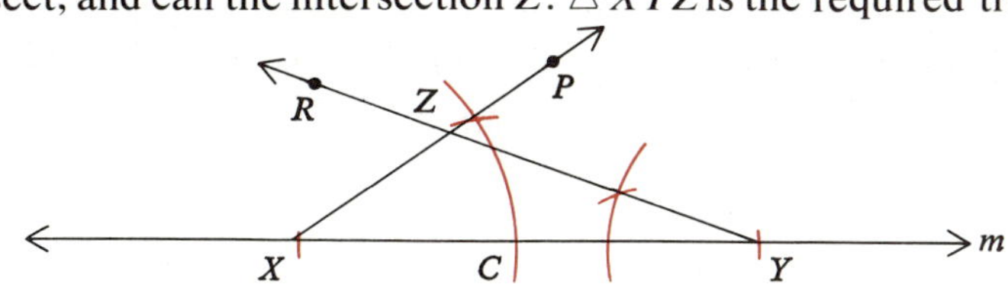

EXERCISES

A Constructions are to be performed with compass and straightedge only. Trace each angle below. Then construct an angle congruent to the given angle.

1. 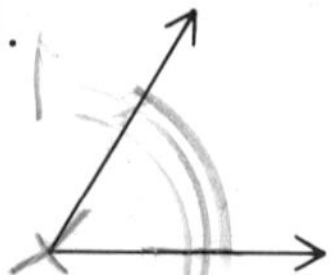

2. 

3. 

Refer to the angles for Exercises 1–3 and construct an angle with a measure equal to the sum of the measures of the two angles in each exercise below.

4. $\angle 1$, $\angle 3$ 5. $\angle 1$, $\angle 2$ 6. $\angle 3$, $\angle 2$

Redraw the figures for Exercises 7–8. Then construct a line parallel to each of the given lines through the external points indicated.

7. 

8. 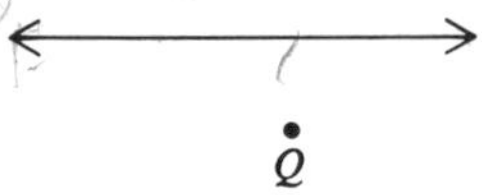

Construct the triangle with the sides and included angle given in each of the following exercises:

9. 

10. 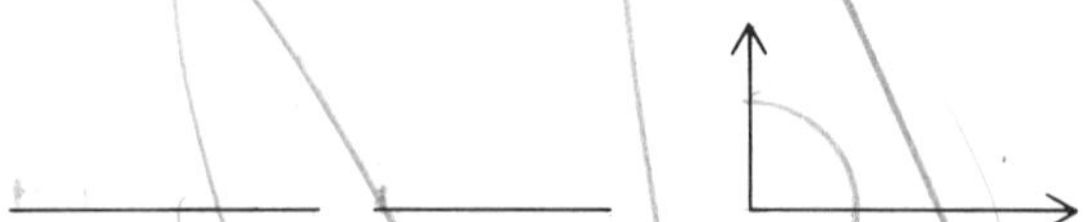

11. 

B Construct the triangle with the angles and included sides given in each of the following exercises:

12. 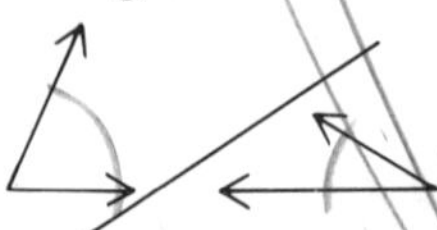

13. 

14. 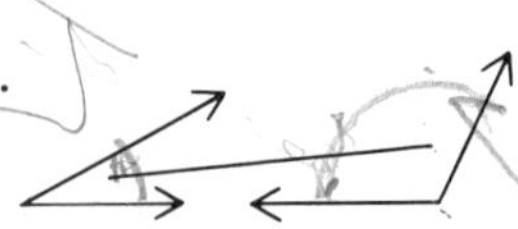

15. Construct $\angle A$ such that m$\angle A = 135$.
16. Construct an isosceles triangle given an altitude of 3 units and the base angle you used in Exercise 9. Use ____ as the unit of linear measure.
17. Construct a parallelogram for any given angle, side, and altitude to the side.
18. Draw an obtuse angle and divide it into four congruent angles.

C 19. Bisect the angles of any acute $\triangle ABC$.
20. Bisect the angles of any right $\triangle PQR$.

21. Bisect the angles of any obtuse $\triangle MNO$.
22. What conclusion can you draw from Exercises 19–21?

11-5 PARTITIONING CONSTRUCTIONS

Many other constructions follow from the basic constructions we demonstrated in the last two sections. The segment-partitioning constructions follow from the construction of parallel lines.

Construction 9 Divide any segment into three congruent segments.

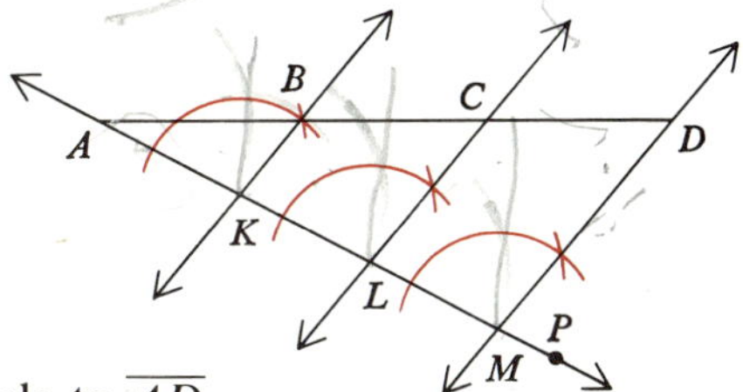

GIVEN $\overline{AD}$

WANTED The partition of $\overline{AD}$ into three congruent segments

CONSTRUCTION
1. Draw $\overleftrightarrow{AP}$ through A at any angle to $\overline{AD}$.
2. Using any appropriate radius of your compass, construct three consecutive congruent segments on $\overline{AP}$ starting at A, such that $AK = KL = LM$.
3. Draw $\overleftrightarrow{MD}$.
4. Construct $\overleftrightarrow{KB} \parallel \overleftrightarrow{MD}$ and $\overleftrightarrow{LC} \parallel \overleftrightarrow{MD}$.
5. If three or more parallel lines intercept congruent segments on one transversal, then they intercept congruent segments on any other transversal. (Theorem 7-6.1.) Thus, $AB = BC = CD$.

Construction 10 Partition any segment into parts proportional to given segments.

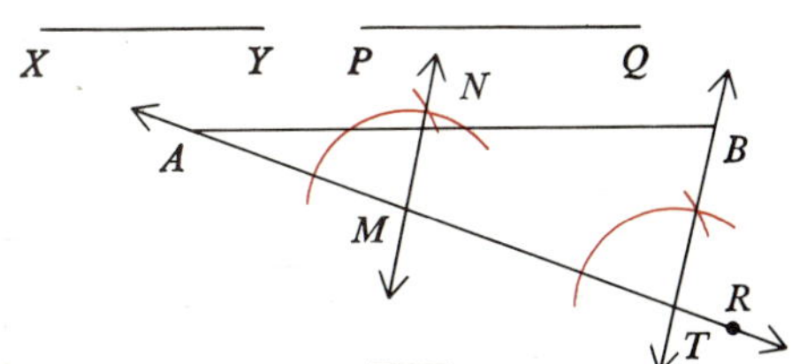

GIVEN $\overline{AB}$, $\overline{XY}$, and $\overline{PQ}$

WANTED Segments proportional to $\overline{XY}$ and $\overline{PQ}$ on $\overline{AB}$

CONSTRUCTION
1. Draw $\overleftrightarrow{AR}$ through A at any convenient angle to $\overline{AB}$.
2. On $\overrightarrow{AR}$, construct $\overline{AM} \cong \overline{XY}$ and $\overline{MT} \cong \overline{PQ}$ so that AMT.
3. Draw $\overline{TB}$.
4. Construct $\overline{MN}$ parallel to $\overline{TB}$ through M.
5. In a triangle ($\triangle ATB$ in this case) a line through two sides and parallel to the third side divides the two sides proportionally, by Corollary 8-2.1a.

$$\frac{AN}{NB} = \frac{AM}{MT} \quad \text{or} \quad \frac{AN}{NB} = \frac{XY}{PQ}$$

Construction 10 provides us with a technique for construction proportionals of given segments.

Construction 11 Construct the fourth proportional of three given segments.

GIVEN $\overline{AB}$, $\overline{CD}$, and $\overline{EF}$

WANTED $\overline{PQ}$ such that $\dfrac{AB}{CD} = \dfrac{EF}{PQ}$

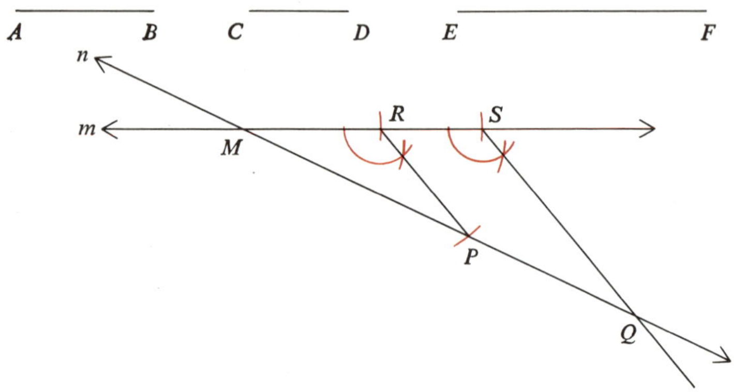

CONSTRUCTION

1. Construct any two intersecting lines m and n intersecting at M.
2. On m construct $\overline{MR}$ and $\overline{RS}$ congruent to $\overline{AB}$ and $\overline{CD}$, respectively.
3. On n construct $\overline{MP} \cong \overline{EF}$.
4. Draw $\overline{RP}$.
5. Through S construct $\overline{SQ}$ parallel to $\overline{RP}$. By Corollary 8-2.1a, $\overline{PQ}$ is such that $\dfrac{AB}{CD} = \dfrac{EF}{PQ}$.

Construction 12 Construct the mean proportional between two given segments.

GIVEN $\overline{AB}$ and $\overline{CD}$

WANTED $\overline{PQ}$ such that $\dfrac{AB}{PQ} = \dfrac{PQ}{CD}$

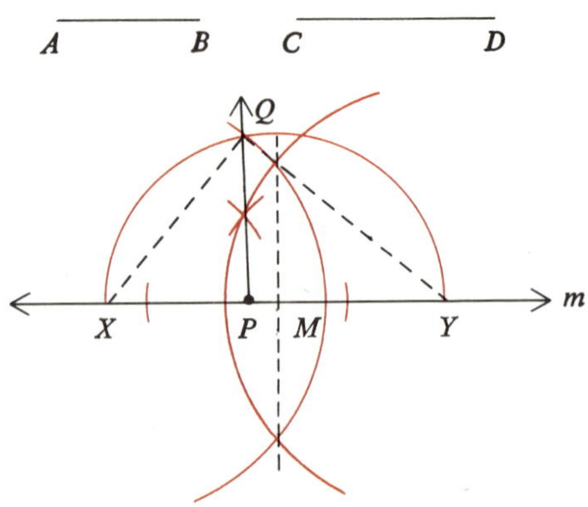

CONSTRUCTION

1. Choose a point P on any line.
2. On each side of P, construct segments $\overline{PX}$ and $\overline{PY}$ such that $PX = AB$ and $PY = CD$.
3. Bisect $\overline{XY}$ in order to find the midpoint M.
4. Construct a semicircle above m using M as center and MX as radius.
5. At P, construct a line perpendicular to m, intersecting the semicircle at Q. PQ is the mean proportional between AB and CD since $\overline{PQ}$ is the altitude to the hypotenuse $\overline{XY}$ of right $\triangle XQY$. (C8-7.1b)

Example 1 Construct the segment of length equal to $a \cdot b$, where a and b are real numbers.

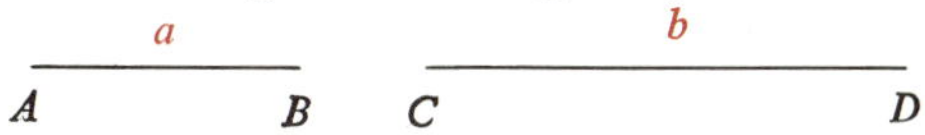

Solution Let m be any horizontal line and n be any line intersecting it. Choose any unit length and let $OU = OU' = 1$. Let $AB = a$ and $CD = b$.

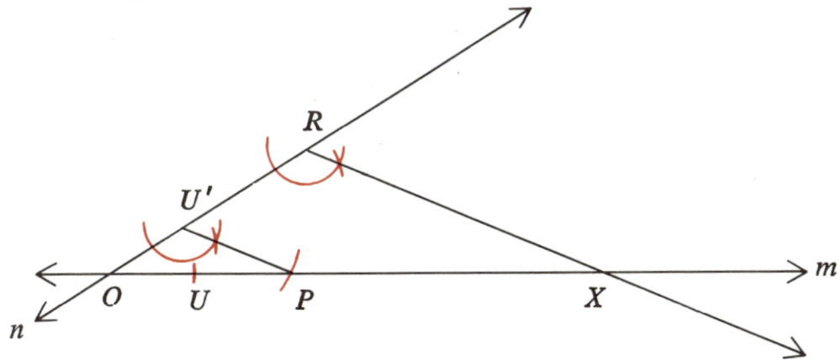

On m, construct $OP = a$, and on n, construct $OR = b$. Draw $\overline{PU'}$ and construct segment $\overline{RX}$ through R parallel to $\overline{PU'}$. By Corollary 8-2.1a, we have $OP : OX = OU' : OR$ or $\frac{a}{OX} = \frac{1}{b}$.
Thus, $ab = OX$

Example 2 Construct the segment of length $\frac{a}{b}$, where a and b are real numbers and $b > a$.

Solution Let m be any horizontal line and n be any line intersecting it. Choose any unit length and let $OU = OU' = 1$.
On m, construct $OP = a$ and $OQ = b$.

Draw $\overline{QU'}$ and construct a line parallel to $\overline{QU'}$ through P intersecting n at X.
By Corollary 8-2.1a,

$$\frac{a}{b} = \frac{OX}{1}$$

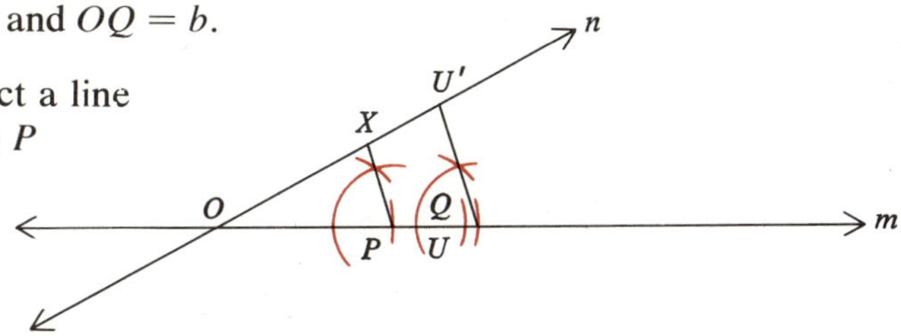

EXERCISES

A Draw four segments congruent to $\overline{XY}$. Partition each one into a different number of congruent segments, as listed below.

1. three 2. six 3. five

Draw four segments congruent to $\overline{XY}$ above. Partition each one into two segments with lengths in the following ratios:

4. $\frac{3}{2}$ 5. $\frac{5}{1}$ 6. $\frac{1}{2}$ 7. $\frac{3}{5}$

Construct a triangle similar to each of the triangles below, letting $\overline{XY}$, the base of the similar triangle, correspond to $\overline{AB}$.

8. 9. 10. 11.

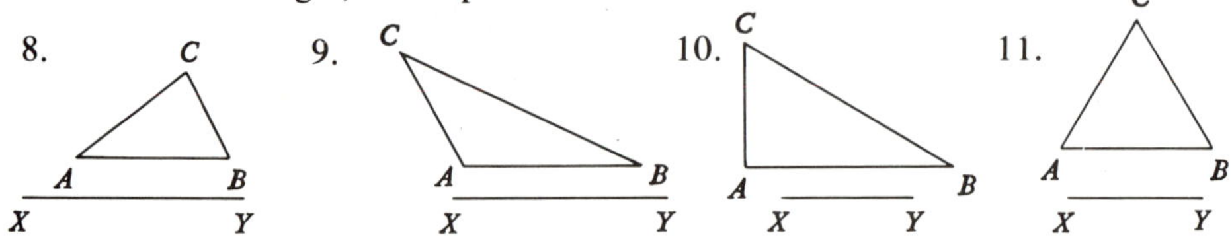

Given the following three segments, construct the segment with measure x in each of the following exercises:

12. $a : b = c : x$ 13. $a : x = x : c$ 14. $b : c = a : x$ 15. $x : b = a : c$

16. $\frac{x}{a} = \frac{c}{x}$ 17. $x = \frac{2a}{3b}$ 18. $x = \sqrt{ab}$ 19. $x = \sqrt{2}b$

B

20. Construct a square with a perimeter equal to XY in Exercise 1.
21. Construct an equilateral triangle with XY from Exercise 1 as perimeter.
22. Construct the equilateral triangle with perimeter $a + b + c$ where a, b, and c are the lengths for Exercises 12–19.
23. Construct a segment of length $\sqrt{2}a$, where a is one-half the length of $\overline{EF}$ for Exercises 12–19.

C

24. Construct an isosceles trapezoid with bases of lengths a and b and altitude of length c. Use the segments from Exercises 12–19 for a, b, and c.
25, Partition a segment into lengths in the ratio 3 : 4 : 5. Then construct the triangle with these lengths as sides.

SOMETHING TO THINK ABOUT

The Radical Spiral

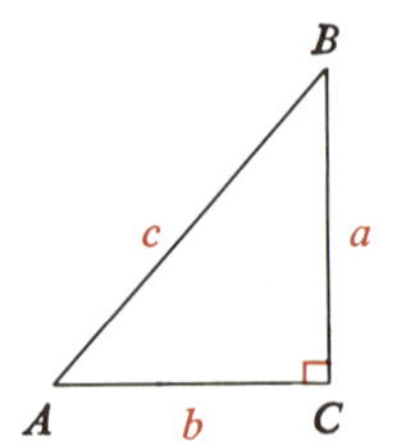

Because of the Pythagorean Theorem, we know that in any right triangle such as the one at right, $c^2 = a^2 + b^2$. If $\triangle ABC$ is an isosceles right triangle such that $a = b = 1$, we know that $c^2 = 2$ or $c = \sqrt{2}$. Thus c is an irrational number. We can use this fact to construct a length of $\sqrt{2}$ or any other length in the radical form $\sqrt{a}$, where a is a positive number.

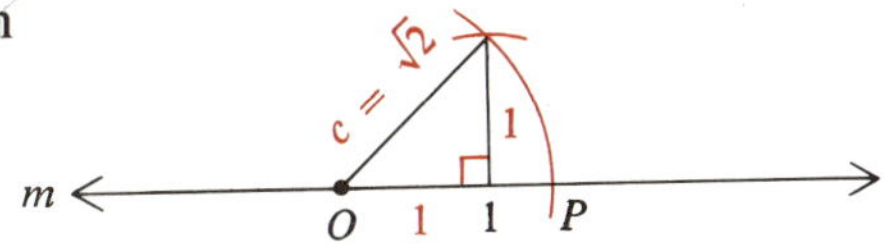

On a number line m, construct an isosceles right triangle with each leg equal to 1 unit. With the origin as center and hypotenuse c as radius, construct an arc intersecting m at P. Then

$$OP = c = \sqrt{2}.$$

We can use this length to construct a segment of length $\sqrt{3}$. Construct a right triangle with legs of lengths $\sqrt{2}$ and 1. The hypotenuse d, then, is

$$d^2 = (\sqrt{2})^2 + 1^2$$
$$d^2 = 2 + 1$$

Thus, $$d = \sqrt{3}$$

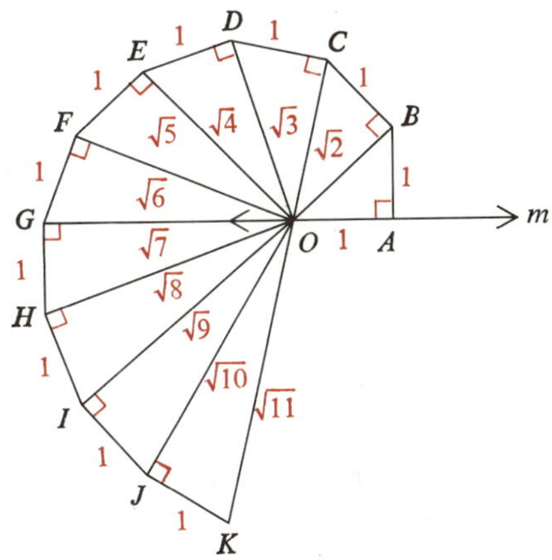

We can continue this process indefinitely. The construction at right incorporates the two constructions above. Each hypotenuse becomes a leg for the next right triangle, with the other leg a unit length. Why do we call this the radical spiral?

11-6 CONSTRUCTIONS INVOLVING CIRCLES

Many of the constructions we would like to make involve circles. We shall present a few of these in this section.

Example 1 Bisect $\overset{\frown}{AB}$

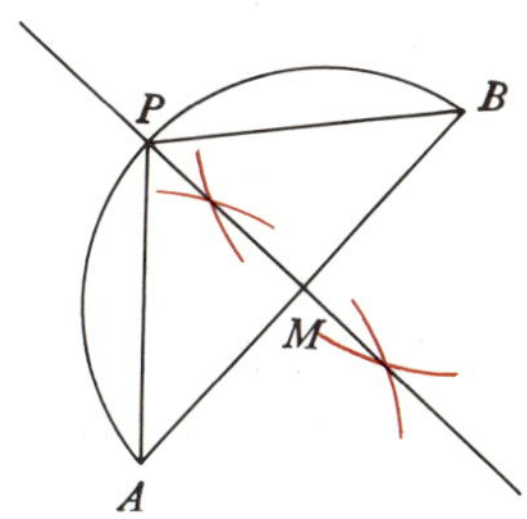

Solution To bisect $\overset{\frown}{AB}$ we construct the perpendicular bisector of chord $\overline{AB}$. This bisector intersects $\overset{\frown}{AB}$ at P. Since all points on $\overleftrightarrow{PM}$ are equidistant from A and B, $\overline{AP} \cong \overline{PB}$. Congruent chords have congruent arcs. Thus, P is the midpoint of $\overset{\frown}{AB}$.

We can see from Example 1 that constructions in circles use some of our other basic constructions.

Construction 13 Locate the center of a given circle.

GIVEN Any circle

WANTED The center, O, of the circle

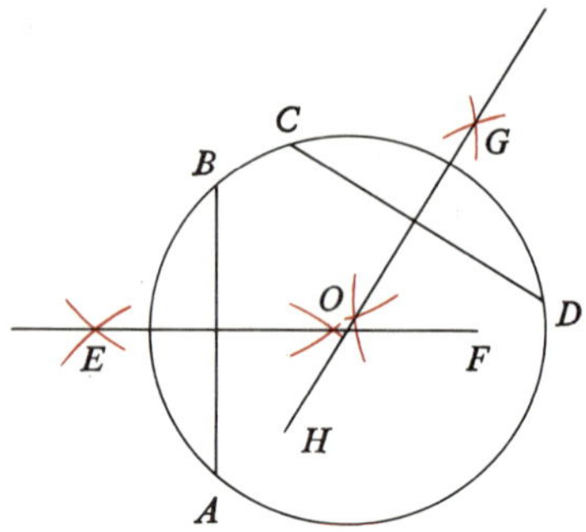

CONSTRUCTION

1. Draw any two nonparallel chords of the circle and label them $\overline{AB}$ and $\overline{CD}$.
2. Construct the perpendicular bisector of each chord. Name them $\overleftrightarrow{EF}$ and $\overleftrightarrow{GH}$, respectively.
3. The intersection of $\overleftrightarrow{EF}$ and $\overleftrightarrow{GH}$ is O, the center of the circle. We know this since the center is in the perpendicular bisector of any chord and O is the only point common to both bisectors.

We can also use some of our basic constructions to construct tangents to circles. The construction of a line tangent to a circle at a point of the circle is similar to the construction of the perpendicular to a segment at its endpoint.

Construction 14 Construct a line tangent to any circle at any point of the circle.

GIVEN Point B of $\odot A$

WANTED A tangent to $\odot A$ at B

The method used here is from Example 2 of Section 11-3.

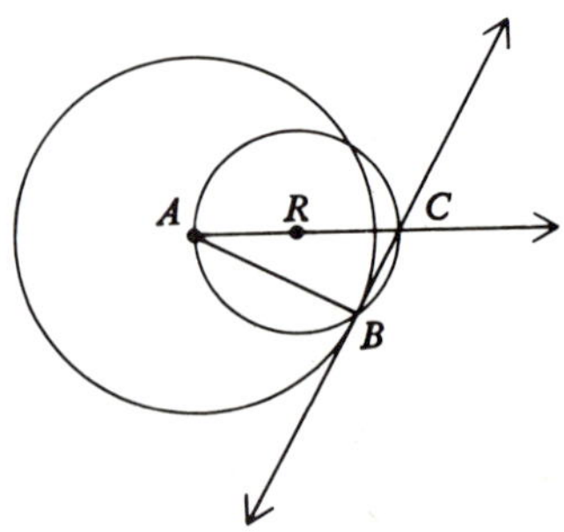

CONSTRUCTION

1. Draw radius $\overline{AB}$.
2. Construct the perpendicular to $\overline{AB}$ at B. Justify the construction marks shown.
3. $\overline{CB}$ is the required tangent since a tangent is perpendicular to the radius at the point of tangency.

Construction 15 Construct a line tangent to a circle through a point external to the circle.

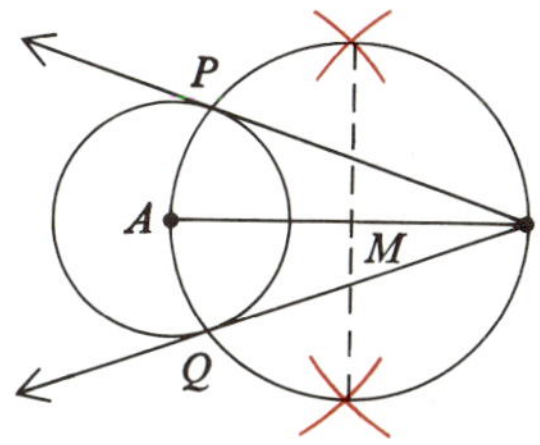

GIVEN Point B in the exterior of $\odot A$

WANTED The line through B tangent to $\odot A$

CONSTRUCTION

1. Draw and bisect $\overline{AB}$. Label the midpoint of M.
2. With M as center and $\overline{MB}$ as radius construct the circle through A and B. Circle M, with $\overline{AB}$ as diameter, intersects $\odot A$ at P and Q.
3. Draw $\overrightarrow{BP}$ and $\overrightarrow{BQ}$. By Corollary 9-6.1c, angles inscribed in a semicircle with sides through the ends of a diameter are right angles. Thus, $\angle APB$ and $\angle AQB$ are right angles. Thus, $\overline{AP} \perp \overleftrightarrow{PB}$ at P and $\overline{AQ} \perp \overleftrightarrow{QB}$ at Q. $\overleftrightarrow{PB}$ and $\overleftrightarrow{QB}$ are the required tangents.

Miscellaneous Constructions

Our basic constructions also allow us to inscribe various regular polygons in given circles.

D **Definition 11-2** The **center of a regular polygon** is the center of its circumscribed or inscribed circle. Segments drawn from the center of the polygon to the vertices of the polygon are radii of the circumscribed circle. Segments drawn from the center of the polygon perpendicular to the sides of the polygon are radii of the inscribed circle.

Construction 16 Inscribe a square in a circle.

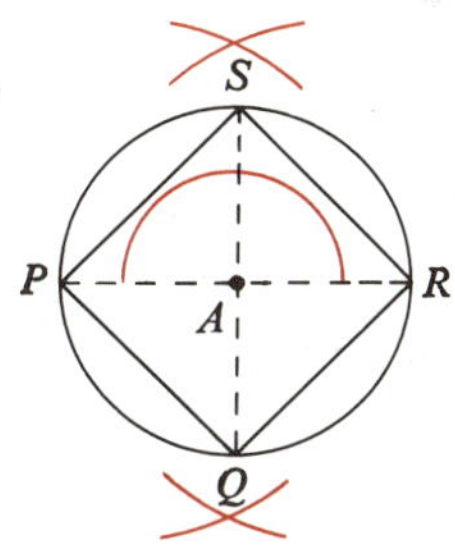

GIVEN Circle A

WANTED A square inscribed in $\odot A$

CONSTRUCTION

1. Draw any diameter $\overline{PR}$.
2. Construct the diameter $\overline{QS}$ perpendicular to $\overline{PR}$. Thus $PA = RA = QA = SA$.
3. Because the diagonals of a square are congruent and perpendicular to each other, $\overline{PQ}$, $\overline{QR}$, $\overline{RS}$, and $\overline{SP}$ form the sides of the square $PQRS$.

From Theorem 9-5.2 we know that if two chords in the same circle are congruent, then their arcs are congruent. Thus, by inscribing a regular hexagon in a circle we form six congruent arcs. Since the arcs are congruent so are the six central angles formed. Because each central angle has a measure of 60, each triangle formed by a side of the hexagon and the radii is equilateral. Referring to this discussion, we can construct a regular hexagon by inscribing it in a circle.

Construction 17 Inscribe a regular hexagon in a circle.

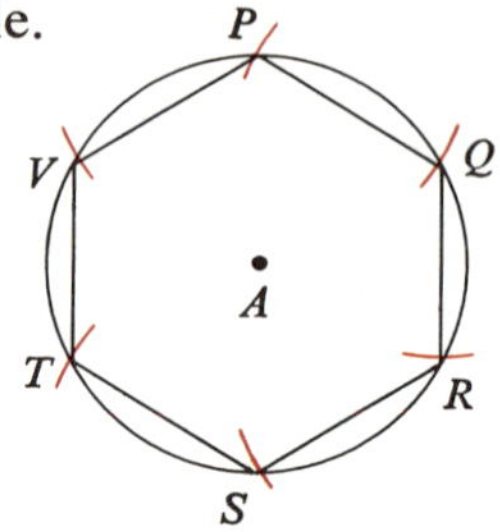

GIVEN Circle A.

WANTED A regular hexagon inscribed in $\odot A$

CONSTRUCTION

1. Using the radius of $\odot A$ and any point P of the circle as center, construct an arc intersecting the circle at Q.
2. Using Q as center and the same radius as in step 1, construct an arc intersecting the circle at P.
3. Continue this process to locate points S, T, and V.
4. Join these points to form $\overline{PQ}$, $\overline{QR}$, $\overline{RS}$, $\overline{ST}$, $\overline{TV}$, and $\overline{VP}$. From our discussion above, can you see why there are exactly six such chords? $PQRSTV$ is the desired regular hexagon.

Joining alternate points found using the last construction will produce an inscribed equilateral triangle.

We have inscribed a square in a circle. We can also circumscribe a square about a circle. Is this the same as inscribing a circle in a square?

EXERCISES

A Copy the illustration at right and bisect each of the following arcs.

1. $\overset{\frown}{AB}$ 2. $\overset{\frown}{BC}$ 3. $\overset{\frown}{AC}$

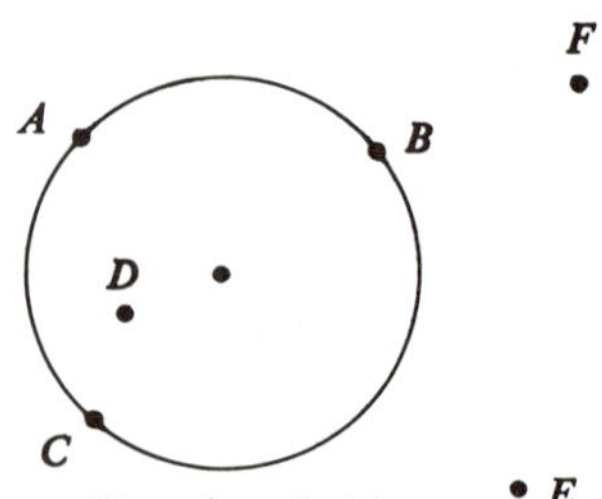

Exercises 1–14

4. Show how to locate the center of the circle.
5. Draw any circle through C and E, first locating the center.

Construct lines tangent to the circle above through each of the following points.

6. A 7. B 8. C 9. E 10. F

Construct four lines tangent to the circle above, each with one of the following added conditions:

11. parallel to the tangent through A
12. parallel to $\overline{AB}$
13. perpendicular to the radius through D
14. parallel to $\overline{EF}$

B

15. Given an acute angle and the hypotenuse, construct a right triangle.
16. Construct internally tangent circles A and B, with radii a and b, respectively.
17. Construct the common tangent of two externally tangent circles.
18. Construct the external tangent common to two nonintersecting circles.
19. Construct the interior tangent common to two nonintersecting circles.

C

20. Construct the angle bisectors of two angles of an acute triangle. With the point of intersection of the bisectors as center, and the perpendicular distance from the center to any of the sides as radius, construct a circle.
21. Repeat the constructions in Exercise 20, using an obtuse triangle.
22. What conclusion can you draw from Exercises 20–21?

11-7 CONCURRENCE

Occasionally several lines constructed in a given geometric figure pass through a common point. Such lines are said to be concurrent.

D

Definition 11-3 If a single point is common to three or more lines, then the lines are **concurrent**. The common point is the **point of concurrence**.

The lines l_1, l_2, l_3 in the figure for the proof of Theorem 11-7.1 are concurrent at Q; Q is the point of concurrence.

Our work with segments and angle bisectors provides the conjectures that form the basis for the theorems in this section.

T

Theorem 11-7.1 The Perpendicular Bisector Theorem for Concurrence The perpendicular bisectors of the sides of a triangle are concurrent at a point equidistant from the vertices of the triangle.

GIVEN Lines l_1, l_2, l_3 are the perpendicular bisectors of the sides of $\triangle ABC$.

PROVE l_1, l_2, and l_3 have point Q in common such that $AQ = BQ = CQ$.

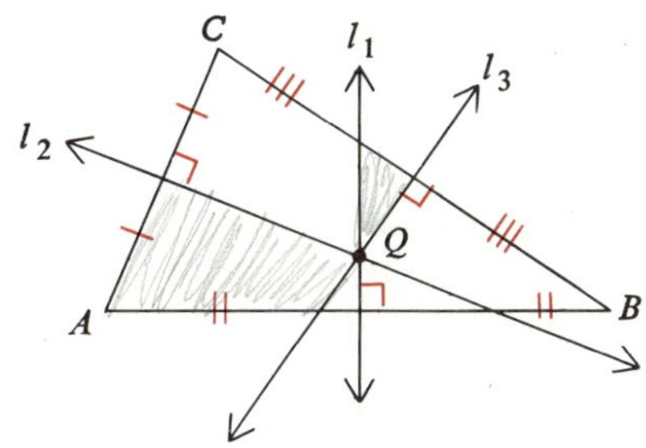

PROOF OUTLINE

Lines l_1, l_2, and l_3 are the perpendicular bisectors of $\overline{AB}$, $\overline{AC}$, and $\overline{CB}$, respectively. Consider l_1 and l_2. They either intersect or they do not. If they do not intersect, then they are parallel. This means that $\overline{AC}$ would have to be parallel to $\overline{AB}$ since, in a plane, lines perpendicular to parallel lines are also parallel. But $\overline{AC}$ and $\overline{AB}$ are adjacent sides of $\triangle ABC$. Thus, our assumption above must be false. Therefore, l_1 and l_2 do intersect at Q.

Since Q is in both perpendicular bisectors, it is equidistant from the endpoints of both segments. That is,

$$AQ = BQ \quad \text{and} \quad AQ = CQ$$

Thus,

$$AQ = BQ = CQ$$

Since $BQ = CQ$, we can say that Q is in the perpendicular bisector of $\overline{BC}$. Therefore, l_1, l_2, and l_3 are concurrent at point Q.

The point of concurrence of the perpendicular bisectors of the sides is called the **circumcenter** of the triangle. If we use the circumcenter as center and use the distance from this point to any of the three vertices as radius, then the vertices are points of the circle drawn. This leads us to the following corollaries. Their proofs are left as exercises.

C **Corollary 11-7.1a** Any three noncollinear points are points of one and only one circle.

C **Corollary 11-7.1b** Two nonconcentric circles intersect in at most two points.

Theorem 11-7.1 and the construction of some auxiliary segments enable us to prove the next concurrence theorem.

T **Theorem 11-7.2 The Altitude Theorem for Concurrence** The lines containing the three altitudes of a triangle are concurrent.

GIVEN The three altitudes of $\triangle ABC$.

PROVE l, m, and n are concurrent.

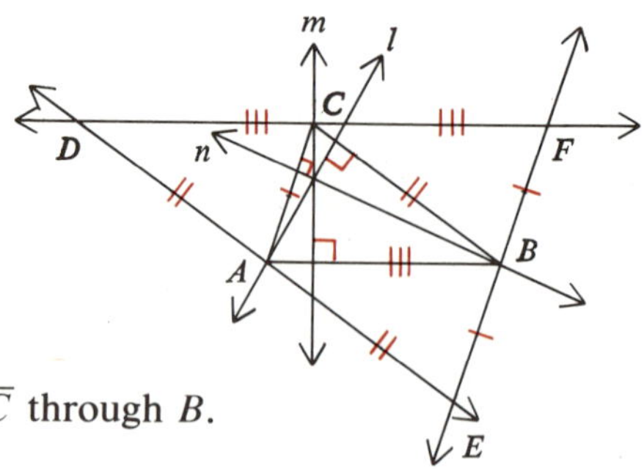

CONSTRUCTION
Construct $\overleftrightarrow{DF}$ parallel to $\overline{AB}$ through C, $\overleftrightarrow{DE}$ parallel to $\overline{BC}$ through A, and $\overleftrightarrow{EF}$ parallel to $\overline{AC}$ through B.

PROOF OUTLINE

Through each vertex of $\triangle ABC$ we have constructed a line parallel to the opposite side. These lines intersect (they cannot be parallel since the sides of $\triangle ABC$ are not parallel) to form $\triangle DEF$.

Three distinct parallelograms are thus formed. ▱$ABFC$ is one of them. Name the other two.

$$AC = BF = BE \quad \text{and} \quad B \text{ is the midpoint of } \overline{FE};$$
$$AB = DC = CF \quad \text{and} \quad C \text{ is the midpoint of } \overline{DF};$$
$$BC = DA = AE \quad \text{and} \quad A \text{ is the midpoint of } \overline{DE}.$$

This means that altitudes l, m, and n are the perpendicular bisectors of the sides of $\triangle DEF$ and are thus concurrent.

We stress that we proved this theorem for the lines containing the altitudes of the triangle and not simply for the altitude segments themselves. Draw an obtuse $\triangle PQR$. Where is the point of concurrence of the lines containing the altitudes of $\triangle PQR$? The point of concurrence of the altitudes of a triangle is called the **orthocenter**.

The medians of a triangle are interior segments whose endpoints are a vertex and the midpoint of the side opposite the vertex. Thus, any two medians would intersect at a point Q in the interior of the triangle. If we can show that the ray from the third vertex through Q passes through the midpoint of the side opposite the third vertex, then we have proved the following theorem:

T **Theorem 11-7.3 The Median Theorem for Concurrence** The medians of a triangle are concurrent at a point of each median located two-thirds of the way from the vertex to the opposite side.

GIVEN The medians of $\triangle ABC$

PROVE The medians are concurrent at a point of any median two-thirds of the way from the vertex to the opposite side.

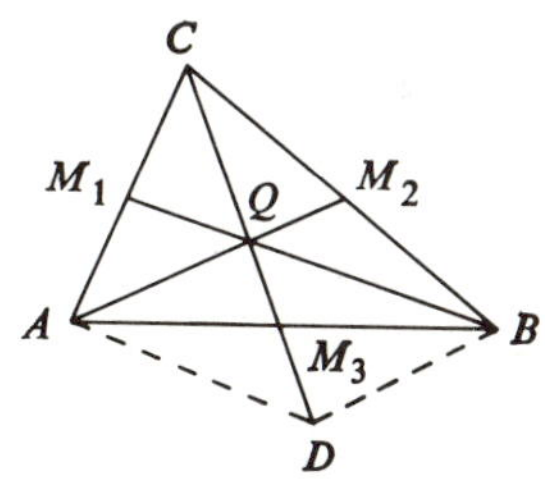

PROOF OUTLINE

The medians $\overline{AM_2}$ and $\overline{BM_1}$ intersect in an interior point of $\triangle ABC$. Call this point Q. Draw $\overrightarrow{CQ}$ from vertex C through point Q. Label the intersection of $\overrightarrow{CQ}$ and $\overline{AB}$ point M_3. Extend $\overrightarrow{QM_3}$ and construct $\overline{QD}$, such that $\overline{CQ} \cong \overline{QD}$. If we draw $\overline{BD}$, $\overleftrightarrow{AQM_2}$ contains the midline of $\triangle DCB$. Thus, by Theorems 7-7.2 and 7-6.3

$$\overleftrightarrow{AQM_2} \,//\, \overline{DB} \quad \text{and} \quad QM_2 = \tfrac{1}{2}DB$$

If we draw $\overline{AD}$ by a similar argument, we can conclude that

$$\overleftrightarrow{BQM_1} \,//\, \overline{DA} \quad \text{and} \quad QM_1 = \tfrac{1}{2}AD$$

Since $\overline{AD} \parallel \overline{QB}$ and $\overline{AQ} \parallel \overline{DB}$, $AQBD$ forms a parallelogram. The diagonals of a parallelogram bisect each other. Thus, $AM_3 = M_3B$ and we can say that M_3 is the midpoint of $\overline{AB}$. This means that $\overline{CM_3}$ is also a median. Therefore, the three medians of a triangle are concurrent.

Since $AQBD$ is a parallelogram,

$$AD = QB \qquad \text{and} \qquad AQ = DB$$

But since $QM_1 = \frac{1}{2}AD$, we can conclude that $QB = 2QM_1$. Thus, $BQ = \frac{2}{3}BM_1$. Likewise, $QM_2 = \frac{1}{2}DB$. Thus $DB = 2QM_2$; or $AQ = \frac{2}{3}AM_2$. In similar fashion we can conclude that $CQ = \frac{2}{3}CM_3$.

The point of concurrence of the medians is called the **centroid** or the **center of gravity**.

Example 1 C is the centroid of $\triangle PQR$. If for median PCM, $PCM = 12$, find the value of CM.

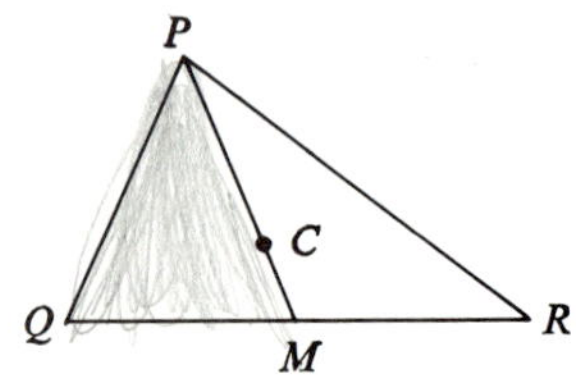

Solution The centroid of $\triangle PQR$ is located two-thirds of the distance from P to M, or a third of the distance from M to P. That is, $CM = \frac{1}{3}PM$. Since $PM = 12$, $CM = \frac{1}{3}(12)$ or $CM = 4$.

T **Theorem 11-7.4 The Angle Bisector Theorem for Concurrence** The angle bisectors of a triangle are concurrent at a point equidistant from the sides of the triangle.

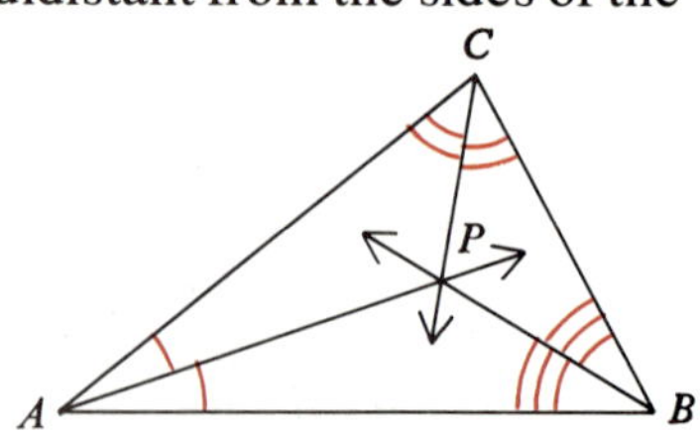

GIVEN The angle bisectors of $\triangle ABC$.

PROVE The bisectors are concurrent at a point equidistant from the sides.

PROOF OUTLINE

In $\triangle ABC$, the bisectors of $\angle A$ and $\angle B$ intersect in a point we shall call P. Since P is in the interior of these two angles, it is in the interior of $\angle C$. From Theorem 11-1.1, we know that P is equidistant from $\overline{AB}$ and $\overline{AC}$ and P is equidistant from $\overline{AB}$ and $\overline{BC}$. Thus, P is equidistant from $\overline{AC}$ and $\overline{BC}$. Therefore, P is also a point of the bisector of $\angle C$.

The point of concurrence of the angle bisectors is called the **incenter** of the triangle. We shall use the incenter later to inscribe a circle in a given triangle.

Constructions

You are now familiar with the circumcenter, the orthocenter, the center of gravity, and the incenter. *Center* is the name usually given to a point of concurrence in a geometric figure. Each center is the center of infinitely many circles. Two of the centers of a triangle, however, are centers of circles of special interest.

Construction 18 Construct a circle circumscribed about a triangle.

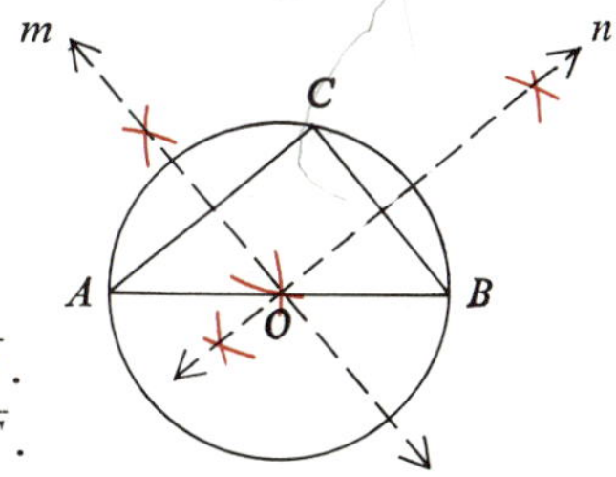

GIVEN $\triangle ABC$

WANTED A circle circumscribed about $\triangle ABC$

CONSTRUCTION
1. Construct n, the perpendicular bisector of $\overline{CB}$.
2. Construct m, the perpendicular bisector of $\overline{AC}$.
3. Lines n and m intersect at O.
4. Construct a circle using O as the center and $\overline{OC}$ as the radius.
5. Circle O is the desired circle.

Construction 19 Construct a circle inscribed in a triangle.

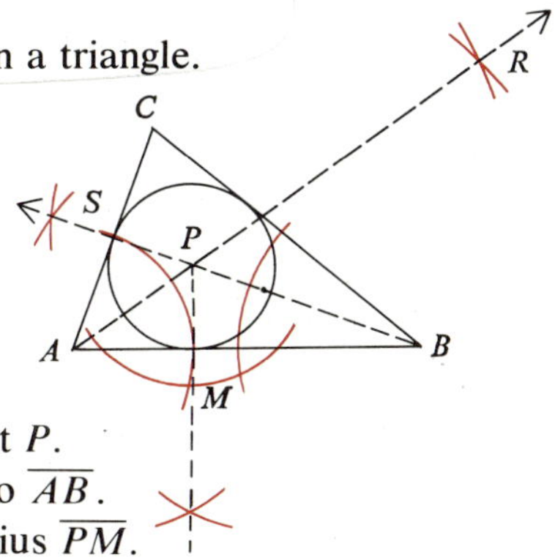

GIVEN $\triangle ABC$

WANTED A circle inscribed in $\triangle ABC$

CONSTRUCTION
1. Construct $\overrightarrow{AR}$, the bisector of $\angle A$.
2. Construct $\overrightarrow{BS}$, the bisector of $\angle B$.
3. Label the intersection of $\overrightarrow{AR}$ and $\overrightarrow{BS}$ point P.
4. Construct the perpendicular $\overline{PM}$ from P to $\overline{AB}$.
5. Construct a circle, using center P and radius $\overline{PM}$.
6. Circle P is the desired circle.

Can you see that center P is the incenter described earlier?

EXERCISES

A

1. Construct four copies of each of the following triangles, or larger but similar triangles, on your paper. Use a different copy to locate the circumcenter, orthocenter, centroid, and incenter of each triangle, by constructing all three appropriate lines in each case.

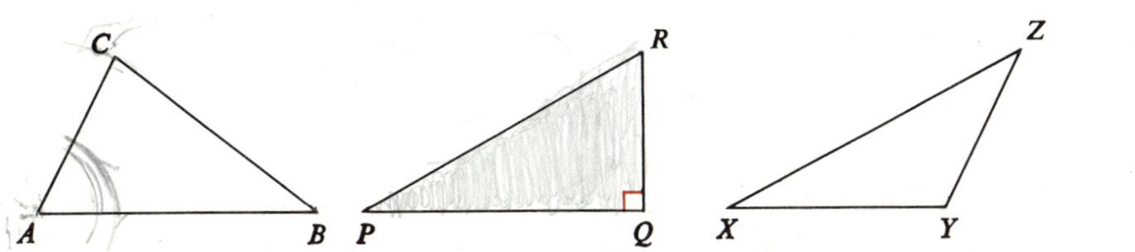

2. Circumscribe a circle about a copy of each of the triangles in Exercise 1.

3. Inscribe a circle in a copy of each of the triangles in Exercise 1.
4. Prove Corollary 11-7.1a
5. Prove Corollary 11-7.1b

B If in $\triangle ABC$, $\overline{AM}$, $\overline{BN}$, and $\overline{CO}$ are medians intersecting in P, find the missing value for each of the following.

6. If $AM = 12$, then $PM =$ ____.
7. If $PM = 4$, then $AP =$ ____.
8. If $BP = 5$, then $BN =$ ____.
9. If $CP = 7$, then $PO =$ ____.
10. If $\triangle ABC$ has vertices $A(0, 0)$, $B(0, 8)$, and $C(10, 0)$, construct $\triangle ABC$.

Using $\triangle ABC$ of Exercise 10, find the coordinates of each of the following:

11. circumcenter
12. orthocenter
13. incenter
14. centroid

C Construct an acute triangle, $\triangle ABC$, on your paper. Extend sides $\overline{AB}$ through B to D and $\overline{AC}$ through C to E. Use this triangle for Exercises 15–18.

15. Bisect external angles CBD and BCE. Label the point of intersection of the bisectors X. X is called the excenter of a triangle.
16. How many excenters does $\triangle ABC$ have?
17. Find the distance from X to $\overline{BC}$.
18. Construct a circle tangent to $\overline{BC}$ using the excenter X as center.

HOW IS YOUR MATHEMATICAL VOCABULARY?

The following words and phrases are introduced in this chapter. How many do you know and understand?

center of gravity (11-7)
centroid (11-7)
circumcenter (11-7)
compass (11-2)
compound loci (11-1)
concurrent lines (11-7)
divider (11-2)
incenter (11-7)
locus (11-1)
orthocenter (11-7)
point of concurrence (11-7)
radius of a compass (11-3)
straightedge (11-2)
Two-Circle Postulate (11-3)

REVIEW EXERCISES

11-1 Locus

Draw and describe the locus of points in a plane satisfying these conditions:

1. The points equidistant from the x-axis.
2. The points equidistant from a pair of perpendicular lines.
3. The points in the interior of a circle with a radius of 3.
4. The points above $y = 2$ or to the right of $x = -3$.
5. The points 6 units from both axes.

11-2 Compass and straightedge

Use the following segments for Exercises 6–16.

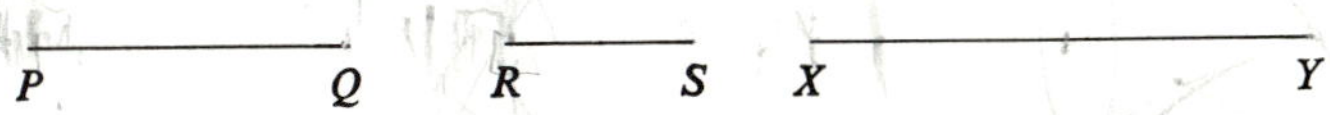

Construct a segment with a length equal to each of the following:

6. $PQ + XY$ 7. $XY - PQ$ 8. $2(XY)$

Construct a circle with each of the following radii:

9. PQ 10. $3(RS)$

11-3 Basic constructions

Construct each of the segments indicated below and the perpendicular bisector of each.

11. $\overline{PQ}$ 12. $\overline{RS}$ 13. $\overline{XY}$

14. Construct the perpendicular to $\overline{XY}$ at Y.
15. Construct the isosceles triangle with base congruent to $\overline{PQ}$ and the sides congruent to $\overline{XY}$ above.
16. Construct a square using $\overline{PQ}$ as a side.
17. Using ___ as unit a, construct an isosceles triangle with a base length a and an altitude length $6a$.

11-4 Angle constructions

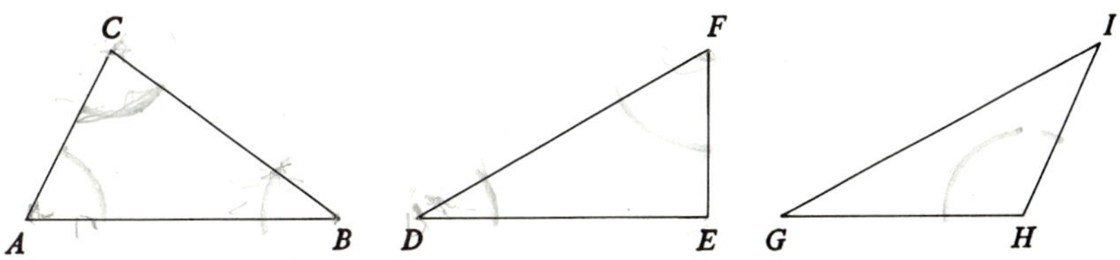

Copy each of the above triangles.

18. Construct an angle congruent to $\angle H$.
19. Construct an angle congruent to $\angle F$.
20. Construct an angle with a measure twice that of $\angle A$.
21. Construct an angle with a measure equal to the sum of the measures of $\angle B$ and $\angle D$.
22. Construct a line parallel to $\overline{AB}$ through C.
23. Construct the triangle with sides congruent to $\overline{AB}$ and $\overline{DF}$ and the included angle congruent to $\angle C$.
24. Bisect the angles of $\triangle DEF$.

11-5 Partitioning constructions

Exercises 25–36

Partition c into the number of congruent segments requested below.
25. four 26. five

Partition c above into segments with lengths in the following ratios:
27. 1 : 3 28. 2 : 5 29. 3 : 4 : 5

30. Construct a triangle similar to $\triangle GHI$, shown above Exercise 18, with c as the base.

Use the segments shown above to construct a segment of measure x in each of the following equations:

31. $a : b = c : x$ 32. $x = \sqrt{ac}$ 33. $x = \sqrt{2}\, a$

34. $x = ab$ 35. $x = \dfrac{3a}{2b}$ 36. $x = \frac{1}{3}(a + c)$

11-6 Constructions involving circles

37. Choose any three noncollinear points and construct the circle through them.

Inscribe the following figures in any given circle.
38. square 39. equilateral triangle 40. regular hexagon

11-7 Concurrence

Construct a triangle congruent to $\triangle ABC$ at right. Then construct the lines necessary to locate the following points:
41. orthocenter 42. centroid
43. circumcenter 44. incenter

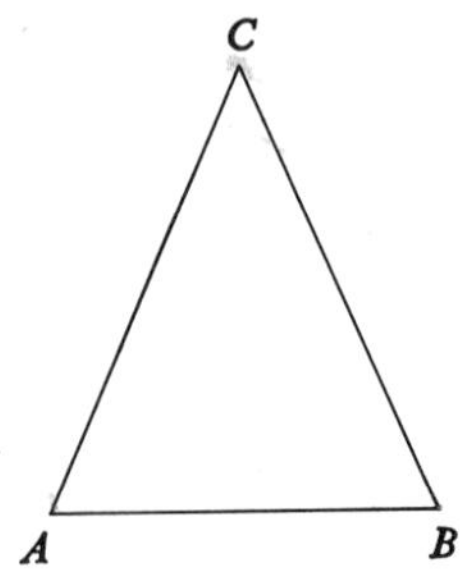

Exercises 41-44

For the following exercises, let $\overline{AM}$ be a median of $\triangle ABC$, with P as centroid.
45. If $AM = 10$, find AP.
46. If $PM = 3$, find AM. 47. If $AP = 6$, find PM.

CHAPTER TEST

1. Draw and describe the locus of points in a plane equidistant from the x-axis and the positive y-axis.
2. Draw and describe the locus of points in a plane that are the centers of circles that are tangent to the x-axis and that pass through the point (0, 5).

3. Draw the locus determined by a spot on the cog of a wheel as this wheel rotates around a larger cogged wheel.
4. Write the equation for the locus in Exercise 1.
5. Write the equation for the locus of points equidistant from $(0, a)$ and $(a, 0)$.
6. Write an equation for the locus three units from (a, b).
7. Draw an obtuse triangle and construct an altitude, angle bisector, and perpendicular bisector on one side.
8. Draw any acute triangle and construct a right triangle with the same base and altitude as the acute triangle.
9. Construct a triangle given any two angles and the included side.
10. Construct a triangle given any two sides and the included angle.

SUGGESTED RESEARCH

1. Read "The Three Famous Problems" in *History of Mathematics*, vol. 2, by D. E. Smith (New York: Dover Publications, 1953), pp. 297-303. Prepare a report for class presentation.
2. Prepare a report on the consequences of Menelaus's and Ceva's Theorems. See *Challenging Problems in Geometry,* by Alfred S. Posamentier and Charles T. Salkind (New York: Macmillan Co., 1970).
3. Prepare a report on the geometry of the fixed compass.
4. Prepare a report on Euclidean constructions.

MATHEMATICAL EXCURSION

Criterion of Constructibility

Construction problems have always played an important and interesting role in the study of geometry. The "rules of the game" were established by the Greeks during the time of Plato. In construction problems, the only permissible instruments are an unmarked straightedge and a compass. Hence, a construction is possible if it can be obtained by drawing straight lines between pairs of points and by drawing circles with a given center and a given radius.

The trisection of an angle, the construction of a square with the same area as a given circle, and the duplication of a cube are three famous construction problems, all of which can be solved easily if the restriction on permissible tools is lifted. These problems are impossible with compass and straightedge. This was finally shown more than 2,000 years after the problems were first proposed.

After mathematicians had tried these constructions for 2,000 years without success, it was reasonable to conclude that the constructions could not be accomplished. But this is not proof. The proof that these constructions cannot be performed is essentially algebraic.

We showed earlier in the chapter how to find a segment whose measure is the sum of the measures of two given segments and how to find a segment whose measure is the difference of the measures of two given segments. If we use directed lengths of segments so that $AB = -BA$, there is no need to subtract. We can see from the figure at right that for any three collinear points, $AB + BC = AC$. The sign of AB will differ from that of AC according to how $|BC|$ compares with $|AB|$. In multiplication the segment whose measure is the product of the measures of two segments depends entirely upon the choice of the segment of unit length. This is not true in addition; the segment representing the sum is independent of the choice of segment of unit length.

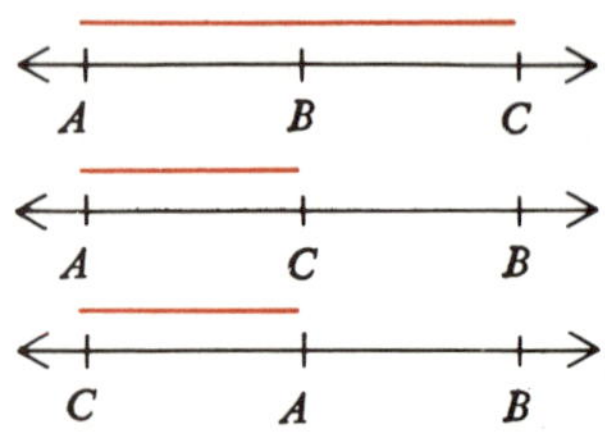

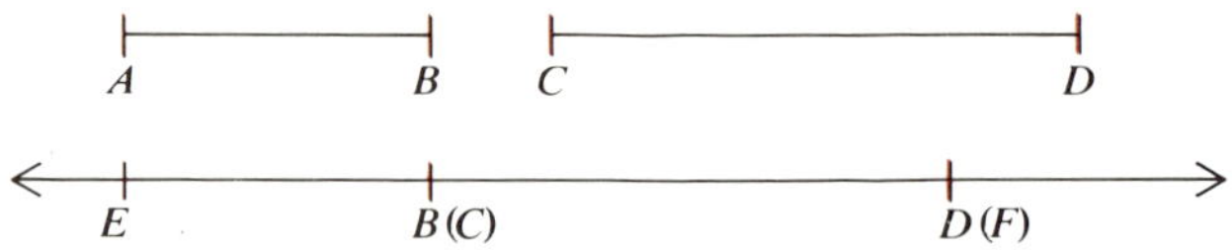

For given lengths AB and CD, $AB + CD = EF$, no matter what segment is used to find the unit length. For example, if the unit length is doubled, the measures of $\overline{AB}$ and $\overline{CD}$ are each cut in half. But so is the measure of $\overline{EF}$.

On the other hand, if we multiply AB by CD, each measure is cut in half, but the product is only one-fourth as great.

Study the following figure to determine how the choice of unit affects the product.

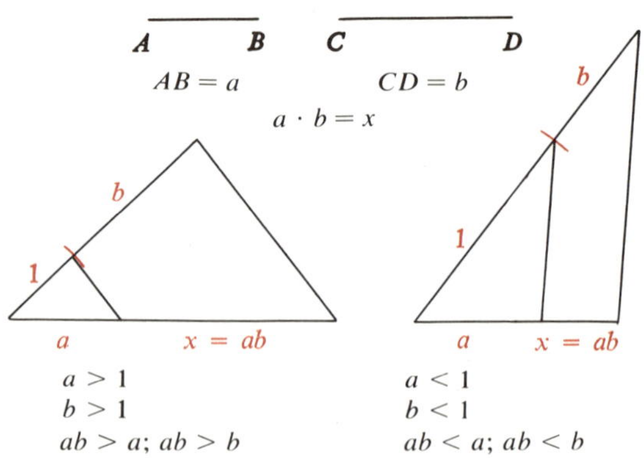

The quotient of two segments can be obtained by treating division as the inverse of multiplication. Again, the choice of a unit plays an important role in the outcome.

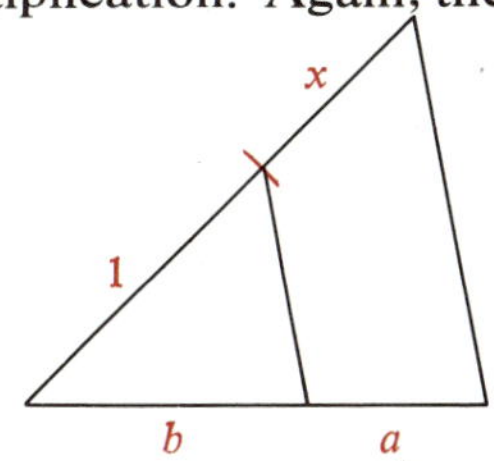

$b \cdot x = a$ or $\frac{a}{b} = x$
$a < b < 1$
$a < x < b < 1$

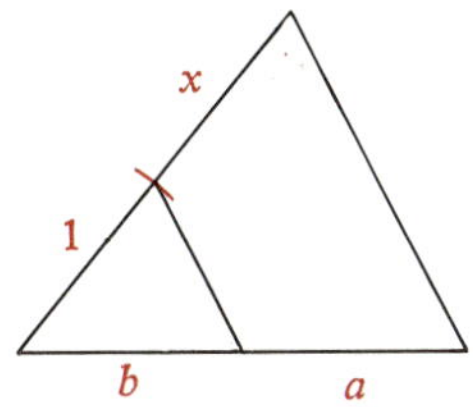

$b \cdot x = a$ or $\frac{a}{b} = x$
$b < a = 1$
$x > a > b$

We shall now define multiplication so that we can multiply with positive and negative numbers. Consider the two figures below as number lines, horizontal and oblique, intersecting at O. To find $a \cdot b$, locate A on the horizontal axis so that $OA = a$, and locate B on the oblique axis so that $OB = b$. Draw the line through the 1 on the oblique axis and A. Through B draw a line parallel to the first line, intersecting the horizontal axis in a point C. Thus, the segment $OC = ab$.

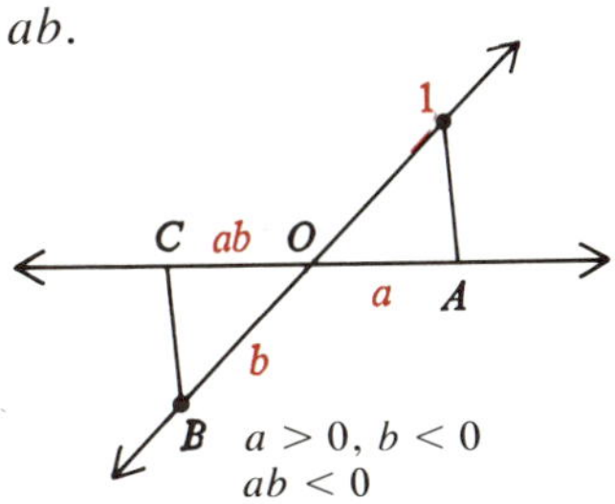

$a > 0, b < 0$
$ab < 0$

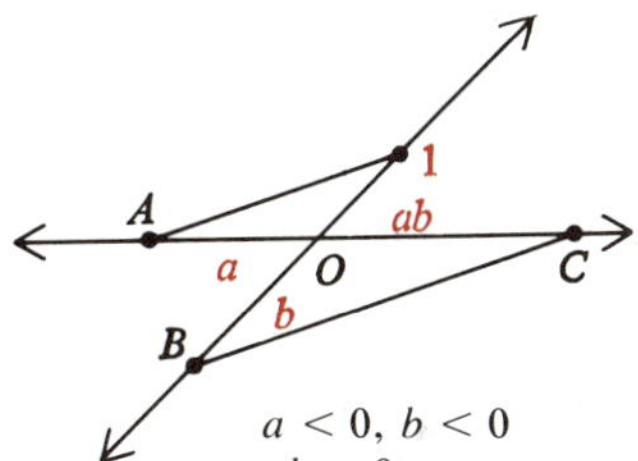

$a < 0, b < 0$
$ab > 0$

As before, we shall find a quotient by considering division as the inverse of multiplication. To find $\frac{a}{b}$, we find x such that $b \cdot x = a$.

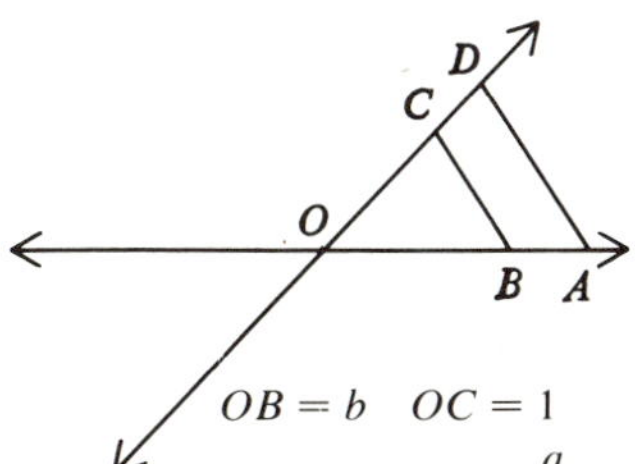

$OB = b \quad OC = 1$
$OA = a \quad OD = \frac{a}{b}$

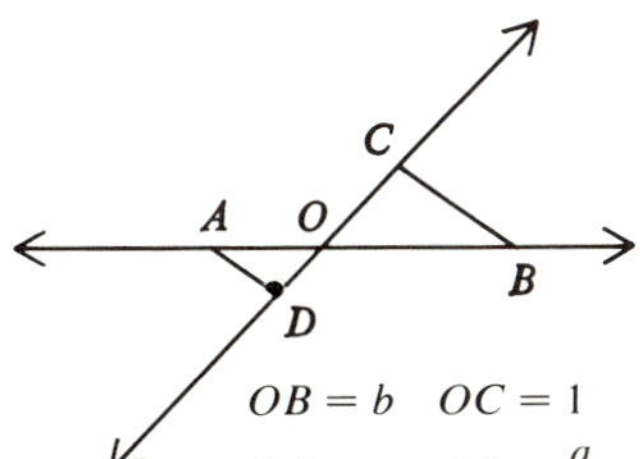

$OB = b \quad OC = 1$
$OA = a \quad OD = \frac{a}{b}$

Do you see why division by zero is undefined when we use this method? If $b = 0$, where is B? What line passes through the 1 and B? Will a line through A, parallel to this line, intersect the oblique axis?

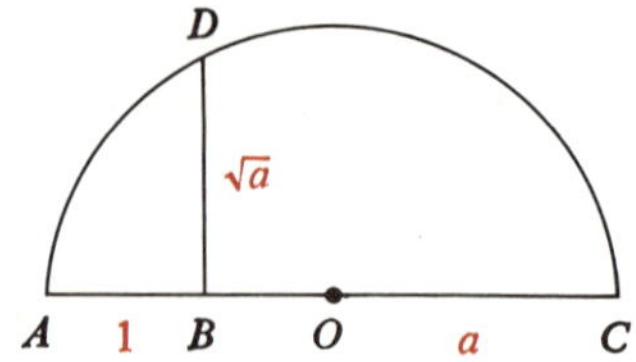

In the chapter, we also showed how to find the square root of a segment. As with multiplication and division, we must use the unit segment. On a line, mark off the unit segment $\overline{AB}$. From B, mark off segment $\overline{BC}$ with length a and collinear with $\overline{AB}$. Use the segment $\overline{AC}$ as a diameter and draw a semicircle. The perpendicular to the diameter at the point B cuts the semicircle in point D. $BD = \sqrt{a}$. This is the mean proportional construction described on page 480. The mean proportional of 1 and a is $\sqrt{a}$.

The solution to a construction problem may be expressed as a root of an equation. For example, consider the problem of duplication of a cube. We must find the edge of a cube whose volume is twice that of a given cube. We can take the edge of the given cube as a unit segment. The construction then becomes a problem of finding x such that $x^3 = 2$.

If we can obtain the solution by a finite number of applications of addition, subtraction, multiplication, division, and extractions of square root, using the given segments and an arbitrary unit of length, then the construction is possible.

Conversely, if the construction is possible, then we can obtain it by a finite number of applications of addition, subtraction, multiplication, division, and extractions of square root, using the given segments and an arbitrary unit of length. We know that the straight lines and circles that we construct are determined either by given segments or those obtained from the intersections of two straight lines, a straight line and a circle, or two circles. To show the converse above, we must show that these intersections can be obtained from the coefficients of the equations by a finite number of applications of the operations of addition, subtraction, multiplication, division, and extraction of square root.

Two straight lines

$$y = mx + b$$
$$y = m'x + b' \qquad m \neq m'$$

have as their point of intersection the point (x, y) with

$$x = \frac{b' - b}{m - m'} \qquad y = \frac{mb' - m'b}{m - m'}$$

These relationships are obtained from the equations by applying the above operations.

An equation for a circle with radius r and center (c, d) is

$$(x - c)^2 + (y - d)^2 = r^2$$

To find the intersection of the circle with the line $y = mx + b$, we can substitute for y in the equation for the circle

$$(x - c)^2 + (mx + b - d)^2 = r^2$$

This forms a quadratic equation in x. Since the solution of the quadratic $ax^2 + bx + c = 0$ is

$$x = \frac{-b \pm \sqrt{b^2 - 4ac}}{2a}$$

we know that the quadratic $(x - c)^2 + (mx + b - d)^2 = r^2$ has a root that can be obtained from the known constants by applying the above five operations.

The intersection of two circles is the same as the intersection of one circle with the common chord. Thus, this case can be reduced to finding the intersection of a circle and a line.

We have now established the Criterion of Constructibility.

Criterion of Constructibility A proposed geometric construction is possible with straightedge and compass alone if and only if the numbers that define algebraically the required geometric elements can be derived from those defining the given elements by a finite number of rational operations and extractions of square root.

We stated earlier that the duplication of a cube is an impossible construction. The segment x, such that $x^3 = 2$, cannot be constructed, using the five operations of the criterion. Solving the equation $x^3 - 2 = 0$, we get

$$(x - \sqrt[3]{2})(x^2 + x\sqrt[3]{2} + \sqrt[3]{4}) = 0$$

Letting each factor equal zero and solving, we have

$$x = \sqrt[3]{2}$$

$$x = \frac{-\sqrt[3]{2} \pm \sqrt{\sqrt[3]{4} - 4\sqrt[3]{4}}}{2} = \frac{-\sqrt[3]{2} \pm \sqrt{-3\sqrt[3]{4}}}{2}$$

The first root $\sqrt[3]{2}$ is neither rational nor the extraction of a square root. The other two roots are not real numbers. They require the square root of a negative number. Hence, the construction is impossible.

The trisection problem requires that a method exists that will trisect any angle.

From a given angle A, we can construct a segment with length equal to $\cos A$. If we take an arbitrary unit-length segment as hypotenuse, $\cos A$ is the length of the segment

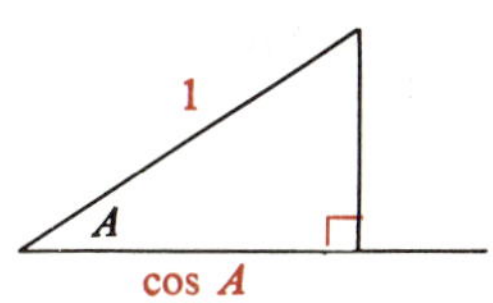

cut off on the other side of angle A by the perpendicular from the opposite end of the hypotenuse. If we can trisect $\angle A$, we can also find the cosine of one-third the angle, $\cos\frac{A}{3}$. If we can show that $\cos\frac{A}{3}$ cannot be constructed, then we have shown that $\angle A$ cannot be trisected. Using proof by counterexample, we shall prove an angle of measure 120 cannot be trisected. We must, however, accept the identity

$$\cos A = 4\cos^3\frac{A}{3} - 3\cos\frac{A}{3}$$

If we multiply through by 2 and replace $2\cos\frac{A}{3}$ with x, we have

$$2\cos A = x^3 - 3x$$

Since $\cos 120 = -\frac{1}{2}$ we have

$$x^3 - 3x + 1 = 0$$

From the Criterion of Constructibility, we know that constructible roots must be of the form

$$a + b\sqrt{c}$$

where a and b are rational and c is constructible.

First, then, we must show that $x^3 - 3x + 1 = 0$ has no rational roots. To do this, we assume that there is a rational root, $\frac{p}{q}$, where p and q have no common factor greater than 1. Substituting for $\frac{p}{q}$, we have

$$\left(\frac{p}{q}\right)^3 - 3\left(\frac{p}{q}\right) + 1 = 0$$

$$p^3 - 3pq^2 + q^3 = 0$$

$$q^3 = 3pq^2 - p^3$$

$$q^3 = p(3q^2 - p^2)$$

This means that q^3, and hence q, has the factor p. Therefore p must equal ± 1. Also, solving for p^3

$$p^3 = 3pq^2 - q^3$$

$$p^3 = q^2(3p - q)$$

This means p and q must have a common factor, and hence $q = \pm 1$. We can conclude from this that the only rational root of $x^3 - 3x + 1 = 0$ is $r = \pm 1$. By substitution, we can show that neither $+1$ nor -1 is a root.

Next, assume $x^3 - 3x + 1 = 0$ has a constructible root $a + b\sqrt{c}$. By substitution in the equation $x^3 - 3x + 1 = 0$, we can show that if $a + b\sqrt{c}$ is a root, then its conjugate, $a - b\sqrt{c}$, is also a root. (See Exercise 7 at the end of this Excursion.) The sum of the roots of the polynomial equation

$$x^n + a_1x^{n-1} + a_2x^{n-2} + \cdots + a_n = 0$$

is $r_1 + r_2 + r_3 + \cdots + r_n = -a_1$. (See Exercise 8 at the end of this Excursion.) It follows from this that the sum of the roots of $x^3 - 3x + 1 = 0$ is zero. If two roots are $a + b\sqrt{c}$ and $a - b\sqrt{c}$, with the third root r, we have

$$a + b\sqrt{c} + a - b\sqrt{c} + r = 0$$
$$r = -2a$$

But a is rational and hence r is rational, and we have a contradiction. Hence the angle whose measure is 120 cannot be trisected.

The third famous problem, that of constructing a square with an area equal to that of a circle, is more difficult to prove impossible. Essentially, the problem requires the construction of a segment π units long, beginning only with the unit segment.

EXERCISES

1. Given $\underline{\quad a \quad}$ and $\underline{\quad b \quad}$. Find the arbitrary unit segment such that $a \cdot b = b$.
2. Show by our definition of multiplication of segments that $a \cdot 0 = 0$.
3. Show that by our definition of multiplication of segments, if $a \cdot b = 0$ and $a \neq 0$, then $b = 0$.
4. Show that by our definition of division of segments, division by zero is impossible.
5. Prove geometrically that $a \cdot \dfrac{1}{a} = 1$ for any segment a.
6. Prove that a regular polygon of 9 sides cannot be constructed with straightedge and compass. (*Hint*: Find the angle subtended at the center by a side.)
7. If $a + b\sqrt{c}$ is substituted for x in $x^3 - 3x + 1 = 0$, some terms will contain $\sqrt{c}$ and some will not. The only way the sum of the terms with $\sqrt{c}$ and those without $\sqrt{c}$ can equal zero is that the sum of the terms without $\sqrt{c}$ equal zero, and the sum of the coefficients of $\sqrt{c}$ equal zero. Show by substitution that if $a + b\sqrt{c}$ satisfies the equation $x^3 - 3x + 1 = 0$, then $a - b\sqrt{c}$ must also satisfy it.
8. If the roots of the equation $x^3 + ax^2 + bx + c = 0$ are r_1, r_2, r_3, then

$$x^3 + ax^2 + bx + c = (x - r_1)(x - r_2)(x - r_3)$$

must be true for all the values of x. This will be true if and only if each power of the variable has the same coefficient. Expand the right member of the above equation and show that the sum of the roots of the equation is equal to $-a$.

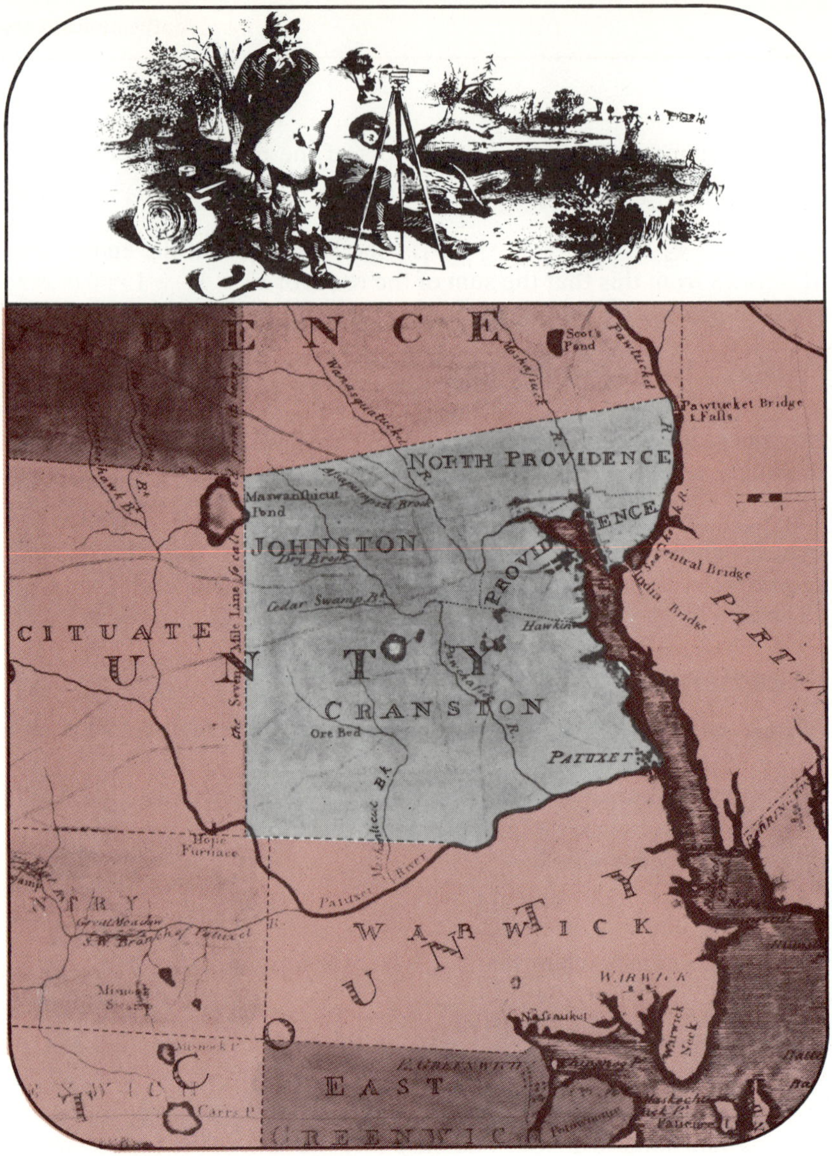

12 Areas of Plane Regions

So far in our study of measurement, we have been concerned with line segments, angles, and arcs of a circle. We can also measure the size of the interior region of a polygon or circle. In this chapter we shall develop formulas for measurement of the areas of the circle and of various polygons. Then we shall establish methods of comparing the areas of similar polygons and circles.

12-1 AREAS OF POLYGONAL REGIONS

Before we begin studying areas, we must define the new terms that we shall need. The two definitions below are important, and you should study them carefully. They define two kinds of sets of points.

D **Definition 12-1** A **triangular region** is the union of a triangle and its interior.

D **Definition 12-2** A **polygonal region** is the union of a finite number of coplanar triangular regions that intersect in either a line segment or a point.

In other words, a polygonal region is a region which can be partitioned (that is, cut up) into triangular regions.

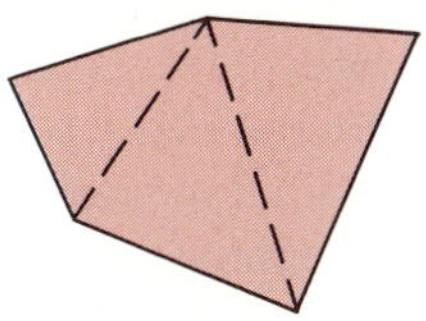
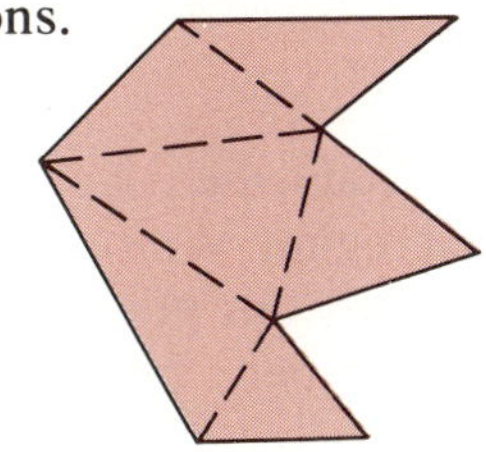
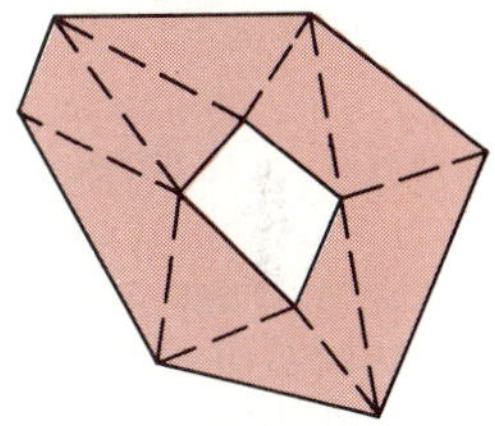

The figures above are examples of some polygonal regions. The figure at right above illustrates that a polygonal region may have a hole in it and still satisfy the definition.

Most of the polygonal regions we shall study have names. When we wish to refer to one without a name, we shall refer to it as a shaded region.

In order to measure the size of a polygonal region we need the following postulates:

P **Postulate 12-1 The Area Postulate** To each polygonal region there corresponds a unique positive real number.

The number referred to in Postulate 12-1 is called the **area** of the polygonal region. The area is the positive real number assigned to a given region.

Two polygons that have the same size and shape should also have the same area. Therefore, we can state the following postulate:

P **Postulate 12-2 The Congruence Postulate for Areas** If two polygons are congruent, then their polygonal regions have equal areas.

For example, if $\triangle ABC \cong \triangle DEF$, then $\mathscr{A}\triangle ABC = \mathscr{A}\triangle DEF$. The symbol $\mathscr{A}$ means "the area of." Thus, the last phrase above is read "the area of $\triangle ABC$ equals the area of $\triangle DEF$." When we refer to the area of a triangle, we mean the area of the triangular region of the triangle.

P

Postulate 12-3 The Area Addition Postulate If a polygonal region R is the union of nonoverlapping polygonal regions R_1 and R_2, then $\mathscr{A}R = \mathscr{A}R_1 + \mathscr{A}R_2$.

This postulate means that we can add areas that intersect in either a line segment, a point, or the empty set. For example, in each pair of figures below, $\mathscr{A}R = \mathscr{A}R_1 + \mathscr{A}R_2$

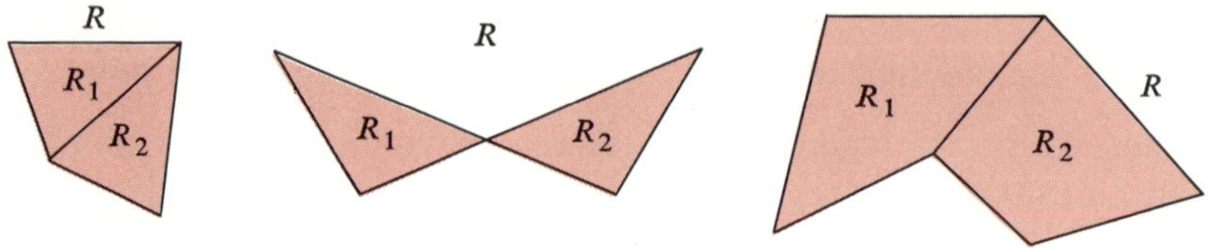

With these three postulates established, we are ready to discuss the actual measurement of a polygonal region. We stated earlier that to every polygonal region there corresponds exactly one positive number. What we need now is a method of determining that number. In other words, we must establish some standard unit of measure.

We could conceivably attempt to find out how many equilateral triangles or how many circles of a certain size would "fit" into a given polygonal region. However, you may recall from your earlier mathematics courses that the area of a polygonal region is measured not in triangle units or circle units, but in **square units**.

Sometimes—on a rectangle, for instance—we can mark off square units on a figure, so that it looks like a piece of graph paper. We can then count the units to find the area. More often, however, we cannot mark off entire square units. This is the case with triangles, irregular quadrilaterals, and many other polygons. Thus, we must rule out determining area by counting. We need to establish some formulas.

We shall begin by postulating a formula for the area of a rectangle and then use this formula to derive formulas for finding the areas of other polygons. Remember, area is measured in square units, such as the square inch, square foot, and square meter.

P

Postulate 12-4 Area of a Rectangle Postulate The area of a rectangle equals the product of the length of its base and its altitude. $\mathscr{A}$ rectangle $= b \cdot h$.

The proof of the following theorem follows directly from the preceding postulate and will be left as an exercise.

T

Theorem 12-1.1 The area of a square equals the square of the length of a side. $\mathscr{A}$ square $= s^2$.

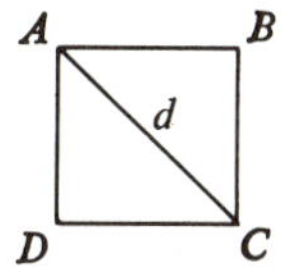

Example 1 Find the area of square $ABCD$, where $AC = d$.

Solution Using Corollary 8-9.la, $AB = \dfrac{d\sqrt{2}}{2}$.

Applying Theorem 12-1.1, we get $(AB)^2 = \left(\dfrac{d\sqrt{2}}{2}\right)^2 = \frac{1}{2}d^2$.

The area of square $ABCD = \frac{1}{2}d^2$.

Example 1 leads us to the following corollary.

C

Corollary 12-1.1a The area of a square equals one-half the square of the length of one of its diagonals. $\mathscr{A}$ square $= \frac{1}{2}d^2$.

Summary. Area Formulas for the Rectangle and Square

1. $\mathscr{A}$rectangle $= b \cdot h$. 2. $\mathscr{A}$square $= s^2$. 3. $\mathscr{A}$square $= \frac{1}{2}d^2$.

EXERCISES

A

Indicate which of the following are true and which are false. Correct the false statements.

1. A triangular region is a polygonal region.
2. Any polygonal region can be separated into triangular regions.
3. The area of a given polygonal region may vary.
4. The area of rectangle $PQRS = PQ \cdot RS$.
5. There exists a square whose area is $\sqrt{19}$.
6. The triangular region includes only the interior of a triangle.
7. A pentagon is a polygonal region.

8. Prove Theorem 12-1.1. 9. Prove Corollary 12-1.1a.

Trace each figure below and partition it into triangular regions to show that it represents a polygonal region.

10.

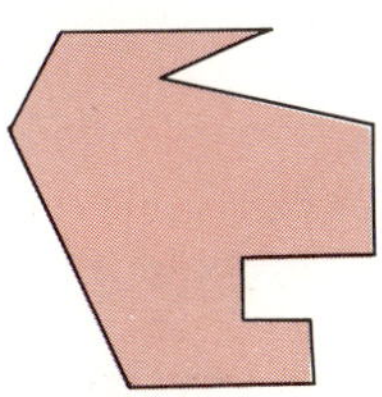

11.

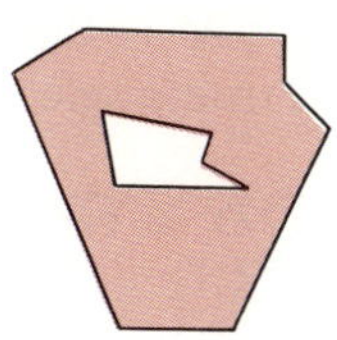

12.

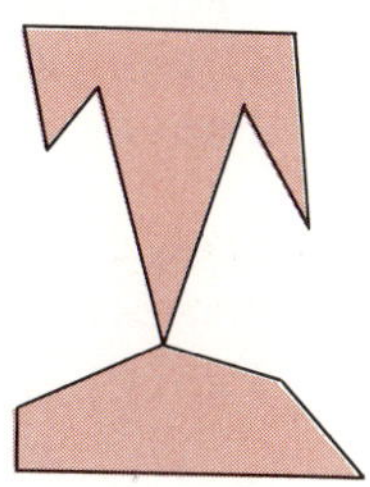

Refer to the figure at right when answering Exercises 13-21.

13. $\mathscr{A}\triangle ABC = \mathscr{A}(R_1) + \mathscr{A}(\quad) + \mathscr{A}(\quad) + \mathscr{A}(\quad)$.

14. $\mathscr{A}$ trapezoid $AFEC = \mathscr{A}(\quad) + \mathscr{A}(\quad)$.

15. $\mathscr{A}$ trapezoid $AFEC = \mathscr{A}$ square $AFED - \mathscr{A}(\quad) - \mathscr{A}(\quad)$.

16. $\mathscr{A}$ rectangle $AGHL = \mathscr{A}(\quad) + \mathscr{A}(\quad) + \mathscr{A}(\quad)$.

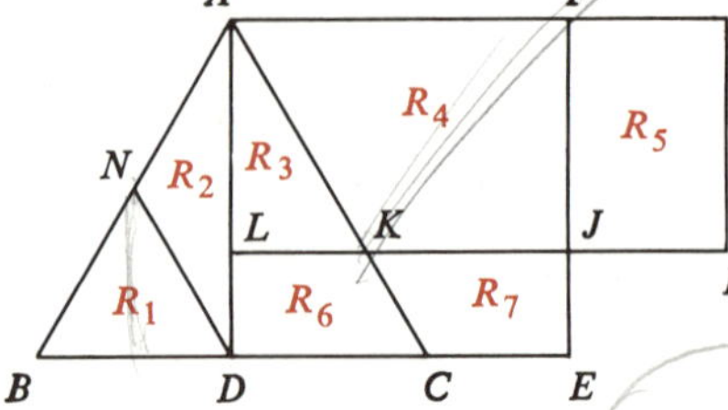

Exercises 13–21

17. Is $\mathscr{A}$ rectangle $AFJL + \mathscr{A}$ trapezoid $AKHG$ equal to $\mathscr{A}(R_3) + \mathscr{A}(R_4) + \mathscr{A}(R_5)$?

18. Is $\mathscr{A}\triangle ADC + \mathscr{A}$ rectangle $LJED$ equal to $\mathscr{A}(R_3) + \mathscr{A}(R_6) + \mathscr{A}(R_7)$?

19. Is $\mathscr{A}\triangle ABD + \mathscr{A}\triangle ALK$ equal to $\mathscr{A}(R_1) + \mathscr{A}(R_2) + \mathscr{A}(R_3)$?

20. If $\overline{AD}$ is an altitude of equilateral $\triangle ABC$, why is $\mathscr{A}\triangle ABD = \mathscr{A}\triangle ACD$?

21. If $\mathscr{A}$ rectangle $AGHL = 15$, $\mathscr{A}(R_3) = 3$, $\mathscr{A}(R_5) = 5$, and $\mathscr{A}(R_7) = 2$, find the area of trapezoid $AFEC$.

Find the area of a rectangle with altitude h and base b, given that:

22. $h = 7, b = 8$ 23. $h = 4\frac{3}{4}, b = 5\frac{4}{15}$ 24. $h = \frac{1}{2}x, b = 2x$

Find the area of each of the squares with the following side lengths:

25. 3 26. $3\frac{1}{2}$ 27. $\frac{1}{2}xy$ 28. $2x^2$

Find the area of each of the squares with the following lengths of a diagonal:

29. 12 30. $7\frac{1}{4}$ 31. $\frac{1}{4}xy$ 32. $4x^2$

B

33. If the length of the base of a rectangle is doubled and the length of the altitude is tripled, then the area is ________.

34. If the length of the altitude of a rectangle is doubled and the length of the base is divided by 2, then the area is ________.

35. If the length of each side of a square is tripled, then the area is ________.

36. If the length of each diagonal of a square is quadrupled, then the area is ________.

37. GIVEN Rectangle $ABCD$; M is the midpoint of $\overline{AB}$.
PROVE $\mathscr{A}\triangle AMD = \mathscr{A}\triangle BMC$

38. GIVEN Rectangle $ABCD$; M is the midpoint of $\overline{AB}$.
PROVE $\mathscr{A}\triangle DMC = \frac{1}{2}\mathscr{A}$ rectangle $ABCD$

39. GIVEN $\overline{BD} \cong \overline{CE}$ and $\overline{GD} \cong \overline{GC}$; $\angle B = \angle E$; $\overline{BDCE}$
PROVE $\mathscr{A}$quadrilateral $AGDB = \mathscr{A}$quadrilateral $GFEC$

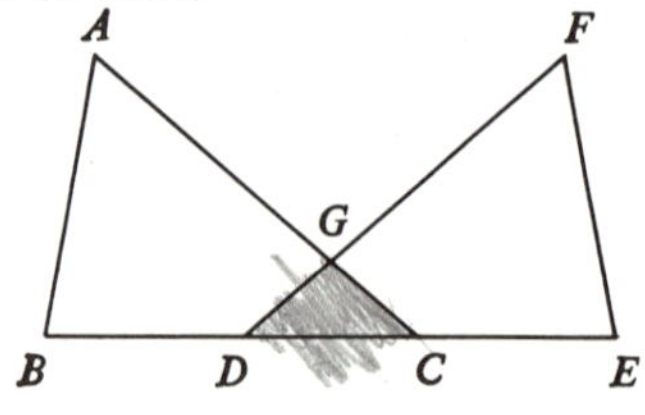

The perimeter of a rectangle is 30. Find the area of the rectangle for each of the following lengths of one side:

40. 4 41. 10 42. 12

43. Find the length of the altitude of a rectangle whose area is 48, if the length of the base is 16.
44. Find the length of the diagonal of a rectangle if the area is 120 and the length of one of the sides is 15.

The length of a diagonal of a rectangle is 20. Find the area of the rectangle if the measure of the angle the diagonal makes with a side is:

45. 30 46. 45 47. 60

48. Find the lengths of the sides of a rectangle if the length of the altitude is represented by $y + 4$, the length of the base is represented by $y - 2$, and the area is represented by $y^2 + 8$.

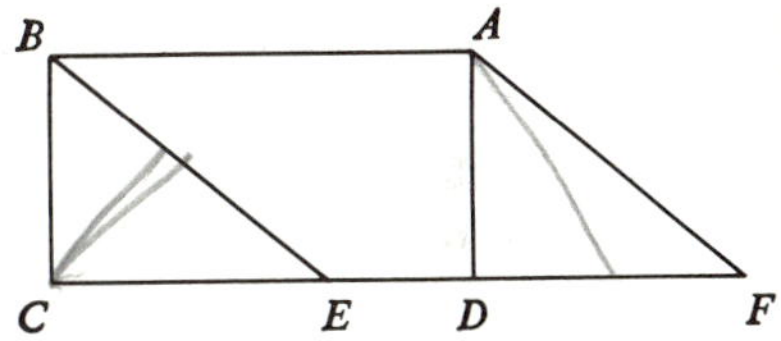

C

49. **GIVEN** Rectangle $ABCD$ and parallelogram $ABEF$; $\overline{FDEC}$
PROVE $\mathscr{A}$ rectangle $ABCD$ $= \mathscr{A}$ rectangle $ABEF$

50. **GIVEN** M, N, K, and L are the respective midpoints of sides $\overline{AB}$, $\overline{BC}$, $\overline{CD}$, and $\overline{DA}$ of square $ABCD$.
PROVE $\mathscr{A}$ quadrilateral $MNKL = \frac{1}{2}\mathscr{A}$ square $ABCD$

12-2 THE AREA OF A TRIANGLE

We shall use our formula for finding the area of a rectangle to derive formulas for finding the areas of various triangles.

Class Exercises

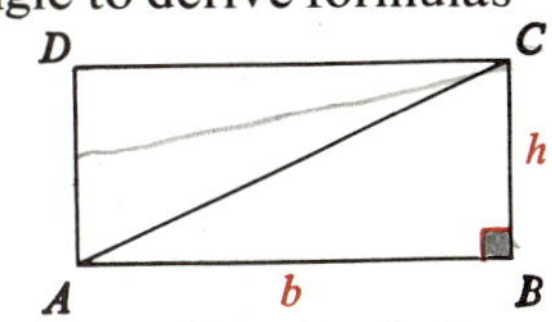

Exercises 1–4

1. What is the area of rectangle $ABCD$ at right?
2. What relationship exists between $\triangle ACB$ and $\triangle CAD$?
3. How does the area of right $\triangle ABC$ compare to the area of rectangle $ABCD$?
4. Therefore, $\mathscr{A}\triangle ABC = \frac{1}{2}(\_\_\_\_)(\_\_\_\_)$.

Exercises 5–10 refer to acute $\triangle ABC$ with altitude $\overline{CD}$.

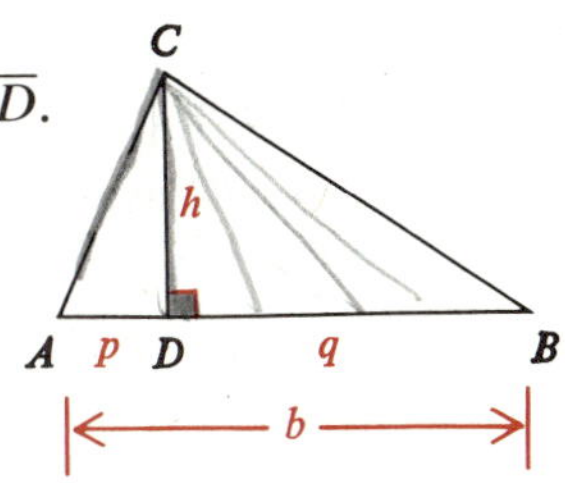

Exercises 5–10

5. What type of triangles are $\triangle ADC$ and $\triangle BDC$?
6. We can infer from Exercise 3 that $\mathscr{A}\triangle ADC = (\_\_\_\_)(h)(\_\_\_\_)$.
7. $\mathscr{A}\triangle BDC = (\_\_\_\_)(h)(\_\_\_\_)$.
8. $\mathscr{A}\triangle ABC = \mathscr{A}\triangle\_\_\_\_ + \mathscr{A}\triangle\_\_\_\_$. Why?
9. $\mathscr{A}\triangle ABC = (\_\_\_\_)(h)(\_\_\_\_) + (\_\_\_\_)(h)(\_\_\_\_)$.
10. $\mathscr{A}\triangle ABC = \frac{1}{2}(h)(\_\_\_\_ + \_\_\_\_) = \frac{1}{2}(h)(\_\_\_\_)$.

Exercises 11–13 refer to obtuse $\triangle ABC$ with altitude $\overline{CD}$, and $m\angle CAB > 90$.

11. $\mathscr{A}\triangle ABC = \mathscr{A}\triangle$____ $- \mathscr{A}\triangle$____.
12. $\mathscr{A}\triangle ABC =$ (____)(h)(____) $-$ (____)(h)(____).
13. $\mathscr{A}\triangle ABC = \frac{1}{2}(h)$ (____ $-$ ____) $= \frac{1}{2}(h)$(____).

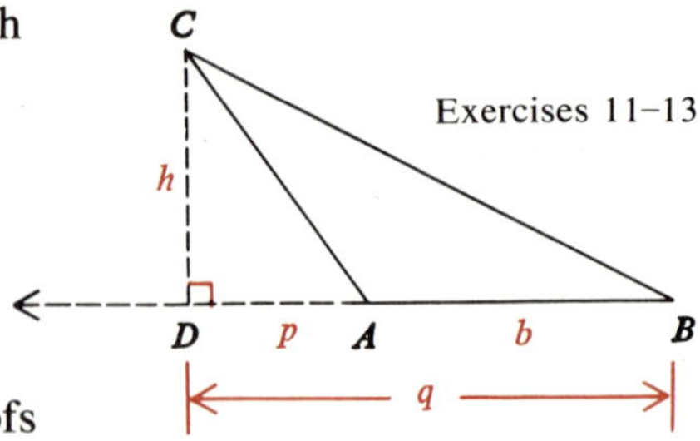

Exercises 11–13

The Class Exercises above outline the proofs for the following theorems. The complete proofs will be left as exercises.

Theorem 12-2.1 The area of a right triangle equals one-half the product of the lengths of its legs. $\mathscr{A}\text{right } \triangle = \frac{1}{2}(l_1 \cdot l_2)$.

Theorem 12-2.2 The area of any triangle equals one-half the product of the lengths of its base and the altitude to that base. $\mathscr{A}\text{triangle} = \frac{1}{2}b \cdot h$.

Referring to the figures in the Class Exercises above, describe the possible positions of the altitudes of an acute and an obtuse triangle.

Example 1 For the figure at right, show that $h_1 b_1 = h_2 b_2 = h_3 b_3$.

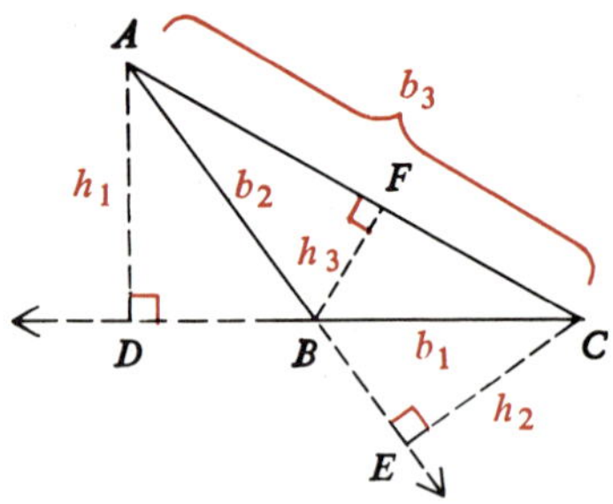

Solution Using Theorem 12-2.2, we find that
$\mathscr{A}\triangle ABC = \frac{1}{2}h_1 b_1$;
$\mathscr{A}\triangle ABC = \frac{1}{2}h_2 b_2$;
$\mathscr{A}\triangle ABC = \frac{1}{2}h_3 b_3$. Therefore,
$\frac{1}{2}h_1 b_1 = \frac{1}{2}h_2 b_2 = \frac{1}{2}h_3 b_3$,
or $h_1 b_1 = h_2 b_2 = h_3 b_3$.

What is true about the product of the lengths of any base of a triangle and its altitude as compared with the product of the lengths of any other base of the triangle and its altitude?

Corollary 12-2.2a Two triangles have equal areas if their bases have the same length and the altitudes to their bases have the same length.

Example 2 Prove that the median of any triangle separates the triangle into two regions of equal area.

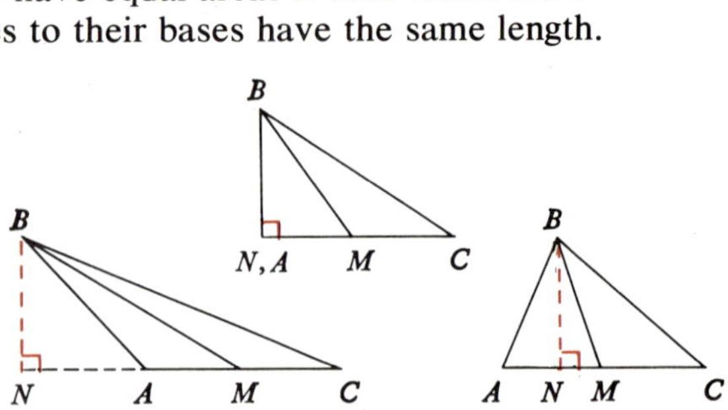

Solution

GIVEN $\overline{BM}$ is a median of $\triangle ABC$.

PROVE $\mathscr{A}\triangle ABM = \mathscr{A}\triangle CBM$

PROOF

STATEMENTS	REASONS
1. $\overline{BM}$ is a median of $\triangle ABC$.	1. Given.
2. M is the midpoint of $\overline{AC}$.	2. Definition of a median of a triangle. (3-9)
3. $AM = MC$	3. Definition of a midpoint of a line segment. (1-15)
4. Locate N in $\overleftrightarrow{AC}$ so that $\overline{BN} \perp \overleftrightarrow{AC}$.	4. Through a point external to a line, there is one and only one line perpendicular to the given line. (4-4.6)
5. $\overline{BN}$ is an altitude of both $\triangle ABM$ and $\triangle CBM$.	5. Definition of an altitude of a triangle. (3-10)
6. $\mathscr{A}\triangle ABM = \mathscr{A}\triangle CBM$	6. Corollary 12-2.2a.

C

Corollary 12-2.2b Triangles that share the same base and have their vertices in a line parallel to the base have equal areas.

In the figure at right, $\overleftrightarrow{AB} \parallel \overleftrightarrow{CDE}$. What is the relationship among the altitudes perpendicular to base $\overline{AB}$ of triangles ACB, ADB, and AEB? Why do these three triangles have equal areas? Your answers to these questions will serve as a guide to the proof of Corollary 12-2.2b, which is left as an exercise.

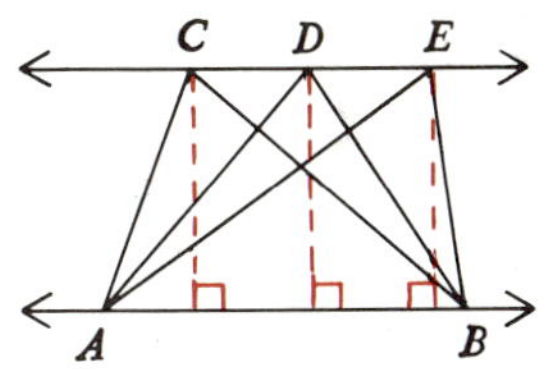

Example 3 How is the area of a triangle affected if the length of the altitude to the base is tripled but the length of the base remains the same?

Solution Suppose the lengths of the base and altitude of the original triangle are b and h, respectively. Then the area of the triangle equals $\frac{1}{2}bh$. If we triple the length of the altitude, the area of the new triangle is $\frac{1}{2}b \cdot 3h = 3(\frac{1}{2}bh)$. Therefore, the area of the new triangle is three times the area of the original triangle.

Example 3 suggests the following theorems:

T **Theorem 12-2.3** If two triangles have congruent bases, then the ratio of their areas equals the ratio of the lengths of their altitudes.

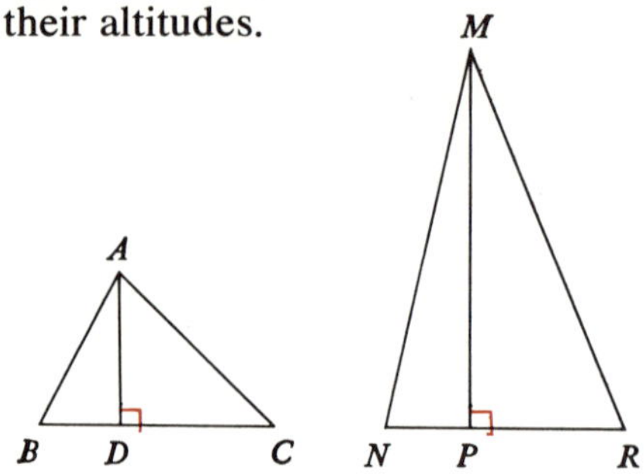

GIVEN $\overline{AD}$ and $\overline{MP}$ are altitudes of $\triangle ABC$ and $\triangle MNR$, respectively; $\overline{BC} \cong \overline{NR}$

PROVE $\dfrac{\mathscr{A}\triangle ABC}{\mathscr{A}\triangle MNR} = \dfrac{AD}{MP}$

PROOF

STATEMENTS	REASONS
1. $\overline{AD}$ and $\overline{MP}$ are altitudes of $\triangle ABC$ and $\triangle MNR$, respectively; $\overline{BC} \cong \overline{NR}$	1. Given
2. $\mathscr{A}\triangle ABC = \frac{1}{2}BC \cdot AD$; $\mathscr{A}\triangle MNR = \frac{1}{2}NR \cdot MP$	2. Theorem 12-2.2
3. $\dfrac{\mathscr{A}\triangle ABC}{\mathscr{A}\triangle MNR} = \dfrac{AD}{MP}$	3. Multiplication property

T **Theorem 12-2.4** If two triangles have congruent altitudes, then the ratio of their areas equals the ratio of the lengths of their bases.

The proof of this theorem is left as an exercise.

Example 4 P is in $\overline{EF}$ so that $EP = 4$ and $PF = 5$. Find $\mathscr{A}\triangle DPF$, if $\mathscr{A}\triangle DEF = 45$.

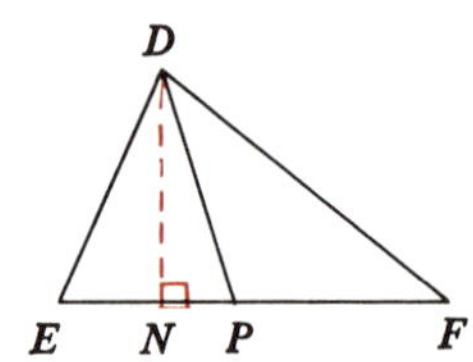

Solution Since $\triangle DPF$ and $\triangle DEF$ share the same altitude, $\overline{DN}$, $\dfrac{\mathscr{A}\triangle DPF}{\mathscr{A}\triangle DEF} = \dfrac{PF}{EF}$, by Theorem 12-2.4. Substituting the given values, we get $\dfrac{\mathscr{A}\triangle DPF}{45} = \dfrac{5}{9}$. Therefore, $\mathscr{A}\triangle DPF = 25$.

Although the formula $\mathscr{A} = \frac{1}{2}bh$ may be used for all types of triangles, there are times when other formulas are more easily applied.

T **Theorem 12-2.5** The area of any triangle equals one-half the product of the lengths of any two sides multiplied by the sine of the included angle. $\mathscr{A}\text{triangle} = \frac{1}{2}ab \cdot \sin \angle C$.

The next two theorems present additional ways of finding the area of an equilateral triangle.

T **Theorem 12-2.6** The area of an equilateral triangle equals $\frac{\sqrt{3}}{4}$ times the square of the length of a side. $\mathscr{A}$ equilateral triangle $= \frac{s^2\sqrt{3}}{4}$.

T **Theorem 12-2.7** The area of an equilateral triangle equals $\frac{\sqrt{3}}{3}$ times the square of the length of an altitude. $\mathscr{A}$ equilateral triangle $= \frac{h^2\sqrt{3}}{3}$.

The proofs of Theorems 12-2.5, 12-2.6, and 12-2.7 are left as exercises.

Summary. Formulas for Finding Areas of Triangles

1. $\mathscr{A}$ right triangle $= \frac{1}{2} l_1 \cdot l_2$.
2. $\mathscr{A}$ triangle $= \frac{1}{2} bh$.
3. $\mathscr{A}$ triangle $= \frac{1}{2} ab \cdot \sin \angle C$.
4. $\mathscr{A}$ equilateral triangle $= \frac{s^2\sqrt{3}}{4}$.
5. $\mathscr{A}$ equilateral triangle $= \frac{h^2\sqrt{3}}{3}$.

EXERCISES

A

Find the area of a triangle with base b and altitude h, where:

1. $b = 3$, $h = 8$
2. $b = 2\frac{1}{4}$, $h = 5\frac{2}{3}$
3. $b = x + 2$, $h = 3x$

Find the altitude to the base b of a triangle of area K, where:

4. $K = 10$, $b = 2$
5. $K = 15$, $b = 45$
6. $K = x - 5$, $b = 2x$

Find the area of a right triangle if the lengths of the three sides are:

7. 3, 4, 5
8. 11, 60, 61
9. $x, y, z; z > x, y$

Find the area of $\triangle ABC$ in each of the following exercises:

10. $AC = 5\sqrt{3}$, $BC = 6$, $m\angle C = 60$
11. $AB = 8$, $BC = 16$, and $m\angle B = 90$

Find the area of an equilateral triangle if the length of a side is:

12. 3
13. $6\sqrt{3}$
14. $8\sqrt{6}$
15. $12x$

Find the area of an equilateral triangle if the length of an altitude is:

16. 2
17. $9\sqrt{3}$
18. $6\sqrt{6}$
19. $12x$

20. GIVEN $\square ABCD$; $\overline{AC}$ and $\overline{BD}$ meet at E
    PROVE $\mathscr{A}\triangle AED = \mathscr{A}\triangle AEB$ $\quad$ $\mathscr{A}\triangle CEB = \mathscr{A}\triangle DEC$
21. GIVEN Right $\triangle ABC$; $m\angle C = 90$; $\overline{CD}$ is an altitude of $\triangle ABC$.
    PROVE $AC \cdot BC = CD \cdot AB$ (*Hint*: Use Theorems 12-2.1 and 12-2.2.)

22. Side $\overline{QMNR}$ of $\triangle PQR$ is partitioned by points M and N so that $QM = 3$, $MN = 2$, and $NR = 1$. If $\mathscr{A}\triangle PQR = 24$, find $\mathscr{A}\triangle MPN$ and $\mathscr{A}\triangle MPQ$.

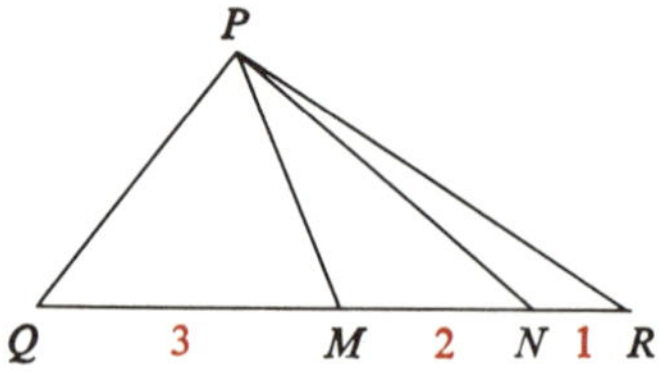

Prove each of the following:

23. Theorem 12-2.1
24. Theorem 12-2.2
25. Corollary 12-1.2b
26. Theorem 12-2.4
27. Theorem 12-2.5
28. Theorem 12-2.6
29. Theorem 12-2.7

30. In the figure at right, $\overline{AE}$ is an altitude of $\triangle ABC$, and $\overline{DF}$ is an altitude of $\triangle DBC$. If $AE = 8$ and $DF = 3$, find the ratio $\dfrac{\mathscr{A}\triangle DBC}{\mathscr{A}\triangle ABC}$.

31. What part of $\mathscr{A}\triangle ABC$ is the area of the shaded region in the figure for Exercise 30?

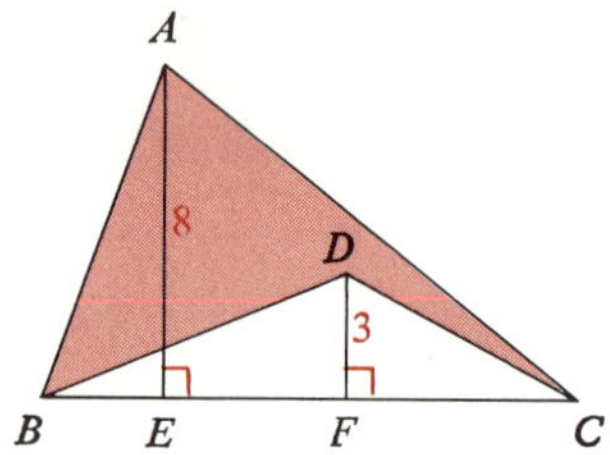

Exercises 30–31

B

32. **Given** $\overline{AM}$ is a median of $\triangle ABC$; $\overline{AM} \cong \overline{BM}$
Prove $\mathscr{A}\triangle ABC = \frac{1}{2}\cdot AB\cdot AC$

Using Theorem 12-2.5, find the ratio of the area of the smaller triangle to the area of the larger triangle in each of the following figures.

33.

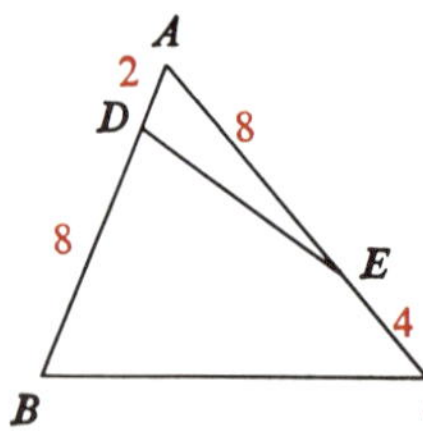

34.

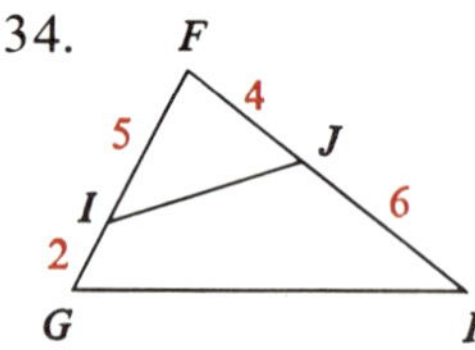

35.

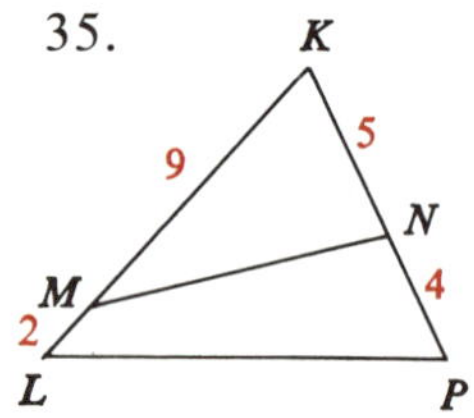

36. $\overline{BD}$ is an altitude of isosceles $\triangle ABC$, where $AB = AC = 13$, and $BC = 10$. Find BD. (*Hint*: Consider altitude $\overline{AE}$.)

37. **Given** P is any point of diagonal $\overline{AC}$ of parallelogram $ABCD$.
Prove $\mathscr{A}\triangle ADP = \mathscr{A}\triangle ABP$

38. Prove that if the diagonals of a quadrilateral separate it into four triangles of equal area, the quadrilateral is a parallelogram.

39. What part is $\mathscr{A}\triangle AFB$ of $\mathscr{A}\triangle ABC$ if the segments have lengths as indicated in the figure at right.

C

40. If, in $\triangle ABC$ of Exercise 39, $BD = 1$ and $DC = 5$, what part is $\mathscr{A}$ quadrilateral $ECDF$ of $\mathscr{A}\triangle ABC$.

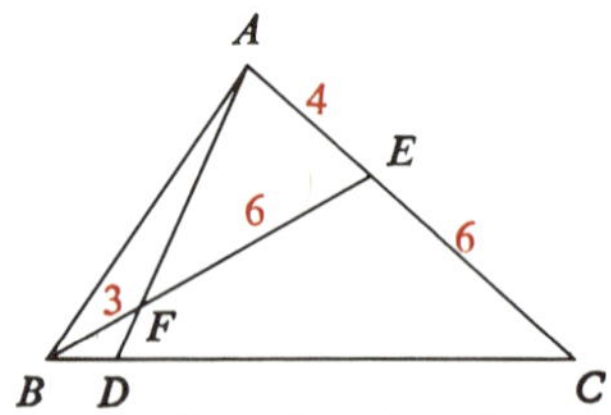

Exercises 39–40

41. For the figure at right, find $\frac{\mathscr{A}\triangle QKR}{\mathscr{A}\triangle PQR}$.

42. If $PM = 8$ and $QM = 6$ in Exercise 41, find $\frac{\mathscr{A}\text{ quadrilateral } PMKN}{\mathscr{A}\triangle PQR}$.

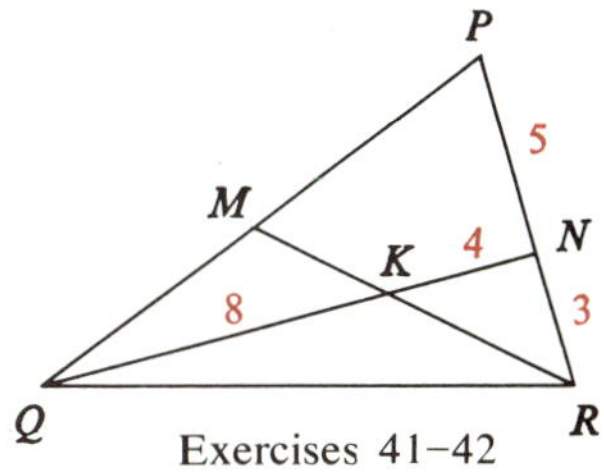

Exercises 41–42

43. Find the radius of a circle if the difference between the areas of the inscribed and circumscribed equilateral triangles is 25.

44. Prove that the three medians of a triangle separate the triangle into six triangles with equal areas.

45. Medians $\overline{CD}$ and $\overline{BE}$ of isosceles $\triangle ABC$ intersect in G. $\overline{AB} \cong \overline{AC}$ and $m\angle A = 30$. Find BA, if $\mathscr{A}\triangle BGC = 48$.

46. Prove that the areas of two triangles inscribed in the same circle or in congruent circles are proportional to the products of the lengths of their sides. (*Hint*: Use the result of Exercise 43 in Section 9-6.)

47. Prove the Proportional Line Segments Postulate by applying Theorem 12-2.4 and Corollary 12-2.2b.

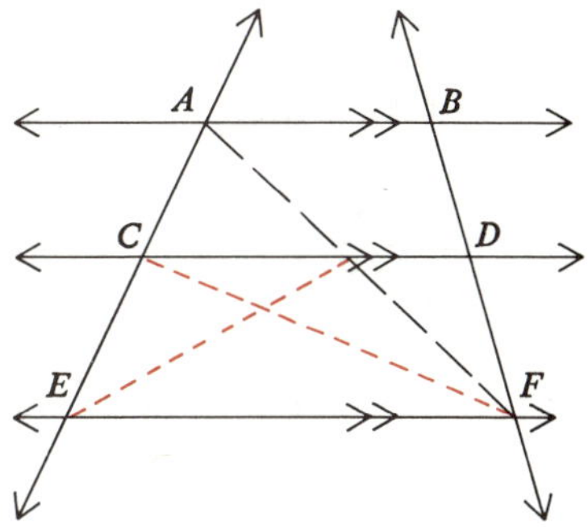

A LOOK AT THE PAST

Heron's Formula

Heron of Alexandria (about 60 A.D.) compiled an encyclopedia on such topics as geometry, computations, and mechanics. In his best-known work on geometry, called *Metrics*, he discusses two methods of finding the area of a triangle with sides of lengths 13, 14, and 15. The first method involves finding the length of one of its altitudes and then applying our Theorem 12-2.2. The second method avoids having to find the length of an altitude. By this method, the area of the given triangle is calculated as follows:

$$13 + 14 + 15 = 42$$

$$\frac{42}{2} = 21$$

$$21 - 13 = 8$$

$$21 - 14 = 7$$

$$21 - 15 = 6$$

$$21 \cdot 8 \cdot 7 \cdot 6 = 7056$$

$$\sqrt{7056} = 84, \text{ the area of the triangle.}$$

In general terms this calculation may be written as $\mathscr{A} = \sqrt{s(s-a)(s-b)(s-c)}$, where a, b, and c are the lengths of the sides of a triangle and $s = \dfrac{a+b+c}{2}$.

For many years, scholars attributed this formula to Heron because he cited no source for it. However, more recently, sufficient evidence has been found to indicate that this celebrated formula was actually derived by Archimedes.

12-3 AREAS OF SOME SPECIAL QUADRILATERALS

In the first section of this chapter, we presented methods of finding areas of rectangles and squares. To complete our study of the areas of quadrilaterals, we shall develop methods of finding the areas of parallelograms, rhombuses, and trapezoids.

T

Theorem 12-3.1 The area of a parallelogram equals the product of the lengths of a base and the altitude to that base. $\mathscr{A}$parallelogram $= b \cdot h$.

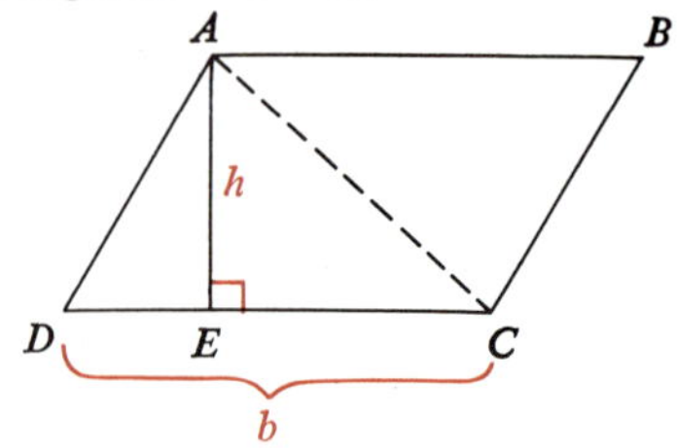

GIVEN $\overline{AE}$ is an altitude of $\square ABCD$; $AE = h$ and $DC = b$

PROVE $\mathscr{A}\square ABCD = bh$

PROOF

STATEMENTS	REASONS
1. $\overline{AE}$ is an altitude of $\square ABCD$; $AE = h$; $DC = b$	1. Given.
2. $\triangle ADC \cong \triangle CBA$	2. A diagonal of a parallelogram divides the parallelogram into two congruent triangles. (7-1.1).
3. $\mathscr{A}\triangle ADC = \mathscr{A}\triangle CBA$	3. Congruence Postulate for Areas. (12-2)
4. $\mathscr{A}\square ABCD = \mathscr{A}\triangle ADC + \mathscr{A}\triangle CBA$	4. Area Addition Postulate. (12-3)
5. $\mathscr{A}\square ABCD = 2\mathscr{A}\triangle ADC$	5. Substitution Postulate. (2-1)
6. $\mathscr{A}\triangle ADC = \frac{1}{2}bh$	6. Theorem 12-2.2.
7. $\mathscr{A}\square ABCD = bh$	7. Substitution Postulate. (2-1)

Since a rhombus is a parallelogram, we can apply the formula from Theorem 12-3.1 to find the area of a rhombus. However, we shall also develop a special formula for the area of a rhombus.

T

Theorem 12-3.2 The area of a rhombus equals one-half the product of the lengths of its diagonals. $\mathscr{A}\text{rhombus} = \frac{1}{2}(d_1 \cdot d_2)$.

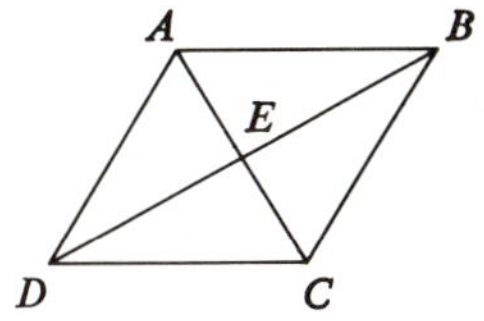

GIVEN Rhombus $ABCD$; $AC = d_1$ and $BD = d_2$

PROVE $\mathscr{A}$ rhombus $ABCD = \frac{1}{2}d_1 d_2$

PROOF

STATEMENTS	REASONS
1. Rhombus $ABCD$; $AC = d_1$ and $BD = d_2$	1. Given.
2. $\overline{AC} \perp \overline{BD}$	2. The diagonals of a rhombus are perpendicular to each other. (7-4.3)
3. $\overline{AE}$ is an altitude of $\triangle ABD$; $\overline{CE}$ is an altitude of $\triangle CDB$.	3. Definition of an altitude. (3-10)
4. $\mathscr{A}$ rhombus $ABCD = \mathscr{A}\triangle ABD + \mathscr{A}\triangle CDB$	4. Area Addition Postulate. (12-3)
5. $\mathscr{A}\triangle ABD = \frac{1}{2}BD \cdot AE$; $\mathscr{A}\triangle CDB = \frac{1}{2}BD \cdot CE$	5. Theorem 12-2.2.
6. $\mathscr{A}$ rhombus $ABCD = \frac{1}{2}BD \cdot AE + \frac{1}{2}BD \cdot CE$	6. Substitution Postulate. (2-1)
7. $\mathscr{A}$ rhombus $ABCD = \frac{1}{2}BD\,(AE + CE)$	7. Distributive property.
8. $\mathscr{A}$ rhombus $ABCD = \frac{1}{2}BD \cdot AC$ or $\frac{1}{2}(d_1 \cdot d_2)$	8. Point Betweenness Postulate. (2-4)

Example 1 Find the altitude AF of a rhombus if the lengths of the diagonals are 10 and 24.

Solution Since the diagonals of a rhombus bisect each other, $EC = 5$ and $DE = 12$. By the Pythagorean Theorem, $DC = 13$.

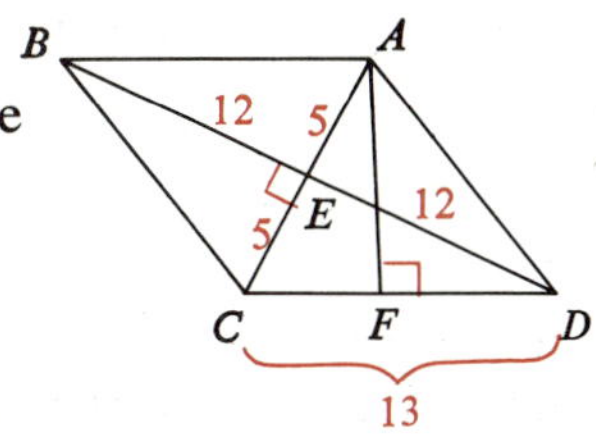

By Theorem 12-3.2, $\mathscr{A}$ rhombus $ABCD = \frac{1}{2}10 \cdot 24 = 120$. However, by Theorem 12-3.1, $\mathscr{A}\text{rhombus } ABCD = 13\,AF$. Therefore, $13\,AF = 120$, and $AF = \frac{120}{13}$.

T

Theorem 12-3.3 The area of a trapezoid equals one-half the product of the length of the altitude and the sum of the lengths of the bases. $\mathscr{A}\text{trapezoid} = \frac{1}{2}h(b_1 + b_2)$.

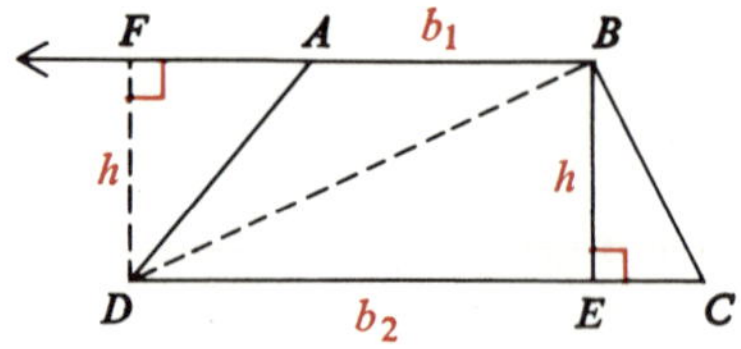

GIVEN Trapezoid $ABCD$; $\overline{AB} \parallel \overline{DC}$; altitude $\overline{BE}$; $BE = h$; $AB = b_1$; and $DC = b_2$

PROVE $\mathscr{A}$ trapezoid $ABCD = \frac{1}{2}h(b_1 + b_2)$

PROOF

STATEMENTS	REASONS
1. Trapezoid $ABCD$ has $\overline{AB} \parallel \overline{DC}$ and altitude $\overline{BE}$; $BE = h$; $AB = b_1$; $DC = b_2$	1. Given.
2. Consider $\overline{DF} \perp \overleftrightarrow{BA}$ at F.	2. Through a point external to a line, there is one and only one line perpendicular to the given line. (4-4.6)
3. $DF = BE = h$	3. If two lines are parallel, then all the points of one line are equidistant from the other line. (7-1.6)
4. $\mathscr{A}$ trapezoid $ABCD = \mathscr{A}\triangle ABD + \mathscr{A}\triangle BDC$	4. Area Addition Postulate. (12-3)
5. $\mathscr{A}\triangle ABD = \frac{1}{2}hb_1$; $\mathscr{A}\triangle BDC = \frac{1}{2}hb_2$	5. Theorem 12-2.2.
6. $\mathscr{A}$ trapezoid $ABCD = \frac{1}{2}hb_1 + \frac{1}{2}hb_2$	6. Substitution Postulate. (2-1)
7. $\mathscr{A}$ trapezoid $ABCD = \frac{1}{2}h(b_1 + b_2)$	7. Distributive property.

Example 2 Find the area of an isosceles trapezoid if the measure of one angle is 135 and the lengths of the bases are 10 and 18.

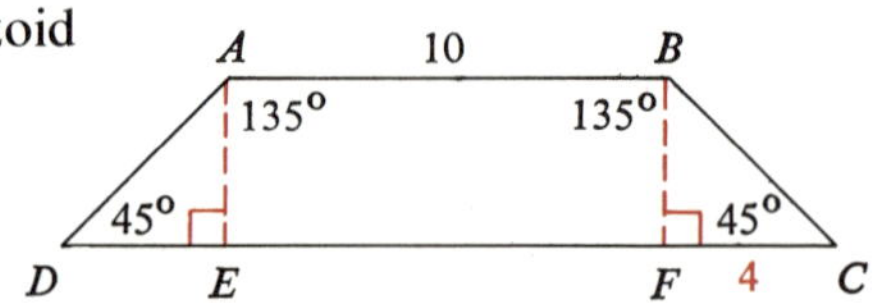

Solution Consider the isosceles trapezoid at right, where $\overline{AB} \parallel \overline{DC}$, $\overline{AD} \cong \overline{BC}$, and $\overline{AE}$ and $\overline{BF}$ are altitudes. Quadrilateral $ABFE$ is a rectangle; therefore, $EF = AB = 10$.

We can easily prove that since $\triangle AED \cong \triangle BFC$, $DE = CF = 4$. Since $m\angle D = 45$, $\triangle AED$ is isosceles (Why?) and $AE = 4$. Therefore, $\mathscr{A}$ trapezoid $ABCD = \frac{1}{2}(AE)(AB + DC) = \frac{1}{2}(4)(10 + 18) = 56$.

Class Exercises

Find the area of each of the quadrilaterals below.

1. 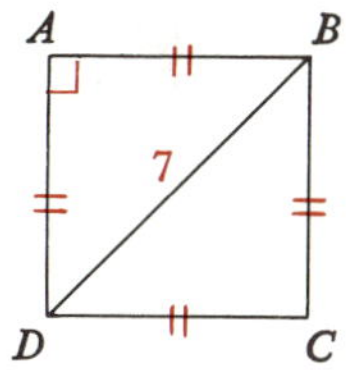

2. 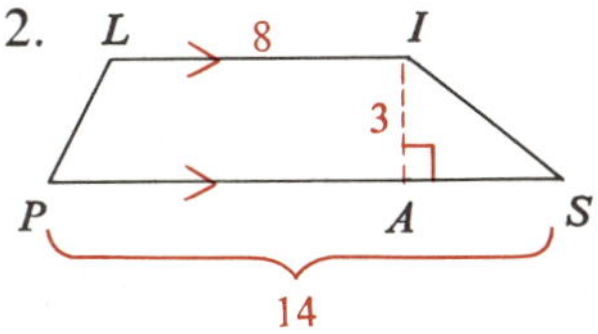

3. 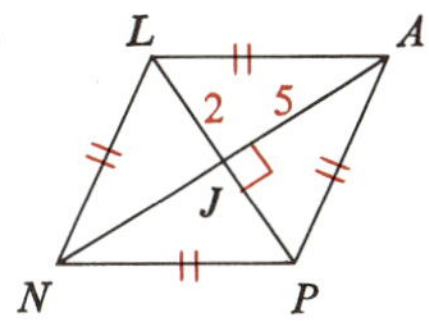

Summary. Formulas for Finding Areas of Quadrilaterals

1. $\mathscr{A}$ parallelogram $= bh$.
2. $\mathscr{A}$ rhombus $= \frac{1}{2}d_1d_2$.
3. $\mathscr{A}$ trapezoid $= \frac{1}{2}h(b_1 + b_2)$.

EXERCISES

A Find the area of a parallelogram whose base b and altitude h have the following lengths:

1. $b = 3$, $h = 8$ 2. $b = 7\frac{1}{2}$, $h = 10$ 3. $b = 2x$, $h = 5x$

Find the altitude to the base b of a parallelogram if the area and base lengths, given in respective order, are:

4. 20, 4 5. 15, 60 6. $18x^2$, $6x$

Find the area of a rhombus if its diagonals have the following lengths:

7. 12, 16 8. $3\frac{1}{2}$, $5\frac{1}{4}$ 9. $7x$, $14x$

Find the area of the trapezoid whose altitude h and bases b_1 and b_2 have the following lengths:

10. $h = 3$, $b_1 = 5$, $b_2 = 7$ 11. $h = 8$, $b_1 = 8$, $b_2 = 11$

Find the area of a parallelogram if two sides have lengths 8 and 12 and include an angle of the following measure:

12. 30 13. 45 14. 60 15. 90

Find the area of an isosceles trapezoid whose bases have lengths 9 and 17, where the measure of an angle is:

16. 135 17. 150 18. 120 19. 45

20. Find the area of an isosceles trapezoid whose longer base has length 20 and whose altitude has length 4, if the measure of one angle is 135.
21. The area of a trapezoid is 144 and the length of its altitude is 16. Find the lengths of the bases, if the length of the smaller base is half that of the longer base.

22. The area of a trapezoid is 27 and length of its altitude is 3. The length of the longer base exceeds the length of the shorter base by 1. Find the lengths of the bases.

23. Prove that the area of a trapezoid equals the product of the lengths of its altitude and median.

Find the length of the median of a trapezoid if the area and length of its altitude, given in respective order, are:

24. 20, 5 25. 32, 10 26. 17, $3\frac{1}{3}$ 27. $24\frac{1}{2}$, $7\frac{3}{4}$

28. Prove that the area of a parallelogram equals the product of the lengths of two adjacent sides and the sine of the included angle.

Find the area of a parallelogram with sides of lengths 10 and 15 where the measure of the included angle is:

29. 150 30. 135 31. 120 32. 90

33. Show how Theorem 12-3.2 may be used to derive Corollary 12-1.1a.

B Find x in each of the following figures:

34.
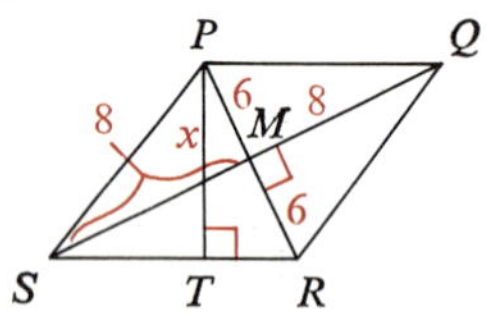

35.
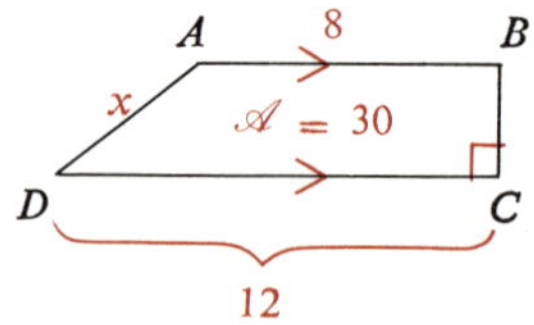

36.
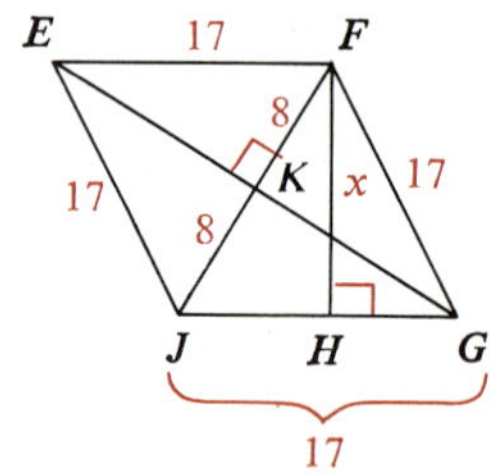

37. Given P is a point in side $\overline{AB}$ of $\square ABCD$.
Prove $\mathscr{A}\ \square ABCD = 2\mathscr{A}\triangle DPC$; $\mathscr{A}\triangle APD + \mathscr{A}\triangle BPC = \mathscr{A}\triangle DPC$

38. Given Diagonals $\overline{AC}$ and $\overline{BC}$ of $\square ABCD$ meet at E.
Prove $\mathscr{A}\ \triangle DEA = \frac{1}{4}\mathscr{A}\ \square ABCD$

39. Given Trapezoid $ABCD$ with median $\overline{MN}$; P is the midpoint of $\overline{MN}$; $\overline{AEB}$ and $\overline{DFC}$
Prove $\mathscr{A}$ trapezoid $AEFD$ $= \mathscr{A}$ trapezoid $CFEB$

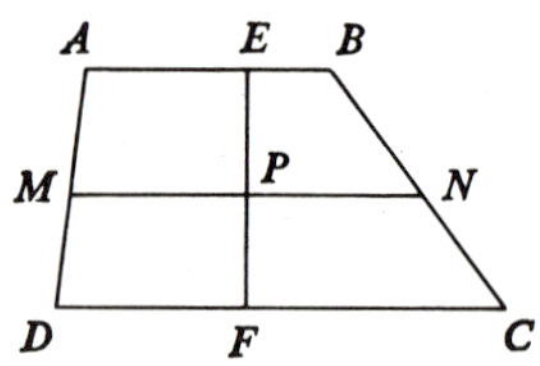

40. Find the area of an isosceles trapezoid whose bases have lengths 15 and 23 and whose legs have length 8.

41. Given P is any point of diagonal $\overline{DB}$ of $\square ABCD$; $\overline{GPH} \parallel \overline{AED}$; $\overline{EPF} \parallel \overline{AFB}$; $\overline{DHC}$; $\overline{BFC}$
Prove $\mathscr{A}\ \square AGPE = \mathscr{A}\ \square CFPH$

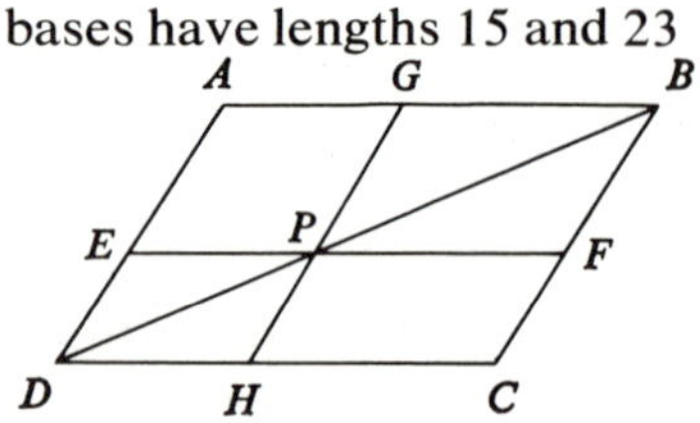

42. GIVEN P is in the interior of ▱$ABCD$.
PROVE $\mathscr{A}\,\triangle APD + \mathscr{A}\,\triangle BPC = \frac{1}{2}\mathscr{A}$ ▱$ABCD$

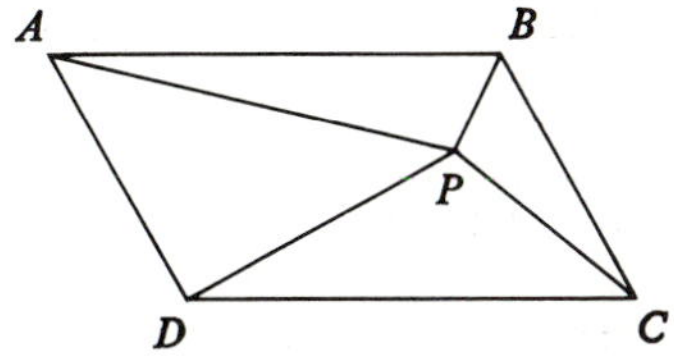

43. Prove that the area of a quadrilateral with perpendicular diagonals equals one-half the product of the lengths of the diagonals.

C

44. Prove that the area of any triangle equals one-half the product of its perimeter and the radius of the inscribed circle.

45. GIVEN M is the midpoint of side $\overline{DC}$ of ▱$ABCD$.
PROVE $\mathscr{A}$ trapezoid $MCBA = 3\ \mathscr{A}\triangle ADM$.

46. GIVEN Trapezoid $ABCD$ with median $\overline{MN}$; $\overline{APB}$
PROVE $\mathscr{A}\ \triangle MPN = \frac{1}{4}\mathscr{A}$ trapezoid $ABCD$

47. GIVEN $\triangle ABC$; M is the midpoint of $\overline{AB}$; N is the midpoint of $\overline{AC}$; $\overline{BDEC}$; $\overline{MD} \parallel \overline{NE}$
PROVE $\mathscr{A}$ quadrilateral $MNED = \frac{1}{2}\mathscr{A}\triangle ABC$

48. GIVEN Trapezoid $ABCD$; M is the midpoint of $\overline{BC}$.
PROVE $\mathscr{A}\triangle AMD = \frac{1}{2}\mathscr{A}$ trapezoid $ABCD$

49. GIVEN $\triangle TQC$; ▱$TIAC$ and ▱$TJBQ$; $\overleftrightarrow{BJ}$ meets $\overleftrightarrow{AI}$ at point L; ▱$QCPN$, where $\overline{QN} = \overline{LT}$ and $\overline{QN} \parallel \overline{LT}$
PROVE $\mathscr{A}$ ▱$TIAC + \mathscr{A}$ ▱$TJBQ = \mathscr{A}$▱$QCPN$
(*Hint*: $\mathscr{A}$▱$TJBQ = \mathscr{A}$ ▱$LTQH = \mathscr{A}$▱$QERN$.)

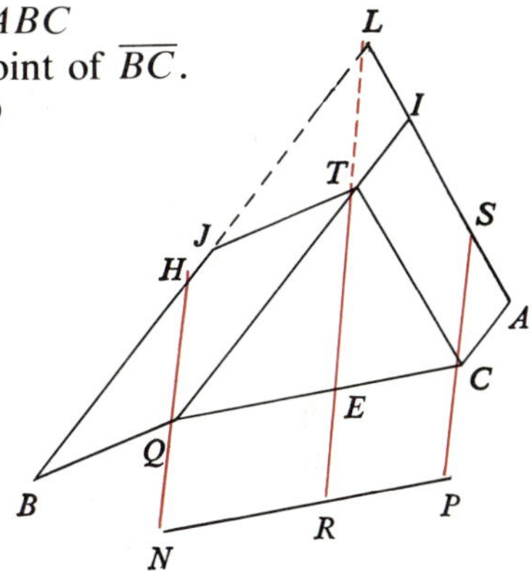

Exercise 49

50. GIVEN ▱$ABCD$ and ▱$EFGD$; $\overline{EAF}$; $\overline{BGC}$
PROVE $\mathscr{A}$▱$ABCD = \mathscr{A}$▱$EFGD$
(*Hint*: Consider $\overline{AG}$.)

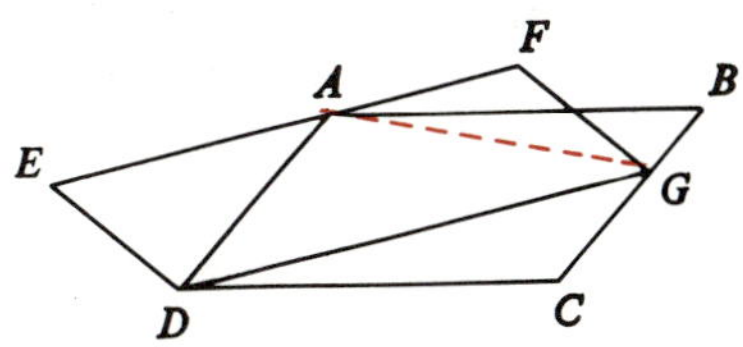

SOMETHING TO THINK ABOUT

The Pythagorean Theorem Revisited

What did President James A. Garfield have in common with Euclid?

In 1876, while still a member of the House of Representatives, James A. Garfield published an original proof of the Pythagorean Theorem. In 300 B.C.,

Euclid also published an original proof. In the discussion that follows, we shall briefly consider these two proofs.

Euclid's Proof

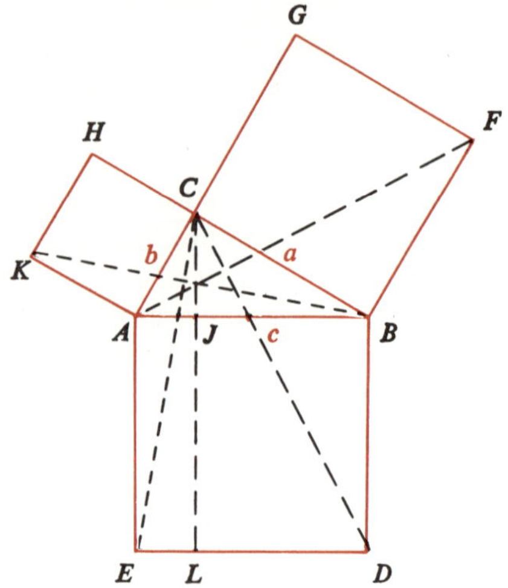

GIVEN Right $\triangle ABC$ with $m\angle C = 90$; $BAED$, $CAKH$, and $CBFG$ are squares.

PROVE $\mathscr{A}$square $CBFG$ $+ \mathscr{A}$square $CAKH$ $= \mathscr{A}$ square $BAED$ (that is, $a^2 + b^2 = c^2$)

PROOF OUTLINE

Consider $\overline{FA}$ and $\overline{CD}$. Why is $\triangle FBA \cong \triangle CBD$? It then follows that $\mathscr{A}\triangle FBA = \mathscr{A}\triangle CBD$. Draw $\overline{CL} \parallel \overline{BD}$. Since $\triangle CBD$ and rectangle $BJLD$ share the same base and altitude, $\mathscr{A}\triangle CBD = \frac{1}{2}\mathscr{A}$ rectangle $BJLD$; likewise $\mathscr{A}\triangle FBA = \frac{1}{2}\mathscr{A}$square $CBFG$. Therefore, $\mathscr{A}$ rectangle $BJLD$ $= \mathscr{A}$rectangle $CBFG$.

Why is $\triangle BAK \cong \triangle EAC$? It follows that $\mathscr{A}$ rectangle $JAEL = \mathscr{A}$ square $CHKA$. Therefore, $\mathscr{A}$ square $CHKA + \mathscr{A}$ square $CBFG = \mathscr{A}$ rectangle $BJLD$ $+ \mathscr{A}$ rectangle $JAEL$. But, $\mathscr{A}$ rectangle $BJLD + \mathscr{A}$ rectangle $JAEL$ $= \mathscr{A}$ square $BAED$. Thus, $\mathscr{A}$ square $CBFG + \mathscr{A}$ square $CHKA$ $= \mathscr{A}$ square $BAED$ or $a^2 + b^2 = c^2$.

The formal proof is left for you to complete.

President James A. Garfield's Proof

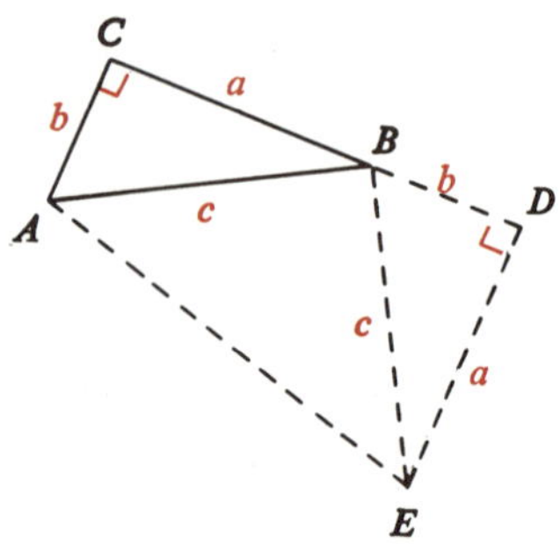

GIVEN Right $\triangle ABC$ with $m\angle C = 90$; $AC = b$; $BC = a$; and $AB = c$

PROVE $a^2 + b^2 = c^2$

PROOF OUTLINE

Select D in $\overleftrightarrow{BC}$ so that $BD = AC$ and $\overline{CBD}$. Consider $\overline{DE} \perp \overline{CBD}$ so that $DE = BC$. What type of quadrilateral is quadrilateral $ACDE$? Why is $\mathscr{A}\triangle ABC = \mathscr{A}\triangle BED$? Why is $AB = BE$?

By Theorem 12-3.3, $\mathscr{A}$ trapezoid $ACDE = \frac{1}{2}CD(AC + DE)$ $= \frac{1}{2}(a + b) \cdot (a + b) = \frac{1}{2}(a + b)^2$.

By Theorem 12-2.1, $\mathscr{A}\triangle ABE = \frac{1}{2}AB \cdot BE = \frac{1}{2}c^2$.

Also by Theorem 12-2.1, $\mathscr{A}\triangle ABC = \frac{1}{2}AC \cdot BC = \frac{1}{2}ab.$

However, $\mathscr{A}$ trapezoid $ACDE = \mathscr{A}\triangle ABE + 2\mathscr{A}\triangle ABC$. Substituting, we get

$$\frac{1}{2}(a+b)^2 = \frac{1}{2}c^2 + 2(\tfrac{1}{2}ab)$$

$$(a+b)^2 = c^2 + 2ab$$

$$a^2 + 2ab + b^2 = c^2 + 2ab$$

$$a^2 + b^2 = c^2$$

The formal proof is left for you to complete.

12-4 THE AREA OF A REGULAR POLYGON

We began studying the area of regular polygons in the first section of this chapter, when we developed methods of finding the area of a square, which is a regular quadrilateral. In the second section, we derived formulas for finding the area of a regular triangle, the equilateral triangle.

In this section we shall develop a formula for finding the area of any regular polygon. To simplify the development of the formula for the area of a regular polygon, we shall first consider a regular pentagon and then generalize our findings to any regular n-gon.

D **Definition 12-3** The **center of a regular polygon** is the center of its circumscribed (or inscribed) circle.

D **Definition 12-4** A segment joining any vertex of a regular polygon with the center of that polygon is a **radius of the polygon**.

Consider the figure for Theorem 12-4.1. If P is the center of the regular pentagon $ABCDE$, then $\overline{PA}$, $\overline{PB}$, $\overline{PC}$, $\overline{PD}$, and $\overline{PE}$ are the radii of the regular pentagon. Are all radii of regular polygons congruent? Why can a regular polygon be inscribed in a circle? How do the radii of the regular polygon and the radii of the circumscribed circle compare?

If $\overline{PN}$ is an altitude of $\triangle DPC$, what is the area of $\triangle DPC$? What is the relationship betweeen the five triangles formed by the radii of the pentagon? Why is the sum of the areas of these five triangles equal to the area of the pentagon? Your answers lead to the next theorem. First, however, we must redefine altitude $\overline{PN}$ in terms of a regular polygon.

D **Definition 12-5** The **apothem** of a regular polygon is the segment from the center of the polygon perpendicular to a side of the polygon.

Are all the apothems of a regular polygon congruent? How many apothems are in a square?

T

Theorem 12-4.1 The area of a regular polygon equals one-half the product of the lengths of the apothem and the perimeter. $\mathscr{A}$ regular polygon $= \frac{1}{2} a \cdot p$.

We shall prove that Theorem 12-4.1 holds true for a pentagon. However, the proof can be extended to any n-gon. This is left as an exercise.

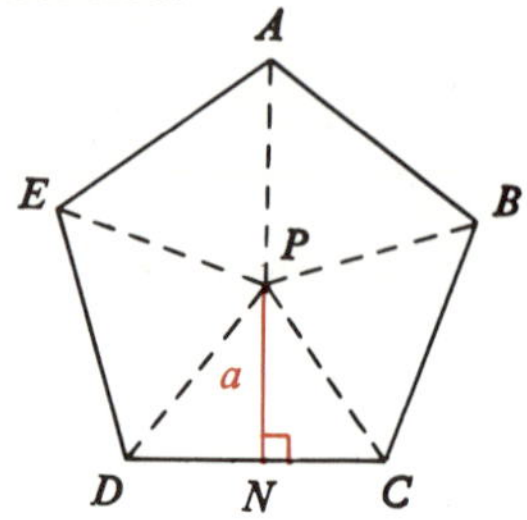

GIVEN Regular pentagon $ABCDE$ with center P and apothem $\overline{PN}$; $\overline{PN} = a$ and perimeter $= p$

PROVE $\mathscr{A}$ pentagon $ABCDE = \frac{1}{2} ap$

PROOF

STATEMENTS	REASONS
1. Regular pentagon $ABCDE$ has center P and apothem $\overline{PN}$; $PN = a$ and perimeter $= p$	1. Given.
2. $AB = BC = CD = DE = EA$	2. Definition of a regular polygon. (6-4)
3. $AP = BP = CP = DP = EP$	3. Definition of a radius of a regular polygon. (12-4)
4. $\triangle APE \cong \triangle BPA \cong \triangle CPB \cong \triangle DPC \cong \triangle EPD$	4. SSS Postulate. (3-3)
5. $\mathscr{A} \triangle APE = \mathscr{A} \triangle BPA = \mathscr{A} \triangle CPB = \mathscr{A} \triangle DPC = \mathscr{A} \triangle EPD$	5. The Congruence Postulate for Areas. (12-2)
6. $\mathscr{A} \triangle DPC = \frac{1}{2} a \cdot DC$	6. Theorem 12-2.2
7. $\mathscr{A}$ pentagon $ABCDE = 5(\frac{1}{2} a \cdot DC)$	7. Area Addition Postulate. (12-3)
8. $\mathscr{A}$ pentagon $ABCDE = \frac{1}{2} a \cdot 5DC$	8. Associative and commutative properties.
9. $5\ DC = p$	9. Definition of perimeter.
10. $\mathscr{A}$ pentagon $ABCDE = \frac{1}{2} ap$	10. Substitution Postulate. (2-1)

Example 1 Find the area of the regular polygon whose perimeter is 40 and whose apothem is 6.

Solution From Theorem 12-4.1, we find that the area of the regular polygon $= \frac{1}{2} ap = \frac{1}{2}(6 \cdot 40) = 120$.

Example 2 Find the area of a regular hexagon if one side has length 6.

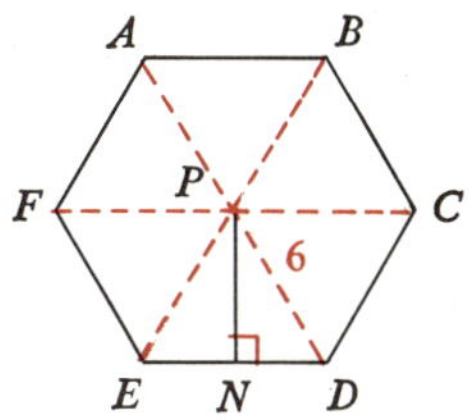

Solution Since $\triangle PED$ is an equilateral triangle (Why?), $PD = 6$, and in 30-60-90 $\triangle PND$, $PN = 3\sqrt{3}$. The perimeter of the hexagon is 36. Therefore, by Theorem 12-4.1, $\mathscr{A}$ hexagon $ABCDEF = \frac{1}{2}(3\sqrt{3} \cdot 36) = 54\sqrt{3}$.

Alternate solution Since a regular hexagon is composed of six congruent equilateral triangles, by Theorem 12-2.6, $\mathscr{A}$ hexagon $ABCDEF = 6\left(\frac{6^2\sqrt{3}}{4}\right) = 54\sqrt{3}$.

EXERCISES

A

Write a formula for the area of polygon R with apothem a, if the length of a side is s, letting polygon R be each of the following figures in turn:

1. pentagon 2. octagon 3. decagon 4. dodecagon

Find the area of a regular polygon with apothem and perimeter, respectively,

5. 5 and 20 6. 7 and 48 7. 9 and 37 8. $4x$ and $17x$

Find the apothem of a regular polygon if the area and perimeter, given in order, are:

9. 16 and 6 10. 8 and 18 11. 17 and 7 12. $15x^2$ and $10x$

Find the apothem and area of a regular 4-gon if the length of a side is:

13. 4 14. 6 15. 7 16. $12x$

Find the apothem and area of a regular hexagon whose perimeter is:

17. 36 18. 24 19. 30 20. 50

Find the apothem and area of a regular 3-gon whose perimeter is:

21. 18 22. 12 23. 15 24. 25

Find the apothem and perimeter of a regular 4-gon, given that the area is:

25. 16 26. 25 27. 36 28. 32

Find the apothem and perimeter of a regular hexagon if the area is:

29. $18\sqrt{3}$ 30. $54\sqrt{3}$ 31. $150\sqrt{3}$ 32. 36

Find the area of a regular hexagon whose perimeter equals that of a square if the area of the square is:

33. 36 34. 16 35. 64 36. 32

Find the area of a regular hexagon if the area of an equilateral triangle with the same perimeter is:

37. $9\sqrt{3}$ 38. $24\sqrt{3}$ 39. $3\sqrt{3}$ 40. $36\sqrt{3}$

B Find the area of a regular pentagon, correct to the nearest square inch, if the length of a side is:

41. 2 in. 42. 6 in. 43. 20 in. 44. 13 in.

Find the area of a regular dodecagon, correct to the nearest square inch, if its perimeter is:

45. 12 in. 46. 36 in. 47. 360 in. 48. 30 in.

Find the area of a regular decagon, correct to the nearest square foot, if the length of the apothem is:

49. 2 in. 50. 7 in. 51. 10 in. 52. 15 in.

53. Find the area of a regular pentagon, correct to the nearest square inch, if the radius of its circumscribed circle is 10 in.
54. Prove that the area of a regular hexagon is twice the area of the equilateral triangle inscribed in the circumscribed circle of the hexagon.
55. What is the ratio of the area of a square to the area of a regular hexagon if both are inscribed in the same circle?
56. What is the ratio of the area of an equilateral triangle to the area of the regular hexagon inscribed in the inscribed circle of the equilateral triangle?
57. The radii of two regular hexagons are 8 and 6. Find the ratio of the areas of these two polygons.
58. Find the radius of a regular dodecagon whose area is 3.
59. Prove Theorem 12-4.1 for an *n*-gon.

C

60. If the hexagon formed by joining the midpoints of a regular hexagon has an area of $\frac{225\sqrt{3}}{2}$, what is the area of the larger hexagon?
61. If the area of an equilateral triangle inscribed in a circle is K, find the area of the regular dodecagon inscribed in the same circle in terms of K.
62. A regular octagon is inscribed in a circle of radius r. Find the area of the octagon in terms of r.
63. Prove that the area of any polygon equals one-half the product of its perimeter and the radius of its inscribed circle.
64. Prove that the sum of the lengths of the segments perpendicular to the sides of a regular polygon from any point in the interior of the polygon equals the product of the length of the apothem and the number of sides.

12-5 THE AREA OF A CIRCLE

Before we can develop a formula for the area of a circle, we must state some definitions.

Definition 12-6 A **circular region** is the union of a circle and its interior.

Consider the regular polygons inscribed in circle Q below. How does the area of the 3-gon compare to that of the 4-gon? Can you prove your answer? How does the area of each polygon below compare to the area of another polygon having a greater number of sides?

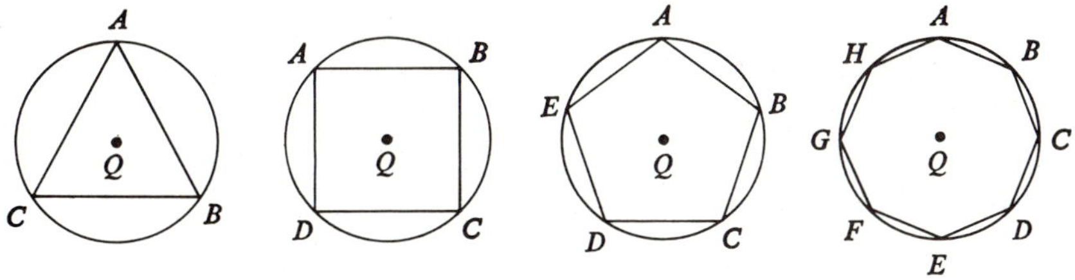

You should be aware that as the number of sides of the regular polygon inscribed in circle Q increases, the area of the circular region which is not in the polygonal region is decreasing. How do you think the size of the polygonal region will compare to the size of the circular region as the number of sides of the regular polygon increases?

In our definition of the circumference of a circle, we indicated that as the number of sides of regular polygons inscribed in the same circle increases, the perimeters of the polygons approach the circumference of the circumscribed circle as a limit. By now you should have anticipated the following definition:

Definition 12-7 The **area of a circular region** is the limit of the areas of the regular polygons inscribed in the circle, as the number of sides increases without bound.

In Section 12-1 we agreed to refer to the area of a polygonal region as the area of the polygon. Similarly, when we speak of the area of a circle, we are actually referring to the area of its circular region.

We shall now examine the various parts of a regular n-gon, with apothem length a and radius r.

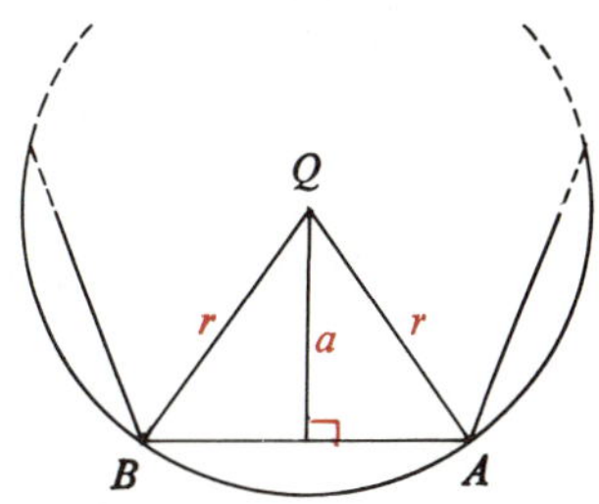

Why is $a < r$? As n increases, $a \to r$ (read "a approaches r as a limit").

What does the perimeter p of the polygon approach as n increases? From the definition of the circumference c of a circle, we can say that $p \to c$ as n increases. Combining these limits, we find that by the formula for the area, K_n, of the polygon,

$$\tfrac{1}{2}ap \text{ approaches } \tfrac{1}{2}rc$$

However, since $c = 2\pi r$,

$$\tfrac{1}{2}ap \text{ approaches } \tfrac{1}{2}r(2\pi r)$$

or

$$\tfrac{1}{2}ap \text{ approaches } \pi r^2$$

From the definition of the area, K, of a circle,

$$K_n \text{ approaches } K$$

That is,

$$\tfrac{1}{2}ap \text{ approaches } K$$

And so we have the familiar formula for the area of a circle, $K = \pi r^2$. This is our next theorem.

T **Theorem 12-5.1** The area of a circle with radius r equals πr^2.

The reasoning in the following example is intuitive rather than deductive.

Example 1 Find the area of the shaded region in the figure at right, where the radius of circle Q is 12, and $m\angle AQB = 90$.

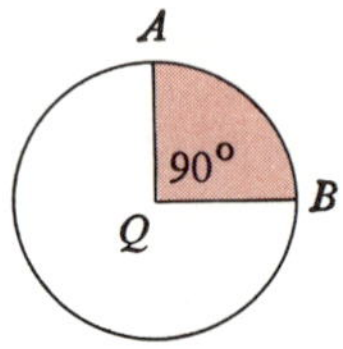

Solution Since $m\overset{\frown}{AB} = 90$, the length of $\overset{\frown}{AB}$ is $\frac{90}{360}$ or one-fourth of the circumference of circle Q. Similarly, it seems reasonable that the area of the shaded region is one-fourth of the area of circle Q. Therefore, the area of the shaded region is $\frac{1}{4}\pi(12)^2 = \frac{1}{4}\pi(144) = 36\pi$. This shaded region is called a sector of circle Q.

D **Definition 12-8** A **sector of a circle** is a region bounded by an arc of the circle and the two radii which contain the endpoints of the arc.

The solution of Example 1 suggests the following theorem:

T **Theorem 12-5.2** The area of a sector with radius r and a central angle of measure n equals $\frac{n}{360} \cdot \pi r^2$.

The proof of this theorem is similar to the one we used to establish a formula for finding the area of a circle. We shall omit the proof here.

Example 2 In the figure at right, $\triangle PQR$ is equilateral and the radius of circle Q is 6. Find the area of the shaded region.

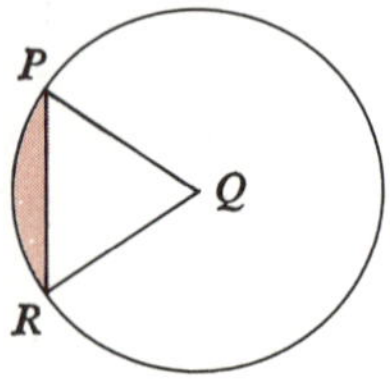

Solution To find the area of the shaded region we must subtract the area of $\triangle PQR$ from the area of sector PQR. The area of sector $PQR = \frac{60}{360} \cdot \pi 6^2 = 6\pi$ (Theorem 12-5.2). The area of equilateral $\triangle PQR = \frac{6^2\sqrt{3}}{4} = 9\sqrt{3}$ (Theorem 12-2.6). Therefore the area of the shaded region equals $6\pi - 9\sqrt{3}$. (*Note*: There is no need to find the decimal approximation of an expression like this unless you are told to do so.)

The shaded region in Example 2 is called a segment of a circle.

D **Definition 12-9** A **segment of a circle** is a region bounded by a minor arc of a circle and the chord containing the endpoints of the arc.

Summary. Formulas for Finding Areas of a Circle and Sector

1. $\mathscr{A}$ circle $= \pi r^2$.
2. $\mathscr{A}$ sector $= \frac{n}{360} \cdot \pi r^2$.

EXERCISES

A

Find the areas of the circles with these diameters.

1. 4 2. 20 3. 15 4. $5\sqrt{2}$

Find the areas of the circles with these circumferences.

5. 10π 6. 36π 7. 25π 8. $9\sqrt{2}$

Find the radii of the circles with the following areas:

9. 25π 10. 81π 11. 24π 12. $8\sqrt{2}$

Find the circumference of the circle whose area is:

13. 49π 14. 144π 15. 18π 16. 16

In a circle of radius 6, find the areas of the sectors whose central angles have the following measures:

17. 30 18. 45 19. 72 20. 84

Find the area of a circle if 24π is the area of a sector whose central angle has a measure of:

21. 90 22. 120 23. 144 24. 80

In a circle whose radius is 12, find the central angle of a sector whose area is:

25. 16π 26. 36π 27. 9π 28. 12π

Find the diameter of a circle, where a sector of area 24π has a central angle with a measure of:

29. 30 30. 60 31. 120 32. 50

Find the area of a circle circumscribed about a square, the length of whose sides is:

33. 8 34. $4\sqrt{2}$ 35. 10 36. $8\sqrt{3}$

Find the area of a circle circumscribed about an equilateral triangle whose altitude has a length of:

37. 6 38. 9 39. $6\sqrt{3}$ 40. 10

Find the area of a circle circumscribed about a regular hexagon whose sides have a length of:

41. 2 42. 6 43. 20 44. $9\sqrt{2}$

Find the area of the shaded region in each of the following figures:

45.

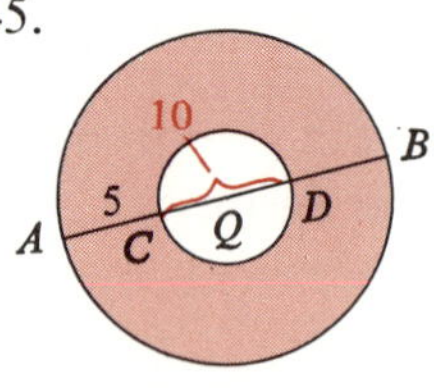

46.

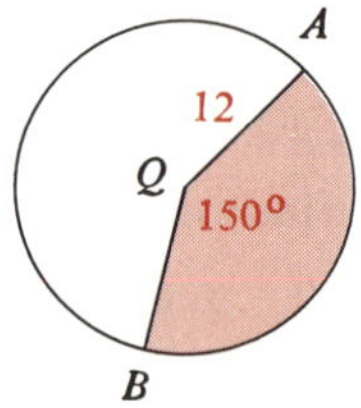

47.

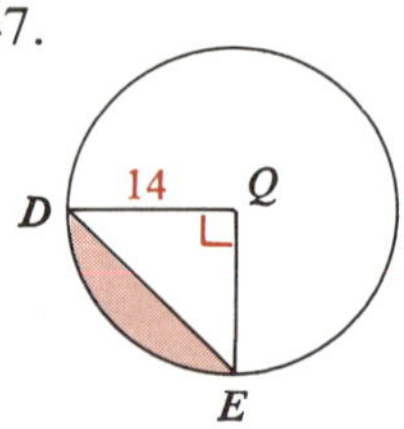

B 48. Show how Theorem 12-5.2 leads to the formula $K = \frac{1}{2}rL$, where K is the area of a sector, r is the radius, and L is the length of the arc of the sector.

49. Find the radius of a circle if the area of a sector is 9 and the length of its arc is 9π.

50. In a circle of radius $\frac{8}{\pi}$, what part of the area of the circle is the area of a sector with an arc of length 4?

Find the area of the shaded region in each of the following figures. The heavy dots represent the centers of circles.

51.

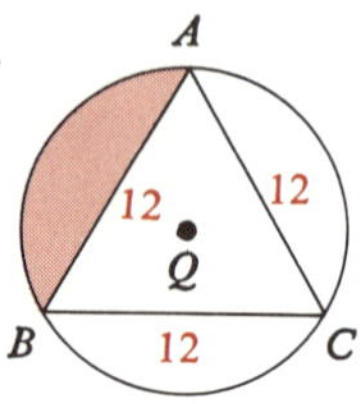

52.

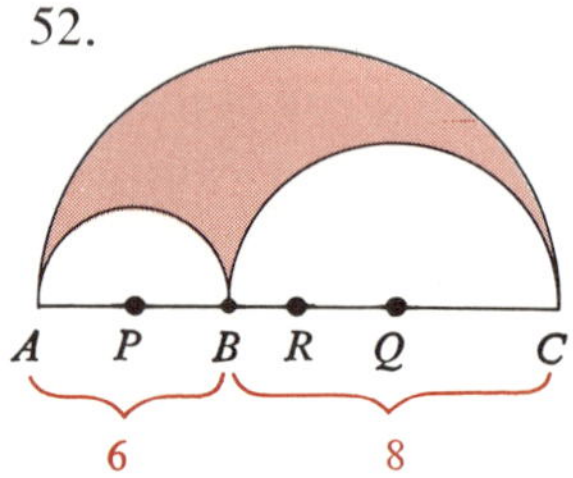

53.

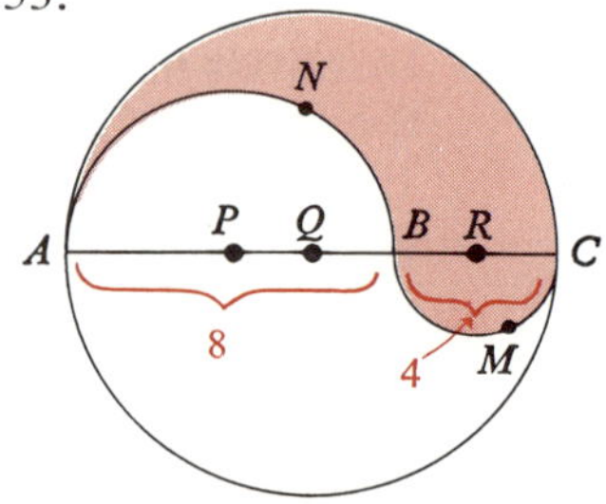

C 54.

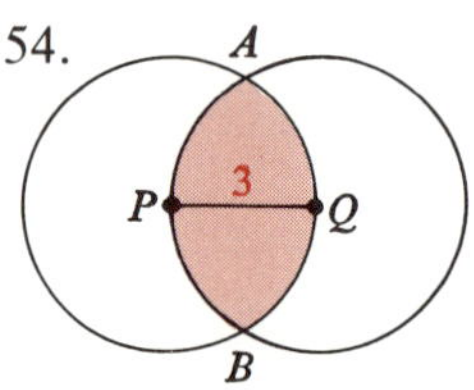

55.

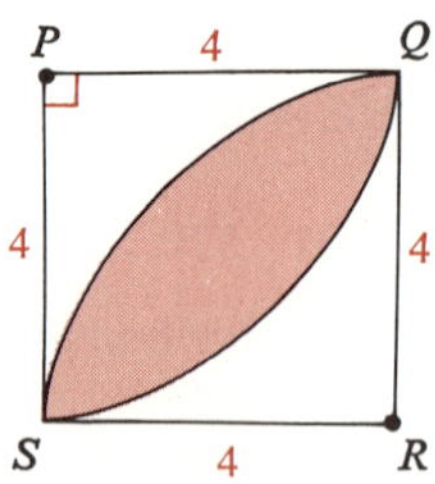

56.

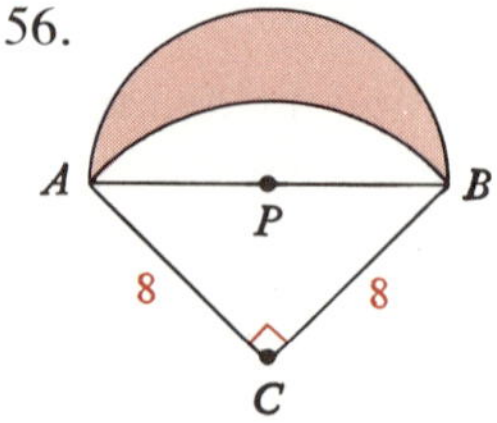

57. GIVEN Right $\triangle ABC$; $m\angle C = 90$; three semicircles with the sides of $\triangle ABC$ as diameters.

PROVE $\mathscr{A}$ semicircle APC + $\mathscr{A}$ semicircle CQB = $\mathscr{A}$ semicircle ARB.

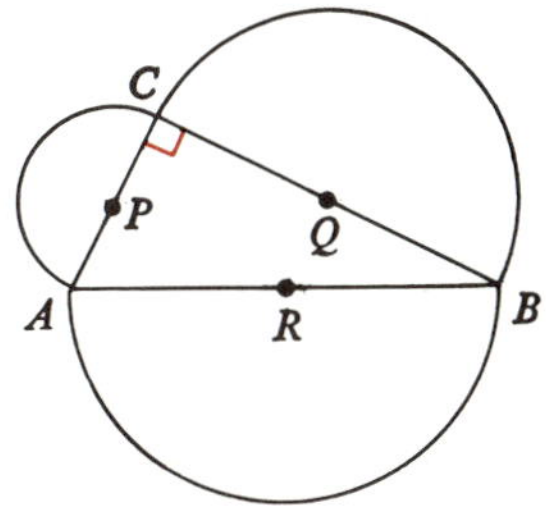

58. In the figure at right, semicircles are drawn with the sides of $\triangle ABC$ as diameters. Prove that $\mathscr{A}R_1 + \mathscr{A}R_2 = \mathscr{A}\triangle ABC$.

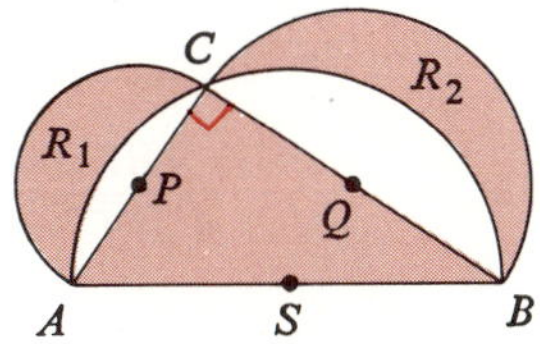

59. In the figure for Exercise 53, $\overline{AB}$ is a diameter of semicircle $\widehat{ANB}$, $\overline{BC}$ is a diameter of semicircle $\widehat{BMC}$. $\overline{ABC}$ is a diameter of circle Q. Prove that the ratio of the area of the shaded region to the area of the nonshaded region is $\frac{BC}{AB}$.

60. Prove that the area of the shaded region bounded by the two concentric circles below equals one-fourth the square of the length of the chord of the larger circle that is tangent to the smaller circle multiplied by π; that is, prove that the area of the shaded region $= \frac{1}{4}\pi(AB)^2$.

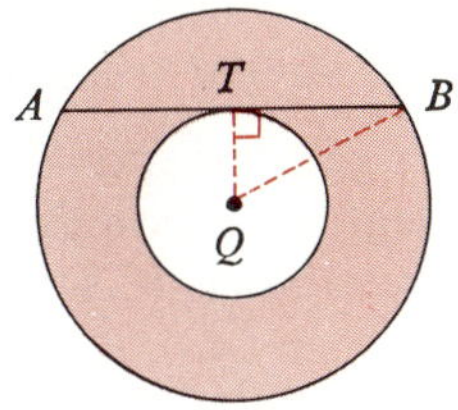

61. GIVEN P is a point of $\widehat{AB}$ of circle Q, so that $\overline{AP} \cong \overline{BP}$; $\widehat{ANB}$ is an arc of circle P; $\overline{AQB}$

PROVE The area of the shaded region $= (PQ)^2$.

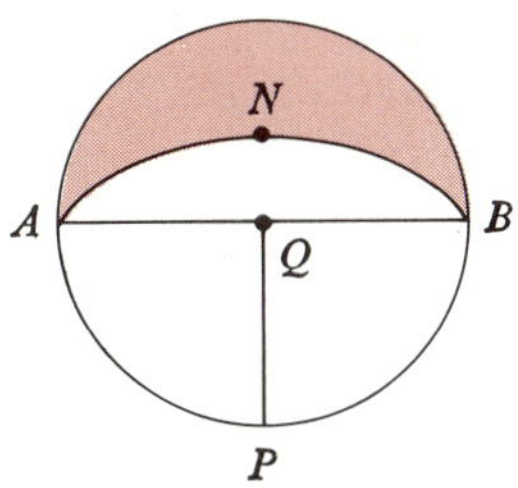

12-6 COMPARING THE AREAS OF SIMILAR POLYGONS

Now that we have developed methods of finding the areas of a variety of regions, we will be able to compare the areas of similar polygons.

T **Theorem 12-6.1** The ratio of the areas of two similar triangles equals the square of their ratio of similitude.

GIVEN $\triangle ABC \sim \triangle A'B'C'$; $\overline{BD}$ and $\overline{B'D'}$ are altitudes of their respective triangles. The lengths of segments are indicated in the figure.

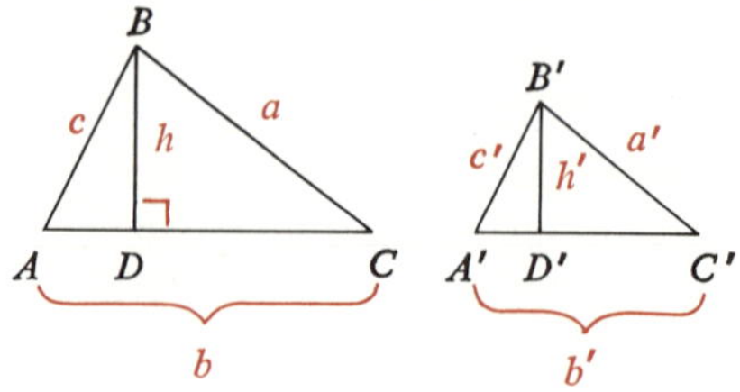

PROVE $\dfrac{\mathscr{A}\triangle ABC}{\mathscr{A}\triangle A'B'C'} = \left(\dfrac{b}{b'}\right)^2$

PROOF

STATEMENTS	REASONS
1. $\triangle ABC \sim \triangle A'B'C'$; $\overline{BD}$ and $\overline{B'D'}$ are altitudes of their respective triangles.	1. Given.
2. $\dfrac{a}{a'} = \dfrac{b}{b'}$	2. Definition of similar triangles. (8-6)
3. $\mathscr{A}\triangle ABC = \frac{1}{2}bh$ $\mathscr{A}\triangle A'B'C' = \frac{1}{2}b'h'$	3. Theorem 12-2.2.
4. $\dfrac{\mathscr{A}\triangle ABC}{\mathscr{A}\triangle A'B'C'} = \dfrac{\frac{1}{2}bh}{\frac{1}{2}b'h'} = \dfrac{bh}{b'h'}$	4. Multiplication property.
5. $\angle BDC \cong \angle B'D'C'$	5. All right angles are congruent. (3-1.1)
6. $\angle C \cong \angle C'$	6. Definition of similar triangles. (8-6)
7. $\triangle BDC \sim \triangle B'D'C'$	7. If two pairs of corresponding angles of two triangles are congruent, then the triangles are similar. (C8-5.1a)
8. $\dfrac{a}{a'} = \dfrac{h}{h'}$	8. Definition of similar triangles. (8-6)
9. $\dfrac{h}{h'} = \dfrac{b}{b'}$	9. Transitive property.
10. $\dfrac{\mathscr{A}\triangle ABC}{\mathscr{A}\triangle A'B'C'} = \dfrac{b^2}{(b')^2} = \left(\dfrac{b}{b'}\right)^2$	10. Substitution Postulate. (2-1)

You should recall that the ratio of similitude of two similar triangles is the ratio of any pair of corresponding sides, altitudes, angle bisectors, or medians.

Example 1 If the ratio of a pair of corresponding altitudes of two similar triangles is $\frac{3}{5}$, and the area of the larger triangle is 75, find the area of the smaller triangle.

Solution Since the ratio of similitude of the two triangles is $\frac{3}{5}$, the ratio of their areas is $\frac{9}{25}$. If K is the area of the smaller triangle, $\frac{9}{25} = \frac{K}{75}$, and $K = 27$. Therefore the area of the smaller triangle is 27.

Example 2 The areas of a pair of similar triangles are 75 and 147. If the length of an angle bisector of the smaller triangle is 15, find the length of the corresponding angle bisector of the larger triangle.

Solution The ratio of the areas of the two triangles is $\frac{75}{147} = \frac{25}{49}$. If x is the length of the corresponding angle bisector of the larger triangle, then $\frac{25}{49} = \left(\frac{15}{x}\right)^2$, by Theorem 12-6.1. By taking the square root of both sides of the above equation, we get

$$\frac{5}{7} = \frac{15}{x}$$
$$x = 21$$

The solution of Example 2 suggests the following corollary to Theorem 12-6.1.

C **Corollary 12-6.1a** The ratio of similitude of any pair of similar triangles equals the square root of the ratio of their areas.

The proof of Corollary 12-6.1a is left as an exercise.

Theorem 12-6.2 extends Theorem 12-6.1 to apply to a pentagon.

T **Theorem 12-6.2** The ratio of the areas of two similar polygons equals the square of their ratio of similitude.

We shall accept the theorem for an n-gon without proving it.

Remember that the ratio of similitude of a pair of similar polygons is the ratio of a pair of corresponding sides, apothems, radii, diagonals, or perimeters.

C **Corollary 12-6.2a** The ratio of similitude of any pair of similar polygons equals the square root of the ratio of their areas.

The proof of Corollary 12-6.2 is left as an exercise.

If we consider a circle as a regular polygon with a boundless number of sides, then we may extend Theorem 12-6.2 to the circle.

C **Corollary 12-6.2b** The ratio of the areas of two circles equals the square of their ratio of similitude.

C **Corollary 12-6.2c** The ratio of similitude of two circles equals the square root of the ratio of their areas.

Example 3 If the areas of two circles are 48 and 75, find the ratio of their circumferences.

Solution The ratio of the areas of the two circles is $\frac{48}{75} = \frac{16}{25}$.

Therefore, from Corollary 12-6.2c, we know that their ratio of similitude is $\sqrt{\frac{16}{25}} = \frac{4}{5}$. Hence, the ratio of the circumferences is $\frac{4}{5}$.

Remember that if polygon $R_1 \sim$ polygon R_2, where $\frac{s_1}{s_2}$ is their ratio of similitude, then

$$\frac{\mathscr{A} R_1}{\mathscr{A} R_2} = \left(\frac{s_1}{s_2}\right)^2 \quad \text{and} \quad \frac{s_1}{s_2} = \sqrt{\frac{\mathscr{A} R_1}{\mathscr{A} R_2}}$$

EXERCISES

A Find the ratios of similitude of the pairs of similar triangles whose areas are in these ratios.

1. $\frac{1}{4}$ 2. $\frac{16}{25}$ 3. $\frac{25}{144}$ 4. $\frac{2}{9}$

Find the ratio of the areas of two regular decagons if the ratio of their apothems is:

5. $\frac{1}{4}$ 6. $\frac{16}{25}$ 7. $\frac{5}{7}$ 8. $\frac{2}{13}$

Find the ratio of the areas of a pair of similar triangles if the lengths of two corresponding angle bisectors are:

9. 3 and 10 10. 15 and 24 11. $12\sqrt{2}$ and $16\sqrt{2}$

For Exercises 12-14, the lengths of a pair of corresponding altitudes of two similar triangles are 15 and 20.

12. Find the ratio of a pair of corresponding medians.
13. Find the ratio of their perimeters.
14. Find the area of the smaller triangle, if the area of the larger triangle is 80.
15. If the area of one equilateral triangle is four times the area of another equilateral triangle, then each side of the larger equilateral triangle is ________ times as long as a side of the smaller equilateral triangle.
16. When a pentagon is enlarged, the angles remain constant, while the lengths of the sides are quadrupled. How does the area change?
17. The ratio of a pair of corresponding medians of two similar triangles is $\frac{5}{9}$. Find the area of each triangle if the sum of their areas is 636.

18. Prove Corollary 12-6.1a
19. Prove Corollary 12-6.2a
20. Prove Corollary 12-6.2b
21. Prove Corollary 12-6.2c
22. A pair of corresponding sides of two similar triangles have lengths 7 and 21. If the area of the larger triangle exceeds the area of the smaller triangle by 320, find the area of the smaller triangle.
23. If the area of one circle is 25 times the area of another circle, and the diameter of the larger circle is 12 more than the diameter of the smaller circle, find the diameter of the smaller circle.
24. If the length of each side of a square is doubled, the area is increased by ____ percent.

B

25. Prove that the midline of a triangle separates the triangle into two regions whose areas are in the ratio $\frac{1}{3}$.
26. The length of a side of an equilateral triangle is s. Another equilateral triangle is drawn with each side of length $s + x$. The area of the second triangle is twice the area of the first triangle. Find x.
27. Prove that the square of the ratio of the lengths of the medians of two similar trapezoids equals the ratio of their areas.
28. GIVEN $\triangle ABC$; $\overline{DE} \parallel \overline{BC}$; $\overline{ADB}$ and $\overline{AEC}$; $\frac{AD}{DB} = \frac{5}{2}$

    PROVE $\frac{\mathscr{A} \triangle ADE}{\mathscr{A} \text{ quadrilateral } DECB} = \frac{25}{24}$

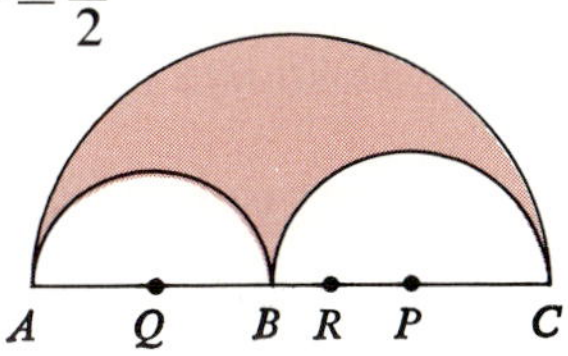

29. In the figure at right, $\overline{AB}$, $\overline{BC}$, and $\overline{AC}$ are the diameters of the semicircles. If $\frac{AB}{BC} = \frac{2}{3}$, find the ratio of the area of the shaded region to the area of the largest semicircle.
30. Points B and C are points of two different but concentric circles with center at A. If $\overline{ABC}$ and $\overline{AB} \cong \overline{BC}$, what part of the area of the larger circle is the area bounded by the two circles?

What part of the entire figure is the shaded region in each of the following exercises?

31.

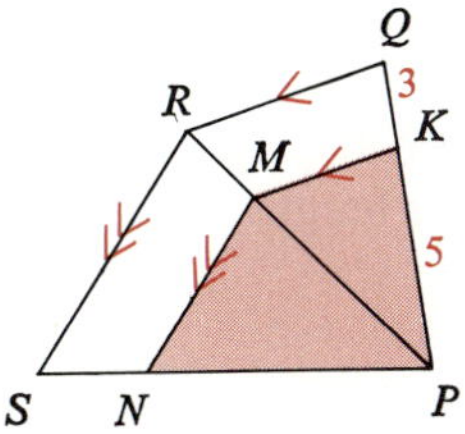

32.

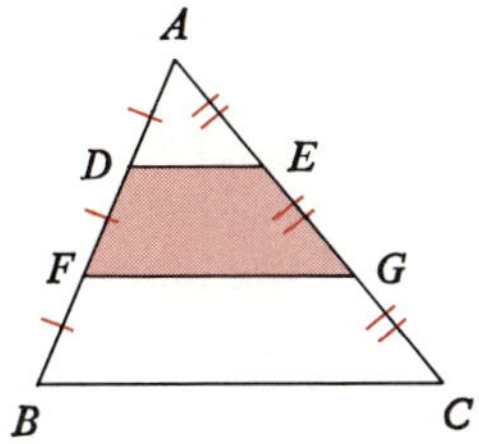

C

33. GIVEN In $\triangle ABC$, $\overline{DE} \parallel \overline{BC}$; $\overline{EF} \parallel \overline{AB}$; $\overline{ADB}$; $\overline{AEC}$; $\overline{BFC}$; $BD = 3(AD)$

    PROVE $\mathscr{A} \ \square DEFB = \frac{3}{8} \mathscr{A} \triangle ABC$

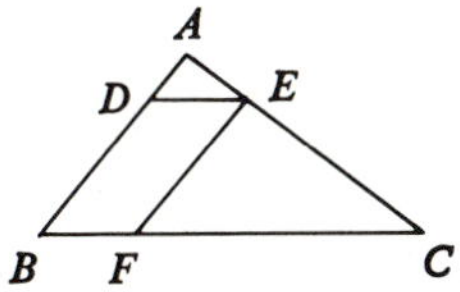

34. Given $\overline{PT}$ is tangent to $\odot Q$ at T; $\overline{ABP}$ is a secant segment; $\overline{TA}$ and $\overline{TB}$ are chords.

Prove $\dfrac{\mathscr{A}\triangle ATP}{\mathscr{A}\triangle TBP} = \dfrac{AP}{BP}$

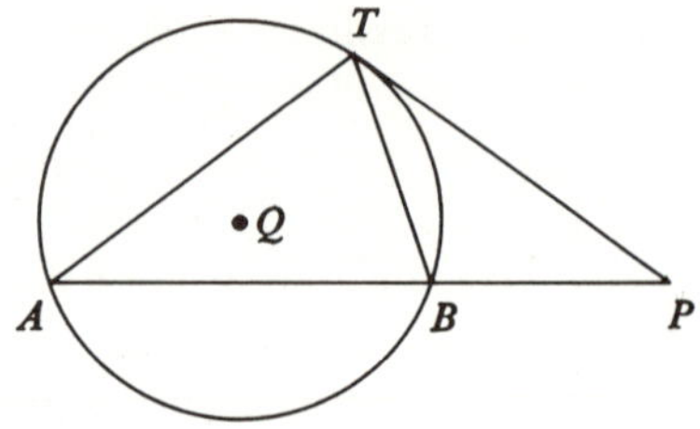

HOW IS YOUR MATHEMATICAL VOCABULARY?

The key words and phrases introduced in this chapter are listed below. How many do you know and understand?

apothem (12-4)
area (12-1)
Area Addition Postulate (12-1)
area of a circular region (12-5)
Area of a Rectangle Postulate (12-1)
Area Postulate (12-1)
center of a regular polygon (12-4)
circular region (12-5)
Congruence Postulate for Areas (12-1)
polygonal region (12-1)
radius of a regular polygon (12-4)
region (12-1)
sector of a circle (12-5)
segment of a circle (12-5)
square unit (12-1)
triangular region (12-1)

REVIEW EXERCISES

12-1 Areas of polygonal regions

1. Find the area of a rectangle whose base and altitude have lengths, respectively, of $3\sqrt{2}$ and $5\sqrt{3}$.
2. Find the area of the square whose diagonal has length 6.

How would the area of a rectangle be affected if each of the alterations below were accomplished?

3. The length of the base were doubled and the length of the altitude were divided by 2.
4. The lengths of the base and altitude were both tripled.
5. The length of the base were quadrupled and the length of the altitude remained the same.
6. Given $\overline{AC} \perp \overline{CEBD}$; $\overline{FD} \perp \overline{CEBD}$; $CE = BD$; $\overline{AB}$ meets $\overline{EF}$ at G; $\overline{EG} \cong \overline{BG}$

   Prove $\mathscr{A}$ quadrilateral $AGEC$ $= \mathscr{A}$ quadrilateral $FGBD$

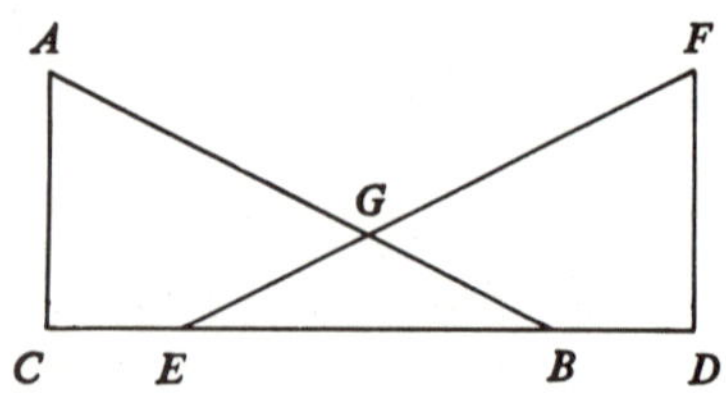

12-2 The area of a triangle

7. Find the area of the right triangle whose legs have lengths 8 and 6.
8. Find the length of the base of a triangle if the length of the altitude to that base is 6 and the area of the triangle is $38\sqrt{2}$.
9. Find the area of a triangle if the lengths of two sides are 6 and 4 and the measure of the included angle is 45.
10. Find the area of an equilateral triangle if the length of a side is $8x$.
11. Find the area of an equilateral triangle if the length of an altitude is 6.
12. In $\triangle ABC$, $\overline{AD}$ and $\overline{BE}$ are altitudes. If $AD = 4$, $BE = 3$, and $AC = 8$, find BC.
13. GIVEN Medians $\overline{AM}$ and $\overline{BN}$ of $\triangle ABC$ meet at G.
    PROVE $\mathscr{A}\triangle BGM = \frac{1}{6}\mathscr{A}\triangle ABC$; $\mathscr{A}$ quadrilateral $GNCM = \frac{1}{3}\mathscr{A}\triangle ABC$

12-3 Areas of some special quadrilaterals

14. Find the length of the altitude of a parallelogram whose area is 48 and whose base has length $8\sqrt{2}$.
15. Find the area of a rhombus whose diagonals have lengths 13 and 7.
16. Find the altitude of the rhombus whose diagonals have lengths 16 and 30.
17. Find the area of a trapezoid whose altitude is 6 and whose bases have lengths 12 and 17.
18. Find the area of an isosceles trapezoid, one of whose angles has measure 135, and whose bases have lengths 15 and 27.
19. GIVEN $\square ABCD$; P is any point of $\overline{AD}$; Q is any point of $\overline{DC}$.
    PROVE $\mathscr{A}\triangle BPC = \mathscr{A}\triangle AQB$

12-4 The area of a regular polygon

20. Find the area of a regular polygon whose perimeter is 30 and whose apothem has length 8.
21. Find the area of a regular hexagon, each of whose sides has length $10\sqrt{2}$.
22. Find the apothem of a regular polygon whose area and perimeter are 60 and $5\sqrt{2}$, respectively.
23. Find the area of a regular pentagon, correct to the nearest square inch, if the length of a side is 50 in.
24. Find the radius of a regular dodecagon whose area is 72.
25. Prove that the area of a regular hexagon, each of whose sides has length s, is $\frac{3}{2}s^2\sqrt{3}$.

12-5 The area of a circle

26. Find the diameter of a circle whose area is 60π.
27. Find the area of the sector of a circle whose radius is 10 and whose central angle has measure 40.

28. Find the radius of a circle if a sector of area 72π has a central angle of measure 270.
29. Find the area of the circle inscribed in an equilateral triangle whose altitude has length 15.

Find the area of the shaded region of each of the following figures with Q the centers of circles.

30.

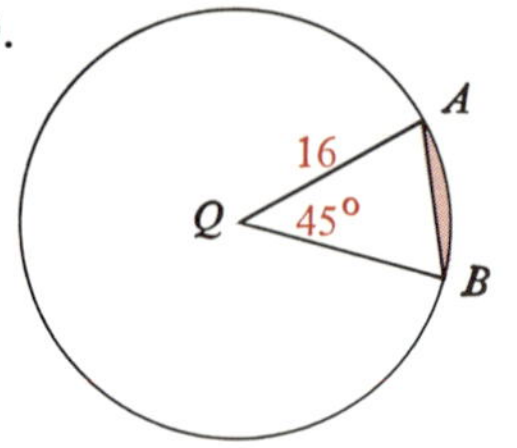

31.

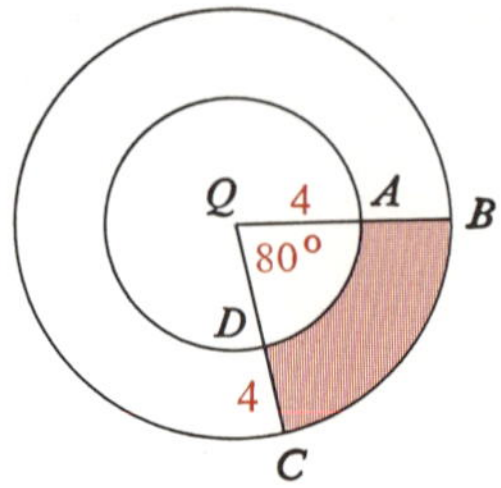

32.

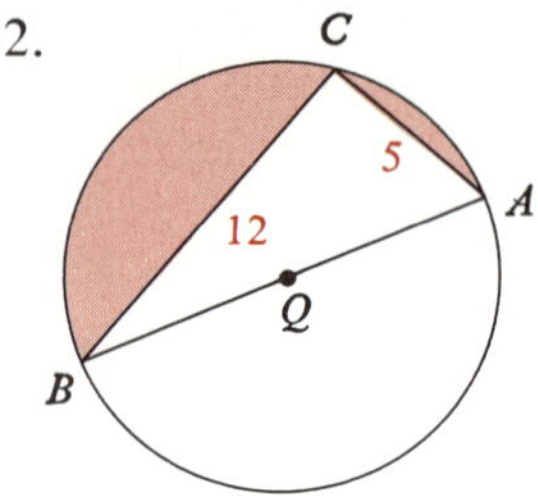

12-6 Comparing the areas of similar polygons

33. If the areas of a pair of similar triangles are in the ratio of $\frac{1}{9}$, what is their ratio of similitude?
34. If a pair of corresponding altitudes of two similar triangles are in the ratio of $\frac{9}{16}$, what is the ratio of their areas?
35. If the apothems of two regular octagons are in the ratio of $\frac{2}{7}$, and the area of the smaller octagon is 16, find the area of the larger octagon.
36. The areas of a pair of similar triangles are 20 and 45, respectively. If the length of a median of the smaller triangle is 14, find the length of the corresponding median of the larger triangle.
37. If the circumferences of two circles are in the ratio of $\frac{9}{25}$, find the ratio of their areas.
38. If the length of the diagonals of a regular pentagon is tripled, how is the area changed in the resulting regular pentagon?
39. K, M, and N are the midpoints of the sides of $\triangle ABC$. What part of $\mathscr{A}\triangle ABC$ is $\mathscr{A}\triangle KMN$?

CHAPTER TEST

Indicate which of the following statements are true and which are false. Correct the false statements.

1. The area of a rhombus equals the product of the lengths of its base and altitude.
2. The area of a square equals the square of the length of its diagonal.
3. If the apothem of a regular pentagon is doubled, then the area is also doubled.
4. A median of a triangle always separates the triangle into two triangles of equal areas.

5. Find the area of a square whose diagonal is $2\frac{1}{2}$ feet long.
6. Find the altitude of a rhombus whose diagonals have lengths of 16 and 12.
7. Find the altitude of an equilateral triangle whose area is $9\sqrt{3}$.
8. Find the area of a triangle that has two sides of lengths 4 and 5, where the measure of the included angle is 30.
9. Find the area of the isosceles trapezoid whose bases have lengths of 18 and 32 and whose legs have length 25.
10. Find the area of the regular hexagon whose sides have length 10.
11. In a circle, find the measure of the central angle of a sector with an area of 12π and a radius of 4.
12. The lengths of a pair of corresponding sides of two similar triangles are 5 and 20. Find the area of the smaller triangle, if the area of the larger triangle is 48.

Find the area of each shaded region in Exercises 13–15.

13.

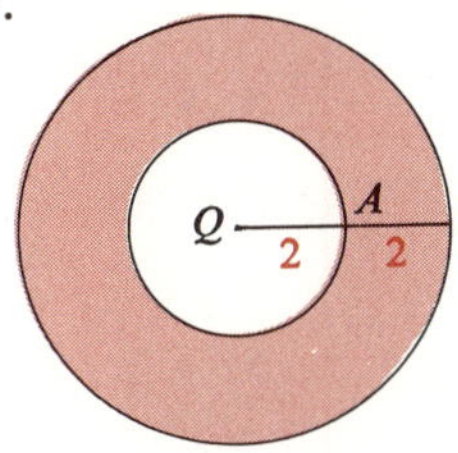

14.

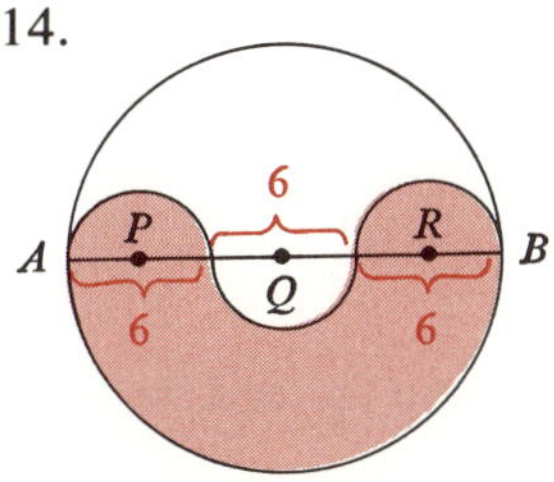

15.

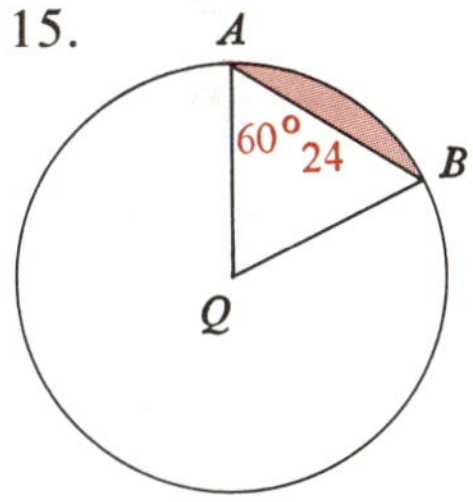

16. GIVEN $\square ABCD$; $\overline{ACP}$ PROVE $\mathscr{A}\triangle BPC = \mathscr{A}\triangle DPC$

17. GIVEN M is the midpoint of side $\overline{BC}$ of $\triangle ABC$. $\overline{DM} \parallel \overline{AC}$; $\overline{EM} \parallel \overline{AB}$; $\overline{ADB}$; $\overline{AEC}$ PROVE $\mathscr{A}\ \square ADME = \frac{1}{2}\mathscr{A}\triangle ABC$

SUGGESTED RESEARCH

1. Present an algebraic and a geometric proof of Heron's formula. The algebraic proof can be found in a number of sources. However, because of its length, a geometric proof of this formula is not too popular. One source for a geometric proof is *Challenging Problems in Geometry*, vol. 2, by A.S. Posamentier and C.T. Salkind (New York: Macmillan Co., 1970).

2. Look up Brahmagupta's formula for the area of a cyclic quadrilateral. How does this formula compare to Heron's formula? Present a proof of Brahmagupta's formula. One source is the book mentioned above.

3. Prepare a report about figures with the same perimeter, but varying areas. One of many sources is *The Enjoyment of Mathematics*, by H. Rademacher and O. Toeplitz (Princeton: Princeton University Press, 1957).

CUMULATIVE REVIEW

Chapters 10, 11, 12

These exercises are designed so that you can check your understanding of some of the principal concepts studied so far. The numbered items in the right column are the correct answers. To complete the review, first cover the answers, read and answer the questions, and check each response by uncovering the corresponding answer.

1. The points (x, y) such that $x < 0$ and $y > 0$ lie in the ____ quadrant.	1. second
2. The distance between (5, 1) and (−4, 7) is ____.	2. $\sqrt{117}$
3. (3, 6) is the midpoint of a segment with (4, 9) and ____ as endpoints.	3. (2, 3)
4. The radius of a circle with center (1, −6) where (5, 3) is a point of the circle is: a. $\sqrt{86}$ b. $\sqrt{45}$ c. $\sqrt{97}$ d. 5 e. none of these	4. c
5. The slope for the segment formed by (−1, 3) and (1, −1) is ____.	5. −2
6. Find the slope of the line perpendicular to the line through (1, 1) and (8, 5).	6. $-\frac{7}{4}$
7. The slope of the line parallel to the line $3x - 2y = 1$ is ____.	7. $\frac{3}{2}$
8. For $5x + 3y = 30$, $a =$ ____, $b =$ ____, and $m =$ ____.	8. $a = 6, b = 10,$ $m = -\frac{5}{3}$
9. Write the equation for a circle with center at the origin and radius 4.	9. $x^2 + y^2 = 16$
10. Write the equation of the circle that has a diameter whose endpoints are (3, 9) and (1, 5).	10. $(x - 2)^2 + (y - 7)^2 = 5$
11. To represent any parallelogram on coordinate axes, use points (0, 0), $(a, 0)$, (b, c), and ____.	11. $(a + b, c)$
12. The equation for the locus of points bisecting the first and second quadrants is ____.	12. $y = \lvert x \rvert$
13. True or false: (4, 3) is on the locus of points lying 5 units from (0, 0).	13. true

14. We can construct a triangle given two sides and the ____.	14. included angle
15. True or false: To construct a triangle similar to a given triangle on a given base, we construct two angles congruent to two angles of the given triangle.	15. true
16. True or false: Bisecting the chord of a segment locates the midpoint of the arc formed by the chord.	16. true
17. In $\triangle ABC$, P is the centroid. If $\overline{AM}$ is a median such that $AM = 9$, then $AP =$ ____.	17. 6
18. True or false: If two triangles have equal areas, then the triangles are congruent.	18. false
19. If the lengths of the altitude and base of a rectangle are doubled, then the area is ____.	19. four times as great
20. Find the area of a rectangle if the length of a diagonal is 26 and the length of one side is 10.	20. 240
21. Find the area of $\triangle ABC$ if $AB = 10$, $BC = 8$, and $m\angle B = 30$.	21. 20
22. If $AB = 3$, the area of equilateral $\triangle ABC$ is ____.	22. $\frac{9\sqrt{3}}{4}$
23. The area of $\square ABCD$ with base of $7\frac{1}{2}$ and altitude of 10 is ____.	23. 75
24. Find the area of trapezoid $DEFG$ with altitude $5\frac{1}{2}$ and bases $2\frac{1}{4}$ and $7\frac{1}{8}$.	24. $\frac{825}{32}$
25. Find the area of a rhombus whose sides have length 6 and where one angle measures 120.	25. $18\sqrt{3}$
26. Find the area of a regular polygon whose apothem and perimeter have respective lengths of 5 and 20.	26. 50
27. The area of a circle with a circumference of 25π is ____.	27. $\frac{625}{4}\pi$
28. Find the radius of a circle whose circumference is numerically equal to its area.	28. 2
29. If the lengths of a pair of corresponding altitudes of two similar triangles are 15 and 20, the ratio of their areas is ____.	29. $\frac{9}{16}$
30. If the ratio of the circumferences of two circles is $\frac{1}{4}$, then the area of the larger circle is ____ times the area of the smaller circle.	30. 16

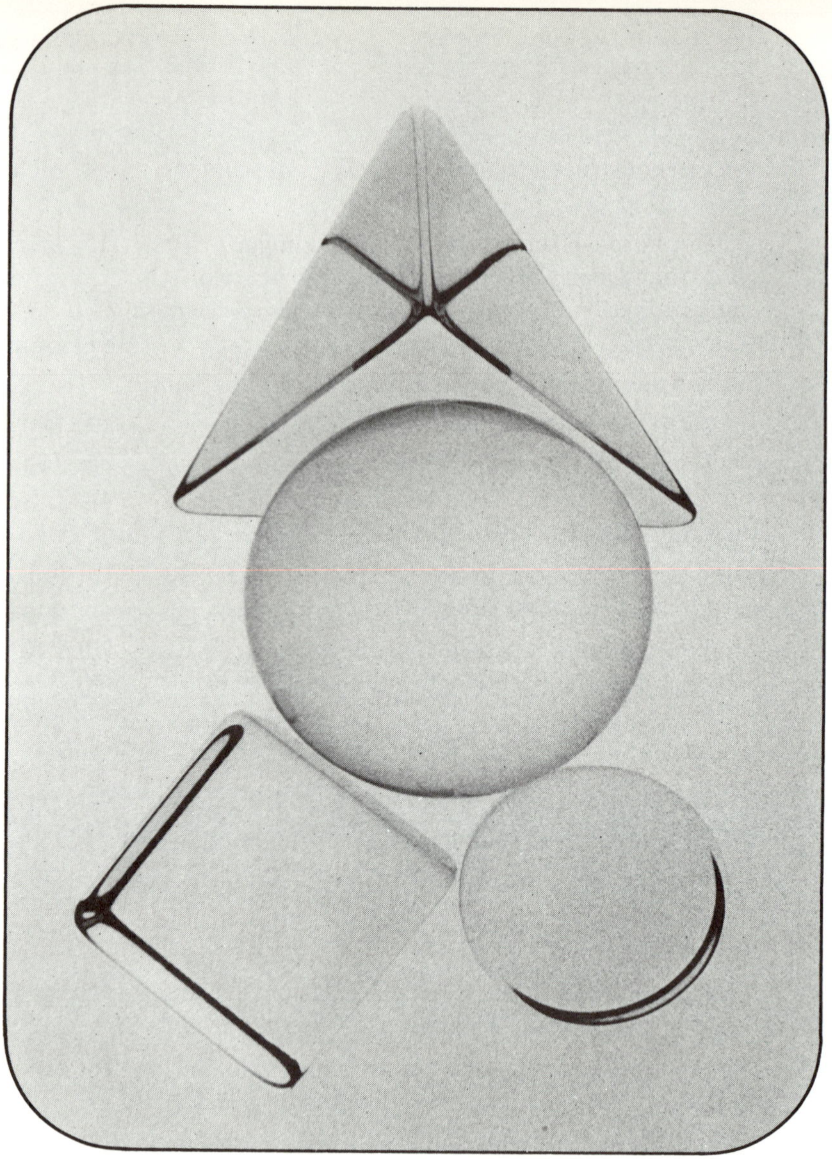

13 Surface Areas and Volumes of Solids

In Chapter 12 we were concerned with plane figures. Although the world we live in is filled with representations of planes, the physical objects of our experience are not planes. They are three-dimensional and may or may not have plane surfaces. In this chapter we shall study the properties of these three-dimensional figures.

13-1 POLYHEDRONS

Earlier we defined a dihedral angle as the union of two intersecting half-planes and their common edge. If a third plane cuts the edge of a dihedral angle, a trihedral angle is formed. The trihedral angle belongs to the general class of polyhedral angles.

D

Definition 13-1 A **polyhedral angle** is the figure formed by three or more planes that intersect in one point.

The point of intersection of all the planes is called the **vertex** of the polyhedral angle. The planes are called **faces**; the intersections of the faces, **edges**; and the plane angles formed by the two edges of each face, **face angles**. We measure a polyhedral angle by finding the sum of the measures of its face angles.

If the intersection of a polyhedral angle with a plane not through the vertex is a convex polygon, then the polyhedral angle is convex. We shall restrict our study to convex polyhedral angles.

A trihedral angle has three face angles that are related somewhat like the sides of a triangle.

T

Theorem 13-1.1 The sum of the measures of any two face angles of a trihedral angle is greater than the measure of the third face angle.

GIVEN Trihedral $\angle V\text{-}ABC$; $\angle AVC$ is the face angle of greatest measure.

PROVE $m\angle AVB + m\angle BVC > m\angle AVC$

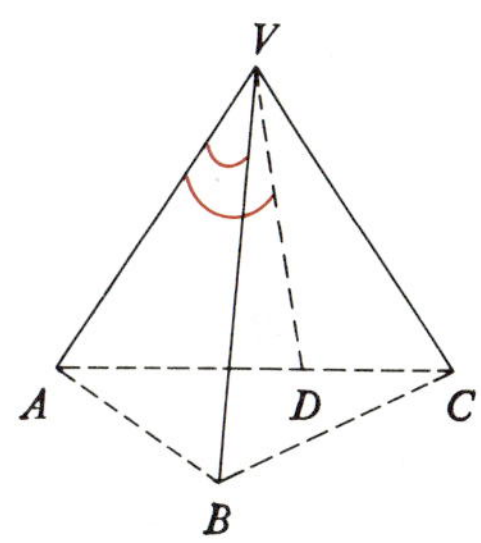

PROOF OUTLINE Draw $\overline{AC}$. In face VAC construct $\angle AVD \cong \angle AVB$. Determine B so that $\overline{VD} \cong \overline{VB}$. Thus, $\triangle VAB \cong \triangle VAD$, with $\overline{AB} \cong \overline{AD}$. $\overleftrightarrow{AC}$ and B determine a plane and $AB + BC > AC$. Hence $BC > DC$. In △s VBC and VDC, we have $m\angle BVC > m\angle DVC$. It follows that $m\angle AVB + m\angle BVC > m\angle AVD + m\angle DVC$; that is, $m\angle AVB + m\angle BVC > m\angle AVC$.

The formal proof is left as an exercise.

Theorem 13-1.1 enables us to establish the following fundamental theorem:

T **Theorem 13-1.2** The sum of the measures of the face angles of any convex polyhedral angle is less than 360.

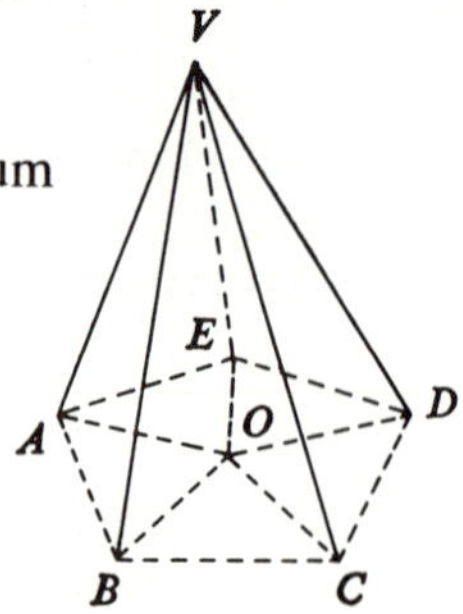

GIVEN Polyhedral $\angle V\text{-}ABCDE \ldots$ with n faces; the sum of the measures of the face angles equals S.

PROVE $S < 360$

PROOF Let a plane cut the edges of the polyhedral angle in $A, B, C, D, E \ldots$. Select any point O in polygon $ABCDE \ldots$ and draw the segments $\overline{OA}$, $\overline{OB}$, $\overline{OC}$, ... Call the n triangles with common vertex O the O-triangles. Likewise, the n triangles whose common vertex is V will be called V-triangles. In trihedral $\angle B\text{-}AVC$, $m\angle ABV + m\angle VBC > m\angle ABC$ by Theorem 13-1.1. Also, $m\angle ABC = m\angle ABO + m\angle OBC$ (Postulate 2-10). Applying the Substitution Postulate (2-1), we have $m\angle ABV + m\angle VBC > m\angle ABO + m\angle OBC$ (similarly for the remaining vertices of polygon $ABCDE \ldots$). Hence, the sum of the vertices of the V-triangles is greater than the sum of the base angles of the O-triangles by the addition property of inequality. Using Theorem 6-4.2 and the multiplication property of equality, we know that the sum of all the angles of the V-triangles is equal to the sum of all the angles of the O-triangles. Combining the last two statements, we have that the sum of the measures of the angles at V must be less than the sum of the measures of the angles at O (subtraction property of inequality). By Definition 9-17, the sum of the measures of the angles at O is 360. Hence, the sum of the measures of the angles at V is less than 360 (P 2-1).

The distinction between a polyhedral region and a polyhedron is similar to the distinction between a polygonal region and a polygon.

D **Definition 13-2** A **polyhedral region** is a solid completely bounded by portions of intersecting planes.

D **Definition 13-3** A **polyhedron** is the union of the bounding plane regions of a polyhedral region.

The smallest number of planes that can enclose a polyhedral region is four. We can think of the polyhedron enclosing this region as a trihedral angle cut by a plane. This polyhedron, which we call a **tetrahedron**, has four faces, four vertices, and six edges.

A **hexahedron** has six faces, eight vertices, and twelve edges. The figures at right show a tetrahedron and a hexahedron.

Tetrahedron

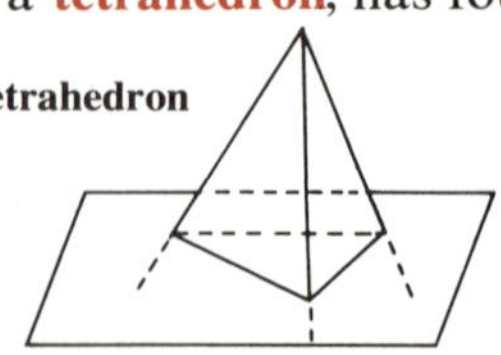

Hexahedron

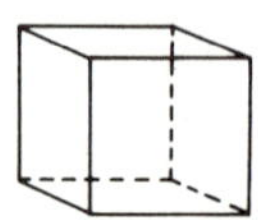

D **Definition 13-4** A **diagonal of a polyhedron** is a segment joining two vertices that are not in the same face.

How does this definition compare with the definition of a diagonal of a polygon? The triangle is the polygon with the fewest number of sides. The tetrahedron is the polyhedron with the fewest number of faces. How many diagonals does a triangle have? How many diagonals do you think a tetrahedron has?

D **Definition 13-5** A polyhedron is a **regular polyhedron** if and only if all its faces are congruent regular polygons and all its polyhedral angles are congruent.

The edges of the faces of a regular tetrahedron form four congruent equilateral triangles. Although the faces are actually triangular regions, it is often more convenient to say simply that the faces are triangles. The faces of a regular hexahedron are six congruent squares. What is another name for a regular hexahedron?

Class Exercises

1. Can each face angle of a trihedral angle be 60°? 90°? 108°? 120°?
2. Find the number of degrees in each angle of a regular polygon of three sides, four sides, five sides, and six sides.
3. Can each face angle of a tetrahedral angle be 60°? 90°? 108°?
4. Can each face angle of a polyhedral angle of five faces be 60°?
5. Can each face angle of a polyhedral angle of six faces be 60°?
6. Show that no more than three regular polyhedrons with triangular faces are possible.
7. Show that no more than one regular polyhedron with square faces is possible.
8. Show that no more than one regular polyhedron with pentagonal faces is possible.

The above Class Exercises do not prove the existence of regular polyhedrons. What they do prove, however, is that there cannot be more than five such polyhedrons. We shall not attempt to prove it here, but all five that are implied by the Class Exercises do exist. These are regular polyhedra of four, eight, and twenty triangular faces; one polyhedron with six square faces; and one with twelve pentagonal faces.

D **Definition 13-6** The **total area of a polyhedron** is the sum of the areas of all its faces.

In the case of a regular polyhedron, the total area is the area of one face multiplied by the number of faces.

Example Find the total area of a regular tetrahedron whose edge is 1 unit.

Solution Since each face is an equilateral triangle with side 1, by Theorem 12-2.6, the area of each face is $\frac{\sqrt{3}}{4}$. Since there are four faces, the total area of the regular tetrahedron is $4 \cdot \frac{\sqrt{3}}{4} = \sqrt{3}$.

EXERCISES

A

1. Is it possible for all the face angles of a tetrahedral angle to be obtuse? Justify your answer.
2. The polyhedron for Theorem 13-1.2 has how many diagonals?
3. If two face angles of a trihedral angle have measures 30 and 70, the measure of the third angle must be greater than what value and less than what value?

The measures of three plane angles are given in each exercise below. Can each set of measures form a trihedral angle? Why?

4. 90, 100, 150
5. 120, 130, 150
6. 20, 100, 150
7. What is the locus of points equidistant from the faces of a trihedral angle? Describe how to find it.
8. A polyhedron consists of a square and 4 congruent isosceles triangles that have a common vertex. If the side of the square is 6 units and the altitude of one of the triangles from the common vertex is 4, find the total area.
9. Find the total area of a regular tetrahedron if an edge is 4 units.

B

10. Prove Theorem 13-1.1.
11. In Exercise 8, if the measure of a side of the square is 8 and the measure of a common edge of each pair of congruent triangles is 5, find the total area.
12. How many planes determine a point in space? Justify your answer.
13. Two straight lines are perpendicular to each other. Can they both be parallel to a plane? Can they both be perpendicular to a plane? Justify your answers.
14. Can three planes determine a polyhedron? Justify your answer.

C

15. Prove that the planes that bisect the dihedral angles of a trihedral angle concur in a line.
16. If the face angles of a trihedral angle are equal, are the dihedral angles equal? Prove your answer.
17. Will three guy-wires of equal length, fastened at the same height on a flagpole, hold the pole rigid? Explain.

SOMETHING TO THINK ABOUT

Euler's Polyhedron Theorem

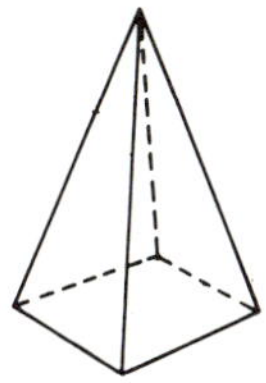
Pentahedron

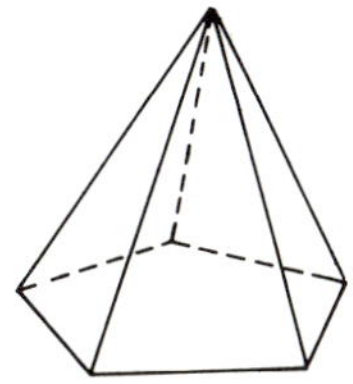
Hexahedron

Consider the pentahedron and the hexahedron illustrated at right. Can you discover a relationship between the number of faces, edges, and vertices that can be generalized to all polyhedrons? In the pentahedron there are five vertices, five faces, and eight edges. In the hexahedron there are six faces, six vertices, and ten edges.

Leonhard Euler, a great Swiss mathematician of the eighteenth century, proved that for any simple polyhedron $V - E + F = 2$, where V is the number of vertices, F the number of faces, and E the number of edges. Verify that this is true of the polyhedrons we have examined. Draw polyhedrons with other numbers of faces and see if this formula is true for them also.

13-2 PRISMS

Before we begin our study of prisms, we need to extend our concepts of parallel lines and planes. In Chapter 1 we introduced planes; in Chapter 4 we discussed perpendicular lines and planes; in Chapter 6 we examined parallel lines. Now we shall direct our attention to parallel lines and planes.

D **Definition 13-7** Two **planes**, or a line and a plane, are **parallel** if they do not intersect.

In a plane, two lines either coincide, intersect, or are parallel. Likewise, two planes either coincide, intersect, or are parallel. However, two lines in two separate parallel planes do not have to be parallel; they may be skew lines. Only if two lines determine a third plane are they parallel.

The proofs of the following theorems and corollaries are left as exercises.

T **Theorem 13-2.1** If a plane intersects two parallel planes, then it intersects them in two parallel lines.

T **Theorem 13-2.2** If a line is perpendicular to one of two parallel planes, then it is perpendicular to the other.

Theorem 13-2.2 is analogous to Corollary 6-1.1b. If we continue the process of analogy, many of the theorems we proved in Chapter 6 for parallel lines suggest corresponding theorems for parallel planes. The following theorem, for example, is analogous to Theorem 6-1.1:

Theorem 13-2.3 If two planes are perpendicular to the same line, then they are parallel.

The following corollary is analogous to Corollary 6-1.1c. The proof of the corollary is left as an exercise.

Corollary 13-2.3a If two planes are both parallel to a third plane, then they are parallel to each other.

From Theorem 4-5.8, we know that two lines perpendicular to the same plane are coplanar. You can use this fact in the proof of Theorem 13-2.4, which is left as an exercise.

Theorem 13-2.4 If two lines are perpendicular to the same plane, then the lines are parallel.

Corollary 13-2.4a If a plane is perpendicular to one of two parallel lines, then the plane is perpendicular to the other line.

The next theorem is the last theorem for parallel planes we shall need before beginning our study of prisms. Its proof is also left as an exercise.

Theorem 13-2.5 Parallel planes are everywhere equidistant.

We are now ready to examine the polyhedron known as a prism. In the figure at right, the pentagonal surfaces are congruent and lie in parallel planes, with their corresponding sides parallel. The other polygons are quadrilaterals. What kind of quadrilaterals are they if they have a pair of congruent and parallel sides?

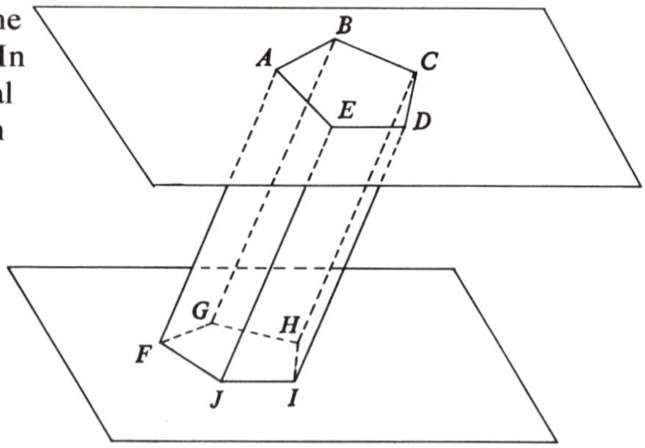

We call the congruent parallel pentagons the **bases** of the figure, and the parallelograms the **lateral faces**. The intersections of the lateral faces are called **lateral edges**. Name the lateral edges, bases, and lateral faces of the prism above.

D **Definition 13-8** A **prism** is a polyhedron whose faces consist of two parallel and congruent polygons, called bases, and the parallelograms, called lateral faces, formed by connecting pairs of corresponding vertices of the parallel polygons.

D **Definition 13-9** The **altitude of a prism** is the perpendicular segment between the parallel planes of the bases, or the length of that segment.

We classify prisms according to the number of sides in their bases. A prism with a triangular base is a triangular prism; a prism with a quadrilateral base is a quadrangular prism, and so on.

D **Definition 13-10** The polygonal region formed by the intersection of a polyhedron and a plane passing through it is a **section of the polyhedron**. A **right section of a prism** is a section formed by a plane which cuts all the lateral edges of the prism and is perpendicular to one of them.

In the definition of a prism, we stated that the lateral faces of a prism are parallelograms. This is justified by Theorem 7-2.2, since a pair of opposite sides, which are also corresponding sides of the bases, are congruent and parallel. We can use this fact to establish Theorem 13-2.6.

T **Theorem 13-2.6** The lateral edges of a prism are congruent and parallel.

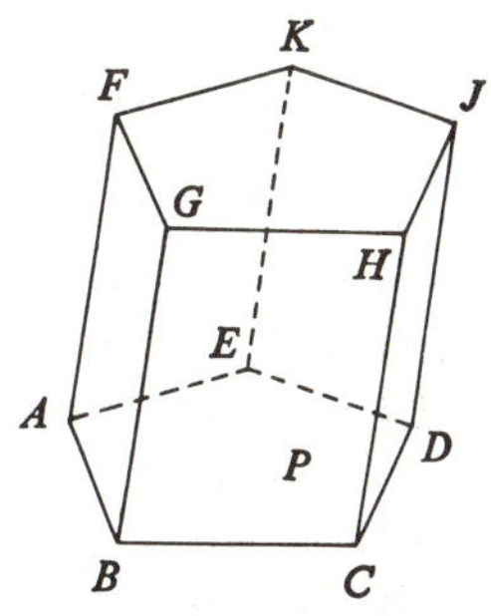

GIVEN Prism P with lateral edges $\overline{AF}$, $\overline{BG}$, $\overline{CH}$, $\overline{DJ}$, $\overline{EK}$, . . .

PROVE $\overline{AF} \cong \overline{BG} \cong \overline{CH} \cong \overline{DJ} \cong \overline{EK} \cong \ldots$; $\overline{AF} \parallel \overline{BG} \parallel \overline{CH} \parallel \overline{DJ} \parallel \overline{EK} \parallel \ldots$

PROOF
By Definition 13-7, quadrilaterals $FGBA$, $GHCB$, . . . are parallelograms. Thus, $\overline{AF} \parallel \overline{BG}$, $\overline{BG} \parallel \overline{CH}$, . . . (Definition 7-1) and $\overline{AF} \cong \overline{BG}$, $\overline{BG} \cong \overline{CH}$, . . . by Theorem 7-1.2. Consequently, the transitive property assures us that $\overline{AF} \cong \overline{BG} \cong \overline{CH} \cong \ldots$ and $\overline{AF} \parallel \overline{BG} \parallel \overline{CH} \parallel \ldots$.

Using the definition of a right section and Theorem 13-2.6, you should be able to present a formal proof of the following corollary as an exercise:

C **Corollary 13-2.6a** The plane of a right section of a prism is perpendicular to all its lateral edges.

D **Definition 13-11** A **right prism** is a prism whose lateral edges are perpendicular to the bases of the prism. A prism that is not a right prism is an **oblique prism**.

D **Definition 13-12** The **lateral area of a prism** is the sum of the areas of the lateral faces.

Theorem 13-2.7 presents a convenient way to find the lateral area of any prism.

T **Theorem 13-2.7** The lateral area of a prism is equal to the product of the perimeter of a right section and the length of a lateral edge.

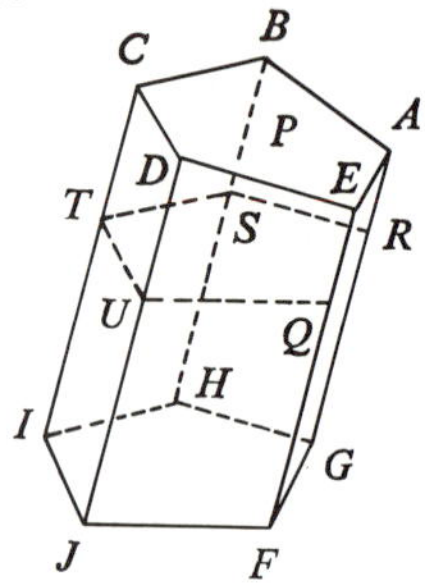

GIVEN Prism P with right section $QRSTU\ldots$; e the length of a lateral edge; $\mathscr{A}$ the lateral area; p the perimeter of the right section.

PROVE $\mathscr{A} = e \cdot p$

PROOF $\overline{UQ}, \overline{QR}, \overline{RS}, \ldots$ are altitudes of the parallelogram faces $DEFJ, EAGF, ABHG, \ldots$ of the prism P by Corollary 13-2.6a and Definition 7-4. Theorem 13-2.6 tells us $AG = BH = \ldots = e$. Also, $\mathscr{A}DEFJ = e \cdot UQ$, $\mathscr{A}EAGF = e \cdot QR, \ldots$ by Theorem 12-3.1. Hence, $\mathscr{A} = e(UQ + QR + \cdots)$ (Definition 13-11), or $\mathscr{A} = e \cdot p$ (Postulate 2-1).

How do we know that the base of a right prism is perpendicular to each lateral edge? Because of this, the base is a right section of the right prism. This fact justifies the following corollary:

C **Corollary 13-2.7a** The lateral area of a right prism is equal to the product of the perimeter of one of its bases and its altitude.

Another useful corollary is suggested by the statement that a section of the prism parallel to the bases cuts the prism into two prisms.

C **Corollary 13-2.7b** Every section of a prism made by a plane parallel to the bases is congruent to the bases.

The proofs of the two corollaries above are left as exercises.

The Volume of a Polyhedral Region

Polyhedral regions not only have surface area, they also occupy a certain amount of space. The measure of this space is the volume of the polyhedral region. We express the volume of such a region in cubic units.

P **Postulate 13-1 The Volume Postulate** To each polyhedral region there corresponds a unique positive real number.

The number we referred to in the Volume Postulate is called the volume of the polyhedral region. Since the polyhedral region is determined by a polyhedron, we often refer to the volume of a polyhedral region as the volume of a polyhedron. If two polyhedrons, P_1 and P_2, have equal volumes, we denote this by $\mathscr{V}\ P_1 = \mathscr{V}\ P_2$.

D **Definition 13-13** A **parallelepiped** is a prism in which the bases are parallelograms.

If all faces of a parallelepiped are rectangles, it is a **rectangular parallelepiped**. The union of a rectangular parallelepiped and its interior is a **rectangular solid**.

P **Postulate 13-2 The Volume Postulate for Rectangular Solids** The volume of a rectangular solid (that is, the volume of a rectangular parallelepiped) equals the product of its length, width, and height. $\mathscr{V} = lwh$.

Class Exercises

1. Each card in a deck has a specific thickness and area. As they are originally packaged, the cards determine a rectangular parallelepiped. Disregarding any space between cards, how can the volume of the rectangular solid be expressed in terms of the volume of a single card?

2. If the end of the deck is pushed so that it slants, as shown at right above, how is the altitude of the parallelepiped affected?
3. Since there is the same number of cards, each with the same thickness and area, how do the two volumes pictured above compare?
4. Suppose we use cards of another shape, but with the same thickness. If this deck consists of the same number of cards as the first deck, under what conditions will it have the same volume?

The above Class Exercises suggest a principle that we shall accept as a postulate.

P **Postulate 13-3 Cavalieri's Principle** If two solid regions have equal altitudes and if sections made by planes parallel to the base of each solid and at the same distance from each base are always equal in area, then the volumes of the solid regions are equal.

T

Theorem 13-2.8 Two prisms have equal volumes if their bases have equal areas and their altitudes are equal.

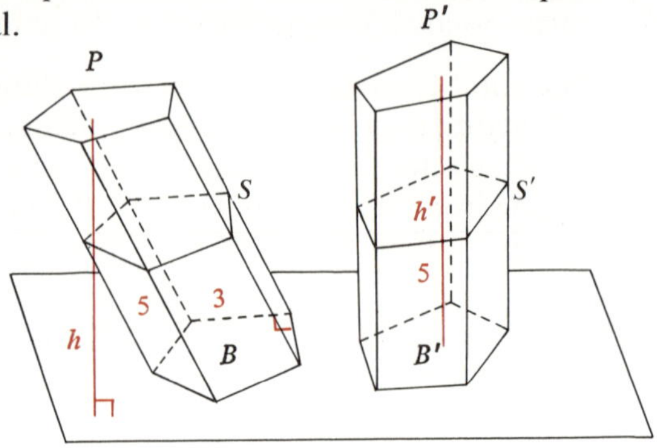

GIVEN Prisms P and P' with bases B and B' such that $\mathscr{A}B = \mathscr{A}B'$ and altitudes h and h' such that $h = h'$

PROVE $\mathscr{V}(P) = \mathscr{V}(P')$

PROOF Consider two planes parallel to the bases of P and P' making sections S and S' at equal distances d and d' from the bases B and B'(Definition 13-9). Corollary 13-2.7b and Postulate 12-2 tell us $S \cong B$, $S' \cong B'$, $\mathscr{A}S = \mathscr{A}B$, and $\mathscr{A}S' = \mathscr{A}B'$. By Postulate 2-1, $\mathscr{A}S = \mathscr{A}S'$. Hence, $\mathscr{V}P = \mathscr{V}P'$ (Postulate 13-3).

Corollary 13-2.8a follows directly from Theorem 13-2.8. Its proof is left as an exercise.

C

Corollary 13-2.8a The plane passing through two diagonally opposite edges of a parallelepiped divides it into two triangular prisms of equal volume.

The next theorem establishes the formula for finding the volume of a prism.

T

Theorem 13-2.9 The volume of a prism is the product of the area of a base and the altitude. $\mathscr{V} = \mathscr{A}Bh$.

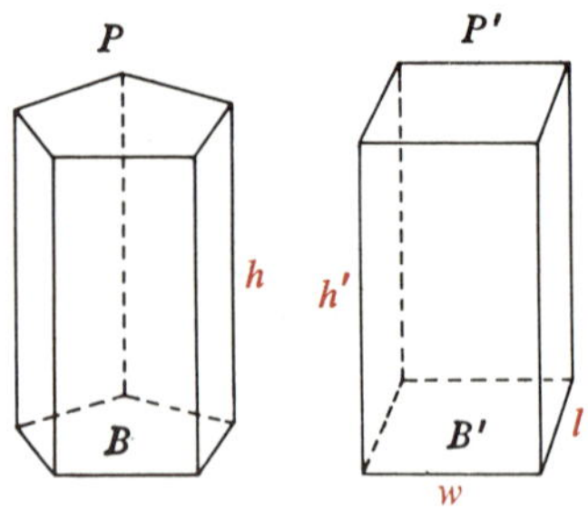

GIVEN Prism P with base B and altitude h.

PROVE $\mathscr{V}P = \mathscr{A}B \cdot h$

PROOF OUTLINE Construct a rectangular parallelepiped P' with base B' such that $\mathscr{A}B = \mathscr{A}B'$, and with an altitude h' equal to that of the prism. Use Cavalieri's Principle to show that the two figures have equal volume ($\mathscr{V}P = \mathscr{V}P'$). Then P and P' have equal volume, bases of equal area, and equal altitudes. Since the volume of the rectangular parallelepiped is the product of the area of a base ($l \times w$) and the altitude (h'), the same is true of the prism.

Since a parallelepiped is a prism, we can use Theorem 13-2.9 to establish the formula for the volume of any parallelepiped.

C

Corollary 13-2.9a The volume of a parallelepiped is the product of the area of any face and the length of the altitude to that face.

Summary. Area and Volume of a Prism

1. $\mathscr{A}$ prism $= ep$ 2. $\mathscr{A}$ right prism $= ap$
3. $\mathscr{V}$ rectangular solid $= lwh$ 4. $\mathscr{V}$ prism $= \mathscr{A}Bh$
5. The volume of a parallelepiped equals the product of the area of a face and the length of the altitude to that face.

EXERCISES

A

1. If a right section of a prism whose lateral edge is 8 is an equilateral triangle with side 5, find the lateral area of the prism.
2. A right prism has a square base and a lateral edge 8. Find the area of the base if the lateral area is 96.
3. The perimeter of a right section of a prism is 30 and the lateral area is 360. What is the length of the lateral edge?
4. Find the diagonal of a cube whose edge is 4.
5. Find the total area of a cube whose diagonal is $8\sqrt{3}$.
6. A rectangular parallelepiped has edges 8, 10, and 15. Find its total area.
7. A rectangular parallelepiped has edges 9, 10, and 15. Find its volume.
8. If each edge of one parallelepiped is three times as long as the corresponding edge of another parallelepiped, how do their total areas and volumes compare?
9. A parallelepiped has a base that is a parallelogram with sides 8 units apart and 12 units long. If the altitude to this base is 20, find the volume.
10. A right section of an oblique prism is a hexagon whose sides are 3, 4, 4, 5, 6, and 7. The lateral edge is 7. Find the lateral area.
11. If each side of a base of a hexagonal right prism is 8, and the altitude is 10, find the total area.
12. The dimensions of a classroom are 25 ft by 18 ft by 12 ft. How many cubic feet of air space does the room contain?
13. The volume of a rectangular parallelepiped is 50 cu in. The area of a base is 10 sq in. What is its altitude?
14. If the surface area of a cube is numerically equal to its volume, find its edge.
15. Find the volume of a prism if a base is 25 sq in. and its altitude is 10 in.

B

16. Find the total area of a right prism whose altitude is $2b$ and whose bases are regular hexagons with each side b.
17. A triangular prism has an isosceles right triangle as a base. The measure of each of the equal sides of a base is a and the measure of the altitude is b. What is the volume?
18. If the prism described in Exercise 17 is a right prism, find its total area.

Prove the following:

19. Theorem 13-2.1
20. Theorem 13-2.2
21. Theorem 13-2.3
22. Corollary 13-2.3a
23. Theorem 13-2.4
24. Corollary 13-2.4a
25. Theorem 13-2.5
26. Corollary 13-2.6a
27. Corollary 13-2.7a
28. Corollary 13-2.7b
29. Corollary 13-2.8a
30. Theorem 13-2.9
31. Corollary 13-2.9a

C

32. A ton of coal occupies 35 cu ft. How high must a bin 10 ft long and 8 ft wide be if it is to hold 10 tons?
33. Find the volume of a right prism each of whose bases is a regular hexagon a inches on a side, and whose altitude is b inches.
34. Each base of a right prism is a rhombus. The diagonals of a base are 12 and 6, and the altitude of the prism is 18. Find the volume and the total area.
35. The total area of a cube is 36. Find the measure of its diagonals.
36. Find the surface area and the diagonal of a cube if its volume is 729.
37. Prove that the diagonals of a parallelepiped bisect each other.

A LOOK AT THE PAST

"Indivisibles"

Class Exercises 1 through 4 in Section 13-2 imply that a solid consists of a stack of planes. This is the basis for rationalizing Cavalieri's Principle. Francesco Bonaventura Cavalieri (1598-1647) was an Italian mathematician whose work in finding areas was a forerunner of modern calculus.

Cavalieri's method was based on the notion that a solid is made up of surfaces, a surface is made up of lines, and a line is made up of points. According to his theory, a line can be subdivided until the ultimate "indivisible" point is reached. The length of a segment is the sum of these "indivisibles." Although this is contrary to the accepted notion that a line is infinitely divisible, Cavalieri used his "indivisibles" in establishing many correct results.

Cavalieri's theory can lead to contradictions, however. If the length of a segment is the sum of the lengths of its "indivisible" points, it can be shown that any two segments have the same length. Let $\overline{AB} \parallel \overline{CD}$ and $AB > CD$.

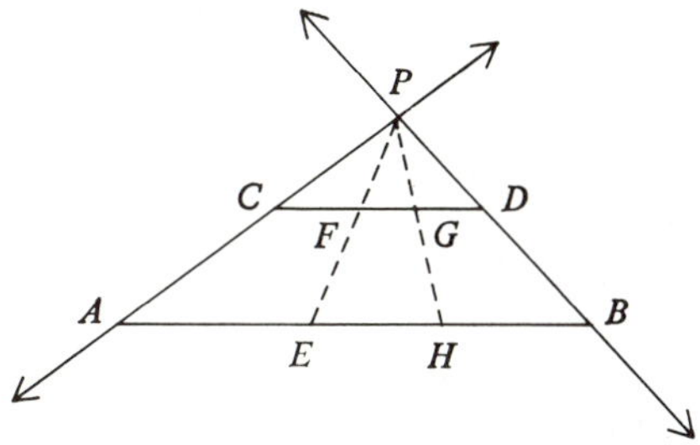

Draw $\overleftrightarrow{AC}$ and $\overleftrightarrow{DB}$, intersecting in P. Corresponding to any point E of $\overline{AB}$ there is a line $\overleftrightarrow{PE}$ that intersects $\overline{CD}$ in a unique point, F. Corresponding to any point G of $\overline{CD}$ there is a line $\overleftrightarrow{PG}$ that intersects $\overline{AB}$ in a unique point, H. We have thus shown there is a one-to-one correspondence between the points of $\overline{AB}$ and $\overline{CD}$. If the length of a segment is found by finding the sum of its points, we have $AB = CD$.

13-3 PYRAMIDS

A polygon and a point not in the plane of the polygon determine a special kind of polyhedron, the pyramid.

D **Definition 13-14** A **pyramid** is a polyhedron formed by joining each point in the sides of a polygonal region to a common point not in the plane of the polygonal region.

The polygonal region is called the **base** and the point is called the **vertex**.

The triangular plane regions that meet at the vertex of a pyramid are called **lateral faces**. The intersections of the lateral faces are the **lateral edges**. The sum of the areas of the lateral faces is the **lateral area** of the pyramid. The length of the perpendicular segment from the vertex to the plane of the base is the **altitude** of the pyramid. We also refer to the segment itself as an altitude.

Like prisms, pyramids are classified according to their bases.

D **Definition 13-15** A **regular pyramid** is a pyramid the sides of whose base form a regular polygon whose center coincides with the projection of the vertex onto the base.

In the figure for Theorem 13-3.1, the projection of P onto the base is O, which is also the center of regular polygon $ABCDE$. When no confusion results, we shall refer to a pyramid by its vertex. The figure on page 554 is pyramid P.

T **Theorem 13-3.1** The lateral edges of a regular pyramid are congruent.

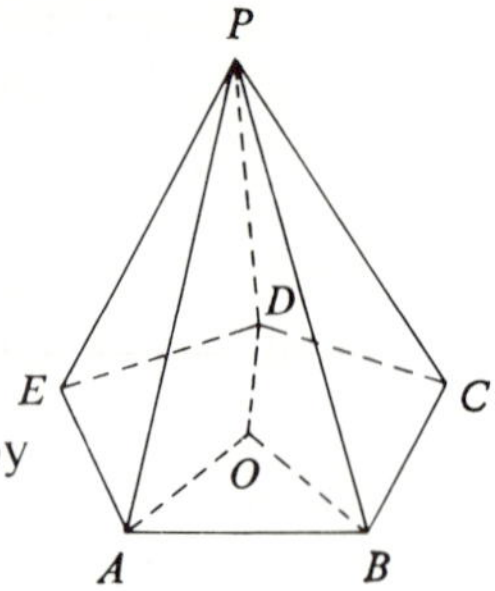

GIVEN Regular pyramid P with base $ABCDE$ . . . and center of base O.

PROVE $\overline{PA} \cong \overline{PB} \cong \overline{PC} \cong \overline{PD} \cong \overline{PE} \cong \ldots$

PROOF In regular pyramid P, $\overline{OA} \cong \overline{OB}$ by Definitions 13-14 and 11-2. Also, $\angle POA \cong \angle POB$ by Definition 4-6 and Theorem 3-1.1. By Theorem 3-1.6, we know that $\overline{OP} \cong \overline{OP}$. Consequently, $\triangle AOP \cong \triangle BOP$ (SAS), and $\overline{PA} \cong \overline{PB}$ by Definition 3-3. Repeating this procedure for $\overline{OC}$, $\overline{OD}$, . . . and applying the Substitution Postulate (2-1), we have $\overline{PA} \cong \overline{PB} \cong \overline{PC} \cong \overline{PD} \cong \ldots$.

The proof of the following corollary is left as an exercise:

C **Corollary 13-3.1a** The lateral edges of a regular pyramid form congruent isosceles triangles.

The **slant height of a regular pyramid** is the length of the altitude to the base of any one of its lateral faces.

T **Theorem 13-3.2** The lateral area of a regular pyramid is equal to one-half the product of its slant height and the perimeter of its base. $\mathscr{L} = \frac{1}{2} s \cdot p$

The proof of this theorem is left as an exercise.

Example 1 Find the lateral area of a regular hexagonal pyramid if both the slant height and the length of a side of the base equal 4.

Solution The perimeter of the base is $6 \cdot 4 = 24$. The slant height is 4. Thus, the lateral area is $\mathscr{L} = \frac{1}{2}(4)(24) = 48$.

We can think of a pyramid as being formed by a plane intersecting all faces of a polyhedral angle. Then a plane parallel to the base of a pyramid cuts off another pyramid. The base of this new pyramid is a section of the first pyramid. This suggests Theorem 13-3.3.

T **Theorem 13-3.3** If a pyramid is cut by a plane parallel to its base, the section is a polygon similar to the base, and the lateral edges and altitude are divided proportionally, with the ratio of their lengths to the lengths of the segments cut off between the section and the vertex equal to the ratio of similitude of the base and the section.

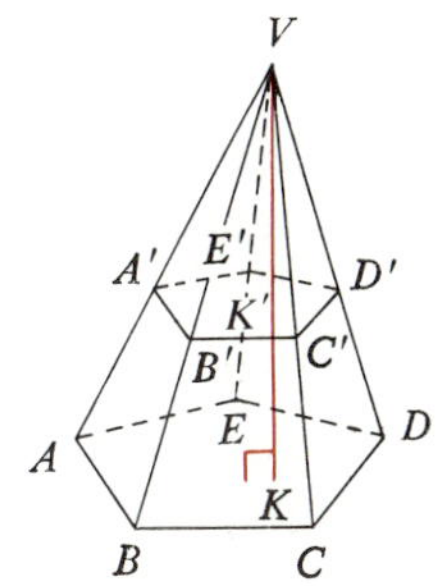

GIVEN Pyramid $V\text{-}ABCDE$ cut by a plane parallel to its base, intersecting the lateral edges in A', B', C', D', E' and the altitude $\overline{VK}$ in K'

PROVE (1) $\dfrac{VA'}{VA} = \dfrac{VB'}{VB} = \ldots = \dfrac{VK'}{VK}$.

(2) $\dfrac{A'B'}{AB} = \dfrac{VA'}{VA} = \ldots = \dfrac{VK'}{VK}$

(3) Polygon $A'B'C'D'E' \sim$ polygon $ABCDE$

PROOF Consider the plane containing $\overline{VA'A}$ and $\overline{VK'K}$ which forms $\triangle VAK$ (P2-5, P2-7, and D1-31). By Theorem 13-2.1, $\overline{A'K'} \parallel \overline{AK}$. Thus $\dfrac{VA'}{VA} = \dfrac{VK'}{VK}$ by Corollary 8-2.1a. Similarly, $\dfrac{VB'}{VB} = \dfrac{VK'}{VK}$, $\dfrac{VC'}{VC} = \dfrac{VK'}{VK}$, ... Since $\angle A'BV' \cong \angle AVB$ and $\dfrac{VA'}{VA} = \dfrac{VB'}{VB}$, we have $\triangle VA'B' \sim \triangle VAB$. By definition of similar triangles (8-6), $\dfrac{A'B'}{AB} = \dfrac{VA'}{VA} = \dfrac{VB'}{VB}$. Similarly, $\dfrac{B'C'}{BC} = \dfrac{VB'}{VB}$. The plane of $\overleftrightarrow{VA}$ and $\overleftrightarrow{VC}$ cuts the base in $\overline{AC}$ and the section in $\overline{A'C'}$. Therefore, by Theorem 13-2.1, $\overline{AC} \parallel \overline{A'C'}$. Since $\overline{AB} \parallel \overline{A'B'}$, and $\overline{BC} \parallel \overline{B'C'}$, we have $\triangle ABC \sim \triangle AB'C'$. Hence, $\angle B \cong \angle B'$. In a similar way, we can show the other corresponding angles congruent. Hence, the base and the section are similar.

C

Corollary 13-3.3a If two pyramids have congruent altitudes and bases with equal areas, sections parallel to the bases at equal distances from the vertices have equal areas.

GIVEN Pyramids $V\text{-}ABC$ and $U\text{-}DEFG$, where $\overline{VK}$ is an altitude of $V\text{-}ABC$ and $\overline{UL}$ is an altitude of $U\text{-}DEFG$; $\overline{VK} \cong \overline{UL}$; $\mathscr{A}\triangle ABC = \mathscr{A}DEFG$; plane $A'B'C' \parallel$ plane ABC and meets $\overline{VK}$ at K'; plane $D'E'F'G' \parallel$ plane $DEFG$ and meets $\overline{UL}$ in L'; $\overline{VK'} \cong \overline{UL'}$.

PROVE $\mathscr{A}\triangle A'B'C' = \mathscr{A}$quadrilateral $D'E'F'G'$

PROOF OUTLINE

$$\frac{\mathscr{A}ABC}{\mathscr{A}A'B'C'} = \frac{(AB)^2}{(A'B')^2} = \frac{(VK)^2}{(VK')^2}$$

Also,

$$\frac{\mathscr{A}DEFG}{\mathscr{A}D'E'F'G'} = \frac{(DE)^2}{(D'E')^2} = \frac{(UL)^2}{(UL')^2}$$

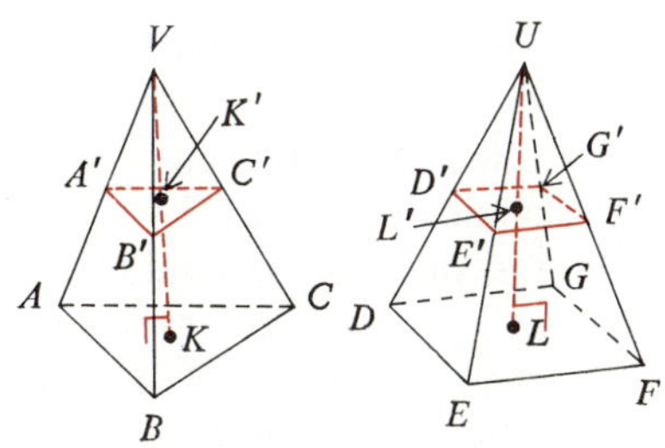

Since $\mathscr{A}ABC = \mathscr{A}DEFG$, $VK = UL$, and $VK' = UL'$, we can conclude that $\mathscr{A}A'B'C' = \mathscr{A}D'E'F'G'$.

Example 2 Two pyramids have bases with equal areas and congruent altitudes. One has a square base and the other a pentagonal base. Their bases are contained in the same plane. A plane parallel to their bases determines a section on the first pyramid that is a square, 2 ft on a side. Find the area of the section determined on the other pyramid.

Solution By Corollary 13-3.3a, the sections have equal areas. The area of the square section is $2 \times 2 = 4$ sq ft. Hence the area of the section of the other pyramid is also 4 sq ft.

Corollary 13-3.3b suggests an important relationship involving the volumes of two pyramids with congruent altitudes and bases that have equal areas.

C **Corollary 13-3.3b** If two pyramids have congruent altitudes and bases with equal areas, then they have equal volumes.

Before Corollary 13-3.3b can be of any value, we must be able to determine the volume of a pyramid. A method for this is provided in the next two theorems.

T **Theorem 13-3.4** The volume of a triangular pyramid equals one-third the product of the area of its base and the altitude. $\mathscr{V} = \frac{1}{3} b \cdot a$.

GIVEN Triangular pyramid O-ABC with volume $\mathscr{V}$, base area b, and altitude length a.

PROVE $\mathscr{V} = \frac{1}{3} b \cdot a$

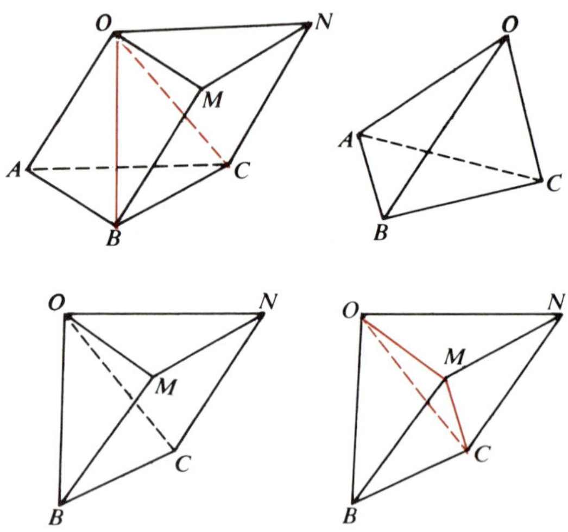

PROOF OUTLINE
Consider a triangular pyramid O-ABC and quadrangular pyramid O-$MNCB$ with common face $\triangle OBC$, and face $\triangle OMN$ congruent and parallel to base $\triangle ABC$. The two pyramids thus compose a triangular prism with bases $\triangle ABC$ and $\triangle OMN$, and lateral sides $\overline{OA}$, $\overline{MB}$, and $\overline{NC}$. The plane determined by $\overleftrightarrow{OM}$ and $\overleftrightarrow{OC}$ divides pyramid O-$MNCB$ into two triangular pyramids, O-MNC and O-MBC. Since face $MNCB$ of the prism is a parallelogram, $\triangle MNC \cong \triangle MBC$. Thus, by Corollary 13-3.3b, pyramids O-MNC and O-MBC have equal volumes, since

their bases have equal areas and they have the same altitude from common vertex O.

But O-MNC is the same as C-MNO which has the same altitude as O-ABC. Also C-MNO and O-ABC have bases of equal area. Therefore, C-MNO and O-ABC have equal volumes. Since the prism consists of three pyramids with equal volumes, the original pyramid, O-ABC, has volume $\mathscr{V} = \frac{1}{3}b \cdot a$.

Example 3 Find the volume of a pyramid whose base is an equilateral triangle with side 10 and whose altitude is 20.

Solution The area of the base is $\dfrac{10^2\sqrt{3}}{4} = 25\sqrt{3}$. Thus the volume is

$V = \frac{1}{3}(25\sqrt{3} \cdot 20) = \dfrac{500\sqrt{3}}{3}$.

We can generalize Theorem 13-3.4 and obtain the formula for the volume of any pyramid. It is necessary merely to divide the pyramid into triangular pyramids with a common vertex and altitude.

T

Theorem 13-3.5 The volume of any pyramid is equal to one-third the product of the area of its base and altitude.

GIVEN Any pyramid P whose base is not a triangle

PROVE $V = \frac{1}{3}b \cdot a$, where b is the area of the base and a is the altitude.

PROOF OUTLINE Consider the planes through P and the diagonals from a single vertex of the base b. These planes divide the pyramid into triangular pyramids with a common altitude a. The sum of the areas of bases of the triangular pyramids, $b_1, b_2, b_3, \ldots$, is equal to the area of the base of the original pyramid. The volume of the original pyramid is the sum of the volumes of the triangular pyramids.

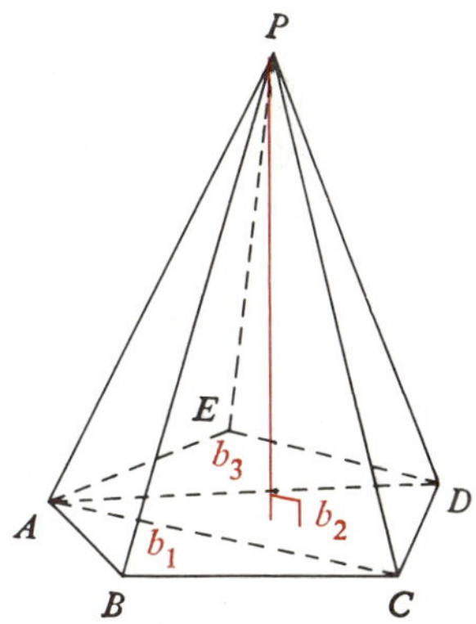

$$\mathscr{V} = \left(\frac{1}{3}b_1 \cdot a\right) + \left(\frac{1}{3}b_2 \cdot a\right) + \left(\frac{1}{3}b_3 \cdot a\right) + \cdots$$
$$\mathscr{V} = \frac{1}{3}a(b_1 + b_2 + b_3 + \cdots) = \frac{1}{3}b \cdot a$$

Class Exercises

1. Find the lateral area and the total area of a regular hexagonal pyramid if each side of the base is 8 and the slant height is 10.
2. The volume of a pyramid is 240. If the area of its base is 120, what is the length of its altitude?

3. The center of a cube is connected with each vertex. How many pyramids are formed? Find the volume of each.
4. The base of a regular pyramid is a pentagon with side 5. If the slant height is 10 find the lateral area.
5. Find the edge of a cube whose volume is equal to that of a pyramid with a square base with side 2 and altitude 8.

Summary. Area and Volume of a Pyramid

1. $\mathscr{L}$regular pyramid $=\frac{1}{2}s \cdot p$. 2. $\mathscr{V}$pyramid $=\frac{1}{3}b \cdot a$.

EXERCISES

A

1. A regular pyramid has for its base an equilateral triangle with side 6. The altitude of the pyramid is 4. Find the lateral area.
2. Find the volume of a pyramid whose altitude is 4 in. and whose base is a right triangle with legs 3 in. and 5 in. long.
3. The altitude of a regular pyramid with a square base is 12 and its slant height is 13. What is its volume?
4. The lateral edge of a regular tetrahedron is 6. What is the altitude?
5. The altitude of a regular quadrangular pyramid is 10 and the sides of its base are 8. Another pyramid has a base which is an isosceles right triangle whose hypotenuse is 16. If the two pyramids have equal volumes, what is the altitude of the second one?
6. Find the lateral area and volume of a regular hexagonal pyramid if each lateral edge is 7 and each side of the base is 5.
7. The center of a cube, which is equidistant from the vertices, is connected with each vertex. How many pyramids are formed? If the edge of the cube is 2, find the volume of one of the pyramids.
8. Find the total area of a regular triangular pyramid if each lateral edge is 10 and the altitude of the base is 5.
9. Find the volume of the pyramid in Exercise 8.
10. How do the volumes of two pyramids compare if they have equal altitudes and the base of one is 25, while the base of the other is 36.

B

11. The area of the base of a pyramid is 48 sq ft. What is the area of a section parallel to the base and midway between the vertex and the base?
12. How far from the base of a pyramid of altitude 10 is a section parallel to the base and equal in area to half of the base?

Prove the following:
13. Corollary 13-3.1a 14. Theorem 13-3.2
15. Theorem 13-3.3 (part 3) 16. Corollary 13-3.3a
17. Corollary 13-3.3b 18. Theorem 13-3.4 19. Theorem 13-3.5

20. Find the volume of a regular quadrangular pyramid if its altitude is 4 and its slant height is 5.
21. Find the volume of a pyramid whose faces are equilateral triangles, each of which has an area of 64 sq ft.
22. If the total surface area of a regular tetrahedron is 144 sq in., find its volume.

C

23. Show that a parallelepiped can be divided into six triangular pyramids of equal volume.
24. Prove that the section of a triangular pyramid made by a plane parallel to two of its edges which do not meet is a parallelogram.

13-4 CYLINDERS AND CONES

In the first three sections of this chapter, we studied areas and volumes of solid figures with plane surfaces. In this section, we shall begin to study figures with other surfaces.

Cylinders

D

Definition 13-16 A **cylindrical surface** is the set of all lines parallel to a given line and intersecting a given curve in a plane that does not contain the given line.

The given curve in the definition is called the **directrix** of the cylindrical surface. Each line in the set of lines referred to in the definition is called an **element of the cylindrical surface.**

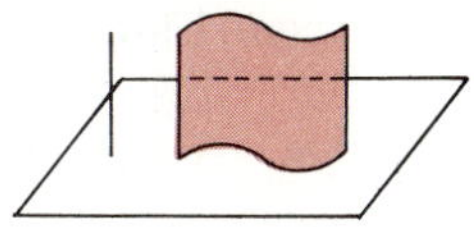

The directrix may be either an open or closed curve. The shape of the directrix determines the shape of the cylindrical surface, but different directrixes can determine the same cylinder. The bottom figure at right is determined by a circular directrix contained in a plane perpendicular to the elements of the surface.

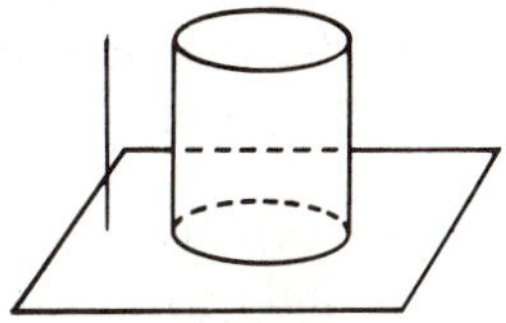

A cylindrical surface is named after its **right sections**, that is, sections contained in planes perpendicular to the elements.

D

Definition 13-17 A **cylinder** is that portion of a closed cylindrical surface between two parallel planes, together with the portions of the planes enclosed by the surface.

The plane regions are called the **bases** of the cylinder. The portion of the cylindrical surface between the planes is called the **lateral area** of the cylinder. The parts of the elements of the cylindrical surface between the bases are the **elements** of the cylinder. Generally, when we use the term *cylinder*, we are referring to a right circular cylinder.

D

Definition 13-18 A **right circular cylinder** is the portion of a circular cylindrical surface lying between two parallel planes that are perpendicular to the elements of the surface, together with the two circular regions of the planes enclosed by the surface.

Many of the terms used with prisms, pyramids, and cylindrical surfaces also apply to cylinders. The elements of a cylinder are the parts of the elements of the surface included between the bases.

The region determined by the intersection of a plane with a cylinder is called a **section** of the cylinder. If the plane cuts all the elements and is perpendicular to them, the section is a **right section**.

The **axis** of a circular cylinder is the segment joining the centers of the bases. In a right circular cylinder the axis is perpendicular to the bases and is thus an altitude of the cylinder.

We shall need the following definition in the proof of Theorem 13-4.1:

D

Definition 13-19 Two **figures are congruent** if and only if every dimension of one is congruent to the corresponding dimension of the other.

T

Theorem 13-4.1 The bases of a cylinder are congruent.

GIVEN A cylinder with bases M and N.

PROVE $M \cong N$

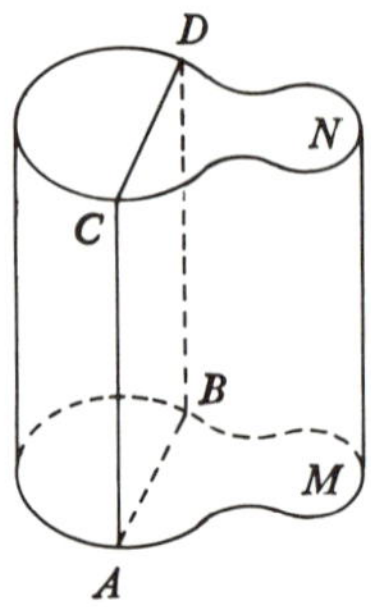

PROOF Let $\overline{AB}$ be any dimension of M and $\overline{CD}$ be the corresponding dimension of N (Definition 1-13). $\overline{AC} \parallel \overline{BD}$ by Theorem 13-2.1. From Definition 13-16 and Theorem 13-2.1, we conclude $\overline{AB} \parallel \overline{CD}$. Thus, $ABCD$ is a parallelogram (Definition 7-1). By Theorem 7-1.2, $\overline{CD} \cong \overline{AB}$. Hence, $M \cong N$ by Definition 13-19.

The proof of the following corollary will be left as an exercise.

C **Corollary 13-4.1a** Every section of a cylinder made by a plane parallel to the bases is congruent to the bases.

We can find the volume of a cylinder by using the formula for the volume of a prism and Cavalieri's Principle (Postulate 13-3). The proof of Theorem 13-4.2 applies to all cylinders, not just those that are circular.

T **Theorem 13-4.2** The volume of a cylinder is the product of the area of the base and the altitude. $\mathscr{V} = \mathscr{A}B \cdot a$.

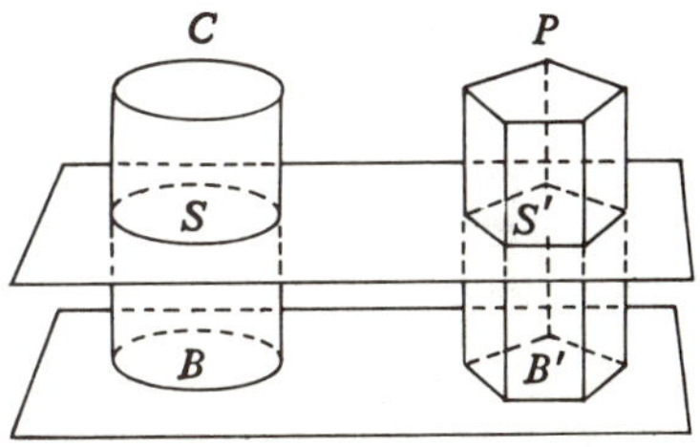

GIVEN Cylinder C with volume $\mathscr{V}$, altitude a, and base B.

PROVE $\mathscr{V} = \mathscr{A}B \cdot a$

PROOF Let P be a prism in the same plane as B with base B' such that a is the altitude and $\mathscr{A}B = \mathscr{A}B'$. Consider a plane parallel to the plane of the bases cutting the prism and cylinder in S' and S, respectively. By Corollary 13-4.1a, $S \cong B$, which implies $\mathscr{A}S = \mathscr{A}B$ (Postulate 12-2). Corollary 13-2.2b tells us that $S' \cong B'$, so that $\mathscr{A}S' = \mathscr{A}B'$. Applying the transitive property, $\mathscr{A}S = \mathscr{A}B' = \mathscr{A}S'$. Thus, $\mathscr{V} = \mathscr{V}'$ by Postulate 13-3 and $\mathscr{V}' = \mathscr{A}B' \cdot a$ (Theorem 13-2.4). Consequently, $\mathscr{V} = \mathscr{A}B \cdot a$ (Postulate 2-1).

If a cylinder is a right circular cylinder, we can use the formula for the area of a circle to determine the area of the base of the cylinder.

C **Corollary 13-4.2a** The volume $\mathscr{V}$ of a right circular cylinder with radius of base r and altitude h equals $\pi r^2 h$. $\mathscr{V} = \pi r^2 h$.

Example 1 Find the volume of a cylindrical can if the diameter of its circular base is 4 in. and its height is 6 in.

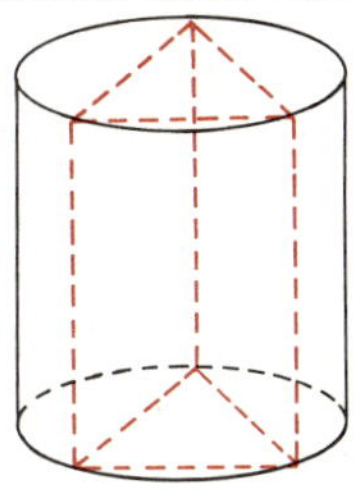

Solution Since the diameter is 4 in., the radius is 2 in.

$$\mathscr{V} = \pi(2^2)(6) = 24\pi$$

If a prism with an equilateral triangle as base is inscribed in a circular cylinder, the volume and surface area of the prism are less than those of the cylinder. If we double the number of sides of the inscribed prism, its volume and surface area will still be less than those of the cylinder, but the differences between them will be smaller. If the number of sides is again doubled, the area and volume of the resulting prism will be even closer to those of the cylinder. We can continue this process until the differences between the area and volume of the regular inscribed prism and the cylinder are less than any amount we choose. The fact that the prism approaches the cylinder as a limit suggests the following postulate, based on Theorem 13-2.2.

P

Postulate 13-4 The Lateral Area Postulate The lateral area of a circular cylinder is equal to the product of an element and the perimeter of a right section.

The following theorem and corollary are immediate consequences of this postulate:

T

Theorem 13-4.3 The lateral area of a right circular cylinder equals the product of its altitude and circumference.

C

Corollary 13-4.3a The total area, $\mathcal{T}$, of a right circular cylinder with altitude h and radius of base r is $2\pi r^2 + 2\pi rh$ or $2\pi r(r + h)$.

Example 2 A cylinder is formed by rotating a 9-by-4 rectangle about a side whose measure is 9. Find the volume and total area of the cylinder.

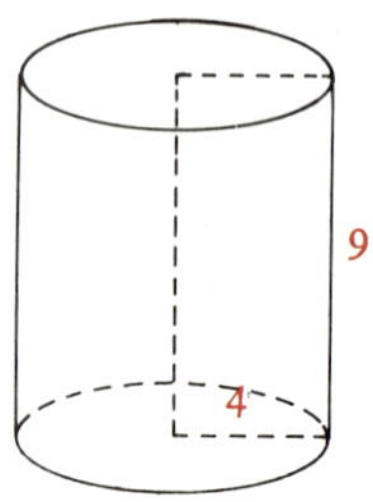

Solution The radius of the base of the cylinder is 4 and the height is 9. Therefore the volume $\mathcal{V} = \pi(4^2)(9) = 144\pi$. The total area $\mathcal{T} = 8\pi(4 + 9) = 104\pi$.

Cones

A prism has parallel bases, while a pyramid has a base and a vertex. A cylinder also has parallel bases. The cone, which we shall define below, is to the cylinder as the pyramid is to the prism.

D

Definition 13-20 A **conical surface** is the set of all lines intersecting a given plane curve and passing through a fixed point that is not in the plane of the curve.

Like the curve of a cylindrical surface, the curve in the above definition is called the **directrix**, and each line is an **element** of the conical surface. The point is called the **vertex**. The parts on each side of the vertex are called **nappes** of the conical surface.

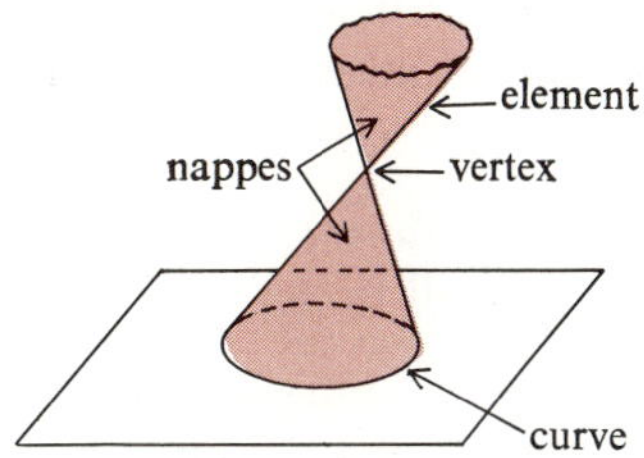

D **Definition 13-21** If a plane intersects one nappe of a closed conical surface, then that part of the surface between the vertex and the plane, together with the region of the plane enclosed by the surface, is a **cone**.

The region of the plane is the **base** of the cone; the area of the portion of the conical surface contained in the cone is the **lateral area** of the cone; the vertex of the conical surface is the **vertex** of the cone; and the parts of the elements of the conical surface between the vertex and the base are the **elements** of the cone.

The **altitude** of a cone is the measure of the perpendicular segment from the vertex to the plane of the base. The **axis** of a circular cone is the segment from the vertex to the center of the base. If the axis forms an angle with the plane of the base other than a right angle, we have an **oblique circular cone**. We shall be primarily concerned with right circular cones, where the length of the axis is the same as the altitude.

D **Definition 13-22** A **right circular cone** is a circular cone whose axis is perpendicular to the plane of the base.

The length of any element of a right circular cone is its **slant height**.

Let us inscribe a regular tetrahedron in a circular cone. Both the volume and total area of the tetrahedron are less than those of the cone. If we increase the number of sides of the base, the volume and total area of the resulting inscribed pyramid remain less than those of the cone, but the differences become even smaller as we continue this process. This relationship between a many-sided pyramid and a circular cone suggests the following postulates. Can you state the theorems to which they are analogous?

P **Postulate 13-5 The Area of a Cone Postulate** The lateral area of a right circular cone is equal to one-half the product of its slant height and the circumference of its base.

P **Postulate 13-6 The Volume of a Cone Postulate** The volume of a circular cone is equal to one-third the product of the area of its base and its altitude. $\mathscr{V} = \frac{1}{3}ba$.

From Postulate 13-5, we derive the following theorem. The proof of the theorem will be left as an exercise.

T **Theorem 13-4.4** If the lateral area of a right circular cone is $\mathscr{L}$, the total area is $\mathscr{T}$, the radius of the base is r, and the slant height is s, then $\mathscr{L} = \pi rs$ and $\mathscr{T} = \pi r^2 + \pi rs = \pi r(r + s)$.

Postulate 13-6 suggests the following theorem concerning the volume of a cone. Its proof will also be left as an exercise.

T **Theorem 13-4.5** The volume $\mathscr{V}$ of a circular cone whose altitude is h and whose base has radius r equals $\frac{1}{3}\pi r^2 h$.

Example 3 A cone is generated by rotating a right triangle with sides 3, 4, and 5 about the leg whose measure is 4. Find the total area and volume of the cone.

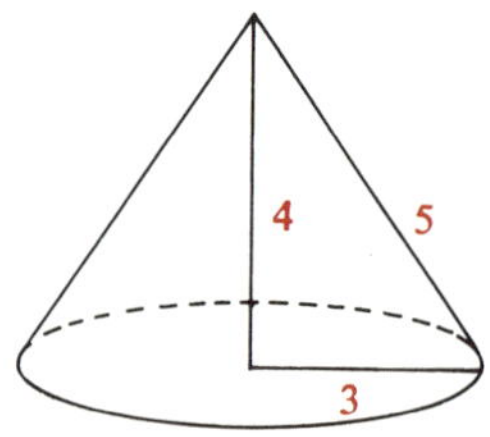

Solution The radius of the base of the resulting cone is 3, the altitude 4, and the slant height 5. Therefore the total area $\mathscr{T} = 3\pi(3 + 5) = 24\pi$. The volume $\mathscr{V} = \frac{1}{3}\pi(3^2 \cdot 4) = 12\pi$.

Summary. Area and Volume of a Cylinder and a Cone

1. $\mathscr{V}$ cylinder $= \mathscr{A}Ba$
2. $\mathscr{V}$ right circular cylinder $= \pi r^2 h$
3. $\mathscr{L}$ right circular cylinder $= 2\pi rh$.
4. $\mathscr{T}$ right circular cylinder $= 2\pi r(r + h)$
5. $\mathscr{L}$ right circular cone $= \pi rs$
6. $\mathscr{T}$ right circular cone $= \pi r(r + s)$
7. $\mathscr{V}$ circular cone $= \frac{1}{3}\pi r^2 h$

EXERCISES

A

1. Find the lateral area and total area of a right circular cylinder with base radius 5 in. and altitude 8 in.
2. Solve the formula $\mathscr{L} = 2\pi rh$ for r and for h.
3. If the radius of the base of a right circular cylinder is 20 ft and the lateral area is 183 sq ft, what is the altitude? the total area?

4. Find the volume of a right circular cylinder if its base radius is 6 and its altitude is 10.
5. If the radius of the base of a right circular cylinder is 26 and the volume is 108, what is the altitude?
6. In order to find its volume, an irregularly shaped object was submerged in water in a cylindrical tank 6 in. in radius. If the water rose 2 in., what was the volume of the object?
7. A right triangle with sides of lengths 5, 12, and 13 revolves about the side whose measure is 5. Find the lateral area of the right circular cone determined by this rotation.
8. What kind of triangle is a section made by any plane containing the axis of a right circular cone?
9. The radius of the base of a right circular cone is 3.2 and the lateral area is 40.4. Find the slant height.
10. Find the volume of a circular cone 5 ft high and with a base 12 sq ft in area.
11. If the diameter of the base of a right circular cone is 5 and its volume is 75, find its altitude.
12. A pile of wheat has the shape of a cone; it is 3 ft high and the circumference of the base is 15 ft. Assuming that a bushel is $\frac{5}{4}$ cu ft, how many bushels of grain are in the pile?

B

13. A right circular cone whose altitude is 12 is cut by a plane parallel to the base and 8 units from the vertex. If the radius of the base is 5, what are the lateral areas and volumes of the two parts of the cone determined by the plane?
14. Describe the locus of the set of all points 6 units from a given line.
15. Find the altitude of a circular cylinder whose diameter is 5 and whose volume is equal to the volume of a cube whose edge is 6.
16. A regular triangular pyramid is circumscribed about a circular cone with altitude 8 and radius of base 3. Compare the volumes of the pyramid and the cone.

Prove the following:

17. Corollary 13-4.1a
18. Corollary 13-4.2a
19. Theorem 13-4.3
20. Corollary 13-4.3a
21. Theorem 13-4.4
22. Theorem 13-4.5

C

23. A regular hexagonal pyramid is circumscribed about a cone with altitude 10 and radius of base 4. Compare the volumes of the pyramid and the cone.
24. If the radius of the base of a right circular cone is r units and the slant height is s units, find the length of the altitude.
25. One cubic foot is approximately 7.5 gallons. If an ice-cream cone is 2 in. in diameter and 2 in. high, how many ice-cream cones can be filled from a gallon of ice cream?

26. A right circular cylinder is 10 in. in diameter and has an altitude of 5 in. A right circular cone has the same base and the same volume as the cylinder. Compare the lateral areas of the two.
27. A cone with radius of base 5 is inscribed in a regular triangular pyramid with altitude 12. Find the volume of the space between the two surfaces.
28. If a cone is cut into two parts of equal volume by a plane parallel to the base, where must the plane be located?

13-5 FRUSTUMS AND SPHERES

Before we can begin to consider the surface area and volume of a sphere, we need to discuss some of the properties of a frustum of a cone.

Frustums

D **Definition 13-23** The **frustum of a cone** is the figure formed by the base of the cone, a section of the cone parallel to the base, and the surface of the cone between the base and the section. The **slant height of the frustum** is that part of the slant height of the cone that is between the parallel planes. The plane areas determined by the conical surface are the **bases** of the frustum.

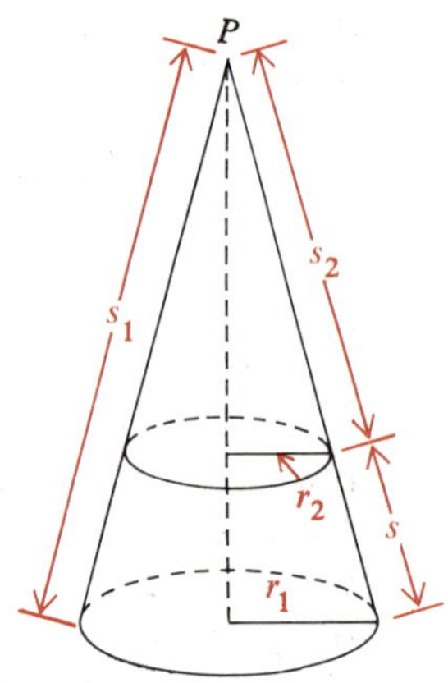

The portion of the cone that is between the frustum and the vertex of the cone is itself a cone. To find the **lateral area** of a frustum, we could subtract the lateral area of the smaller cone from that of the larger cone.

Class Exercises

The figure at right shows two right circular cones. Let the radii of the bases be r_1 and r_2 and the slant heights of the cones be s_1 and s_2.

1. The lateral area, $\mathscr{L}_1$, of the entire cone is ____.
2. The lateral area of the upper cone, $\mathscr{L}_2$, is ____.
3. The lateral area of the frustum is $\mathscr{L}_1 - \mathscr{L}_2 =$ (____ − ____).

Actually, we prefer to express the lateral area of the frustum in terms of the slant height. The **slant height** of the frustum, s, equals $s_1 - s_2$ and the lateral area, $\mathscr{L}$, equals $\mathscr{L}_1 - \mathscr{L}_2$. If you examine the illustration above, you can see that the slant heights are hypotenuses of right triangles.

Thus, $$\frac{s_2}{s_1} = \frac{r_2}{r_1} \quad \text{and} \quad s_2 = \frac{s_1 r_2}{r_1}$$

Substituting for s_2 in the formula from Exercise 3 in the Class Exercises above, and simplifying, we have

$$\mathscr{L} = \pi\left(r_1 s_1 - \frac{r_2{}^2 s_1}{r_1}\right) = \frac{\pi}{r_1}(r_1{}^2 s_1 - r_2{}^2 s_1)$$

$$= \frac{\pi s_1}{r_1}(r_1 - r_2)(r_1 + r_2) = \left(\pi s_1 - \frac{\pi s_1 r_2}{r_1}\right)(r_1 + r_2)$$

Substituting s_2 in the last equation,

$$\mathscr{L} = \pi(s_1 - s_2)(r_1 + r_2)$$

Substituting s and multiplying by $\frac{1}{2} \cdot 2$,

$$\mathscr{L} = \tfrac{1}{2}s(2\pi r_1 + 2\pi r_2)$$

This is the basis for Theorem 13-5.1.

T

Theorem 13-5.1 The lateral area of a frustum of a right circular cone is one-half the product of the slant height and the sum of the circumferences of the two bases.

A frustum may be formed from any polyhedron. The example below shows how to find the volume of a frustum.

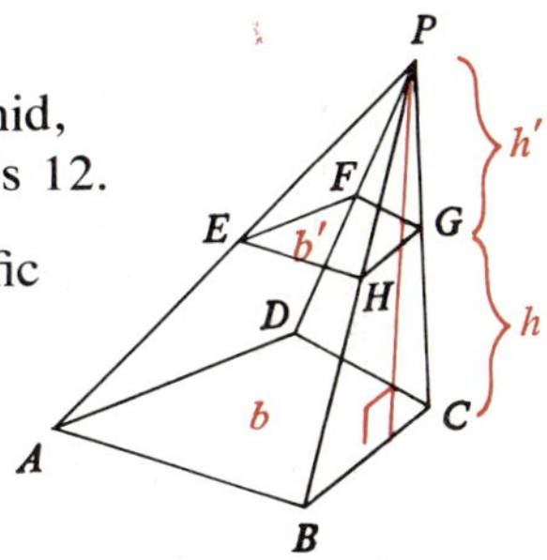

Example 1 Find the volume of a frustum of a pyramid, if the areas of the bases are 2 and 8 and the altitude is 12.

Solution Before beginning our solution of this specific problem, we must develop a general formula for finding the volume of any frustum. Let b and b' represent the areas of the bases, and h, the altitude. Now consider the pyramid formed when the lateral edges are extended to intersect in P. The volume of pyramid $P\text{-}ABCD = \frac{1}{3}b(h + h')$, where h' is the altitude of pyramid $P\text{-}EFGH$. The volume of pyramid $P\text{-}EFGH = \frac{1}{3}b'h'$. Hence, the volume of the frustum is

$$\tfrac{1}{3}b(h + h') - \tfrac{1}{3}b'h' = \tfrac{1}{3}[bh + h'(b - b')]$$

Since each triangular face of pyramid $P\text{-}ABCD$ is similar to the corresponding face of pyramid $P\text{-}EFGH$, by the definition of similar triangles we get $\dfrac{h'}{h + h'} = \dfrac{EH}{AB}$. But by Corollary 12-6.1a, $\dfrac{EH}{AB} = \dfrac{\sqrt{b'}}{\sqrt{b}}$.

Therefore

$$h' = \frac{h\sqrt{b'}}{\sqrt{b} - \sqrt{b'}}$$

By substituting in the above expression for the volume of the frustum, we get

$$\mathscr{V} = \tfrac{1}{3}\left[bh + \frac{h\sqrt{b'}}{\sqrt{b} - \sqrt{b'}}(b - b')\right]$$

$$= \tfrac{1}{3}h\left[b + \frac{\sqrt{b'}}{\sqrt{b} - \sqrt{b'}} \cdot \frac{\sqrt{b} + \sqrt{b'}}{\sqrt{b} + \sqrt{b'}} \cdot (b - b')\right] = \tfrac{1}{3}h[b + b' + \sqrt{bb'}\,]$$

To find the solution to the specific example above, we merely substitute the appropriate values in this formula. Therefore,

$$\mathscr{V} = \tfrac{1}{3}(12)(8 + 2 + \sqrt{2 \cdot 8}\,) = (4)(10 + \sqrt{16}) = 56$$

The volume of the frustum of the pyramid is 56.

The following theorem applies to the lateral area of three figures: the frustum of a right circular cone, the right circular cylinder, and the right circular cone.

T

Theorem 13-5.2 The area of the surface generated by a line segment revolving about an axis in its plane, but not perpendicular to it nor crossing it, is equal to the product of the projection of the segment onto the axis and the circumference of the circle whose radius is the perpendicular to the segment drawn from its midpoint to the axis.

Proof Outline There are three cases to consider.

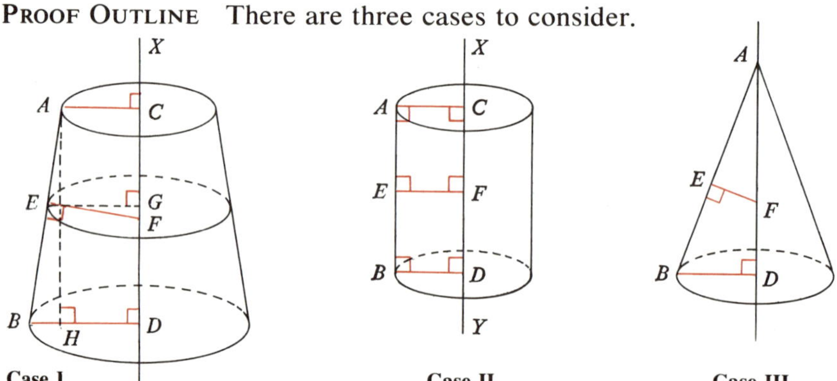

Case I **Case II** **Case III**

Case I The revolving segment is oblique to and does not touch the axis. In this case, the segment generates the frustum of a right circular cone. The segment is $\overline{AB}$, the axis is $\overleftrightarrow{XY}$. The projection of $\overline{AB}$ onto $\overleftrightarrow{XY}$ is $\overline{CD}$. The perpendicular bisector of $\overline{AB}$ is $\overline{EF}$. Draw $\overline{EG} \perp \overline{XY}$ and $\overline{AH} \perp \overline{BD}$. If $\mathscr{S}$ is the surface area generated by $\overline{AB}$, then all the following are true: $EG = \frac{1}{2}(AC + BD)$, $\mathscr{S} = AB \cdot 2\pi EG$, $\triangle ABH \sim \triangle EFG$, $\dfrac{AB}{EF} = \dfrac{AH}{EG}$, $AB \cdot EG = EF \cdot AH$, $AB \cdot EG = EF \cdot CD$, and $\mathscr{S} = CD \cdot 2\pi\, EF$.

Case II The revolving segment is parallel to the axis. In this case, the segment generates the lateral area of a right circular cylinder. Since $\overline{EF} \cong \overline{BD}$, the area $\mathscr{L} = CD \cdot 2\pi \cdot EF$.

Case III The revolving segment meets the axis. In this case, $\mathscr{L}$ is the lateral area of a right circular cone. If E is the midpoint of $\overline{AB}$ and if $\overline{AD}$ is the projection of $\overline{AB}$ onto the axis, then $\mathscr{L} = \frac{1}{2}AB \cdot 2\pi BD = AE \cdot 2\pi BD$. Draw $\overline{EF} \perp \overline{AB}$. Since $\triangle AEF \sim \triangle ADB$, we have $\dfrac{BD}{AD} = \dfrac{EF}{AE}$. Thus, $AE \cdot BD = AD \cdot EF$. Substituting, $\mathscr{L} = AD \cdot 2\pi \cdot EF$.

Theorem 13-5.2 suggests an area postulate for spheres. As the number of chords placed in a semicircle is increased, the area generated by rotation of the chords about the diameter approaches the area of the sphere with the semicircle as an arc of a great circle. The distance from the center of the semicircle to the midpoint of the chord approaches the radius.

P **Postulate 13-7 The Area Postulate for Spheres** The area of a sphere is the product of 2π, the diameter, and the radius. $\mathscr{L} = 2\pi \cdot 2r \cdot r = 4\pi r^2$.

If we restate the postulate, viewing the diameter as two radii, we see that the area of a sphere is four times the area of a great circle, that is, $\mathscr{L} = 4\pi r^2$.

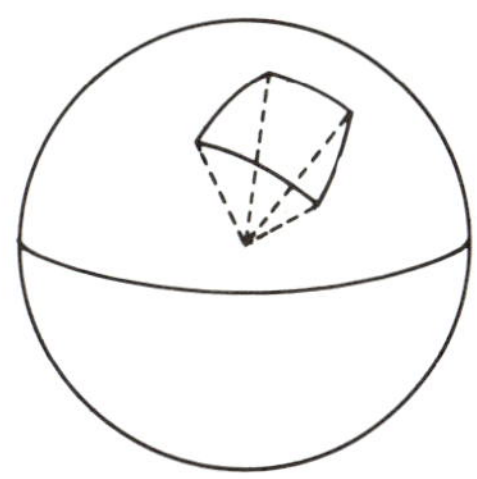

We can provide background explanation of the volume postulate for spheres more easily than was the case with the area postulate. Consider any polygon drawn on the surface of a given sphere and the radii of the sphere drawn to the vertices of the polygon. The figure will resemble a pyramid with its vertex at the center of the sphere, and its altitude equal to the radius of the sphere. The base of the pyramid, however, is a region with a spherical surface instead of a plane surface.

However, the more we reduce the length of the sides of the polygon we draw in the surface of the sphere, the closer the polygon becomes to a plane region. If a spherical region is divided into many of these pyramid-like figures, the volume of the sphere is the sum of the volumes of the pyramid-like figures, the sum of whose bases is the total area of the sphere. From the formula for the volume of a pyramid we derive $\mathscr{V} = \frac{1}{3} r \cdot 4\pi r^2 = \frac{4}{3}\pi r^3$.

P **Postulate 13-8 The Volume Postulate for Spheres** The volume $\mathscr{V}$ of a sphere with radius r is $\frac{4}{3}\pi r^3$.

Example 2 A sphere has a radius of 2.5 in. Find its surface area and its volume.

Solution The surface area $\mathscr{S} = 4\pi(2.5)^2 = 25\pi$. The volume $\mathscr{V} = \frac{4}{3}\pi(2.5)^3 = \frac{125}{6}\pi$.

Summary. Area and Volume of a Frustrum and Sphere

1. $\mathscr{L}\text{frustum} = \frac{1}{2}S(2\pi r_1 + 2\pi r_2)$
2. $\mathscr{L}\text{sphere} = 4\pi r^2$
3. $\mathscr{V}\text{frustum} = \frac{1}{3}h(b + b' + \sqrt{bb'})$
4. $\mathscr{V}\text{sphere} = \frac{4}{3}\pi r^3$

EXERCISES

A

1. Find the area of a sphere with radius 4.
2. The areas of two spheres are 16π and 100π. Find the ratio of the circumferences of their great circles.
3. Find the surface of a sphere that is inscribed in a cube with an edge 8 in. long.
4. The radius of a sphere is 6 ft. What is the maximum height of a right circular cylinder inscribed in the sphere?
5. If the diameter of a sphere is 10 ft, find the lateral area of a right circular cylinder with an 8-ft altitude that is inscribed in the sphere.
6. Find the volume of a spherical shell whose inner radius is 4 and whose outer radius is 6.

B

7. How do the radii of two spheres compare if the volume of one is three times that of the other?
8. A right circular cone has a radius equal to the radius of a sphere and an altitude equal to the diameter of the sphere. What is the ratio of their volumes?
9. A sphere fits exactly into a cube. Compare the surface area of the sphere with the total area of the cube.
10. Find the radius of a sphere if its surface area and volume are numerically equal—that is, the number of square units of area equals the number of cubic units of volume.

C

11. Prove Theorem 13-5.1
12. Prove Theorem 13-5.2
13. A sphere with a 10-in. radius is cut by a plane 4 in. from the center of the sphere. Find the area of an equilateral triangle inscribed in the circle of intersection.
14. A sphere has the same surface area as a cube. Find the ratio of the volume of the sphere to the volume of the cube.

HOW IS YOUR MATHEMATICAL VOCABULARY?

The key words and phrases introduced in this chapter are listed below. How many do you know and understand?

altitude of a cone (13-4)
altitude of a prism (13-2)
altitude of a pyramid (13-3)
Area of a Cone Postulate (13-4)
Area Postulate for Spheres (13-5)
axis of a circular cone (13-4)
axis of a circular cylinder (13-4)
base of a cone (13-4)
base of a cylinder (13-4)
base of a frustum (13-5)
base of a prism (13-2)
base of a pyramid (13-3)
Cavalieri's Principle (13-2)
cone (13-4)
congruent solid figures (13-4)
conical surface (13-4)
cubic unit of volume (13-2)
cylinder (13-4)
cylindrical surface (13-4)
diagonal of a polyhedron (13-1)

directrix of a conical surface (13-4)
directrix of a cylindrical surface (13-4)
edge of a polyhedral angle (13-1)
element of a cone (13-4)
element of a conical surface (13-4)
element of a cylinder (13-4)
element of a cylindrical surface (13-4)
face angle of a polyhedral angle (13-1)
face of a polyhedral angle (13-1)
frustum of a cone (13-5)
hexahedron (13-1)
lateral area of a cone (13-4)
lateral area of a cylinder (13-4)
lateral area of a frustum (13-5)
lateral area of a prism (13-2)
lateral area of a pyramid (13-3)

Lateral Area Postulate (13-4)
lateral edge of a prism (13-2)
lateral edge of a pyramid (13-3)
lateral face of a prism (13-2)
lateral face of a pyramid (13-3)
nappe of a conical surface (13-4)
oblique circular cone (13-4)
oblique prism (13-2)
parallel planes (13-2)
parallelepiped (13-2)
polyhedral angle (13-1)
polyhedral region (13-1)
polyhedron (13-1)
prism (13-2)
pyramid (13-3)
rectangular parallelepiped (13-2)
rectangular solid (13-2)
regular polyhedron (13-1)
regular pyramid (13-3)
right circular cone (13-4)

right circular cylinder (13-4)

right prism (13-2)

right section of a cylinder (13-4)

right section of a cylindrical surface (13-4)

right section of a prism (13-2)

section of a cylinder (13-4)

section of a polyhedron (13-2)

slant height of a right circular cone (13-4)

slant height of a frustum (13-5)

slant height of a regular pyramid (13-3)

tetrahedron (13-1)

total area of a polyhedron (13-1)

vertex of a cone (13-4)

vertex of a conical surface (13-4)

vertex of a polyhedral angle (13-1)

vertex of a pyramid (13-3)

Volume of a Cone Postulate (13-4)

Volume Postulate (13-2)

Volume Postulate for Rectangular Solids (13-2)

Volume Postulate for Spheres (13-5)

REVIEW EXERCISES

13-1 Polyhedrons

1. Can the face angles of a trihedral angle be 100°, 125°, and 150°? Justify your answer.
2. Find the total area of a regular tetrahedron with an edge of length 5.
3. What is the minimum number of planes required to determine a polyhedron?
4. Describe how to locate a point 2 in. from one face of a trihedral angle, 4 in. from a second face, and 6 in. from the third face.

13-2 Prisms

5. The lateral edge of a prism is 10 and a right section is a square with side 3. Find the lateral area.
6. If the diagonal of a cube is $6\sqrt{3}$, find its total area.
7. Each base of a prism is an isosceles triangle with sides 4, 4, and 6. The altitude of the prism is 10. Find its total area and volume.
8. The total area of a cube is 64. Find its volume.
9. A right section of a prism has a perimeter 13. Find the measure of its edge if its lateral area is 360.

13-3 Pyramids

10. Find the lateral area and total area of a regular pyramid whose altitude is 10 and whose base is a square with side 5.

11. The altitude of a regular pyramid with a square base is 5 and its slant height is 13. Find its volume.
12. The measure of a lateral edge of a regular tetrahedron is 9. What is the altitude?
13. The altitude of a regular tetrahedron is 6. What is its total area?
14. Find the total area of a regular triangular pyramid if the lateral edge is 12 and the altitude of the base is 6.
15. Find the volume of a regular quadrangular pyramid if its altitude is 6 and its slant height is 8.
16. If the total surface area of a regular tetrahedron is 400 sq in., find its volume.

13-4 Cylinders and cones

17. Find the lateral area and total area of a right circular cylinder with a 6-in. radius and a 10-in. altitude.
18. The radius of a right circular cylinder is 13 and the volume is 250. Find the altitude.
19. Find the volume of a circular cone that is 6 ft high and has a base area of 10 sq ft.
20. Find the altitude of a cylinder whose diameter is 10 and whose volume is equal to the volume of a cube whose edge is 10.
21. The lateral area of a right circular cone is 64. The slant height is 8. Find the altitude.

13-5 Frustums and spheres

22. Find the surface area of a sphere that is inscribed in a cube whose volume is 64.
23. If the volume of one sphere is twice that of another, how do their radii compare?
24. Find the volume of a spherical shell 1 in. thick if the inner radius is 4 in.
25. A right circular cone has a base equal to the great circle of a sphere whose radius is 6. Find the altitude of the cone if its volume is equal to the volume of the sphere.
26. Find the lateral area of a right circular cylinder with altitude 7 that is inscribed in a sphere with radius 5.

CHAPTER TEST

1. Two face angles of a tetrahedral angle are 70° and 100°. Find the range of possible values for the third face angle.
2. Find the total area of a regular tetrahedron if an edge is 10.

3. An oblique prism has as a right section a pentagon with sides 3, 5, 4, 6, 4, and a lateral edge 12. Find the lateral area.
4. A right prism has triangular bases with sides 5, 12, and 13, and an altitude of 10. Find its volume.
5. What is the altitude of a regular quadrangular pyramid if a side of its base is 8 and its volume is 32.
6. Find the total area of a regular triangular pyramid if the lateral edge is 8 and the altitude is 6.
7. Find the total area of a right circular cylinder if the diameter of its bases is 12 and the altitude is 8.
8. Find the volume of a right circular cone with altitude 6 and slant height 10.
9. The slant height of a conical tent is 12 ft and the radius of its base is 6 ft. How many square yards of canvas does it contain?
10. Find the volume of a sphere that is inscribed in a cube with edge 5.

SUGGESTED RESEARCH

1. The five regular polyhedrons have been called "Platonic bodies." Prepare a report on Plato's contribution to the study of polyhedrons.
2. Investigate the possible kinds of intersections of a plane with both nappes of a conical surface. Summarize your findings.
3. Present a report to your class on how symmetric polyhedral angles differ from congruent polyhedral angles.
4. Prepare a report on how to find the areas of spherical polygons.

MATHEMATICAL EXCURSION

Topology and a Proof of Euler's Formula

A simple polyhedron is one without any "holes" in it. The polyhedrons we have been studying are convex polyhedrons. Every convex polyhedron is simple, but not every simple polyhedron is convex.

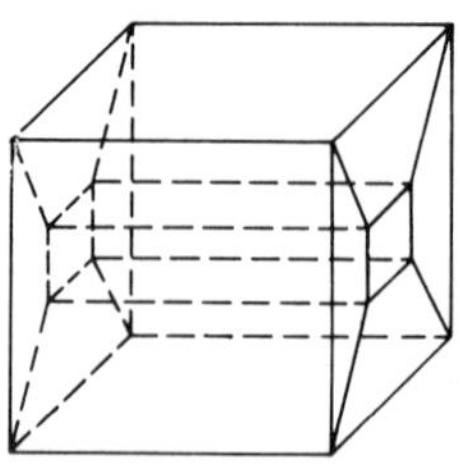

Nonsimple

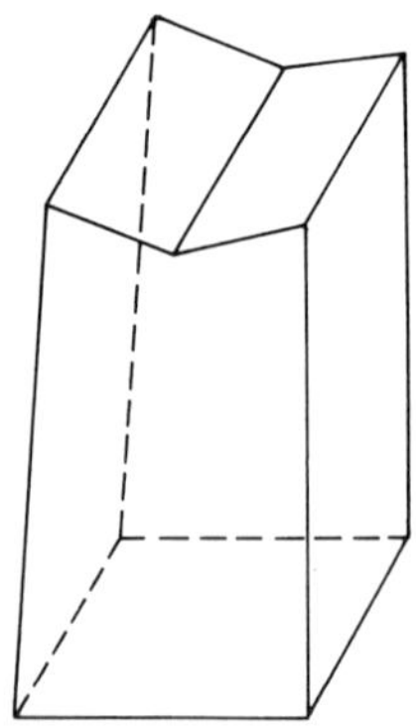

Simple, nonconvex

Euler's Formula, $V - E + F = 2$ is true for all simple polyhedrons, but not for all nonsimple ones. In the figure at left above, there are 16 faces, 16 vertices, and 32 edges. $16 - 32 + 16 = 0$. In the polyhedron at right above, there are 7 faces, 10 vertices, and 15 edges. $10 - 15 + 7 = 2$.

If we pass a plane cutting all the edges of a trihedral angle of a polyhedron, we separate one of the vertices from the rest of the polygon. But, in the process, we add to the polyhedron 1 face, 3 edges, and 3 new vertices. If V is increased by 2, F increased by 1, and E increased by 3, then $V - E + F$ remains unchanged.

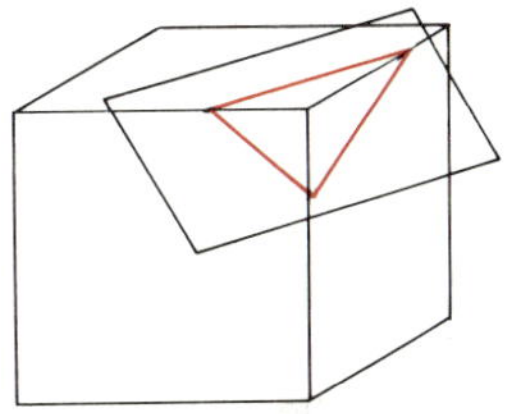

We can obtain a similar result for any polyhedral angle. The new polyhedron will have a new face with the same number of vertices as edges. Since we lose one vertex but gain one face, there is no change in the expression $V - E + F$.

We know the Euler Formula applies to a tetrahedron. From the above argument, we can conclude that it applies to any polyhedron that can be derived by passing a plane that cuts off a vertex of a tetrahedron a finite number of times. However, we would like it to apply to all simple polyhedrons. In the proof we need to show that in regard to the value of the expression $V - E + F$, any polyhedron agrees with the tetrahedron. To do this we need to discuss a new branch of mathematics called topology.

Topology is a very general type of geometry. Establishment of Euler's Formula is a topological problem. Two figures are topologically equivalent if one can be made to coincide with the other by distortion, shrinking, stretching, or bending, but not by cutting or tearing. A teacup and a doughnut are topologically equivalent. The hole in the doughnut becomes the inside of the handle of the teacup.

Topology has been called rubber-sheet geometry. If a face of a polyhedron is removed, the remaining figure is topologically equivalent to a region of a plane. We can deform the figure until it stretches flat on a plane. The resulting figure does not have the same shape or size, but its boundaries are preserved. Edges will become sides of polygonal regions. There will be the same number of edges and vertices in the plane figure as in the polyhedron. Each face of the polyhedron, except the one that was removed, will be a polygonal region in the plane. Each polygon not a triangle can be cut into triangles, or triangular regions, by drawing diagonals. Each time a diagonal is drawn, we increase the number of edges by 1 but we also increase the number of faces by 1. Hence, the value of $V - E + F$ is undisturbed.

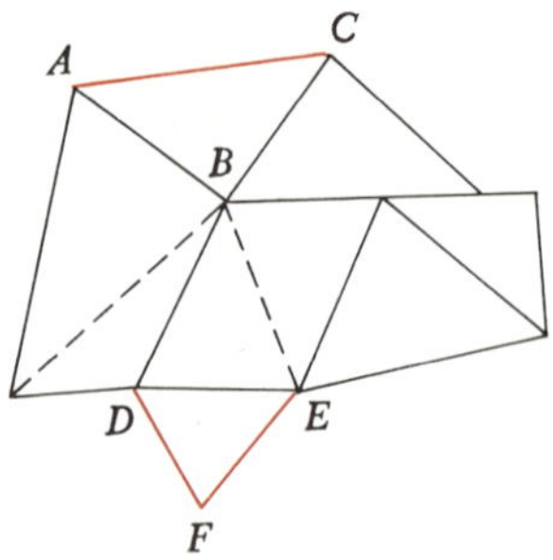

Triangles on the outer edge of the region will have either 1 edge on the boundary of the region, as $\triangle ABC$ in the figure at right, or have 2 edges on the boundary, as $\triangle DEF$. We can remove triangles like $\triangle ABC$ by removing the one boundary side. In the figure, this is $\overline{AC}$. This decreases the faces by 1 and the edges by 1. Still $V - E + F$ is unchanged. If we remove the other kind of boundary triangle, such as $\triangle DEF$, we decrease the number of edges by 2, the number of faces by 1, and the number of vertices by 1. Again, $V - E + F$ is unchanged. This process can be continued until one triangle remains.

The single triangle has 3 vertices, 3 edges, and 1 face. Hence, $V - E + F = 1$. Consequently, $V - E + F = 1$ in the plane figure obtained from the polyhedron by distortion. Since one face had been eliminated, we conclude that for the polyhedron

$$V - E + F = 2$$

This procedure applies to any simple polyhedron, even if it is not convex. Can you see why it cannot be applied to a nonsimple polyhedron?

An alternate to the approach of distorting the polyhedron to a plane after a face has been eliminated can be named "shrinking a face to a point." If a face is replaced by a point, we lose the n edges of the face and the n vertices of the face, and we lose a face and gain a vertex (the point that replaces the face). This leaves $V - E + F$ unchanged. This process can be continued until only 4 faces remain. Then any polyhedron has the same value for $V - E + F$ as does a tetrahedron. The tetrahedron has 4 faces, 4 vertices, and 6 edges: $4 - 6 + 4 = 2$.

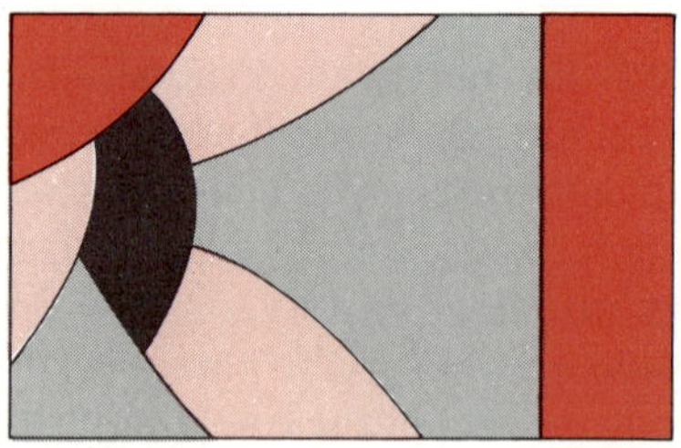

Another topological problem that sounds very simple, but is not, is known as the four-color problem, or the mapmaker's problem. Have you ever paid any attention to the number of colors used to color a map? The rule in this problem is to use different colors for two regions with a common boundary. If two regions have at most a point in common, they are not considered to have a common boundary. The problem is: How many colors are needed to color a map according to our rules? No one has ever produced a map that required more than four colors, but it has never been proved that the number needed is four. This is an open topological problem. In this sense, it differs from the problem of trisecting any angle. It has been shown that the trisection problem cannot be done with straightedge and compass. The trisection issue is closed.

In this chapter we have studied figures composed of plane surfaces and curved surfaces. They enclosed a region. These figures have an inside and an outside. Is it possible to have a surface with only one side? You might enjoy making one and studying its properties. Cut a strip of paper about 12 in. long and $1\frac{1}{2}$ in. wide. Draw a line down the middle of the strip on both sides. In the center, put an A on one side and a B on the other side. Hold the strip

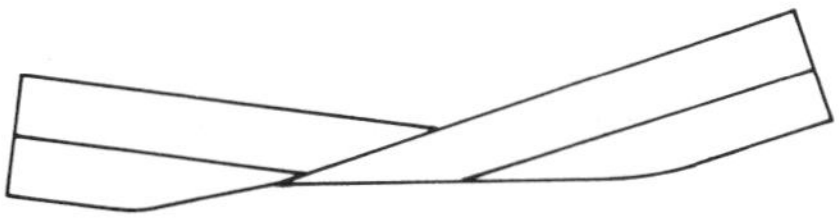

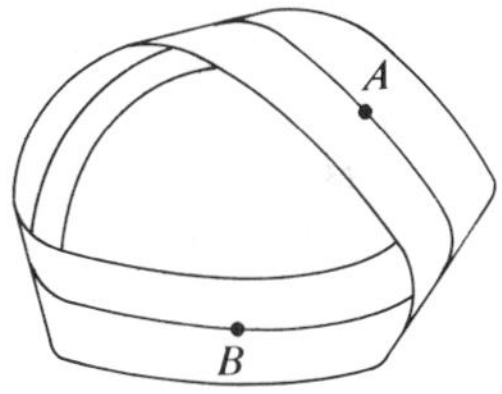

at both ends, one end in each hand. Turn one end through a half twist. Bring the ends together, making sure the dividing lines coincide. Paste the ends together. Now if you will locate A and move along the middle line, you will ultimately get to B and, continuing, back to A. A figure like this is called a Möbius strip.

EXERCISES

1. Alter the nonsimple polyhedron at the beginning of this excursion so that at the right-hand end of the "hole" there is a single face rather than four. Find the value of $V + F - E$ in the resulting nonsimple polyhedron.
2. Construct a "map" that cannot be colored with less than four colors if regions with a common boundary are not colored the same color.
3. Show that the "map" on page 576 can be colored with fewer than four colors.
4. Make a Möbius strip. Try to visualize what will happen if it is cut all the way around down the center line. Cut it as described and compare the results with what you visualized.
5. Follow up Exercise 4 by cutting the resulting strip down the middle.
6. Make another Möbius strip and cut it along a line $\frac{1}{3}$ of the way in from the edge. Describe your results.

14 Vectors

Many quantities, such as 5 pounds, 20 feet, and $3\frac{1}{2}$ gallons, can be represented with only a number and the name of a unit. Others, such as a wind of 15 miles per hour in a northeasterly direction and a velocity of 30 miles per hour south 10° west, have direction as well as magnitude. These are called vectors. Although the concept of vectors originated in physics, the mathematical theory is very extensive. In this chapter, we shall show how the concept of vectors can be used to establish some geometric theorems.

14-1 MATHEMATICAL REPRESENTATION OF VECTORS

We can use a directed segment to represent a vector quantity. Suppose we have two segments, $\overline{AB}$ and directed $\overline{AB}$. In both cases we are interested in the distance between A and B. In the case of $\overline{AB}$, the locations of A and B are not important. But in the case of the directed segment, position, as well as the distance between the two points, is very important. The directed segment from A to B is not the same as the directed segment from B to A.

Sometimes an arrow is used to indicate a directed segment—that is, $\overrightarrow{AB}$ may denote directed $\overline{AB}$. But we already use this symbol to indicate a ray. We cannot use a bar, as in $\overline{AB}$, because we use this to denote the undirected segment. We shall use a modified arrow. Thus, $\overrightharpoon{AB}$ denotes the segment directed from A to B, or **vector** AB.

D **Definition 14-1** **Vector** AB is the directed segment from A to B.

We can also use a single letter to represent a vector, printing the letter in boldface to distinguish it from the name of a point. The use of a single letter is particularly helpful when several vectors begin at the same point.

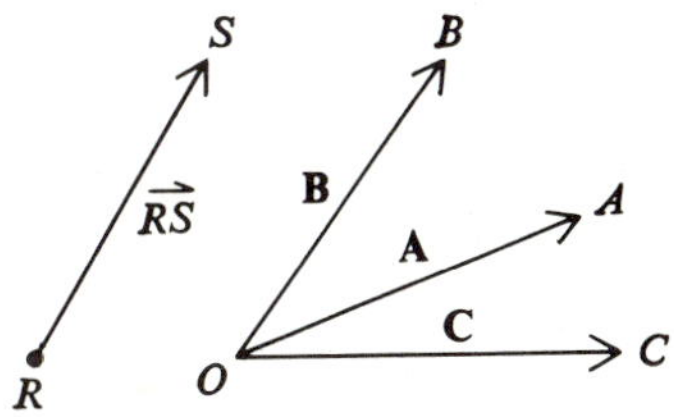

In the figure, the length of $\overrightharpoon{RS}$ is the same as the length of $\overline{RS}$, and can thus be represented by RS. However, we can also indicate the length by $|\overrightharpoon{RS}|$.

Like any three segments, three vectors do not have to be coplanar. We shall confine our consideration, however, to vectors in a plane.

It can be useful to know the differences between the coordinates of the endpoints of a vector. Consider points $A(3, 5)$ and $B(4, 9)$ in the Cartesian plane. $\overrightharpoon{AB}$ is the directed segment from (3, 5) to (4, 9). The difference between the x coordinates of A and B, which we shall call x, is $4 - 3$. The difference between the y coordinates, which we shall call y, is $9 - 5$. (Since the vector goes from A to B, we subtract the coordinate of A from the corresponding coordinate of B to find the change from A to B.) The ordered pair (x, y) is $(4 - 3, 9 - 5)$ or (1, 4). Using this ordered pair, we can determine the direction and length of $\overrightharpoon{AB}$ even if we do not know the coordinates of points A and B. We say that $\overrightharpoon{AB} = (1, 4)$. But many other vectors have the same direction and length, also determined by (1, 4). Since a vector is a directed segment, all of these vectors are equal.

D **Definition 14-2** **Equal vectors** are vectors having equal lengths and the same direction.

D **Definition 14-3** The set of all vectors equal to a particular vector is called a **class of equal vectors.** The ordered pair of numbers (x, y) defines this class.

The term *class* is sometimes used interchangeably with *set*. The class defined in Definition 14-3 is an equivalence class. For a discussion of equivalence relations and equivalence classes, see the Mathematical Excursion in Chapter 3.

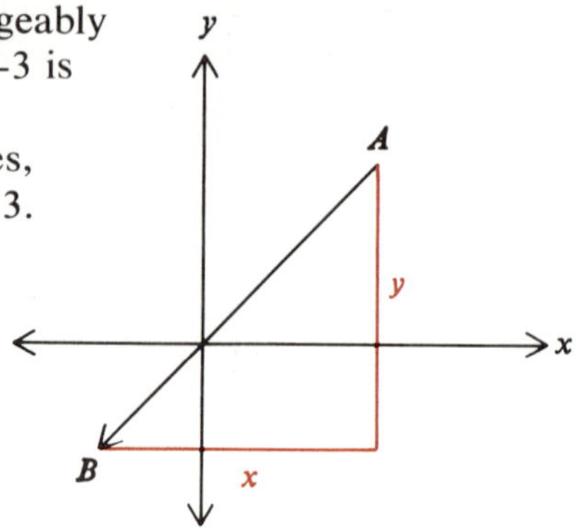

Since the ordered pair (x, y) gives the changes in the x and y coordinates in going from A to B, in the figure at right, you can see that if we apply the Pythagorean Theorem, then the length of $\overrightarrow{AB}$ is $\sqrt{x^2 + y^2}$.

D **Definition 14-4** The **length of a vector** $\overrightarrow{AB}$ belonging to the class defined by (x, y) equals $\sqrt{x^2 + y^2}$.

Example 1 Given $A(5, 2)$ and $B(-4, -3)$, find $|\overrightarrow{AB}|$.

Solution $x = -4 - 5 = -9$ $\quad y = -3 - 2 = -5$

$$|\overrightarrow{AB}| = \sqrt{(-9)^2 + (-5)^2} = \sqrt{106}$$

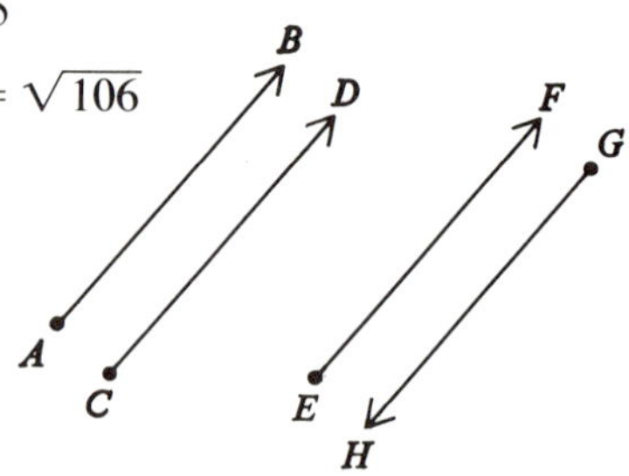

In the diagram at right, $\overrightarrow{AB}$ and $\overrightarrow{CD}$ are equal. $\overrightarrow{EF}$ and $\overrightarrow{GH}$ are opposite.

D **Definition 14-5** **Opposite vectors** are vectors having equal lengths, but opposite directions.

The opposite of a given vector $\mathbf{V} = (x, y)$ is $\mathbf{V}' = (-x, -y)$.

Example 2 Given $A(2, -1)$, $B(3, 4)$, $C(5, 12)$, and $D(4, 7)$, show that $\overrightarrow{AB}$ and $\overrightarrow{CD}$ are opposite vectors.

Solution $\overrightarrow{AB} = (3 - 2, 4 + 1) = (1, 5)$
$\overrightarrow{CD} = (4 - 5, 7 - 12) = (-1, -5)$

EXERCISES

A Determine the vector formed by each of the following pairs of points:

1. $A(2, 5)$ to $B(1, 7)$ 2. $M(3, -1)$ to $N(-6, 5)$ 3. $C(-3, -4)$ to $D(8, -6)$
4. $E(0, 0)$ to $B(1, 7)$ 5. $E(0, 0)$ to $F(3, 9)$ 6. $F(3, 9)$ to $B(1, 7)$

Compare $\overrightarrow{AB}$ with $\overrightarrow{BA}$, given the following coordinates of A and B:

7. $A(-6, -2)$, $B(5, -1)$ 8. $A(4, 0)$, $B(5, 0)$ 9. $A(2, 6)$, $B(4, 12)$

If quadrilateral $ABCD$ is a parallelogram, $\overrightarrow{AB} = (3, -4)$, and $\overrightarrow{BC} = (-2, 3)$, find the following:

10. $\overrightarrow{CD}$ 11. $|\overrightarrow{AB}|$ 12. $\overrightarrow{CB}$ 13. $\overrightarrow{DA}$ 14. $|\overrightarrow{DA}|$ 15. $-\overrightarrow{CD}$

B Given $\overrightarrow{AB} = (-2, 3)$, find the following:

16. The coordinates of C if $\overrightarrow{CD} = \overrightarrow{AB}$ and the coordinates of D are $(-5, 3)$.
17. The coordinates of D if $\overrightarrow{CD} = \overrightarrow{AB}$ and the coordinates of C are $(-5, 3)$.
18. $|\overrightarrow{AB}|$ 19. $|\overrightarrow{BA}|$
20. The coordinates of M if the coordinates of N are $(5, -6)$ and $\overrightarrow{MN}$ is the opposite of $\overrightarrow{AB}$.

C Given $A(4, 3)$, $B(2, 5)$, $C(-4, 1)$, and $D(0, 0)$, find the following:

21. $\overrightarrow{AB}$ 22. $\overrightarrow{BC}$ 23. $\overrightarrow{AC}$ 24. $\overrightarrow{DA}$ 25. $\overrightarrow{DB}$ 26. $\overrightarrow{DC}$
27. Compare x for $\overrightarrow{AB}$ and $\overrightarrow{BC}$ with x for $\overrightarrow{AC}$.
28. Compare y for $\overrightarrow{AB}$ and $\overrightarrow{BC}$ with y for $\overrightarrow{AC}$.
29. Compare the coordinates of a point with the vector from the origin to the point.

A LOOK AT THE PAST

Vectors in Physics

The late Albert Einstein was essentially a physicist, yet he had to invent new mathematical models in order to carry out his research. His laboratory equipment was the pencil and paper which he used to express his ideas as mathematical equations.

The vectors invented by the physicists are restricted to the plane and to space. In the former case, vectors may have two coordinates (x, y), and in the latter case, three (x, y, z). Mathematicians conceive of vectors much more broadly. They see the physicists' applications as special cases of a general notion. A mathematical vector may have any number of coordinates $(x_1, x_2, x_3, \ldots, x_n, \ldots)$. We provide a glimpse of the generalized idea of vectors in the Mathematical Excursion at the end of this chapter.

14-2 VECTOR OPERATIONS

We can show experimentally that if two forces $\mathbf{F}_1$ and $\mathbf{F}_2$ act simultaneously upon a point P, the net result is a force $\mathbf{R}$. We shall represent force $\mathbf{R}$ by the diagonal from P of a parallelogram whose sides from P are the representations of the forces $\mathbf{F}_1$ and $\mathbf{F}_2$. The force $\mathbf{R}$ is called the **resultant** of the forces $\mathbf{F}_1$ and $\mathbf{F}_2$. This is shown in the first figure on page 582.

The figure suggests the definition of the addition of two vectors. Since the opposite sides of a parallelogram are parallel and congruent, we can also start $\mathbf{F}_2$ at the end of $\mathbf{F}_1$, thus forming a triangle.

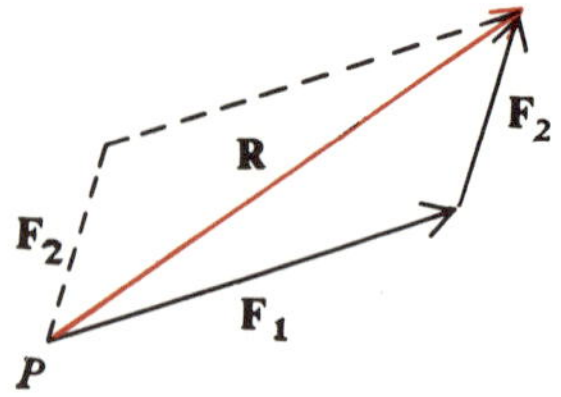

D **Definition 14-6** The **sum A + B of two vectors A and B** is the vector from the initial point of **A** to the terminal point of **B**, when the initial point of **B** is at the terminal point of **A**.

In the figure above, $\mathbf{R} = \mathbf{F}_1 + \mathbf{F}_2$.

In the drawing above, the vector $\mathbf{F}_2$ is equivalent to the dotted segment $\mathbf{F}_2$ even though they are distinct directed segments. This is analogous to the representation of a rational number by an equivalent fraction. Any one of the equivalent fractions may be used to represent the number. A vector which is represented by a set of directed segments is called a *free vector*. Vector $\mathbf{F}_2$ above is a free vector. This means we may place the vector anywhere we wish, as long as its length and direction are fixed.

Example 1 Find the sum of $\mathbf{A} = (3, 5)$ and $\mathbf{B} = (2, -1)$

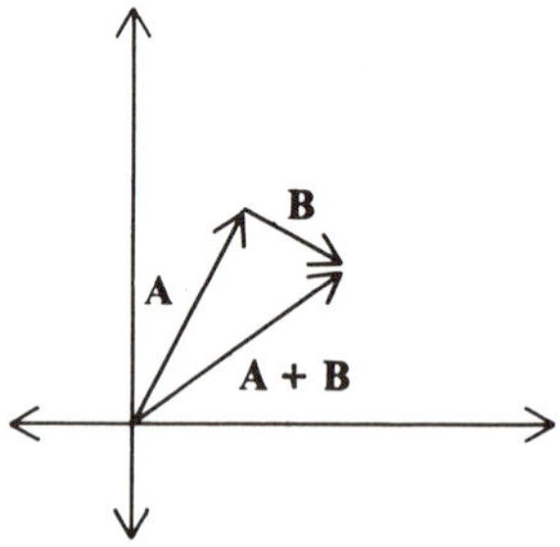

Solution For convenience, we let **A** begin at the origin and terminate at (3, 5). If we place the initial point of **B** at the terminal point of **A**, then **B** will terminate at $(3 + 2, 5 - 1)$ or $(5, 4)$. Then the sum is the vector from (0, 0) to (5, 4). It is the vector defined by $(5 - 0, 4 - 0)$ or (5, 4).

In the example, it is not necessary to place **A** with its origin at (0, 0). Suppose we wish to start it at $(3, -4)$. The terminal point of **A** will be $(3 + 3, -4 + 5)$ or (6, 1). With the initial point of **B** at (6, 1), the terminal point of **B** is (8, 0). Thus, the sum is the vector from $(3, -4)$ to (8, 0) or the vector $(8 - 3, 0 + 4)$, or (5, 4).

If **A** and **B** are designated by pairs of numbers, we can find their sum as a pair of numbers. Let $\mathbf{A} = (x_1, y_1)$ and $\mathbf{B} = (x_2, y_2)$. Since **A** and **B** are free vectors, we shall let the initial point of **B** be at the terminal point of **A**. Then the x value of $\mathbf{A} + \mathbf{B}$ is $x_1 + x_2$ and the y value is $y_1 + y_2$.

D **Definition 14-7** The sum $\mathbf{A} + \mathbf{B}$ of $\mathbf{A} = (x_1, y_1)$ and $\mathbf{B} = (x_2, y_2)$ is the vector $(x_1 + x_2, y_1 + y_2)$.

Example 2 Given $A(3, -2)$, $B(5, 6)$, and $C(4, -3)$, find $\overrightarrow{AB}$, $\overrightarrow{BC}$, $\overrightarrow{AB} + \overrightarrow{BC}$ and compare them with $\overrightarrow{AC}$.

Solution

$$\overrightarrow{AB} = (5 - 3, 6 + 2) = (2, 8)$$
$$\overrightarrow{BC} = (4 - 5, -3 - 6) = (-1, -9)$$
$$\overrightarrow{AB} + \overrightarrow{BC} = (2 - 1, 8 - 9) = (1, -1)$$
$$\overrightarrow{AC} = (4 - 3, -3 + 2) = (1, -1)$$
$$\overrightarrow{AB} + \overrightarrow{BC} = AC$$

EXERCISES

A

Given points $A(3, 2)$, $B(-4, 1)$, $C(5, -3)$, $D(1, 4)$, $E(0, 5)$, find the following:

1. $\overrightarrow{AB}$
2. $\overrightarrow{BC}$
3. $\overrightarrow{AB} + \overrightarrow{BC}$
4. $\overrightarrow{CA}$
5. $\overrightarrow{ED}$
6. $\overrightarrow{DE}$
7. $\overrightarrow{BC} + \overrightarrow{CD} + \overrightarrow{DE}$
8. $\overrightarrow{BE}$
9. $\overrightarrow{AC} + \overrightarrow{CE}$
10. $\overrightarrow{EA}$

B

Use vectors to determine whether each quadrilateral determined by the following vertices is a parallelogram.

11. $A(2, 3)$, $B(5, 6)$, $C(4, 9)$, $D(1, 6)$
12. $A(-1, 2)$, $B(3, -5)$, $C(6, 2)$, $D(9, 5)$

C

13. Show that if A, B, and C are not collinear, then $\overrightarrow{AB} + \overrightarrow{BC} + \overrightarrow{CA} = 0$, on the assumption that $\overrightarrow{AC} + \overrightarrow{CA} = 0$.
14. Show that the relationship in Exercise 13 holds even when A, B, C, are collinear. Does the order of the points make any difference?

14-3 VECTOR RELATIONSHIPS

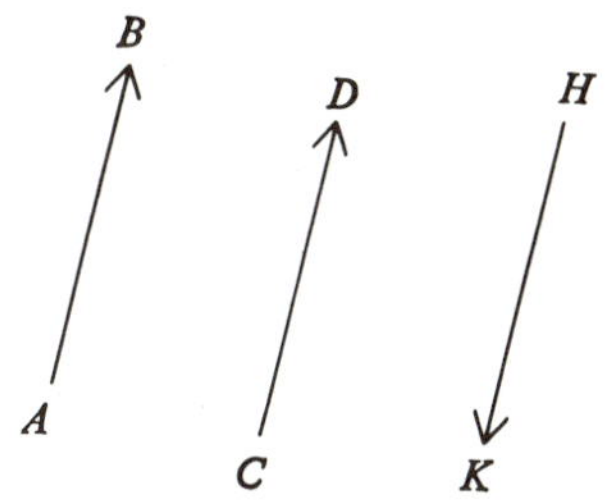

We saw in Section 14-1 that two equal vectors have the same direction. If you consider two parallel one-way streets, you can see why we do not establish the direction of a vector by saying that it is parallel to a particular directed segment. Two segments can be parallel and go in opposite directions. We must also establish the direction of the vectors along the parallel. The lines $\overleftrightarrow{AB}$, $\overleftrightarrow{CD}$, and $\overleftrightarrow{HK}$ at right are parallel. Also, the measures of $\overline{AB}$, $\overline{CD}$, and $\overline{HK}$ are equal. But because of their directions, $\overrightarrow{AB} = \overrightarrow{CD}$, while $\overrightarrow{AB} \neq \overrightarrow{HK}$.

By definition, any two equal vectors are parallel. Are any two parallel vectors equal?

D

Definition 14-8 Two **vectors are parallel** if and only if they lie in the same line or in parallel lines.

Class Exercises

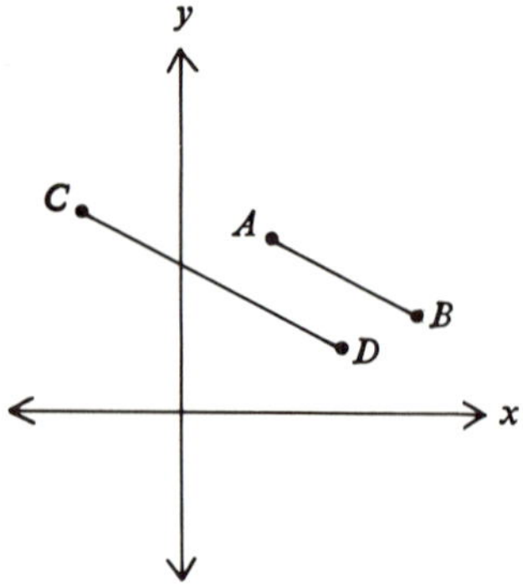

Points $A\,(3, 5)$, $B\,(7, 3)$, $C\,(-3, 6)$, and $D\,(5, 2)$ are shown in the figure at right.

1. Find the slope of $\overleftrightarrow{AB}$.
2. Find the slope of $\overleftrightarrow{CD}$.
3. Are $\overleftrightarrow{AB}$ and $\overleftrightarrow{CD}$ parallel? Why?
4. Find $\overrightarrow{AB}$.
5. Find $\overrightarrow{CD}$.
6. How does x in $\overrightarrow{CD}$ compare with x in $\overrightarrow{AB}$?
7. How does y in $\overrightarrow{CD}$ compare with y in $\overrightarrow{AB}$?
8. How do the components, that is, x and y, compare in two parallel vectors?

The above Class Exercises suggest the following definition:

Definition 14-9 $\mathbf{A} = (x_1, y_1)$ and $\mathbf{B} = (x_2, y_2)$ are parallel vectors if and only if $x_2 = kx_1$ and $y_2 = ky_1$, where $k \neq 0$.

Perpendicular vectors are vectors that lie in perpendicular lines. Since, by Theorem 10-3.1, the product of the slopes of perpendicular lines that are not parallel to the axes is -1, the vectors $\mathbf{A} = (x_1, y_1)$ and $\mathbf{P} = (-y_1, x_1)$ are perpendicular. The vector $(-ky_1, kx_1)$ is parallel to $\mathbf{P}$, where $k \neq 0$. Hence, if we have vectors of the form (x_1, y_1) and $(-ky_1, kx_1)$, they are perpendicular.

Definition 14-10 **Perpendicular vectors** are of the form (x_1, y_1) and $(-ky_1, kx_1)$, where $k \neq 0$, or of the form $(x, 0)$ and $(0, y)$.

Describe the positions of the vectors in the latter case.

Example 1 Show that $\mathbf{A} = (3, 6)$ and $\mathbf{B} = (-2, -4)$ are parallel vectors. Show that $\mathbf{A} = (3, 6)$ and $\mathbf{C} = (-2, 1)$ are perpendicular vectors.

Solution To show that $\mathbf{A}$ and $\mathbf{B}$ are parallel vectors, we shall apply Definition 14-9. Since $-2 = -\frac{2}{3} \cdot 3$ and $-4 = -\frac{2}{3} \cdot 6$, $\mathbf{A} = (3, 6)$ and $\mathbf{B} = (-2, -4)$ are parallel.

To show that $\mathbf{A}$ and $\mathbf{C}$ are perpendicular, we shall apply Definition 14-10. Since $-2 = -\frac{1}{3} \cdot 6$ and $1 = \frac{1}{3} \cdot 3$, $\mathbf{A} = (3, 6)$ and $\mathbf{C} = (-2, 1)$ are perpendicular.

In addition to being equal, opposite, parallel, or perpendicular, vectors may have equal lengths. The length of a vector is its absolute value.

Definition 14-11 The absolute values of $\mathbf{A} = (x_1, y_1)$ and $\mathbf{B} = (x_2, y_2)$ are equal if and only if $\sqrt{x_1^2 + y_1^2} = \sqrt{x_2^2 + y_2^2}$. We denote the equality of the absolute values of $\mathbf{A}$ and $\mathbf{B}$ by $|\mathbf{A}| = |\mathbf{B}|$.

Example 2 Show that $\mathbf{A} = (3, 4)$ and $\mathbf{B} = (0, 5)$ are neither parallel nor perpendicular, but $|\mathbf{A}| = |\mathbf{B}|$.

Solution The two vectors are not parallel because there is no k such that $3 \cdot k = 0$ and $4 \cdot k = 5$. They are not perpendicular because there is no k such that $3k = 5$ and $4(-k) = 0$. $|\mathbf{A}| = |\mathbf{B}|$ because $\sqrt{3^2 + 4^2} = \sqrt{0^2 + 5^2}$ or $5 = 5$.

EXERCISES

A Find a vector parallel to, but not equal to, each of the following vectors:

1. $\mathbf{A} = (3, 7)$ 2. $\mathbf{B} = (-4, 5)$ 3. $\mathbf{C} = (0, 13)$ 4. $\mathbf{D} = (\frac{1}{4}, -\frac{2}{3})$
5. $\mathbf{E} = (5, 0)$ 6. $\mathbf{G} = (1, 1)$ 7. $\mathbf{H} = (3, -\frac{3}{4})$ 8. $\mathbf{J} = (-3, -5)$

Find a vector perpendicular to each of the following vectors:

9. $\mathbf{V}_1 = (4, 7)$ 10. $\mathbf{V}_2 = (5, -3)$ 11. $\mathbf{V}_3 = (-2, 5)$ 12. $\mathbf{V}_4 = (-6, -3)$
13. $\mathbf{V}_5 = (3, 1)$ 14. $\mathbf{V}_6 = (4, -1)$ 15. $\mathbf{V}_7 = (0, -3)$ 16. $\mathbf{V}_8 = (\frac{1}{2}, -3)$

B Find x in each of the following vectors:

17. $|(4, 0)| = |(2, x)|$ 18. $|(3, -4)| = |(5, x)|$ 19. $|(2, 2)| = |(x, 1)|$
20. $|(x, 5)| = |(-4, -3)|$ 21. $|(12, 5)| = |(0, x)|$ 22. $|(3, 1)| = |(1, x)|$

Consider the sides of parallelogram $ABCD$ as the vectors $\overrightarrow{AB}$, $\overrightarrow{BC}$, $\overrightarrow{CD}$, $\overrightarrow{DA}$.

23. If $\overrightarrow{AB} = (2, 7)$, what is $\overrightarrow{CD}$?
24. If $\overrightarrow{AB} = (2, 7)$ and $\overrightarrow{BC} = (-\frac{7}{2}, 1)$, what kind of parallelogram is $ABCD$?
25. If $\overrightarrow{BC} = (-\frac{3}{2}, 11)$, then what is $\overrightarrow{DA}$?
26. If $\overrightarrow{AB} = (3, 4)$ and $\overrightarrow{BC} = (-4, -3)$, what kind of parallelogram is $ABCD$?
27. If $\overrightarrow{AB} = (3, 4)$ and $\overrightarrow{BC} = (-4, 3)$, what kind of parallelogram is $ABCD$?
28. If $ABCD$ is a square and $\overrightarrow{AB} = (1, 1)$, express the remaining three sides as vectors.

14-4 PROPERTIES OF VECTOR ADDITION

We defined the addition of vectors $\mathbf{V}_1 = (x_1 + y_1)$ and $\mathbf{V}_2 = (x_2 + y_2)$ as $\mathbf{V}_1 + \mathbf{V}_2 = (x_1 + x_2, y_1 + y_2)$. If we add two vectors, will the result always be a vector? At first glance, it would seem so. If x_1 and x_2 are real numbers, then their sum, $x_1 + x_2$, is also a real number. But, to be completely sure, we must mention the vector $\mathbf{V} = (0, 0)$. Do you think this represents a directed segment? We frequently think of a point as a segment whose measure is zero. Such a segment can also be considered as having any desired direction. We say that its direction is indeterminate. These considerations suggest the following:

D **Definition 14-12** The **zero vector**, $\mathbf{0}$, equals $(0, 0)$.

Since the direction of the zero vector is indeterminate, we consider it both parallel and perpendicular to any vector.

Since the sum of two vectors is always a vector, we can say that addition of vectors is closed. The zero vector is the **additive identity vector**. If $\mathbf{V}_1 = (x_1, y_1)$, then $\mathbf{V}_1 + \mathbf{0} = (x_1 + 0, y_1 + 0) = (x_1, y_1) = \mathbf{V}_1$.

Every vector $\mathbf{V}$ has an additive inverse, its opposite $\mathbf{V}'$.

Using vector addition, we can show that any vector $\mathbf{V} = (x, y)$ has an inverse $-\mathbf{V} = \mathbf{V}' = (-x, -y)$.

$$\mathbf{V} + (-\mathbf{V}) = (x + (-x), y + (-y)) = (0, 0) = \mathbf{0}$$

Example Show that if $\mathbf{V}_1 = (-1, 3)$ and $\mathbf{V}_2 = (5, -4)$ then $\mathbf{V}_1 + \mathbf{V}_2 = \mathbf{V}_2 + \mathbf{V}_1$.

Solution
$$\mathbf{V}_1 + \mathbf{V}_2 = \quad (-1) + 5, 3 + (-4) \quad = (4, -1)$$
$$\mathbf{V}_2 + \mathbf{V}_1 = (5 + (-1), (-4) + 3) = (4, -1)$$

A generalization of this example will show that vector addition is commutative.

T **Theorem 14-4.1** Addition of vectors is commutative. $(\mathbf{V}_1 + \mathbf{V}_2 = \mathbf{V}_2 + \mathbf{V}_1)$

T **Theorem 14-4.2** Addition of vectors is associative.
$(\mathbf{V}_1 + \mathbf{V}_2) + \mathbf{V}_3 = \mathbf{V}_1 + (\mathbf{V}_2 + \mathbf{V}_3)$

The proofs of Theorems 14-4.1 and 14-4.2 are left as an exercise.

Do vectors follow the same laws of addition that real numbers follow?

EXERCISES

A

1. $\mathbf{V}_1 = (3, -1)$, $\mathbf{V}_2 = (-4, 1)$, $\mathbf{V}_1 + \mathbf{V}_2 =$ ____.
2. $\mathbf{V}_3 = (2, 5)$, $\mathbf{V}_4 = (4, 6)$, $\mathbf{V}_3 + \mathbf{V}_4 =$ ____.
3. $\mathbf{V}_5 = (-2, -7)$, $\mathbf{V}_6 = (4, -1)$, $\mathbf{V}_5 + \mathbf{V}_6 =$ ____.

Find the additive inverse of each of the following:

4. $\mathbf{V}_1 = (3, -2)$ 5. $\mathbf{V}_2 = (-4, -1)$ 6. $\mathbf{V}_3 = (-2, 6)$
7. $\mathbf{V}_4 = (5, 0)$ 8. $\mathbf{V}_5 = (0, -6)$ 9. $\mathbf{V}_6 = (3, -\frac{1}{3})$

Given $\mathbf{A} = (3, -4)$, $\mathbf{B} = (2, -3)$, $\mathbf{V}_1 = (-3, -5)$, $\mathbf{V}_2 = (3, -5)$, and $\mathbf{V}_3 = (-3, 4)$, find the following:

10. $\mathbf{A} + \mathbf{B}$ 11. $\mathbf{B} + \mathbf{A}$ 12. $\mathbf{V}_1 + \mathbf{V}_2$ 13. $\mathbf{V}_2 + \mathbf{V}_1$
14. $\mathbf{A} + \mathbf{V}_3$ 15. $\mathbf{V}_3 + \mathbf{A}$ 16. $(\mathbf{A} + \mathbf{B}) + \mathbf{V}_1$ 17. $\mathbf{A} + (\mathbf{B} + \mathbf{V}_2)$
18. $(\mathbf{V}_1 + \mathbf{V}_2) + \mathbf{V}_3$ 19. $\mathbf{V}_1 + (\mathbf{V}_2 + \mathbf{V}_3)$

B

A given $\triangle ABC$ has vertices $A(2, 3)$, $B(5, 7)$, $C(-3, 4)$.

20. Show that $\overrightarrow{AB} + \overrightarrow{BC} = \overrightarrow{AC}$. 21. Show that $(\overrightarrow{AB} + \overrightarrow{BC}) + \overrightarrow{CA} = \mathbf{0}$.
22. Show that $\overrightarrow{AB} + (\overrightarrow{BC} + \overrightarrow{CA}) = \mathbf{0}$. 23. Show that $\overrightarrow{AB} + \overrightarrow{BA} = \mathbf{0}$.

C

24. Prove that vector addition is commutative and associative.

14-5 MULTIPLICATION OF A VECTOR BY A SCALAR

In the introduction to this chapter, we indicated that some quantities—such as 5 pounds or 20 feet—can be represented by a number. In a discussion of vectors, numbers such as these are called scalars. A **scalar** is a quantity having only magnitude, which can thus be described by a single number.

We cannot add a scalar and a vector, but we can multiply vectors by scalars. When we do, the product is always a vector. As long as the scalar is a natural number, multiplication is similar to repeated addition. For example,

$$3\mathbf{V}_1 = \mathbf{V}_1 + \mathbf{V}_1 + \mathbf{V}_1$$

But what does $3\mathbf{V}_1$ mean? If $\mathbf{V}_1 = (x_1, y_1)$, then $\mathbf{V}_1 + \mathbf{V}_1 + \mathbf{V}_1 = (x_1 + x_1 + x_1, y_1 + y_1 + y_1) = (3x_1, 3y_1)$. The x and y of the vector $\mathbf{V}$ are called its **components**. In the expression $(3x_1, 3y_1)$, which is equal to $3\mathbf{V}_1$, the scalar is distributive over the components of a vector.

D **Definition 14-13** If $\mathbf{V}_1 = (x_1, y_1)$ and a is a scalar, then $a\mathbf{V}_1 = (ax_1, ay_1)$.

Example 1 Find $\frac{2}{3}\mathbf{V}_1$, if $\mathbf{V}_1 = (6, 5)$.

Solution $\frac{2}{3}\mathbf{V}_1 = (\frac{2}{3} \cdot 6, \frac{2}{3} \cdot 5) = (4, \frac{10}{3})$.

Is the product $(4, \frac{10}{3})$ a vector?

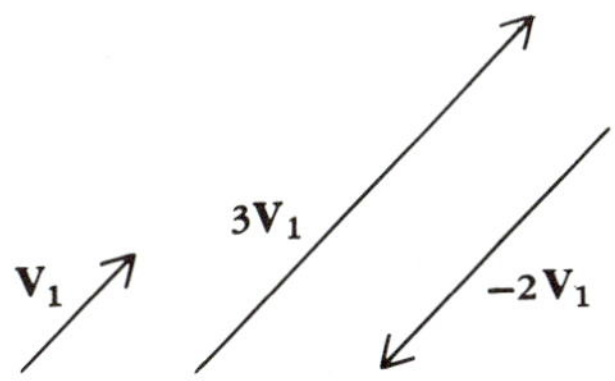

Graphically, if we multiply a vector by a scalar, we increase or decrease the length of the vector the number of times indicated by the scalar.

Note carefully that Definition 14-13 does not allow us to say that $\mathbf{V}_1 a$ is equal to $a\mathbf{V}_1$. Under this definition, what properties of multiplication of vectors by scalars can we derive?

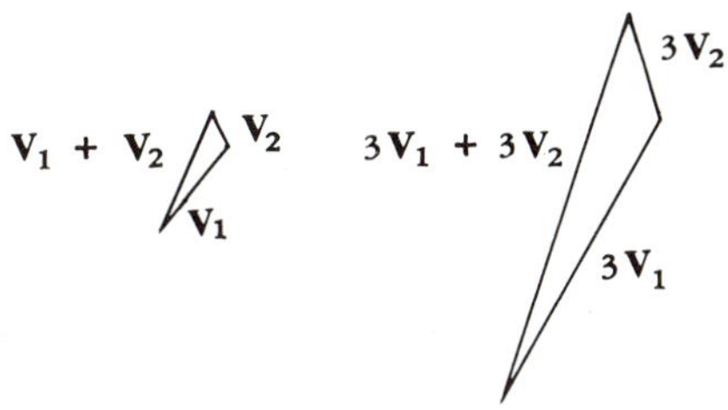

We shall begin to answer this question by investigating the possibility that a scalar is distributive over the sum of two vectors. Does $3(\mathbf{V}_1 + \mathbf{V}_2) = 3\mathbf{V}_1 + 3\mathbf{V}_2$? Are the expressions on each side of the equal sign defined? The two triangles in the diagram at right appear to be similar triangles with the ratio of similitude 3. Does the diagram prove anything? In this case, $3(\mathbf{V}_1 + \mathbf{V}_2) = 3\mathbf{V}_1 + 3\mathbf{V}_2$. Theorem 14-5.1 is a general statement of the property which this case illustrates.

T

Theorem 14-5.1 If a is a scalar and $\mathbf{V}_1$ and $\mathbf{V}_2$ are vectors, then $a(\mathbf{V}_1 + \mathbf{V}_2) = a\mathbf{V}_1 + a\mathbf{V}_2$.

GIVEN $\mathbf{V}_1 = (x_1, y_1)$; $\mathbf{V}_2 = (x_2, y_2)$; a is a scalar.

PROVE $a(\mathbf{V}_1 + \mathbf{V}_2) = a\mathbf{V}_1 + a\mathbf{V}_2$

PROOF

$a(\mathbf{V}_1 + \mathbf{V}_2) = a[(x_1, y_1) + (x_2, y_2)]$	by Postulate 2-1.
$= a[(x_1+x_2, y_1 + y_2)]$	by Definition 14-7.
$= [a(x_1 + x_2), a(y_1 + y_2)]$	by Definition 14-13.
$= [(ax_1 + ax_2), (ay_1 + ay_2)]$	(Distributive property.)
$a\mathbf{V}_1 + a\mathbf{V}_2 = a(x_1, y_1) + a(x_2, y_2)$	by Definition 14-7.
$= (ax_1, ay_1) + (ax_2, ay_2)$	by Definition 14-13.
$= (ax_1, ay_1) + (ax_2, ay_2)$	by Definition 14-7.
$a(\mathbf{V}_1 + \mathbf{V}_2) = a\mathbf{V}_1 + a\mathbf{V}_2$	by Postulate 2-1.

Example 2 Show that $9\mathbf{V} = 4\mathbf{V} + 5\mathbf{V}$.

Solution Let $\mathbf{V} = (x, y)$.

$9(x, y) = (9x, 9y) = 9\mathbf{V}$
$4(x, y) = (4x, 4y) = 4\mathbf{V}$
$5(x, y) = (5x, 5y) = 5\mathbf{V}$

Adding the vectors $4\mathbf{V} + 5\mathbf{V}$, we have $(4x, 4y) + (5x, 5y) = (9x, 9y) = 9\mathbf{V}$.

What property is illustrated by this example?

T

Theorem 14-5.2 If a and b are scalars and $\mathbf{V}$ is a vector, then $(a + b)\mathbf{V} = a\mathbf{V} + b\mathbf{V}$.

GIVEN Scalars a and b and vector $\mathbf{V} = (x, y)$

PROVE $(a + b)\mathbf{V} = a\mathbf{V} + b\mathbf{V}$

PROOF

$(a + b)\mathbf{V} = (a + b)(x, y)$	by Postulate 2-1.
$= (a + b)x, (a + b)y$	by Definition 14-13.
$= (ax + bx, ay + by)$	(Distributive property.)
$a\mathbf{V} + b\mathbf{V} = a(x, y) + b(x, y)$	by Postulate 2-1.
$= (ax, ay) + (bx, by)$	by Definition 14-13.
$= (ax + bx), (ay + by)$	by Definition 14-7.
$(a + b)\mathbf{V} = a\mathbf{V} + b\mathbf{V}$	by Postulate 2-1.

Since we have not introduced multiplication of a vector by a vector, the only associative property of multiplication we can examine is for a scalar times a vector.

Example 3 Show that $3(5\mathbf{V}) = 15\mathbf{V}$.

Solution
Let $\mathbf{V} = (x, y)$
$5\mathbf{V} = (5x, 5y) \qquad 3(5\mathbf{V}) = 3(5x, 5y) = (15x, 15y) = 15\mathbf{V}$

T **Theorem 14-5.3** If a and b are scalars and $\mathbf{V}$ is a vector, then $a(b\mathbf{V}) = (ab)\mathbf{V}$.

The proof of this theorem is similar to the two preceding proofs. It is left as an exercise.

EXERCISES

A

Find the following products:

1. $3(-2, 5)$
2. $-1(4, 3)$
3. $6(-5, -9)$
4. $0(6, -1)$
5. $13(1, -3)$
6. $4(0, 0)$
7. $2(-1, 1)$
8. $-1(2, -2)$

Express each of the following as a single vector:

9. $4[(3, 1) + (2, -3)]$
10. $4(5, -2)$
11. $-3[(2, 0) + (4, 1)]$
12. $-3(6, 1)$
13. $7[(5, -3) + (-5, 3)]$
14. $2(3, 5) + 5(3, 5)$

In the following exercises, let $\mathbf{V}_1 = (3, 1)$, $\mathbf{V}_2 = (-4, 2)$, $\mathbf{V}_3 = (0, 3)$, $\mathbf{V}_4 = (-1, -5)$, and $\mathbf{V}_5 = (2, -3)$. Express each quantity as a single vector in component form.

15. $2\mathbf{V}_1 + 3\mathbf{V}_2$
16. $3\mathbf{V}_1 - 2\mathbf{V}_2 - \mathbf{V}_3$
17. $\frac{1}{2}(\mathbf{V}_1 + \mathbf{V}_2) + \frac{1}{3}\mathbf{V}_3$
18. $20(\mathbf{V}_5 - \mathbf{V}_4) + 10\mathbf{V}_3$
19. $6\mathbf{V}_4 - 5(\mathbf{V}_3 - \mathbf{V}_2)$
20. $4(2\mathbf{V}_3 + 3\mathbf{V}_5) + 2(\mathbf{V}_4 - 3\mathbf{V}_1)$
21. $2(\mathbf{V}_3 + \mathbf{V}_4) - 3(\mathbf{V}_5 + \mathbf{V}_1) + 4\mathbf{V}_2$

B

22. Let $a = 3$, $\mathbf{V}_1 = (-1, 2)$, $\mathbf{V}_2 = (3, 1)$. Illustrate geometrically that $a(\mathbf{V}_1 + \mathbf{V}_2) = a\mathbf{V}_1 + a\mathbf{V}_2$.
23. Let $a = 2$, $b = 3$, and $\mathbf{V} = (3, 4)$. Illustrate geometrically that $(a + b)\mathbf{V} = a\mathbf{V} + b\mathbf{V}$.
24. Let $a = 2$, $b = 3$, and $\mathbf{V} = (3, 4)$. Illustrate geometrically that $a(b\mathbf{V}) = (ab)\mathbf{V}$.

C

25. Prove Theorem 14-5.3.

SOMETHING TO THINK ABOUT

Vector Space

In Section 14-6 we shall see how vectors can be useful in proving geometric theorems. But it would be a mistake to leave the impression that this is their only mathematical importance.

If we were studying physics, we would be interested in vectors of only two $[\mathbf{V} = (x, y)]$ or three $[\mathbf{V} = (x, y, z)]$ dimensions. But mathematicians have extended the concept of vectors so that vectors may possess any number of dimensions. The resulting system has many varied mathematical applications. The set of all vectors that have n components

$$\mathbf{V} = (x_1, x_2, x_3, \ . \ . \ . \ , x_n)$$

is said to belong to an n-space. The properties that we have developed for two-dimensional vectors can be generalized to any number of dimensions. The set of all n-dimensional vectors satisfying these properties of addition and scalar multiplication is called a *vector space*.

You might enjoy trying your hand at verifying the addition and scalar multiplication properties for three-dimensional vectors.

14-6 USE OF VECTORS IN PROVING THEOREMS

Some prominent mathematicians and mathematics educators think all of elementary geometry should be studied through the medium of vectors. Obviously, that is not the point of view of this book. However, it is true that some theorems can be proved rather neatly by utilizing vectors. In this section, we shall examine a few.

First, however, we must discuss a few important ideas. If two vectors **A** and **B** have a common origin, we can join their terminal points to form a vector triangle. As shown by the arrows in the figure at left below, the third side of the triangle is $\mathbf{B} - \mathbf{A}$. If we rotate the triangle clockwise, we have $\mathbf{A} + (\mathbf{B} - \mathbf{A}) - \mathbf{B} = \mathbf{0}$. In $\triangle ABC$ below, if A' is the midpoint of $\overline{BC}$, $\overrightarrow{BA'} = \frac{1}{2}\overrightarrow{BC}$; $\overrightarrow{BA'} = -\overrightarrow{CA'}$; $\overrightarrow{BA} + \overrightarrow{AA'} + \overrightarrow{A'B} = \mathbf{0}$.

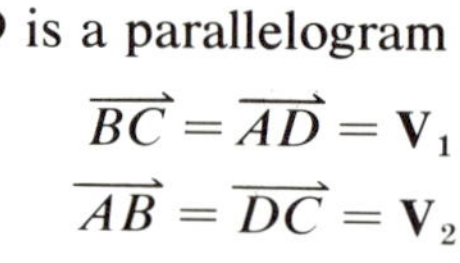

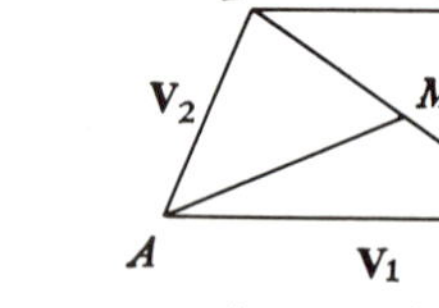

For our first vector proof, we shall prove the theorem that the diagonals of a parallelogram bisect each other.

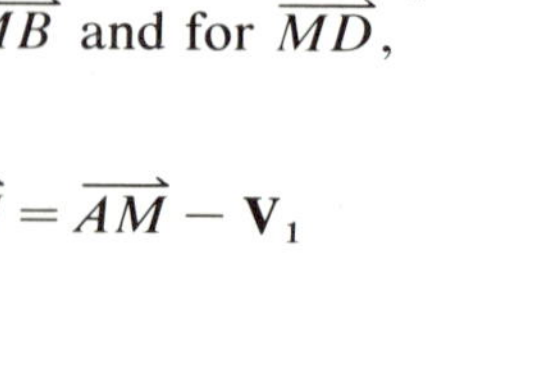

Since $ABCD$ is a parallelogram

$$\overrightarrow{BC} = \overrightarrow{AD} = \mathbf{V}_1$$
$$\overrightarrow{AB} = \overrightarrow{DC} = \mathbf{V}_2$$

Draw $\overline{BD}$, with midpoint M. In $\triangle AMB$, we have $\overrightarrow{AM} + \overrightarrow{MB} = \mathbf{V}_2$ and in $\triangle AMD$ we have $\overrightarrow{AM} + \overrightarrow{MD} = \mathbf{V}_1$. Solving for $\overrightarrow{MB}$ and for $\overrightarrow{MD}$,

$$\overrightarrow{MB} = \mathbf{V}_2 - \overrightarrow{AM}$$
$$\overrightarrow{MD} = \mathbf{V}_1 - \overrightarrow{AM} \qquad \text{or} \qquad \overrightarrow{DM} = \overrightarrow{AM} - \mathbf{V}_1$$

Since M is the midpoint of $\overline{BD}$, $\overrightarrow{DM} = \overrightarrow{MB}$.

Substituting, $\quad \overrightarrow{AM} - \mathbf{V}_1 = \mathbf{V}_2 - \overrightarrow{AM}$

$$2\overrightarrow{AM} = \mathbf{V}_2 + \mathbf{V}_1$$

Referring back to the parallelogram we have $\mathbf{V}_2 + \mathbf{V}_1 = \overrightarrow{AC}$ and hence, $2\overrightarrow{AM} = \overrightarrow{AC}$ or $\overrightarrow{AM} = \frac{1}{2}\overrightarrow{AC}$. Thus M bisects both diagonals $\overline{BD}$ and $\overline{AC}$.

Another theorem that lends itself to a vector proof is the theorem that the line segment joining the midpoints of two sides of a triangle is parallel to the third side and is one-half the length of the third side.

Let C' and B' be the midpoints of $\overline{AB}$ and $\overline{AC}$, respectively.

$$\overrightarrow{C'B'} + \overrightarrow{B'C} + \overrightarrow{CB} + \overrightarrow{BC'} = \mathbf{0}$$

$$\overrightarrow{C'B'} = \overrightarrow{BC} - \overrightarrow{B'C} - \overrightarrow{BC'}$$

$$2\overrightarrow{C'B'} = 2\overrightarrow{BC} - \overrightarrow{AC} - \overrightarrow{BA}$$

$$2\overrightarrow{C'B'} = \overrightarrow{BC} + \overrightarrow{BC} + \overrightarrow{CA} + \overrightarrow{AB}$$

$$2\overrightarrow{C'B'} = \overrightarrow{BC} + \mathbf{0}$$

$$\overrightarrow{C'B'} = \frac{1}{2}\overrightarrow{BC}$$

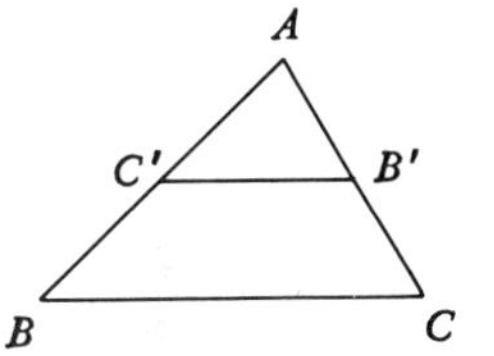

Thus $\overline{C'B'}$ is parallel to $\overline{BC}$ and equal to $\frac{1}{2}$ $\overline{BC}$.

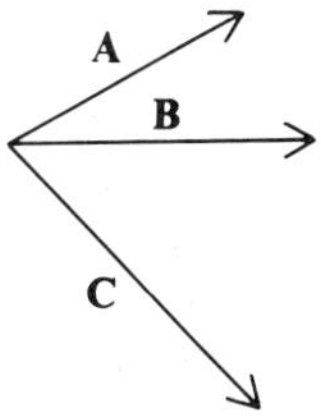

EXERCISES

A

Reproduce the figure at right and construct the following vectors:

1. $\mathbf{A} - \mathbf{C}$ 2. $\mathbf{B} + \mathbf{C}$ 3. $\mathbf{B} - \mathbf{C}$
4. $-\mathbf{B} - \mathbf{C}$ 5. $\mathbf{C} - \mathbf{A}$ 6. $\mathbf{A} + \mathbf{B}$
7. The sum of the forces (vectors) acting on a point in equilibrium is the zero vector. Forces of 30 lb and 40 lb act on a point at right angles. Describe the third force necessary to keep the point in equilibrium.
8. Using vectors, show that the midpoints of consecutive sides of a quadrilateral are vertices of a parallelogram.
9. If the diagonals of a quadrilateral bisect each other, prove that the quadrilateral is a parallelogram, using vectors.

B

10. A', B', and C' are the midpoints of $\overline{BC}$, $\overline{CA}$, and $\overline{AB}$, respectively, in $\triangle ABC$. If P is any point, prove that $\overrightarrow{PA} + \overrightarrow{PB} + \overrightarrow{PC} = \overrightarrow{PA'} + \overrightarrow{PB'} + \overrightarrow{PC'}$.
11. Prove that for any vector $\mathbf{V}$ and the scalar 0 the product $0 \cdot \mathbf{V} = \mathbf{0}$. Prove that the zero vector, $\mathbf{0}$, is unique. (*Hint*: Assume another zero vector $\mathbf{0}'$ and show that $\mathbf{0} = \mathbf{0}'$.)
12. In the figure at right, $ABCD$ is a parallelogram. In diagonal $\overline{AC}$, $\overline{AE} = \overline{CF}$; $\overleftrightarrow{DE}$ and $\overleftrightarrow{BF}$ meet sides $\overline{AB}$ and $\overline{DC}$ in X and Y. Prove that $\overleftrightarrow{DX}$ is parallel to $\overleftrightarrow{BY}$, using vectors.

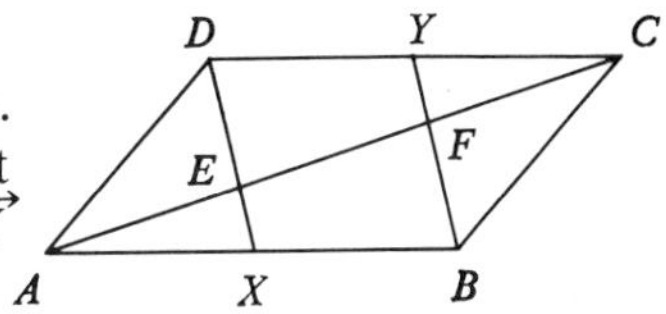

HOW IS YOUR MATHEMATICAL VOCABULARY?

The key words and phrases introduced in this chapter are listed below. How many do you know and understand?

additive identity vector (14-4)
class of equal vectors (14-1)
components of a vector (14-5)
equal vectors (14-1)
free vector (14-2)
length of a vector (14-2)
multiplication of a vector by a scalar (14-5)
opposite vectors (14-1)
parallel vectors (14-3)
perpendicular vectors (14-3)
resultant force (14-2)
scalar (14-5)
sum of two vectors (14-2)
vector (14-1)
zero vector (14-4)

REVIEW EXERCISES

14-1 Mathematical representation of vectors

Express as an ordered pair each vector described below.

1. From $A(5, -2)$ to $B(-3, 1)$
2. From $C(-2, -5)$ to $D(-4, 2)$
3. From $E(3, 4)$ to $F(8, 6)$
4. From $G(-3, 6)$ to $H(-2, -3)$
5. If $\overrightarrow{AB} = (5, -3)$, find $\overrightarrow{BA}$.
6. If $\overrightarrow{CD} = (-4, 5)$, find $\overrightarrow{DC}$.
7. Show that $|\overrightarrow{AB}| = |\overrightarrow{BA}|$ for any vector $\overrightarrow{AB}$.

14-2 Vector operations

Given points $A(0, 3)$, $B(-2, 1)$, and $C(3, -5)$, find the following:

8. $\overrightarrow{AB}$ 9. $\overrightarrow{BC}$ 10. $\overrightarrow{CA}$ 11. $\overrightarrow{AB} + \overrightarrow{BC}$ 12. $\overrightarrow{AC}$
13. If $A(3, 2)$, $B(6, 5)$, and $C(9, 4)$ are three vertices of a parallelogram, find the fourth vertex D.
14. There are three answers to Exercise 13. Explain why. Find the other two.

14-3 Vector relationships

15. If $\mathbf{V}_1 = (3, 7)$ is parallel to $\mathbf{V}_2 = (x, 10)$, find x.
16. If $\mathbf{V}_1 = (6, 9)$ is parallel to $\mathbf{V}_2 = (3, x)$, find x.

Find x in the following vector equations:

17. $|(3, 4)| = |(5, x)|$ 18. $|(7, 15)| = |(x, 5)|$ 19. $|(x, 0)| = |(0, 7)|$
20. Find a vector perpendicular to and whose absolute value is equal to $\mathbf{V}_1 = (1, 3)$.
21. If in parallelogram $ABCD$, $\overrightarrow{AB} = (5, 3)$, find $\overrightarrow{CD}$.

14-4 Properties of vector addition

If $\mathbf{V}_1 = (-3, 2)$, $\mathbf{V}_2 = (5, 7)$, and $\mathbf{V}_3 = (4, -3)$, find the following:

22. $\mathbf{V}_1 + \mathbf{V}_2$ 23. $\mathbf{V}_2 + \mathbf{V}_3$ 24. $\mathbf{V}_3 + \mathbf{V}_1$ 25. $\mathbf{V}_1 + \mathbf{V}_1$
26. $\mathbf{V}_1 + \mathbf{V}_3$ 27. $\mathbf{V}_1 + (\mathbf{V}_2 + \mathbf{V}_3)$ 28. $(\mathbf{V}_1 + \mathbf{V}_2) + \mathbf{V}_3$

29. Find the additive inverse of $\mathbf{V}_1 = (-3, 2)$.
30. Prove that $\mathbf{0} = (0, 0)$ is the additive identity vector.

14-5 Multiplication of a vector by a scalar

Find the products of the following scalars and vectors:

31. $5(-2, 3)$ 32. $-4(6, -2)$ 33. $12(0, 0)$
34. $0(9, 5)$ 35. $6(2, -2)$ 36. $7(-5, -8)$

Express each of the following as a single vector:

37. $-2[(-1, 4) + (1, -4)]$ 38. $3[4(2, 3)]$
39. $5[(0, 2) + (3, 1)]$ 40. $-1[3(2, 5)]$

In the following exercises, let $\mathbf{V}_1 = (2, -3)$, $\mathbf{V}_2 = (-4, 2)$, and $\mathbf{V}_3 = (3, -1)$. Express each quantity as a single vector.

41. $3\mathbf{V}_1 + 5\mathbf{V}_2$ 42. $2\mathbf{V}_1 - 5\mathbf{V}_3$ 43. $\frac{2}{3}(\mathbf{V}_1 - 2\mathbf{V}_2)$
44. $2\mathbf{V}_2 - (2\mathbf{V}_1 + \mathbf{V}_3)$ 45. $6(\mathbf{V}_1 + \mathbf{V}_3)$ 46. $6\mathbf{V}_1 + 6\mathbf{V}_3$

14-6 Use of vectors in proving theorems

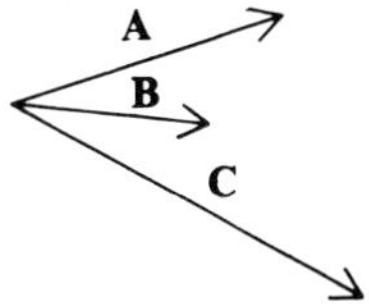

Reproduce the figure at right. Then construct the following vectors.

47. $\mathbf{A} - \mathbf{C}$ 48. $\mathbf{B} + \mathbf{C}$ 49. $\mathbf{B} - \mathbf{C}$ 50. $\mathbf{C} - \mathbf{A}$
51. Prove by vectors that the diagonals of a rectangle are equal and that they bisect each other.

CHAPTER TEST

1. Distinguish between a segment and a vector.
2. Explain what a free vector is.
3. Given points $A(3, 5)$, $B(-1, 3)$, and $C(4, 3)$, find a point D such that $\overrightarrow{AB} = \overrightarrow{CD}$.

Given points $A(3, 5)$, $B(-1, 3)$, and $C(4, 3)$, evaluate the following:

4. $|\overrightarrow{AB}|$ 5. $|\overrightarrow{BC}|$ 6. $|\overrightarrow{CA}|$
7. Given points $A(3, 5)$ and $B(-1, 3)$, find a vector $\mathbf{V}$ parallel to $\overrightarrow{AB}$ such that $|\overrightarrow{AB}| \neq |\mathbf{V}|$.
8. Plot points $A(3, 2)$, $B(-1, 5)$, and $C(1, -2)$ and demonstrate that $\overrightarrow{AB} + \overrightarrow{BC} + \overrightarrow{CA} = \mathbf{0}$.

If $\mathbf{V}_1 = (-1, 5)$, $\mathbf{V}_2 = (3, 1)$, $\mathbf{V}_3 = (1, 4)$, find:

9. $3\mathbf{V}_1 - 5\mathbf{V}_2 + \mathbf{V}_3$ 10. $3(\mathbf{V}_1 + \mathbf{V}_2)$
11. Given $\mathbf{V}_1$ and $\mathbf{V}_2$, show how to find their sum geometrically.
12. Show how to find the difference of $\mathbf{V}_1$ and $\mathbf{V}_2$ geometrically.
13. If the diagonals of a quadrilateral are equal, is the quadrilateral a rectangle? Prove your answer.

SUGGESTED RESEARCH

1. Sir Isaac Newton was a famous physicist as well as a mathematician. Many great mathematicians were also accomplished in other areas. Prepare a short report on the lives of Descartes, Leibniz, and Fermat, discussing their contributions to other fields.

2. Vectors are used extensively in physics to represent forces. Find the meaning of *moment of force* and *equilibrium of a point* and relate the meaning of the first term to the second term.

MATHEMATICAL EXCURSION

Inner Product of Vectors

In this chapter we did not introduce the concept of the product of two vectors. There are two kinds of vector multiplication. The inner product of two vectors, also called dot product, is a scalar. The vector product, or cross product, results in a vector.

The operation of cross product is essentially a geometric concept. If we have $A \times B = C$, then A and B lie in a plane and C is always perpendicular to that plane. Thus, if $C \neq 0$, we introduce three-dimensional space, and vectors take the form $\mathbf{V} = (x, y, z)$.

The inner product of vectors does not have these limitations. There is a dot product of any two vectors, provided the two vectors are of the same dimension, regardless of what that dimension might be.

The dot product of $\mathbf{V}_1 = (3, 4)$ and $\mathbf{V}_2 = (1, 2)$ is $\mathbf{V}_1 \cdot \mathbf{V}_2 = 3 \cdot 1 + 4 \cdot 2 = 11$, which is a scalar.

In the general case, if
$$\mathbf{V}_1 = (x_1, x_2, x_3, \ldots, x_n)$$
$$\mathbf{V}_2 = (y_1, y_2, y_3, \ldots, y_n)$$
then
$$\mathbf{V}_1 \cdot \mathbf{V}_2 = (x_1y_1 + x_2y_2 + x_3y_3 + \cdots + x_ny_n)$$

This product is the sum of the products of the correspondingly placed components. Since the components are real numbers, the scalar product is a real number.

Of what value is the inner product? First, we shall examine the value relative to a plane, that is, vectors with two components.

$$\mathbf{V}_1 \cdot \mathbf{V}_1 = (x_1, x_2) \cdot (x_1, x_2) = x_1^2 + x_2^2$$

But the length of $\mathbf{V}_1$ or $|\mathbf{V}_1| = \sqrt{x_1^2 + x_2^2}$. So we have that $\sqrt{\mathbf{V}_1 \cdot \mathbf{V}_1} = |\mathbf{V}_1|$.

This concept can be extended easily to three dimensions. First, recall that in a plane $\mathbf{V}_1 = (x, y)$ means that the horizontal distance between the initial and terminal points of the vector is x and the vertical distance is y. In three dimensions, we have a similar situation.

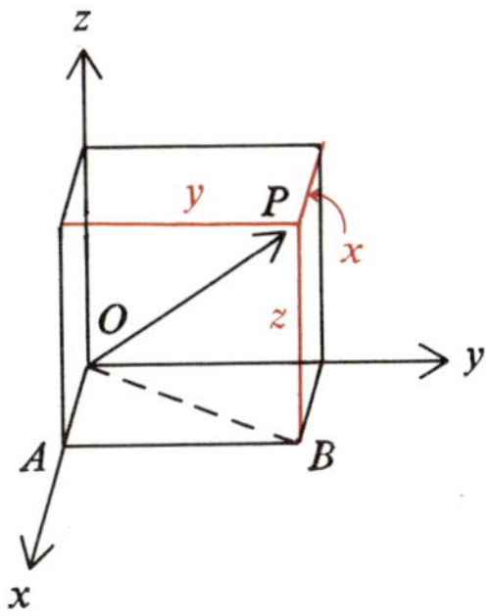

In the figure at right, let the vector $\overrightarrow{OP} = (x, y, z)$ with its initial point at the origin in a three-dimensional coordinate system. Its terminal point is a distance z from the xy-plane, a distance y from the xz-plane, and a distance x from the yz-plane. We can find the length of $\overrightarrow{OP}$ by first finding $OB = \sqrt{x^2 + y^2}$. Then, in $\triangle OPB$,

$$OP^2 = OB^2 + z^2 \qquad OP^2 = x^2 + y^2 + z^2$$

Here,

$$OP = \sqrt{x^2 + y^2 + z^2}$$

$$OP = \sqrt{\overrightarrow{OP} \cdot \overrightarrow{OP}}$$

Using analogy, we can define the length of a vector of n dimension,

$$\mathbf{V}_n = (x_1, x_2, x_3, \ldots, x_n),$$

as

$$|\mathbf{V}_n| = \sqrt{\mathbf{V}_n \cdot \mathbf{V}_n} = \sqrt{x_1^2 + x_2^2 + x_3^2 + \cdots + x_n^2}.$$

It should be clear that if $n > 3$ we are not dealing with "length" in the sense of physical space.

In Section 14-3, we defined perpendicular vectors in two dimensions,

$$\mathbf{V}_1 \perp \mathbf{V}_2 \qquad \text{if } \mathbf{V}_1 = (x, y) \text{ and } \mathbf{V}_2 = (-ky, kx)$$

Unfortunately, this definition cannot be extended to the general case of n dimensions. However, let us examine the inner product $\mathbf{V}_1 \cdot \mathbf{V}_2$.

$$\mathbf{V}_1 \cdot \mathbf{V}_2 = x(-ky) + y(kx) = -kxy + kxy = 0$$

From this, we can define perpendicularity.

D **Definition** Two vectors are perpendicular if the inner product is equal to zero. If

$$\mathbf{V}_1 = (x_1, x_2, x_3, \ldots, x_n) \qquad \text{and} \qquad \mathbf{V}_2 = (y_1, y_2, y_3, \ldots, y_n)$$

$$\mathbf{V}_1 \cdot \mathbf{V}_2 = (x_1y_1 + x_2y_2 + x_3y_3 + \cdots + x_ny_n) = 0$$

then $\mathbf{V}_1$ and $\mathbf{V}_2$ are perpendicular.

It should be fairly easy to show that the inner product is commutative, that is, that

$$\mathbf{V}_1 \cdot \mathbf{V}_2 = \mathbf{V}_2 \cdot \mathbf{V}_1$$

Is the inner product distributive over addition of vectors? That is, is

$$\mathbf{V}_1 \cdot (\mathbf{V}_2 + \mathbf{V}_3) = \mathbf{V}_1 \cdot \mathbf{V}_2 + \mathbf{V}_1 \cdot \mathbf{V}_3$$

The proof of this statement for two-dimensional vectors is left as an exercise.

We said earlier that if the inner product equals zero, the vectors are perpendicular. But, for any vector $\mathbf{V}_1$, $\mathbf{V}_1 \cdot \mathbf{0} = 0$.

The above statement is consistent with the concept of a zero vector. Why? Notice that the inner product of vectors is the same as the product of real numbers in that if there is a zero factor, the product must be zero. But unlike the product of real numbers, the inner product can be zero without either factor being zero.

The inner product can be used to prove theorems involving perpendiculars.

Prove that the diagonals of a rhombus are perpendicular to each other.

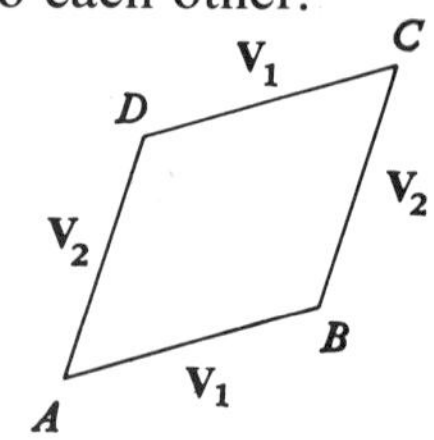

In the figure, quadrilateral $ABCD$ is a rhombus.

$\overrightarrow{AB} = \overrightarrow{DC} = \mathbf{V}_1$ $\overrightarrow{AD} = \overrightarrow{BC} = \mathbf{V}_2$

$\overrightarrow{DB} = \mathbf{V}_1 - \mathbf{V}_2$ $\overrightarrow{AC} = \mathbf{V}_1 + \mathbf{V}_2$

If we show that $(\mathbf{V}_1 - \mathbf{V}_2) \cdot (\mathbf{V}_1 + \mathbf{V}_2) = 0$, then since neither diagonal represents the zero vector, it will follow that the diagonals are perpendicular. Using the distributive property

$$(\mathbf{V}_1 - \mathbf{V}_2) \cdot (\mathbf{V}_1 + \mathbf{V}_2) = (\mathbf{V}_1 - \mathbf{V}_2) \cdot \mathbf{V}_1 + (\mathbf{V}_1 - \mathbf{V}_2) \cdot \mathbf{V}_2$$

Using the commutative property

$$(\mathbf{V}_1 - \mathbf{V}_2) \cdot \mathbf{V}_1 + (\mathbf{V}_1 - \mathbf{V}_2) \cdot \mathbf{V}_2 = \mathbf{V}_1 \cdot (\mathbf{V}_1 - \mathbf{V}_2) + \mathbf{V}_2 (\mathbf{V}_1 - \mathbf{V}_2)$$

And the distributive property again yields

$$\mathbf{V}_1 \cdot (\mathbf{V}_1 - \mathbf{V}_2) + \mathbf{V}_2 (\mathbf{V}_1 - \mathbf{V}_2) = \mathbf{V}_1 \cdot \mathbf{V}_1 - \mathbf{V}_1 \cdot \mathbf{V}_2 + \mathbf{V}_2 \cdot \mathbf{V}_1 - \mathbf{V}_2 \cdot \mathbf{V}_2$$

Since $\mathbf{V}_1 \cdot \mathbf{V}_1 = |\mathbf{V}_1|^2$, we have

$$(\mathbf{V}_1 - \mathbf{V}_2) \cdot (\mathbf{V}_1 + \mathbf{V}_2) = |\mathbf{V}_1|^2 - |\mathbf{V}_2|^2 = 0 \quad \text{Why?}$$

Therefore, the diagonals are perpendicular.

EXERCISES

1. Prove that $\mathbf{V}_1 \cdot \mathbf{V}_2 = \mathbf{V}_2 \cdot \mathbf{V}_1$ using two-dimensional vectors.
2. If $\mathbf{V}_1 = (1, 3, 5)$, $\mathbf{V}_2 = (2, 0, 3)$, $\mathbf{V}_3 = (-2, 1, -3)$, then verify that
$$\mathbf{V}_1 \cdot (\mathbf{V}_2 + \mathbf{V}_3) = \mathbf{V}_1 \cdot \mathbf{V}_2 + \mathbf{V}_1 \cdot \mathbf{V}_3$$
3. Find the length of $\mathbf{V}_1 = (3, -2, -4)$.
4. Show that $\mathbf{V}_1 = (1, 4, 3)$ and $\mathbf{V}_2 = (4, 2, -4)$ are perpendicular.
5. If $\mathbf{V}_1 \cdot \mathbf{V}_2 = \mathbf{V}_1 \cdot \mathbf{V}_3$, does $\mathbf{V}_2$ equal $\mathbf{V}_3$?
   Try $\mathbf{V}_1 = (1, 1, 1)$, $\mathbf{V}_2 = (3, 2, 1)$, $\mathbf{V}_3 = (2, 3, 1)$.
6. Prove that the altitudes of a triangle are concurrent.
7. Prove that $\mathbf{V}_1 \cdot (\mathbf{V}_2 + \mathbf{V}_3) = \mathbf{V}_1 \cdot \mathbf{V}_2 + \mathbf{V}_1 \cdot \mathbf{V}_3$ is true for two-dimensional vectors.

Non-Euclidean Geometry

For more than 2,100 years after Euclid first introduced the Parallel Postulate, many mathematicians expended considerable time and energy attempting a proof. Most believed that they had found valid proofs, but they had always assumed an equivalent statement. It was not until the nineteenth century that mathematicians began seeking alternatives to the postulate. In this discussion, we shall examine some of these attempts and the consequences of several of the alternative assumptions.

Attempts to Prove the Parallel Postulate

In Chapter 6, we stated the Parallel Postulate in a form known as Playfair's Axiom. Euclid did not use our Postulate 6-1 as his Parallel Postulate. You can see from the following statement of Euclid's postulate that he did not even mention parallel lines.

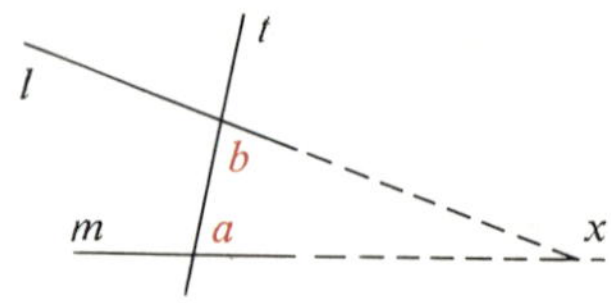

Euclid's Fifth Postulate *If a transversal cuts two lines and if the sum of the measures of the interior angles on one side of the transversal is less than 180, then the two lines meet on this side of the transversal.* The figure shows the condition described by this postulate. Playfair's Axiom and Euclid's Fifth Postulate are equivalent since we can assume either and prove the other. While Playfair's Axiom states that there cannot be more than one line parallel to a given line through a point not on the line, it does not establish the existence of parallel lines. We stated and proved the existence of parallels in Theorem 6-1.1. Since we did not use either Playfair's Axiom or Euclid's Parallel Postulate or any other equivalent statement to prove Theorem 6-1.1, we can use it to show that the two assumptions are equivalent.

Any statement that is equivalent to either Euclid's Fifth Postulate or Playfair's Axiom is automatically equivalent to the other. Why is this true? There are many statements equivalent to the Parallel Postulate. Some of them may seem surprising. For example, "Similar noncongruent triangles exist"; "Three noncollinear points always determine a circle that passes through them"; "The angle-measure sum of a triangle is 180." In fact, the opposite of Theorem 6-1.1, "If one of two lines is perpendicular to a third line but the other is not, then the two lines are not parallel," is equivalent to the Fifth Postulate. If we use any one of these as an assumption in place of the assumptions of either Euclid or Playfair, and if we retain all our other assumptions, such as the Substitution Postulate, we can develop the same geometry.

Ptolemy, a contemporary of Euclid, made one of the earliest attempts to prove the Parallel Postulate. He tried to prove that the sum of the measures of the interior angles on either side of a transversal cutting two parallel lines equals 180. Ptolemy reasoned that parallel lines must be parallel in both directions. However, when he concluded that angles on opposite sides of a transversal must have equal measures, he assumed a statement equivalent to the Parallel Postulate. Using this equivalent assumption, Ptolemy proved the contrapositive of Euclid's postulate, and thus the postulate itself.

Nasîr ed-din, a Persian mathematician of the thirteenth century, made another interesting attempt to prove the Parallel Postulate. In his proof, Nasîr ed-din tried to show that the sum of the measures of the angles of a quadrilateral with two right angles is 360. In so doing, he assumed one of the equivalent forms of the Fifth Postulate which we mentioned earlier. Can you guess which one?

The main interest of Nasîr ed-din's attempted proof lies in its relation to later developments made by an Italian monk, Girolamo Saccheri (1667–1773). Saccheri was an extremely competent logician who was not prone to making unwarranted assumptions. His method was to deny the Parallel Postulate while retaining the remainder of Euclid's assumptions. In doing this he expected to develop a geometry which contradicted itself, since he had no doubt that the Parallel Postulate was a necessary truth. In his proof, Saccheri used a quadrilateral similar to that of Nasîr ed-din.

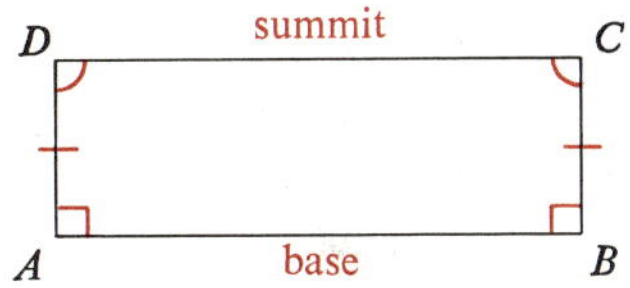

Such a quadrilateral, with two right angles and two congruent sides, one adjacent to each angle, is called a Saccheri quadrilateral. Saccheri called $\overline{AB}$ the base of the quadrilateral and $\overline{CD}$ its summit. The angles ACD and BDC were called the summit angles. He tried to prove that $m\angle C = m\angle D = 90$, a statement which is equivalent to Euclid's Parallel Postulate. He succeeded in proving that the two summit angles are congruent without using a parallel assumption. He hoped to complete his proof by showing, through a process of elimination, that the summit angles had to be right angles. Although he completed many new theorems in the effort, Saccheri never succeeded in proving that the summit angles could not be acute. His persistent attempt failed because the assumption that the summit angles of a Saccheri quadrilateral may be acute is perfectly valid. Had he recognized this, non-Euclidean geometry might have been developed a century earlier than it was.

Even though it might be appropriate to call any geometry which differs from Euclid's non-Euclidean, we generally reserve that term for those geometries which are based on an assumption contrary to the Parallel Postulate. For more than two thousand years, the rules which Euclid had laid down for geometric relations in space were assumed to be as inviolate as the multiplication table. This sovereignty was undermined by the work of several outstanding mathematicians of the nineteenth century, among them Gauss, Lobachevski, Bolyai, Helmholtz, Riemann, and Clifford. They discovered that, while Euclid's system was perfectly adequate as a representation of ideal space and as an exercise in logic, non-Euclidean geometries were not only logically possible, but might serve as better descriptions of physical space.

The distinguished German mathematician Karl Friedrich Gauss (1777–1855) was apparently the first to realize that it was possible to develop a logically consistent geometric system based upon an assumption contrary to the Parallel Postulate. However, realizing what a shock this would be to the mathematical world and possibly fearing the ridicule of his colleagues, Gauss kept silent about his discovery. It remained for John Bolyai (1802–1860) and Nikolai Lobachevski (1793–1856) to independently develop the same non-Euclidean geometry. It may seem strange to you that two mathematicians, living in Europe at the same time, could develop the same system independently. However, this is fairly common in the history of mathematics and, in this case, was not due entirely to chance. Bolyai's father, Wolfgang, and one of Lobachevski's teachers were classmates of Karl Gauss. Undoubtedly, the three students discussed the parallel problem and exchanged ideas about it. The apparently unconnected successes of Bolyai and Lobachevski probably had a common beginning in the sessions of Gauss and his classmates.

The assumed omnipotence of Euclidean geometry was not the only reason that mathematicians did not question the accuracy of the Parallel Postulate. Early mathematicians were considerably inhibited by the theory of space of Immanuel Kant, a German philosopher. It was Kant's contention that some knowledge of space existed in the mind, independent of experience, as a sort of inborn understanding.

Attempts to prove the Parallel Postulate in the early nineteenth century centered around one of two assumptions: (1) there are no lines parallel to a given line through a point not in the line, or (2) there is more than one line parallel to a given line through a point not in the line. Bolyai and Lobachevski used the second assumption in order to develop the non-Euclidean geometry known as **hyperbolic geometry**.

Non-Euclidean Geometries

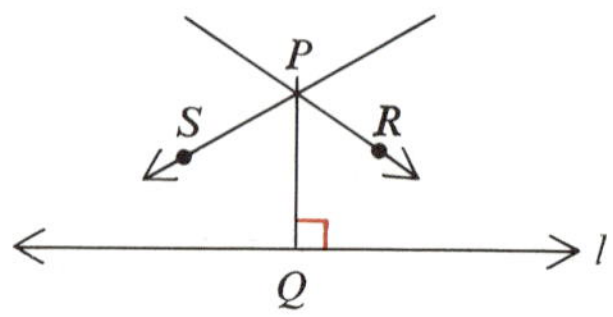

In place of the Parallel Postulate, Lobachevski assumed that through a given point not in a line, more than one line can be drawn not meeting the given line. This is illustrated in the figure, where $\overleftrightarrow{PS}$ and $\overleftrightarrow{PR}$ are two such lines. It is natural to ask whether there are more than two lines which do not intersect line l. Lobachevski proved that there are an infinite number of lines passing through P which do not intersect l, and, furthermore, that these lines lie in the exterior of $\angle SPR$. The two lines, $\overleftrightarrow{PS}$ and $\overleftrightarrow{PR}$ are called parallel lines. All other nonintersecting lines are referred to as transparallel or hyperparallel lines. Lobachevski proved further that $\angle SPQ$ and $\angle QPR$ in the figure are congruent and must be acute angles. These correspond to the acute summit angles of the Saccheri quadrilateral.

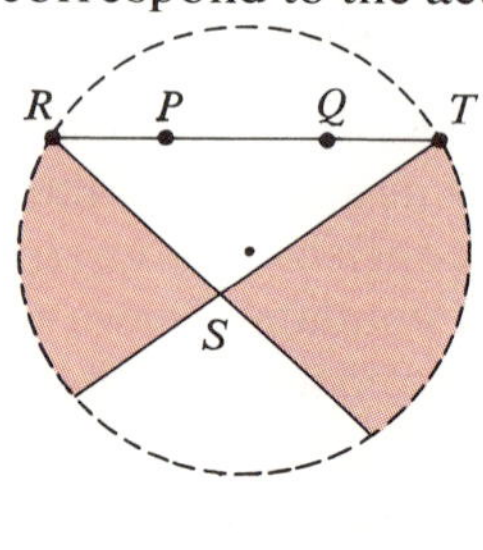

One of the famous models for hyperbolic geometry is due to Felix Klein (1849–1925). His model for the hyperbolic plane corresponds to the interior of a circle. Two points within the circle determine a chord. In the figure, if chord

$\overline{PQ}$ meets the circle at points R and T and if S is any other point within the circle, then $\overleftrightarrow{RS}$ and $\overleftrightarrow{ST}$ do not meet $\overleftrightarrow{PQ}$ since points on the circumference do not belong to the model. Consequently, the lines $\overleftrightarrow{RS}$ and $\overleftrightarrow{ST}$ are models of the parallels to $\overleftrightarrow{PQ}$, and any line containing S and lying within the shaded region corresponds to a transparallel.

Klein's model, in addition to illustrating the new theory of parallels, served a more useful purpose: it demonstrated the inherent impossibility of proving the Parallel Postulate from Euclid's other axioms. If such a proof were possible, the postulate would be a theorem in the geometry based upon Klein's model since Euclid's other postulates are valid in hyperbolic geometry. It can be shown, however, that the Parallel Postulate is not valid in this model. The model also demonstrated that hyperbolic geometry is as consistent (free of contradictions) as Euclidean geometry.

You may wish to explore another famous model of Lobachevskian geometry, which was developed by Henri Poincaré (1854–1912). It was obtained by stereographic projection, that is, the projection of a sphere onto a horizontal plane.

The developers of hyperbolic geometry assumed that all of Euclid's postulates were valid except the Parallel Postulate. One of Gauss' famous pupils, Georg Bernard Riemann (1826–1866), developed another type of non-Euclidean geometry by assuming that: (1) there are no parallel lines, and (2) while lines are boundless, they are not infinite. What does it mean for a line to be boundless and finite? The concept is analogous to walking around the earth from pole to pole in a straight path without ever stopping. Your trip is finite in that you are always tracing the same set of points, but it is boundless since you never stop. We call this type of non-Euclidean geometry **elliptic geometry**.

You may well be wondering why Riemann bothered to make the second assumption, since it led to many conclusions which were completely contrary to both Euclidean and hyperbolic geometry. He made this assumption, which seemed totally fantastic to mathematicians of his time, because he recognized that parallels will always exist as long as a line is considered infinite in extent.

If we consider the points and lines of elliptic geometry to be the points and great circles of a sphere, then we have a natural model for Riemann's system. You will recall from the Mathematical Excursion following Chapter 9 that great circles always intersect. Consequently, lines must always intersect and there are no parallel lines. Furthermore, because infinitely many great circles pass through any two diametrically opposite points, we can no longer assert that two points determine a unique line. This is unfortunate in one sense because it makes the transition from Euclidean to elliptic geometry more complicated than that from Euclidean to hyperbolic. There are actually two forms of elliptic geometry, only one of which is due to Riemann. The form we are considering is called double elliptic geometry. The sphere is a model for this geometry because it has an interior and an exterior. Furthermore, 2 lines always have 2 points in common. Another elliptic geometry, developed by Klein, is called single elliptic geometry. Its model, a Möbius strip, has only one surface. Furthermore, 2 lines always have 1 point in common. If you are not acquainted with the properties of such a strip, you may wish to investigate them.

We mentioned earlier that the developers of non-Euclidean geometry hoped that their systems might better describe physical space. This is certainly true with double elliptic geometry since it is the mathematical representation of what sailors observe when they sail direct courses over the curved oceans of the globe. What do you think is true about the sum of the degree measures of a triangle in elliptic geometry? Riemann proved that this sum is greater than 180. In hyperbolic geometry we observed that the summit angles of a Saccheri quadrilateral were equal and acute. From this it follows that the sum of the angle measures of a triangle is less than 180°.

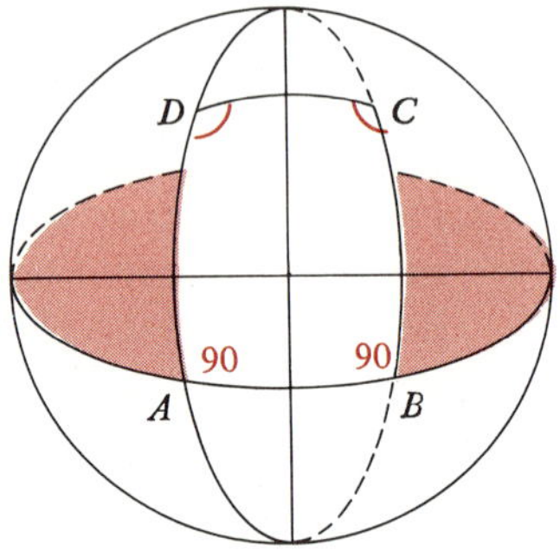

In elliptic geometry it can be shown that the summit angles of a Saccheri quadrilateral are equal and obtuse. We can observe this fact in the figure, where $ABCD$ is a Saccheri quadrilateral and $\angle C$ and $\angle D$ are the summit angles. Several interesting theorems can be proved in both hyperbolic and elliptic geometry once we have established the properties of a Saccheri quadrilateral. Two which you may wish to prove on your own are "There are no squares in hyperbolic or elliptic geometry," and "Any two similar triangles are congruent."

Riemann, who died at the age of 40, did far more for mathematics and physics than develop a non-Euclidean geometry. He proved that all three forms of geometry, Euclidean, hyperbolic, and elliptic, were instances of a still more general geometry. Riemann developed a purely mathematical theory with no concern for its practical application. Albert Einstein found in Riemann's system the mathematical tool he needed to develop his Theory of Relativity. This serves to illustrate that so-called pure mathematical research is not necessarily without practical application.

Neutral Geometry

Our discussion thus far has centered around the number of parallels to a given line. Do you think it is possible to form a geometric system in which the notion of parallels is not even mentioned? There are many theorems in geometry which do not depend upon any particular concept of parallelism. In fact, the theorems we proved in the first five chapters did not make use of a parallel assumption. However, if we drop the idea of parallel completely, some problems result. We cannot, for example, prove or disprove Theorem 6-4.2 until we make some assumption about parallel lines. When we remove a postulate which is required for the proof of a theorem in a system, logicians call the system incomplete. An incomplete system is formed in geometry when we choose to ignore the concept of parallel lines. Because it does not take a stand on the nature of parallelism, this system is known as **neutral geometry**. A theorem belongs to the body of neutral geometry if the proof of the theorem does not require any assumption about parallel lines. Neutral geometry contains all theorems which are common to Euclidean and hyperbolic geometry. For example, it is possible to prove the following statement in neutral geometry: "If the sum of the measures of the angles of any triangle is 180, then the sum of the measures of the angles of all triangles is 180."

Concluding Remarks

The development of non-Euclidean geometry had a profound effect on the subsequent development of much of mathematics in general. Prior to the nineteenth century, mathematicians were concerned with finding unique and undeniable truths about the real world. The postulates and theorems of geometry were thought to constitute an accurate description of the real world. The arrival of non-Euclidean geometry caused mathematicians to adopt a new point of view toward their work. Obviously, Euclidean, hyperbolic, and elliptic geometry could not all be perfect models of physical space since they were contradictory. However, it is impossible to say that only one of these systems is correct. In order to establish this, we would have to use measurement. As you know, measurement is an approximation process which always involves some error. It is this factor of error which prevents us from asserting that only one geometry is correct. Consequently, mathematicians today recognize that any mathematical system consists of relative rather than absolute truths. Albert Einstein summarized this position as follows: "As far as the laws of mathematics refer to reality, they are not certain; and as far as they are certain, they do not refer to reality."

SUGGESTED READINGS

Eves, Howard, *A Survey of Geometry* (Vol. 1, Boston: Allyn and Bacon, Inc., 1963). (Particularly Chapter 7)

Klein, Felix, *Elementary Geometry from an Advanced Standpoint* (New York: The Macmillan Co., 1932), pages 174–188.

Moise, E., *Elementary Geometry from an Advanced Viewpoint* (Reading, Mass.: Addison-Wesley Publishing Company, Inc., 1963). (Chapter 9)

Wolfe, H. E., *Introduction to Non-Euclidean Geometry* (New York: The Dryden Press, Inc., 1945).

SYMBOLS

$\lvert x \rvert$	absolute value of x
$\angle ABC$	angle ABC
$\approx$	approximately equal to
$\overset{\frown}{ABC}$	arc ABC
$\mathscr{A}$	area
$\odot$	circle
$\cong$	congruent to
$\ncong$	not congruent to
$\wedge$	conjunction
$\leftrightarrow$	correspondence
$A \times B$	A cross B
$\angle A\text{-}BC\text{-}D$	dihedral angle
$\vee$	disjunction
$\in$	element of
$\varnothing$	empty set
$=$	equal to
$\neq$	not equal to
$\leftrightarrow$	equivalence
$>$	greater than
$\ngtr$	not greater than
$\geq$	greater than or equal to
$\overset{\circ\!\!\rightarrow}{AB}$	half-line AB
$\rightarrow$	implication
$\cap$	intersection
$\mathscr{L}$	lateral area
AB	length of segment AB
$<$	less than
$\nless$	not less than
$\leq$	less than or equal to
$\overleftrightarrow{AB}$	line AB
$\overline{AB}$	line segment AB
$m\angle ABC$	measure of angle ABC
$m\overset{\frown}{ABC}$	measure of $\overset{\frown}{ABC}$
$\sim p$	not p
(x, y)	ordered pair containing x and y
$//$	parallel to
$\not\parallel$	not parallel to
$\perp$	perpendicular
π	pi
$a:b, \frac{a}{b}$	ratio of a to b
$\overrightarrow{AB}$	ray AB
$\mathscr{R}$	relation
$\sim$	similar
$\sqrt{}$	square root
$\subseteq, \subset$	subset of
$\mathscr{S}$	surface area
$\therefore$	therefore
$\mathscr{T}$	total area
$\triangle$	triangle
$\triangle\!\!\!\!\text{s}$	triangles
$\cup$	union
$\overrightarrow{AB}$	vector AB
$\mathbf{A}$	vector A
$\mathscr{V}$	volume

The following properties are those properties that the students will need to know but are not listed anywhere else in the book.

EQUALITY PROPERTIES

1. **Dichotomy property** For all real numbers a and b, either $a = b$ or $a \neq b$.
2. **Reflexive property** For any real number a, $a = a$. (This is sometimes referred to as the *identity property*.)
3. **Symmetric property** For all real numbers a and b, if $a = b$, then $b = a$.
4. **Transitive property** For all real numbers a, b, and c, if $a = b$ and $b = c$, then $a = c$.

5. **Addition property** For all real numbers a, b, and c, if $a = b$, then $a + c = b + c$.
6. **Multiplication property** For all real numbers a, b, and c, if $a = b$, then $ac = bc$.

ORDER PROPERTIES

1. **Trichotomy property** For all real numbers a and b, either $a > b$, $a = b$, or $a < b$.
2. **Transitive property** For all real numbers a, b, and c, if $a > b$ and $b > c$, then $a > c$.
3. **Addition property** For all real numbers a, b, and c, if $a > b$, then $a + c > b + c$.
4. **Multiplication property** For all real numbers a, b, and c, if $a > b$ and $c > 0$, then $ac > bc$; and if $a > b$ and $c < 0$, then $ac < bc$.

FIELD PROPERTIES

1. **Closure property of addition** The sum of any two real numbers is a real number.
2. **Commutative property of addition** For any two real numbers a and b, we have $a + b = b + a$.
3. **Associative property of addition** For any three real numbers, a, b, and c, we have $(a + b) + c = a + (b + c)$.
4. **Identity property for addition** The set of real numbers includes an element 0, such that for all real numbers n, $n + 0 = 0 + n = n$. The number zero is called the *additive identity element*.
5. **Additive inverse property** For every real number n, there is another real number $-n$, such that $n + (-n) = 0$.
6. **Closure property of multiplication** The product of any two real numbers is a real number.
7. **Commutative property of multiplication** For any two real numbers a and b, we have $a \cdot b = b \cdot a$.
8. **Associative property of multiplication** For any three real numbers a, b, and c, we have $(ab)c = a(bc)$.
9. **Identity property for multiplication** The set of real numbers includes an element 1, such that for all real numbers n, $n \cdot 1 = 1 \cdot n = n$. The number one is called the *multiplicative identity element*.
10. **Multiplicative inverse property** For every nonzero real number n, there is another real number $\frac{1}{n}$, such that $(n)(\frac{1}{n}) = 1$.
11. **Distributive property** For any three real numbers a, b, and c, we have $a(b + c) = ab + ac$.

INDEX

3 4 5 6 7 8 9 10 KPKP 81 80 79 78 77 76 75 74